GUIDANCE AND CONTROL 2000

CONFERENCE COMMITTEE OFFICERS

Chairperson: Eileen Dukes
Lockheed Martin Astronautics

Co-Chairperson: Doug Wiemer
Ball Aerospace & Technologies Corp.

Secretary: Joyce Idler
Lockheed Martin Astronautics

Treasurer: Pam Marlette
Lockheed Martin Astronautics

AAS National Chairperson: James McQuerry
Ball Aerospace & Technologies Corp.

AAS Rocky Mtn. Chairperson: Larry Germann
Left Hand Design Corp.

Conference Liaison: Ron Rausch
Lockheed Martin Astronautics

ADDITIONAL COMMITTEE MEMBERS

Robert Culp	University of Colorado
Ed Euler	Lockheed Martin Astronautics
Arlo Gravseth	Delfin Systems
Dave Igli	Lockheed Martin Astronautics
Joanne Kitlen	Ball Aerospace & Technologies Corp.
James Medbery	Ball Aerospace & Technologies Corp.
Donald Meyers	Honeywell, Inc.
Lou Morine	Schafer Corp.
Marv Odefey	Left Hand Design Corporation
Charlie Schira	Ball Aerospace & Technologies Corp.
Joe Vellinga	Lockheed Martin Astronautics

AAS PRESIDENT
    Dr. Wesley T. Huntress, Jr.   Carnegie Institution of Washington

VICE PRESIDENT - PUBLICATIONS
    Dr. Roger D. Launius   NASA Headquarters

EDITORS
    Dr. Robert D. Culp   University of Colorado
    Eileen M. Dukes   Lockheed Martin Astronautics

SERIES EDITOR
    Robert H. Jacobs   Univelt, Incorporated

Front Cover Illustration:

    This is an artist's conception of the Mars Surveyor 2001 Orbiter scheduled for launch in April, 2001. It will carry a set of instruments designed to determine the composition of surface materials, detect shallow buried ice, and study the radiation environment.

Frontispiece:

    This is a picture of the Mars Global Surveyor spacecraft in final assembly prior to launch. The MGS has been operating in orbit around Mars since 1997, gathering high resolution images and other science data to further our understanding of the Red Planet.

GUIDANCE AND CONTROL 2000

Volume 104
ADVANCES IN THE ASTRONAUTICAL SCIENCES

Edited by
Robert D. Culp
Eileen M. Dukes

Proceedings of the Annual AAS Rocky Mountain Guidance and Control Conference held February 2-6, 2000, Breckenridge, Colorado.

*Published for the American Astronautical Society by
Univelt, Incorporated, P.O. Box 28130, San Diego, California 92198*

Copyright 2000

by

AMERICAN ASTRONAUTICAL SOCIETY

AAS Publications Office
P.O. Box 28130
San Diego, California 92198

Affiliated with the American Association for the Advancement of Science
Member of the International Astronautical Federation

First Printing 2000

Library of Congress Card No. 57-43769

ISSN 0065-3438

ISBN 0-87703-468-0 (Hard Cover)
ISBN 0-87703-469-9 (Soft Cover)

*Published for the American Astronautical Society
by Univelt, Incorporated, P.O. Box 28130, San Diego, California 92198*

Printed and Bound in the U.S.A.

FOREWORD

HISTORICAL SUMMARY

The Annual Rocky Mountain Guidance and Control Conference began as an informal exchange of ideas and reports of achievements among local guidance and control specialists. Since most area guidance and control experts participate in the American Astronautical Society, it was natural to gather under the auspices of the Rocky Mountain Section of the AAS.

In 1977, Bud Gates, Don Parsons, and Bob Culp organized the conference formally and began the annual series of meetings the following winter. In March, 1978, the Annual Rocky Mountain Guidance and Control Conference met at Keystone, Colorado. It met there for eighteen years. The last five Conferences have been held at Breckenridge, Colorado.

The first Conference was chaired by Bud Gates, co-chaired by Bob Culp, with arrangements by Don Parsons. Bob Culp was Section Chairman. The local chairmen were Bob Barsocchi, Carl Henrikson, and Lou Morine; session chairmen were Sherm Seltzer, Pete Kurzhals, and Lou Herman. Other members of the original organizing committee were Ed Euler, Joe Spencer, and Tom Spencer. This year was the twenty-third Conference.

The Rocky Mountain Section of the Society established a broad-based Conference Committee, the Rocky Mountain Guidance and Control Committee, chaired ex-officio by the next Conference Chairman, to run the Annual Conference. The Conference has been a success from the start. Currently, the meeting attracts about 200 of the nation's top specialists in space guidance and control:

Year	Conference Chairman	Attendance
1978	Robert L. Gates	83
1979	Robert D. Culp	109
1980	Louis L. Morine	130
1981	Carl Henrikson	150
1982	W. Edwin Dorroh, Jr.	180
1983	Zubin Emsley	192
1984	Parker S. Stafford	203
1985	Charles A. Cullian	200
1986	John C. Durrett	186
1987	Terry Kelly	201
1988	Paul Shattuck	244
1989	Robert A. Lewis	201
1990	Arlo Gravseth	254
1991	James McQuerry	256

Year	Conference Chairman	Attendance
1992	Dick Zietz	258
1993	George Bickley	220
1994	Ron Rausch	182
1995	Jim Medbery	169
1996	Marv Odefey	186
1997	Stuart Wiens	192
1998	David Igli	189
1999	Doug Wiemer	188
2000	Eileen Dukes	199

In recent years, the Rocky Mountain Section of the AAS has strongly supported students interested in the aerospace program. The Section every year gives $2,000 to two scholarships at the University of Colorado: one in Aerospace Engineering Sciences and one in Electrical and Computer Engineering. The Rocky Mountain Section has made a commitment to continue this support through a $50,000 endowment, which will assure the existence of these scholarships indefinitely.

In 1983, Don Parsons, Joe Spencer, Ed Dorroh, and Bob Culp, at the request of the National Board of the American Astronautical Society, prepared a charter which established the Annual Conference as the AAS Rocky Mountain Guidance and Control Conference. The Conference is sponsored by the National Organization of the AAS, meets annually in Colorado, and is run by the Rocky Mountain Section.

The AAS Guidance and Control Technical Committee, with its national representation, provides oversight to the local conference committee. W. Edwin Dorroh, Jr., was the first chairman of the AAS Guidance and Control Technical Committee. At the 1985 Conference, Robert L. Gates took over as the chairman of the National AAS Guidance and Control Committee. In 1995, James McQuerry assumed the chairmanship of this committee. This year the chairmanship passed to Larry Germann. This committee meets every year at the Conference, and also at either the summer Guidance and Control Conference, or the fall AAS Annual Meeting.

Thus, the AAS Rocky Mountain Guidance and Control Conference continues as the premier conference of its type anywhere. As a National Conference sponsored by the AAS, it promises to be the preferred idea exchange for guidance and control experts for years to come.

The Conference Committee extends congratulations to the 2000 Conference Chairman, Eileen Dukes, for an excellent conference and a job well done. Her report follows.

Robert D. Culp
University of Colorado
Boulder, Colorado

PREFACE

The 23rd Annual AAS Rocky Mountain Guidance and Control Conference was held in Breckenridge, Colorado at the Beaver Run Resort on February 2-6, 2000. The conference began with the Wine and Cheese Social and Registration on Wednesday evening. It provided an opportunity for conference attendees to get acquainted amid some welcoming remarks and a presentation of extracurricular activities available in the Breckenridge area.

The technical sessions began on Thursday morning with *Advances in Guidance and Control*. This session is a traditional one offering papers on innovative ideas in guidance and control and is designed to cover a broad range of topics. The session this year covered enhanced fault-tolerance, a two-step optimal estimator, an automated rendezvous system, all-stellar attitude determination in theory and in practice, Genesis attitude dynamics, and a fire control system. The *Advances* session was co-chaired this year by Dr. Michael Polites of NASA Marshall Space Flight Center and Lt. Colonel Carlee Bishop of the United States Air Force Academy.

The Thursday afternoon session, *Formation Flying and Constellations*, covered the guidance and control aspects of low thrust hardware and explored different options for control of spacecraft formations. Trajectory determination and generation options for formation control were also explored. This session was co-chaired by Garry Burdick of the NASA Jet Propulsion Laboratory and Dr. Russell Carpenter of the NASA Goddard Space Flight Center.

The third session, *Guidance and Control Issues for the International Space Station*, was held on Friday morning. This session provided an excellent survey of the issues related to the incremental build and the associated interim control problems of a large, diverse space structure. Discussions of the interdependency of the U.S. and Russian GN&C systems in the early construction phases and the assembly of the Propulsion module and Crew Return Vehicle were most informative. Both of the co-chairs for this session, Karen Frank of NASA Johnson Space Center and Thomas Russell of Boeing Space and Defense, were deeply involved in their subject.

Friday evening is always the time for the popular *Storyboards* session, co-chaired this year by Dr. Ian Gravseth of Ball Aerospace and Rick Jackson of Lockheed Martin. This session allows individuals to present their innovations and achievements in a posterboard format and have direct discussion with conference attendees. This format also provides an opportunity to see the latest advances in space-related guidance and control hardware. The many displays covered a wide range of topics including student rocket experiments to the latest sensor technologies and everything in between. Many thanks to the corporate sponsors who help make Storyboards possible every year.

The fifth session, *GNC Technology for Micro/Nano Spacecraft*, addressed the opposite end of the spectrum from the third session by covering new technologies in sensors, actuators, and system design for the new generation of smaller, cheaper satellites. The presenters provided a cross-section of the current work as they came from universities, government, and industry, including an international perspective. This session was chaired by Dr. Don Mackison of the University of Colorado.

Saturday evening was the conference banquet which provides an opportunity for conference attendees and their families to enjoy a nice dinner and an informative and entertaining presentation. This year's banquet speaker was Dr. Richard Zurek, the Project Scientist for the Mars '98 mission, on *Exploring the Climate of Mars: Mars Polar Lander in the Land of the Midnight Sun*. Dr. Zurek described the science mission for the Mars Polar Lander (MPL) and how important the south polar data is to understanding Mars. He also described the design of the instruments and the spacecraft. Unfortunately, he wasn't able to include results from the MPL mission as he had intended. He reminded us that space exploration is a risky business because we don't have all the answers about the places we explore. Dr. Zurek has not lost his enthusiasm for Mars exploration and hopes that another mission will someday capture the science lost when MPL was lost.

The last session of the conference on Sunday morning was *Recent Experiences*. This session is always popular as it provides a forum to share flight experiences, gained through successes and failures. Lessons learned through experience prove most valuable when shared with others in the G&C community. This year's session covered some of the headliner missions of the year including experiences with the Chandra X-ray Observatory, Mars Global Surveyor, QuickScat, UoSat-12, TOMS-EP, and Globalstar. The session was co-chaired by Marv Levenson of the Naval Research Laboratory and Angela Buckley of The Aerospace Corporation.

Conference attendance was close to two hundred attendees again this year indicating a continued interest and support of this conference by the G&C community. I was able to renew many old acquaintances and meet some new associates. The support of the research laboratories, the universities, the many corporations, and the G&C engineers and researchers from around the country and the world continues to make this conference a great success.

Sincere appreciation and congratulations are expressed to the conference planning committee, the local session chairs, the session co-chairs, authors, presenters, and Beaver Run Resort in Breckenridge. I want to express my personal thanks to everyone for making this year's conference a success and I look forward to seeing you at next year's conference, January 31 – February 4, at Beaver Run Resort.

Eileen Dukes
Conference Chairperson
2000 AAS Guidance & Control Conference

CONTENTS

	Page
FOREWORD	vii
PREFACE	ix

SECTION I
ADVANCES IN GUIDANCE AND CONTROL   1

Enhanced Fault-Tolerant Attitude Control for Lockheed Martin's A2100 Spacecraft (AAS 00-001)
    N. Goodzeit, M. Patel and H. Weigl 3

The Two-Step Optimal Estimator and Example Applications (AAS 00-002)
    N. Jeremy Kasdin 15

Test Results for the Automated Rendezvous and Capture System (AAS 00-003)
    Craig A. Cruzen, James J. Lomas and Richard W. Dabney 35

Lost-in-Space: A Star Pattern Recognition and Attitude Estimation Approach for the Case of No Prior Attitude Information (AAS 00-004)
    Gwanghyoek Ju, Hye-Young Kim, Thomas C. Pollock, John L. Junkins, Jer-Nan Juang and Daniele Mortari 57

Design, Implementation, and Flight Results for All-Stellar Attitude Determination (AAS 00-005)
    Jim D. Chapel, Stephen M. Micciche and Richard Kiessig 73

Attitude Dynamics of the Genesis Spacecraft (AAS 00-006)
    Carl Hubert 91

Fire Control Design for High Energy (Laser) Systems (AAS 00-007)
    Timothy J. Schneeberger, S. M. Seltzer and Robert Van Allen 105

SECTION II
FORMATION FLYING AND CONSTELLATIONS   117

A Projection Approach to Spacecraft Formation Attitude Control (AAS 00-011)
    Jonathan Lawton, Randal W. Beard and Fred Y. Hadaegh 119

Gravitational Perturbations, Nonlinearity and Circular Orbit Assumption Effects on Formation Flying Control Strategies (AAS 00-012)
    Kyle T. Alfriend, Hanspeter Schaub and Dong-Woo Gim 139

	Page

Nonlinear Dynamics, Trajectory Generation, and Adaptive Control of Multiple Spacecraft in Periodic Relative Orbits (AAS 00-013)
    Qiguo Yan, Guang Yang, Vikram Kapila and Marcio S. de Queiroz . . . 159

Validating a Formation Flying Control System Design: The GRACE Project Experience (AAS 00-014)
    H. D. Stevens, Jack Rodden, Phil Morton and Matthias Fehrenbach 175

A Tethered Formation Flying Concept for the SPECS Mission (AAS 00-015)
    David A. Quinn and David C. Folta 183

Project Orion: Carrier Phase Differential GPS Navigation for Formation Flying (AAS 00-016)
    Franz D. Busse, Gokhan Inalhan and Jonathan P. How 197

Mode and Logic-Based Switching for the Formation Flying Control of Multiple Spacecraft (AAS 00-017)
    Mehran Mesbahi and Fred Y. Hadaegh 213

SECTION III
GUIDANCE AND CONTROL ISSUES FOR THE INTERNATIONAL SPACE STATION   241

Evolution of International Space Station GN&C System Across ISS Assembly Stages (AAS 00-021)
    Roscoe Lee 243

International Space Station Assembly and Operation Control Challenges (AAS 00-022)
    Nazareth Bedrossian 259

Studies on the Attitude Control System Design for the Crew Return Vehicle (X38) (AAS 00-025)
    K. Abdel-Motagaly, O. Rombout, R. Gonzalez, K. Berrier, D. Hasan, D. Strack and B. Rishikof 281

Thermal Radiator Pointing for International Space Station (AAS 00-026)
    Scott A. Green 297

Command Level Maneuver Optimization for the International Space Station (AAS 00-027)
    Gregory E. Chamitoff, Adam L. Dershowitz and Amy L. Bryson 311

SECTION IV
GUIDANCE AND CONTROL STORYBOARD DISPLAYS   327

Adaptation of a Spaceborne Geolocation System to Airborne Experiments (AAS 00-031)
    Alan S. Hope, Henry M. Pickard and Jay W. Middour 329

Nanosol – A Next Generation Sun Sensor (AAS 00-032)
    John Glaberson 347

	Page
The TERMA Star Tracker for the NEMO Satellite (AAS 00-034) L. Maresi, T. Paulsen, R. Noteborn, O. Mikkelsen and R. Nielsen	355
A Cost Effective, High Reliable RWA Solution to a Commercial Market Demand (AAS 00-035) Terance Marshall, Mitchell Fletcher and Joseph Zuckerbrow	365
Redundant Launch Vehicle Guidance (AAS 00-036) R. Joe Wright	379
Radiation Hardened Power PC 603e™ Space Processor (AAS 00-037) Robert D. Campbell, Richard F. Elmhurst and Gary R. Brown	387
ARU Architectural Solutions for Long Life Satellite Missions (AAS 00-038) Edward C. Moulton, Robert H. Fall and Thomas G. Stottlar	403
Turbo-Charged Torqwheels™ (AAS 00-041) Bill Bialke	415
Progress of Fiber Optic Gyroscope Development for Space Applications (AAS 00-043) James Goodwin, Pei-hwa Lo, Mark Mariak and Ming Yu	425
SOHO: Loss and Recovery 1998, Gyroless 1999 (AAS 00-044) A. van Overbeek	433
A Novel New GEO Earth Sensor Provides High Accuracy (AAS 00-045) George Rullman, Richard Burton and Len Anderson	447
The Design and Operation of a COTS Space GPS Receiver (AAS 00-046) Martin J. Unwin and Michael K. Oldfield	465
Rocket Sounding Balloon Experimental Section 2000 (AAS 00-049) Kenneth Dalton, Gretchen England, Justin Eisenach, Chris Kilzer, Thanh Tran and Echezona Onwuatuegwu	475
In Flight Performance of the ZARM Magnetic Torquers MT80-1/MT140-2 Flown on the ABRIXAS Mission (AAS 00-050) Matthias Wiegand, Oliver Matthews, Peter Offterdinger and H. J. Rath	483

SECTION V
GNC TECHNOLOGY FOR MICRO/NANO SPACECRAFT 497

Formation Flying and Relative Navigation – A Nanosatellite Research Mission (AAS 00-061) Frank R. Chavez and David K. Schmidt	499
MEMS Rate Sensors for Space (AAS 00-062) Joel Gambino	515
Dynamics and Control of Nanosatellite ASUSat1 (AAS 00-063) Brian K. Underhill, Assi Friedman and Helen L. Reed	523

	Page
Attitude Control – An Afterthought: The MightySat I Experience (AAS 00-064)	
Jeff Benton and Thomas Itchkawich	541
MEMS-Based GN&C Sensors and Actuators for Micro/Nano Satellites (AAS 00-065)	
J. Connelly, N. Dennehy, P. Hattis, W. Johnson, D. Sargent and M. Socha	561
Digital Reaction Wheel Assembly RSI 01-5 for Small and Low-Cost Spacecraft (AAS 00-066)	
Armin Landes and Stephen Böttcher-Arff	577
An Attitude Control System for an ASAP5-Launched Interplanetary Spacecraft (AAS 00-067)	
Tobin Anthony and Bhavesh Patel	589

Section VI
RECENT EXPERIENCES IN GUIDANCE AND CONTROL — 601

Autonomous Orbit Control: Initial Flight Results From UoSAT-12 (AAS 00-071)
  James R. Wertz, Jeffrey L. Cloots, John T. Collins, Simon D. Dawson, Gwynne Gurevich, Brian K. Sato and L. Jane Hansen 603

Fuel Optimization During Mars Global Surveyor Aerobraking (AAS 00-072)
  Stuart R. Spath and Dave F. Eckart 615

QuikSCAT Attitude Control System Initialization and Early On-Orbit Operations (AAS 00-073)
  Dan Hegel and Scott Mitchell 629

Chandra X-Ray Observatory Pointing Control System Performance During Transfer Orbit and Initial On-Orbit Operations (AAS 00-074)
  Peter Quast, Frank Tung, John Wider and Mark West 647

The Recovery of TOMS-EP (AAS 00-076)
  Brent Robertson, Phil Sabelhaus, Todd Mendenhall and Lorraine Fesq . . . 665

Operational Experiences With Globalstar Constellation Control and Stationkeeping Using Solar Radiation Pressure (AAS 00-077)
  Lee Barker and Benjamin Lange 687

APPENDICES — 697
Publications of the American Astronautical Society 698
  Advances in the Astronautical Sciences 699
  Science and Technology Series 706
  AAS History Series 713

INDEX — 715
Numerical Index 717

Author Index 720

Section I
ADVANCES IN GUIDANCE AND CONTROL

SESSION I

Joint Session Chairperson: Dr. Michael Polites
NASA Marshall Space Flight Center

Joint Session Charperson: Lt. Col. Carlee Bishop
U.S. Air Force Academy

Local Session Chairperson: Eric Lander
Lockheed Martin Astronautics

The following paper numbers were not assigned:
AAS 00-008 to -010

ENHANCED FAULT-TOLERANT ATTITUDE CONTROL FOR LOCKHEED MARTIN'S A2100 SPACECRAFT

N. Goodzeit, M. Patel and H. Weigl[*]

To provide increased service reliability, autonomy, and ease of operations, the Lockheed Martin A2100 geosynchronous communications spacecraft features a redundancy management system that operates to automatically detect and correct attitude control component failures. Additionally, this system can reconfigure the attitude control system to operate without certain critical components while failure conditions are present. As an example of this capability, the system can simultaneously reconfigure the gyros, power supply, and processor of the internally redundant inertial measurement unit and transition to a gyroless backup control mode. This backup control mode provides high-performance attitude control that prevents disruption in payload operations while gyro data is unavailable. The gyroless system features an innovative control structure that combines elements of zero momentum, momentum-bias, and disturbance rejection control designs. When failure conditions are absent, the system automatically resumes the normal gyro-based control. All aspects of component switching, backup control, and the return to normal control are completely automatic. The paper describes the structure of this gyroless control system and gives actual A2100 spacecraft flight performance results. In addition, extensions to the system are described that allow long-term gyroless control operation while meeting mission life and performance requirements. These extensions include gyroless earth re-acquisition and gyroless stationkeeping attitude control.

---

[*] Lockheed Martin Commercial Space Systems.

INTRODUCTION

To provide increased service reliability, autonomy, and ease of operations, the Lockheed Martin A2100 geosynchronous communications spacecraft features a redundancy management system (REDMAN) that operates to automatically detect and correct hardware failures. This system, shown in Figure 1, includes failure detection tests and hardware switching macros specific to each attitude control component.

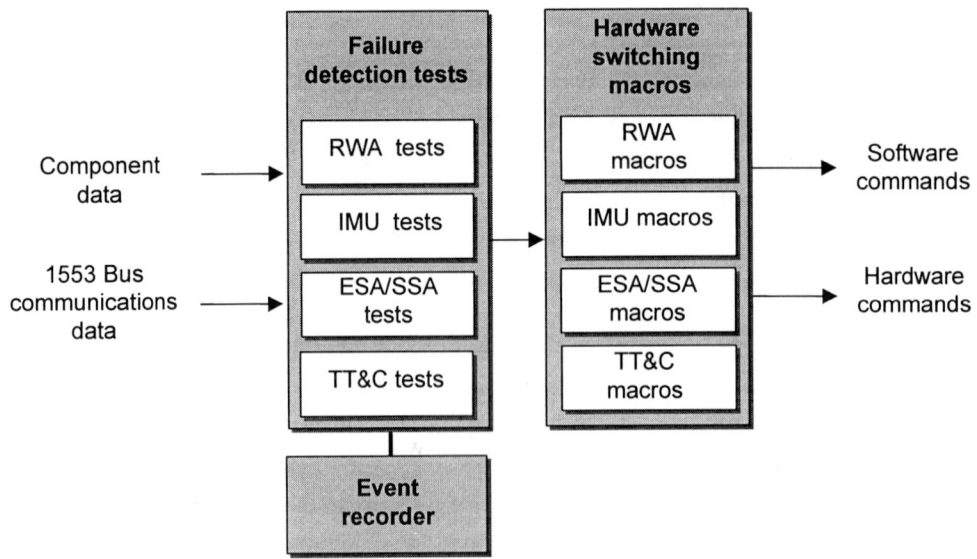

Figure 1. A2100 Spacecraft Redundancy Management System

The REDMAN tests monitor performance data for each component to detect failures. The hardware switching macros are command lists stored in the spacecraft on-board processor that switch to redundant components. When a failure is detected, the macros rapidly re-establish a working hardware configuration to minimize attitude transients and prevent any disruption in communications service. The system is designed for flexibility and ease of operations. The failure detection tests can be individually enabled and disabled by ground command. The failure correction macros can be modified as required to tailor the system's failure response based on customer operational preferences and the available redundant hardware. A full complement of telemetry provides complete information on all failure detection tests and component switching actions. An event recorder logs REDMAN state data to capture failure and recovery events if telemetry is not available.

To add robustness and further insulate the spacecraft from the effects of component failures, alternate control modes that do not rely on data from the failed component may be temporarily engaged. This approach is taken for the Inertial Measurement Unit (IMU), which provides continuous three-axis rate data for three-axis attitude determination and control. The A2100 spacecraft IMU is a high-reliability, internally redundant unit featuring four hemispherical resonator gyros, two power supplies and two processors. If REDMAN detects an IMU failure, one of five IMU failure correction macros executes to switch the IMU internal configuration. What automatic switching action is taken depends on which failure detection tests have triggered and on evaluation of the extensive built-in-test information supplied by the IMU. In addition to the failure correction response, the system automatically switches to a gryoless backup attitude control mode that provides earth-pointing accuracy sufficient to allow continued payload operations while valid gyro data is unavailable. When failure conditions are absent, the gyroless control system is automatically disabled and normal gyro-based control resumes. All aspects of IMU configuration switching, the switch to gyroless control, and the return to gyro-based control are completely automatic.

The gyroless control system uses data from the earth sensor (ESA) and sun sensor (SSA) to provide high-accuracy three-axis attitude control using reaction wheels (RWAs). This system uses the nominal gyro-based zero-momentum control structure with several modifications. To reduce RWA torquing in response to ESA measurement noise, a novel filtering scheme combines RWA speed information with the angular rates derived from ESA data. When SSA data is unavailable for direct yaw control, yaw is indirectly controlled through the roll/yaw gyroscopic coupling provided by pitch momentum stored in the RWAs. The momentum bias yaw controller includes gyroscopic torque cross feeds that circulate the yaw/roll RWA momentum at orbit rate and compensation that adds damping to the resulting yaw/roll coupling mode. Yaw performance is enhanced by feeding forward yaw and roll disturbance torque estimates. Unlike standard disturbance estimation approaches, the gyroless system disturbance estimator does not rely on knowledge of the spacecraft mass properties or the spacecraft pitch momentum. To facilitate the switch to gyroless yaw control, a small positive pitch momentum from 80 to 200 in-lb-sec is maintained in the RWAs at all times. This pitch momentum is easily accommodated with no less than one-week between RWA momentum adjust maneuvers.

Although the gyroless system is designed to accommodate single IMU failures, the system can be expanded to provide long-term gyroless system operation. The gyroless stationkeeping system allows north/south maneuvers to be executed at any orbital position by continuous firing of high-efficiency arcjets and east/west maneuvers to be performed by pulsing reaction engine assemblies (REAs). The yaw reference for these maneuvers is provided either by the SSA or is derived on-board the spacecraft from payload transponder channel telemetry data. In the event of loss of earth lock, the gyroless safe mode first acquires the Sun using only information from the solar arrays and SSA. Once the Sun is acquired, the Earth is located in the ESA field-of-view by rotating about the sunline. The approach features innovative IMU replacement software (Cyber-MU™) that derives three-axis rate information from solar array telemetry, SSA data, and control torque demands.

The sections below present the details of the gyroless backup control system. Gyroless system performance is illustrated with in-orbit test data and simulation results for a high-power A2100 spacecraft. Finally, the extensions needed for total gyroless control are presented.

GYROLESS SYSTEM OVERVIEW

The standard A2100 operational orbit control system uses an ESA, SSA, and IMU for three-axis attitude determination and control. Three-axis body rates are measured directly by the IMU and used to propagate the spacecraft inertial attitude. The inertial attitude is continuously updated using the roll and pitch angles measured by the ESA. The SSA data is used to update the yaw attitude when the sun is present in the SSA field-of-view (FOV). The spacecraft attitude is controlled using a proportional-integral-derivative (PID) controller that computes control torque demands based on the sensed attitude and rate errors. Control torques are applied to the spacecraft using thrusters or reaction wheels (RWAs).

When an IMU failure is detected by REDMAN, gyroless control is activated to maintain spacecraft pointing and prevent disruption of payload service while gyro data is unavailable. Although the REDMAN system is designed to quickly correct an IMU failure, the gyroless control system can maintain spacecraft pointing for as long as necessary to provide time for IMU troubleshooting and manual reconfiguration if necessary. The gyroless control system is fully integrated into the standard A2100 flight software and reuses most of the normal control logic. If an IMU fails in operational orbit, orbit adjust and momentum adjust maneuvers are automatically terminated. The REDMAN logic reconfigures the IMU while the spacecraft switches to gyroless control using the RWAs. Once the REDMAN logic declares that the IMU failure is corrected, gyroless control is autonomously disabled and the spacecraft switches back to normal gyro-based control.

The block diagram of Figure 2 illustrates how gyroless three-axis control is provided when gyro data is unavailable. High accuracy roll and pitch pointing is achieved using high-bandwidth control directly referenced to the ESA roll and pitch angles. Because the IMU body rate measurements are unavailable, the roll and pitch rates are estimated by differencing the measured ESA angles. Roll and pitch proportional-derivative (PD) control torque demands are computed based on the ESA angles and derived rates. The impact of differencing noisy ESA angles on the control torque demand is minimized using an innovative filtering scheme. The ESA derived rates are combined with RWA rate information to provide noise attenuation and minimize phase lag into the control loop. As indicated in Figure 2, the roll and pitch control torques are applied to the spacecraft by the RWAs.

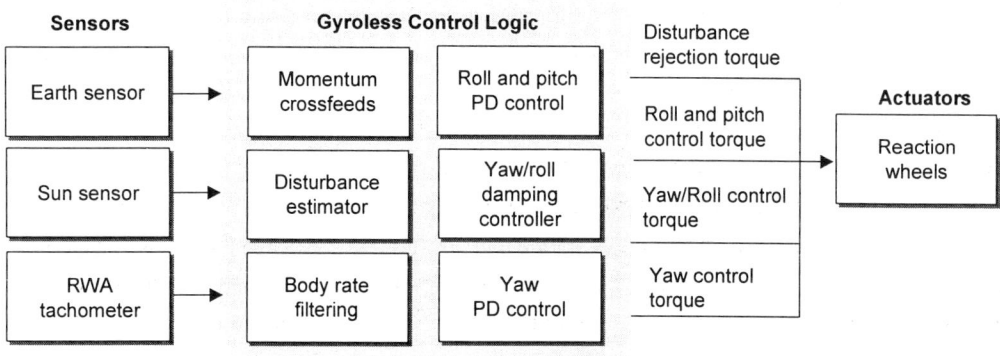

Figure 2. A2100 Gyroless Control System Block Diagram

Unlike the roll and pitch attitude, which is continuously measured by the ESA, yaw attitude measurements are only available when the sun is within the SSA FOV. During this time, the gyroless attitude control system provides high-bandwidth PD yaw control using the SSA measurements. The yaw body rate is estimated by differencing the SSA yaw measurements. The gyroless attitude control system continues to maintain yaw pointing performance when SSA measurements are unavailable. In this case, the yaw attitude is controlled using the coupling between the yaw and roll axes introduced by a positive pitch momentum bias as described below.

Because the gyroless control system torques are applied internally using RWAs, the RWA momentum vector magnitude and inertial direction are uncontrolled. The gyroless control system accounts for this by applying crossfeeds to rotate the RWA momentum in the yaw/roll plane at orbit rate. An RWA torque is applied to the yaw axis that is equal to the product of orbit rate and the roll RWA momentum. This torque cancels the gyroscopic torque on the yaw axis created by the orbit rate rotation along the pitch axis and the roll axis momentum stored in the RWAs. Similarly, a torque is applied to the RWAs along the roll axis that is equal to the negative product of orbit rate and the measured yaw RWA momentum. These momentum crossfeeds force the yaw/roll momentum to interchange between the yaw and roll axes at orbit rate.

The component of RWA momentum along the pitch axis is used to control yaw when no yaw measurements are available (recall that high-bandwidth roll and pitch control is applied continuously based on the ESA measurements). Because of the pitch momentum bias, a torque applied to the yaw axis couples into roll motion that is sensed by the ESA. This coupling is the basis for the yaw controller described below.

The control design and analysis is based on the linearized attitude dynamics with the roll and pitch PD control loops closed, momentum crossfeeds applied to the yaw and roll axes, and a positive pitch momentum bias. Table 1 lists the modal frequencies and damping ratios for a typical A2100 spacecraft with 150 in-lb-sec of pitch momentum and without the momentum bias yaw control loop closed. The mode at the origin corresponds to the pitch momentum that is unconstrained. The oscillatory mode at orbit rate corresponds to the yaw/roll momentum, the magnitude of which is unconstrained and that interchanges between the yaw and roll axes at orbit rate. The highly damped modes at 0.03 and 0.08 rad/sec are introduced by the roll and pitch PD control. The neutrally stable mode at 0.0027 rad/sec (with a period of 6.5 hours) is the yaw/roll coupling mode that results from the pitch momentum bias, the roll PD control, and the momentum crossfeeds.

Table 1. System Modes without Yaw/Roll Control

Mode	Natural Frequency (rad/sec)	Damping Ratio
Pitch momentum	0	1
Yaw/roll momentum	0.00007	0
Yaw/roll coupling	0.00027	0
Roll PD control	0.03	0.7
Pitch PD control	0.08	0.8

To maintain yaw pointing, the gyroless attitude control system applies a yaw control torque based on the measured ESA roll angle to damp the yaw/roll coupling mode. The closed loop system modes with the additional yaw/roll damping controller are listed in Table 2. The roll and pitch PD control modes as well as the momentum modes are unchanged from Table 1. However, there is an additional mode associated with the damping controller. In addition, the frequency of the yaw/roll coupling mode is shifted. Both of these modes have a damping ratio close to 0.4. The damping controller is designed to ensure that the damping ratio of these two modes will be greater than 0.2 over a wide range of pitch momentum.

Table 2. Closed-Loop System Modes with Yaw/Roll Control

Mode	Natural Frequency (rad/sec)	Damping Ratio
Pitch momentum	0	1
Yaw/roll momentum	0.00007	0
Yaw/roll coupling	0.00016	0.4
Yaw/roll controller	0.00046	0.4
Roll PD control	0.03	0.7
Pitch PD control	0.08	0.8

When SSA measurements are available, high-bandwidth yaw PD control is applied that provides good yaw disturbance rejection. However, yaw axis disturbance rejection is poor outside of the SSA control region, since the yaw control is then dependent on the low frequency coupling between the yaw and roll axes introduced by the pitch bias momentum. Outside of the SSA window, accurate yaw pointing cannot be provided for rapid momentum adjust or stationkeeping. System enhancements that provide this capability are described below.

One major source of yaw disturbances that affects performance without SSA yaw control is the inertial environmental disturbance torque resulting from solar pressure on the solar arrays. To improve long-term gyroless yaw control performance, a disturbance torque estimator is incorporated into the gyroless control architecture. The estimator computes the diurnal yaw and roll disturbance torques from the measured RWA momentum. Including the estimator in the control system does not significantly affect the gyroless control modes given in Table 2. The formulation of the disturbance estimator is independent of the spacecraft mass properties and the pitch momentum bias. The estimated diurnal torques are applied to the yaw and roll axes to cancel the actual disturbances. Higher harmonics of the environmental disturbances are not estimated, but have a relatively small effect on yaw pointing.

As described above, the A2100 gyroless control system relies on momentum crossfeeds to cancel the gyroscopic torques due to the transverse momentum stored in the RWAs. However, the momentum interchange control is not exact due to uncertainties in the actual RWA spin axis alignments and moments of inertia. These errors act as additional diurnal and constant disturbances that further degrade yaw pointing performance. Worst-case estimates of these disturbances and their effect on yaw pointing performance have been computed by Monte Carlo analysis with the individual error sources varied according to their expected probability distributions.

IN-ORBIT PERFORMANCE RESULTS

Figure 3 shows the spacecraft attitude errors during an on-orbit test of the A2100 spacecraft gyroless control system. The entire test is conducted while the Sun is present in the SSA FOV so that the yaw pointing error can be directly measured. As indicated in the figure, the gyroless control system is enabled at the beginning of the test. The ground-commanded transition to gyroless control illustrates the autonomous action that would be taken if the REDMAN logic were to detect an IMU failure. During the test, the roll and pitch attitude errors are less than 0.01° at all times. The small additional variation in the sensed roll and pitch errors while under gyroless control is due to ESA noise. The rate filtering included in the gyroless system design reduces the RWA torque noise (not shown) to less than 0.05 in-lb peak-to-peak.

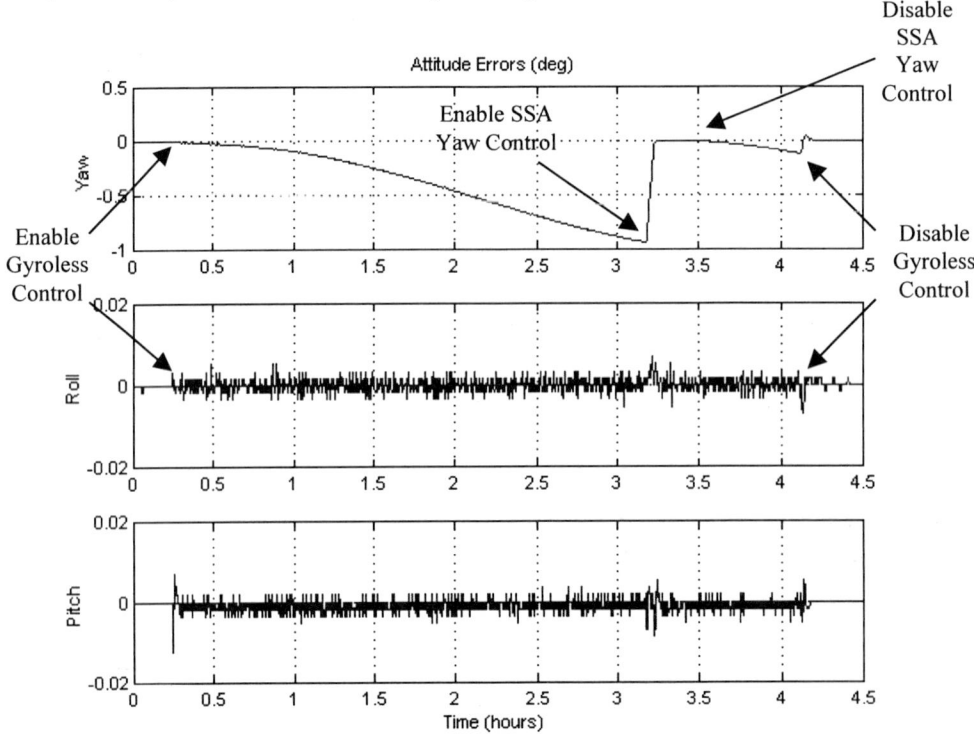

Figure 3. On-Orbit Gyroless Control System Performance

During the first three hours of gyroless control, the SSA yaw control is disabled in order to measure the pointing performance without the PD yaw control. The peak yaw transient during this time is less than 1°. This yaw error is a result of disturbance torques due to RWA misalignments and inertia uncertainties as well as inexact cancellation of the environmental disturbance torque (the disturbance estimator was not fully converged at the start of the test). As shown in Figure 3, the yaw attitude error is reduced to less than

0.01° when SSA yaw control is enabled later in the test. When the SSA control is again disabled, the yaw error increases to roughly 0.15° during the final half an hour of the test. At the end of the test, the gyroless system is disabled by ground command and the system executes a transient-free return to gyro-based operation.

GYROLESS SYSTEM EXTENSIONS

The system described above can be enhanced with features necessary to provide long-term gyroless attitude control. A block diagram of the enhanced gyroless system is shown in Figure 4. This system can be used to perform high-efficiency stationkeeping maneuvers at any orbital position. In addition, the system can automatically acquire a power-safe attitude and re-acquire the Earth if earth-lock is lost.

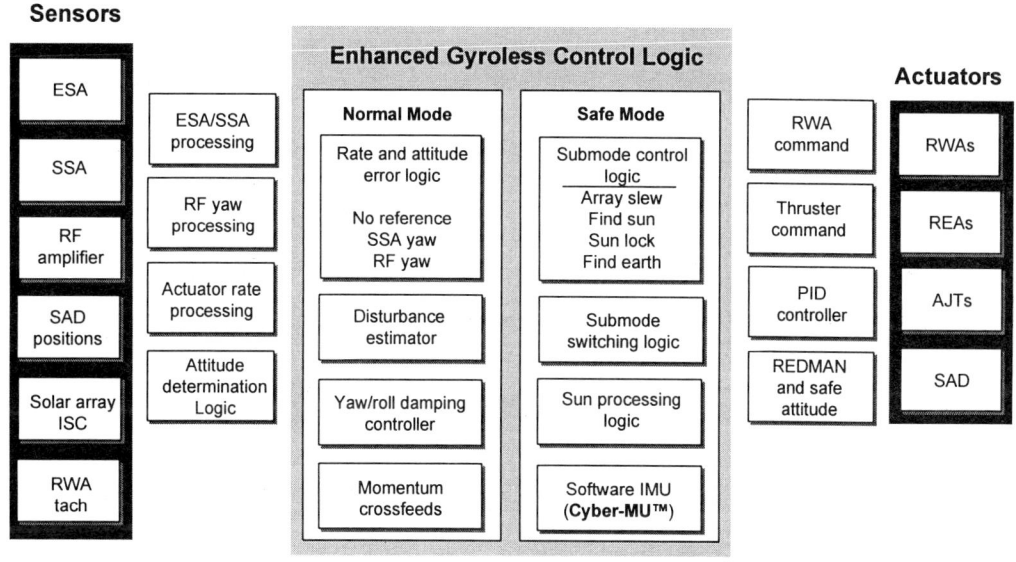

Figure 4. Enhanced Gyroless Control System

The yaw reference for stationkeeping attitude control is provided either by the SSA or is derived on-board the spacecraft from payload transponder channel telemetry data. A direct yaw reference is necessary for high accuracy attitude control while arcjets are fired continuously for north/south stationkeeping or REAs are pulsed repetitively for east/west stationkeeping. The payload yaw reference is generated using a carrier signal from the ground that is input to a vertically polarized transponder channel. The polarization of the carrier signal is selected to maximize the sensitivity of the RF amplifier drive current to changes in spacecraft yaw angle. In response to changes in yaw angle the vertical component of the carrier signal at the receive antenna increases or decreases, causing proportional changes in the amplifier drive current. Prior to the start of a maneuver the mean value of the amplifier drive current is determined by the on-board software and used as the yaw control reference value during the maneuver. Because the RF-derived yaw error provides only a short-term yaw reference while stationkeeping is performed, the long-term stability of the signal due to atmospheric, temperature, and other effects is unimportant. Figure 5 shows simulated attitude responses for a north/south stationkeeping maneuver using the RF yaw reference, and Figure 6 shows the simulated control thruster pulse width demands.

A filtering scheme using thruster and RWA information attenuates the noise in the yaw rate signal that is derived from the noisy RF yaw data (standard deviation of 0.28 degrees). This allows a control system bandwidth sufficient for good disturbance rejection, yet prevents thruster pulsing in response to noise that wastes propellant.

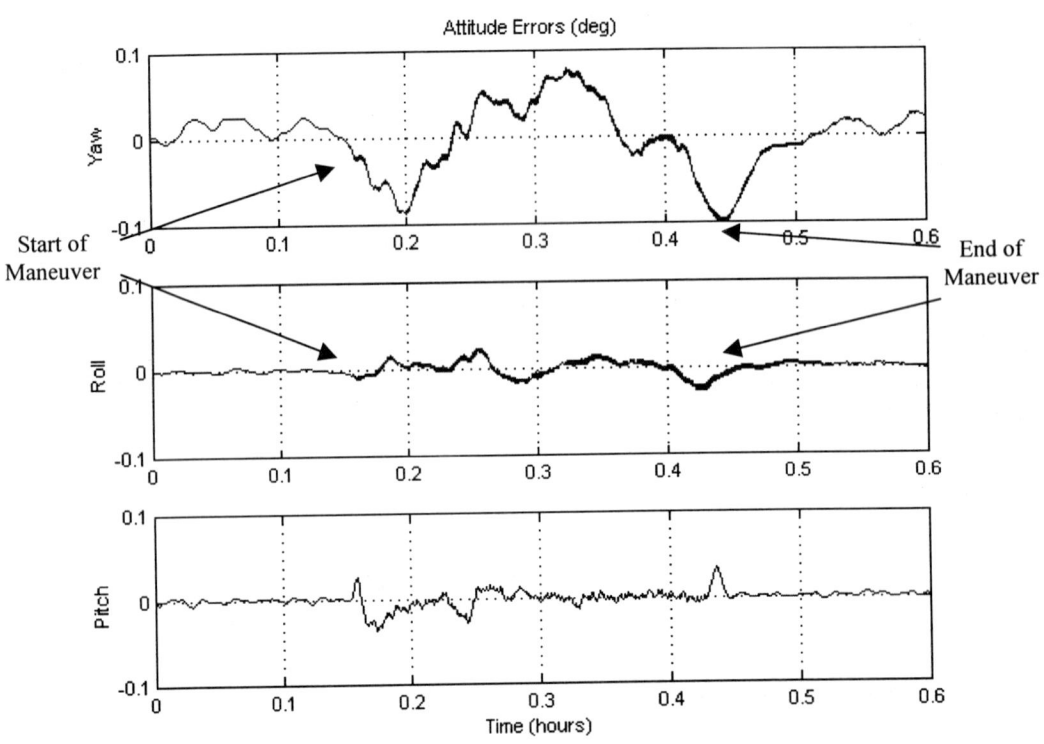

Figure 5. Gyroless North/South Stationkeeping Simulated Attitude Errors

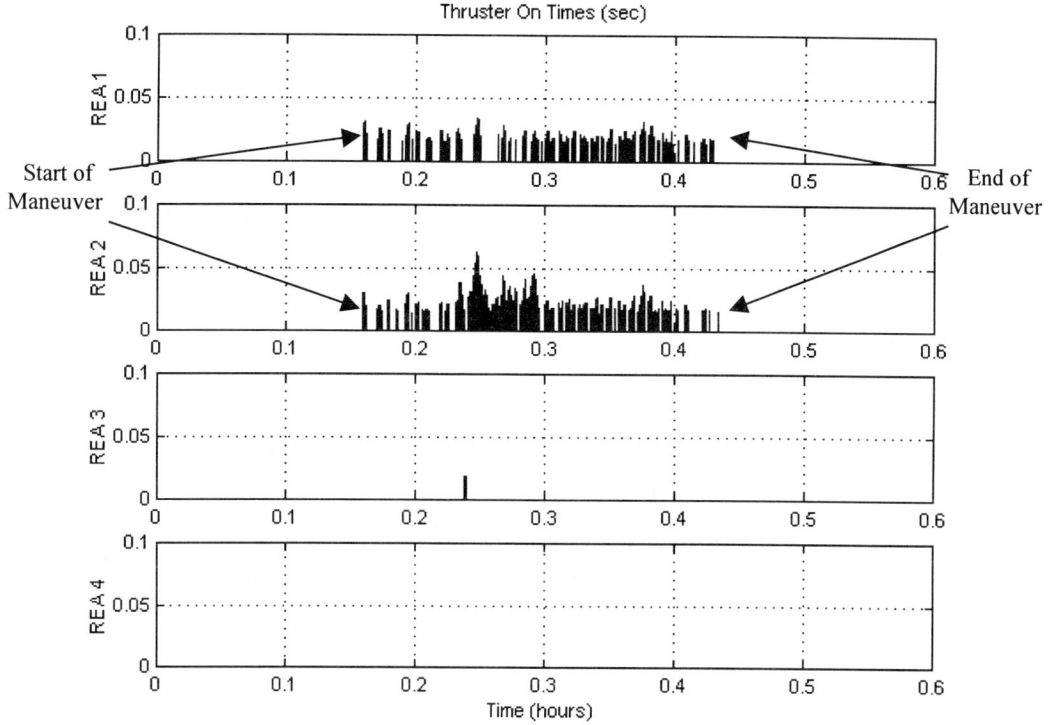

Figure 6. Gyroless Stationkeeping Simulated Thruster On-Times

In the unlikely event that earth-lock is lost, the gyroless safe mode uses attitude information from the solar arrays and SSA to place the spacecraft in a power-safe attitude. Once the Sun is acquired, earth-lock is automatically re-established by rotating about the sunline until the Earth enters the ESA FOV. To acquire the Sun, the system first separates the north and south arrays to form a wide FOV sun sensor that brackets the SSA FOV. Based on the shorted solar cell current (ISC) data from the north and south arrays, thruster control is applied to rotate the spacecraft so that the Sun enters the SSA FOV. Once the Sun is aligned with a target vector in the body frame, the rate about the sunline is estimated and nulled. Gyroless safe mode uses Cyber-MU™ software to estimate the spacecraft three-axis rates from ISC, SSA, and actuator data. The ambiguity in the ISC-derived sun vector is resolved by checking for consistency between the derived rates and the rate changes expected from thruster firing. The sunline rates are estimated based on the RWA momentum perpendicular to the sunline and the gyroscopic component of the control torque demand. Earth acquisition is timed to occur when the Sun is nearly coincident with the SSA boresight vector. When the Earth is located, the system transitions to three-axis gyroless control using ESA and SSA data. The RWA pitch momentum is then adjusted to the value for normal gyroless operation.

CONCLUSION

As customers demand higher levels of spacecraft autonomy and reliability, standard approaches to redundancy management will need to be enhanced with the addition of alternate control modes. Such control modes are needed to provide high-accuracy earth-pointing control in the event that automatic component switching actions are not immediately successful at restoring a working hardware configuration. For the A2100 spacecraft, a gyroless backup control system is provided to maintain attitude pointing during periods when valid gyro data is unavailable from the IMU. The gyroless control mode is automatically activated if an IMU failure is detected and remains in operation until normal IMU function is restored. The A2100 spacecraft gyroless system architecture promotes expansion to incorporate features that allow long-term gyroless operation. The ability to provide higher levels of robustness and redundancy with a single-fault tolerant hardware configuration is expected to become increasingly important in the future as the cost and revenue-generating value of communications spacecraft continue to increase.

THE TWO-STEP OPTIMAL ESTIMATOR AND EXAMPLE APPLICATIONS

N. Jeremy Kasdin[*]

The two-step filter is a new approach for nonlinear recursive estimation that substantially improves the estimate error relative to the extended Kalman Filter (EKF) or the iterated extended Kalman filter (IEKF). Historically, when faced with an optimal estimation problem involving a set of nonlinear measurements, designers have been forced to choose between optimal, but off-line, iterative batch techniques or sub-optimal, approximate techniques, typically the EKF or IEKF. These techniques linearize the measurements and dynamics to take advantage of the well known Kalman filter equations. While broadly used, these filters typically result in sub-optimal and biased estimates and often can go unstable. The two-step estimator, introduced in 1996, provides a dramatic improvement over these filters for situations with nonlinear measurements. It accomplishes this by dividing the estimation problem (a quadratic minimization) into two-steps – a linear first step and a non-linear second step. The result is a filter that comes much closer to minimizing the desired cost, virtually eliminating any biases and dramatically reducing the mean-square error relative to the EKF. This paper presents an overview of the two-step estimator, outlining the derivation of the two-step measurement update and cost function minimization. It also presents the newest time update, resulting in a robust and accurate estimation technique. This presentation is followed by several simple aerospace examples to illustrate the utility of the filter and its improvement over the EKF and IEKF. These include both open loop estimation and closed loop control applications.

INTRODUCTION

The problem of optimally estimating the state of a system given a series of measurements is an old one. A least squares solution for the static estimation of constant state variables from a sequence of noisy measurements has long been available [see, e.g., Refs. 1,2,3]. The Wiener filter[4] generalizes this to random processes by finding the optimal frequency weighted filter for measurements of time varying signals. Kalman, in 1960, developed the famous time domain solution bearing his name for the optimal estimation of stochastic differential systems driven by noise from noisy measurements of a subset of the states.

While the Kalman filter proved to be revolutionary and enabling for many technologies, including the Apollo moon landings, and while it remains ubiquitous today, its utility is still limited to linear

---

Professor, Mechanical and Aerospace Engineering Department, Princeton University, Princeton, New Jersey 08544.

problems. The solution is only optimal for linear dynamic systems and linear measurements. Optimal solutions to general nonlinear problems have proved elusive. Most nonlinear estimation algorithms force a choice between an optimal, but batch, solution and a suboptimal recursive formulation. If the estimation can be done offline, various nonlinear search and iterative batch algorithms can be used to minimize the desired cost function. However, many applications require the real-time state estimates provided by a recursive formulation of the estimation problem.

For the most part, designers needing a recursive algorithm have been limited to the extended Kalman filter (EKF), a somewhat *ad hoc* approach to linearizing the Kalman filter. While the EKF has proven effective for many systems, it is known to be a biased and inconsistent estimator and, in some cases (particularly where there is poor initial knowledge), it can be wildly divergent.

Alternatives to the EKF exist. In particular, the iterated extended Kalman filter (IEKF) is often used where more accuracy is required. Similar to the EFK, the IEKF improves upon the accuracy by iterating the measurement update (more on this later). Unfortunately, the IEKF is still suboptimal and also frequently results in biased estimates. A number of other nonlinear filtering approaches have been discussed in recent years (such as predictive filtering, the pseudo-linear Kalman filter (PLKF), the SDRE, etc.) and research continues on techniques and applications. Each of these approaches has its own drawbacks. All are suboptimal and approximate in some sense and some only apply to certain types of problems. Still, by far the most common filter for nonlinear applications is the EKF.

This paper reviews the development of the two-step filter, first introduced in 1996 by Haupt, Kasdin, et. al.[6] The two-step estimator is an attempt to find a closer to optimal estimation algorithm for the most general nonlinear estimation problem and thus provide an alternative to the EKF or IEKF. It accomplishes this by dividing the estimation problem into steps – a linear first step and a nonlinear second step. The linear first-step problem can be solved optimally and recursively using a standard Kalman filter. The second step states are then found by treating the first step state estimates as measurements and using an iterative Gauss-Newton like algorithm to perform a nonlinear least-squares fit. The result is a filter that comes much closer to minimizing the desired cost and dramatically reduces the biases and mean-square error of the EKF.

The following section will review the derivation and past work on the two-step filter. Following this is a discussion of initialization approaches and the presentation of a new, more robust and accurate method for setting the initial conditions of the filter. This is followed by a number of simplified aerospace examples to demonstrate the utility of the filter, its performance improvements over the EKF, and some of its limitations. The paper is concluded with some comments on current research being applied to further improve the filter.

THE TWO-STEP ESTIMATOR

Nonlinear Least Squares Estimation

The dynamic estimation problem can be formulated as the optimal solution that minimizes the following quadratic cost function (least-squares estimation):

$$J = \frac{1}{2}(x_1 - \bar{x}_1)^T M_{x_1}^{-1}(x_1 - \bar{x}_1) + \frac{1}{2}\sum_{k=1}^{N-1} w_k^T Q_k^{-1} w_k$$
$$+ \frac{1}{2}\sum_{k=1}^{N}(z_k - F(x_k,t_k))^T R_k^{-1}(z_k - F(x_k,t_k))$$

(1)

subject to the dynamic constraint on the desired $n \times 1$ state vector

$$\dot{x} = A(x,t) + B(x,t)w(t)$$ (2)

with the $\ell \times 1$ measurement vector z_k given by

$$z_k = F(x_k, t_k) + v_k \tag{3}$$

As usual, $w(t)$ represents a vector, Gaussian white noise stochastic process with covariance matrix Q and v represents a vector of Gaussian i.i.d. measurement noise with covariance matrix R. The subscript k denotes the value at time step t_k. The vector x_1 represents the *a priori* state estimate with error covariance M_{x_1}. Note that for this case of Gaussian noise the least squares solution to Eq. (1) is also the maximum likelihood solution and approaches the minimum variance solution for large numbers of measurements.

It is also worth noting that there is a large body of work on the meaning of white noise and the stochastic differential equation (s.d.e.) in Eq. (2). As white noise is not a *bona fide* process (it does not exist in practice), special definitions of Eq. (2) are necessary. The reader is referred to Refs. 7, 8, & 9 for extensive discussions of the various definitions and properties of s.d.e.'s. The reader should also note that for linear systems these distinctions don't apply – this only becomes an issue when we consider nonlinear problems. For the purposes of this work, Eq. (2) is considered to be a stochastic differential equation in the sense of Ito. Current research by the author is directed at the properties of Eq. (2), at simulation methods, and at the implications to nonlinear filtering.

For linear measurement problems – that is, problems where the measurement function F_k is linear in the states ($z_k = H_k x_k + v_k$) and the dynamics functions A and B in Eq. (2) are independent of the state– the well known recursive Kalman filter can be derived as the global minimum solution using, for example, the sweep method[3]. Unfortunately, there is no such general solution to the stated *nonlinear* problem.

The typical alternative approach is to exploit the fact that the linear Kalman filter can be derived in a number of different but equivalent ways and to derive the nonlinear filter in a similar fashion. For instance, one common approach for the linear problem is to minimize a static quadratic cost function and then manipulate the resulting solution to derive a recursive form. In other words, the linear Kalman filter can be derived by first minimizing the following linear cost function for a "batch" solution of the static state:

$$J = \frac{1}{2} \sum_{k=1}^{N} (z_k - H_k x)^T R^{-1} (z_k - H_k x) \tag{4}$$

The resulting solution provides an optimal estimate of the states x in terms of H and z_k, represented as usual by $\hat{x}$. This batch solution can then be manipulated to find a recursive equation for the estimate $\hat{x}$ (that is, an equation for an update to the estimate, given an additional measurement, that does not require reformulating the batch solution). The resulting formula is the same as the Kalman filter measurement update. The time update is then added using the equations for stochastic process propagation[2]. The time update is independent of the measurement and thereby provides a new *a priori* estimate for the recursive measurement update.

This observation, i.e., that the static problem with time update is equivalent to the batch dynamic problem, is exploited in the derivation of the two-step estimator.[6,10] The nonlinear measurement problem is recast as a static batch problem and solved for the best recursive estimate of the states. This static nonlinear cost function is given by:

$$J = \frac{1}{2} \sum_{k=1}^{N} (z_k - F(x, t_k))^T R^{-1} (z_k - F(x, t_k)) \tag{5}$$

The optimal solution is a stationary point of the cost function in Eq. (5), i.e., a point that solves:

$$\frac{\partial J}{\partial x} = -[\xi - \mathcal{F}(x, t)]^T \mathcal{R}^{-1} \frac{\partial \mathcal{F}(x, t)}{\partial x} = 0 \tag{6}$$

where we have streamlined the notation by replacing the summation with large matrices. Note that a batch approach is one where a numerical search is performed to find the solution to Eq. (6).

The two-step filter formulation rewrites the cost function in Eq. (5) as a linear measurement in an $m \times 1$ intermediate (first-step) state vector y

$$J_y = \frac{1}{2}\sum_{k=1}^{N}(z_k - Hy_k)^T R^{-1}(z_k - Hy_k) \qquad (7)$$

where $y_k = f_k(x_k)$.

The resulting linear cost function can be solved exactly for $\hat{y}_k$ and also reformulated as a recursive update. The desired state estimate $\hat{x}_k$ is found in a second-step minimization with cost function

$$J_x = \frac{1}{2}(\hat{y}_k - f_k(x_k))^T P_y^{-1}(\hat{y}_k - f_k(x_k)) \qquad (8)$$

where P_y is the first-step state estimate error covariance. This optimization is performed using any convenient least squares search algorithm, though most commonly the Gauss-Newton or Levenberg-Marquardt algorithms are used[6,10,11]. It is straightforward to show that this two-step approach *exactly* minimizes the desired static cost function in Eq. (5).[6] That is, the resulting estimate, $\hat{x}$, is the optimal solution to the original static problem.

This reformulation as a two-step problem now greatly simplifies the derivation of a recursive form. The linear first step estimation can be converted to a recursive filter using the standard approach (see, e.g., Refs. 1 & 2). In other words, given a previous estimate, $\bar{y}_k$, and covariance M_{y_k}, perhaps from a batch least squares fit on previous measurements, a new estimate, $\hat{y}_k$, and covariance P_{y_k}, can be found from the following equations:

$$\hat{y}_k = \bar{y}_k + P_{y_k} H_k^T R_k^{-1}(z_k - H_k\bar{y}_k)$$
$$P_{y_k} = \left(M_{y_k}^{-1} + H_k^T R_k^{-1} H_k\right)^{-1} \qquad (9)$$

or, equivalently,

$$K_k = M_k H_k^T \left[H_k M_k H_k^T + R_k\right]^{-1}$$
$$\hat{y}_k = \bar{y}_k + K_k(z_k - H_k\bar{y}_k) \qquad (10)$$
$$P_{y_k} = (I - K_k H_k)M_k$$

These equations are the Kalman filter measurement update. The second step state estimates, $\hat{x}_k$, are again found by numerically minimizing the cost function in Eq. (8) at each time step. Once again, the estimate from these equations is *optimal* – it minimizes the exact cost function in Eq. (5).

Before completing the filter presentation by introducing the time update, it is enlightening to compare the two-step estimator derivation at this point to the extended and iterated extended Kalman filter. Such a comparison vividly shows the improvement obtained by use of the two-step formalism.

Comparison to EKF & IEKF

The above approach of starting with a static, batch problem and then finding a recursive form does not work for the EKF and IEKF. There is no convenient way to find a recursive form based on linearizing the cost in Eq. (5). A direct comparison can be made, however, by examining another approach to deriving the Kalman filter. In this approach, a recursive form is found directly by minimizing a recursive cost:

$$J_k = \frac{1}{2}(x_k - \bar{x}_k)^T M_{x_k}^{-1}(x_k - \bar{x}_k) + \frac{1}{2}(z_k - H_k x_k)^T R_k^{-1}(z_k - H_k x_k) \qquad (11)$$

Here, $\bar{x}_k$ and M_{x_k} are the *a priori* estimate and covariance prior to the update. In the static case discussed so far, these are estimates from a previous recursion or the result of a batch fit on all previous measurements.

For linear measurements, it is possible to show that the minimizing recursive solution is the same as the batch solution of the original cost function in Eq. (4). In fact, it is straightforward to show that the cost in Eq. (4) evaluated at the minimizing solution for the first k-1 measurements is the same as the first term in the recursive cost in Eq. (11), proving the optimality of the solution.

The problem is more complex for nonlinear measurements. The EKF approach is to use a recursive cost as in Eq. (11) but for the nonlinear measurements:

$$J_k = \frac{1}{2}(x_k - \bar{x}_k)^T M_{x_k}^{-1}(x_k - \bar{x}_k) + \frac{1}{2}(z_k - F(x_k, t_k))^T R_k^{-1}(z_k - F(x_k, t_k)) \quad (12)$$

and to linearize the function $F(x_k, t_k)$ about the previous estimate, $\bar{x}_k$. This cost function can then be easily minimized by treating it as a linear problem to find the updated estimate $\hat{x}_k$:

$$\hat{x}_k = \bar{x}_k + P_{x_k} H_k^T R_k^{-1}(z_k - F(\bar{x}_k, t_k))$$

$$P_{x_k} = \left(M_{x_k}^{-1} + H_k^T R_k^{-1} H_k\right)^{-1} \quad (13)$$

$$H_k = \left.\frac{\partial F(x,t)}{\partial x}\right|_{\bar{x}_k}$$

The IEKF improves upon this estimate by repeatedly linearizing about the estimate from previous iterations[1]. It can be shown that the IEKF is equivalent to a Gauss-Newton search on the cost function in Eq. (12).[12] It thus results in the exact minimum of that cost function. Why then is it not an optimal, static estimator for nonlinear problems?

The answer lies in the relationship between the cost in Eq. (12) and the original, batch cost function in Eq. (5). Unlike for the linear case, these cost functions are no longer equivalent. Thus, minimizing the cost in Eq. (12) does not result in the minimum of the original, desired cost function and thus is not an optimal solution. The result is a biased estimate.

In contrast, the two-step estimator can be found by minimizing the modified recursive cost function:

$$J_k = \frac{1}{2}(y_k - \bar{y}_k)^T M_{y_k}^{-1}(y_k - \bar{y}_k) + \frac{1}{2}(z_k - H_k y_k)^T R_k^{-1}(z_k - H_k y_k) \quad (14)$$

This function *is* equivalent to the batch cost at the minimum as it is linear in the first step states y. This can be minimized exactly and results in the same Kalman filter measurement update for the first step states as in Eq. (9). Again, the result is an *optimal* estimate of the static states.

One additional caveat needs to be mentioned. Because of the least-squares minimization necessary to determine the second-step states (Eq. (8)), the number of first-step states must equal or exceed the number of second-step states (m ≥ n). Violation of this requirement would result in a non-unique minimum for the second-step optimization. For situations where other means of satisfying this requirement are not practical, the suggested approach is to simply augment the choice of first-step states with the second-step state vector x.[6,10]

Time Update for Dynamic Systems

The above discussion was only for static systems, that is, systems where $A(x,t)$ and $B(x,t)$ are zero. Even for this simple case the EKF and IEKF were shown to be suboptimal while the two-step estimator was a minimizing solution of the desired cost function. In this section a time update is presented for the two-step filter.

For traditional problems, including the EKF and IEKF, the time update is simply a method for propagating the estimate $\hat{x}_k$ and covariance P_{x_k} to the next time step, k+1, based on the dynamics in Eq. (2). The resulting estimate and covariance is then used as the *a priori* information in the cost function in Eq. (12). For the linear case this is particularly easy. When $A(x,t)$ and $B(x,t)$ are independent of x, the dynamics can be solved exactly:

$$x_{k+1} = \Phi_k x_k + \Gamma_k w_k \qquad (15)$$

where Φ is the state transition matrix and w is an i.i.d. noise process with covariance Q_d. The estimates and covariance are then propagated using the standard formulas:

$$\bar{x}_{k+1} = \Phi_k \bar{x}_k$$
$$M_{x_{k+1}} = \Phi_k P_{x_k} \Phi_k^T + \Gamma_k Q_d \Gamma_k^T \qquad (16)$$

For nonlinear problems this is more difficult. The best method for updating is still an outstanding problem and the subject of current research by the author. For now, the most common approach is to use a standard Runge-Kutta integration technique on the estimate and to propagate the covariance via Eq. (16) but with the transition matrices found by linearizing the dynamics about the previous estimate:

$$P_y(k+1) = \Phi(t_{k+1}, t_k) P_y(k) \Phi^T(t_{k+1}, t_k) + Q_d(k) \qquad (17)$$

where the state transition matrix is the solution of the linear differential equation:

$$\frac{d}{dt}\Phi(t, t_o) = \left.\frac{\partial A(x,t)}{\partial x}\right|_{\hat{x}} \Phi(t, t_o) \qquad (18)$$

and the discrete process noise covariance matrix is the solution of the integral:

$$Q_d(k) = \Phi_k \left\{ \int_{t_{k-1}}^{t_k} \Phi^{-1}(\tau) B(\hat{x}, \tau) Q B^T(\hat{x}, \tau) \Phi^{-T}(\tau) d\tau \right\} \Phi_k^T \qquad (19)$$

Equations (18) and (19) are most easily solved using the algorithm due to Van Loan[13].

The time update for the two-step estimator is more complicated. Here, both $\hat{y}$ and $\hat{x}$ must be propagated. A propagated value of $\bar{y}$ is necessary in order to substitute into the cost in Eq. (14) (or the resulting measurement update equations). Even for linear dynamics, the problem is difficult because of the nonlinear relationship between x and y. It is worth noting that were it possible to find an *exact* update formula for y then the resulting filter would be truly optimal, minimizing the cost function in Eq. (1).

The original approach to time updating y is presented in Refs. 6 & 10. In those works, the dynamics were assumed linear and thus the second step states were easily updated using Eq. (16). The first step state update was obtained by using the identities[6,10]

$$y_k = f_k(x_k, t_k) \qquad (20)$$

and

$$y_{k+1} = y_k + f_{k+1}(x_{k+1}) - f_k(x_k) \qquad (21)$$

where x is the desired n × 1 state vector. The time update for the first step states y was found by taking the expectation of Eq. (21). Unfortunately, the nonlinear measurement, $f(x)$, makes an exact expectation impossible in practice. The expectation is made tractable by expanding the last two terms to first order in the second-step estimation error,[1,2]

$$f_k(x_k) \approx f_k(\hat{x}_k) + \left.\frac{\partial f_k}{\partial x}\right|_{\hat{x}_k} (x_k - \hat{x}_k) \qquad (22)$$

and likewise for $f_{k+1}(x_{k+1})$ about $\bar{x}_{k+1}$. By taking the expected value of this expanded version and assuming the distribution of estimation error is unbiased, the first-step time update was found:

$$\bar{y}_{k+1} \approx \hat{y}_k + f_{k+1}(\bar{x}_{k+1}) - f_k(\hat{x}_k) \qquad (23)$$

The covariance is likewise found by subtracting Eq. (23) from Eq. (21), expanding to first order, squaring, and taking the expected value. After much algebra, the result is:

$$M_{y_{k+1}} \approx P_{y_k} + \left.\frac{\partial f_{k+1}}{\partial x}\right|_{\bar{x}_{k+1}} M_{xk+1} \left.\frac{\partial f_{k+1}}{\partial x}\right|_{\bar{x}_{k+1}}^T - \left.\frac{\partial f_k}{\partial x}\right|_{\hat{x}_k} P_{xk} \left.\frac{\partial f_k}{\partial x}\right|_{\hat{x}_k}^T \qquad (24)$$

These equations have the desirable property that they are updates for the first-step states and covariance. That is, they are small corrections on the previous estimates as in the traditional Kalman filter. However, it is not difficult to see that the covariance update in Eq. (24) has potential problems. A true covariance is always positive definite and symmetric. Because of the subtraction in Eq. (24), though, the covariance of the first step states can potentially drop rank or become negative. This could be catastrophic for the filter.

Unfortunately, such behavior was seen in simulations of some systems. Garrison, Axelrad, and Kasdin[11] showed that this eventuality is predictable and depends upon the system, measurement, and trajectory. This calls into question the value of the original two-step time update. Ref. 14 presents an alternative time update that eliminates this potential problem. This time update is guaranteed to be symmetric and positive definite. This is done by finding a differential equation directly for the first step states, y, and propagating as above for the EKF and IEKF. Here, we turn to the details of stochastic calculus and rely on the interpretation of Eq. (2) as an Ito s.d.e. We can then use the modified chain rule of the Ito calculus to find a differential expression for each component of $y = f(x)$:

$$\dot{y}_i = \frac{\partial f_i}{\partial t} + \frac{\partial f_i}{\partial x} A(x,t) + \frac{1}{2} Tr\left\{\frac{\partial^2 f_i}{\partial x^2} B(x,t) Q B^T(x,t)\right\} + \frac{\partial f_i}{\partial x} B(x,t) w(t) \qquad (25)$$

where $Tr\{\}$ refers to the trace of the enclosed matrix expression. This equation is now used along with Eq. (2) to jointly numerically propagate x and y. Recalling the earlier discussion regarding the requirement that there be at least as many first step states as second step states (m ≥ n), it becomes straightforward to develop a propagation algorithm. In that discussion it was proposed that the first step states be augmented by the second step states to guarantee satisfaction of that requirement. Now, this augmentation is required as the states x and y and their covariance must be simultaneously numerically integrated. Thus, the first step states are always going to be given by an expression that looks like,

$$y = f(x) = \begin{bmatrix} \hat{F}(x,t) \\ x(t) \end{bmatrix} \qquad (26)$$

where $\hat{F}(x,t)$ may be just $F(x,t)$ as previously proposed or it may be some other convenient combination.

The first step states are now propagated using any available nonlinear propagation algorithm. The nonlinearities are expanded to first order in the estimate error and assumed central. The expected value is taken to find the deterministic differential equations for the estimates,

$$\dot{\bar{y}} = \begin{bmatrix} \dfrac{d}{dt}\hat{F}(\bar{x},t) \\ \dot{\bar{x}} \end{bmatrix} = \begin{bmatrix} \left.\dfrac{\partial f_1}{\partial t}\right|_{\hat{x}} + \left.\dfrac{\partial f_1}{\partial x}\right|_{\hat{x}} A(\hat{x},t) + \dfrac{1}{2}Tr\left\{\left.\dfrac{\partial^2 f_1}{\partial x^2}\right|_{\hat{x}} B(\hat{x},t)QB^T(\hat{x},t)\right\} \\ \vdots \\ \left.\dfrac{\partial f_\ell}{\partial t}\right|_{\hat{x}} + \left.\dfrac{\partial f_\ell}{\partial x}\right|_{\hat{x}} A(\hat{x},t) + \dfrac{1}{2}Tr\left\{\left.\dfrac{\partial^2 f_\ell}{\partial x^2}\right|_{\hat{x}} B(\hat{x},t)QB^T(\hat{x},t)\right\} \\ A(\hat{x},t) \end{bmatrix} \equiv \tilde{A}(\hat{x},t) \quad (27)$$

In the work done so far this equation is propagated using a classical, third or fourth order Runge-Kutta formula to find $\bar{y}_{k+1}$. The covariance is propagated as described above for the EKF but with A replaced by $\tilde{A}$. As mentioned, current research is directed at more accurate methods for performing this nonlinear propagation.

Performance with this new approach for the time update has proven to be far more robust than the original method. Its only drawback arises in systems with linear dynamics. Whereas in a standard filter the time update for linear dynamics is straightforward, this new approach for the two-step estimator results in the need for complex nonlinear propagation methods for all systems. However, the benefits of exact optimization of the measurement step far outweigh this drawback. In fact, this points out the key advantage of the two step estimator. For problems with nonlinear measurements, the two-step technique moves all of the approximations to the time-update while the EKF and IEKF are suboptimal in the measurement update. The two-step estimator makes better use of the measurement information and results in a more accurate filter.

Summary

Using the derivations described above, the two-step filter can be summarized as follows:

Measurements:
$$\mathbf{z}_k = F(\mathbf{x}_k, t_k) + \mathbf{v}_k = H_k \mathbf{y}_k + \mathbf{v}_k \quad k=1\ldots N$$

Dynamics:
$$\dot{\mathbf{x}} = A(\mathbf{x},t) + B(\mathbf{x},t)\mathbf{w}(t)$$

Nonlinearity: $\quad \mathbf{y} = \mathbf{f}(\mathbf{x}) = \begin{bmatrix} \hat{F}(\mathbf{x},t) \\ \mathbf{x}(t) \end{bmatrix}$

Initial Conditions: $\bar{\mathbf{x}}_1, M_{x_1}$

where $E[\mathbf{v}_k] = 0 \quad E[\mathbf{v}_k \mathbf{v}_k^T] = R_k$

$$E[\mathbf{w}(t)] = 0 \quad E[\mathbf{w}(t)\mathbf{w}(t+\tau)^T] = Q\delta(\tau)$$

First - Step Optimization:

Measurement Update:
$$\hat{\mathbf{y}}_k = \bar{\mathbf{y}}_k + P_{y_k} H_k^T R_k^{-1}(\mathbf{z}_k - H_k \bar{\mathbf{y}}_k) \quad k=1\ldots N$$
$$P_{y_k} = (M_{y_k}^{-1} + H_k^T R_k^{-1} H_k)^{-1}$$

Time Update:
$$\dot{\bar{y}} = \begin{bmatrix} \dfrac{d}{dt}\hat{F}(\bar{x},t) \\ \dot{\bar{x}} \end{bmatrix} = \begin{bmatrix} \left.\dfrac{\partial f_1}{\partial t}\right|_{\hat{x}} + \left.\dfrac{\partial f_1}{\partial x}\right|_{\hat{x}} A(\hat{x},t) + \dfrac{1}{2}Tr\left\{\left.\dfrac{\partial^2 f_1}{\partial x^2}\right|_{\hat{x}} B(\hat{x},t)QB^T(\hat{x},t)\right\} \\ \vdots \\ \left.\dfrac{\partial f_\ell}{\partial t}\right|_{\hat{x}} + \left.\dfrac{\partial f_\ell}{\partial x}\right|_{\hat{x}} A(\hat{x},t) + \dfrac{1}{2}Tr\left\{\left.\dfrac{\partial^2 f_\ell}{\partial x^2}\right|_{\hat{x}} B(\hat{x},t)QB^T(\hat{x},t)\right\} \\ A(\hat{x},t) \end{bmatrix} \equiv \tilde{A}(\hat{x},t)$$

$$\dot{P}_y = \left.\frac{\partial \tilde{A}(x,t)}{\partial x}\right|_{\hat{x}} P_y \left.\frac{\partial \tilde{A}(x,t)}{\partial x}\right|_{\hat{x}}^T + \tilde{B}(\hat{x},t)Q\tilde{B}^T(\hat{x},t)$$

where,

$$\tilde{B}(x,t) = \left[\frac{\partial f_1}{\partial x}B(x,t) \quad \ldots \quad \frac{\partial f_\ell}{\partial x}B(x,t) \quad B(x,t)\right]^T$$

Second - Step Optimization:

Measurement Update (Gauss - Newton iteration on i):

$$\hat{x}_{k,i+1} = \hat{x}_{k,i} - H_{Gk,i}^{-1} q_{k,i}^T \quad k = 1\ldots N$$

$$H_{Gk,i} = \left.\frac{\partial f_k}{\partial x}\right|_{\hat{x}_{k,i}}^T P_{y_k}^{-1} \left.\frac{\partial f_k}{\partial x}\right|_{\hat{x}_{k,i}}$$

$$q_{k,i} = -(\hat{y}_k - f_k(\hat{x}_{k,i}))^T P_{y_k}^{-1} \left.\frac{\partial f_k}{\partial x}\right|_{\hat{x}_{k,i}}$$

$$P_{xk,i} = H_{Gk,i}^{-1}$$

INITIALIZATION OF THE TWO-STEP FILTER

One of the trickier issues associated with the two-step filter not mentioned above is the problem of initializing the first step states. In standard Kalman filter problems some initial estimate of the states, $\bar{x}_1$, is assumed, with an initial error covariance, M_{x_1}. Typically, this initial covariance is taken very large to reflect the initial lack of knowledge regarding the true states of the system.

To implement the two-step estimator an initial guess at the first step states and covariance, $\bar{y}_1$ and M_{y_1}, is needed. There are a number of possible approaches, one of which is discussed in Ref 10. In the original two-step estimator examples[6,10], these initial conditions were chosen *a priori* based on *ad hoc* analysis of the relationship between x and y. While this was successful, it lacked robustness and simulation studies showed two-step estimator performance to be very sensitive to selection of initial conditions. A more rigorous approach was desired.

One complicating factor is again the requirement that there be more first step states than second step states. A simple first order expansion of $f(x,t)$ would result in a covariance for y that is of rank n rather than m. Unfortunately, it is also necessary that M_y be invertible in order to perform the weighted least squares search to find the second step states. One possible solution is to use higher order terms in the expansion[10]. While this was shown to work, it requires a much more complex analysis with many higher order derivatives. Another drawback of the expansion approach is its limited accuracy. Underlying such an expansion is an assumption that the initial error in $\bar{x}_1$ is small, in direct contradiction to the usual practice of using large initial covariance.

While other approaches are possible, such as direct calculation of the expected values by numerical integration, it has been found that the most efficient, robust, and accurate method is to use a Monte-Carlo generation technique. Here, the initial value for y is simply:

$$\bar{y}_1 = f(\bar{x}_1, t_o) \tag{28}$$

Since we only have one available sample of the initial condition distribution, then there is no better estimate for the initial first step states than that given by Eq. (28). To find the initial covariance, we generate a random sample from a multivariable Gaussian distribution with mean zero and covariance M_{x_1}, written as x_i, and compute a sample first step state:

$$y_i = f(x_i, t_o) \qquad (29)$$

The first step covariance is found by averaging N samples:

$$M_{y_1} = \frac{1}{N}\sum_{i=1}^{N}(y_i - \bar{y}_1)(y_i - \bar{y}_1)^T \qquad (30)$$

The number of averages, N, should be chosen very large to ensure a good estimate of the initial covariance, particularly for large initial covariance on x. For the simulations in this paper, N was typically chosen as large as 50,000. While this seems like a large computational price to pay, it should be recalled that this step is only performed once prior to operation of the filter. It imposes no burden on the efficiency of the estimation routine itself.

BENCHMARK EXAMPLE

The two-step filter was first used for a static estimation problem taken from the Gravity Probe B mission.[6] In that problem, as series of constants were estimated from a nonlinear, time varying series of measurements. It's performance was shown to be optimal and to far exceed that of the EKF or IEKF on the same problem. In this paper, however, we are focusing on dynamic problems, so we present the second example from Ref. 6, what we call our benchmark problem. This problem is to estimate the horizontal position, a, horizontal velocity, $\dot{a}$, and altitude, b, of an aircraft flying straight and level from only a single range measurement taken at discrete intervals (see Fig. 1). It can be thought of as a two-dimensional simplification of a dynamic GPS measurement problem with fewer than 4 satellites. This problem is characterized by a highly nonlinear measurement but linear dynamics. It is an effective benchmark because of the reliance on the dynamics to make the system observable. Since there is only a single ranging site, the states can not be discerned from only one measurement. The dynamic model must be used to determine the system state. This exercises the estimation algorithm to the maximum extent and, as we shall see, the EKF and IEKF do not perform very well.

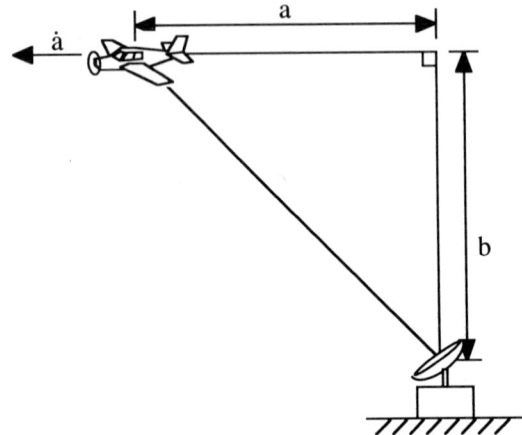

Fig. 1: Radar Ranging of on Airplane flying overhead at constant altitude

This system is characterized by the state vector:

$$x = \begin{bmatrix} a & \dot{a} & b \end{bmatrix}^T \qquad (31)$$

and measurement:

$$z(t) = \sqrt{a^2 + b^2} + v \qquad (32)$$

where v is the measurement noise. The dynamics are given by:

$$\dot{x} = \begin{bmatrix} 0 & 1 & 0 \\ 0 & 0 & 0 \\ 0 & 0 & 0 \end{bmatrix} x + \begin{bmatrix} 0 \\ 1 \\ 0 \end{bmatrix} w \qquad (33)$$

or, in discrete form,

$$x_{k+1} = \Phi x_k + \Gamma w_k = \begin{bmatrix} 1 & T & 0 \\ 0 & 1 & 0 \\ 0 & 0 & 1 \end{bmatrix} x_k + \begin{bmatrix} T^2/2 \\ T \\ 0 \end{bmatrix} w_k \qquad (34)$$

where T is the time step between measurements. To find the first step states we augment the measurement as described earlier:

$$y_k = f(x_k) = \begin{bmatrix} \sqrt{a_k^2 + b^2} & a_k & \dot{a} & b \end{bmatrix}^T \qquad (35)$$

These equations are now used in the two-step estimator as described in the previous section with a Levenberg-Marquardt algorithm for finding the second step states. For the first example simulation, we repeat the same parameter values used in Ref. 6. These are given in Table 1. Note that there are several differences in this simulation from that in Ref. 6. First, we are using the new method where the differential equation for y is used for the time update. Second, we are establishing the initial conditions for the first step states using the Monte-Carlo approach described earlier rather than the *ad hoc* technique of Ref. 6. Lastly, the initial covariance has been increased from 10^4 to 10^6. The mean error of 100 ft. previously used is representative of very good initial knowledge. It was felt that 1000 ft. error was slightly more realistic of common practice and better exercised the filters.

Table 1 Parameter values for the radar ranging problem

Given:
T = 0.1 sec   sampling time
R_k = 1 ft^2 measurement noise variance
Q_k = 1 ft^2 / sec^4/Hz   process noise variance

States:
$$\mu_x = \begin{bmatrix} a \\ \dot{a} \\ b \end{bmatrix} \qquad M_{x_1} = \begin{bmatrix} 1000^2 & 0 & 0 \\ 0 & 1000^2 & 0 \\ 0 & 0 & 1000^2 \end{bmatrix}$$

where:
a = 1000 ft, $\dot{a}$ = 500 ft/sec, b = 3000 ft

The results of the simulation are shown in Figs. 2 and 3. In Fig. 2 we display the bias in altitude estimate for the two-step estimator, the EKF, and the IEKF after 250 runs of the filter. While the two-step estimator is still a biased one, the dramatic improvement over the previous techniques is apparent. In Fig. 3 is shown the estimator bias for the velocity estimates. Note that while the two-step estimator has effectively zero bias, the iterated extended Kalman filter never was able to improve on the initial error and converge to the correct bias. This is extremely important as this large bias in the velocity resulted in a *divergent* error in the horizontal range estimate. In Ref. 6 the filter did not diverge for the smaller initial covariance. The EKF and IEKF are very sensitive to initial conditions and tend to only perform when there is good initial information. The two-step estimator performs well for any initial error.

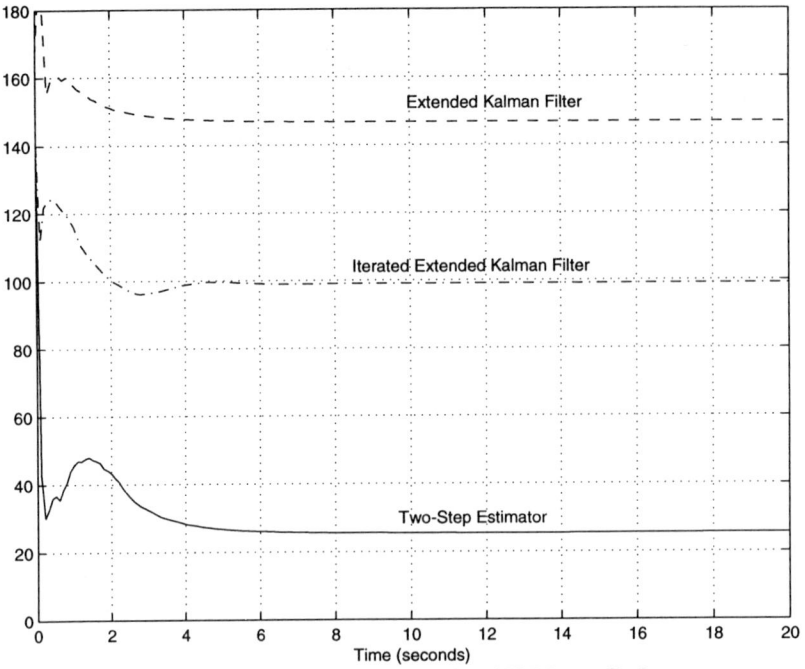

Fig. 2: Altitude estimate error bias after 250 Monte-Carlo runs

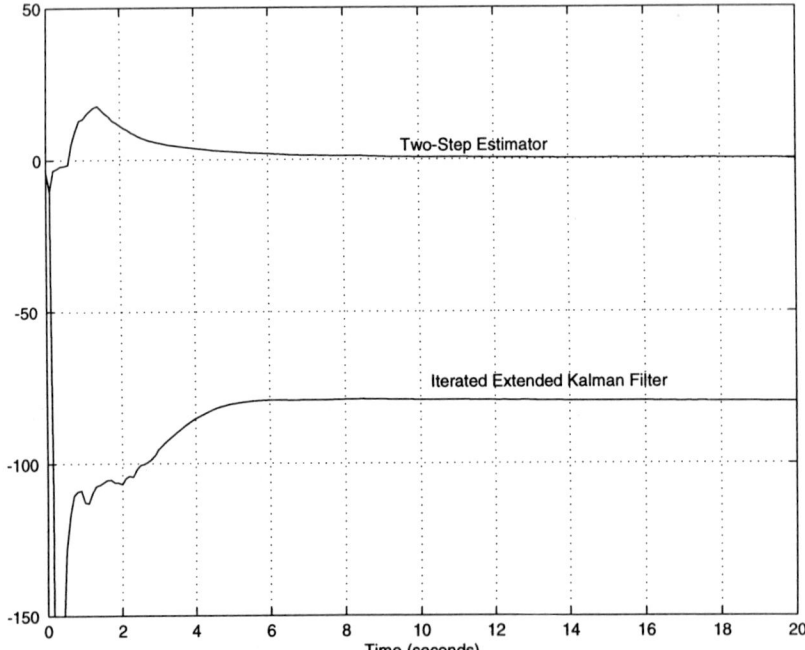

Fig. 3: Velocity estimate error bias after 250 Monte-Carlo runs

While the measurement error of 1 ft. is very low, the comparable process noise, along with the unobservability based on single measurements, means the filters must rely heavily on the time update. As all of the two-step filter approximations are in the time update, this reliance produces the resulting bias.

Because the benchmark problem is representative of the most difficult estimation tasks but is simple and quick to simulate, it is an instructive to use it to investigate the estimator behavior for different parameter values. For example, Fig. 4 shows the bias performance for very large measurement noise relative to process noise ($R = 1000$ ft^2 and $Q = 1$ ft^2/s^4/Hz). The important observation here is that the steady state bias error is roughly the same. As one might expect from standard, linear Kalman filter analysis, the ratio of R to Q determines the time constant of the filter (as well as the steady state covariance), with larger R and smaller Q resulting in a slower convergence. The steady state biases and convergence of the nonlinear filters is largely determined by the initial conditions and the nature of the approximations. It is worth noting that this second example is fairly realistic of actual systems, with range measurement error on the order of 30 ft. In this example, again, the estimate error for the horizontal position diverged in the EKF.

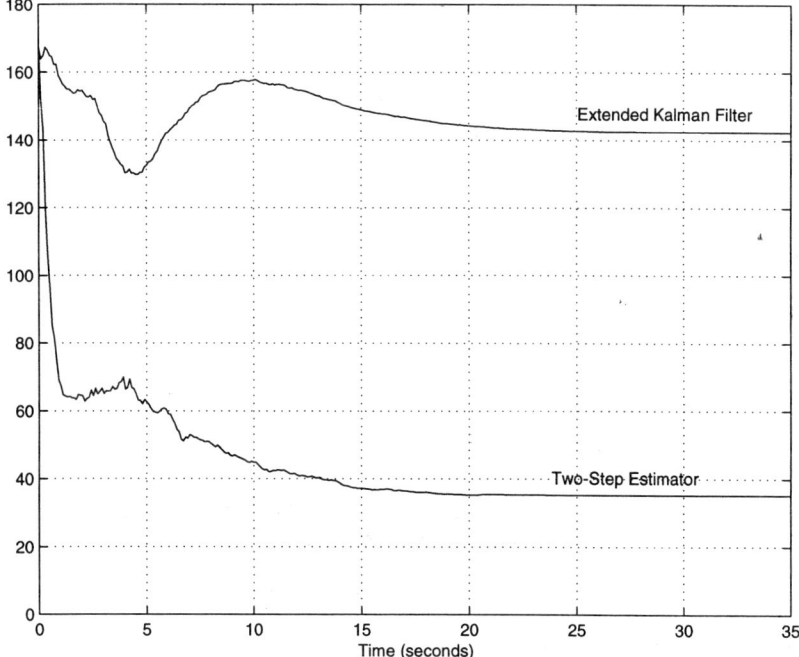

Fig. 4: Altitude estimate error bias for 250 Monte-Carlo runs
$R = 1000$ ft^2, $Q = 1$ ft^2/s^4/Hz

OTHER SIMPLE APPLICATIONS

The next example is taken from Refs. 11, 14, and 15. Recall that it was shown in Ref. 11 that the original two-step time update failed catastrophically on some systems. Garrison[15] was using the two-step filter for an orbit determination problem – trying to estimate the relative orbits of two satellites from range measurements between them. To explore the nature of the estimator failure, he developed a simple, 2-D analog of this problem. While dissimilar in the dynamics, the trajectory and measurements are representative of the larger space mechanics problem. And, as in the benchmark problem above, the simplicity of the example allows for rapid simulation and ease of understanding. This example also introduces a system with nonlinear dynamics.

This same example was used to demonstrate the effectiveness of the new, differential time update technique[14]. Here, we will not reproduce the failed results from the original approach but merely show the performance of the new two step method as a further validation of the technique.

The example is a purely kinematic motion of a particle on a spiral trajectory defined by a constant angular velocity ω_o and a constant radial velocity v_o (see Fig. 5). The desired (second-step) state vector consists of the two-dimensional position of the particle, (x_1, x_2). The motion of the particle in a Cartesian coordinate system is described by two first-order, non-linear differential equations,

$$\frac{dx_1}{dt} = \frac{x_1 v_0}{\sqrt{x_1^2 + x_2^2}} - x_2 \omega_0 + w_1$$
$$\frac{dx_2}{dt} = \frac{x_2 v_0}{\sqrt{x_1^2 + x_2^2}} + x_1 \omega_0 + w_2 \tag{36}$$

Here, w_1 and w_2 are process noise terms with diagonal covariance matrix Q.

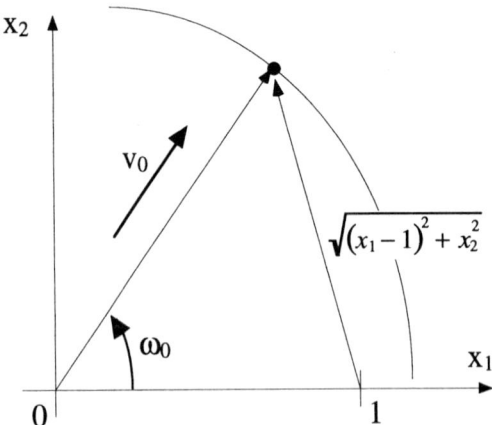

Fig. 5: Kinematic Particle on a Spiral Trajectory

The objective of the filter is to find estimates of the states x_1 and x_2 given only range measurements from a fixed point located at (1,0). The measurement equation is thus given by,

$$z = \sqrt{(x_1 - 1)^2 + x_2^2} \tag{37}$$

The two-step filter is formed using the augmented first-step state vector,

$$y = \begin{bmatrix} \sqrt{(x_1 - 1)^2 + x_2^2} \\ x_1 \\ x_2 \end{bmatrix} \tag{38}$$

with measurement matrix $H = \begin{bmatrix} 1 & 0 & 0 \end{bmatrix}$.

Table 2 lists the values used in the simulation.[11] The particle starts out at the position (2.583, 0.313) and propagates for 6 seconds with $\omega_o = 1$ and $v_o = 1$. While a process noise term is used in the filter model, no process noise is added to the simulation of the truth model. A plot of the reference trajectory is shown in Fig. 6.

Table 2 Parameter values for second example problem, spiral motion

Time Step, Δt	0.0002
Filter initial state, $\hat{x}_o$	(2.0, 0.0)
True initial state, $x(0)$	(2.583, 0.313)
Measurement covariance, R	10^{-4}
Filter initial second-step state covariance, P_{x_o}	diag{0.25, 0.25}
Second-step continuous process noise, Q	diag{10^{-12}, 10^{-12}}

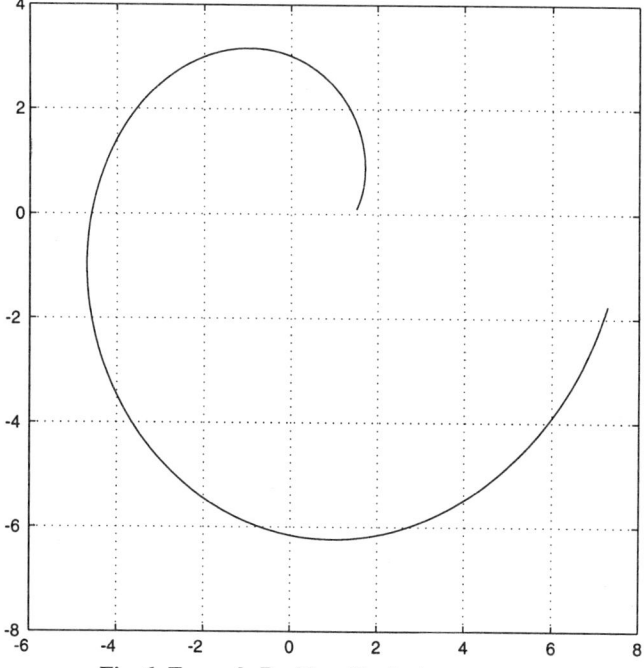

Fig. 6: Example Problem Trajectory, x_1 vs. x_2

The two-step estimator was run on this problem. The new approach was used with the following differential equation for y:

$$\dot{y} = \begin{bmatrix} \dot{r} \\ \dot{x}_1 \\ \dot{x}_2 \end{bmatrix} = \begin{bmatrix} \frac{\sqrt{x_1^2 + x_2^2}}{r} v_o - \frac{x_1 v_o}{r\sqrt{x_1^2 + x_2^2}} + \frac{x_2 \omega_o}{r} + \frac{q}{2r} \\ \frac{x_1 v_o}{\sqrt{x_1^2 + x_2^2}} - x_2 \omega_o \\ \frac{x_2 v_o}{\sqrt{x_1^2 + x_2^2}} + x_1 \omega_o \end{bmatrix} + \begin{bmatrix} \frac{x_1 - 1}{r} & \frac{x_2}{r} \\ 1 & 0 \\ 0 & 1 \end{bmatrix} \begin{bmatrix} w_1 \\ w_2 \end{bmatrix} \quad (39)$$

where $r = \sqrt{(x_1 - 1)^2 + x_2^2}$ and q is the magnitude of each diagonal element of Q. The results are shown in Figs. 7 and 8, comparing performance to the EKF and IEKF. Note that the EKF, even with fairly good initial conditions, has a very large bias. While the IEKF has improved performance, it is still very biased compared to the two-step filter.

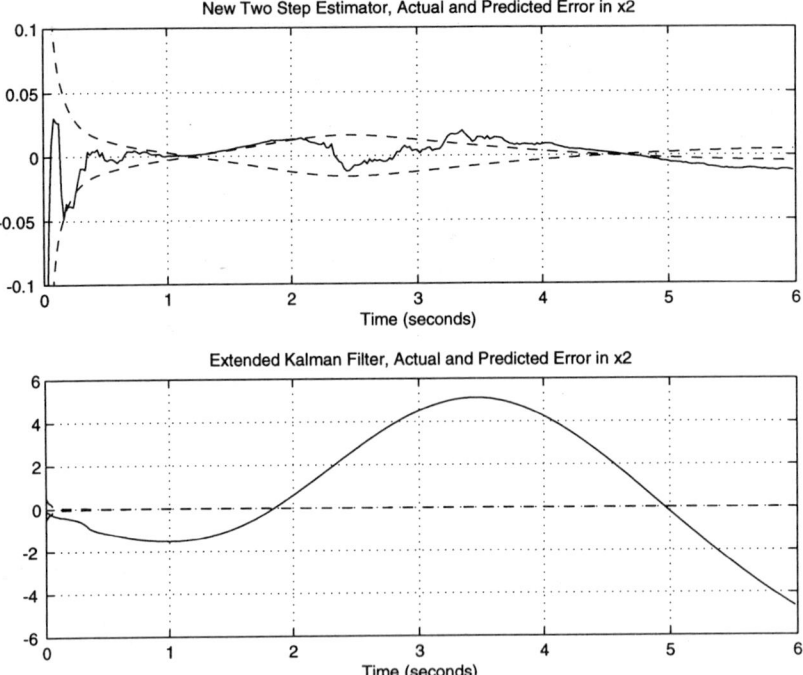

Fig. 7: Comparison of Two-Step Filter with Extended Kalman Filter

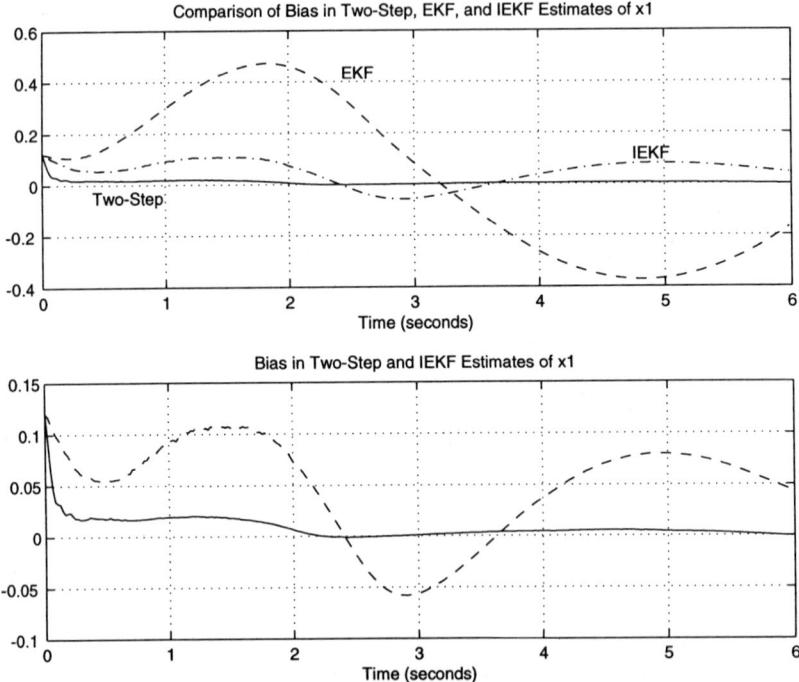

Fig. 8: Comparison of Bias in EKF, IEKF, and Two-Step Filter

The final example is a simple attitude control problem. This problem allows us to use the two-step estimator in a closed loop control system. The problem is again taken from the Gravity Probe B mission. In this problem we consider two axis pointing control of a rolling rigid spacecraft at an inertially fixed stellar reference.[16] Control is achieved via measurements from a precise telescope and performed using continuous thrusters. For brevity, we don't show the dynamic equations of motion. The nonlinearity is in the telescope output. Rather than being directly proportional to the pitch and yaw of the satellite, the telescope instead measures:

$$\mathbf{z}_k = \begin{bmatrix} \theta_p \left(1 + C_1(\theta_p/\theta_o) + C_2(\theta_p/\theta_o)^2 + C_3(\theta_p/\theta_o)^3\right) \\ \theta_y \left(1 + C_1(\theta_y/\theta_o) + C_2(\theta_y/\theta_o)^2 + C_3(\theta_y/\theta_o)^3\right) \end{bmatrix} + v_k \quad (40)$$

where θ_p and θ_y are the pitch and yaw angles of the satellite respectively and all angles are measured in milliarcsec. It is assumed that a prior calibration has provided values for the constants coefficients. The problem is to find the best estimate of the pitch, yaw, and rates from the nonlinear and noisy measurements. One could of course simply invert the equations, but this ignores the measurement noise. It is assumed that the measurement noise is large enough that such an inversion would produce unacceptable error.

We choose arbitrary values for the constants to produce a significant nonlinearity. The measurement function used for the simulations is shown in Fig. 9. The nonlinearity is large enough that ignoring it and using a simple, linear, full state feedback controller does not stabilize the system. However, we can use the two step estimator with states:

$$y = \begin{bmatrix} \theta_p & \theta_p^2 & \theta_p^3 & \theta_y & \theta_y^2 & \theta_y^3 & \dot{\theta}_p & \dot{\theta}_y \end{bmatrix}^T \quad (41)$$

where for numerical reasons we have only included the first two terms of the polynomial.

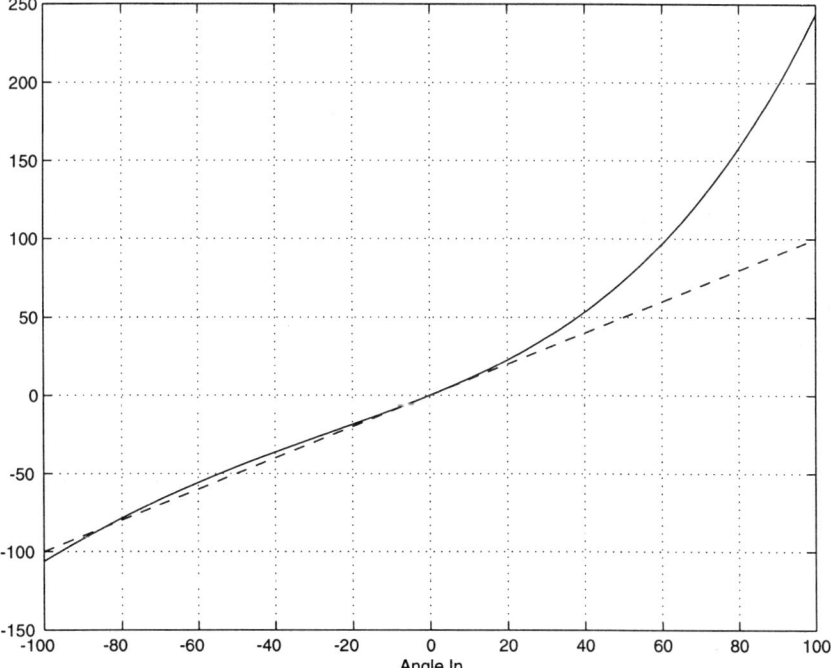

Fig. 9: Polynomial Telescope Output

For the measurement relationship in Eq. (40), even an IEKF wasn't stable. However, the two-step filter did result in a stable implementation with excellent performance as shown in Fig. 10. It should be pointed out that with the EKF or IEKF outside the control loop, the estimator performance in the steady state was essentially the same as the two-step filter. This is because as the controller locks onto the star, deviations are small enough that the nonlinear terms of the measurement no longer contribute.

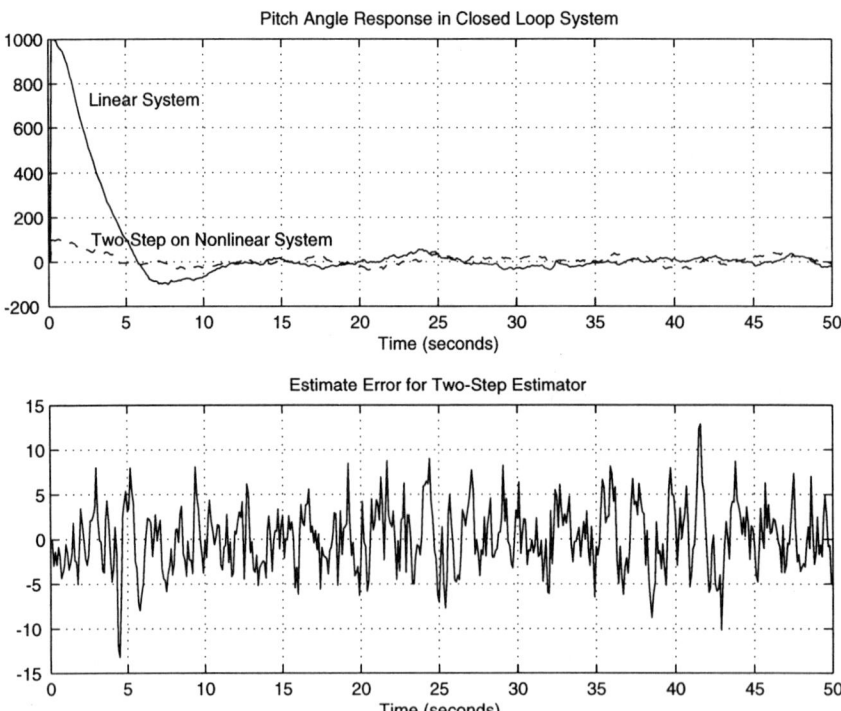

Fig. 10: Performance of Satellite Pointing Control with Two-Step Estimator

CURRENT AND FUTURE RESEARCH

There are many areas of current or potential research on the two-step estimator. Of particular interest to the author is improvements in the time update. While the algorithms presented here have been shown to work quite effectively, it was also stated that the remaining error in the two-step is primarily due to inaccuracies in the time update for nonlinear dynamics. Current research by the author is directed at more exact methods for numerically propagating nonlinear stochastic differential equations with applications to Kalman filter time updating.

Also of merit are investigations into the numerical properties and robustness of the two-step estimator. Some discussion of this can be found in Ref. 10. In particular, a square root implementation of the original filter is presented. One of the main advantages of the two-step is its conversion of the measurement update into a standard, linear Kalman filter form. This allows the use of previously introduced methods for numerical robustness, such as square root forms or sequential measurement processing. However, research is warranted on the second step update in the two-step filter. While both the Gauss-Newton and Levenberg-Marquardt algorithms have been utilized successfully, some cases have been seen in simulation with less than optimal performance because the search algorithm converged to false minimums.

Ref. 14 has a brief discussion of other implementation considerations with the two-step filter, particularly regarding execution speed. Since all of the studies presented in this paper were MATLAB™

based simulations, it is difficult to draw definitive conclusions regarding performance, though the two step estimator typically performed as fast or faster than the IEKF. Future work is directed at a wider variety of applications and at a real time implementation of the filter.

Finally, little has been said about how to go about selecting the first step states. Very little work has been applied to that problem to date. It is certainly true that one potential drawback of the technique is that it can dramatically increase the size of the problem for certain choices of first step states (see, for instance, example three above). It has been felt that observability studies of the filter may shed light on more rigorous or optimal approaches to selection of the first step states.

CONCLUSIONS

This paper presented a review of past work on the two-step optimal estimator. It was intended to provide a thorough description of the filter and its implementation. It also introduced a new, Monte-Carlo technique for initializing the filter. The two step filter has been shown to be closer to optimal for all problems than either the extended Kalman filter or the iterated extended Kalman filter. Three simple examples, two of which were taken from previous works, were presented to highlight the performance of the filter and study different aspects of it. In all studies to date the two-step filter has been seen to either outperform or perform at least as well the EKF and IEKF. The two-step filter is believed to be a significant improvement over these techniques and should see increased applications as further studies confirm these results.

REFERENCES

1. Gelb, Arthur, Ed., *Applied Optimal Estimation,* The M.I.T. Press, Cambridge, MA, 1974.
2. Stengel, R. F., *Stochastic Optimal Control, Theory and Application,* John Wiley & Sons, New York, 1986.
3. Bryson, A. E., and Ho, Y.C., *Applied Optimal Control,* Hemisphere Publishing Corp., Washington, 1975.
4. Kailath, T., *Lectures on Wiener and Kalman Filtering,* Springer Verlag, New York, 1981.
5. Kalman, R. E., "A New Approach to Linear Filtering and Prediction Problems," *Transactions of the ASME,* Series D: Journal of Basic Engineering, Vol. 82, 1960, 35-45.
6. Haupt, G. T., Kasdin, N. J., Keiser, G. M., and Parkinson, B. W., "An Optimal Recursive Iterative Algorithm for Discrete Nonlinear Least-Squares Estimation," *Journal of Guidance, Control, and Dynamics,* May-June, 1996, Vol. 19, No. 3, pp. 643-649.
7. Kasdin, N. J. "Discrete Simulation of Colored Noise and Stochastic Processes and $1/f^\alpha$ Power Law Noise Generation," *Proceedings of the IEEE,* Vol. 83, No. 5, May 1995, pg. 802-827.
8. Soong, T.T. *Random Differential Equations in Science and Engineering.* New York: Academic Press, 1973.
9. Gard, T. C. *Introduction to Stochastic Differential Equations.* New York: Marcel Dekker, Inc., 1988.
10. Kasdin, N. J., Haupt, G. T., "Second-Order Correction and Numerical Considerations for the Two-Step Optimal Estimator," *Journal of Guidance, Control, and Dynamics,* March-April, 1997, Vol. 20, No. 2, pp. 362-369.
11. Garrison, J. L., Axelrad, P., Kasdin, N. J., "Ill-Conditioned Covariance Matrices in the First-Order Two-Step Estimator," *Journal of Guidance, Control, and Dynamics,* Sept.-Oct., 1998, Vol. 21, No. 5, pp. 754-760.
12. Bell, B. M., and Cathey, F. W., "The Iterated Kalman Filter Update as a Gauss-Newton Method," *IEEE Transactions on Automatic Control,* Vol. 38, No. 2, 1993, pp. 294-297.
13. Van Loan, C. F., "Computing Integrals Involving the Matrix Exponential," *IEEE Transactions on Automatic Control,* June, 1978, vol. AC-23, pp. 395-404.
14. Kasdin, N. J., "A New, Guaranteed Positive Time Update for the Two-Step Optimal Estimator," upcoming issue of *Journal of Guidance, Control, and Dynamics.*

15. Garrison, J. L., "Recursive Nonlinear Estimation for Relative Navigation in Elliptical Orbits," Ph. D. Dissertation, Dept. of Aerospace Engineering Sciences, Univ. of Colorado, Boulder, CO, Dec. 1997.
16. Kasdin, N. J., "Precision Pointing Control of the Spinning Gravity Probe B Spacecraft," Ph.D. dissertation, Dept. of Aeronautics and Astronautics, Stanford University, Stanford, CA, March 1991, SUDAAR 606.

AAS 00-003

TEST RESULTS FOR THE AUTOMATED RENDEZVOUS AND CAPTURE SYSTEM

Craig A. Cruzen, James J. Lomas and Richard W. Dabney[*]

The Automated Rendezvous and Capture (AR&C) system was designed and tested at NASA's Marshall Space Flight Center (MSFC) to demonstrate technologies and mission strategies for automated rendezvous and docking of spacecraft in Earth orbit. The system incorporates some of the latest innovations in Global Positioning System (GPS) space navigation, laser sensor technologies and automated mission sequencing algorithms. The system's initial design and integration was completed in 1998 and underwent testing in 1999. This paper describes the major components of the AR&C system and presents results from the official system tests performed in MSFC's Flight Robotics Laboratory with digital simulations and hardware in the loop tests. The results show that the AR&C system can safely and reliably perform automated rendezvous and docking missions in the absence of system failures. When system failures were included, the system used its automated collision avoidance logic to recover in a safe manner. The primary objective of the AR&C project is to prove that by designing a safe and robust automated system, mission operations cost can be reduced by decreasing the personnel required for mission design, preflight planning and training required for crewed rendezvous and docking missions.

INTRODUCTION

Since the 1960's, the National Aeronautics and Space Administration (NASA) has been performing rendezvous and docking missions between two spacecraft. Recent examples of how this has become commonplace in space operations include the servicing of the Hubble Space Telescope, Space Shuttle/MIR dockings and International Space Station (ISS) construction missions. One common thread that remains between the current NASA mission philosophy and the very first Gemini docking mission is that at least one of these spacecraft has always been piloted by astronauts and supported by a virtual army of ground personnel. When the Russian space program developed an automated docking system, it was seen as a way to decrease costs of space flight by reducing the amount of support personnel required for docking operations. However, the near fatal accident that occurred when a docking attempt ended in collision between an unmanned, tele-operated controlled Soyuz supply ship and MIR showed that a great deal of safety and redundancy must be designed into any docking system. NASA has several missions on the horizon that will require an Automated Rendezvous and Capture (AR&C) capability[1]. In support of these mission requirements, engineers at NASA's George C. Marshall Space Flight Center (MSFC) have designed and tested an AR&C system which, along with the capability to lower mission operation costs, also has a great deal of safety, redundancy and reliability designed into it. The system incorporates some of the latest innovations in Global Positioning System (GPS) space navigation, laser sensor technologies and automated mission planning algorithms as well as the continuous capability for ground monitoring and intervention. This paper presents results from the official system tests performed in MSFC's Flight Robotics Laboratory (FRL) with digital simulations and hardware-in-the-loop-tests. The test cases in this paper, which are only a subset of the entire test profile, cover the operating ranges from 40 km to dock.

---

[*] Aerospace Engineers, NASA Marshall Space Flight Center, Huntsville, Alabama 35812.

The results show that the AR&C system can safely and reliably perform automated rendezvous and docking missions in the absence of system failures. When failures were included, the system used its automated collision avoidance logic to recover safely.

AR&C PROJECT BACKGROUND

The objective of the AR&C project is to advance rendezvous and docking technologies from manual to automated capabilities. This goal is seen as essential for two reasons: to reduce the recurring cost of routine docking missions; and for missions that require automated operations due to long communication delays, (i.e. robotic missions to Mars[1]). To that end, hardware, software, documentation and test facilities were developed to support the design of an AR&C system. Specifically, the project objectives were defined as follows: provide design criteria, procedures and simulation techniques that will influence standardization of AR&C systems; establish test facilities and procedures to support the development and verification of future systems prior to flight; demonstrate relevant technologies for future AR&C systems; establish functional performance capabilities of subsystem elements for an AR&C system; demonstrate spacecraft automated rendezvous, proximity operations, station keeping, capture and collision avoidance maneuvers in a controlled ground simulation and in space; demonstrate the capability to dock with 100 percent success in the absence of system failures; demonstrate the safety of AR&C including recovery from anomalous situations; and contribute to the future capability to conduct robotic spacecraft operations with the ISS and other space platforms[2]. Most of these objectives were demonstrated through flight experiments, detailed 6 Degree-of-Freedom (6 DoF) Hardware-In-The-Loop (HITL) simulations and digital tests.

AR&C SYSTEM DESCRIPTION

The AR&C system consists of several components, each of which is necessary in one or more phases of an automated docking mission. Table 1 lists the assumptions made about the Chaser and Target Vehicles during the design and testing process[3]. Figure 1 gives a pictorial overview of the elements included in the AR&C system. On the Chaser Vehicle (CV), an on-board computer performs hardware commanding, telemetry, guidance, navigation and control, collision avoidance maneuvers (CAMs) and system monitoring functions. Long-range absolute and relative navigation is accomplished using a 12 channel, L1, CA code GPS receiver in combination with an Inertial Measurement Unit (IMU). Short-range (100 meters to dock) relative navigation and attitude information is provided by the Video Guidance Sensor (VGS)[4]. The Three-Point Docking Mechanism (TPDM) performs the actual physical latching of the two spacecraft. The Target Vehicle (TV) is assumed to be stabilized in attitude and equipped with a set of trunions that align with the TPDM latches, a set of passive reflectors that serve as the VGS target, a 12 channel, L1, CA GPS receiver and a short-range transmitter that sends GPS data to the CV. All of this hardware was integrated together and tested in the MSFC FRL. It is significant to note that the AR&C project has already tested one element of the system in space on two separate occasions. The VGS was flown on STS-87 and STS-95. The purpose of these flight experiments was to verify the operational characteristics of the VGS in the low earth orbit environment. Both were "very successful"[4].

Table 1 CHASER AND TARGET VEHICLE ASSUMPTIONS

Chaser Vehicle Assumptions	Target Vehicle Assumptions
One 4,000 N thrust, main engine, Isp = 260 sec	TV has a passive VGS target mounted on it.
RCS thrust available in each body direction: 50 N, Isp = 220 sec, Full 6 DoF Control	TV transmitting raw GPS data for CV relative navigation
Total CV mass at Arrival in Orbit: 34,000 kg	TV GPS data transmission range : 7 km
Propellant for main engine and RCS: 4,500 kg	TV Orbit: 407 km, Circular, 51.6° Inclination
CV Payload delivered to TV: 25,000 kg	TV attitude stabilized to ±1° in each axis.

The AR&C system mission scenarios include each of the following functions: autonomous phasing and rendezvous with a target spacecraft after the CV's arrival in orbit, automated approach and departure maneuvers and automated "soft dock" with the TV. This system is able to meet all of these requirements without ground intervention while providing real time ground monitoring capability and intervention[3]. The results presented in this paper only cover proximity operations. Automated orbit transfer and rendezvous test results will be presented in a later paper.

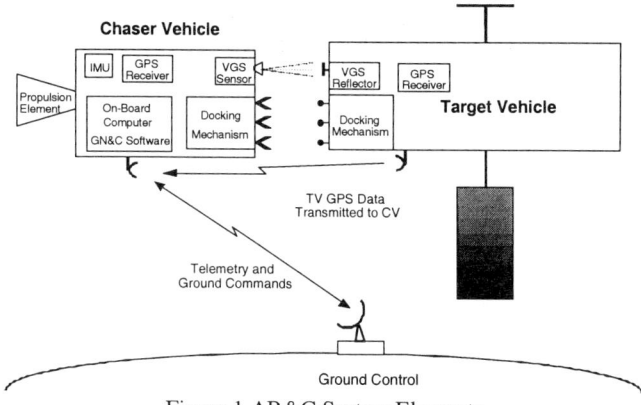

Figure 1 AR&C System Elements

AR&C SYSTEM TEST OVERVIEW

The AR&C tests were performed in the MSFC FRL (see Figure 2). For all test cases, "truth" dynamics were calculated on a Harris Night Hawk, which served as the central simulation computer. This computer stimulated a 20-channel GPS simulator, a high fidelity IMU math model, thruster models as well as all environment models in real time. The 30 m to docking tests were 6 DoF HITL cases using the Dynamic Overhead Target Simulator (DOTS), docking mechanisms, VGS sensor and target hardware. The relative motion of the two vehicles was calculated on the Night Hawk and modeled by moving the DOTS.

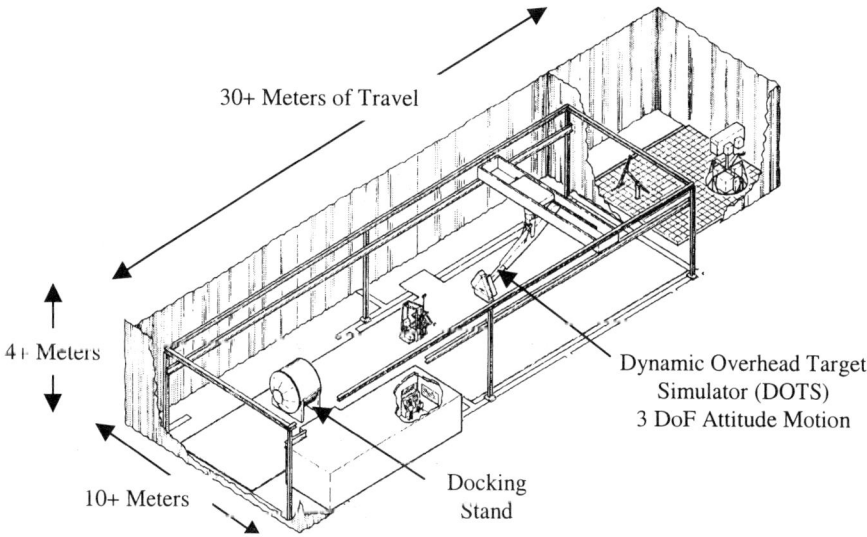

Figure 2 MSFC Flight Robotics Laboratory

The DOTS was outfitted with the passive elements of the system (VGS target and docking trunions). Although the DOTS represented the Target vehicle, usually thought of as stationary, it was easier to mount passive elements on the DOTS rather than elements that require power and data connections. Due to the large distances in the rendezvous to 30 m cases, the DOTs and VGS hardware were not used.

Instead, these were digital tests where the Night Hawk computer drove a functionally equivalent VGS digital model. In all cases, the AR&C flight computer, a Motorola Power PC with Vx Works operating system, executed the Guidance, Navigation and Control (GN&C) software as well as the command, telemetry handling and housekeeping functions. Commands issued by the GN&C were sent to the system elements (GPS receiver, VGS, TPDM) via communications lines. Thrust commands were sent to the Night Hawk computer to be included in the dynamics calculations.

Rendezvous Profile

The AR&C system has been tested throughout all phases of a rendezvous mission starting from orbit insertion, through orbit transfer and phasing, proximity operations, docking and also undock and back away. Due to space constraints, this paper discusses only test results concerning proximity operations and dock. Proximity operations for the AR&C system begin with the Chaser in a coelliptic orbit with the Target at a point 40 km behind and 5 km below. A diagram of the nominal motion of the Chaser vehicle relative to the Target is shown in Figures 3 and 4. Included in the test cases are approaches along the +V-Bar, +R-Bar and –R-Bar axes of the Target. As shown in Figure 3, +V-Bar and –R-Bar type missions approach the TV from in front (direction of orbital velocity). +R-Bar profiles approach from behind and underneath (nadir direction) the Target vehicle[5].

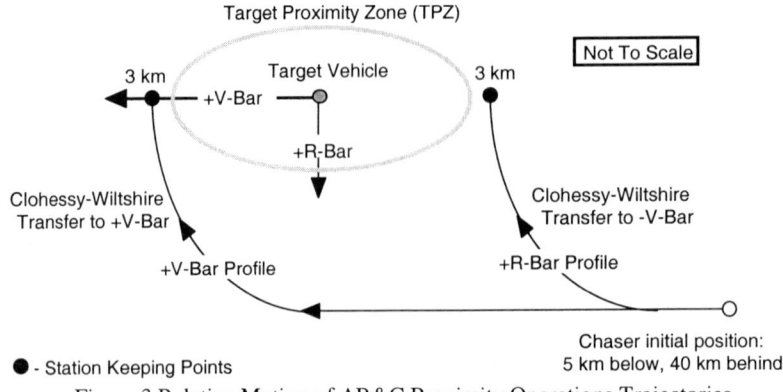

Figure 3 Relative Motion of AR&C Proximity Operations Trajectories

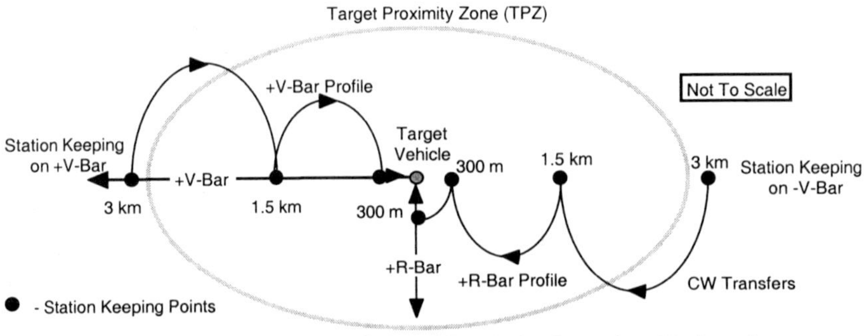

Figure 4 Relative Motion of AR&C Proximity Operations Trajectories

Seven test cases are presented that included nominal approaches as well as commanded aborts and hardware failures that resulted in automatic CAMs. The test cases are listed in Table 2 and are described in detail in the results section.

Table 2 AR&C PROXIMITY OPERATIONS TEST CASES

Case Number	Test Cases	Starting Position	Ending Position	Duration (Hr:Min)	Commands Sent	Faults
1	Rendezvous to 30 m, +V-Bar	40 km Behind, 5 km Below	30 m on +V-Bar	4:20	None	None
2	Rendezvous to 30 m, +R-Bar	40 km Behind, 5 km Below	30 m on +R-Bar	4:30	None	None
3	Rendezvous to 30 m, +V-Bar	40 km Behind, 5 km Below	30 m on +V-Bar	5:10	None	Loss of TV GPS Data, Loss of VGS Data
4	Rendezvous to 30 m, –R-Bar	40 km Behind, 5 km Below	30 m on -R-Bar	6:40	TPZ Abort, New Event Sequence	None
5	30 m to Dock, +V-Bar	30 m on +V-Bar	Docked	0:27	None	None
6	30 m to Dock, +R-Bar	30 m on +R-Bar	Docked	0:27	None	None
7	30 m to Dock, +V-Bar	30 m on +V-Bar	Docked	0:41	Wave Off	Loss of VGS Data, TPDM Fault

AR&C SYSTEM TEST RESULTS

Test 1: Nominal Rendezvous to 30 m, +V-Bar

The Chaser began at the initial rendezvous point with a relative position of 40 km behind and 5 km below the Target vehicle. The Chaser was in controlled drift mode and used GPS navigation to determine its own state and a propagated state vector for the Target to calculate its relative position. During the entire test, the Chaser's control system maintained a local vertical, local horizontal attitude. The controlled drift lasted for an hour during which time the range between the two vehicles decreased due to the differences in orbital rates. At point 2 in Figure 5, the CV initiated a Clohessy-Wiltshire (CW) transfer to the 3 km point on the +V-Bar of the Target. Note that just prior to the CW transfer, there is a large deviation between the navigated and true relative position. This was caused by a timing error in the CV GPS receiver tracking loops. This error only lasted for a few minutes but had a dramatic effect on the nav error. At point 3 on Figure 5, the CV entered the broadcast range of the Target's GPS data (assumed to be 7 km). Once within this range the navigation mode switched to using the two vehicles' raw GPS data to determine their relative state. Point 4 on Figure 6 shows the CV entering into a station keeping event 3 km on the +V-Bar. The allowable station keeping limits are illustrated on the relative motion plots and decrease in size as the CV approaches the TV. The size of the station keeping "boxes" are based upon ISS approach requirements. If the CV were to go outside of these "boxes", an automatic CAM would be triggered, causing the CV to retreat to the previous station keeping point. Points 5 and 6 on Figure 6 show CW transfers to the 1.5 km and 300 m points. Inside of 300 m, the CV used straight line, forced motion transfers to approach the TV. This method and the approach corridor limits are also based upon ISS requirements. From 300 m, the CV approaches to the 100 m point on the +V-Bar and transitions to using the VGS as its primary navigation sensor. The VGS has an approximate maximum tracking range of 150 m and experience has shown while approaching the TV, transitioning from relative GPS to VGS is best accomplished when the CV is station keeping. Figure 7 shows the CV's approach to 100 m with good agreement between the navigated and true states. Figure 8 shows the 100 m station keeping event as well as the transition to VGS nav. After the station keeping event, the CV approached to the 30 m point along the +V-Bar and entered a final station keeping.

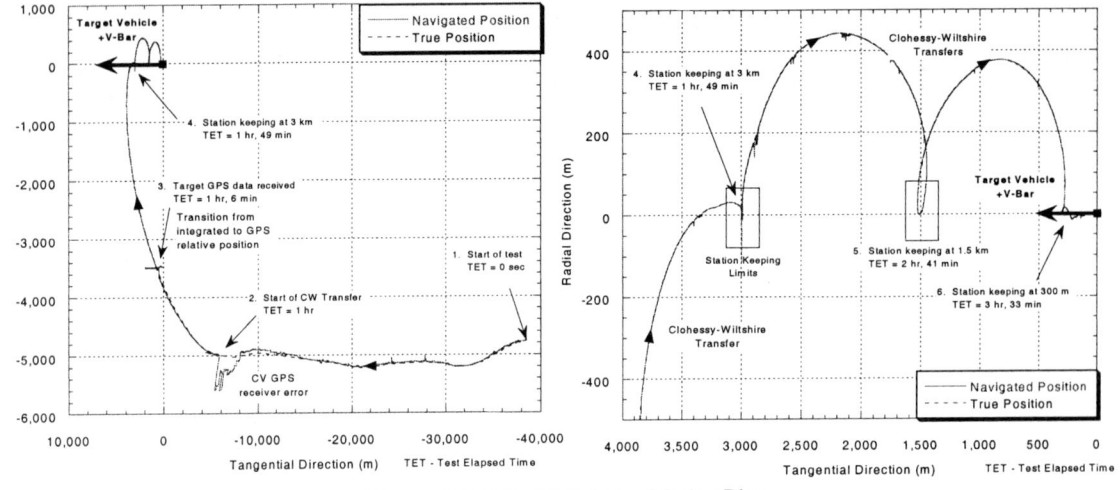

Figures 5 & 6 Test 1 Relative Motion Plots

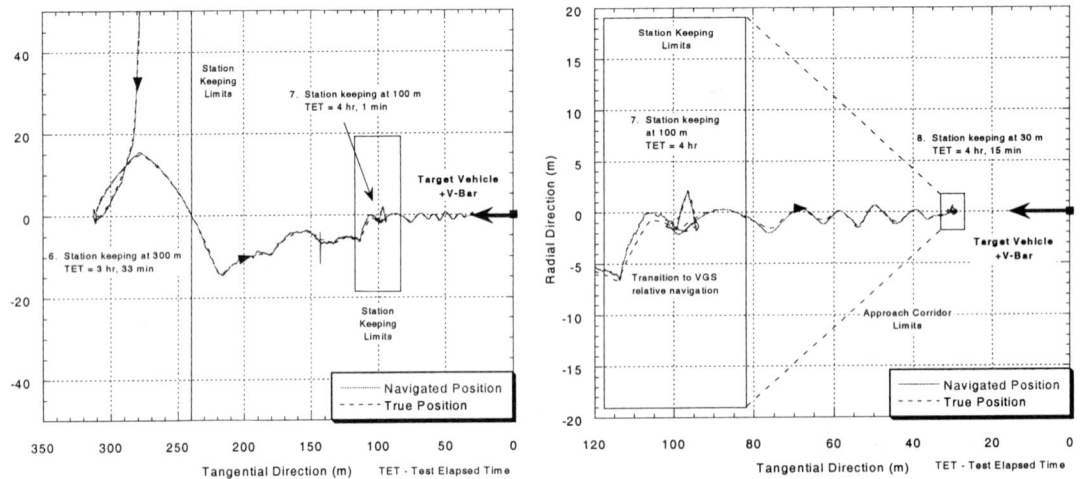

Figures 7 & 8 Test 1 Relative Motion Plots

Table 3 lists the navigation error statistics for the portion of the mission that used relative GPS (40 km to 100 m). Table 4 shows the error statistics while the VGS was used as the primary navigation sensor (100 m to 30 m). The point of maximum error during the approach from 100 m to dock occurred during the 100 m station keeping event. The errors then steadily decreased with range. This is illustrated in Figure 8. Table 5 gives the duration, amount of propellant used and the calculated delta velocity (Delta-V) for each event and the totals.

Table 3 TEST 1 RELATIVE POSITION ERROR USING GPS (7 km to 100 m)

	Radial Error (m)	Tangential Error (m)	Normal Error (m)
Maximum	1.47	2.15	1.34
Mean	-0.37	0.15	0.07
Std Deviation	0.96	1.35	0.36

Table 4 TEST 1 RELATIVE POSITION ERROR USING VGS (100 m to 30 m)

	Radial Error (cm)	Tangential Error (cm)	Normal Error (cm)
Maximum	110	142	110
Mean	-11	21	-24
Std Deviation	30	33	28

Table 5 TEST 1 TIME, PROPELLANT AND DELTA-V BUDGET

Event	Duration (min)	Delta Mass (kg)	Delta-V (m/s)
Controlled Drift	60.9	122.95	8.18
CW Transfer to 3 km	48.78	117.76	7.88
SK at 3 km	5	2.99	0.21
CW Transfer to 1.5 km	46.5	41.25	2.76
SK at 1.5 km	5	9.33	0.64
CW transfer to 300 m	46.73	26.64	1.79
SK at 300 m	5	14.22	0.96
Approach to 100 m	23.75	18.35	1.23
SK at 100 m	5	8.70	0.58
Approach to 30 m	9	35.18	2.36
SK at 30 m	5	13.33	0.90
Totals	**4 hrs, 20 min**	**410.70 kg**	**27.49 m/s**

Test 2: Nominal Rendezvous to 30 m, +R-Bar

Test 2 began with identical initial conditions as Test 1 and the on-board computer was loaded with a +R-Bar approach sequence. The primary difference between the +V-Bar and +R-Bar profiles is that for the latter, the CV approaches along the –V-Bar before transferring to the +R-Bar axis. The reason for this is to allow the CV to have stable station keeping points during the approach. Station keeping points on the ± R-Bar are not stable because to maintain a relative position to the Target, the Chaser must continuously thrust in the direction of orbital velocity to null the difference in mean orbital rates. For Test 2, the Chaser began in controlled drift mode then executed a CW transfer to 3 km on the -V-Bar. Calculated relative navigation was used until the CV came within 7 km of the Target and then switched to relative GPS, shown at point 3 in Figure 9. Upon reaching 3 km on the –V-Bar, the CV station kept for a short time. This was followed by two more CW transfers and two station keeping events at 1.5 km and 300 m (see Figure 10). After the station keeping event at 300 m on the –V-Bar, the CV transferred to 300 m point on the +R-Bar (see point 7 in Figure 10). It then continued the approach to the 100 m point. At 100 m, the Chaser switched to VGS relative navigation and approached up the +R-Bar to 30 m. The test successfully concluded with the Chaser at the 30 m point on the +R-Bar (see point 9 in Figure 12). Tables 6 and 7 list the navigation error statistics for the relative GPS and VGS portions of the mission. Again note that the largest VGS errors occurred during the 100 m station keeping event on +R-Bar (see Figure 12). Table 8 gives the duration, propellant used and Delta-V for each event and the totals. One interesting point to note is the fuel required for the controlled drift portion of Test 1 versus Test 2. Although the two drift distances were essentially the same, Test 1 used almost twice as much propellant. The reason lies in the fact that during Test 1, there was a GPS receiver timing error which caused the relative state to be in error thereby causing the guidance to command a large radial (upward) thrust. Since this timing error did not occur in Test 2, the navigated relative position stayed very close to the true position and no extraneous thrust commands were given.

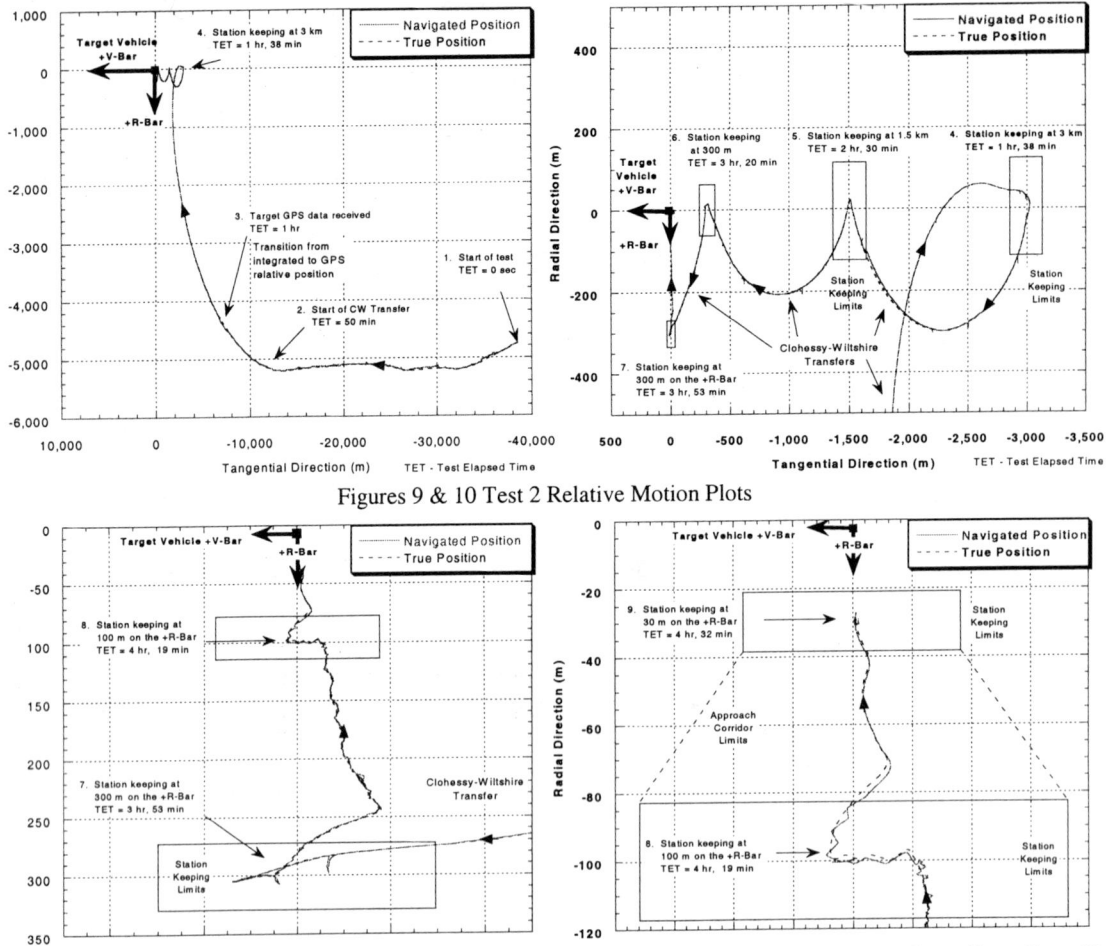

Figures 9 & 10 Test 2 Relative Motion Plots

Figures 11 & 12 Test 2 Relative Motion Plots

Table 6 TEST 2 RELATIVE POSITION ERROR USING GPS (7 km to 100 m)

	Radial Error (m)	Tangential Error (m)	Normal Error (m)
Maximum	9.10	8.92	1.93
Mean	0.55	-0.21	0.10
Std Deviation	1.79	1.18	0.41

Table 7 TEST 2 RELATIVE POSITION ERROR USING VGS (100 m to 30 m)

	Radial Error (cm)	Tangential Error (cm)	Normal Error (cm)
Maximum	223	97	19
Mean	-124	-2	5
Std Deviation	32	28	8

Table 8 TEST 2 TIME, PROPELLANT AND DELTA-V BUDGET

Event	Duration (min)	Delta Mass (kg)	Delta-V (m/s)
Controlled Drift	49.57	63.05	4.19
CW to 3 km -Vbar	49	127.56	8.51
3 km SK	5	4.03	0.28
CW to 1.5 km -Vbar	46.67	35.46	2.37
1.5 km SK	5	3.21	0.22
CW to 300 m -Vbar	46.43	34.60	2.32
300 m SK	5	6.44	0.44
CW to 300 m +Rbar	26.18	33.52	2.25
300 m SK	5	16.70	1.13
Approach to 100 m +Rbar	20.85	25.85	1.74
100 m SK	5	10.30	0.69
Approach to 30 m +Rbar	7.92	9.98	0.67
30 m SK	5	10.23	0.69
Totals	4 hrs, 30 min	380.93 kg	25.50 m/s

Test 3: Rendezvous to 30 m, +V-Bar with GPS and VGS Faults

This test was identical to Test 1 except for four induced faults. The first occurred when the Chaser was approaching from 3 km to 1.5 km. During this phase of the mission, the CV was using relative GPS navigation supplied by its GPS receiver as well as the Target's raw GPS data. The fault involved disrupting the data link from the Target for 5 seconds. This did not cause a CAM, as the AR&C flight rules allow for up to a 30 second loss of relative GPS data in this phase of the mission. Then the TV GPS signal was cut completely and the CV performed an automatic CAM back to the 3 km point on the +V-Bar (point 5, Figure 14). At 3 km, the Target GPS signal was reestablished and the Chaser continued its approach to 100 m on the +V-Bar. The Chaser then used the VGS to continue its approach to 30 m. The third fault was induced by disrupting VGS data for 5 seconds when the CV was between 70 and 80 m from the target. This short outage did not cause the Chaser to CAM. The VGS data was then terminated, causing the CV to CAM back to the 100 m point (point 10, Figure 16). When the Chaser arrived back at the 100 m point, VGS data was restored and the test concluded with the Chaser at the 30 m point on the +V-Bar. Tables 9 and 10 list the navigation error statistics for the relative GPS and VGS portions of the mission. Table 11 gives the duration, propellant used and Delta-V for each event and the totals. The total propellant required for this test was essentially the same as Test 1. This was because although the GPS timing error did not occur in the controlled drift, a significant amount of fuel was required to execute the CAM back to the 3 km point.

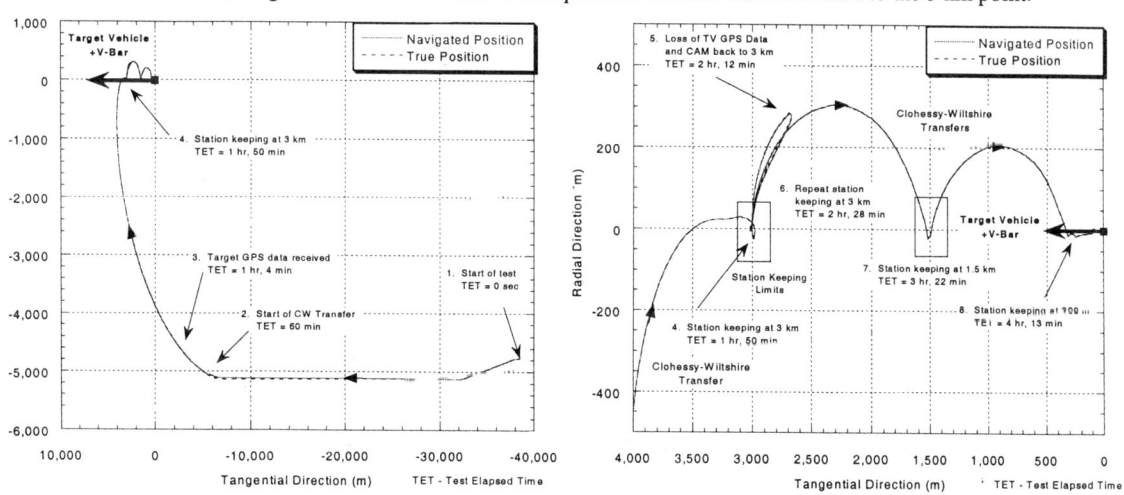

Figures 13 & 14 Test 3 Relative Motion Plots

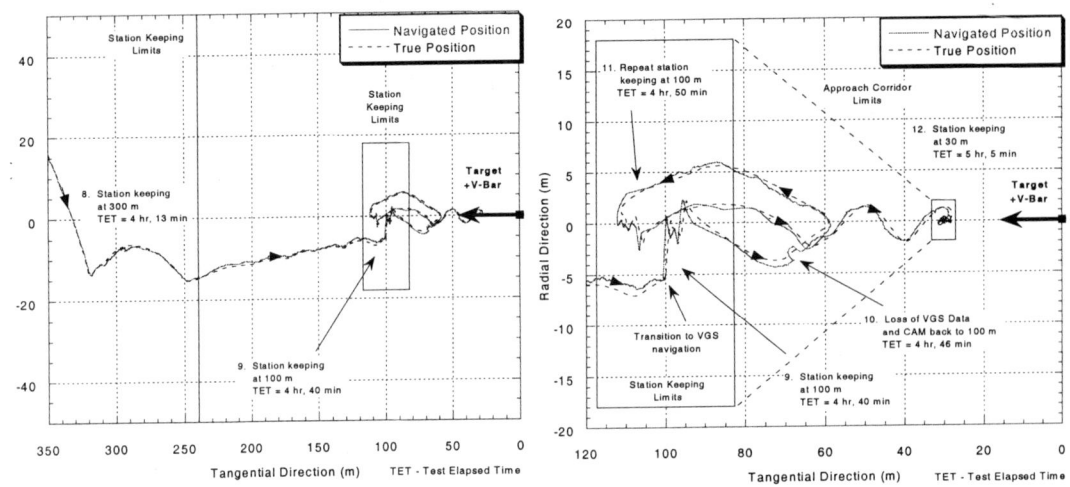

Figures 15 & 16 Test 3 Relative Motion Plots

Table 9 TEST 3 RELATIVE POSITION ERROR USING GPS (7 km to 100 m)

	Radial Error (m)	Tangential Error (m)	Normal Error (m)
|Maximum|	6.15	2.81	1.63
Mean	-0.33	0.01	0.03
Std Deviation	1.36	0.76	0.40

Table 10 TEST 3 RELATIVE POSITION ERROR USING VGS (100 m to 30 m)

	Radial Error (cm)	Tangential Error (cm)	Normal Error (cm)
|Maximum|	115	198	116
Mean	-15	113	-27
Std Deviation	34	25	33

Table 11 TEST 3 TIME, PROPELLANT AND DELTA-V BUDGET

Event	Duration (min)	Delta Mass (kg)	Delta-V (m/s)
Free Drift	61.72	61.90	4.11
CW to 3 km +Vbar	48.95	89.12	5.94
3 km SK	5.02	4.15	0.28
CW to 1.5 km/CAM	34.92	86.05	5.75
3 km SK	5.02	6.19	0.43
CW to 1.5 km	46.60	31.28	2.09
1.5 km SK	5.02	3.63	0.22
CW to 300 m +Vbar	46.30	28.65	1.92
300 m SK	5.02	3.63	0.25
Approach to 100 m +Vbar	20.60	15.42	1.04
100 m SK	5.02	10.12	0.68
Approach to 30 m/CAM	6.58	20.38	1.37
100 m SK	5.02	14.19	0.95
Approach to 30 m +Vbar	9.62	16.93	1.14
30 m SK	5.02	18.80	1.27
Totals	5 hrs, 10 min	410.44 kg	27.44 m/s

Test 4: Rendezvous to 30 m, -R-Bar with TPZ Abort and New Sequence Command

This test was identical to Test 1 except that during the mission two non-standard commands were issued. This test began at the initial rendezvous point (40 km behind and 5 km below the Target) and the mission progressed nominally until the Chaser was approaching from 1.5 km to 300 m. At this point, a TPZ Abort command was issued from the command console, which caused the CV to retreat immediately to 3 km (point 6, Figure 18). Upon arriving at the 3 km point, the Chaser entered station keeping and waited for further instructions. Then a new mission sequence was sent to the Chaser that changed the approach type from +V-Bar to -R-Bar. This new approach is similar to the +V-Bar however at the 300 m position on the +V-Bar the CV executed a CW transfer to the 300 m point on the -R-Bar (point 9, Figure 19). From there, the Chaser approached along the -R-Bar to the 100 m point (with 5 minute station keeping events at 300 m and 100 m). At the 100 m point, the Chaser transitioned to the VGS navigation mode (point 11, Figure 20) and approached to the 30 m point. The test concluded with the Chaser vehicle at 30 m on the -R-Bar. Tables 12 and 13 list the navigation error statistics for the relative GPS and VGS portions of the mission. Table 14 gives the duration, propellant used and Delta-V for each event and the totals. From this table it's easy to see that the TPZ abort maneuver is very expensive in terms of propellant. The CAM maneuver took approximately 146 kg of propellant as opposed to only 21 kg for the nominal CW transfer.

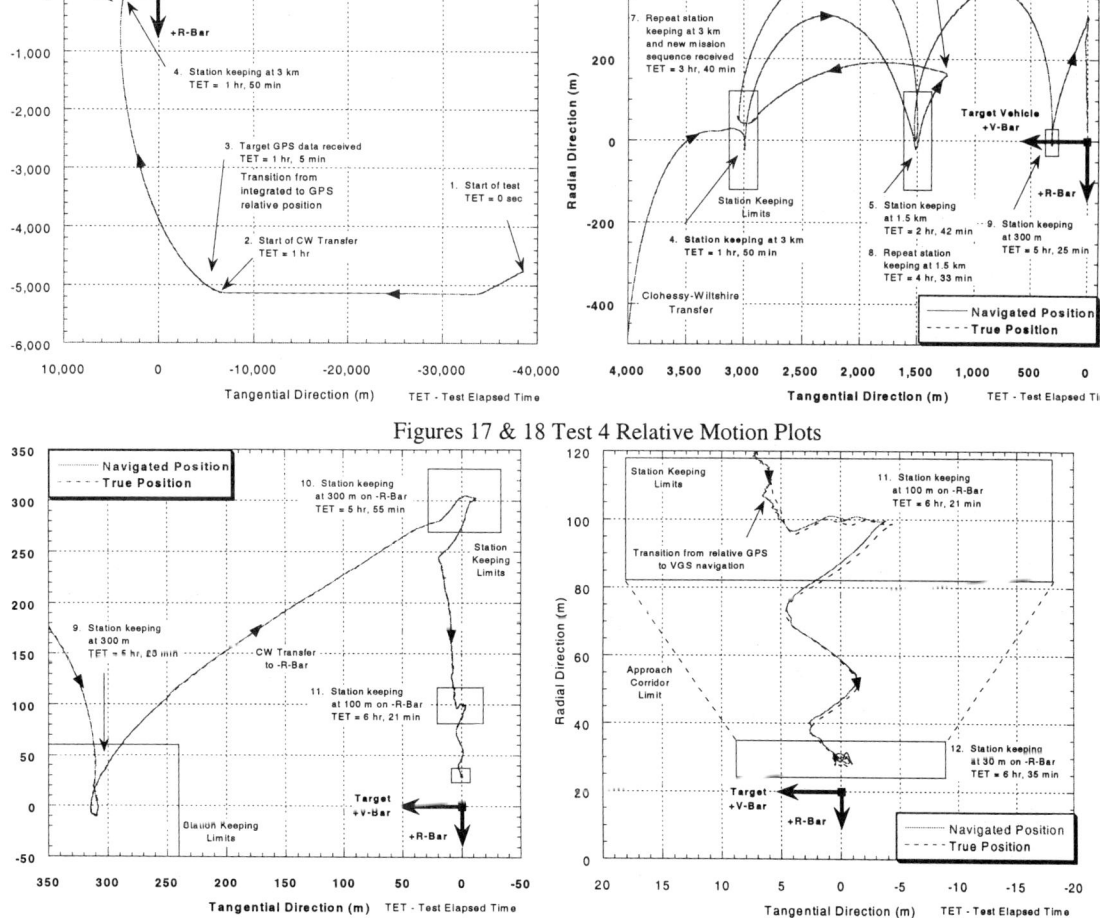

Figures 17 & 18 Test 4 Relative Motion Plots

Figures 19 & 20 Test 4 Relative Motion Plots

Table 12 TEST 4 RELATIVE POSITION ERROR USING GPS (7 km to 100 m)

	Radial Error (m)	Tangential Error (m)	Normal Error (m)
Maximum	3.41	1.20	0.96
Mean	-0.05	-0.02	0.04
Std Deviation	0.60	1.09	0.38

Table 13 TEST 4 RELATIVE POSITION ERROR USING VGS (100 m to 30 m)

	Radial Error (m)	Tangential Error (m)	Normal Error (m)
Maximum	2.08	1.08	0.42
Mean	1.12	0.18	0.31
Std Deviation	0.22	0.31	0.05

Table 14 TEST 4 TIME, PROPELLANT AND DELTA-V BUDGET

Event	Duration (min)	Delta Mass (kg)	Delta-V (m/s)
Free Drift	61.77	60.92	4.05
CW to 3 km	48.65	89.23	5.96
3 km SK	5.02	4.98	0.34
CW to 1.5 km	46.57	33.15	2.21
1.5 km SK	5.02	3.50	0.24
CW to 300 m/TPZ ABORT	53.80	146.46	9.81
3 km SK	5.03	0.37	0.03
CW to 1.5 km	47.05	25.42	1.71
1.5 km SK	5.02	1.93	0.14
CW to 300 m +Vbar	46.95	21.46	1.44
300 m SK	5.02	2.14	0.15
CW to 300 m -Rbar	25.47	25.88	1.74
300 m SK	5.02	15.02	1.02
Approach to 100 m -Rbar	21.20	29.94	2.02
100 m SK	5.02	12.29	0.83
Approach to 30 m -Rbar	8.40	14.90	1.00
30 m SK	5.02	19.60	1.32
Totals	**6 hrs, 40 min**	**507.19 kg**	**34.01 m/s**

Test 5: Nominal 30 m to Dock, +V-Bar

This was a HITL test that used the DOTS overhead crane, VGS and TPDM hardware. It began with the Chaser 30 m away from the Target vehicle on its +V-Bar. The Chaser started in station keeping mode and after acquiring lock with the VGS, performed a forced motion approach along the +V-Bar to the 10 m point (point 2, Figure 21). After station keeping at 10 m, the Chaser transitioned to terminal autopilot mode and proceeded into dock. The test concluded with the TPDM hardware achieving dock and the Chaser entering standby mode. Figure 21 shows the navigated and true relative motion profile for the test. Tables 15 and 16 give the relative error statistics for the run and the final error at dock. These errors show a very good agreement between navigated and true relative states.

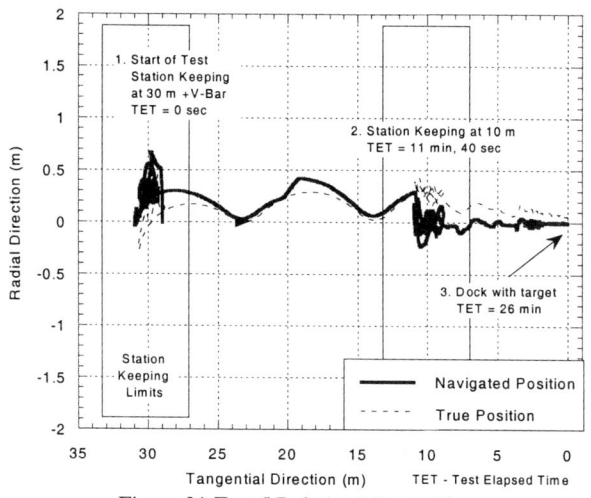

Figure 21 Test 5 Relative Motion Plot

Table 15 TEST 5 RELATIVE POSITION ERROR USING VGS

	Radial Error (cm)	Tangential Error (cm)	Normal Error (cm)
|Maximum|	53	63	26
Mean	-6	20	-5
Std Deviation	17	18	13

Table 16 TEST 5 NAVIGATED POSITION ERROR AT DOCK

Radial Error (cm)	Tangential Error (cm)	Normal Error (cm)
5	3	9

Test 6: Nominal 30 m to Dock, +R-Bar

This was also a HITL test that used the DOTS. It began with the Chaser 30 m away from the Target along its +R-Bar. The Chaser started in a station keeping event and after locking on with the VGS, performed a forced approach up the +R-Bar to the 10 m point (point 2, Figure 22). After another station keeping event, the Chaser transitioned to terminal autopilot mode and proceeded into dock. During the final approach, the VGS lost lock on the target for 6 seconds (point 3, Figure 22). During that time the relative state "jumped" approximately 2 meters as the primary navigation sensor changed from the VGS to the GPS filter. Had the dropout exceeded 10 seconds, an automatic CAM would have been triggered and the CV would have retreated back to the 10 m point. However the VGS regained lock and the approach continued. The test concluded with the TPDM hardware achieving dock and the CV entering standby mode. Figure 22 shows the navigated and true relative motion profile for the test. Tables 17 and 18 give the relative error statistics for the run and the final error at dock. The maximum error values in Table 17 represent the jump in state when the VGS lost lock. One obvious improvement to the system would be to filter both the VGS and GPS relative states to avoid this type of behavior during data drop outs.

Table 17 TEST 6 RELATIVE POSITION ERROR

	Radial Error (cm)	Tangential Error (cm)	Normal Error (cm)
|Maximum|	365	48	41
Mean	-18	8	9
Std Deviation	22	19	19

Table 18 TEST 6 NAVIGATED POSITION ERROR AT DOCK

Radial Error (cm)	Tangential Error (cm)	Normal Error (cm)
-2	-5	0.5

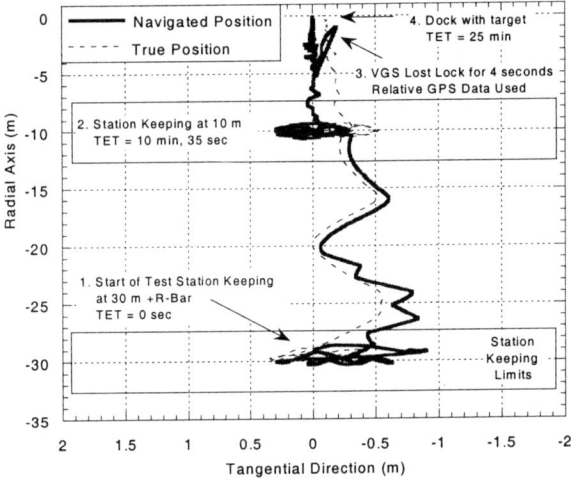

Figure 22 Test 6 Relative Motion Plot

Test 7: 30 m to Dock, +V-Bar with Wave Off Command and VGS, TPDM Faults

This final case was also a HITL test that used the DOTS in the FRL. It began with the same initial conditions as Test 5, which had the Chaser station keeping at the 30 m point on the +V-Bar of the Target vehicle. After locking on with the VGS, the CV performed a forced approach to the 10 m point and started station keeping. At the end of the station keeping event, the CV transitioned to terminal autopilot mode and started its final approach to dock. When the Chaser was between 10 and 6 m away from the target, a planned VGS fault was introduced which stopped the VGS data to the OBC for 3 seconds (point 3, Figure 23). This did not cause a CAM as the AR&C flight rules state that the system can handle a loss of VGS data in this range for up to 10 seconds without issuing a CAM. A second fault was then introduced which reported a TPDM "bad" status to the flight computer. This fault caused the CV to CAM back to the 10 m point and station keep until the fault was resolved (point 4, Figure 23). When the Chaser achieved station keeping at 10 m, a TPDM "good" status indication was restored and after station keeping for 5 minutes, the chaser again approached to dock. At the 5 m point, a Wave Off command was issued from the simulation control console which again caused the Chaser to CAM back to the 10 m point and station keep (point 5, Figure 23). After this had been verified, the approach continued and the test concluded with Chaser docking with the Target and entering standby mode.

Table 19 TEST 7 RELATIVE POSITION ERROR

	Radial Error (cm)	Tangential Error (cm)	Normal Error (cm)
Maximum	81	58	69
Mean	-28	13	5
Std Deviation	28	17	27

Table 20 TEST 7 NAVIGATED POSITION ERROR AT DOCK

Radial Error (cm)	Tangential Error (cm)	Normal Error (cm)
-8	5	-3

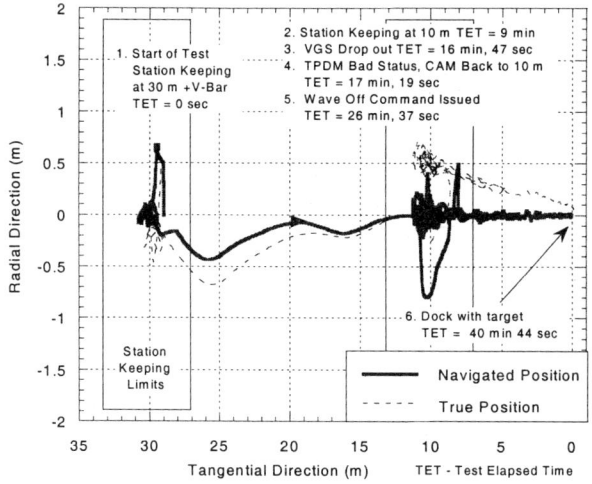

Figure 23 Test 7 Relative Motion Plot

GPS NAVIGATION TESTING AND PERFORMANCE

The use of GPS receivers has become the standard in modern orbital navigation. Rendezvous and proximity operations of two orbital vehicles have taken advantage of this navigation sensor to replace or augment radar as a method to determine relative ranging between vehicles. This can be accomplished by several methods. One method is the state vector differencing of two GPS receiver's navigational solutions. This method is robust but uncommon satellite geometries can induce larger than desired errors. A second method of using GPS receivers for relative navigation can be accomplished when line of sight data from the two receivers is input into a Kalman filter to generate a more accurate relative state between vehicles even in cases of minimal common GPS satellites. The use of a GPS relative navigation filter for long and medium range navigation was tested in the MSFC FRL as part of the AR&C program. A 20-channel Nortel GPS simulator was used to stimulate the receivers being tested. The simulator incorporated error models for GPS satellite Selective Availability (SA), multipath, almanacs, ephemerides, ionospheric and tropospheric models, and receiver antenna patterns. SA and multipath errors were not modeled in the results presented here but were tested. A future paper will go into detail on the filter design and performance in the presence of all these relevant error sources. Testing of both the receiver's and the filter's performance was accomplished prior to the AR&C system test described earlier.

GPS Receiver Testing and Performance

The first tests of the GPS receiver covered the International Traffic in Arms Regulations (ITAR) limits. US manufacturers of GPS receivers have these limits imposed on them so that foreign powers could not use this hardware for military applications. If justification for need is approved, manufacturers can disable this software function. Orbital GPS receivers must have these limits removed or the unit will not operate in the space environment. The first test to see if these software limits are in place requires that the receiver be stimulated with a vertical rise from the Earth's surface to a height above the ITAR altitude limit of 60,000 ft. The second test trajectory is similar to that of an aircraft traveling at a rate greater than the ITAR velocity limit of 1000 knots. The third trajectory combines these two limits in a low Earth circular orbit. In each case the trajectories were long enough to allow almanac and ephemeride downloads and satellite switching to occur. Functionality, robustness and performance were then analyzed.

Tests were also conducted on the receivers to verify their ability to correctly account for the weekly rollover of GPS satellite almanacs and the 4-6 hour change in satellite ephemerides. Tests included the modeling of almanac and ephemeride changes to assure that discontinuities in the navigation solutions did not occur. The weekly change in the GPS almanac and ephemerides affects a receiver's ability to lock onto satellites quickly and accurately. There are four possible inputs into a GPS receiver to aid it in acquiring and tracking satellites. The four inputs are current time, current almanac, current ephemerides and current location and velocity. Not all of these inputs are necessary for a GPS receiver to work correctly. An error in the current time of a receiver can cause incorrect propagation of the GPS satellite constellation and the related Doppler shift of each satellite. The acquisition capability of a receiver is affected by not knowing which satellites to look for and what their Doppler shift should be. Propagation of a particular GPS satellite almanac is essential for a receiver to begin tracking of the satellites. Once a receiver tracks a satellite it can download information from it about the current almanac and that satellite's ephemeris. The almanac download takes about 30 seconds per satellite. Some receivers do not use this new information until it has downloaded all satellite information and verified that the complete data set is correct. Once a receiver knows the correct time and orientation of the GPS constellation, it needs to know approximately its own position and velocity. A limited channel receiver uses this information to look for only the GPS satellites that should be visible. The receiver's velocity is used in conjunction with the calculated GPS satellite's velocity to determine the Doppler shift window to search for acquisition of satellites. Terrestrial velocities have little effect on the shift but orbital velocities have a major impact on acquisition.

Several tests were performed to quantify the receivers' acquisition time. The time to first fix (TTFF) is the time required by the receiver to start tracking and produce a navigation solution. The TTFF performance of the receiver was tested in several modes. Table 21 summarizes the performance of the unit being tested.

Table 21 GPS RECEIVER TIME TO FIRST FIX RESULTS

Startup Description	TTFF
Cold – Without Aiding	15 minutes
Cold – With Aiding	3 minutes
Warm – With Aiding	15 seconds

All TTFF tests were performed after the initialization of the receivers with the current GPS time. The first test of TTFF was performed with a stationary position on the Earth, a 200 week old almanac and no information on the receiver's location or velocity. This is considered a "cold" startup. In the first test, after powering up the unit, it went into "sky search" mode. In this mode, the receiver cycles through all satellites and Doppler shifts to acquire a satellite. This mode does not work for orbital receivers due to the high Doppler shift and rapidly changing constellation they experience. Orbital receivers must be given aiding (position and velocity) information to acquire. In the second case the receiver started with a 200 week old almanac and no satellite ephemerides. It was given information on its position and velocity to aid in acquisition. This case was tested in an orbital and terrestrial environment. Knowledge of the receiver's location and the correct time allowed the embedded receiver propagation algorithm to correct the satellite location and determine ranging data accurately enough to acquire satellites quickly. In the third case the receiver started with a current almanac and the most recent set of satellite ephemerides. In this case the receiver could very quickly acquire visible satellites. The results of these tests depend on the number of receiver channels available for tracking of satellites and the fidelity of embedded software in the receiver.

The receiver performance from Test 1 can be seen in Figures 24 and 25. Figure 24 shows the positional error in the Chaser and Target vehicle's receiver solutions. Figure 25 shows the velocity error in the Chaser and Target vehicle's receiver solutions. The time period for this test is more than two and one-half hours long, which is long enough to allow a total switch of GPS satellites during the nearly two orbits. Receiver performance, using CA code and with SA turned off was expected to be 10 meters in position and 1.0 m/sec in velocity. The receivers demonstrated that they could meet this requirement[2].

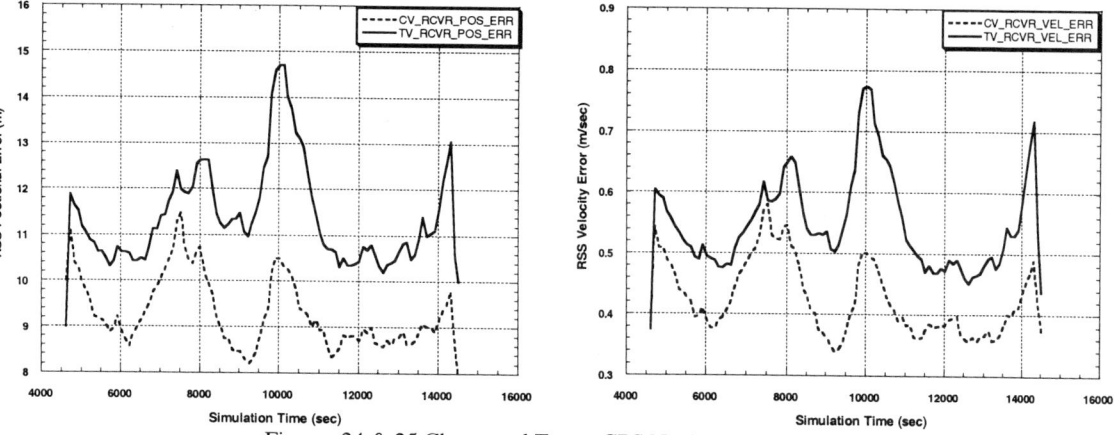

Figures 24 & 25 Chaser and Target GPS Navigation Performance

Relative Navigation Filter Testing and Performance

The GPS Relative Navigation Filter[3,6,7] requests several packages of information from the receivers. The filter requests ephemerides, timing and line of sight data. Part of the timing information the filter requests is the Universal Coordinated Time (UTC) leap seconds. This variable is in the GPS satellite navigation message and is needed due to the coordinate system in which the filter works. GPS receivers put out data in the WGS84 Earth Centered Earth Fixed frame. Since the filter was designed to work in an orbital environment, the filter puts out its solution in the Earth Centered Inertial frame. The UTC leap second correction is used to convert from one frame to the other. Testing of the filter showed that when a UTC leap second change occurred, the filter did not cause a discontinuity in the vehicle's position solution.

The GPS Relative Navigation Filter was designed with an accuracy requirement[2] of 10 meters in position and 0.1 m/sec in velocity (three sigma values.) The filter was to meet these requirements in the presence of realistic error sources and satellite visibility and commonality. The use of 12-channel GPS receivers and 80 degree half-cone angle antenna patterns, increased the likelihood of more than four common GPS satellites. Figure 26 shows the results from Test 1 of the number of common satellites between the two vehicles during the test. When less that four common satellites are used in the filter's solution, performance begins to degrade.

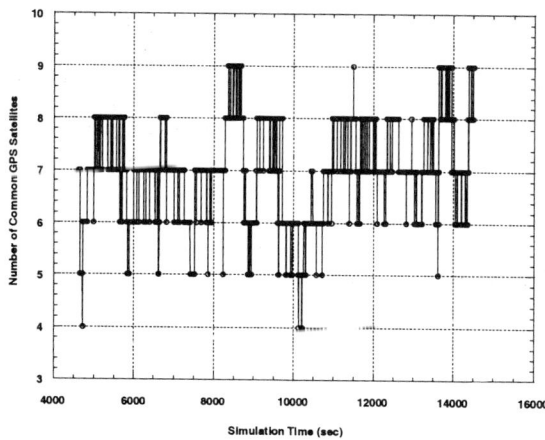

Figure 26 GPS Relative Filter Chaser and Target Common Satellites

The filter developed and tested for the AR&C project was hosted on a Personal Computer (PC) with a real-time operating system and three bi-directional RS232 interfaces. The filter communicated with two GPS receivers and the flight computer. The filter was designed as an asynchronous process external to the flight computer's navigation function. The relative solution that the filter on the PC generated was passed with validity flags to the navigation function on the flight computer. Navigation would take this data and propagate it to current vehicle time. The navigation function was designed to smoothly transition from filter relative solutions to state vector differencing of the receiver states when a filter solution was not available. Testing of the filter generated errors in position as seen in Figure 27 and in velocity as seen in Figure 28.

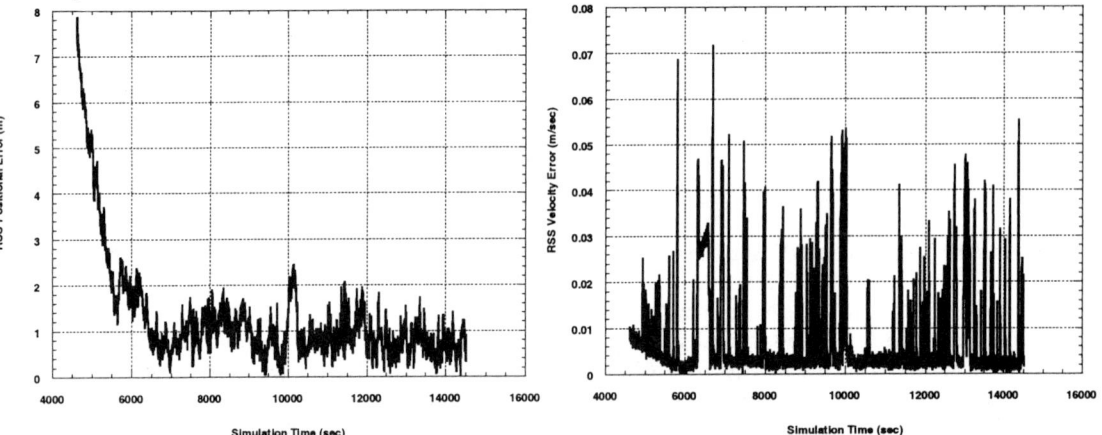

Figures 27 & 28 Relative GPS Navigation Position and Velocity Performance

The positional error plot shows an exponentially decreasing error as the simulation time increases. This can be accounted for from two sources. First, the vehicles are getting closer together and the mathematical local level relative coordinate system, that is assumed by the filter, does not take into account curvilinear affects. At 6,540 seconds simulation elapsed time the Chaser vehicle is within 3 km of the Target and located along the Target's velocity vector. By this time the curvilinear effects are small and the vehicles are relatively close to each other. Second, the filter does take some time to converge on a solution. Tests show this convergence time to be 15-20 seconds if large burns or orbit transfers are not involved. A large positional error can be seen around 6000 seconds on Figure 27 that represents the final burn of a CW transfer to the Vbar of the Target vehicle. Subsequent burns in the proximity zone of the Target vehicle are numerous but small. It can be seen from the figures that once the vehicles are within close proximity of each other the filter produces a very good result. Velocity errors from this type of filter, as shown in Figure 28, are sensitive to the type of data provided to the Kalman filter. The filter tested has three types of inputs that are used for velocity calculations. They are; range-rate, Doppler or carrier phase. The performance results shown are for a carrier phase input run. The accuracy of the GPS receiver to track the carrier phase will have an impact on the performance of the velocity solution by the filter.

The testing results shown here will be presented in much more detail in a future paper that discusses the filter in detail. Also, that paper will give the testing results for the more demanding case where SA errors and multipath errors are modeled.

TERMINAL AUTOPILOT AND ATTITUDE CONTROL PERFORMANCE

The primary function of the Terminal Autopilot is to control the position, velocity and attitude of the Chaser vehicle relative to the Target vehicle from 10 m to dock[8]. It keeps the CV position on the projected docking axis of the TV, and the CV attitude aligned with that of the docking plane. Its goal is to fly the approach which results in the softest possible docking with the mimimal risk of an abort. This is different from the forced approach guidance function, used from 300 m to 10 m, which maintains the CV as close as possible to the V-bar or R-bar as desired, without regard to relative attitude. Figures 29 and 30 provide vertical and horizontal profile views of Test case 5, a typical V-bar approach, starting from a stationkeep hold at 30 m. Figure 29 shows that upon taking control after the 10 m stationkeep, the Terminal Autopilot reduced the four centimeter residual radial misalignment to 0.5 cm before contact occurred. Figure 30 shows that it also held the normal alignment error to under one cm for the entire duration. Both plots show that this was accomplished while holding velocity changes to a minimum, indicating a high degree of system stability.

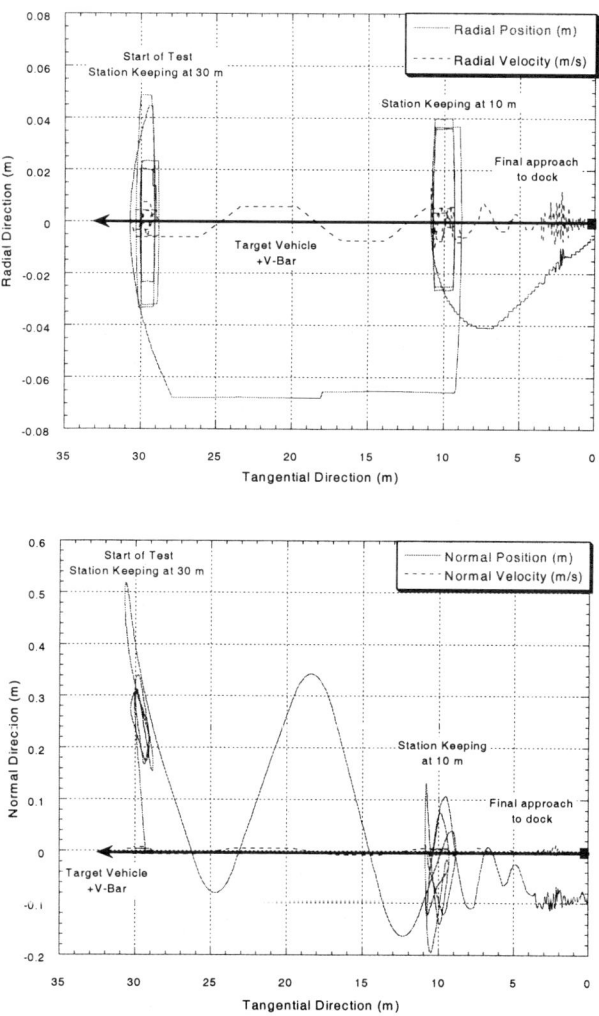

Figures 29 & 30 Vertical and Horizontal Final Aproach Profiles, Test 5

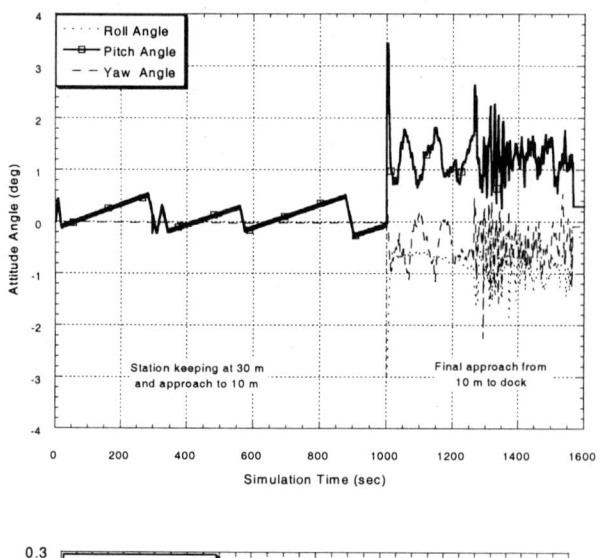
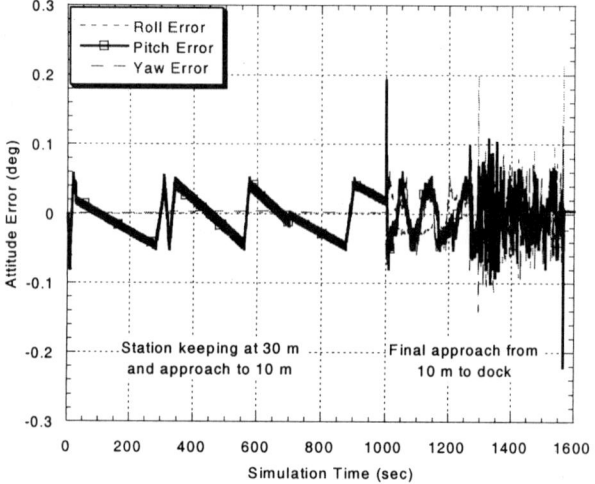

Figures 31 & 32: Vertical and Horizontal Final Aproach Profiles, Test 5

The attitude control function of the autopilot, which is active in both terminal and guidance-directed modes, is demonstrated in figures 31 and 32. Figure 31 indicates the actual attitude of the Chaser, initially relative to the Local Vertical, Local Horizontal reference frame, then (after the transition to Terminal Autopilot mode at 1,000 seconds) relative to the TV. Figure 32 depicts the attitude errors during the same run. Figure 32 clearly shows that during the approach segment preceding the 10 m stationkeep, the autopilot maintained the roll and yaw errors to almost imperceptible values, and pitch to within less than 0.1 degree of the commanded angle. Pitch requires more active control during a forced approach because of the need to track orbital rate, thus the larger error swings. During the terminal portion of the run, it can be seen from Figure 32 that relative attitude in all three axes was maintained to within 0.1 degree of the Target, despite the presence of noise from the VGS, which is typically an order of magnitude higher than that from the IMU. Also, it is evident by the near-zero attitude error values at dock in Figure 32, that the autopilot compensation for the hardware offsets seen in figure 31 (non-zero attitude values at dock) was accomplished sucessfully.

The Terminal Autopilot is also responsible for providing velocity control during the last 10 m of the approach, smoothly slowing the vehicle as much as possible without bringing it to a complete stop prior to contact[3]. Figure 33 illustrates this action and shows that velocity was reduced from 2 cm per second at the beginning of the final approach to about 0.25 cm per second immediately prior to entering the docking mechanism. The burst of noise at about three meters resulted from a software problem in the VGS which has been corrected, and was filtered out by the Terminal Autopilot[4]. Such a low contact velocity has never before been repeatably achieved by either human pilots or other automated systems, and makes possible reductions in the wieght and complexity of docking fixtures because kinetic energy absorbtion is no longer an issue.

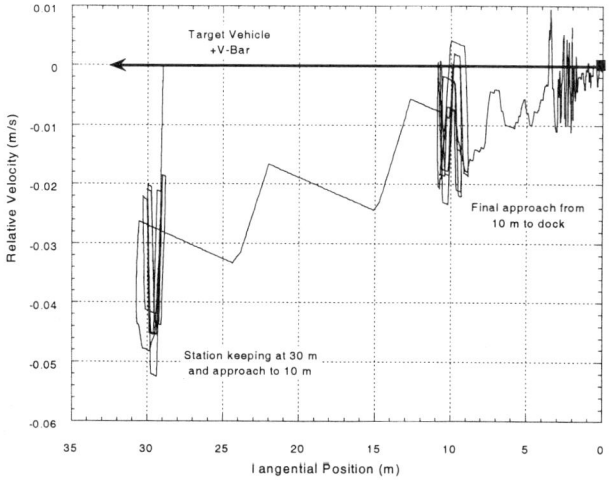

Figure 33 Chase Vehicle Relative Velocity vs. Range, Test 5

CONCLUSION

The results presented in this paper show that the current AR&C system was able to dock with a Target vehicle in the absence of failures. When system anomalies were introduced, the AR&C CAM logic automatically detected them and acted to ensure the safety of the two vehicles while trying to preserve mission success. Analysis of the test data thus far shows that all system and subsystem requirements were met. The AR&C project engineers will continue to scrutinize the test data and make improvements to the system. In the mean time, the MSFC Pathfinder office is working to identify a vehicle or vehicles that could benefit from these technologies. To this end, the joint NASA-Boeing Pathfinder program plans to use the AR&C system in a rendezvous and close approach demonstration in low earth orbit on the second flight of the X-37 vehicle in 2003. One of the primary goals for this flight is to prove that by designing a safe and robust automated system, recurring mission operations costs can be significantly reduced.

ACKNOWLEDGEMENT

The authors would like to acknowledge the following people who contributed to the design of MSFC's AR&C system. Michael Book, Linda Brewster, Thomas Bryan, Geoffrey Burton, Mark Coffman, Greg Dukeman, Tina Ehresman, Richard Howard, Jonathan Lohr, Carrie Olsen, Ronald Phillips, Kenneth Ricks, Margaret Stroud and John Weir. We would also like to thank the AR&C project and MSFC management for their support.

REFERNCES

1. Polites, M.E., "An Assessment of the Technology of Automated Rendezvous and Capture in Space", NASA/TP-1998-208528 July 1998.

2. "Automated Rendezvous and Capture System Requirements Document", MSFC-RQMT-2371 B, 1996.

3. Cruzen, C.A., Lomas, J.J., "Design of the Automated Rendezvous and Capture Docking System", AIAA Paper, AIAA International Space Station Service Vehicles Conference, Houston TX, April 26-28, 1999.

4. Howard, R., Bryan, T., Book, M., Dabney, R., "The Video Guidance Sensor – A Flight Proven Technology", AAS Paper 99-025, Presented at the 22nd AAS Guidance and Control Conference, Breckenridge, CO, Feb 3-7, 1999.

5. Deaton, A.W., Lomas, J.J., Mullins, L.D., "A Plan For Spacecraft Automated Rendezvous", NASA/MSFC Technical Memorandum, NASA TM-10835, October 1992.

6. Galdos, J.I., Upadhyay, T.N., Deaton, A.W., Lomas, J.J., "A GPS Relative Navigation Filter for Automatic Rendezvous and Capture", Mayflower Communications Final Report, Contract No. NAS 8-37852, March 9, 1993.

7. Upadhyay, T.N., Cotterill, S., Deaton, A.W., "Autonomous Reconfigurable GPS/INS Navigation and Pointing System for Rendezvous and Docking", AIAA Paper 92-1390, AIAA Space Programs and Technologies Conference, Huntsville AL, March 24-27, 1992.

8. Calhoun, P., Dabney, R., "A Solution to the Problem of Determining the Relative 6 DoF State for Spacecraft Automated Rendezvous and Docking", *Proceedings of SPIE Space Guidance, Control and Tracking II*, 1995.

LOST-IN-SPACE: A STAR PATTERN RECOGNITION AND ATTITUDE ESTIMATION APPROACH FOR THE CASE OF NO PRIOR ATTITUDE INFORMATION

Gwanghyoek Ju,[*] Hye-Young Kim,[*] Thomas C. Pollock,[†] John L. Junkins,[‡] Jer-Nan Juang[**] and Daniele Mortari[††]

A novel attitude determination approach is presented and results from night sky validation tests are discussed. The central ideas are extensions of the *K-Vector* method recently introduced by Mortari. It involves a construction of a judicious "star pair catalog" prior to launch, wherein all cataloged stars are considered as star pairs and are ordered {k = 1, 2, ... N~O(10^6)}, over the whole sky, sorted in the order of increasing inter-star angle. From this, we have a "searchless" means to access the candidate set of stars for each measured star pair, but this *K*-Vector method may still give 10s and sometimes 100s of candidate stars for each measured pair. We introduce in this paper a method to identify the measured stars in the subset of candidate stars accessed using the *K*-vector. The new method is based upon a logical process which *pivots* about two stars (the first identified pair) to more efficiently identify or ignore the remaining measured stars. Using this Pivot Method, we show that star identification can be reliably and efficiently implemented on-orbit, even for the Lost-In-Space case, and thereby this paper introduces a globally valid process, consistent with real-time, on-board computational constraints, that solves the most fundamental problem associated with star pattern identification. Results from night sky experiments are discussed which support the validity of the analysis and practicality of this approach. Also discussed are plans for on-orbit experiments. The StarNav experiment is planned for Shuttle Mission STS 107, January 2000; this will represent the first ever on-orbit demonstration of a star sensor implementing a Lost-In-Space star identification and attitude determination process.

Introduction

Star observations are widely used by spacecraft as a primary means of precision attitude determination. Star line of sight vectors are usually measured by a CCD (Charge Coupled Device) camera generating a 2D image of a small section of the sky. Since the first star tracker based on the use of CCD area array imaging sensors was developed in the early 1970s at JPL (Jet Propulsion Laboratory), many families of solid state star sensors have been developed to meet the desire for increasingly precise, real-time, on-board attitude estimation during the last two decades. Currently, CCD based star trackers allow us to obtain precise spacecraft attitude estimates by applying star pattern recognition algorithms in an on-board computer, which has been implemented with several existing algorithms.

---

[*] Graduate Research Assistant, Aerospace Engineering Dept., Texas A&M University, College Station, Texas 77843.
[†] Associate Professor, Aerospace Engineering Department, Texas A&M University, College Station, Texas 77843.
[‡] George Eppright Chair Professor, Aerospace Engineering Department, Texas A&M University, College Station, Texas 77843. Fellow AAS.
[**] Principal Scientist, Structural Dynamics Branch, NASA Langley Research Center, Hampton, Virginia 23681. Fellow AAS.
[††] Assistant Professor, Aerospace Engineering School, University of Rome "La Sapienza." Also, Visiting Associate Professor of Aerospace Engineering, Texas A&M University, College Station, Texas 77843. Member AAS.

However, attitude control, pointing control, and navigation systems of future advanced spacecraft will be characterized by requirements for a high degree of autonomy, automatic on-orbit calibration, very high accuracy, efficient commandability, and fast fault recovery. These characteristics are incompatible with the constraints of conventional star trackers and current flight operational algorithms which require at least approximate prior attitude estimates or, in many cases, additional approximate sensors (such as sun, horizon, or magnetic field sensors) and associated algorithms. In addition to being inadequate for implementing ambitious plans for future spacecraft systems, conventional star trackers and algorithms are costly to use because their lack of flexibility/adaptability necessitates the expenditure of significant human, ground software, uplink, redundant sensors, and telemetry resources.

With the advent of more accurate and sensitive CCDs, faster microprocessors, and higher-density memory chips, it has become feasible to build an *autonomous, intelligent* star tracker and software system which is capable of alleviating the above-mentioned problems of existing systems. Accompanying the development of star tracker and microprocessor hardware, many different approaches for star identification and associated algorithms have been devised to solve the attitude determination problem[1-14]. Nevertheless, only a few star pattern identification methods, especially for the *lost-in-space* case, have been developed[5, 11-14]. No algorithm capable of solving the lost-in-space problem has been implemented for an actual space mission. Thus the present algorithm, scheduled to test on-orbit aboard STS-107 in early 2001, will represent a significant milestone.

Van Bezooijen [5] developed an autonomous star identification method for the AST (Autonomous Star Tracker) developed and tested by Lockheed Palo Alto Research Laboratory (Currently, Lockheed Martin's Advanced Technology Center). The algorithm, which was also designed to consider the lost-in-space case, finds the largest group of observed stars that matches a group of guide stars and attempts to maximize its success rate by exploiting the full information content of star patterns including the visual magnitude information. The guide star database was specially designed, which consists of the position and brightness of the acquisition guide stars plus a sorted list of all acquisition guide star pairs that fit in the FOV (Field-Of-View) of AST.

Mortari [11] developed a robust star identification approach, called the SLA (Search-Less Algorithm), for the *lost-in-space* case, which solves a problem common to all star identification techniques. By introducing the *K*-vector technique, Mortari's method [11] eliminates the star catalog searching phase which represents the heaviest computational load of existing star identification procedures. This has enabled new *whole sky* star pattern recognition approaches that *do not require either prior sub-catalogs, or any searching phase to find candidate stars for pattern matching*. The *K*-vector method is a general procedure that directly addresses a global data base that corresponds to, for example, $\sim 10^5$ stars in a mission catalog. The *K*-vector method can be adapted for almost any existing approach and improve it dramatically in terms of speed. The method enables real-time non-iterative solution of the problem: Identify the indices of all the data x_i satisfying a given requirement $x_i \in [x_{min}, x_{max}]$, within any n-long data vector x. The original SLA approach consists of two different identification processes: ***K*-vector star-pair identification** and *Reference-Star* matching identification. The ***K*-vector star-pair identification** process, which is common and essential to the ***K*-vector** based star identification methods, selects an admissible star-pair set from the star-pair catalog whereas the subsequent reference star matching identification process performs the final identification sequence by evaluating angular separation combinations with a subset of all of the admissible star-pairs. In order to find the reference-star, the identification process of this approach mainly relies on two major indices of the highest occurrences by computing a histogram of the overall admissible index vectors. Once the reference star pair is identified, it is possible to proceed with the identification of the other observed stars by the analysis of the index vectors corresponding to the reference vector [11]. In order to increase the efficiency in the identification process for multiple FOV star cameras, the strategy of the SLA has been modified [14], a method that finds common indices among subsets of measured star triads instead of using the reference star as in [11].

Mortari and Junkins [14] introduced the SP-Search (Spherical Polygon-Search) approach to identify the stars as observed by a wide FOV star tracker, which uses more than once the ***K*-vector** technique, does not require any (accurate or not) initial guess of the spacecraft attitude, does not use magnitude information, and is particularly suitable for the *lost-in-space* general case. The method is based upon simultaneously matching interstar angles for three stars; first the two vectors for a reference observed star pair are adopted to create a vector basis with which the star identification process is accomplished, by projecting the vectors locating for all of the other remaining (*n*–2) observed stars, and using the measurement uncertainty to identify the usually unique third star (of any measured triad) in the ***K*-vector** accessed subset of candidate stars. In the unique formulation of [14], this is accomplished is a searchless fashion, however some rules-based logic is required to operate on the ***K*-vector** accessed stars to make the algorithm essentially failsafe. The performance of this method is maximized when interstar angles approach $90°$, so it is attractive for multiple FOVs star trackers.

While the ***K*-Vector** method gives direct searchless access to a candidate set of stars for each measured star pair, this method may still provide 10s up to 100s of candidate stars from the master mission catalog, for each measured pair, and thus a significant number of stars must still be considered in the final star identification logic. In this paper, we introduce a method to deal with multiple measured pairs, an idea based upon *pivot measured stars*. Using the new *Pivot Method*, we show that star identification can be reliably implemented on-orbit, for the *Lost–In–Space* case, and thereby we establish a globally valid process, consistent with real-time, on-board computational constraints, that solves the most fundamental problem associated with star pattern identification. The whole sky star identification procedure is explained in the following sections, which includes the generation of the star pair catalog using K-vector technique, ***K*-vector** star-pair identification, and pivot method based star identification.

Star Pair Catalog Generation using K-Vector Technique

The ***K*-vector** technique has been devised in order to avoid the searching phase characterizing essentially all existing star pair identification algorithms. This vector directly provides the indices of the admissible star-pairs range, as discussed briefly below and in detail in [11]. The method is general, and can be applied to any sorted *n*-long real vector Y, in which we want to identify a subset of elements of Y, specifically, all of the elements falling between (y–δ) and (y+δ). It involves a construction of a judicious "star pair catalog" prior to launch, wherein all catalogued stars are considered as star pairs are ordered {k = 1, 2, ..., N~O(10^6)}, over the whole sky, indexed continuously in the order of increasing inter-star angle. Mortari [11] has shown that, given a measured interstar angle and an associated maximum star centroiding error, then a deterministic, non-iterative access using the K-Vector information gives the k indices of the feasible (*all possible*) star pairs from the catalog, and therefore, the master catalog indices using (i =I(k), j = J(k)) of all possible cataloged star pairs from the whole sky, that could have the measured interstar angle. Thus if the measured star pair is in the catalog, we can be certain that it is contained in the non-iteratively accessed subset of stars.

The master star catalog adopted for this study is the catalog used for the Near Earth Asteroid Rendezvous Mission (NEAR); this NEAR catalog was constructed from the Smithsonian Astrophysical Observatory (SAO) bright star catalog by T. E. Strikwerda et al at the was Johns Hopkins University Applied Physics Laboratory (JHU/APL). This catalog contains 12,620 stars covering the magnitude range from –1.5 up to 6.4 with a magnitude precision of ~0.2 in which each star has stored a unique integer name, its visual magnitude, and three components of direction cosines. The catalog is considered to have worst case errors of a small fraction of an arc second in star positions. This precision can be considered perfect for most practical purposes of star identification, since state of the art star cameras have about one order of magnitude larger errors associated with image centroiding. In our case, we have single star centroiding errors that average about 5 arc seconds, one sigma, and image stars down to about visual magnitude 6.5. Thus the NEAR catalog is well-matched with our sensor precision and sensitivity. Figure 1 shows a histogram of the number of the 12,620 stars in the NEAR master catalog vs. visual magnitude.

Construction of the K-vector star *pair catalog* can be initiated by pre-calculating the interstar sine values of all combinations of catalogued stars within the given cut-off threshold of visual magnitude. For convenience, we use the square of the interstar sine values to avoid calculation with the trigonometric functions as well as to easily use components of direction cosines as given in the master star catalog.

$$\mu_{ij} = \sin^2 \Theta_{ij} = \left|\hat{u}_i \times \hat{u}_j\right|^2 = (u_{yi}u_{zj} - u_{zi}u_{yj})^2 + (u_{zi}u_{xj} - u_{xi}u_{zj})^2 + (u_{xi}u_{yj} - u_{yi}u_{xj})^2$$

$$P = \{\cdots \sin^2 \Theta_{ij} \cdots\}^T, \quad I_P = \{\cdots i \cdots\}^T, \quad J_p = \{\cdots j \cdots\}^T \quad (1)$$

P is a one dimensional array, sorted with increasing values of $\mu_{ij} = \sin^2 \Theta_{ij}$.
This pair catalog can be truncated, deleting pairs with interstar angle ranges that could not possibly be measured due to a particular star camera system's field of view and similar geometric considerations. We typically restrict the pair catalog to ignore stars closer together than θ_{min} and further apart than θ_{max}. For a single $10°$ circular field of view camera, clearly $\theta_{max} = 10°$ and usually we take $\theta_{min} \cong 0.5°$.

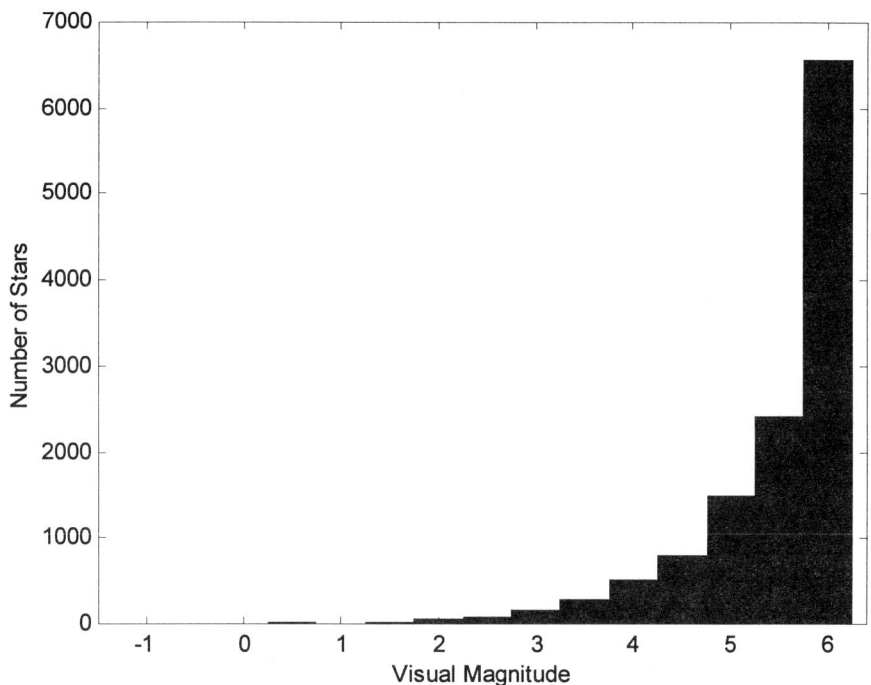

Figure 1 Star Distribution versus Visual Magnitude in the NEAR Master Catalog

Any star-pair that passes the logical truncation: $\sin^2 \theta_{min} \leq |\hat{u}_i \times \hat{u}_j|^2 \leq \sin^2 \theta_{FOV} = \sin^2 \theta_{max}$ is called "admissible" and is retained in the final P array. The size (N) of the resulting *star pair catalog P* is the total number of star pairs in the master catalog that could possibly be imaged by the camera, considering the full sky. Clearly N is mainly dominated by the size of FOV and the cut-off threshold value of visual magnitude (which is in turn dictated by the sensor integration time, CCD quantum efficiency, and lens design). The star pair catalog is constructed from the (say) 12,000 star catalog through a computationally expensive one-time search process over the entire catalog; the stars are ordered in increasing interstar angle (i.e., starpair number 1 will have a separation angle of about θ_{min} and star pair number N will have a separation angle of about θ_{max}). Obviously, this star pair catalog is constructed in a one-time ground-based computation and then stored in the flight computer. With a typical setting of parameters, for our camera, we find $N \sim 10^6$. In order to avoid the cases such that the number of measured stars is not enough to perform the successful pattern identification, the star sensor FOV and integration time are selected carefully to assure a sufficient average number of observed stars. The prototype of the *StarNav* camera used for our current research has a 10 deg x 10 deg FOV and an integration times from 0.01 sec, the cut-off magnitude is about 6.5, extensive tests indicate these choices result in 5 or more of imaged stars in the FOV for virtually all pointing directions, and typically, we image about 20 stars [15-16]. Figure 2 shows the average number of measured stars vs. FOV, the relationship between the size of star pair catalog and the cut-off threshold level of visual magnitude according to the FOV of star tracker.

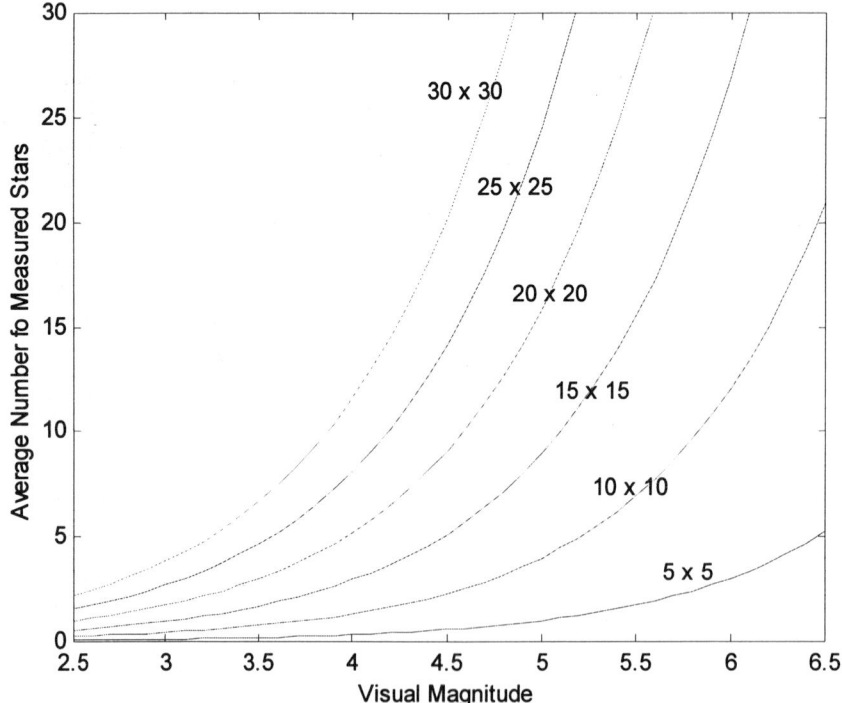

Figure 2 Average Number of Measured Stars within given FOV

When constructing the star pair catalog, in a pair of separate *index pointer arrays* (I, J), a pair of integers are stored for each (the kth) admissible star pair in this sequence, giving the integer indices (I(k) = i, J(k) = j) of the $(i, j)^{th}$ stars in the master star catalog. That is, all the admissible star pairs, with which the associated integer indices maintain the corresponding relationships, are arranged in ascending order as

$$Y = \{\cdots P(k) \cdots P(l) \cdots\}^T, \quad I = \{\cdots I_p(k) \cdots I_p(l) \cdots\}^T, \quad J = \{\cdots J_p(k) \cdots J_p(l) \cdots\}^T \quad (2)$$

So once k is known, we immediately have the master catalog indices of both stars in the pair from the index arrays $\{i= I(k), j = J(k)\}$. These arrays can be built as discussed below.

Following Mortari [11], we find the straight line connecting the two extremes $[1, Y(1)]$ and $[n, Y(n)]$, has, on average, only one element $Y(i)$ for each $D = [Y(n)- Y(1)]/(n-1)$ step. Let us consider the slightly steeper line which connects the two points $[1, Y(1)-D/2]$ and $[n, Y(n)+D/2]$. Because this straight line is found [11] to be an excellent approximation, it simplifies the process by avoiding many index checks and directly locates the desired elements of $Y(i)$. The equation of this line is

$$\sin^2 \Theta_{ij} = a_1 k + a_0, \quad \text{where} \quad \begin{cases} a_1 = nD/(n-1) \\ a_0 = Y(1) - a_1 - D/2 \end{cases} \quad (3)$$

where $i=1\text{-}n$. Starting with $K(1)=0$, an integer vector K is then built as follows

$$K(i) = j \quad \text{where} \quad Y(j) \leq a_1 i + a_0 \leq Y(j+1) \tag{4}$$

From a practical point of view, the i-th element of the K vector represents the number of elements $Y(i)$ below the value $Y(i)=a_1 i+a_0$. Once this vector is built, the evaluation of the two indices identifying, in the Y vector, the range of all the possible catalog star-pairs matching with the observed one, becomes an easy and fast task. The main details of the procedure to find the admissible range is explained in the following section. The specifications of the K-vector star pair catalog generated by this procedure are summarized in the Table 1.

Table 1 K-Vector Star Pair Catalog Specifications

	Circular FOV		Square FOV	
FOV size(deg)	10		10 x 10	
Visual Magnitude Threshold	5.5	6.0	5.5	6.0
Number of Stars	4787	8146	4787	8146
Dimension of K-vectors	186027	541132	369413	~10^6

K-vector based Star-Pair Identification

The K-vector star-pair identification procedure starts by accessing the candidate set of cataloged stars for each measured star pair. For any observed (p, q) star-pair, a number of h admissible star-pairs, which can be arranged to form the h-dimensional index vector arrays $[I, J]$, are obtained from the star pair catalog. The dimension of h is to be determined by how many star pairs exist such that interstar sine angle of each star pair falls within the measured value, plus or minus the sensor accuracy threshold.

Once the angular separation and sensor accuracy are given for any measured star-pair, the indices associated with these values in the Y vector are simply provided as

$$\begin{cases} j_{bot} = bot\{[\sin^2(\theta-\delta) - a_0]/a_1\} \\ j_{top} = top\{[\sin^2(\theta+\delta) \quad a_0]/a_1\} \end{cases} \tag{5}$$

where the function $top\{z\}$ is defined as the larger integer number next to z, and $bot\{z\}$ is the smaller integer number immediately below z. Once j_{bot} and j_{top} are evaluated, it is possible to compute the bounding indices

$$k_{start} = \mathbf{K}(j_{bot}) + 1 \quad \text{and} \quad k_{end} = \mathbf{K}(j_{top}) \tag{6}$$

where k_{start} and k_{end} represent the extremes of the index set

$$k_{start} \leq k \leq k_{end} \tag{7}$$

This set contains the indices of all possible cataloged star pairs that have interstar angles $\theta \pm \gamma$. The **K**-vector can easily be applied to the star pairs identification. The **Y** vector, contains the sorted sine squared of all the observable star pair angles, and the $(y - \delta)$ and $(y + \delta)$ range limits represent $\sin^2 \theta \pm \delta$, where $\delta = \gamma \sin(2\theta)$, respectively, where θ is the angle between two observed stars and γ is the sensor precision (we adopt for γ the 4 - sigma centroiding error for an individual star, in radians). Note δ is the variation of $\sin^2 \theta$ due to a perturbation of θ by γ. The much smaller **K**-vector completely substitutes the vector **Y** in that it contains pointers to only the subset of star pairs that *could possibly* have been measured. Ref. [11] contains a more detailed presentation and discussion of the **K**-vector technique.

The size of the admissible range for the associated admissible star-pair set $[k_{start}, k_{end}]$ in Eq. (7) mainly depends on the precision of the star tracker and the prescribed visual magnitude threshold. According to Mortari's work the average value of the admissible star pair number, that is, the average value of h $(= k_{start} - k_{end} + 1)$ is less than 2 for a reference star tracker having FOV = 30 deg, $M_T = 3$, and the sensor centroiding precision of 10 arcsec. However, since the number of the admissible star-pairs $(k_{end} - k_{start} + 1)$ quickly increase for a more sensitive sensor (lower magnitude treshold), we find that when using a cutoff at visual magnitude 6 typically gives $(k_{end} - k_{start} + 1)$ on the order of 10s up to 100s of candidate star pairs for each measured pair. Thus it is evident that further judicious logic is required to finalize the star identification, given these easily accessed candidate sets of stars for each measured pair.

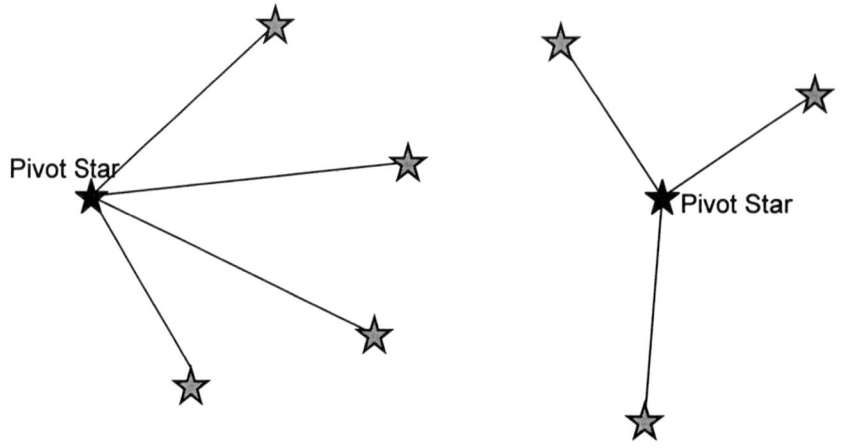

Figure 3 Schematic Diagram of the Pivot Method for Star Identification

Pivot Method for Star Identification

The star pattern recognition approach used in this paper is mainly derived from the modified SLA (Search Less Algorithm) identification method. The main strategy of the modified SLA is as follows: given subsets of measured star triads, three different combinations can be used to obtain the admissible star pair set with **K**-vector method.

Since the true index of each measured star must appear at least twice in the whole set of admissible ranges, star identification can be considered as a success if we have the index set occurring twice in each set of admissible ranges. For sub-ten-arc second centroiding errors, this method has been found to fail occasionally and then only for the case of near double stars, and we have precluded this most likely failure mode by deleting all star pairs closer than θ_{min} from our star pair catalog. By having five or more measured stars, and insisting that all interstar angles between a minimum of four measured stars match the corresponding cataloged stars, we have found this scheme virtually never gives incorrect identfications. However, occasionally this conservative identification approach will fail to identify stars when there are only a small number of measured stars in the FOV. For high frame rate cameras, an occasional failure is generally acceptable, because use of a Kalman Filter to recursively operate on the attitude estimates will provide robustness with respect to occasional dropouts of the star-derived precision attitude estimates. Obviously, some index comparison logic over the small *K*-vectors subset of accessed stars is required to find the matching star indices from two or more *K*-vector accessed subsets of cataloged stars. The Pivot method is designed to accelerate this process by greatly increasing the probability to near-certainty that the same star is contained in the two subsets of star pair indices being compared, and upon location of the index match, we will immediately identify the additional measured stars paired with the pivot stars.

A flowchart of the pivot method for star identification is depicted in Fig. 4. First, it is checked if a sufficient number of stars (at least, 4) is available to enable successful identification. If a sufficient number of measured stars is available for identification, search for the pivot star candidate can be then initiated. To acquire the pivot star, a triad matching procedure from the modified SLA identification method is adopted. Brightest star among the first matching triad subset is normally chosen as a candidate of the pivot star. If even one subset of triad stars is not matched, the star identification is reported as a *failure*. Once the pivot star is chosen, it is possible to proceed with the identification of the additional (that is, the 3rd) measured star connected to the pivot star by checking the common indices among the admissible K-vector index set. If at least 3 stars are confirmed as connected to the pivot star, the star identification is considered to be *successful*. If there are more than 2 ambiguous set of triads, which are not identified by the additional match with the additionally connected stars, star identification can be confirmed by checking whether they are in the same FOV with their direction cosine information.

Figure 4 Flowchart of the Pivot Method for Star Identification

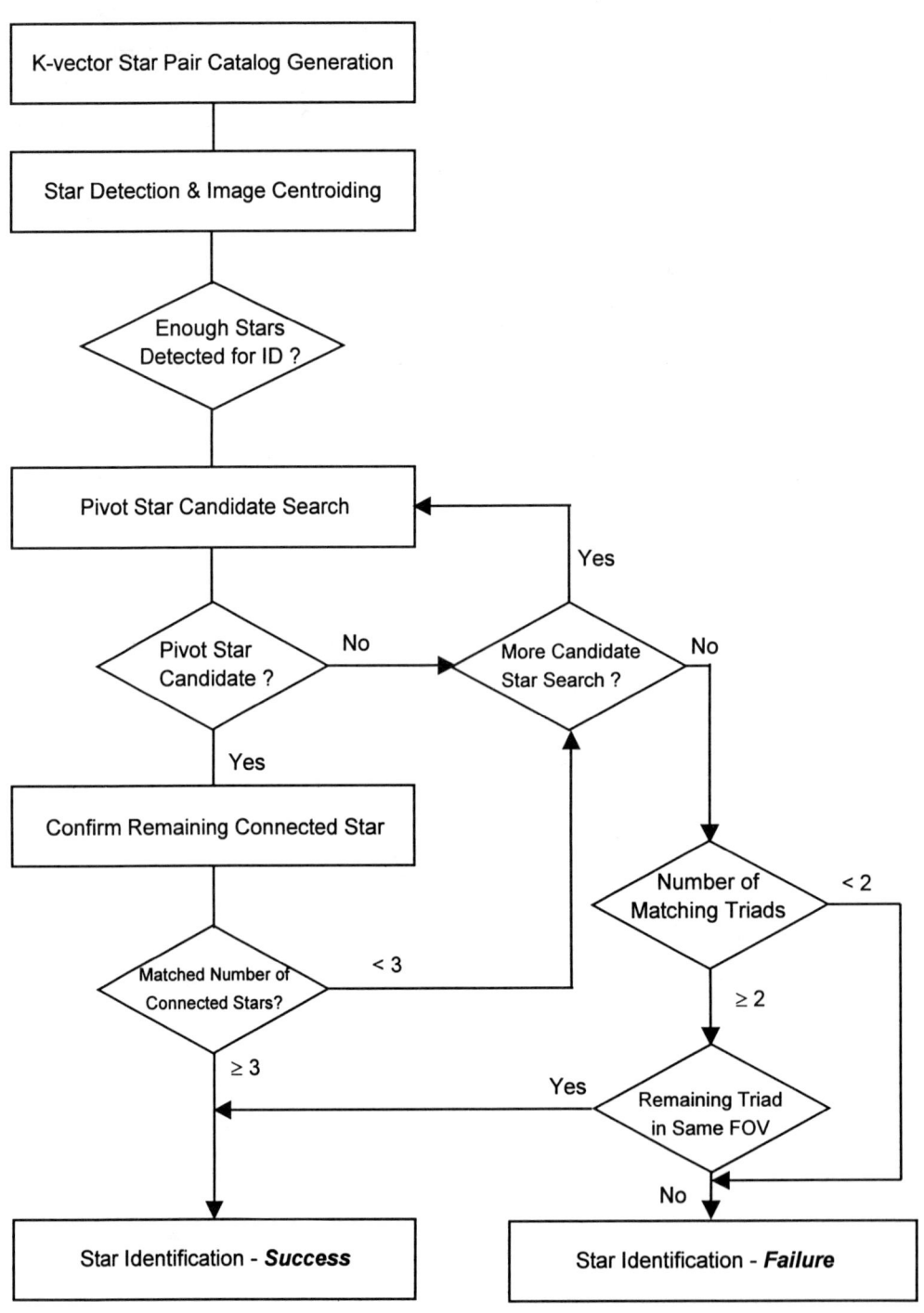

Results from Ground Validation Experiments

To evaluate further the validity and practicality of the pivot method, a family of night sky experiments were performed with the first prototype of StarNav micro star tracker. The first StarNav prototype is a low cost star tracker for micro spacecraft developed at Texas A&M University (TAMU) [15, 16], which has a 10° of single FOV with a magnitude 6.5 sensitivity and has shown excellent performance results from the night sky experiments recently. Its camera head unit has a 70 mm f/1.0 lens, 512 x 512 CCD detector with high quantum efficiency, and a thermoelectric cooler. The imager is an Apogee AP7 CCD camera (Apogee Instruments) which is interfaced to the computer via an ISA bus card that also carries the camera's power supply modules. The Octagon PC500 embeddable single board computer manufactured by Octagon Systems has a Pentium-133 processor with 32Mb RAM and a 40Mb "solid state disk" as well as conventional drives.

This 16-bit grey scale camera was selected primarily because it uses a thinned, back illuminated CCD sensor (SITe) which has an unusually high QE (0.85 peak and averaging >0.7 over the wavelength range $380 < \lambda < 900$ nm). Following laboratory validation and integration with appropriate optics, the camera was configured on an equitorial telescope mount, so it could be easily swept across the sky with known orientation; and the entire system validated end-to-end, under night sky viewing conditions.

Two night sky images were selected for validation of the pivot method, as regards the discussion in this paper. Also, the performance of and differences between star identification using the *pivot method* and the *modified SLA method are* discussed. Each image was taken with 10 ms exposure near Taurus and Lyra, respectively. The two methods were applied to the first star image (Fig. 5), in which include a very distinctively bright star 'Taurus'. The advantage of the triad (or triangle) match based on the modified SLA method is that triangle patterns are more distinct than pairs, thus eliminating many spurious stars, and handling triad groups is much more manageable. Even though the uniqueness probability for matching star triads is several orders of magnitude less likely to result in an invalid match than using a single star pair star pair, occasionally (less than a fraction of a percent of the time) the algorithm fails to correctly match the correct triangle and invokes a totally different catalog star triad in a incorrect sky location; this is more likely when the precision tolerance is too loosely prescribed. The subset of triads in which the 3 brightest stars are included result in an incorrect identification in the totally different location of the sky when the modified SLA was applied with a loose precision tolerance ($\gamma = 3\sigma = 0.003$ deg ~ 11 arc sec), whereas the star identification was successful with a tighter precision ($\gamma = \sigma = 0.001$ deg ~ 4 arc sec). Now we could anticipate that brighter stars normally result in bigger centroid error than medium bright stars, due to image blooming effects on the CCD. Also, the initial match is so critical, and there are usually ~ 20 stars in the field of view; a more conservative approach can be taken. Therefore, a conservative approach can be taken: The tighter precision tolerance ($\gamma = 3\sigma$) was used when searching for an initial candidate pivot star, whereas the looser precision threshold ($\gamma = 3\sigma$) is enforced for the each additional connected star match when the pivot method was implemented. Since the subsequent matches must simultaneously match the previously identified stars, the 3σ tolerance is still leads to a high probability identification. Figure 5 shows a better identification result using the pivot method. As shown in the figure, star 1, 6 and 14 were used as pivot stars.

While the first part of Figure 6 shows the real night-sky image near Lyra and graphical presentation of star identification result, the second part of the figure presents the table for the identification results, including the cataloged number, sensor location, right ascension, declination, and direction cosines of identified stars. Both the pivot method and the modified SLA star identification methods were highly successful in these identifications. Further studies are underway to evaluate efficiency over a large number of night sky experiments.

Figure 5 Night-Sky Image and Star Identification Result: Taurus

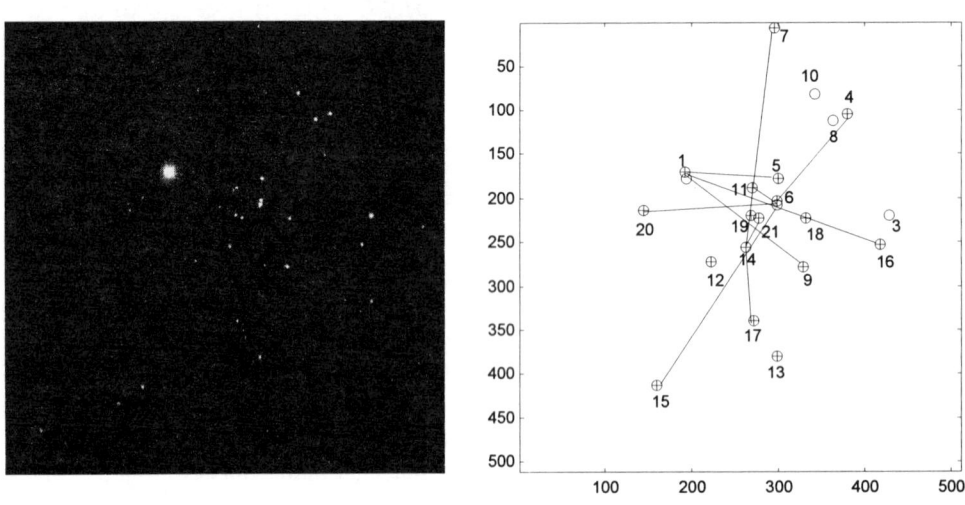

```
 #  cat#      X          Y          RA      DEC      Direction Cosines

 1  2738  192.204468  170.308151  68.9801  16.5095  0.343905  0.894972  0.284174
 4  2716  380.276642  104.873596  65.7336  17.5426  0.391867  0.869247  0.301414
 5  2724  301.455353  178.440033  67.1098  16.3597  0.373218  0.883953  0.281666
 6  2726  299.638000  202.602936  67.1435  15.9621  0.373448  0.885952  0.275002
 7  2725  296.490753    6.685185  67.154   19.1804  0.366703  0.870395  0.328544
 9  2721  329.673584  278.363739  66.6519  14.7139  0.383319  0.888006  0.253992
11  2731  270.547211  188.750610  67.64    16.194   0.365331  0.888118  0.27889
12  2736  222.941879  271.536743  68.462   14.8444  0.354865  0.89913   0.256195
13  2728  298.541382  379.811249  67.2088  13.0475  0.377373  0.898122  0.225759
14  2734  263.493439  255.538467  67.7799  15.1051  0.3651    0.893753  0.260591
15  2740  159.894897  413.726105  69.5393  12.5109  0.341264  0.914665  0.216625
16  2713  417.640564  253.431641  65.1512  15.0957  0.405724  0.876106  0.260433
17  2733  271.536682  339.548248  67.6556  13.7245  0.369318  0.898507  0.237254
18  2720  332.723083  222.907700  66.5864  15.6184  0.382694  0.883777  0.269229
19  2732  269.656769  219.599258  67.6619  15.6918  0.365906  0.890485  0.270462
20  2741  144.871796  213.666672  69.7882  15.8     0.332438  0.902967  0.27228
21  2729  277.480560  222.789368  67.5356  15.6378  0.367965  0.889911  0.269556
```

Figure 6   Night-Sky Image and Star Identification Result: Lyra

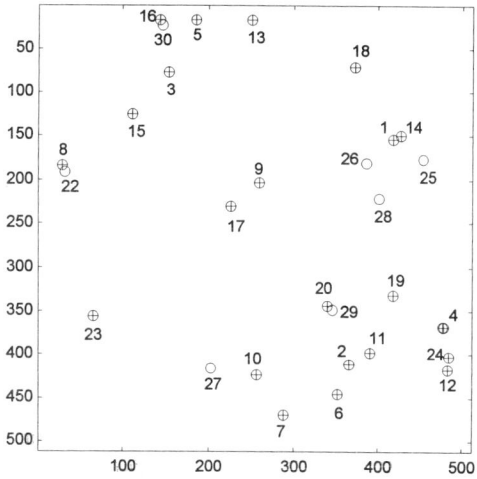

```
 #   cat#      X          Y         RA       DEC       Direction Cosines

 1   1890   416.871704  154.336884  -76.3741  36.8985  0.188395  -0.777192  0.600399
 2   1893   366.374054  411.801910  -75.2641  32.6896  0.214075  -0.813927  0.540087
 3   1904   153.192368   78.159874  -70.9079  38.1337  0.257277  -0.743306  0.617499
 4   1888   478.806030  368.499268  -77.48    33.3627  0.181056  -0.815345  0.549937
 5   1902   185.738876   17.257172  -71.5606  39.146   0.245304  -0.735723  0.631299
 6   1894   352.893890  445.561310  -74.9968  32.1455  0.219188  -0.817836  0.532072
 7   1897   288.823364  470.000000  -73.759   31.7442  0.23784   -0.816469  0.526128
 8   1911    29.430668  183.519440  -68.4619  36.3179  0.295805  -0.749481  0.592265
 9   1898   260.747131  203.904068  -73.1745  36.1002  0.233879  -0.773399  0.589199
10   7921   256.999329  423.763916  -73.1435  32.5016  0.244559  -0.80714   0.537323
11     94   390.847687  398.762360  -75.7436  32.9015  0.206763  -0.813748  0.543197
12   1887   483.555054  417.852600  -77.5296  32.5511  0.182014  -0.823026  0.538052
13   1899   251.643600   17.758486  -72.9449  39.155   0.227429  -0.741339  0.631421
14   1889   426.698334  149.708679  -76.5685  36.9717  0.185579  -0.77708   0.601421
15   1906   111.728462  125.683754  -70.0874  37.3307  0.270817  -0.747609  0.606415
16   1905   143.704849   17.712189  -70.6668  39.1211  0.256842  -0.732065  0.630962
17   1900   227.924942  230.763916  -72.5152  35.661   0.244112  -0.774941  0.582989
18   1892   372.800232   71.841324  -75.4921  38.2662  0.196689  -0.760106  0.619316
19   1891   416.848419  332.852081  -76.2812  33.9686  0.196685  -0.805684  0.558738
20   1895   340.736572  343.991058  -74.7699  33.802   0.218291  -0.80178   0.556325
21   1888   477.937622  368.565369  -77.48    33.3627  0.181056  -0.815345  0.549937
23   1907    65.756310  356.789154  -69.3611  33.5183  0.293864  -0.780203  0.552203
24   1886   484.669586  401.922119  -77.5589  32.8128  0.181062  -0.82071   0.541896
```

The basic approach pursued in both of these identification strategies is to insist that the initial triple of stars be matched with a stringent one σ precision, and form a hypothesis that these stars have been correctly matched with the corresponding cataloged stars accessed using the K-Vector method. If one or more subsequent stars have interstar angles matching all three of these stars to within 3σ, then the hypothesis of a correct match is confirmed, if not, the search is continued until the initial triple is confirmed by additional matches, or until all measured stars have been considered. We have found that when matching 4 or more stars with this precision (< 10 arc seconds), having the correct match between measured and cataloged stars is essentially the certain event (we have found no counter examples). Only for a small fraction of the measured star patterns studied (<< 1%) with this conservative strategy will fail to achieve a match in the catalog. Regarding failure to match an occasional measured star pattern, our philosophy is that it is far preferable to have an occasional pattern match failure, than to incur frequent incorrect star identifications. We wish to reduce the probability of an incorrect identification to essentially zero. It is anticipated that a continuous Kalman filter will be processing the discrete attitudes determined from the identified stars, and a rare data dropout resulting from a star identification failure will not represent a significant operational concern. This is especially true in the most likely event that the spacecraft will have high quality rate gyros to measure angular velocity, which can be accurately integrated for several minutes with minimal accuracy degradation.

Plans for On-Orbit Experiments

In order to represent the on-orbit demonstration of a Lost-In-Space star identification and attitude determination process for the second prototype of the hardware, the StarNav experiment is planned for Shuttle Mission STS 107, January 2001. The second StarNav prototype is also a low cost star tracker for micro spacecraft under being developed at Texas A&M University (TAMU), which has a $7.5 \sim 8°$ of single FOV with a magnitude 6.5 of sensitivity. Its camera head unit has a 70 mm f/1.0 lens, 512 x 512 CCD detector with high quantum efficiency, and thermoelectric cooler. The imager is a Pegasus-XL CCD camera which will be interfaced to the on-board computer via a parallel port that also carries the camera's power supply modules. The PC-104 embeddable single board computer has a Pentium-133 processor with 32Mb RAM and a 40Mb "solid state disk" as well as conventional drives. All of these components have been customized to significantly increase radiation hardness and ability to withstand the launch and on-orbit environment. All software is being implemented in ANSI-C under the real-time environment of QNX operating system. The ground user interface support software, which is written in Microsoft Visual C++, is also being developed to send uplink commands and to acquire the downloaded images during the space mission. A minimum of ten hours of experimentation is planned, however it is hoped that 30 or 40 hours of operation will be carried out. The results of the imaging and attitude estimation results will be available in near-real time over the internet, interested parties should contact the authors to obtain final details of these plans during the fall of 2000.

Conclusions

We have presented an approach to star pattern identification and attitude determination suitable for solving the general *lost-in-space* problem wherein no prior attitude estimate is available. Results from ground test experiments are also reported which support the validity and practicality of the approach. It is planned to test these algorithms in the StarNav star tracker and processor to be flown aboard STS-107 in 2001; this will represent an important milestone toward the autonomous spacecraft envisioned for the new millennium.

References

1. Junkins, J. L., White, C. C. III, and Turner, J. D., "Star Pattern Recognition for Real Time Attitude Determination," *Journal of the Astronautical Sciences*, Vol. XXV, No.3, July-September, 1977, pp. 251-270.

2. Strikwerda, T. E. and Junkins, J. L., "Star Pattern Recognition and Spacecraft Attitude Determination," U.S. Army Engineer Topographic Laboratories, ETL-0260, Fort Belvoir, Virginia, May 1981.

3. Sasaki, T. and Kosaka, M., "A Star Identification Method for Satellite Attitude Determination using Star Sensors," *Proceedings of the 15th International Symposium on Space Technology and Sciences*, Tokyo, Japan, May 1986, pp. 1125-1130.

4. Alveda, P. and San Martin, A. M., "Neural Network Star Pattern Recognition of Spacecraft Attitude Determination and Control," *Advances in Neural Information Processing System I*, Denver, Colorado, 1988, pp. 314-322.

5. Van Bezooijen, R. W. H., "A Star Pattern Recognition Algorithm for Autonomous Attitude Determination," *Preprint of XIth IFAC Symposium on Automatic Control in Aerospace*, Tsukuba, Japan, July 1989, pp. 51-58.

6. Strikwerda, T. E., Fisher, H. L., Kilgus, C. C., and Frank, L. J., "Autonomous Star Identification and Spacecraft Attitude Determination with CCD Star Trackers," *First International Conference on Spacecraft Guidance, Navigation and Control Systems*, Noordwick, The Netherlands, June 1991, pp. 195-200.

7. Anderson, David, *Autonomous Star Sensing and Pattern Recognition for Spacecraft Attitude Determination*, Ph.D. Thesis, Texas A&M University, College Station, TX, May 1991.

8. Liebe, C. C., "Pattern Recognition of Star Constellations for Spacecraft Applications," *IEEE Aerospace Electronics Systems Magazine*, Vol. 7, 1992, pp. 34-41.

9. De Antonio, L., Udomkesmalee, S., Alexander, J. W., Blue, R., Dennison, E., Sevaston, G., and Scholl, M., "Star-tracker Based, All-sky, Autonomous Attitude Determination," *SPIE Proceedings on Space Guidance, Control, and Tracking*, Vol. 1949, 1993, pp. 204-215.

10. Udomkesmalee, S., Alexander, J. W., and Tolivar, A. F., "Stochastic Star Identification," *Journal of Guidance, Control, and Dynamics*, vol. 17, No. 6, November-December 1994, pp. 1283-1286.

11. Mortari, D., "Search-Less Algorithm for Star Pattern Recognition," *The Journal of the Astronautical Sciences*, Vol. 45, No. 2, April-June 1997, pp. 179-194.

12. Mortari, D., "SP-Search: A New Algorithm for Star Pattern Recognition," AAS Paper No. 99-181, AAS/AIAA Space Flight Mechanics Meeting, Breckenridge, CO, Feb. 7-10, 1999.

13. Mortari, D. and Angelucci, M., "Star Pattern Recognition and Mirror Assembly Misalignment for DIGISTAR II and III Multiple FOVs Star Sensors," AAS Paper No. 99-182, AAS/AIAA Space Flight Mechanics Meeting, Breckenridge, CO, Feb. 7-10, 1999.

14. Mortari, D. and Junkins J. L., "SP-Search Star Pattern Recognition for Multiple Fields of View Star Trackers," AAS Paper No. 99-437, AAS/AIAA Astrodynamics Specialist Conference, Girdwood, Alaska, Aug. 16-19, 1999.

15. Ju, G., Kim, H., Pollock, T. C. and Junkins, J. L., "Micro Star Tracker and Attitude Determination System," Third International Symposium on Reducing the Cost of Spacecraft Ground Systems and Operations, Tainan, Taiwan, Mar. 22-24, 1999.

16. Ju, G., Kim, H., Pollock, T. C. and Junkins, J. L., "DIGISTAR: A Low Cost Micro Star Tracker," AIAA Paper No. 99-4603, AIAA 1999 Space Technology Conference & Exposition, Albuquerque, NM, Sep. 28-30, 1999.

AAS 00-005

DESIGN, IMPLEMENTATION, AND FLIGHT RESULTS FOR ALL-STELLAR ATTITUDE DETERMINATION

Jim D. Chapel,[*] Stephen M. Micciche[†] and Richard Kiessig[‡]

An all-stellar attitude determination capability has been developed for both the Mars Climate Orbiter (MCO) and the Stardust spacecraft. This capability was designed to both save power and extend IMU life, thereby enabling the extended missions of these spacecraft. Both spacecraft incorporate a wide field-of-view (WFOV) Star Tracker Camera developed by Corning-OCA under contract to Lockheed Martin. The "lost-in-space" capability provided by this camera and the associated Stellar Compass™ software allows a robust all-stellar solution.

The Stellar Compass™ software uses pattern-matching techniques to provide a quaternion output. In acquisition mode, "candidate" stars are compared with the entire star catalog at 0.1 Hz to achieve a lost-in-space capability. Once an initial attitude measurement is obtained, the software can be operated in a more computationally efficient track mode at 1 Hz. An all-stellar attitude determination capability has been designed around the Stellar Compass™ software, providing an accurate and robust attitude and attitude-rate estimation capability without the benefit of gyro data. High fidelity simulation results are provided showing the attitude determination performance of this design.

Although MCO and Stardust are both 3-axis stabilized spacecraft, the all-stellar control implementations are quite different. MCO uses reaction wheels and Stardust uses small RCS thrusters for attitude control, thereby necessitating different design approaches. Control designs accommodating all-stellar attitude determination are summarized for both spacecraft.

Finally, recent all-stellar flight results from MCO are provided. These results show excellent agreement with the performance predicted in the high-fidelity simulation. Additionally, the All-Stellar performance is compared with more traditional gyro-based Kalman Filter results, which was running simultaneously on the spacecraft during test and check out.

INTRODUCTION

Advances in sensors and processing capability have allowed new paradigms to be considered for spacecraft attitude determination. As addressed in this paper, more capable

---

[*] Senior Staff Engineer, Lockheed Martin Astronautics, P.O. Box 179, Denver, Colorado 80201.
[†] Staff Engineer, Lockheed Martin Astronautics, P.O. Box 179, Denver, Colorado 80201.
[‡] President, Intelligent Decisions Incorporated, Los Altos, California.

star sensing components and expanded processing and software capabilities can improve the reliability of attitude determination for long-duration interplanetary missions. In this paper, an all-stellar attitude determination design is presented whereby attitude sensing is provided by a single wide-field-of-view (WFOV) star camera. Of course, attitude determination performance with this type of scheme is limited by the star sensor performance; the performance improvements provided by Kalman filtering are not available with a single sensor system. However, lower performance is often acceptable if mission requirements can be met with increased reliability and with lower cost.

For various reasons, several other recent programs have examined similar schemes for attitude sensing.[1,2] First, limiting the number of components required for attitude determination enhances reliability. Second, providing an attitude determination mode that does not require the use of other attitude sensing components also enhances functional redundancy. Finally, powering-off the Inertial Measurement Unit (IMU) not only saves power, but also extends the useful life of the IMU, ensuring that it is available for mission critical events such as maneuvers and safing.

The development of the star camera discussed in this paper has been previously described in detail,[3] so only an overview of the component is provided here. Recent flight results of the star camera are presented here which substantiate the previously reported ground test results. To implement an all-stellar attitude determination capability, a number of software algorithm enhancements are required. An overview is presented of the all-stellar attitude determination algorithms, and the associated attitude control algorithms, designed and implemented for the Mars Climate Orbiter (MCO) and the Stardust spacecraft. Finally, flight results of the MCO all-stellar implementation are shown, including favorable comparisons to the more traditional Kalman filter-based attitude determination.

STARDUST AND MARS CLIMATE ORBITER MISSIONS

The Stardust spacecraft, as shown in Figure 1, is one of NASA's Discovery missions. The primary science objective of the Stardust program is to collect comet particles and return them to earth for analysis. Stardust launched in February of 1999. It will use an earth gravity assist in January of 2001, fly through the coma of the comet Wild-2 in January 2004, and return the Sample Return Capsule to earth in January 2006. The total mission duration for Stardust is nearly seven years, and the mission profile takes the spacecraft as far as 2.7 AU away from the sun. The length of the mission (with its implied life requirement for the IMUs), along with the power limitations 2.7 AU away from the sun, led to the all-stellar attitude determination development. For Stardust, the all-stellar attitude determination requirements were derived from the high-gain antenna pointing requirements during cruise. The derived attitude knowledge requirement, which includes misalignments, is 3.5 mrad,, 3-sigma per axis,. Because of the modest accuracy requirements, the Stardust mission is compatible with all-stellar attitude determination performance.

Under the auspices of the MSP '98 program, two spacecraft were built and launched during the Mars 1998/1999 launch opportunity. The Mars Climate Orbiter (MCO), shown in Figure 2, was launched in December 1998. It's original mission required interplanetary cruise to Mars in approximately 10 months, aerobraking over a 2 month period to achieve its science orbit, a science mission over a 2 year period (1 Martian year), and a data relay mission for another 3 years. Total mission duration of the Orbiter was to exceed 6 years. As with Stardust, the length of the mission along with the power limitations led to the all-stellar attitude determination development. The science mission for the Orbiter dictated only modest pointing requirements. The most stringent pointing requirements for

MCO were during the science mission; attitude knowledge and control errors were to be less than 10 mrad.. However, the attitude knowledge requirements were still consistent with the modest pointing requirements derived for Stardust.

Unfortunately, a ground software parameter error produced a significant navigational error for Mars Orbit Insertion (MOI),[4] Subsequently, MCO was lost in September 1999 just prior to orbiting Mars. Thus the all-stellar implementation was never completely validated in flight. However, all-stellar attitude determination was used during cruise with the IMUs powered off. The cruise results are presented in this paper.

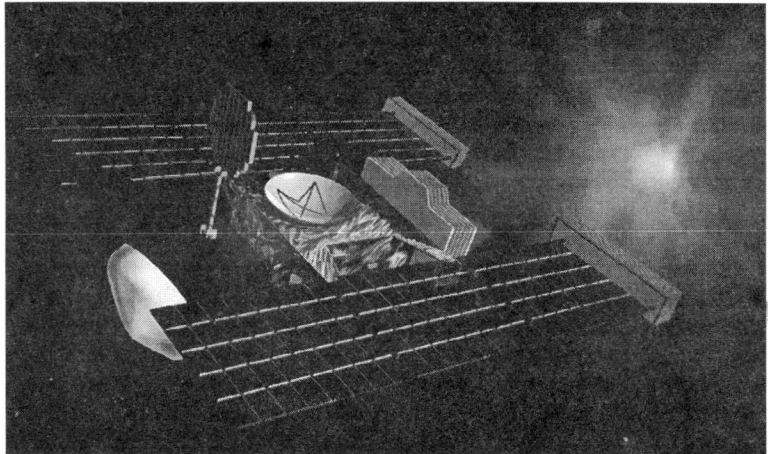

Figure 1 Stardust

Figure 2 Mars Climate Orbiter

Although the missions vary greatly, the two spacecraft share similar avionics architectures. Each spacecraft incorporates two Star Tracker Cameras (see detailed description below) in a block redundant fashion. To simplify both the avionics and the fault protection designs, the star cameras are not cross-strapped between Side A and Side B of the Command and Data Handling (C&DH) System. One star camera is designated as the Side A star camera, and interfaces only to Side A of the C&DH. Similarly, the other star camera is designated as the Side B star camera, and only interfaces to Side B of the C&DH. A failure of either camera would require switching to the other avionics string.

STAR CAMERA TRACKER DESCRIPTION

Corning OCA and Lockheed Martin Astronautics cooperatively developed the Star Tracker Camera shown in Figure 3.[3,6] The design of the Corning OCA Star Tracker Camera drew heavily upon the Clementine star camera, which successfully flew as a technology demonstration funded by the Strategic Defense Initiative (SDIO) and built by the Naval Research Laboratory (NRL).[5]

Figure 3 Corning-OCA Star Tracker Camera (photo courtesy of Corning-OCA)

In the flight configuration, the unit weighs 1.15 kg as shown (including baffle), and consumes approximately 6.7W of power with 1.0 Hz star updates, and 4.7W of power with 0.1 Hz star updates. The approximately 25 degree circular FOV provided by the optics is inscribed in the 512x512 square CCD array. The electronics provides a digital 8-bit image to the spacecraft host computer. The unit, when used in conjunction with the Stellar Compass™ software, provides 3-axis attitude knowledge with errors ≤500 μrad (3σ) in pitch and yaw, and ≤1500 μrad (3σ) in roll (boresight) at end of life. Although the Stellar Compass™ software that produces the attitude update does not reside in the unit itself, it is an integral part of the Star Tracker Camera system, and is critical to providing the robustness needed for all-stellar attitude determination.

STELLAR COMPASS™ SOFTWARE

Acquire Mode

After the camera acquires an image, the Stellar Compass™ Software examines the image and determines the vehicle's orientation. The first step in this process is to identify the objects in the image that are probably stars. The software takes special care to exclude many different types of image artifacts. This includes single hot pixels, objects that are too bright or too large and image streaks. Stray light is handled using an adaptive filter that adjusts to changing background levels through the image. A cellular logic algorithm examines all potential stars to ensure that they have an energy distribution pattern that "looks like a star," based on the point spread function (PSF) of the lens. Working together, the algorithms are able to screen out most non-stellar objects. Stars with a signal-to-noise ratio (SNR) as low as 2.0 can be seen through local clutter. Once objects are identified that pass all of the screening criteria, their centroids and total brightness are calculated. The 2-D coordinates of the star image on the CCD are then back-projected through the lens to obtain the 3-D coordinates on a local unit sphere.

The next step is to pick bright stars three at a time from the list of objects that passed the screening criteria. Bright stars are used preferentially in the matching process because they have a better SNR than dim stars. This allows increased centroiding accuracy, and therefore increased overall system accuracy. A triangle is built from each three stars, and the triangle side lengths are matched against an on-board database. When a match is found, a quaternion is calculated that rotates the triangle formed by the stars from the CCD image onto the one formed by the three corresponding catalog stars. When enough of the resulting quaternions agree, the matching process is complete. It takes at least four stars (and therefore four triangles) to obtain a confident match in acquire mode. The triangle matching process is very effective at filtering out remaining image artifacts that look like stars, but which are not.

The final step in the process is quaternion calculation. The QUEST algorithm[7] is used, with measurement weights assigned to each star as a function of its brightness. QUEST takes as input the measured star image locations and the corresponding star positions from the star catalog. The output is a "best-fit" quaternion that rotates the measured star positions into the catalog star positions, along with information on the quality of the result. The resulting quaternion can be directly used in a Kalman filter attitude determination scheme.[8]

Track Mode

Once the software has processed a sequence of images resulting in high-confidence matches, it can be switched into "track" mode. This mode reduces CPU usage by about a factor of five. It is also more robust against certain kinds of image artifacts, since only part of each image is examined. Given the location of the stars in the current image, and the calculated attitude from the current image and the previous one, the software calculates where it expects to find those same stars in the next image. The software is also aware of stars moving out of the field of view. When a star enters a "boundary zone" near the edge of the FOV, a candidate replacement star (the brightest star in the field that is not currently being tracked) is chosen from the on-board catalog.

When a new image arrives, a small area around each predicted star location is searched. The software implements robustness against multiple missing stars. If they do not appear as expected, alternative candidates are immediately chosen. A minimum of two stars is required. If too many stars are missing from their predicted locations, track is "broken," and subsequent images are again processed in acquire mode. The triangle pattern matching process does not take place in track mode. Measured and catalog star positions are passed directly to QUEST for final quaternion generation.

STAR TRACKER CAMERA/STAR CATALOG SKY COVERAGE

The star catalog used by the software is based on the Hipparcos catalog,[9] which is complete down to visual magnitude 7.0, with measurements taken to magnitude 12.4. In selecting the star catalog used by the Star Tracker Camera, the key discriminators were star brightness and star separation. The resulting catalog contains 933 stars, with a minimum instrument magnitude of 4.85. A histogram of the star catalog, corrected for instrument sensitivity, is shown in Figure 4.

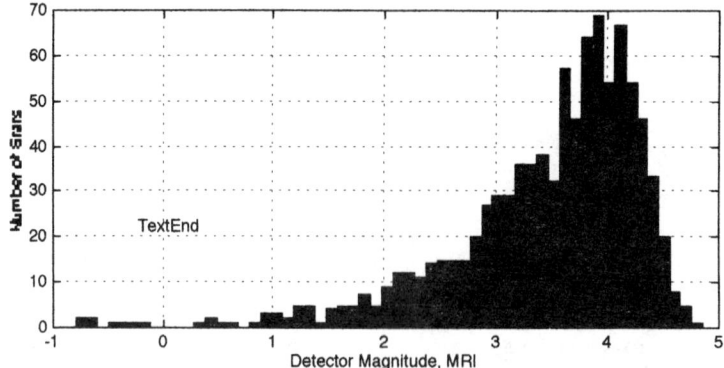

Figure 4 Histogram of Star Magnitudes in the Star Catalog

The selected star catalog was analyzed for "holes" with the Star Tracker Camera FOV. For a minimum sensitivity of MRI=4.85 (instrument magnitude), acquire mode assures full-sky coverage (i.e., 5-star solutions can be obtained for any attitude). Acceptance testing of the flight units has provided end-of-life performance estimates for the Star Tracker Camera. The results indicate that the end-of-life sensitivity will be MRI=4.8 or better. Therefore, the Star Tracker Camera is ideally suited to an all-stellar scheme.

Further examination of the star catalog allows the robustness of an all-stellar solution to be assessed against degraded sensor performance. Because the stars are not uniformly distributed in the sky, "holes" can appear with reduced sensor sensitivity. As a function of right ascension and declination, Figure 5 graphically shows what the sky coverage of the Star Tracker Camera would be if the sensitivity were to degrade to MRI=4.5. The gray-scale image represents the number of stars sensed for a particular camera boresight. White represents 12 or more stars in the field of view, and black represents zero stars. Shades of gray indicate that the number of stars range between 0 and 12. The dark spots show that a sensitivity of MRI=4.5 is not sufficient to provide solutions in some attitudes. Analysis of the data indicates that approximately 0.33% of camera orientations would have 3 stars or less in the field of view. These attitudes would not provide sufficient stars for acquire mode to obtain a solution. This analysis reveals an important point: verification of star sensor sensitivity is crucial to implementing a viable all-stellar attitude determination capability.

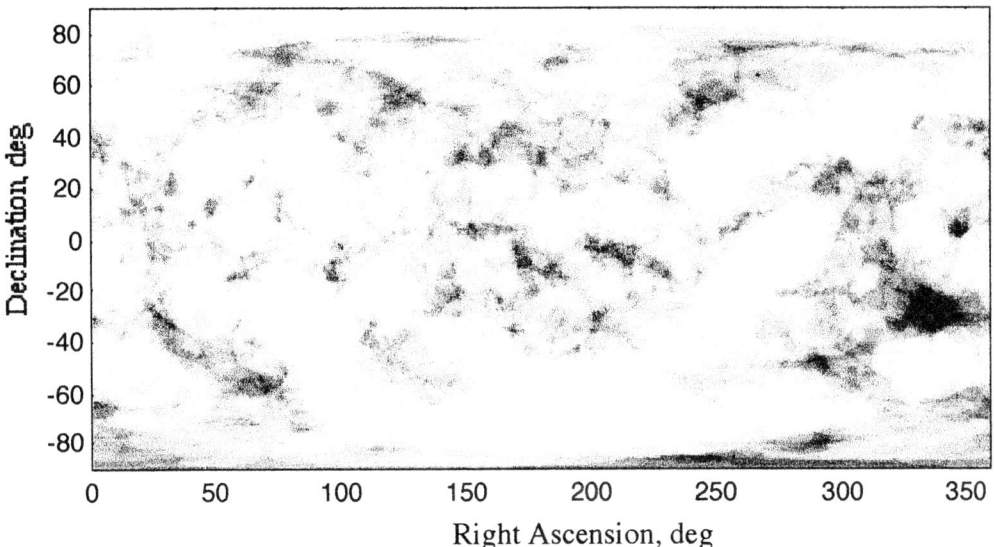

Figure 5 Sky Coverage of the Star Tracker Camera, Sensitivity of MRI=4.5

ATTITUDE DETERMINATION AND CONTROL DESIGN

All-Stellar Attitude Determination

As discussed previously, the derived attitude knowledge requirements for the MCO and the Stardust spacecraft were very similar. Specifically, the most stringent requirement identified for either spacecraft was attitude knowledge of 3.5 mrad per axis, 3-σ. No specific rate knowledge requirement had been identified for either spacecraft, although large rate knowledge errors would complicate the control designs. Because the two spacecraft were being developed nearly simultaneously, and because the Corning OCA Star Tracker Camera was being integrated in both spacecraft, it was desirable to develop a common all-stellar attitude determination scheme to minimize development costs.

The star camera measurement accuracies are consistent with the desired all-stellar performance. The most challenging part of the design was accommodating the different update rates of acquisition mode and track mode, and the transitions between the two. As discussed previously, acquisition mode provides 0.1 Hz attitude updates, whereas track mode provides 1.0 Hz updates. Although an all-stellar design accommodating only track mode would have been much simpler, significantly improved robustness is afforded by the design presented here because of its ability to gracefully handle short duration star outages.

Figure 6 presents a block diagram of the all-stellar attitude determination algorithm.[10] In between star measurements, the previous attitude is propagated forward using the previous attitude rate estimate. Two different methods are used to estimate attitude rate, one for track mode and one for acquisition mode. For acquisition mode, a simple back difference between successive attitude updates provides a reasonable rate estimate. For track mode, a much better rate estimate is obtained by filtering the attitude updates. During an outage period, attitude is propagated forward for the duration of the outage based upon the latest attitude measurement and derived rate estimate. Because the measurement noise of the star camera causes inaccuracies in the rate estimates, the attitude estimate will tend to drift. The current design dictates fault protection will step in after one minute without a valid star update. This limits the attitude knowledge errors to well under one degree.

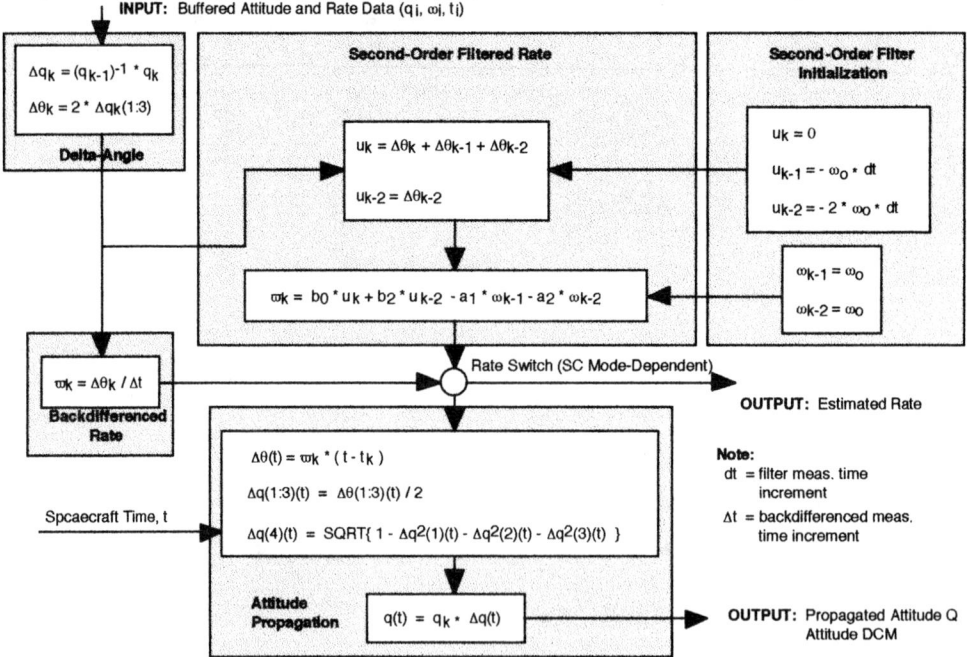

Figure 6 All-Stellar Attitude Determination Block Diagram

Figure 7 shows simulated performance of the algorithm, including performance for acquisition mode, track mode, transitions between modes, and a 60-second star outage. In this simulation, star camera performance parameters were incorporated that represent acceptance test data. As can be seen by the plot, the design meets the derived attitude knowledge requirements with margin, except toward the end of the outage period.

An important feature of the all-stellar implementation is the ability to run a traditional gyro-based attitude determination scheme in parallel with all-stellar attitude determination. Flight data is shown later for this so-called "dual" mode. The gyro-based attitude determination features a full six-state Kalman filter, which estimates quaternion errors and gyro bias errors.[8] The filter propagates gyro data at 10 Hz, and accepts attitude updates from the star camera at 0.1 Hz (even if star updates are available at a higher frequency). In dual mode, the attitude control uses the gyro-based estimates only; the all-stellar data is simply telemetered to the ground for evaluation. Attitude control uses the all-stellar attitude and rate estimates after transition from dual to all-stellar mode. Flight data is also shown later for attitude control using the all-stellar data.

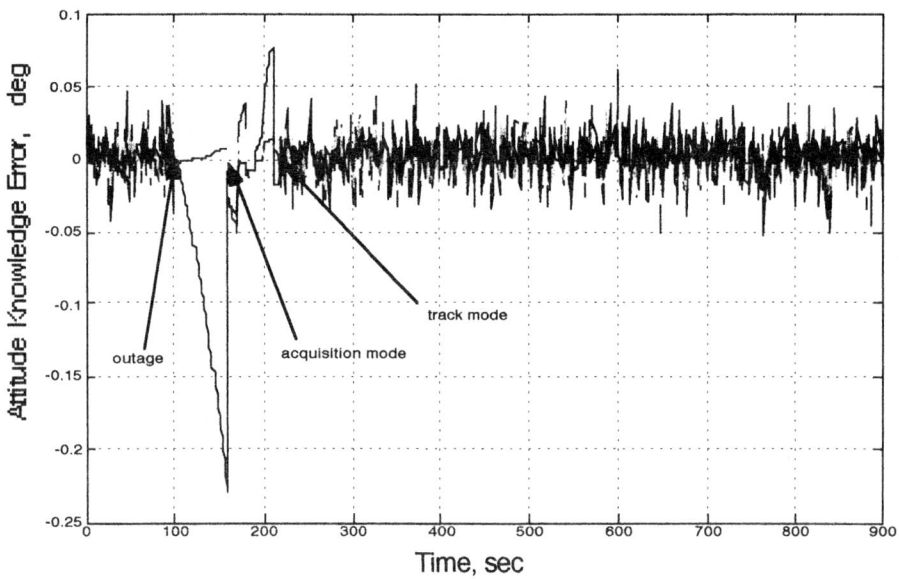

Figure 7 All-Stellar Attitude Determination Simulation with Star Outage

Attitude Control

Unlike the attitude determination design, the attitude control designs for Stardust and MCO are significantly different. Although both are 3-axis spacecraft, the MCO design utilized 3 orthogonally-mounted reaction wheels for attitude control, whereas Stardust incorporated a suite of 0.9 N thrusters for attitude control. Additionally, MCO had two articulable appendages, namely the solar array and the high-gain antenna. Stardust implemented no articulable appendages. In the design of attitude control supporting all-stellar attitude determination, MCO is clearly the more difficult control design problem of the two spacecraft. As discussed previously, the pointing requirements for MCO were most stringent for the orbital mapping mission. During mapping, the RSS of attitude control and attitude knowledge errors was not to exceed 5.8 mrad per axis, 3-σ. Additional requirements were imposed on spacecraft stability of 1.5 mrad over 1 s and 3 mrad over 12 s, both 3-σ. These pointing and stability requirements were relaxed during the propulsive desaturation events.

Several control design constraints are imposed by the all-stellar attitude determination design described above. First, a single control design must accommodate both track mode and acquisition mode. This design constraint eliminates the complexity associated with switching controllers based upon attitude determination state. Second, the control design must account for data latencies of 1 s in track mode and 10 s in acquisition mode. Third, the design must accommodate the phase delay associated with the two rate estimation schemes discussed earlier. Finally, the design must account for attitude disturbances resulting from articulating the high gain antenna and the solar array.

Figure 8 shows the control block diagram for the resulting controller.[11] Track and acquisition modes are both accommodated by limiting the attitude control bandwidth to <0.005 Hz. Reaction wheel dynamics, such as friction, are minimized by implementing a tight control loop on wheel speed, and commanding wheel momentum in the attitude loop. Appendage dynamics and thruster firings are fed forward to minimize attitude disturbances. It should be noted that the control architecture shown was implemented for both Stardust and MCO, with parts omitted in the Stardust implementation that were not applicable.

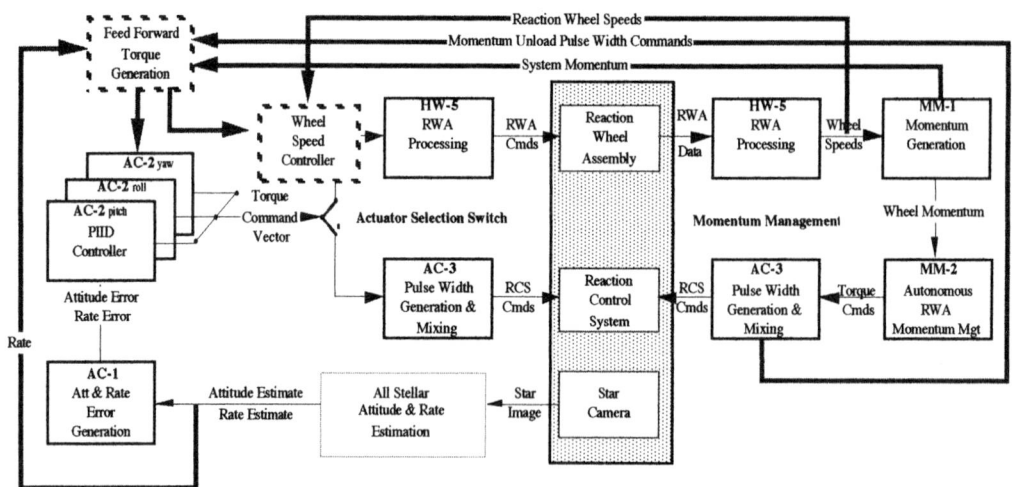

Figure 8 Control Block Diagram for All-Stellar Attitude Determination

Figure 9 shows simulated attitude control performance for MCO. The simulation results incorporate both the all-stellar attitude determination design shown in Figure 6 and the attitude control design shown in Figure 8. As seen from the plot, the overall pointing requirements are met with margin. Close examination of the plot shows slightly degraded performance for acquisition mode. It should be noted that performance during desaturation events depends upon the accuracy of the feed forward torques, which in turn depends upon the accuracy of the thruster model. Although the simulation does account for knowledge errors in all the feed forward terms, flight data revealed that the thruster models were not as accurate as assumed in this simulation. Compared with the simulation results showing ~0.1 deg attitude errors during a desaturation event, desats in flight produced attitude errors of more than one degree. This was deemed acceptable performance for cruise, although an updated model probably would have been loaded onto the spacecraft had it successfully achieved the mapping orbit.

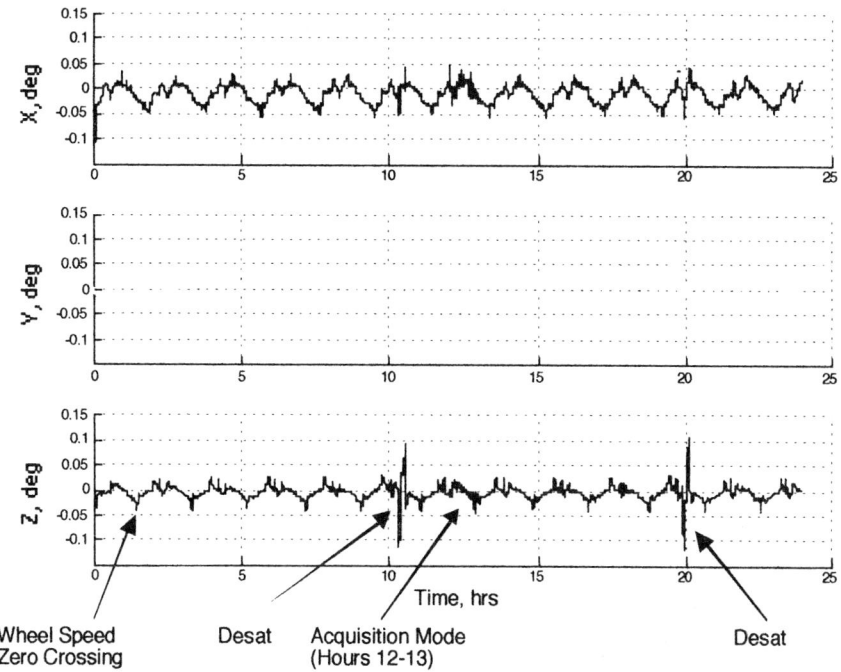

Figure 9 Simulated MCO Attitude Control Performance for All-Stellar Design

FLIGHT RESULTS

Star Imaging Flight Results

Star camera images have been periodically downloaded from both Stardust and MCO during cruise. These images have been used to verify the star camera sensitivity and noise models, as well as to trend the performance over time. An example image from MCO is shown in Figure 10, where the approximate star camera 25-degree circular field of view and the locations of the stars in the star catalog have been added. As shown in the figure, there are 11 stars from the on-board star catalog in this image, with the brightest star being MRI=2.5 and the dimmest being MRI=4.5. The corresponding star map is shown in Figure 11, including the Hipparcos catalog star identifications and the instrument magnitudes of the stars in the catalog. Note the "egg-shaped" projection of the circular field of view onto the right ascension/declination map.

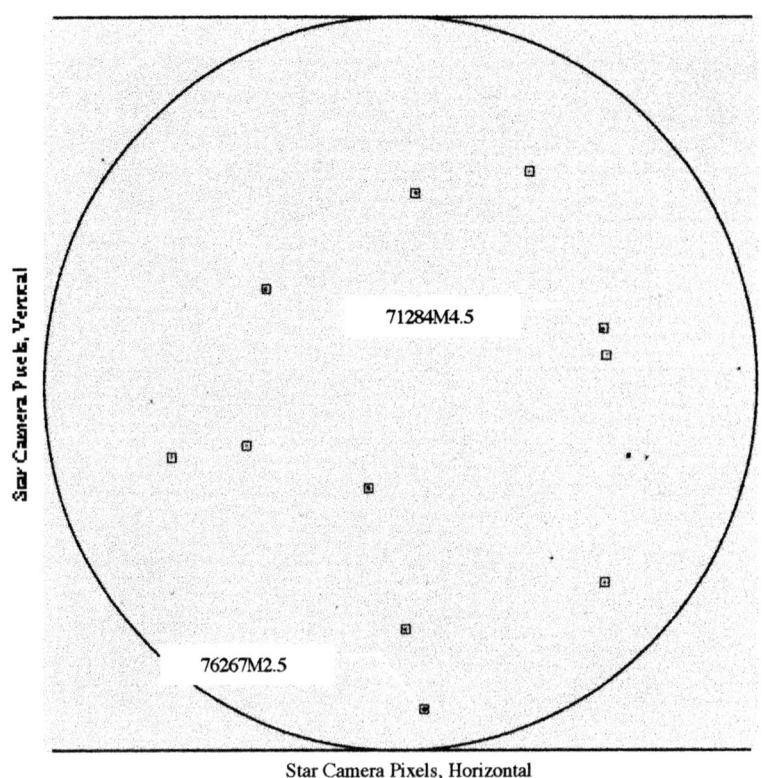

Fig 10 MCO Star Cam Image from Early Cruise

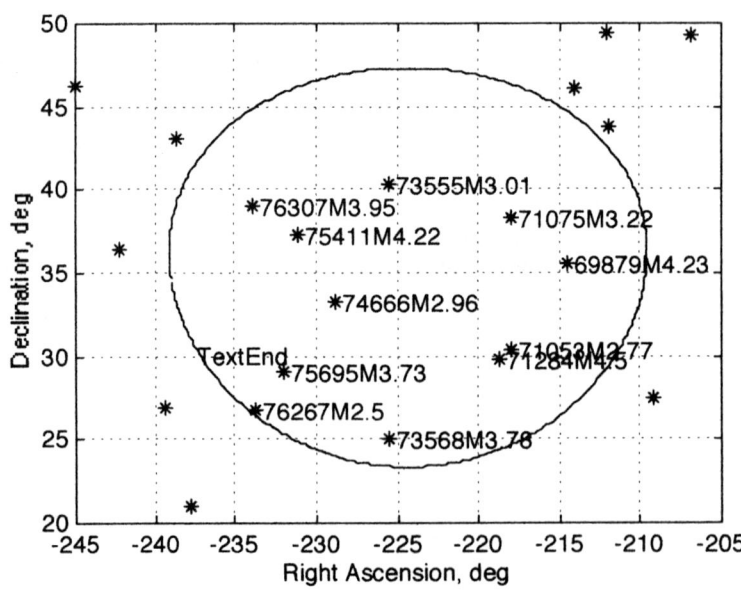

Figure 11 Hipparcos Star Identification Numbers and Star Magnitudes for MCO Image

The star camera sensitivity can be assessed by comparing individual star images between flight data and simulated data. Comparisons are shown in Figure 12 for the MRI=2.5 star, and in Figure 13 for the MRI=4.5 star. As evident from the figures, the peak number of counts is very close between flight data and simulation (within 2 counts for both of the stars shown). However, the sum over a 5x5 set of pixels (25 pixels total) reveals that the simulation sensitivity estimate is very conservative. After subtracting out the background level, the sum over the 25 pixels is more than 30% higher for the flight data when compared with the simulation data. Therefore, the sensitivity assumptions made for the full-sky coverage analysis appear conservative for beginning-of-life performance. As the Stardust mission progresses (as was planned for MCO as well), image data will be periodically gathered to assess sensitivity degradation over life.

The star camera noise characteristics can provide an indication of radiation damage by trending flight data. From the flight image shown in Figure 10, the mean background level is 11.9 counts with a standard deviation of 1.33 counts, whereas the mean background level of the simulation is 11.7 counts with a standard deviation of 1.34 counts. The model shown was established prior to flight. The close proximity of the flight data and model data indicates that radiation damage was insignificant up to that point in the MCO mission.

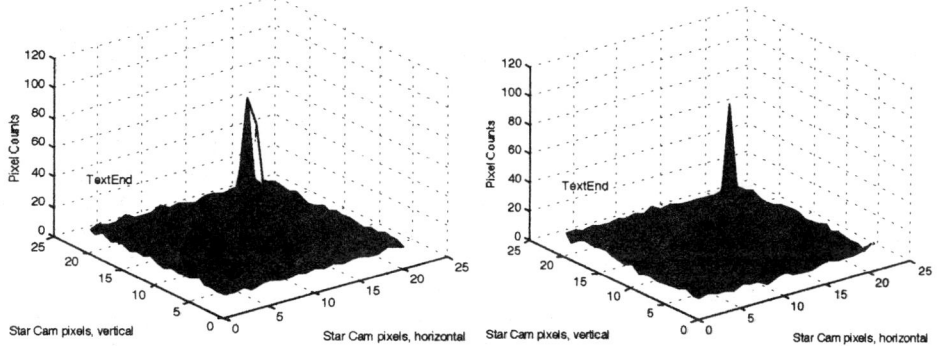

Figure 12 Flight Data (left) vs. Simulation Data (right) for MRI=2.5 Star

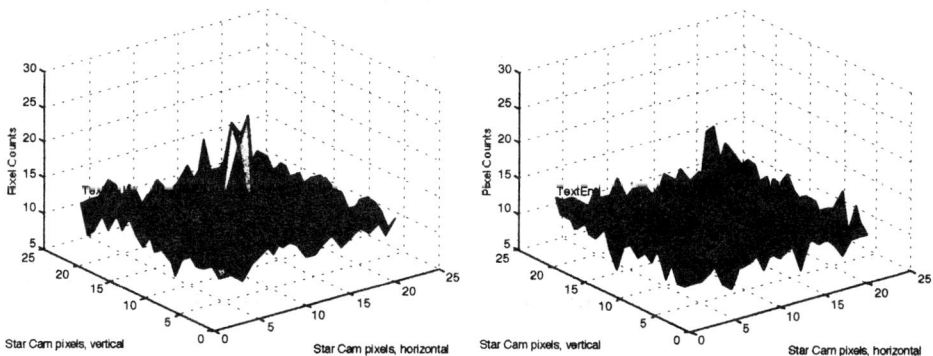

Figure 13 Flight Data (left) vs. Simulation Data (right) for MRI=4.5 Star

All-Stellar Flight Results

In this section, we present some key all-stellar flight results for MCO during cruise. Because the Stardust implementation is nearly identical, including image processing and attitude estimation, the flight results shown characterize the all-stellar design for both spacecraft. MCO performed two demonstrations of all-stellar attitude determination, and then transitioned to long term all-stellar use in June 1999. Gyro-based attitude determination was used three additional times to prepare for various spacecraft events. Table 1 shows the high-level command history for MCO attitude determination state.

Table 1 Command History of MCO Attitude Determination State

Date	Event	PURPOSE
11 Jun 99	Dual State Commanded	Short Term Demonstration and Check Out – Dual State
16 Jun 99	All Stellar Commanded	Short Term Demonstration and Check Out – All Stellar
23 Jun 99	Transition to All Stellar	Initial Long Term Implementation
23 Jul 99	Return To Gyro Based	Preparation for Trajectory Correction Maneuver #3
28 Jul 99	Transition to All Stellar	Return from TCM #3 configuration
20 Aug 99	Return To Gyro Based	Preparation for special Two Axis Gimbal Testing
21 Aug 99	Transition to All Stellar	Return from TAG Testing
02 Sep 99	Return to Gyro Based	Preparation for Calibrations and Mars Orbit Insertion

Transferring image data from the camera during acquisition mode utilizes significant processor throughput. Star search algorithms must also process an entire image at 0.1 Hz during acquisition mode. During track mode, predicted star locations are employed to identify portions of the star camera image for transfer and processing. This allows the image capture frequency to be increased from 0.1 Hz to 1 Hz. Consistent with the reduced data presented to Stellar Compass during Track Mode, the number of events associated with the image processing drops significantly. These swaths comprise between 20 and 25% of a full image, and hence processing events would be expected to decrease correspondingly. Image processing begins with a scan of pixel data looking for local brightness maxima. On each such brightness peak, more detailed pixel comparisons, denoted as impulse filtrations, are applied in the attempt to distinguish stars from CCD noise. A dramatic drop in impulse filtrations was observed during the first dual demo, as shown in Figure 14. Pixels that pass this detailed scrutiny are then considered candidate stars, for subsequent star and triangle identification algorithms. Similar to the decrease in impulse filtrations, Figure 15 shows a dramatic drop in candidate stars upon transition to track mode.

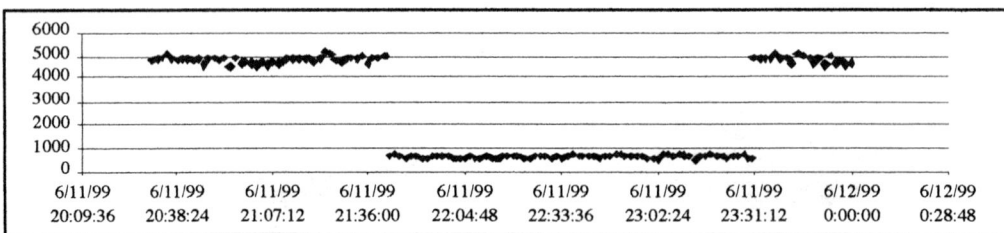

Figure 14 Impulse Filtration Decrease During Track Mode

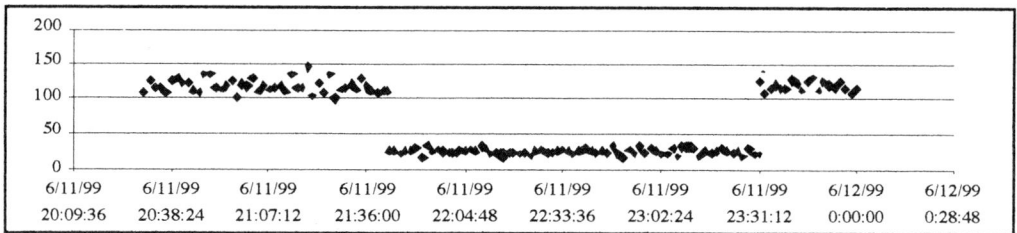
Figure 15 Candidate Star Decrease During Track Mode

Because the entire attitude determination software, including image processing, was hosted on the spacecraft C&DH processor, processor utilization has been closely monitored during flight. Figure 16 shows the processor utilization for the June 16 all-stellar transition, which included a short period of dual operation. Processor utilization increases slightly during dual operation, since both gyro-based and all-stellar attitude determination algorithms are running simultaneously. The transition from gyro-based to dual also allows Stellar Compass to transition from acquire mode to track mode. Even though star updates are computed 10 times faster during track mode, the CPU utilization actually drops slightly once the Kalman filter and the IMU data processing are no longer running. This can be seen in Figure 16 by comparing the data after the all-stellar transition with the data prior to the dual transition.

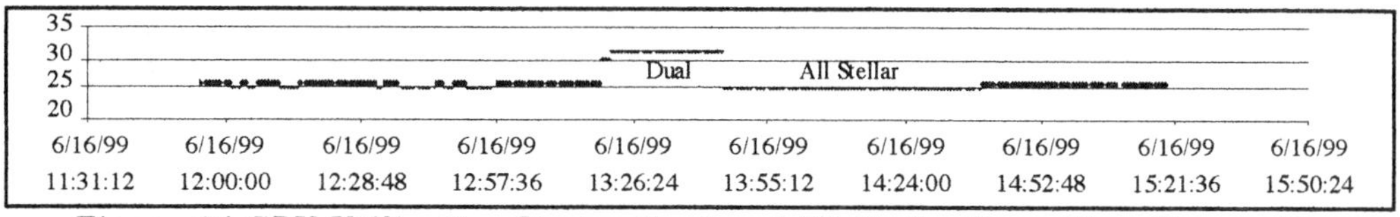

Figure 16 CPU Utilization During Dual and All-Stellar Operation, % capacity

The most important measure of all-stellar performance is the accuracy of the attitude measurements. As discussed earlier, both the Kalman-filter (gyro-based) scheme and the all-stellar scheme ran simultaneously during the dual demonstration. Although only the Kalman-filter based attitude and rate estimates were use in the attitude control loop, a complete set of telemetry was recorded and downlinked for both schemes. Figure 17 shows the resulting attitude error between the Kalman-filter attitude estimate and the all-stellar attitude estimate. The error is computed by comparing the two quaternion solutions and computing the angular error between the two. The results show that total difference between the Kalman-filter attitude and the all-stellar attitude has a mean of 0.0214 deg and a standard deviation of 0.0092 deg. This level of performance allows the attitude knowledge error allocation to be easily met.

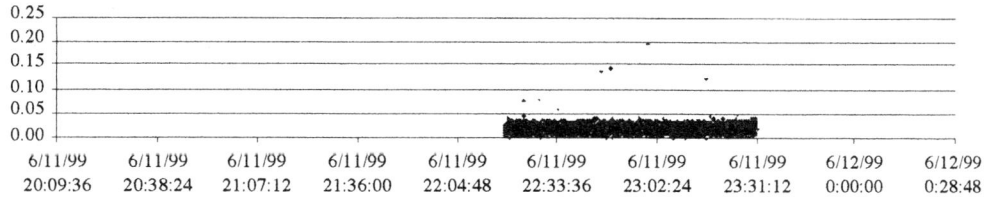
Figure 17 All-Stellar Attitude Estimation Error, deg

The rate estimation accuracy is also key to understanding the performance of all-stellar attitude determination. For the dual demonstration, all-stellar rate estimates have been compared against filtered IMU data, and the results are presented in Figure 18. As expected, the rate error has zero mean in all three axes (the dual demonstration used a fixed inertial attitude to maintain telemetry link). The variation of the rate estimate was very small, approximately 0.002 deg/s, 3-σ for the worst case axis. Also as expected, the worst-case axis is the spacecraft's x-axis, which is the axis most closely aligned with the boresight of the star camera.

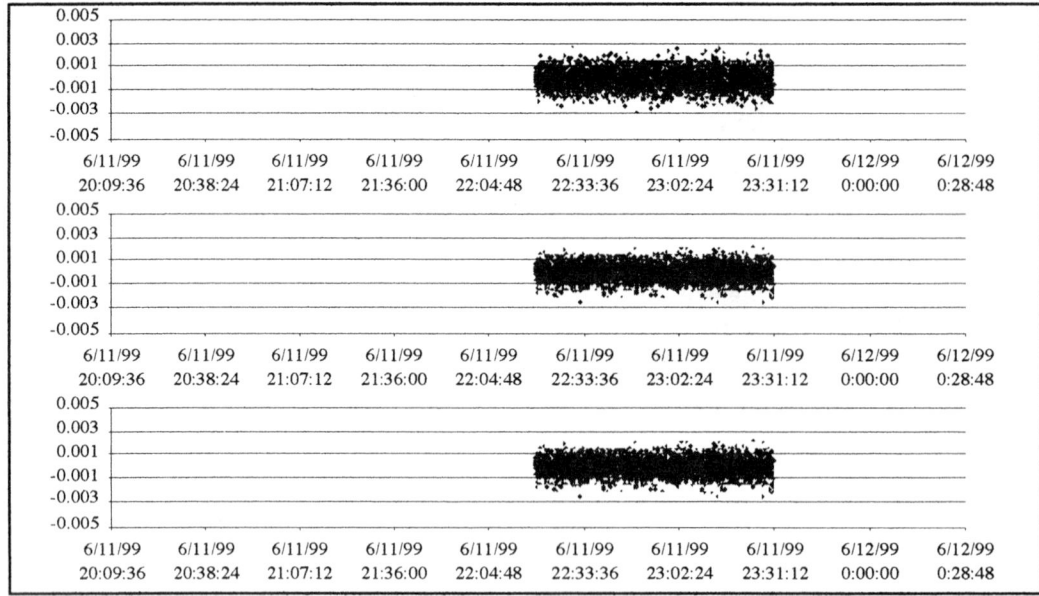

Figure 18 All-Stellar Rate Estimation Error, deg/s (x, y, z axes)

The results presented above characterize the all-stellar attitude estimation performance. As discussed earlier, all-stellar attitude estimation also places constraints on the control design. As part of the dual demonstration, the attitude control algorithms were reinitialized with parameters designed for use with all-stellar attitude estimation. Although the attitude control still used the Kalman-filter derived attitude estimates during the dual demonstration, the commands to the reaction wheels were computed using the all-stellar control law. This approach allowed the attitude estimation validation and the attitude control validation to be performed simultaneously while isolating any problems to either attitude control or estimation. Upon invocation of the all-stellar control parameters, a transient in wheel torque commands was observed, commensurate with a small spike in attitude error. The response is shown in Figure 19. The transient is a consequence of transitioning from a torque-based to a momentum-based control law, and is within expected limits. As seen in the plot, the lower-bandwidth momentum-based control law does result in slightly degraded attitude control performance. However, the attitude control errors are still extremely small, with the maximum excursion still less than 0.003 degrees. Closer examination of the data shows that the control design meets the pointing stability requirements as well.

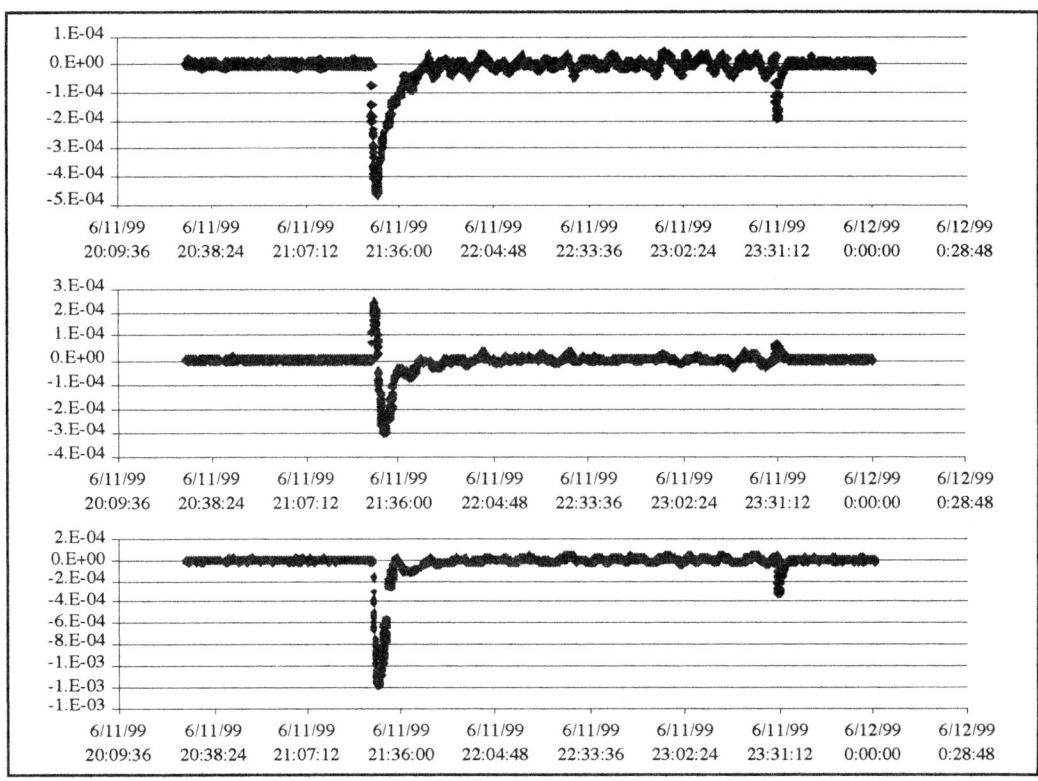

Figure 19 Attitude Control Errors During Dual Demonstration, rad (x, y, z)

CONCLUSION

An all-stellar attitude determination design has been designed, implemented, and validated in flight. The design takes advantage of the full-sky coverage and lost-in-space capabilities afforded by the Star Tracker Camera, developed jointly by Corning OCA and Lockheed Martin. The all-stellar attitude determination design provides flexibility, accepting star updates at either 0.1 Hz or 1 Hz, and accommodating short duration star outages. A robust attitude control law was developed to support the all-stellar attitude determination scheme. The control law accommodates either reaction wheel control or thruster deadband control, and accommodates propulsive desaturation events and articulable appendages. Flight data demonstrates the design provides single-sensor attitude estimation and control while meeting moderate pointing and stability requirements.

ACKNOWLEDGMENT

This work was supported by the Mars Surveyor Program under contract to the Jet Propulsion Laboratory, contract number 960457. Many thanks go to the dedicated analysts at Lockheed Martin who helped develop this design, worked through the details, and made it flightworthy, and to those at JPL for their careful review. Special thanks go to Eric Lander, who developed the overall all-stellar architecture, Tom Kelecy, who developed the all-stellar attitude determination design, Jason Wynn, who developed the control design, Brad Haack and Dale Howell, who debugged and verified the flight software implementation, and Kent

Hoilman, who worked on star camera modeling and sky coverage. We also gratefully acknowledge the assistance provided by Corning-OCA in producing this paper.

REFERENCES

1) C. Heatwole, L. Herman, G. Manke, and C. Voth, "Attitude Determination for the STEX Spacecraft Using Virtual Gyros," accepted for publication, Proceedings of the 23rd Annual AAS Guidance and Control Conference, Breckenridge, CO, Feb 2000.
2) "Sea Winds Mission Status," Media Relations Office, Jet Propulsion Laboratory, Pasadena, CA, Jan 10, 2000.
3) Jim D. Chapel and Richard Kiessig, "A Lightweight, Low-Cost Star Camera Designed for Interplanetary Missions," Proceedings of the 21st Annual AAS Guidance and Control Conference, Breckenridge, CO, Feb 1998.
4) James Oberg, "Why the Mars Probe Went Off Course," IEEE Spectrum, Dec 1999.
5) Paul DeLaHunt, Steve Gates, and Marv Levenson, "Attitude Determination and Control of Clementine During Lunar Mapping," AIAA Journal of Guidance, Control, and Dynamics, May-Jun 1996, Vol 19, No 3, p. 505.
6) "Wide Field-of-View Star Tracker Design Specification," Corning OCA Specification Number OCA 25001, Jan. 1998.
7) M.D. Shuster, S.D. Oh, "Three Axis Attitude Determination from Vector Observations", AIAA Journal of Guidance and Control, Jan-Feb 1981, Vol 4, p. 70.
8) E. J. Lefferts, F. L. Markley, and M. D. Shuster, "Kalman Flitering for Spacecraft Attitude Estimation," Journal of Guidance, Sep-Oct, 1982, Vol 5, No 5, p. 417.
9) M.A.C. Perryman, "The Hipparcos and Tycho Catalogues, Introduction and Guide to the Data," ESA Publications Division, 1997, Vol 1.
10) T. Kelecy, "All-Stellar Attitude Determination, Rev. 2" Mars Surveyor Program Controls Algorithm Notebook, Lockheed Martin, April 1998.
11) "Mars 98 Orbiter All-Stellar Attitude Determination and Control Design & Analysis," Engineering Change Board Briefing Package, Lockheed Martin Astronautics, Oct. 1997.

ATTITUDE DYNAMICS OF THE GENESIS SPACECRAFT

Carl Hubert[*]

The Genesis spacecraft has a science payload that requires a spinning platform but that also requires mechanical reconfigurations of a type that are normally not found on a spin-stabilized vehicle. This paper addresses the dynamics of these reconfigurations and describes the resulting constraints on vehicle operations. The reconfigurations in question include the opening and closing of the reentry capsule's massive cover, which must be done in a way that avoids tumbling the vehicle. Other critical reconfigurations involve exposing different sets of solar wind collectors without violating tight attitude constraints. The various mechanical reconfigurations, attitude maneuvers, and orbital maneuvers all induce nutational motions that must be attenuated in a timely manner. Hence, nutation damping requirements are also addressed. Key damper design trades are discussed and the reasons for selecting the final configuration are outlined. Finally, the design and test of the vehicle's passive nutation dampers are described.

INTRODUCTION

The Genesis mission will be the first since Apollo to return extraterrestrial material to Earth. After its launch in early 2001, Genesis will be maneuvered into a halo orbit about the L1 libration point between Earth and the Sun. This orbit will place Genesis outside Earth's magnetosphere, where it will be exposed to the raw solar wind. Genesis will spend two years collecting samples of the solar wind and will return the samples to Earth in 2003.

Figure 1 shows Genesis in its launch configuration with the solar panels stowed. After the panels are deployed (Fig. 2) they remain fixed for the rest of the mission. Except during orbit change maneuvers, the spacecraft spin axis (perpendicular to the solar panels) will be kept oriented toward the sun. Figure 2 shows the cruise configuration in which Genesis travels out to the L1 point and later returns to Earth. After Genesis enters the halo orbit it is reconfigured for the Science phase of the mission (Fig. 3). First, the sample return capsule's cover is opened,

---

[*] Hubert Astronautics, Inc., 13123 Taylor Court, West Windsor, New Jersey 08550.

exposing a canister that contains the solar wind collector arrays. (The canister has a separate sealed lid that protects the collector arrays from contamination both before launch and after return to Earth.) The canister lid is then opened and the collector arrays are deployed.

At the end of the Science phase of the mission, the collector arrays are stowed, the canister lid is closed, and the sample return capsule's cover is also closed. Once it is back in the cruise configuration the vehicle returns to Earth. Shortly before entry into Earth's atmosphere the sample return capsule is released for a parachute landing at the Pentagon's Utah Test and Training Range. A small propulsive maneuver then diverts the spacecraft bus for reentry over the Pacific Ocean.

The science payload requires a spinning platform but it also requires mechanical reconfigurations of a type that are normally not found on a spin-stabilized vehicle. These reconfigurations include the opening and closing of the sample return capsule's large cover. Other critical reconfigurations involve exposing and shading different sets of solar wind collectors (see Fig. 3) without violating tight attitude constraints.

The science payload doesn't operate during solar panel deployment or during the opening and closing of the sample return capsule cover. Hence, beyond the need to damp residual nutation within a reasonable amount of time, pointing accuracy is not an issue during these events. The only dynamics question of importance is vehicle safety. Specifically, these events must not cause the vehicle to tumble. Here, the word "tumble" refers to a condition in which the spacecraft is inverted and spinning backwards.

SOLAR PANEL DEPLOYMENT

Figures 1 and 2 show Genesis before and after the solar panels are deployed. The deployment mechanism comprises hinge springs, hinge dampers, and a pair of short-stroke push-off springs to separate the two solar panels The deployment occurs with the vehicle initially spinning at 7.0 ± 2.5 rpm.

Since solar panel deployment involves no externally-applied torques, it will not change the magnitude or direction in space of the vehicle's angular momentum vector. From an attitude standpoint, therefore, the most important question is whether it is possible for the deployment to tumble the spacecraft. If a tumble is not possible, the next most important question is whether the worst-case post-deployment nutation will have an acceptable magnitude. Hence, the focus of the deployment analysis was not on expected performance but on whether extreme behavior could create an unacceptable condition. As described in the next few paragraphs, even under improbably extreme conditions, the solar panel deployment is a dynamically safe operation.

If the vehicle has no initial nutation and if the solar panels deploy symmetrically, the reaction torques will be balanced in such a way that there will be relatively little nutation and absolutely no chance of a tumble. However, if there is a large initial nutation and/or if the deployment deviates significantly from its nominal symmetry, a large nutation can result. The extreme worst-case condition involves the unlikely scenario of sequential deployment, in which one panel doesn't start to move until after the other panel is completely deployed. This might happen, for example, if an initially stuck panel was jarred loose by the shock of other panel's end-of-travel impact against its stop. Although this type of deployment is highly improbable, it was examined as bounding case. If an extreme worst-case sequential deployment cannot tumble the vehicle, then the worst-case simultaneous deployment at unequal rates will not be able to cause a tumble either.

Although intuition suggests that a fast single-sided deployment should produce a larger residual nutation than a slow deployment, simulations show that a nearly instantaneous deployment is not the worst case. The true worst-case asymmetric deployment involves an initial movement that requires a non-trivial fraction of a nutation cycle to complete. A simulation study showed that the extreme worst-case involves the minimum initial spin rate (4.5 rpm), the maximum initial transverse rate (6 deg/sec), and the first panel starting to move at a moderate speed at the worst point in the nutation cycle. In the extreme worst-case scenario, the first panel requires approximately 33 sec. to deploy, the second panel starts moving within a few seconds after the first one hits its stop, and the second panel deploys at an improbably high speed (approximately 3 sec. to deploy). Although this extreme — and extremely unlikely — scenario produces a large residual nutation, it does not tumble the spacecraft. The result of this and many other simulations showed that it is impossible for any panel deployment sequence to tumble the vehicle. From the standpoint of vehicle safety, this was a key result.

More realistic than sequential deployment is the situation in which both panels start to move almost simultaneously but at significantly different angular rates. Starting at 4.5 rpm with a 6 deg/sec transverse rate, the worst-case involves one panel achieving full deployment in about 33 seconds with the other panel requiring several minutes to fully deploy. This situation, which is almost as unlikely as a sequential deployment, could be caused by improbably extreme differences in the hinge damper temperatures. With worst-case phasing of the deployment relative to the nutation cycle, the residual nutation is about 33 deg. Although this is high, it is still far below the nutation needed to tumble the vehicle.

The most realistic asymmetric situation is one in which the two panels start to move simultaneously and complete their deployments within a few seconds of each other. Under these circumstances the post-deployment nutation could still be large if the initial spin rate is at its lower bound while the transverse rate is at its upper bound. Depending on the point in the nutation cycle when the panel motion begins, the post-deploy nutation could range between 6 and 21 deg. Although 21 deg. may seem large, the driving requirement is that the residual nutation be reduced to no more than 0.2 deg. within 8 hours after the deployment. As discussed below, this is well within the capabilities of the passive nutation dampers.

SAMPLE RETURN CAPSULE OPENING AND CLOSING

As Figures 1 through 3 clearly show, the Sample Return Capsule (SRC) has a very large hinged cover. During the process of opening this cover, the vehicle's equilibrium spin axis (the maximum moment of inertia axis) deviates more than 30 deg. from its initial orientation relative to the spacecraft bus. This is a temporary condition. Once the cover is completely open, the spin axis is very close to its initial orientation. However, movement of the principal axis during the cover opening (and closing) excites nutation. The more rapid the principal axis motion, the greater the induced nutation. The greatest nutation occurs if the SRC cover moves through a large angle in less than one nutation cycle. Hence, care must be taken during the cover opening and closing to avoid tumbling the spacecraft. Specifically, the cover movements must be smooth, continuous, and slow.

Because Genesis is inertially asymmetric, it is theoretically possible to make it tumble at a nutation angle less than 90 deg. Furthermore, because the inertial asymmetry is greatest during the SRC cover deployment, potential susceptibility to tumbling is greatest during the deployment. Because of this, it was determined that the principal axis angular shift during a single nutation cycle should be much less than the minimum nutation needed to tumble the vehicle. This set a conservative upper bound on the cover deployment rate. Mathematically, this constraint can be expressed as follows:

$$\Omega_C < \frac{\eta}{2\pi} \sqrt{(\sigma_{min}-1)(\sigma_{max}-1)} \ \tan^{-1} \sqrt{\frac{\sigma_{min}-1}{\sigma_{max}-\sigma_{min}}}$$

where

Ω_C = normalized cover deployment rate (the ratio of the cover's angular rate to the spacecraft spin rate)

σ_{min} = minimum principal axis inertia ratio ($I_{max}/I_{intermediate}$)

σ_{max} = maximum principal axis inertia ratio (I_{max}/I_{min})

η = tumble parameter
 = R_S / Ω_P

Ω_P = normalized principal axis offset rate
 = the rate that the spacecraft's major axis shifts relative to body coordinates divided by the cover's angular rate during the deployment. Ω_P varies during the cover deployment. The maximum value is used in order to calculate a conservative value for η.

R_S = a dimensionless safety factor
 = the maximum allowable principal axis shift in one nutation cycle divided by the minimum nutation angle that can tumble the spacecraft

A safe upper bound for Ω_C is established by setting an appropriate upper bound on η. This is done by selecting an appropriate value for the safety factor (R_S), which should typically be no greater than 0.5.

The rate constraint discussed above applies to continuous cover motions. Stopping and starting, if badly phased, can pump-up nutational motion to a level well beyond that possible from a simple continuous motion. The worst-case would be to deploy (or stow) the cover with a series of short movements that are separated by one nutation cycle. To avoid this, the cover is opened and closed in a single stage without stopping.

COLLECTOR ARRAY RECONFIGURATION

The solar wind collector arrays are shown in Fig. 3. Three of these arrays are occasionally slewed between shaded and unshaded positions depending on measured characteristics of the solar wind. An important difference between collector array transitions and the SRC cover opening/closing is that the array transitions occur during science data collection. This imposes much more stringent requirements on the attitude transients than is needed for the SRC cover opening and closing. During collector reconfigurations it is not enough to maintain dynamic stability. Instead, nutation should be kept below 0.5 deg. during and after every reconfiguration.

Genesis has five collector arrays. One of these is fixed inside the lid of the canister that contains the stowed arrays. This fixed array has no effect on transient behavior during the Science phase of the mission. After the canister is opened, the other four collector arrays rotate from the stowed to the deployed state in preparation for the Science phase of the mission. The top array does not move beyond this deployed position. However, the lower three arrays periodically move back and forth between their deployed and "unshaded" positions (Fig. 3). At any given time, one or two of these lower collector arrays can be in the unshaded position. Reconfiguration can involve the movement of one, two, or all three lower collector arrays. For reconfigurations involving multiple collector arrays, the movements are sequential rather than simultaneous and the attitude response depends very much on the order in which the arrays are moved. If we assume that the mission can involve collectors transitioning from any combination of one or two unshaded arrays to any other combination of one or two unshaded arrays, and if we further assume that all sequences are possible, then there are 54 possible collector array transitions.[*] Further complicating any worst-case analysis is the fact that the collector arrays can move at either of two different speeds.

Rather than making an exhaustive study of the various possibilities, the analysis concentrated on the motions of a single collector array. The selected array was chosen because it

---

[*] Moving only one array involves 6 possible transitions (3 arrays, each with 2 possible starting positions). Moving two arrays involves 12 possible transitions (6 reconfigurations, each with 2 possible sequences). Moving three arrays involves 36 possible transitions (6 reconfigurations, each with 6 possible sequences).

is furthest from the vehicle's center of mass and thus has the greatest effect on the attitude motion. The sensitivity to mass properties and array speed was examined. From these analytical results, the worst-case effect of multiple array motions was estimated by superposition.

Simulations were made of forward and backward collector motions at different speeds. Representative behavior is shown in Figure 4, which plots the results of 26 simulations that differed only in the speed with which the collector array moved. The figure shows the expected result that peak nutation gets smaller as the collector motion gets slower. However, it also shows significant variability in the residual nutation, which can be nearly zero or as large as the peak nutation. These wide swings in residual nutation are functions of the relationship between the nutation period and the time required to complete the array transition. With best-case phasing, the collector start-up transient is almost exactly cancelled by the end transient. Worst-case phasing, however, can lead to a substantial residual nutation. Whether the phasing is good or bad depends not only on the speed of the collector motion but also on the nutation frequency, which is a function of the vehicle's moments of inertia.

If solar wind conditions require more than one collector array to be moved, they are relocated sequentially with no pause between the individual relocations. The nutation resulting from these sequential motions depends strongly on their phasing relative to the nutation cycle. In other words, the response depends on the speed of the collector motions, the sequence of the collector motions, and the vehicle mass properties. In the best case, multiple collector movements could lead to near zero residual nutation. In the extreme worst case, the effects of the individual movements could be essentially additive. In other words, the residual nutation from relocating two or three collector arrays could nearly equal the sum of the residual nutations that would be induced by moving each of the collectors individually.

The transition with the potential for an absolute worst-case residual nutation would be one involving two collectors moving from the deployed to the unshaded position with the third collector moving from the unshaded to the deployed position. With the faster of the two collector speeds and worst-case lower bounds on the vehicle moments of inertia, there is a small possibility that the residual nutation could briefly exceed the 0.5 deg. goal. Fortunately, this extreme condition requires an unlikely combination of slew speeds and actual flight mass properties. Furthermore, even if the 0.5 deg. limit was slightly exceeded, the nutation dampers would quickly reduce the nutation to the nominally required level.

NUTATION DAMPING

The Genesis spacecraft spins about its maximum moment of inertia axis during all phases of the mission. Hence, energy dissipation by structural hysteresis and by liquid propellant motion provide inherent nutation damping. However, this inherent damping is poorly defined and unlikely be fast enough to meet the mission requirements; especially toward the end of the mission when little propellant remains. It was therefore necessary to incorporate a nutation damper in the design. Active damping was briefly considered and rejected for a variety of systems considerations, including the fact that it would have required an additional sensor that would have no other function. A trade study of passive damper designs led to the selection of a pair of viscous ring dampers.

Passive Damper Requirements

The worst-case nutation and the maximum allowable damping time vary with mission phase. Hence, the damper performance requirements also vary with mission phase. The key damper time constant requirements are listed in Table 1, which also identifies the nominal number of nutation cycles per time constant for each vehicle configuration. The number of cycles per time constant is important because the fewer cycles available per time constant, the more energy dissipation that is needed per cycle to achieve that time constant. The last column in Table 1 shows that the requirements for operations at 1.6 rpm are more difficult to meet than those for 10.0 rpm. Thus the damper design was optimized for the nutation frequencies associated with 1.6 rpm operations.

TABLE 1

NUTATION DAMPER TIME CONSTANT REQUIREMENTS

Vehicle Configuration	Spin Rate (rpm)	Required Time Constant (hr.)	Nutation Cycles per Time Constant
Pre-Science Cruise	10.0	2.2	785
Science	1.6	6.5	465
Final Entry	1.6	6.3	305
Divert Burn	10.0	3.7	1,200

A key consideration in the passive damper design is the fact that when the spacecraft is mounted on its spinning upper stage, the device becomes a nutation de-damper. The stack comprising the spacecraft and upper stage spins about its minimum moment of inertia axis. In such a configuration, any energy dissipation causes nutation to grow. A design goal for the passive nutation dampers was that the divergent nutation time constant should be greater than 10 minutes while the vehicle is mounted on the spinning upper stage.

In addition to being able to meet its performance goals, the damper had to be testable in a 1-g environment using relatively simple equipment and test procedures. The requirement of testability was a key driver in the selection of the design approach.

Passive Damper Design Trades

A trade study examined several passive damper types before the final design was selected. The damper types that were considered and their advantages and disadvantages are as follows:

Ball-in-Tube Dampers. Devices of this type have a solid ball that rolls or slides along a tube in response to nutational motion. Energy is dissipated by the flow of gas or liquid through the small gap between the ball and the tube wall. This type of damper is optimized for a specific inertia ratio and spin rate by selecting the curvature of the tube and the gap between the ball and wall. Although this "tuning" can provide strong damping at the design point, dampers of this type can be inefficient if the spacecraft's inertia ratio or spin speed differs significantly from the design point. Since the Genesis dampers must be effective over a wide range of spin rates and inertia ratios, sharp tuning is a potentially severe limitation.

Another problem with a ball-in-tube damper is that to get adequate performance at the low Genesis spin rate, the gap between ball and wall would have to be extremely small. This poses a major manufacturing challenge: making a curved tube with a very precise inside diameter. Furthermore, the damping time constant is sufficiently sensitive to error in the gap that ground-based performance testing would be essential. However, the slow operational spin rate makes realistic performance tests impossible in a 1-g environment. During the science phase of the Genesis mission, the centripetal acceleration at the damper is approximately 0.003 g; three orders of magnitude away from the conditions during ground-based testing. Finally, it was unclear whether materials could be found that would maintain the appropriate gap over the full range of thermal conditions.

Pendulum Dampers. These devices typically have a lumped mass suspended at the end of a flexible wire. Energy is dissipated by nutation-induced movement of the mass through a viscous medium (liquid or gas) or via magnetic hysteresis. An advantage over the ball-in-tube damper is the lack of a stiction deadband. In addition, because the pendulum bob isn't confined to a tube or track, it responds to two-axis motion. This increases nutation damping in inertially asymmetric spacecraft such as Genesis. Pendulum dampers, however, are more complex than ball-in-tube dampers because they require a mechanism to hold the pendulum bob in place during launch. Furthermore, like ball-in-tube dampers, pendulum dampers intended for slowly-spinning spacecraft are impossible to test realistically in a 1-g environment.

Fluid Slug Dampers. These devices dissipate energy through viscous effects as liquid is driven through a tube by nutational motion. These dampers can take several forms. One type is a U-shaped tube with a partially-filled reservoir at each end. Another type is a closed, tubular ring that is partially filled with liquid. A distinguishing characteristic of these dampers is that the liquid has a free surface. These dampers are simple to manufacture. The only area where some precision is required is in the amount of liquid used, because the liquid volume affects the tuning.

Fluid slug dampers have many of the same advantages and disadvantages as ball-in-tube dampers. They can be tuned to provide strong damping at the expected nutation frequency, but they can also be inefficient if the spacecraft's inertia ratio differs significantly from the design point. At small nutation angles and low spin rates surface tension might lock the liquid in position, effectively eliminating energy dissipation by the damper. Given the relatively low centripetal accelerations associated with the Genesis operational spin rate, surface tension lock-up was a potential concern. Fluid slug dampers also have the same sensitivity to gravity that makes ball-in-tube dampers for slowly spinning spacecraft difficult to test on the ground.

Partially Filled Ring Damper. This device is a variant of the fluid slug damper. It is simply a tube bent into a circular ring and partially filled with mercury. The ring is oriented perpendicular to the vehicle's spin axis and is centered on the spin axis. The performance of this type of damper is difficult to predict at small nutation angles and low spin rates – the conditions under which Genesis spends most of its mission. Realistic ground-based testing for these conditions is impossible. At large nutation angles the mercury can pool into a "slug" and enter a nutation synchronous mode. In this mode, the liquid has a unidirectional flow around the tube at the nutation frequency. This mode dissipates energy very rapidly, which is an advantage for recovery from large disturbances. However, this mode of rapid energy dissipation is a potentially serious disadvantage when the spacecraft is atop the spinning upper stage.

Completely Filled Viscous Ring Damper. This type of device is similar to the partially-filled ring damper in that it is a closed tubular ring. However, the tube is completely filled and the ring does not need to be circular. As long as the bends aren't too sharp, the ring can be any shape that conveniently fits within the spacecraft's envelope. Another difference is that the damper is oriented parallel to the spin axis instead of being perpendicular to it. To prevent bursting at high temperature and to prevent the formation of voids at low temperature, a bellows mechanism is required to accommodate fluid expansion and contraction. Like fluid slug dampers, filled viscous ring dampers are relatively simple to manufacture. However, the needed bellows can make them heavier than fluid slug dampers.

By their nature, filled viscous ring dampers are less sharply tuned than other nutation dampers. Although this reduces their effectiveness at the nominal design point, it makes them effective over a broader range of spin rates and inertia ratios. Furthermore, because there is no liquid free surface, they cannot lock up due to surface tension. In other words, they have no deadband. The lack of a free surface also makes the performance of these dampers insensitive to gravity. This is a key advantage because it makes realistic, high-fidelity ground-based testing possible.

Silicone fluid is the preferred liquid for this type of damper because its viscosity can be "tuned" to optimize performance at a design point represented by a nominal combination of vehicle mass properties, spin rate, and operating temperature. This tuning maximizes the energy dissipation per unit mass of fluid. Silicone fluid can be made to have virtually any viscosity by blending off-the-shelf fluids with viscosities that are just above and just below the desired value. For a given nutation frequency and tube size, the optimum damper fluid viscosity is given by the following equation:

$$\nu = 0.1581 \, \lambda \, r^2$$

where ν is the kinematic viscosity, λ is the nutation frequency, r is the tube's inside radius, and the coefficient is dimensionless.

Nutation Damper Design

Because they can operate well over a very wide range of inertia ratios and spin rates, and because they are relatively easy to manufacture and test, filled viscous ring dampers were selected for Genesis. Each of the two Genesis nutation dampers is a tubular ring that is completely filled with silicone fluid. Nutation induces a cyclic fluid flow within the tube, and the resulting dissipation of kinetic energy causes the nutation to exponentially decay. As shown in Figure 5, the damper tubes have the shape of rounded squares. Silicone fluid was selected for the reasons discussed in the previous section. The fluid viscosity was chosen to optimize performance during the science phase of the Genesis mission. The dampers were sized to have a factor of 2 performance margin relative to the requirements listed in Table 1.

Damper Testing

The dampers were tested on a horizontal torsional pendulum which had a fundamental frequency that matched the Genesis nutation frequency during the science phase of the mission. The test configuration is diagrammed in Fig. 6. The "instrumentation" was extremely simple: a thermometer (to measure ambient temperature for fluid viscosity calculations), a stop watch (to determine the pendulum's period), a battery-powered laser, several large pieces of paper, and a pen. As is described below, the tests were monitored by manually tracking a small, slowly-moving spot of light on a large piece of paper.

The equipment and balance masses were distributed around the platform so that the center of mass was at the torsion rod. The platform's moment of inertia about the torsion rod (including equipment and balance masses) was large enough that the damper fluid had essentially no effect on the platform dynamics other than energy dissipation.

Because the pendulum had a low frequency (approximately 0.02 Hz), it was expected that the test platform's inherent damping could be sensitive to air currents from room ventilation. Testing of the engineering development unit confirmed this sensitivity and the test platform was therefore enclosed in a plastic tent to minimize the effect of air currents. Subsequent tests within the tent showed that the test platform's damping characteristics still varied with ambient conditions. Each functional test was therefore immediately preceded and followed by tare runs using a dummy damper. In other words, each "test" involved a set of three runs. The first was a tare run with the dummy damper, the second was with an actual damper, and the third again used the dummy device. The results of each damper test were compared with the average of the preceding and following tare runs to determine the unit's performance. By bracketing damper runs with tare runs, the effect of variations in ambient conditions was calibrated out.

The original intention was for the tare runs to be done with a dummy damper that simply matched the mass properties of the real damper. However, shakedown runs that preceded the first damper tests revealed sufficient sensitivity to platform aerodynamics that a cardboard mockup of the damper and the bellows support structure was added to the platform. This mockup increased the damping during dummy runs by about 10%. The cardboard mockup was therefore used for all subsequent tare runs

As shown in Fig. 6, the primary "instrumentation" consisted of a small laser that was mounted on the torsional pendulum. Damping was monitored by projecting the laser beam on a large piece of paper that was located a short distance from the test platform. During each cycle of a test run, the leftmost and rightmost locations of the projected light spot were marked on the paper. These marks were used to determine the peak-to-peak motion of each cycle and these amplitudes were used to determine the damping rate. This inexpensive and seemingly crude method of data collection was effective and accurate because the pendulum's 50 second period made manual tracking quite easy (although a bit tedious).

Each flight damper was tested several times, and all tests showed good correlation with pre-test analytical predictions. In all cases, the tests showed damper performance that was within 8% of the analytically predicted value.

CONCLUSION

The Genesis spacecraft has unusual physical characteristics for a spinning vehicle. Despite this, the vehicle's dynamic behavior is compatible with the mission requirements.

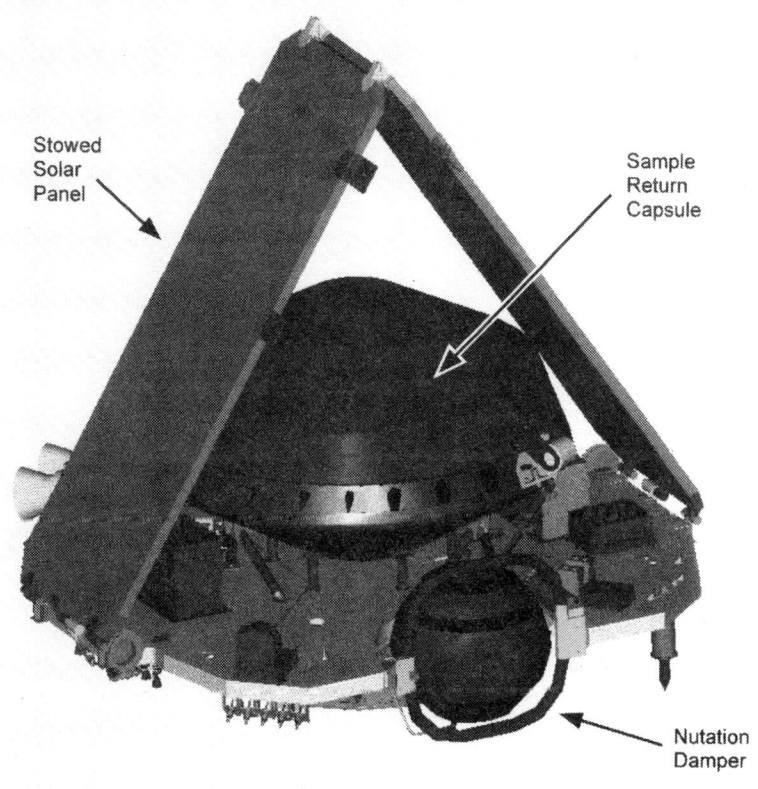

Figure 1. Launch Configuration

Figure 2. Cruise Configuration

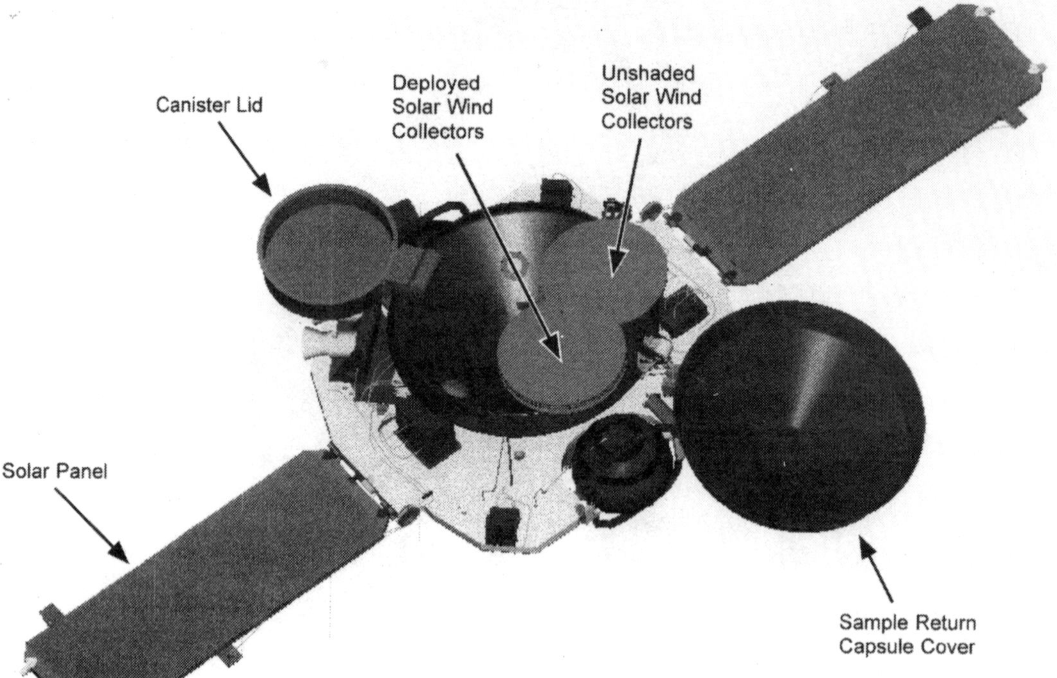

Figure 3. Science Configuration

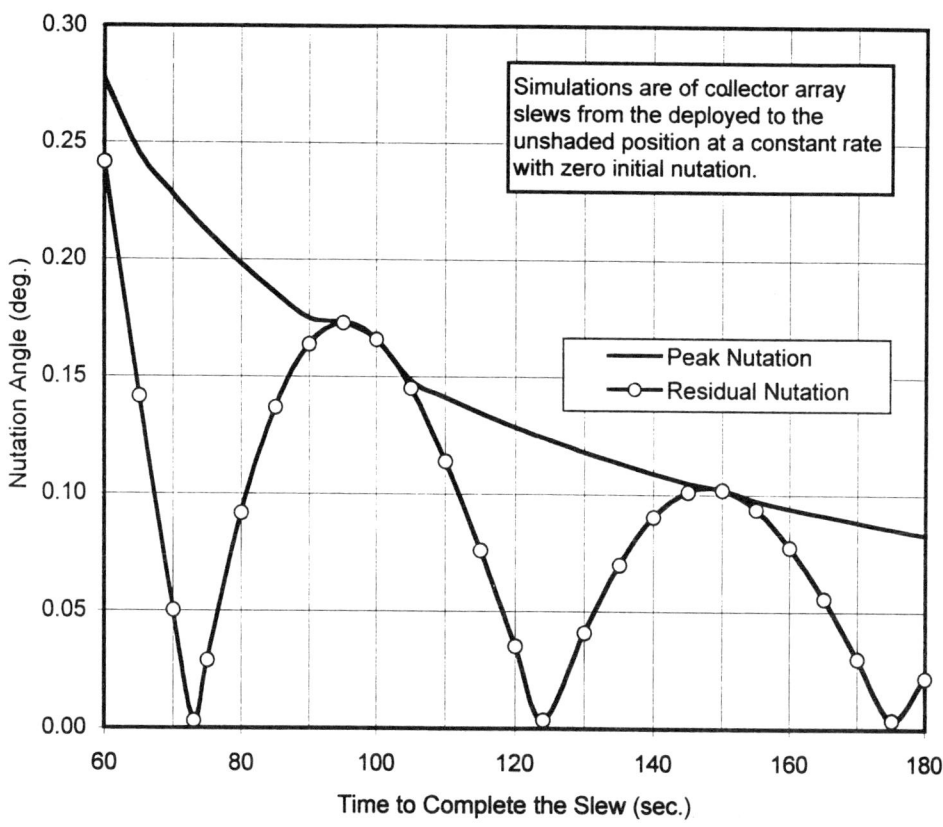

Figure 4. Peak and Residual Nutation vs. Collector Array Speed

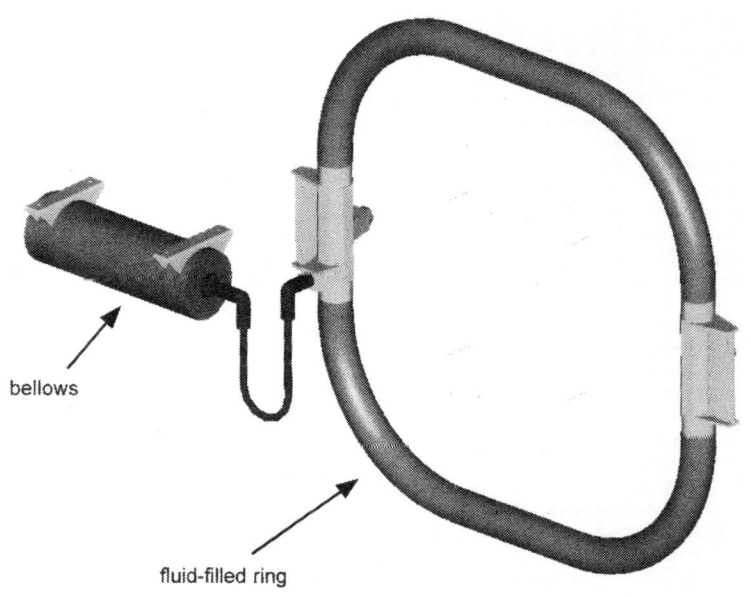

Figure 5. Viscous Ring Nutation Damper

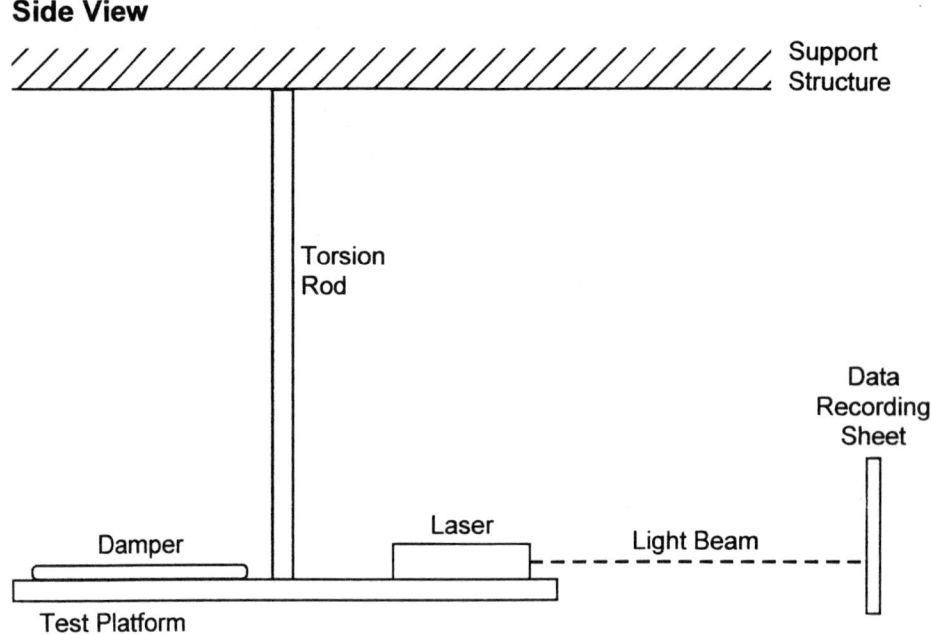
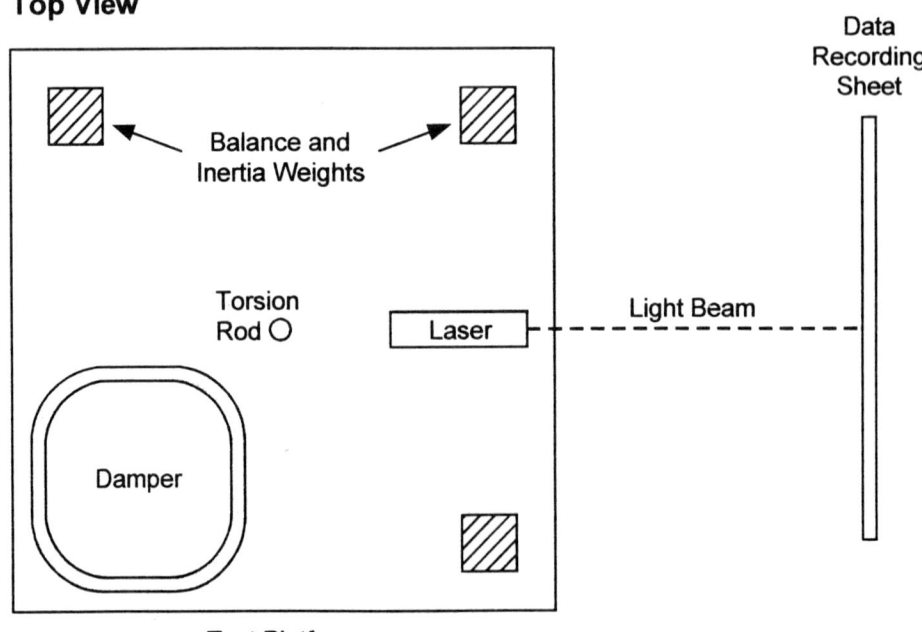

Figure 6. Side and Top Views of the Nutation Damper Test Setup

AAS 00-007

FIRE CONTROL DESIGN FOR
HIGH ENERGY (LASER) SYSTEMS

Timothy J. Schneeberger,[*] S. M. Seltzer[†] and Robert Van Allen[‡]

Fire Control systems for high-energy laser applications have very stringent (and often competing) requirements. For example, precision pointing requires high spatial resolution information while stabilization requires high temporal resolution. Sometimes, complex real-time decisions (turning major subsystems on, declaring the mission has been completed, etc.) use information from low bandwidth imaging sources. These decisions are based on merged high and low bandwidth rates. The engineering process associated with the design of such systems has many unique features and is the subject of this paper. Example top level requirements are given and a systematic flow-down into the various elements of the fire control system described. System design and test validation approaches are also discussed. Finally, a unique approach for generating real-time processor software code from detailed Matlab/Simulink simulations is shown along with examples from the Air Force Airborne Laser (ABL) program.

---

[*] Ph.D., VP for Technology, SVS Inc.
[†] Ph.D., Chief Engineer, SVS Inc.
[‡] Ph.D., COO, SVS Inc.

INTRODUCTION

A boost phase laser-based defensive weapon system must quickly respond to initial threat indicators and transition through a sequence of acquisition, tracking and pointing tasks which culminate with deposition of lethal fluence on the target. The logical constructs, software and hardware, and specific information extraction algorithms that effect this automatic sequencing comprise the Fire Control (FC) Element and its associated mode logic. For example, the FC subsystem for the Airborne Laser System (ABL) manages the target engagement timeline and implements real time trackers for six beam or target position sensors. The FC Element can be thought of as the conductor of an orchestra, responsible for the precise coordination of the weapon system instruments during a target engagement. The FC Element consists of hardware interfaces (from sensed plume scenes, for example), algorithms (velocity vector estimation, for example), and real time software/processors to execute tasks required to successfully complete a target engagement. The ABL subsystem is realized in a single 19-inch rack of flight electronics.

The ABL system provides an excellent example of the importance of FC to a high-energy laser (HEL) pointing system. The operational construct for engaging a single theater ballistic missile target with an ABL is shown in Figure 1. The notional timeline for engaging and negating a target in the boost phase clearly shows that the appropriate units for completion of FC functions are seconds! In addition to rapidly acquiring and tracking the target, the FC subsystem is responsible for calculating the target aimpoint, where accuracies to a fraction of a target width are typically required.

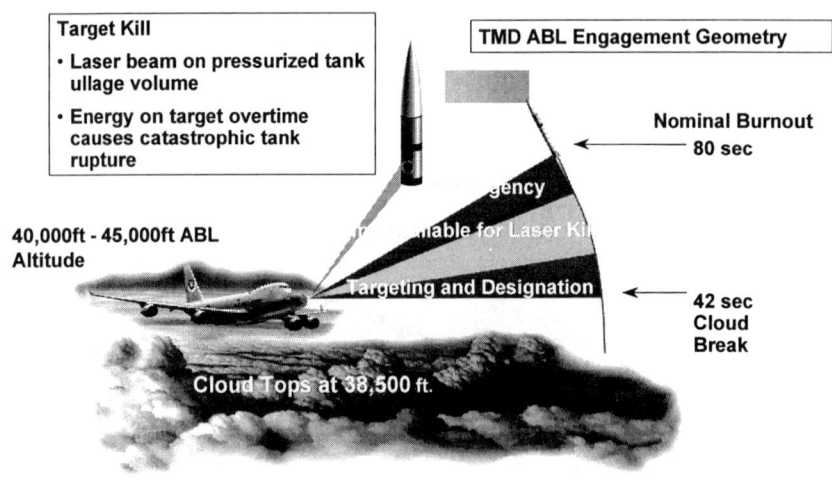

Figure 1: ABL is a boost phase defense system

Design Drivers and Development of FC Functions

Design Drivers identify those Fire Control sub-system requirements that have the most impact on the design process and that present the greatest design challenges to the Fire Control Engineering team. The two key design drivers for the ABL system design are as follows:
1) Engagement Timeline

Trade studies have been completed on the most efficient and time optimal approach to transitioning the ABL "region-of-interest" from the two pi steradians of the typical surveillance systems to the microradian regime required to precisely place the beam. Passive tracking functionality is shown in Figure 2, and active tracking and lasing is shown in Figure 3.

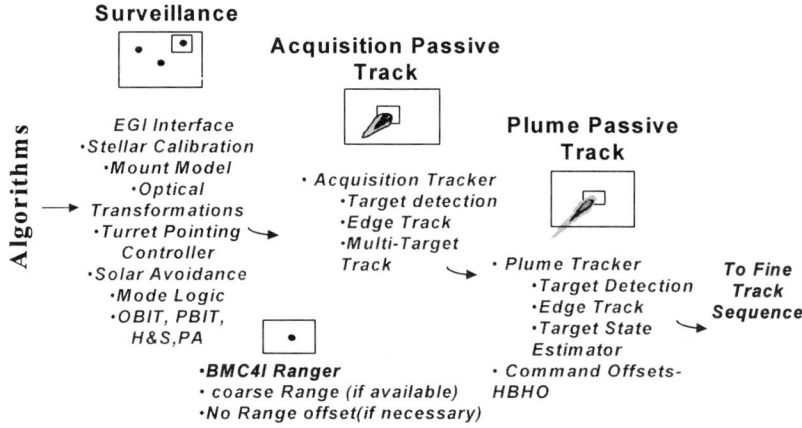

Figure 2: FC functional design for passive tracking.

2) Active Track and Aimpoint

Laser weapon systems require accurately placing the beam on target, in the presence of own-ship disturbances, atmospheric turbulence and target dynamics. For the ABL system the FC element is responsible for active track algorithm implementation and the computation of all offsets that effect transitions in pointing between passive, active and laser pointing. The algorlthms associated with these tasks are indicated in Figure 3.

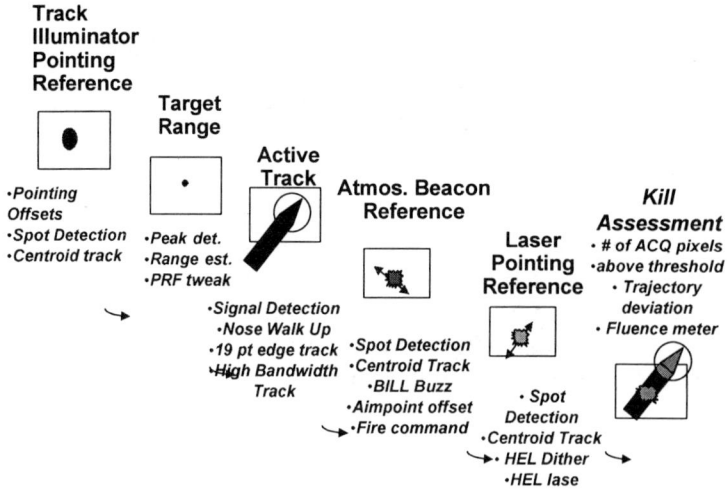

Figure 3:FC functional flow for active tracking and aimpoint.

Requirements Flow-down

FC is a subsystem of the overall Beam Control Fire Control (BCFC) segment on ABL, as it is on many HEL weapon systems. The Beam Control functionality includes implementation of the detailed control laws for precision pointing, and correction for atmospheric and local environmental disturbances. The BCFC segment specification flows from an overall system specification. Requirements for FC then are derived from the segment specification. FC however must have significant interaction with the top-level system requirements, which are typically defined by a set of operational scenarios and engagements. For ABL, a Technical Requirements Document contains details on targets, trajectories and associated engagement parameters. These details are fully incorporated in the FC trade study and design process. In addition, the FC element has significant interactions with data and commands to/from the Battle Management and Laser Segments. Thus a highly interactive design communication activity is essential to a seamless weapon system operation. Figure 4 shows this approach to requirements identification. The functional decomposition of FC into Algorithms, HW and SW, iterated against performance simulations and risk reduction tests is also shown in Figure 4.

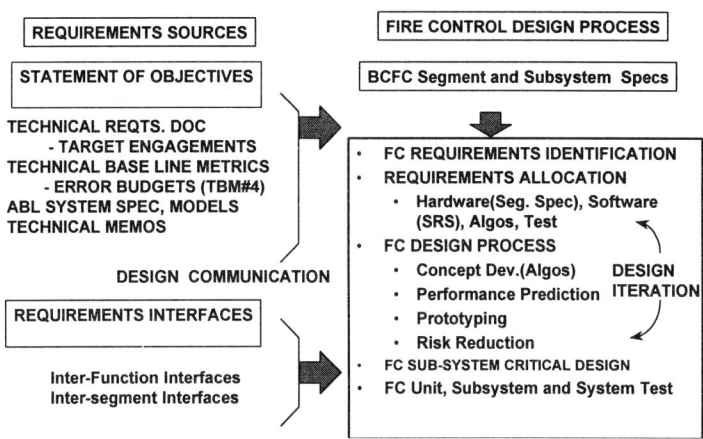

Figure 4: Requirements Flow-down

The design of the FC subsystem (and many other engineering subsystems), given initial definition of requirements and interfaces, can be decomposed into hardware (HW), software (SW), and algorithms (Algos), in union with a build-to-test design philosophy. This set of functional elements can be efficiently integrated using the power of simulation and information management tools. This overall approach to subsystem design is shown schematically in Figure 5. It supplements the well know "waterfall" chart of systems engineering with a concurrent engineering overlay, managed through integrated information tools.

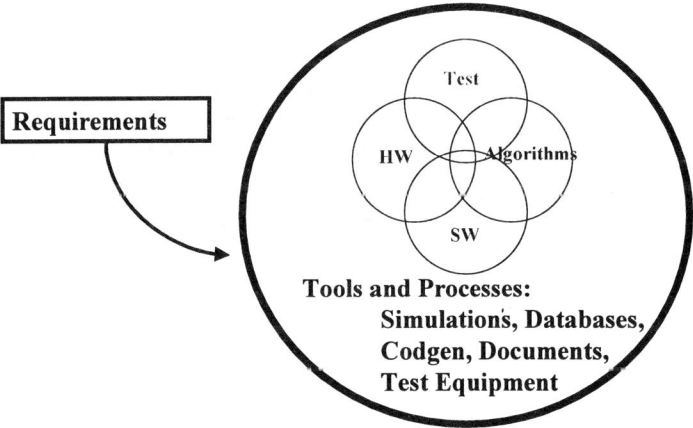

Figure 5: Integrating Test, SW, HW and Algorithms in the Design Process

FC Functions

The system operational flow shown in Figures 2 and 3 is used as a basis to initiate conceptual work on the states of the FC subsystem during an engagement, and the algorithms that operate in each state. In parallel, the sensor interfaces, latencies, and the software requirements document are used to develop the SW and processor architecture for the subsystem.

Figure 6 shows the FC derived algorithms for ABL.

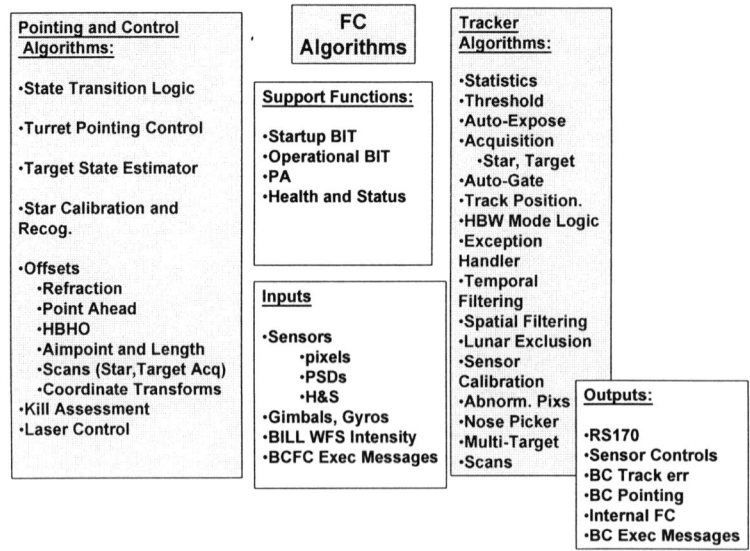

Figure 6: FC Algorithms

Top level SW architecture is shown in Figure 7. The SW decomposes into a number of SW components – Figure 8 shows the engagement software component, and lists the algorithms mapped to this software. The FC design process is able to integrate this software design effort directly with the simulations by using a process of automatic code generation. In this way simulation code can be directly compiled for a real time processor. This assures complete consistency between simulation and flight system.

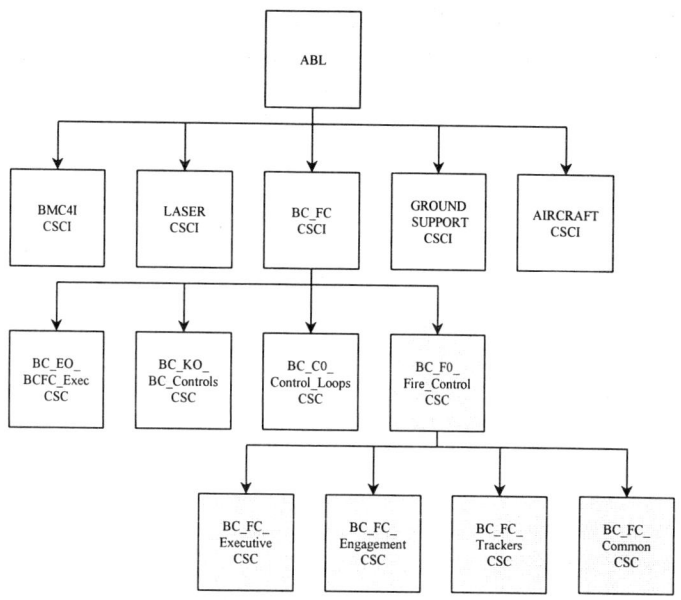

Figure 7: Top Level FC SW

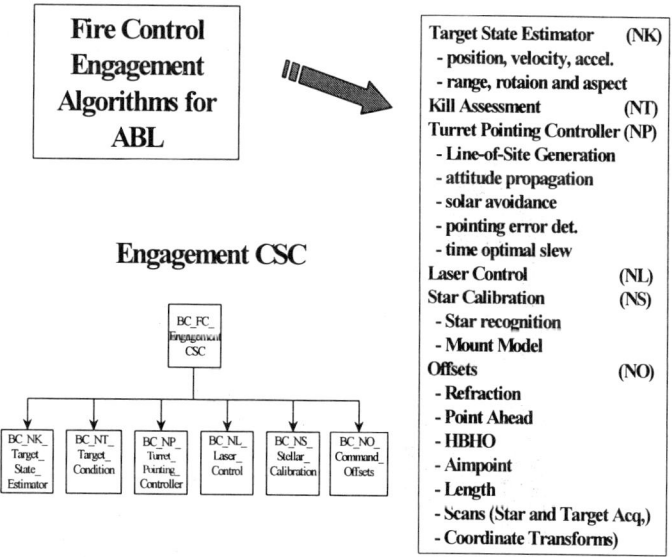

Figure 8: Fc Engagement SW Component

One such algorithm, the State Transition Logic, is particularly important to the overall functionality of the FC design. The approach to this logic is shown in Figure 9. This logic is implemented in the simulation using matrix and vector mathematics, and is easily converted to real time code.

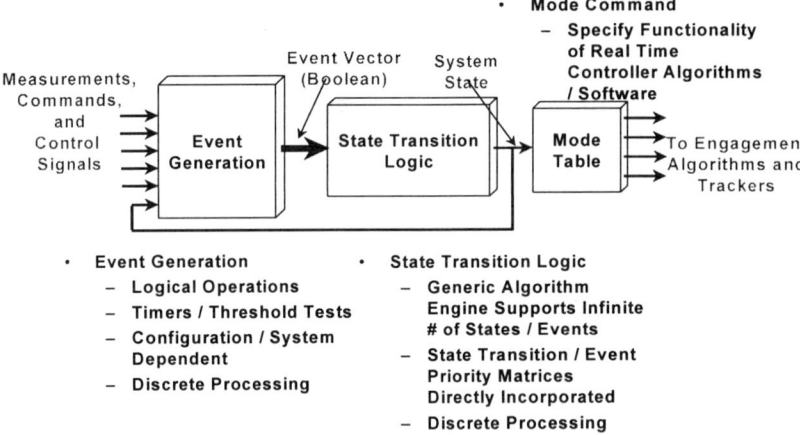

Figure 9: State Transition Logic

The realization of the SW and algorithms into HW is show in Figure 10. The FC HW utilizes a set of Power-PC processors running the VxWorks real-time operating system. Three processing chases map to the high bandwidth tracking, low bandwidth tracking and executive/engagement/performance assessment functions.

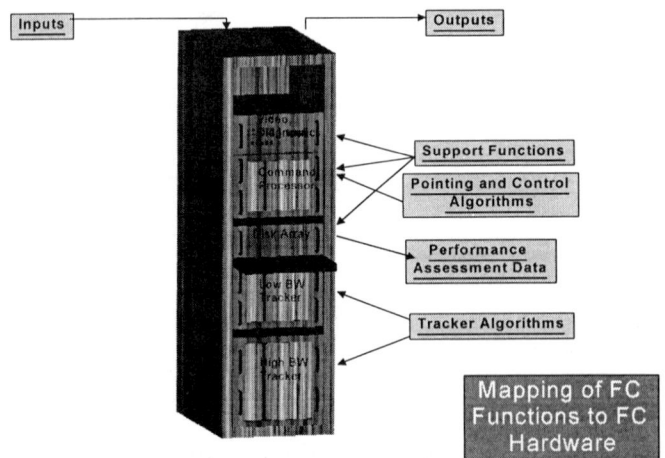

Figure 10: FC Hardware

Managing the Information

Requirements are assessed and analyzed and mapped to the HW, SW and Algorithm functional areas as shown in Figure 11. The requirements are stored in a database that can produce reports detailing, for example, all requirements that track to a given algorithm. Figure 11 shows example analyses and requirements tables, and Figure 12 shows a typical requirement database report.

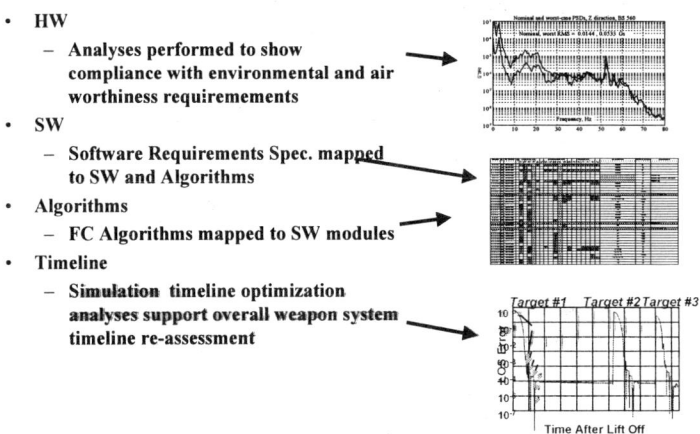

Figure 11: Analyses are mapped to requirements tables

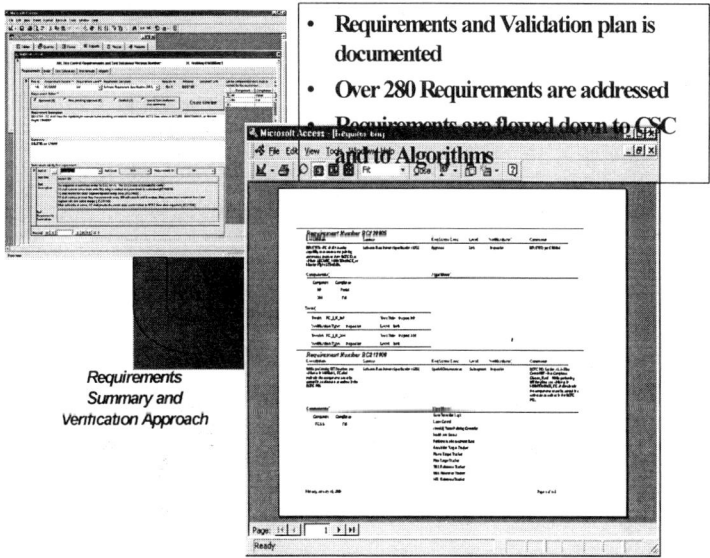

Figure 12 Requirements database tracks all FC requirements

Information Integration

A set of databases has been created to not only map requirements but also map specific signals. The "dataflow" database was created to store all FC signals and identify each signal with a set of descriptors, including origin, destination, path, bandwidth, dynamic range, and data type. This database allows FC to make reports which can show, for example, all the signals that cross two specific SW modules. Thus this signal database becomes a design tool.

This approach has been taken farther by using this signal database to automatically generate definition files for the FC simulation. Thus the FC simulation uses, where appropriate, the exact same signal list as the real time SW will use. Figure 13 shows the FC dataflow database.

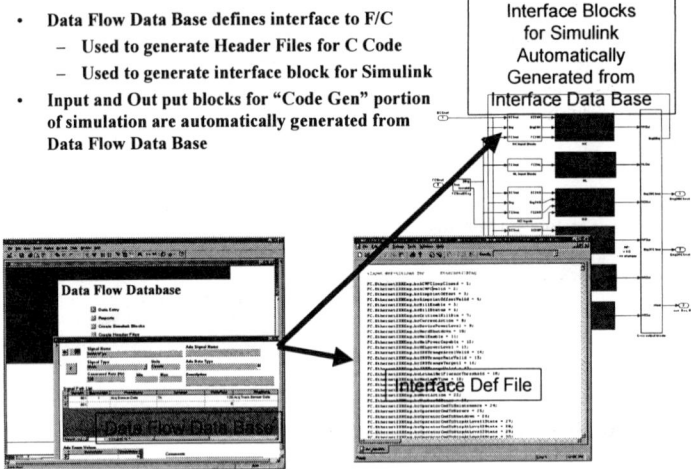

Figure 13: Dataflow Database

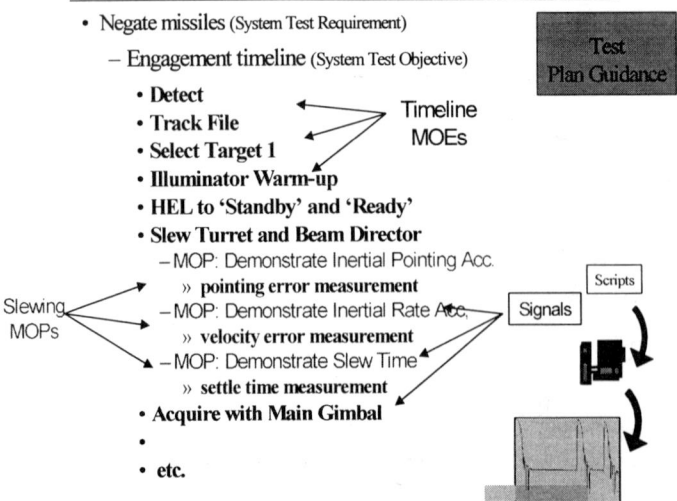

Figure 14: Mapping Signals to top level test objectives and analysis SW.

A further extension of the database in systems engineering is to use the database to configure a test signal analysis system. By designing for test concurrently, the FC subsystem signals are mapped to key measures of performance and measures of effectiveness. For example, for the key timeline performance, specific signals are used to indicate that the FC subsystem has successfully transitioned to the next state in the timeline. These signals are in the database and can be mapped to analysis SW to create a report that documents success or failure on these timeline state transitions. Figure 15 shows how the database ties to both the real time code and the post-test data analysis system.

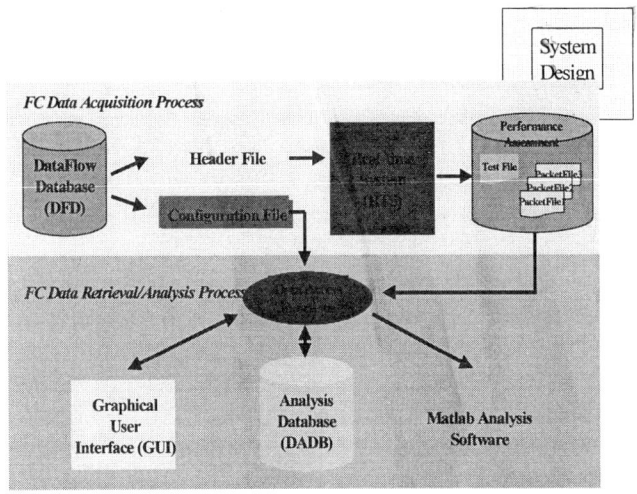

Figure 15: Data Integration shows database driving both real-time and post test analysis system.

Simulation

The FC performance prediction simulation, SABL, is based on Matlab/Simulink software. It includes high fidelity target hardbody and plume signatures in a full 3D, 6 DoF implementation.

The primary purpose of SABL is to validate the weapon system timeline performance by using detailed high fidelity models for targets, engagements, controls, dynamics, sensors, algorithms and mode logic. SABL is a full three-dimensional model incorporating all coordinate systems and transformations currently used in the baseline design. SABL is also used as a design tool. By implementing detailed designs and evaluating performance not only in terms of timeline, but in terms of design performance parameters such as radiometrics, loop bandwidth, and jitter rejection, the stability, performance, robustness, and design margin can be evaluated. Significant portions of SABL have been validated using various data sources. Model validation is considered a critical component in using simulations as both a design tool and performance predictor.

Figure 16 shows an overview of the SABL simulation architecture. The "Environment" block contains target & platform trajectories, plume databases created by the Strategic Scene Generation Model (SSGM), atmospheric transmissions & backgrounds, target hardbody reflectance maps created by the GRC Laser Radar Cross Section (LRCS) model, as well as data for various other target and physical effects. "System Computers" is further divided into the individual computers operating in the weapon system including battle management, fire control, beam control, and the laser device. This block also contains the system mode logic and algorithms, some of which are identified in the figure. "Beam Control & Dynamics" modules contain the dynamics, kinematics, and control laws for all significant elements including the gimbals outer and inner loops, complete with time optimal slew, as well as all major control loops specified by the design. The block includes all optical transformations in 3 dimensions. The "Sensors & Tracker" block contains models for the active & passive sensors, routines to interface with the numerous databases, and specific track algorithms including leading edge, centroid, and peak pixel track. Fire Control functions are simulated with a complete set of algorithms, states, mode logic, events and event priorities.

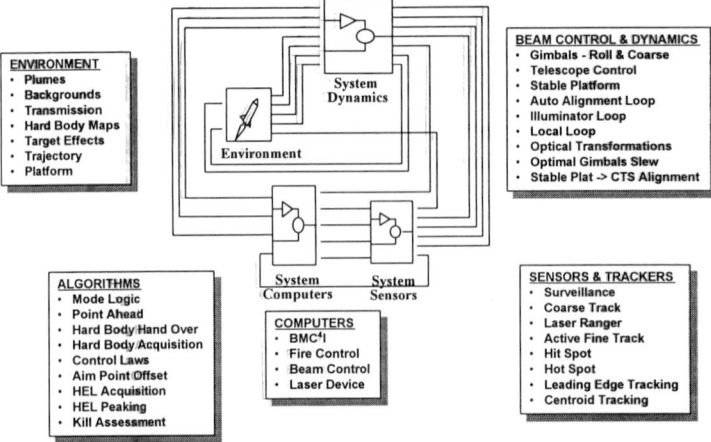

Figure 16: Simulation Architecture

Summary

The integration of simulation, algorithms, HW, SW and test functionality is shown in Figure 17. The use of automated tools results in an efficient, testable and highly integrated design approach. This approach continuously tests the system, validates the simulation and develops the design.

Section II
FORMATION FLYING AND CONSTELLATIONS

SESSION II

Joint Session Chairperson: Garry Burdick
                                                                 Jet Propulsion Laboratory

Joint Session Charperson: Dr. Russell Carpenter
                                                 NASA Goddard Space Flight Center

Local Session Chairperson: Dr. David Sonnabend
                                                       Analytical Engineering

The following paper numbers were not assigned:
AAS 00-018 to -020

AAS 00-011

A PROJECTION APPROACH TO SPACECRAFT FORMATION ATTITUDE CONTROL

Jonathan Lawton,[*] Randal W. Beard[†] and Fred Y. Hadaegh[‡]

This paper considers the problem of rotating a group of spacecraft each about a fixed axis while maintaining relative spacecraft alignment. We decompose the spacecraft attitude, and angular velocity into components in the direction of the desired eigenaxis rotation and "perpendicular" to the desired eigenaxis rotation. After making the decomposition we use leader-following and behavioral-based control to coordinate the spacecraft motion in the direction of the desired eigenaxis maneuver and we use simple P.D. control to regulate the spacecraft motion in the perpendicular direction. Simulation results show the effectiveness of the approach.

1 INTRODUCTION

Travel to neighboring galaxies would require space voyages lasting thousands of years. As a result further space exploration can only be practically achieved by indirect observation of astronomical objects. Much can be determined about a space object from the emitted light. To make such delicate observations, space-based interferometers with baselines on the order of one to ten kilometers would be needed.

Large monolithic space-based interferometers cannot be physically implemented. In [1] and [2] a free-flying three spacecraft interferometer is proposed. Figure 1 shows how a free-flying multiple spacecraft interferometer might be implemented. Two spacecraft would move within an observation plane to sample light from an astronomical body. The light is then reflected to the third spacecraft which observes the interference pattern from the two different light paths (see Figure 1).

Interferometry requires precise distance measurements between spacecraft on the order of a wavelength of light. The ST3 mission proposed by NASA/JPL will use an on-board laser metrology system to make these measurements [3]. There must be a pairwise alignment of spacecraft in order to lock the metrology sensors one to another. Establishing this alignment is a nontrivial process called formation initialization. Once sensor lock is established the spacecraft can monitor their relative orientation and relative position with respect to neighboring spacecraft. Sensor lock being established it is important that sensor lock not be loss or else the whole process of formation initialization must be repeated.

To understand the type of measurements that must be performed consider the van Cittert-Zernike result [4] which is the basis of interferometric imaging. Define the $\nu - \eta$ plane as the observation plane. The location of the two collector spacecraft are (ν_1, η_1) and (ν_2, η_2) respectively. The star is

---

[*] Department of Electrical & Computer Engineering, Brigham Young University, Provo, Utah 84602-4099.
[†] Corresponding Author: Department of Electrical & Computer Engineering, Brigham Young University, Provo, Utah 84602-4099. E-mail: beard@ee.byu.edu.
[‡] Jet Propulsion Laboratory, California Institute of Technology, 4800 Oak Grove Drive, Pasadena, California 91109-8099.

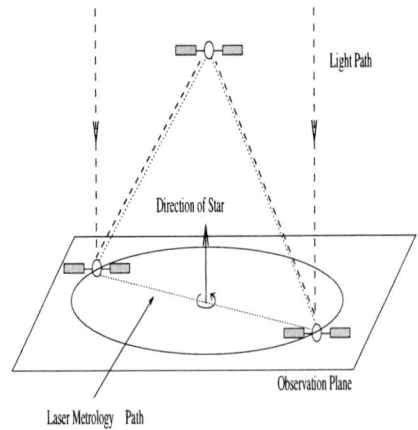

Figure 1: Free-Flying Interferometry problem.

located at (x, y) a distance z away from the observation plane. We can define new coordinates

$$u = \frac{\eta_1 - \eta_2}{z\lambda}$$
$$v = \frac{\nu_1 - \nu_2}{z\lambda},$$

where λ is the wave length of the observed light. These new coordinates define the $U - V$ plane.

Observation of the interference pattern for different values of (u, v) allows for measurement of the complex mutual coherence function $\mu(u, v)$. The van Cittert-Zernike result is that the inverse Fourier transform of $\mu(x, y)$ gives the desired irradiance pattern $I(x, y)$ of the celestial body.

Two basic formation maneuvers are required to sample the irradiance pattern. Namely formation resize and formation reorientation. Motion between stars requires a formation retarget maneuver. Figure 1 shows the position and attitude requirements in order to maintain sensor lock, for resize maneuvers and for reorientation/retarget maneuvers.

A resize maneuver (see Figure 1a) requires that the relative attitude be maintained while the spacecraft position is varied. On the other hand, both reorientation and retarget maneuvers (see Figure 1b) require that the relative attitude between spacecraft be varied at the same rate (but opposite direction) as the formation to maintain sensor lock.

To implement these types of maneuvers there exists three approaches to formation flying found in the literature. These approaches have application to the coordination of spacecraft, multiple robots, and aircraft. They are the leader-following [5, 6], behavioral [7], and virtual structure [8] approaches.

In leader-following, one vehicle is designated as the leader, while the rest of the vehicles are designated as followers. The idea is that the followers track the position and orientation of the leader with some prescribed (possibly time-varying) offset. There are numerous variations on this theme including designating multiple leaders, forming a chain (spacecraft i tracks spacecraft $i-1$), and other tree topologies.

A variation on leader following is the virtual structure approach. This approach creates a virtual leader. The virtual leader traces out the trajectory that the formation is to follow. The remaining vehicles (followers) track the trajectory of the virtual leader at a prescribed distance and orientation separation.

The idea behind the behavioral approach is to prescribe several desired behaviors for each agent, and then to make the control action of each agent a weighted average of the control for each behavior. Possible behaviors include collision avoidance, obstacle avoidance, goal seeking, and formation keeping. There are also numerous variations on the behavioral approach to multi-agent coordination,

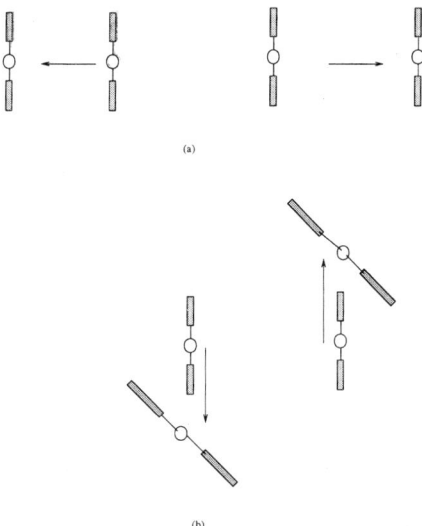

Figure 2: This illustrates the motion angular and translational motion required for a expansions and reorientations/retargeting.

most of which are derived by novel weightings of the behaviors. The disadvantage of the behavioral approach is that analytic convergence proofs are difficult to obtain.

When implementing any of these techniques, much of the information that is shared among the spacecraft is not relevant to the coordination problem. To illustrate this consider the simple problem of two spacecraft moving in a straight line while trying to coordinate their relative position. Figure 1 illustrates that there will be disturbances in the trajectories of both spacecraft perpendicular to the intended direction of motion. It is desirable for spacecraft to jointly coordinate their motion in the intended direction of motion while independently attenuating any motion perpendicular to the intended direction of motion. For linear motion this decomposition idea is straight forward (see [9]). However, implementing a similar concept for the attitude control problem is more difficult.

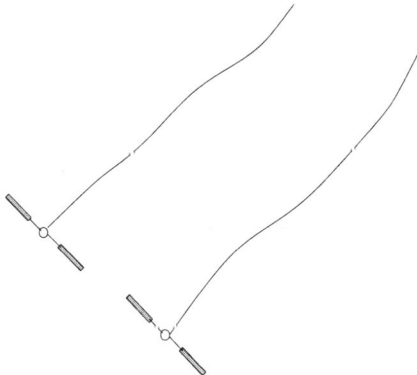

Figure 3: 'In formation control you do not want to try to coordinate disturbances.'

The main contribution of this paper is the factorization of the rotational dynamics into an eigenaxis component and a component perpendicular to the eigenaxis. Then by forcing the perpendicular component to be small the motion of each spacecraft will approximate an eigenaxis rotation. The entire coordination problem is reduced to coordinating the eigenaxis component of the rotation. This coordination can be accomplished by the three above mentioned formation control methods. In this paper we will only consider leader-following and behavioral based control since the virtual structure approach is a special case of leader-following but with a virtual leader.

The paper is organized as follows. In Section 2 the unit quaternion is factored into one component in the direction of the intended rotation and another component perpendicular to that direction. In Section 3 we state the objective of the formation control problem. In Section 4 we present the reduced order leader-following control. In Section 5 we give the reduced order behavioral based control. In Section 6 we present simulations. Section 7 gives our conclusions. Finally in Appendix A we give some properties of quaternions.

2 PROJECTIONS

There are three key objectives to attitude formation control. First, keep formation during the entire maneuver. Second, each spacecraft should converge to its desired attitude. Third, each vehicle should rotate though the smallest angle possible. The shortest rotation is an eigenaxis rotation, [10] i.e. a rotation about a fixed axis. For an eigenaxis rotation from

$$\mathbf{q}(0) = \begin{bmatrix} \sin(\theta/2)\,\mathbf{u} \\ \cos(\theta/2) \end{bmatrix}$$

to

$$\mathbf{q}_f = \begin{bmatrix} 0 \\ 1 \end{bmatrix}$$

the vector component of $\mathbf{q}$ will always be in the direction of $\mathbf{u}$. Theorem 2.1 factors the quaternion $\mathbf{q}$ into the product of two quaternions. One having its vector component in the direction of $\mathbf{u}$ and the other having its vector component perpendicular to this direction. For the convenience of the reader a review of some of the properties of quaternions, is given in the Appendix.

Theorem 2.1 *Let the spacecraft attitude be given by* $\mathbf{q}$. *Define*

$$\mathcal{V}(\mathbf{q}^{\shortparallel}) = \sin\left(\frac{\theta}{2}\right)\mathbf{u}$$
$$\mathcal{S}(\mathbf{q}^{\shortparallel}) = \cos\left(\frac{\theta}{2}\right) \tag{1}$$

$$\mathbf{q}^{\perp} = \mathbf{q}^{\shortparallel *}\mathbf{q}, \tag{2}$$

where

$$\tan\left(\frac{\theta}{2}\right) = \frac{\mathbf{u}^T \mathcal{V}(\mathbf{q})}{\mathcal{S}(\mathbf{q})}$$

and $\mathbf{q}^{\shortparallel *}$ *is the quaternion conjugate of* $\mathbf{q}^{\shortparallel}$, $\mathcal{V}(\mathbf{q})$ *is the vector component of* $\mathbf{q}$ *and* $\mathcal{S}(\mathbf{q})$ *is the Scalar component of* $\mathbf{q}$ *as defined in the Appendix. Then*

1. $\mathbf{q} = \mathbf{q}^{\shortparallel}\mathbf{q}^{\perp}$
2. $\mathcal{V}(\mathbf{q}^{\shortparallel})^T \mathcal{V}(\mathbf{q}^{\perp}) = 0$.

Proof:

1. The decomposition that $\mathbf{q} = \mathbf{q}''\mathbf{q}^\perp$ follows directly from Equation 2 and the fact that $\mathbf{q}''^*$ is the multiplicative inverse of $\mathbf{q}''$.

2. By direct calculation
$$\mathbf{q}^\perp = \mathbf{q}''^* \mathbf{q}$$
$$\begin{bmatrix} \mathcal{V}(\mathbf{q}^\perp) \\ \mathcal{S}(\mathbf{q}^\perp) \end{bmatrix} = \begin{bmatrix} \cos(\frac{\theta}{2})\mathcal{V}(\mathbf{q}) - \mathcal{S}(\mathbf{q})\sin(\frac{\theta}{2})\mathbf{u} - \sin(\frac{\theta}{2})\mathbf{u} \times \mathcal{V}(\mathbf{q}) \\ \cos(\frac{\theta}{2})\mathcal{S}(\mathbf{q}) - \sin(\frac{\theta}{2})\mathbf{u}^T \mathcal{V}(\mathbf{q}) \end{bmatrix},$$

where the $\times$ operator is the vector cross product. Therefore we can calculate the quantity

$$\mathcal{V}(\mathbf{q}'')^T \mathcal{V}(\mathbf{q}^\perp) = \sin\left(\frac{\theta}{2}\right) \mathbf{u}^T \left(\cos\left(\frac{\theta}{2}\right) \mathcal{V}(\mathbf{q}) - \mathcal{S}(\mathbf{q})\mathbf{u} - \sin\left(\frac{\theta}{2}\right) \mathbf{u} \times \mathcal{V}(\mathbf{q}) \right)$$
$$= \sin\left(\frac{\theta}{2}\right) \left(\cos\left(\frac{\theta}{2}\right) \mathbf{u}^T \mathcal{V}(\mathbf{q}) - \mathcal{S}(\mathbf{q})\sin\left(\frac{\theta}{2}\right) \right)$$
$$= \sin\left(\frac{\theta}{2}\right) \cos\left(\frac{\theta}{2}\right) \mathcal{S}(\mathbf{q}) \left(\frac{\mathbf{u}^T \mathcal{V}(\mathbf{q})}{\mathcal{S}(\mathbf{q})} - \tan\left(\frac{\theta}{2}\right) \right)$$
$$= 0$$

∎

We can similarly decompose the angular velocity into two different components.

Lemma 2.1 *If*
$$\omega'' = 2\dot{\mathbf{q}}''\mathbf{q}''^*$$
$$\omega^\perp = 2\mathbf{q}''\dot{\mathbf{q}}^\perp \mathbf{q}^{\perp *}\mathbf{q}''^*$$
$$\omega''_d = 2\dot{\mathbf{q}}''_d \mathbf{q}''^*_d$$

then

1. $\omega'' + \omega^\perp = \omega$
2. $\frac{d}{dt}\rho(\mathbf{q}^\perp, 1) = \omega^{\perp T} \mathbf{q}''^* \mathcal{V}(\mathbf{q}^\perp)\mathbf{q}''$
3. $\frac{d}{dt}\rho(\mathbf{q}'', \mathbf{q}''_d) = (\omega'' - \omega''_d)^T \mathcal{V}(\mathbf{q}''_1 \mathbf{q}''^*_d)$.

Proof:

1. From the product rule of differentiation of quaternions
$$\omega = 2\dot{\mathbf{q}}\mathbf{q}^*$$
$$= 2(\dot{\mathbf{q}}''\mathbf{q}^\perp + \mathbf{q}''\dot{\mathbf{q}}^\perp)\mathbf{q}^*$$
$$= 2\dot{\mathbf{q}}''\mathbf{q}^\perp \mathbf{q}^{\perp *}\mathbf{q}''^* + 2\mathbf{q}''\dot{\mathbf{q}}^\perp \mathbf{q}^{\perp *}\mathbf{q}''^*$$
$$= 2\dot{\mathbf{q}}''\mathbf{q}''^* + 2\mathbf{q}''\dot{\mathbf{q}}^\perp \mathbf{q}^{\perp *}\mathbf{q}''^*$$
$$= \omega'' + \omega^\perp.$$

2. From Lemma A.2 we get that
$$\frac{d}{dt}\rho(\mathbf{q}^\perp, 1) = \mathcal{V}(\mathbf{q}^\perp)^T \mathbf{q}''^* \omega^\perp \mathbf{q}''$$
$$= \mathcal{S}(\mathcal{V}(\mathbf{q}^\perp)\mathbf{q}''^* \omega^{\perp *}\mathbf{q}'')$$
$$= \mathcal{S}(\omega^{\perp *}\mathbf{q}''\mathcal{V}(\mathbf{q}^\perp)\mathbf{q}''^*)$$
$$= \omega^{\perp T}(\mathbf{q}''\mathcal{V}(\mathbf{q}^\perp)\mathbf{q}''^*).$$

3. From Lemma A.2 we get that

$$\frac{d}{dt}\rho(\mathbf{q}'', \mathbf{q}''_d) = \mathcal{V}(\mathbf{q}''\mathbf{q}''^*_d)^T(\omega'' - \omega''_d).$$

∎

Now that we have decomposed the spacecraft attitude and angular velocity into a component in the direction of intended motion and a component perpendicular to the direction of intended motion we will force the perpendicular component to be small will coordinating the parallel component.

3 PROBLEM STATEMENT

In this Section we formalize our formation control objective. In order to present our results in a general framework we present some definitions that will help specify the structure of the given group of spacecraft.

Let $\mathcal{I} = \{1, \ldots, N\}$ be the index set of all spacecraft in the formation. Additionally let the set $\mathcal{L} \subset \mathcal{I}$ to be the set of all vehicle that have knowledge of their final goal (i.e. leaders). We may use graph theoretic techniques to examine the flow of information in the formation. The set $\mathcal{I}$ is the set of vertices contained in the graph. Let $\mathcal{P}$ be the set of edges in the graph. Each edge represents a path through which information can flow in the formation. Note that the graph $(\mathcal{I}, \mathcal{P})$ can either be a connected or a disconnected graph (see Figure 3).

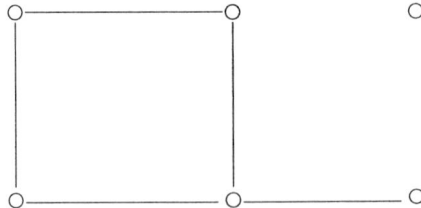

a) The set P for the disconnected case

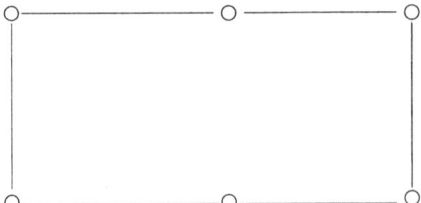

b) The set P for the connected case.

Figure 4: This illustrates the set $\mathcal{P}$ for a connected case and a disconnected case.

Formation flying in the attitude domain has three main purposes. They are

1. Formation keeping,

2. Convergence to the final desired attitude.

3. Executing eigenaxis rotations.

We wish to develop a metric for the formation that will take into account each of these goals. First consider the function E_F given by

$$E_F = \sum_{(i,j)\in \mathcal{P}} \rho(\mathbf{q}''_i, \mathbf{q}''_j). \quad \text{Where } (\mathcal{I}, \mathcal{P}) \text{ is a connected graph}$$

This function expresses how well the group of spacecraft are keeping formation. The term $\rho(\mathbf{q}''_i, \mathbf{q}''_{i+1}) = 0$ only when $\mathbf{q}''_i = \mathbf{q}''_{i+1}$. This reflects that fact that the spacecraft have the same angular distance remaining in the direction about the eigenaxis to reach the final desired attitude.

Next consider the metric E_e given by

$$E_e = \sum_{i \in \mathcal{I}} \rho(\mathbf{q}^\perp_i, \mathbf{1}).$$

$E_e = 0$ only when $\mathbf{q}^\perp_i = \mathbf{1}$. This implies that $\mathbf{q}_i = \mathbf{q}''_i$ which means that all of the spacecraft are rotating about the fixed axis $\mathbf{u}$. Thus E_e measures how well the spacecraft are performing an eigenaxis maneuver.

The last error function E_G given by

$$E_G = \sum_{i \in \mathcal{L}} \rho(\mathbf{q}''_i, \mathbf{1})$$

measures how well the spacecraft are converging to their final goal. $E_G = 0$ only when $\mathbf{q}''_i = \mathbf{1}$. This indicates that each spacecraft would have completed its rotation.

We can now present the objective of our formation control problem.

Definition 1 (Formation Control Problem) *We wish to find a set of controls τ_i for each spacecraft such that E_F, E_e and E_G asymptotically converge to zero.*

The leader-following approach (see Section 4) gives $E_F, E_G, E_e \to 0$ independent of each other. The burden of driving E_F to zero is placed on the followers. The burden of driving E_G to zero is placed on the leaders. The coupled dynamics approach (See Section 5) provides that the weighted average of the errors

$$E = k_G E_G + k_F E_F + k_e E_e.$$

converges to zero. The burden of driving E_F and E_G to zero is shared equally by all of the spacecraft.

4 REDUCED ORDER LEADER-FOLLOWING

Having decoupled the formation motion into a component that is perpendicular to the direction of motion and a component that is in the direction of motion we are prepared to apply it to the familiar leader-following formation control. Often the leader motion as observed by the followers is quite noisy. The decoupling of the previous section reduces the amount of information that each follower must receive from the leader.

For the leader-following problem define $\mathcal{L} = \{1\}$, this means that only the leader is aware of the final goal for the formation. The remaining vehicles simply follow its motion. Further define $\mathcal{P} = \{(1,2),(1,3),\ldots(1,N)\}$. Thus information flows from the leader to every given spacecraft. These two sets are illustrated in Figure 4. In Figure 4a the 'o' represents the members of $\mathcal{I}$ and the circled spacecraft represent the members of $\mathcal{L}$. In Figure 4b the pairs of spacecraft connected by an arrow represent the member of $\mathcal{P}$.

The orientation dynamics of the ith spacecraft are given by

$$J_i \dot{\omega}_i = -\omega_i \times J_i \omega_i + \tau_i,$$

a) For leader-following only the leader knows the final goal.

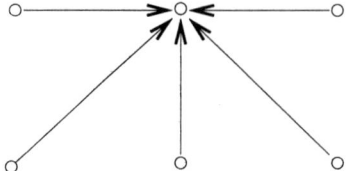

b) All the followers reference themselves to the leader's position.

Figure 5: This illustrates the sets $\mathcal{L}$ and $\mathcal{P}$ for leader following.

where J_i is the moment of inertia of the spacecraft and τ_i is the applied torque. The control

$$\tau_i = -\omega_i \times J_i\omega_i + J_i\mathbf{f}_i,$$

feedback linearizes the system dynamics into

$$\dot{\omega}_i = \mathbf{f}_i \tag{3}$$

Since

$$\omega_i = \omega''_i + \omega^\perp_i,$$

differentiating we get that

$$\mathbf{f}_i = \dot{\omega}''_i + \dot{\omega}^\perp_i.$$

Define

$$\mathbf{f}''_i = \dot{\omega}''_i$$
$$\mathbf{f}^\perp_i = \dot{\omega}^\perp_i.$$

It follows that the feedback linearized force can be decomposed as

$$\mathbf{f}_i = \mathbf{f}''_i + \mathbf{f}^\perp_i.$$

Using the feedback linearized dynamics, Theorem 4.1 will implement a reduced order control for leader-following.

Theorem 4.1 *Let the spacecraft attitude controls be given by*

$$\mathbf{f}^\perp_i = -d_e\omega^\perp_i - k_e\mathbf{q}''_i\mathcal{V}(\mathbf{q}^\perp_i)\mathbf{q}''^*_i$$
$$\mathbf{f}''_1 = -d_G\omega''_1 - k_G\mathcal{V}(\mathbf{q}''_1)$$
$$\mathbf{f}''_i = \dot{\omega}''_1 - d_F(\omega''_i - \omega''_1) - k_F\mathcal{V}(\mathbf{q}''_i\mathbf{q}''^*_1) \text{ for } i > 1,$$

where $d_G, d_e, k_e, k_G, k_F > 0$. *Then*

1. $\dot{\omega}_i^\perp, \omega^\perp{}_i \perp \mathbf{u}$ and $\dot{\omega}^{''}, \omega^{''} \parallel \mathbf{u}$

2. $E_F \to 0$ asymptotically.

3. $E_G \to 0$ asymptotically.

4. $E_e \to 0$ asymptotically.

The significance of the first statement is that the angular velocity terms can be easily calculated. The first component $\omega^\perp$ is the component of ω that is orthogonal to $\mathbf{u}$ and $\omega^{''}$ is the component that is parallel to $\mathbf{u}$.

Proof:

1. Note that the closed loop dynamics of the perpendicular motion of the ith spacecraft is given by

$$\dot{\omega}_i^\perp + d_e \omega^\perp{}_i + k_e \mathbf{q}^{''}_i \mathcal{V}(\mathbf{q}^\perp{}_i) \mathbf{q}^{''*}_i = 0,$$

thus

$$\frac{d}{dt}(e^{d_e t} \omega^\perp{}_i) = -e^{d_e t} k_e \mathbf{q}^{''}_i \mathcal{V}(\mathbf{q}^{''*}_i).$$

Integration gives

$$\omega^\perp{}_i(t) = e^{-d_e t}\omega^\perp{}_i(0) - k_e \int_0^t e^{d_e(\tau_i - t)} \mathbf{q}^{''}_i \mathcal{V}(\mathbf{q}^\perp{}_i) \mathbf{q}^{''*}_i d\tau.$$

Note that $\mathcal{V}(\mathbf{q}^\perp{}_i)$ is perpendicular to $\mathbf{u}$. Hence when $\mathcal{V}(\mathbf{q}^\perp{}_i)$ is rotated about $\mathbf{u}$, it will remain perpendicular to $\mathbf{u}$. Thus $\mathbf{q}^{''}\mathcal{V}(\mathbf{q}^\perp{}_i)\mathbf{q}^{''*}_i$ is perpendicular to $\mathbf{u}$ for all time. Taking $\omega^\perp{}_i(0) \perp \mathbf{u}$ it follows that $\omega^\perp{}_i(t) \perp \mathbf{u}$ for all time. Since $\omega^\perp{}_i(t)$ is confined to the plane perpendicular to $\mathbf{u}$, $\omega^\perp{}_i(t_1) - \omega^\perp{}_i(t_2)$ is also confined to the same plane. Taking the limit as $t_1 \to t_2$ and dividing by $t_1 - t_2$ implies that $\dot{\omega}_i^\perp \perp \mathbf{u}$. Similar reasoning can be used to show that $\omega^{''}_i, \dot{\omega}^{''}_i \parallel \mathbf{u}$.

2. It is sufficient to show that $\mathbf{q}^{''}_i \to \mathbf{1}$ for all i. Consider the Lyapunov function candidate

$$V_i^\perp = \frac{1}{2}\omega^{\perp T}_i \omega^\perp{}_i + k_e \rho(\mathbf{q}^\perp{}_i, \mathbf{1}).$$

Differentiating $V_i^\perp$ using Lemma 2.1 gives

$$\begin{aligned}\dot{V}_i^\perp &= \omega^{\perp T}_i \mathbf{f}^\perp{}_i + k_e(\omega^\perp{}_i)^T \mathbf{q}^{''*}_i \mathcal{V}(\mathbf{q}^\perp{}_i)\mathbf{q}^{''}_i \\ &= \omega_i^T(\mathbf{f}^\perp{}_i + k_i^\perp \mathbf{q}^{''}_i \mathcal{V}(\mathbf{q}^\perp{}_i)\mathbf{q}^{''*}_i) \\ &= -d_e^\perp \omega^{\perp T}_i \omega^\perp{}_i \\ &\leq 0.\end{aligned} \quad (4)$$

Thus $\mathbf{q}^\perp{}_i, \omega^\perp{}_i$ converges to the set $\Omega_i = \{\mathbf{q}^\perp{}_i, \omega^\perp{}_i | \omega^\perp{}_i = 0\}$. Application of LaSalle's invariance principle can be used to show that $\mathbf{q}^\perp{}_i$ converges to the set defined by

$$\mathbf{q}^{''}_i \mathcal{V}(\mathbf{q}^\perp{}_i)\mathbf{q}^{''*}_i = 0. \quad (5)$$

Equation (5) is only true when $\mathbf{q}^\perp{}_i = \pm \mathbf{1}$, which corresponds to the same physical orientation.

3. It is sufficient to show that $\mathbf{q}^{''}_1 \to \mathbf{1}$ asymptotically. Consider the Lyapunov function candidate

$$V_1^{''} = \frac{1}{2}\omega^{''T}_1 \omega^{''}_1 + k_G \rho(\mathbf{q}^{''}_1, \mathbf{1}).$$

Differentiating V_1'' gives

$$\dot{V}_1'' = \omega_1''^T(\mathbf{f}_1'' + k_G\mathcal{V}(\mathbf{q}_1''))$$
$$= -d_G\omega_1''^T\omega_1''$$
$$\leq 0.$$

The convergence follows by applying LaSalle's invariance principle as before.

4. It is sufficient to show that $\mathbf{q}_i'' \to \mathbf{q}_1''$ asymptotically. Consider the Lyapunov function candidate

$$V_i'' = \frac{1}{2}(\omega_i'' - \omega_1'')^T(\omega_i'' - \omega_1'') + k_F\rho(\mathbf{q}_i'', \mathbf{q}_1'').$$

Differentiating V_i'' gives

$$\dot{V}_i'' = (\omega_i'' - \omega_1'')^T(\mathbf{f}_i'' - \dot{\omega}_1'' + k_F\mathcal{V}(\mathbf{q}_i''\mathbf{q}_1''^*))$$
$$= -d_F(\omega_i'' - \omega_1'')^T(\omega_i'' - \omega_1'') \quad (6)$$
$$\leq 0.$$

Again, applying LaSalle's invariance principle will lead to the results that $\mathbf{q}_i'' \to \mathbf{q}_1''$. ∎

5 COUPLED DYNAMICS BASED CONTROL

The basic idea behind a behavior based control is to consider a group of desired behaviors for each spacecraft (i.e. formation keeping, eigenaxis rotation and convergence to the final goal.) Then the desired group dynamics are said to emerge as the result of the desired behaviors. Here we will use a reduced order coupled dynamics control for the coordinated formation reorientation.

The behavior-based control that we implement is the coupled dynamics control introduced in [9]. For this control we take the set $\mathcal{L} = \mathcal{I}$, (i.e. all spacecraft know the final goal). Next we take $\mathcal{P} = \{(1,2), (2,3), \ldots, (N-1, N), (N, 1)\}$. These two sets are illustrated in Figure 5. In Figure 5a the $'o'$ represents the members of $\mathcal{I}$ the circled spacecraft represent the members of $\mathcal{L}$. In Figure 4b the pairs of spacecraft connected by an arrow represent the member of $\mathcal{P}$. Theorem 5.1 states the coupled dynamic based control and its convergence result.

Again we will use the same feedback linearization for each spacecraft. That is

$$\tau_i = -\omega_i \times J_i\omega_i + J_i(\mathbf{f}_i'' + \mathbf{f}_i^\perp),$$

which results in

$$\dot{\omega}_i'' = \mathbf{f}_i''$$
$$\dot{\omega}_i^\perp = \mathbf{f}_i^\perp.$$

Theorem 5.1 *Given the control*

$$\mathbf{f}_i^\perp = -d_e\omega_i^\perp - k_e\mathbf{q}_i''\mathcal{V}(\mathbf{q}_i^\perp)\mathbf{q}_i''^*$$
$$\mathbf{f}_i'' = -d_F\omega_i'' - k_F\mathcal{V}(\mathbf{q}_i''\mathbf{q}_{i+1}''^*) - k_F\mathcal{V}(\mathbf{q}_i''\mathbf{q}_{i-1}''^*) - k_G\mathcal{V}(\mathbf{q}_i'')$$

where $k_F \geq 0$ and $d_G, d_F, k_G, k_e > 0$, then

1. $\dot{\omega}_i^\perp, \omega_i^\perp \perp \mathbf{u}$ *and* $\dot{\omega}_i'', \omega_i'' \parallel \mathbf{u}$,

a) For coupled dynamics all spacecraft know the final goal.

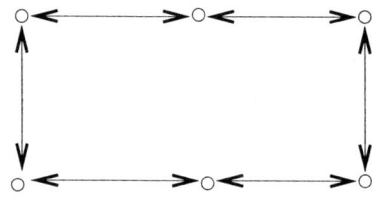

b) All vehicles reference themselves to two neighbors

Figure 6: This illustrates the sets $\mathcal{L}$ and $\mathcal{P}$ for coupled dynamics.

2. $E = k_F E_F + k_G E_G + k_e E_e \to 0$ *asymptotically.*

Proof:

1. Similar to Theorem 4.1 part 1.

2. Consider the Lyapunov function candidate

$$V = E + \frac{1}{2}\sum_{i=1}^{N} \omega_i^T \omega_i$$
$$= k_e \sum_{i=1}^{N} \rho(\mathbf{q}^{\perp}{}_i, \mathbf{1}) + k_F \sum_{i=1}^{N} \rho(\mathbf{q}^{\|}{}_i, \mathbf{q}^{\|}{}_{i+1}) + k_G \sum_{i=1}^{N} \rho(\mathbf{q}^{\|}{}_i, \mathbf{1}) + \frac{1}{2}\sum_{i=1}^{N} \omega_i^T \omega_i.$$

From Lemma 2.1 when we differentiate we get that

$$\dot{V} = k_e \sum_{i=1}^{N} \omega^{\perp}{}_i^T (\mathbf{q}^{\|*}_i \mathcal{V}(\mathbf{q}^{\perp}{}_i) \mathbf{q}^{\|}{}_i) + k_F \sum_{i=1}^{N} (\omega^{\|}{}_i - \omega^{\|}{}_{i+1})^T \mathcal{V}(\mathbf{q}^{\|}{}_i - \mathbf{q}^{\|}{}_{i+1})$$
$$+ k_G \sum_{i=1}^{N} \omega^{\|}{}_i^T \mathcal{V}(\mathbf{q}^{\|}{}_i) + \sum_{i=1}^{N} \omega_i^T \mathbf{f}_i. \quad (7)$$

Substituting

$$\omega_i^T \mathbf{f}_i = (\omega^{\|}{}_i + \omega^{\perp}{}_i)^T (\mathbf{f}^{\|}{}_i + \mathbf{f}^{\perp}{}_i) = \omega^{\perp}{}_i^T \mathbf{f}^{\perp}{}_i + \omega^{\|}{}_i^T \mathbf{f}^{\|}{}_i,$$

into Equation (7) renders

$$\dot{V} = k_e \sum_{i=1}^{N} \omega^{\perp T}_i (\mathbf{q}''^*_i \mathcal{V}(\mathbf{q}^\perp_i) \mathbf{q}''_i) + k_F \sum_{i=1}^{N} \omega''^T_i (\mathcal{V}(\mathbf{q}''_i + \mathbf{q}''^*_{i+1}) \mathcal{V}(\mathbf{q}''_i \mathbf{q}''^*_{i-1}))$$

$$+ k_G \sum_{i=1}^{N} \omega''^T_i \mathcal{V}(\mathbf{q}''_i) + \sum_{i=1}^{N} (\omega''^T_i \mathbf{f}''_i + \omega^{\perp T}_i \mathbf{f}^\perp_i)$$

$$= \sum_{i=1}^{N} \omega^{\perp T}_i (\mathbf{f}^\perp_i + k_i \mathbf{q}''^*_i \mathcal{V}(\mathbf{q}^\perp_i) \mathbf{q}''_i) + \sum_{i=1}^{N} \omega^{\perp T}_i (\mathbf{f}''_i + k_G \mathcal{V}(\mathbf{q}''_i) \quad (8)$$

$$+ k_F \mathcal{V}(\mathbf{q}''_i \mathbf{q}''_{i+1}) + k_F \mathcal{V}(\mathbf{q}''_i \mathbf{q}''_{i-1}))$$

$$= -d_e \sum_{i=1}^{N} \omega^{\perp T}_i \omega^\perp_i - d_G \sum_{i=1}^{N} \omega''^T_i \omega''_i$$

$$\leq 0.$$

To show convergence we will apply LaSalle's invariance principle. Let $\Omega^\perp = \{\mathbf{q}^\perp_i | \omega^\perp_i = 0\}$. On the set $\Omega^\perp$

$$\mathbf{f}^\perp_i = -d_e \omega_i - k_e \mathbf{q}''^*_i \mathcal{V}(\mathbf{q}^\perp_i) \mathbf{q}''_i. \quad (9)$$

The largest invariant set of $\Omega^\perp$ will be when $\mathcal{V}(\mathbf{q}^\perp) = 0$. This will only occur when $\mathbf{q}_i = \pm 1$ which corresponds to the same orientation.

Now let $\Omega'' = \{\mathbf{q}''_i | \omega''_i = 0\}$. On the set Ω''

$$\mathbf{f}''_i = -k_F \mathcal{V}(\mathbf{q}''_i \mathbf{q}''^*_{i+1}) - k_F \mathcal{V}(\mathbf{q}''_i \mathbf{q}''_{i-1}) - k_G \mathcal{V}(\mathbf{q}''_i).$$

Applying Equation (1) gives

$$\mathbf{f}''_i = -(k_G + k_F \cos(\theta_{i-1}/2) + k_F \cos(\theta_{i+1}/2)) \sin(\theta_i/2)$$
$$+ k_F \cos(\theta_i/2) \sin(\theta_{i-1}/2) + k_F \cos(\theta_i/2) \sin(\theta_{i-1}/2).$$

On the largest invariant set of Ω''

$$- (k_G + k_F \cos(\theta_{i-1}/2) + k_F \cos(\theta_{i+1}/2)) \sin(\theta_i/2)$$
$$+ k_F \cos(\theta_i/2) \sin(\theta_{i-1}/2) + k_F \cos(\theta_i/2) \sin(\theta_{i-1}/2) = 0 \quad (10)$$

Let

$$z = (\sin(\theta_1/2), \sin(\theta_2/2), \ldots, \sin(\theta_N/2))^T.$$

and let B be the matrix defined by

$$B_{ii} = -(k_G + k_F \cos(\theta_{i-1}/2) + k_F \cos(\theta_{i+1}/2))$$
$$B_{i,i+1} = k_F \cos(\theta_i/2)$$
$$B_{i,i-1} = k_F \cos(\theta_i/2)$$
$$B_{i,j} = 0 \text{ otherwise.}$$

Equation (10) can be written as

$$Bz = 0.$$

If we assume that $-\pi \leq \theta_i \geq \pi$ then it follows that $z = 0$ if and only if $\theta_i = 0$ for all i and $\cos(\theta_i/2) \geq 0$ for all θ_i. Our desired results will follow from LaSalle's invariance theorem if and only if B is full rank. Since $\cos(\theta_i) \geq 0$ it follows that B^T is strictly diagonally dominant. Therefore B^T is full rank and so is B.

∎

6 SIMULATIONS

In this section simulations are presented to show the performance of the reduced order leader-following and the reduced order coupled dynamics control.

In the case of leader-following we assume that we have one leader and one follower. The initial orientation for each simulation are given by $\mathbf{q}_1 \approx \mathbf{q}_2 \approx = [1,0,0,1]/\sqrt{2}$. The goal is to drive the final formation error to zero while maintaining formation during the course of the maneuver. In each simulation four plots are given. The first plot shows the leader's progression in the parallel direction toward its final goal ($\rho(\mathbf{q}''_1, \mathbf{1})$). The second plot shows the formation keeping in the parallel direction between the leader and the follower ($\rho(\mathbf{q}''_1, \mathbf{q}''_2)$). Finally the remaining plots show the formation error in the perpendicular direction for both spacecraft ($\rho(\mathbf{q}^\perp_1, \mathbf{1})$ and $\rho(\mathbf{q}^\perp_2, \mathbf{1})$).

In the first simulation

$$\mathbf{q}_1 = [10, 0, 0, 9]^T/\sqrt{181}$$
$$\mathbf{q}_2 = [10, 0, 0, 10]^T/\sqrt{200}.$$

In other words the spacecraft begin slightly out of formation, however, the initial formation error is in the parallel direction. Figure 7 shows the results of this simulation. Note that the initial formation error is driven to zero without altering the perpendicular errors. The spacecraft also converge to their final desired positions.

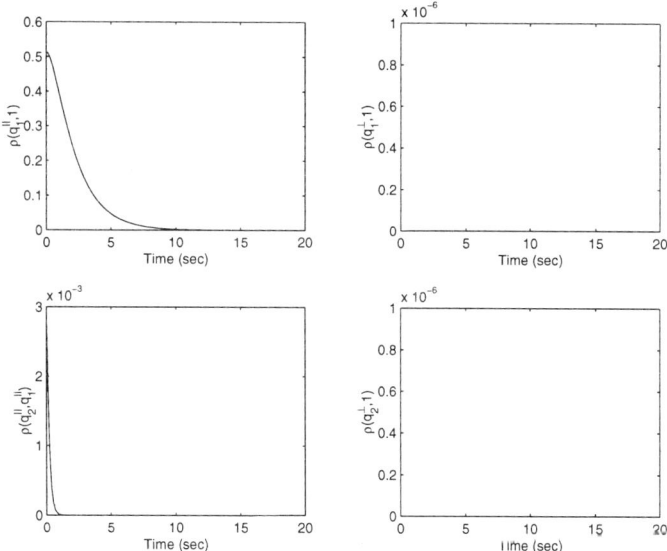

Figure 7: 'Reduced order leader-following when follower starts with and initial parallel error.'

In the next simulation

$$\mathbf{q}_1 = [10, 0, 0, 9]^T/\sqrt{181}$$
$$\mathbf{q}_2 = [10, 0, 1, 10]^T/\sqrt{201}.$$

In this case the spacecraft begin the maneuver out of formation in both the perpendicular and the parallel directions. Figure 8 shows the results of this simulation. Note that the formation keeping, eigenaxis and total convergence error terms all converge to zero.

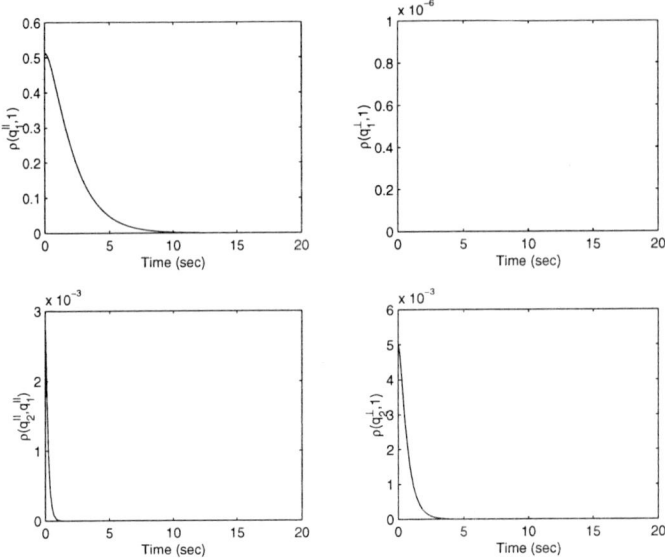

Figure 8: 'Reduced order leader-following when follower starts with an initial parallel and perpendicular error.'

The next set of simulations examine the coupled dynamics control. In this case there are three spacecraft flying in formation. The spacecraft begin the maneuver in formation with $\mathbf{q}_1 \approx \mathbf{q}_2 \approx \mathbf{q}_3$. For each simulation one plot is given to show the overall convergence of one of the spacecraft. Two plots are given to show the pairwise formation keeping between that spacecraft and its two neighbors in terms of the parallel motion. Finally three plots are given to show the motion in the perpendicular direction.

In the first simulation

$$\mathbf{q}_1 = [10, 0, 0, 10]^T / \sqrt{200}$$
$$\mathbf{q}_2 = [10, 0, 0, 9]^T / \sqrt{181}$$
$$\mathbf{q}_3 = [10, 0, 0, 11]^T / \sqrt{221}.$$

The spacecraft begin out of formation, in the parallel direction. Figure 9 shows the results of this simulation. Note that the initial formation error is driven to zero without altering the perpendicular errors. The spacecraft also converge to their final desired positions.

In the next simulation

$$\mathbf{q}_1 = [10, 0, 0, 10]^T / \sqrt{200}$$
$$\mathbf{q}_2 = [10, 0, 1, 9]^T / \sqrt{182}$$
$$\mathbf{q}_3 = [10, 1, 0, 11]^T / \sqrt{222}.$$

In this case the spacecraft begin the maneuver out of formation in both the perpendicular and the parallel directions. Figure 10 shows the results of this simulation. Note that the formation keeping, eigenaxis and total convergence error terms all converge to zero.

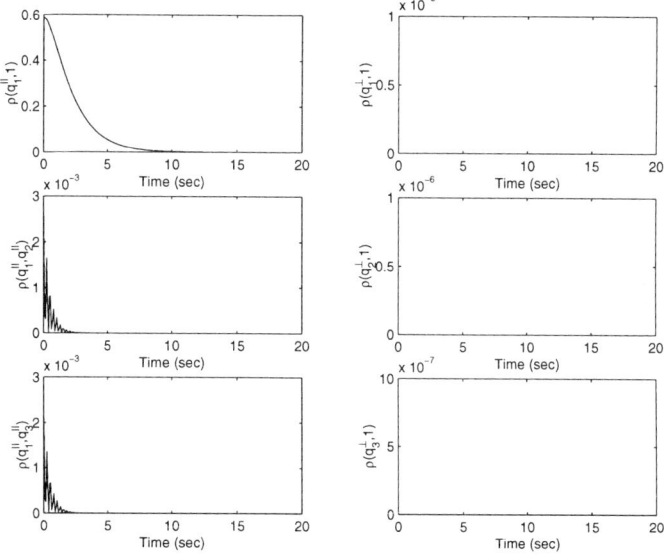

Figure 9: 'Coupled dynamics control where spacecraft begin maneuver out of formation in the parallel direction.'

7 CONCLUSIONS

We found that factoring the attitude of each spacecraft decouples the formation control problem into a lower-order formation control problem and a regulation problem. The formation control problem is solved by using leader-following and behavioral-based control. The regulation problem is solved using P.D. control. Simulation results are given. These illustrate both leader-following and behavioral-based control. In both case the decoupling of the motion is illustrated.

A Useful Properties of Quaternions

We will use the unit quaternion given by

$$\mathbf{q} = \begin{bmatrix} q_1 \\ q_2 \\ q_3 \\ q_4 \end{bmatrix},$$

where $q_1^2 + q_2^2 + q_3^2 + q_4^2 = 1$ to represent spacecraft attitude in this paper.

We can define the vector component of a quaternion by $\mathcal{V}(q) = [q_1, q_2, q_3]^T$ and the scalar component of a quaternion by $\mathcal{S}(q)$. Using these definitions quaternion multiplication can be written as

$$\mathbf{qp} = \begin{bmatrix} \mathcal{V}(\mathbf{qp}) \\ \mathcal{S}(\mathbf{qp}) \end{bmatrix}$$
$$= \begin{bmatrix} \mathcal{S}(\mathbf{q})\mathcal{V}(\mathbf{p}) + \mathcal{S}(\mathbf{p})\mathcal{V}(\mathbf{q}) + \mathcal{V}(\mathbf{q}) \times \mathcal{V}(\mathbf{p}) \\ \mathcal{S}(\mathbf{q})\mathcal{S}(\mathbf{p}) - \mathcal{V}(\mathbf{q})^T \mathcal{V}(\mathbf{p}) \end{bmatrix}.$$

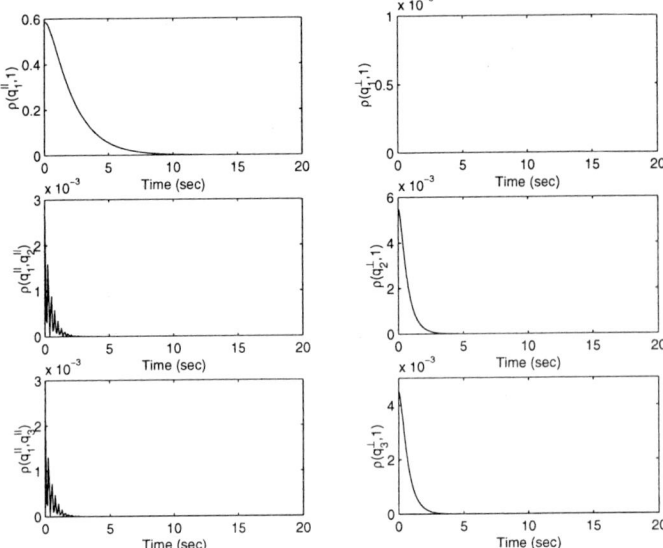

Figure 10: 'Coupled dynamics control where the spacecraft begin the maneuver out of formation in both the parallel and perpendicular direction'

Furthermore define the quaternion identity by

$$\mathbf{1} = \begin{bmatrix} 0 \\ 0 \\ 0 \\ 1 \end{bmatrix}.$$

Given quaternion multiplication and the quaternion identity the following properties are true

1. Unit quaternions are closed under multiplication as defined above,
2. $\mathbf{q1} = \mathbf{1q} = \mathbf{q}$,
3. The product rule of differentiation is

$$\frac{d\mathbf{qp}}{dt} = \dot{\mathbf{q}}\mathbf{p} + \mathbf{q}\dot{\mathbf{p}},$$

4. The conjugate of a quaternion is defined by

$$\mathbf{q}^* = \begin{bmatrix} -\mathcal{V}(\mathbf{q}) \\ \mathcal{S}(\mathbf{q}) \end{bmatrix}$$

and is the unique multiplicative inverse of $\mathbf{q}$ i.e. $\mathbf{qq}^* = \mathbf{q}^*\mathbf{q} = \mathbf{1}$,

5. $\mathbf{q}^T\mathbf{p} = \mathcal{S}(\mathbf{qp}^*) = \mathcal{S}(\mathbf{q}^*\mathbf{p}) = \mathcal{S}(\mathbf{pq}^*) = \mathcal{S}(\mathbf{p}^*\mathbf{q})$,
6. $\mathcal{S}(\mathbf{qp}) = \mathcal{S}(\mathbf{pq})$.

The kinematics of the unit quaternion are given by

$$\frac{d\mathcal{V}(\mathbf{q})}{dt} = \frac{1}{2}\mathcal{S}(q)\omega + \frac{1}{2}\omega \times \mathcal{V}(\mathbf{q})$$
$$\frac{d\mathcal{S}(\mathbf{q})}{dt} = -\frac{1}{2}\mathcal{V}(q)^T\omega, \qquad (11)$$

where $\omega \in \mathbb{R}^3$ is the angular velocity.

If we consider the angular velocity to be a quaternion with the last element identically zero then the kinematic equation can be written as

$$\dot{\mathbf{q}} = \frac{1}{2}\omega\mathbf{q}, \qquad (12)$$

where the product between ω and $\mathbf{q}$ is a quaternion product. It is also useful to mention that we can solve for ω by right multiplying by $2\mathbf{q}^*$ which renders

$$\omega = \dot{\mathbf{q}}\mathbf{q}^*. \qquad (13)$$

The following lemma shows that ω will always be a vector given that $\mathbf{q}$ is a unit quaternion.

Lemma A.1 ω *is a vector if and only if*

$$\frac{d(\mathbf{q}^T\mathbf{q})}{dt} = 0.$$

A special case of the above lemma is when $\mathbf{q}$ is a unit quaternion.

Proof: From Equation (13)

$$\omega \text{ is a vector} \Leftrightarrow \mathcal{S}(\dot{\mathbf{q}}\mathbf{q}^*) = 0$$
$$\Leftrightarrow \dot{\mathbf{q}}^T\mathbf{q} = 0$$
$$\Leftrightarrow \frac{d(\mathbf{q}^T\mathbf{q})}{dt} = 0.$$

∎

Definition 2 *The function $\rho(\mathbf{q}_1, \mathbf{q}_2)$ will establish the error between the quaternions $\mathbf{q}_1$ and $\mathbf{q}_2$.*

$$\rho(\mathbf{q}_1, \mathbf{q}_2) = \|\mathbf{q}_1 - \mathbf{q}_2\|$$
$$= 2(1 - \mathbf{q}_1^T\mathbf{q}_2) \qquad (14)$$
$$= 2(1 - \mathcal{S}(\mathbf{q}_1\mathbf{q}_2^*)).$$

It is apparent that $\rho(\mathbf{q}_1, \mathbf{q}_2) = 0$ if and only if $\mathbf{q}_1 = \mathbf{q}_2$.

The Lemma A.2 gives the derivative of $\rho(\mathbf{q}_1, \mathbf{q}_2)$.

Lemma A.2 *Given*

$$\dot{\mathbf{q}}_1 = \frac{1}{2}\omega_1\mathbf{q}_1$$
$$\dot{\mathbf{q}}_2 = \frac{1}{2}\omega_2\mathbf{q}_2$$

then

$$\frac{d}{dt}\rho(\mathbf{q}_1, \mathbf{q}_2) = \mathcal{V}(\mathbf{q}_1\mathbf{q}_2^*)^T(\omega_1 - \omega_2).$$

Proof:

$$\frac{d}{dt}\rho(\mathbf{q}_1,\mathbf{q}_2) = -2\frac{d}{dt}\mathcal{S}(\mathbf{q}_1\mathbf{q}_2^*)$$
$$= -2\mathcal{S}(\frac{d}{dt}(\mathbf{q}_1\mathbf{q}_2^*))$$
$$= -2\mathcal{S}(\dot{\mathbf{q}}_1\mathbf{q}_2^* + \mathbf{q}_1\dot{\mathbf{q}}_2^*)$$
$$= -2\mathcal{S}(\frac{1}{2}\omega_1\mathbf{q}_1\mathbf{q}_2^* + \frac{1}{2}\mathbf{q}_1\mathbf{q}_2^*\omega_2^*). \tag{15}$$

Since ω_2 is a vector $\omega_2^* = -\omega_2$. From Lemma 2.1

$$\mathcal{S}(\omega_1(\mathbf{q}_1\mathbf{q}_2^*)) = -[\omega_1^T 0]\begin{bmatrix}\mathcal{V}(\mathbf{q}_1\mathbf{q}_2^*)\\ \mathcal{S}(\mathbf{q}_1\mathbf{q}_2^*)\end{bmatrix}$$
$$= -\omega_1^T \mathcal{V}(\mathbf{q}_1\mathbf{q}_2^*) \tag{16}$$

and similarly $\mathcal{S}(\mathbf{q}_1\mathbf{q}_2^*\omega_2) = -\omega_2^T \mathcal{V}(\mathbf{q}_1\mathbf{q}_2^*)$. Therefore

$$\frac{d}{dt}\rho(\mathbf{q}_1,\mathbf{q}_2) = \mathcal{V}(\mathbf{q}_1\mathbf{q}_2)^T(\omega_1 - \omega_2).$$

∎

Acknowledgment

The research was partially supported by the Jet Propulsion Laboratory, California Institute of Technology under contract #96-1245, and by the NASA Rocky Mountain Space Grant Consortium.

References

[1] Decou, "Orbital station-keeping for multiple spacecraft interferometry", *The Journal of the Astronautical Sciences*, vol. 39, no. 3, pp. 283–297, July-September 1991.

[2] A.B. Decou, "Multiple spacecraft optical interferometry, preliminary feasibility assessment", Internal Report D-8811, Jet Propulsion Laboratory, Pasadena California, August 1991.

[3] P. Gorham, "DS-3 descope mission description", Presentation to potential industrial partners detailing the descope of the DS-3 mission found at http://huey.jpl.nasa.gov/rfp_ds3/descope_mission_overview.pdf.

[4] Sanjay S. Joshi, "An informal introduction to sythetic aperature imaging", Interoffice Memorandum 3450-98-0004, Jet Propulsion Laboratory, 4800 Oak Grove Dr., Pasadena, CA 91109, February 1998.

[5] P.K.C. Wang and F.Y. Hadaegh, "Coordination and control of multiple microspacecraft moving in formation", *The Journal of the Astronautical Sciences*, vol. 44, no. 3, pp. 315–355, July-September 1996.

[6] P.K.C. Wang and F.Y. Hadaegh, "Optimal formation-reconfiguration for multiple spacecraft", in *Proceedings of The American Institute of Aeronautics and Astronautics Guidance and Control Conference*, Boston, Mass., 1998, pp. 686–696.

[7] Tucker Balch and Ronald C. Arkin, "Behavior-based formation control for multirobot teams", *IEEE Transactions on Robotics and Automation*, vol. 14, no. 6, pp. 926–939, December 1998.

[8] Randal W. Beard, Jonathan Lawton, and Fred Y. Hadaegh, "A coordination architecture for spacecraft formation control", *IEEE Control Systems Technology*, Submitted, Available at http://www.ee.byu.edu/~beard/papers/cst99.ps.

[9] Jonathan Lawton, Brett Young, and Randal Beard, "A decentralized approach to elementary formation maneuvers", in *Proceedings of The IEEE International Conference on Robotics and Automation*, San Fransico, CA, April 2000.

[10] B. Wie, H. Weiss, and A Arapostathis, "Quaternion feedback regulator for spacecraft eigenaxis rotations", *Journal on Guidance*, vol. 12, no. 3, pp. 375–380, May-June 1989.

AAS 00-012

GRAVITATIONAL PERTURBATIONS, NONLINEARITY AND CIRCULAR ORBIT ASSUMPTION EFFECTS ON FORMATION FLYING CONTROL STRATEGIES

Kyle T. Alfriend,[*] Hanspeter Schaub[†] and Dong-Woo Gim[‡]

Hill's equations have been used in most formation flying studies to determine relative motion orbits and control strategies. Hill's equations assume the Chief satellite orbit is circular, the Earth is spherically symmetric and the nonlinear terms in the relative motion variables can be neglected. This paper presents an approach for determining the effect of these assumptions on the fuel consumption for establishing and maintaining a relative motion orbit. Initial results on the errors in predicting the relative motion using Hill's equations are presented.

Nomenclature
Subscripts
c – refers to Chief satellite

d – refers to deputy satellite

0 – refers to conditions at the initial time

Reference frames
E – Earth centered inertial

C – chief orbit frame with x-axis along the radius vector, the y-axis in the direction of motion and the z-axis perpendicular to the orbit plane. Origin coincides with chief satellite. Unit vectors are $(\vec{e}_{xc}, \vec{e}_{yc}, \vec{e}_{zc})$.

D – deputy orbit frame with u-axis along the radius vector, the v-axis in the direction of motion and the w-axis perpendicular to the orbit plane. Origin coincides with deputy satellite. Unit vectors are $(\vec{e}_{xd}, \vec{e}_{yd}, \vec{e}_{zd})$.

Variables
T^{BA} - transformation matrix for transforming a vector from the A frame to the B frame.

---

[*] Professor and Head, Department of Aerospace Engineering, Texas A&M University, H.R. Bright Building, Room 701, MS 3141 AMU, College Station, Texas 77843-3141. E-mail: alfriend@aero.tamu.edu.

[†] Research Engineer, Sandia National Laboratories, P.O. Box 5800, MS-1003, Albuquerque, New Mexico 87185-1003. E-mail: hschaub@sandia.gov.

[‡] Graduate Student, Department of Aerospace Engineering, Texas A&M University, H.R. Bright Building, Room 701, MS 3141 AMU, College Station, Texas 77843-3141.

$(\vec{R}_c, \vec{R}_d)$ - position vectors of the chief and deputy.

$\vec{r}_d$ - position of the deputy relative to the chief.

$(\vec{V}_c, \vec{V}_d)$ - velocity of the Chief and deputy.

(V_r, V_t) - radial and tangential components of the velocity.

$\vec{v}_d$ - velocity of the deputy relative to the chief.

(v_{dr}, v_{dt}) - radial and tangential components of the relative velocity vector.

a – semi-major axis

e – eccentricity

i – inclination

Ω – right ascension

ω – argument of perigee.

f – true anomaly

θ – argument of latitude, $\theta = f + \omega$

$q_1 - e\cos\omega$

$q_2 - e\sin\omega$

$\delta\alpha$ - variation of the variable α with respect to the chief orbit.

INTRODUCTION

Spacecraft flying in precise formation is a subject drawing considerable attention within NASA and the DoD[1]. O-orbit experiments are planned within the near future[2,3]. Satellites flying in formation is not a new challenge, but flying in precise formation and operating autonomously is a significant challenge. It is important to design the relative motion orbits such that fuel consumption is minimized and lifetime maximized. Most studies[4-6] have used Hill's equations[7] (sometimes called the Clohessy-Wiltshire or CW equations) to describe the relative motion of the satellites. These equations assume that a) the Earth is spherically symmetric, b) the Chief or reference satellite orbit is circular, and c) the equations can be linearized in the relative motion variables. For small formations the effects of the neglected nonlinear terms are probably negligible, but the effects of the ignored gravitational perturbations and the eccentric reference orbit can be significant[7-9]. Formations that will emulate large apertures will require some of the satellites to have out-of-plane motion relative to the reference orbit. This out-of-plane motion is achieved by some combination of small changes in the inclination, δi, and the right ascension, $\delta\Omega$. (see Figure 1) An inclination difference results in the maximum out-of-plane separation occurring at the maximum latitude. In contrast, a right ascension difference results in the maximum separation occurring at the equator. A constellation emulating a large aperture at all times would have satellites with varying combinations of inclination and right ascension differences. A differential inclination has three negative effects, it causes the deputy satellites to have a slightly different nodal precession rate, a slightly different orbit period and a slightly different argument of perigee rate. Since all three of these effects cause the two satellites to slowly separate these effects must be negated, either by control or design of the relative motion orbit. Reference 7 derives initial conditions for minimizing these effects by selection of the

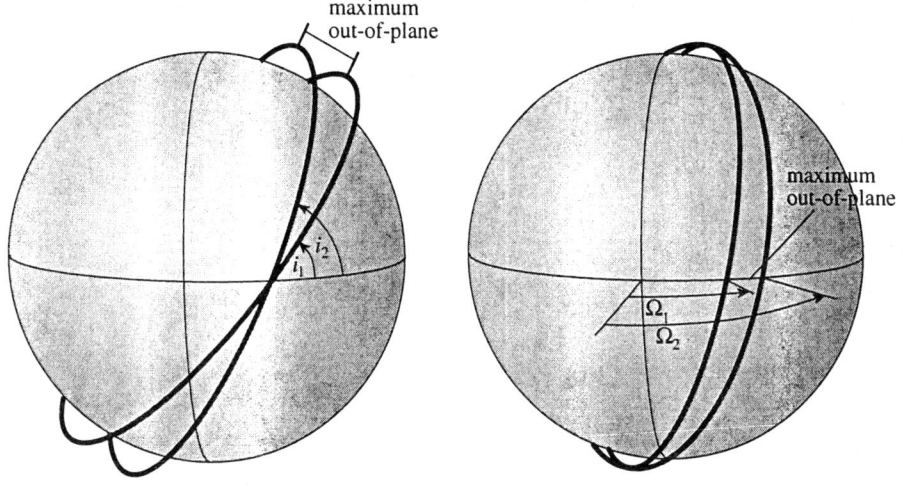

Figure 1 Achieving out-of-plane motion

orbital parameters. However, in most cases some control will still be needed. Some control approaches are presented in references 4,6,8 and 9. An unanswered question is what effect does using Hill's equations for the control have on fuel consumption. The gravitational perturbations create short period oscillations in the orbital elements that then create short period oscillations in the relative motion variables that are not captured by Hill's equations. If these oscillations are outside of the deadband of the control system they cold be interpreted by the control system as a secular rate and the control system would then try to negate these natural motions. The reference orbit eccentricity can have a similar effect. Trying to continually negate these natural motions will waste fuel. How much is the unanswered question. Also, if the model does not include the gravitational perturbation effects then fuel may be wasted trying to negate the differential secular rates. The system will not know how to select the correct orbital parameters to minimize the secular rates. The purpose of this paper is to develop a method for evaluating these effects and to present some results on the errors that occur in estimating the relative motion with Hill's equations. Essentially, we develop a state transition matrix for a system that includes the gravitational perturbations and reference orbit eccentricity. Research is underway on another method for a state transition matrix that will include the effect of the neglected nonlinear terms. One of the methods includes the effects of the nonlinear terms in the relative motion variables. A state transition matrix that includes the reference orbit eccentricity for small eccentricities has been derived by Melton[10].

HILL'S EQUATIONS

Referring to figure 2 the relative motion is described using reference frame O, which is a rotating reference with its origin at the reference satellite, the x-axis is along the radius vector, the y-axis is in the orbit plane in the direction of motion, and the z-axis is perpendicular to the orbit plane. Assuming the reference satellite orbit is circular and there are no gravitational perturbations the linearized relative equations of motion are:

$$\ddot{x} - 2n\dot{y} - 3n^2 x = 0$$
$$\ddot{y} + 2n\dot{x} = 0 \tag{1}$$
$$\ddot{z} + n^2 z = 0$$

n = mean motion of reference orbit

Note that the out-of-plane motion is decoupled from the in-plane motion. The solution is

$$x = 2(2x_0 + \dot{y}_0/n) - (3x_0 + 2\dot{y}_0/n)\cos\psi + (\dot{x}_0/n)\sin\psi$$
$$y = (y_0 - 2\dot{x}_0/n) - 3(2x_0 + \dot{y}_0/n)\psi + (2\dot{x}_0/n)\cos\psi + 2(3x_0 + 2\dot{y}_0/n)\sin\psi \tag{2}$$
$$z = z_0 \cos\psi + (\dot{z}_0/n)\sin\psi$$
$$\psi = nt$$

For periodic motion

$$2x_0 + \dot{y}_0/n = 0 \tag{3}$$

This condition is just the requirement that the semi-major axes of the two satellites must be equal. Also requiring that the center of the relative motion be at the reference satellite periodic relative motion orbits are given by

$$x = x_0 \cos\psi + (y_0/2)\sin\psi = A\sin(\psi + \alpha)$$
$$y = y_0 \cos\psi - 2x_0 \sin\psi = 2A\cos(\psi + \alpha)$$
$$z = z_0 \cos\psi + (\dot{z}_0/n)\sin\psi = B\sin(\psi + \beta) \tag{4}$$
$$A = (x_0^2 + y_0^2/4)^{1/2}, \tan\alpha = 2x_0/y_0$$
$$B = (z_0^2 + \dot{z}_0^2/n^2)^{1/2}, \tan\beta = nz_0/\dot{z}_0$$

Note that the projection of the relative periodic orbits in the x-y (orbit) plane is a 2-1 with the long axis in the y-direction. Two periodic orbits of interest are a) a circular relative motion orbit, and b) an orbit for which the projection of the motion in the horizontal (y-z) plane is a circle. This has application for emulating large circular apertures. The initial conditions for these orbits are:

Circular Relative Orbit

$$B = \sqrt{3}A, \alpha = \beta$$
$$x^2 + y^2 + z^2 = 4A^2 \tag{5}$$

This relative motion orbit is inclined at 30 degrees to the horizontal plane.

Circular Horizontal Plane Orbit

$$B = 2A, \alpha = \beta$$
$$y^2 + z^2 = 4A^2 \tag{6}$$

This relative motion orbit is inclined at 26.56 degrees to the horizontal plane.

The TechSat21 program has three satellites flying in formation. One option for a portion of the program is the circular horizontal plane orbit. In this configuration the three satellites would form an equilateral triangle in the horizontal plane. Thus, the constellation would appear to be a rotating equilateral triangle. For this configuration the differential inclination will be different for each satellite. Thus, the rate that each satellite drifts from the equilateral triangle configuration will be different. Assume that at $t=0$ the chief satellite is on the equator and the constellation is as shown in Figure 2 and is rotating counterclockwise as a circle of radius $2A$. The initial conditions for the three satellites are

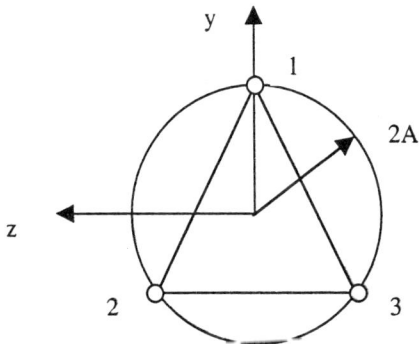

$$y_{10} = 2A, \dot{y}_{10} = 0, z_{10} = 0, \dot{z}_{10} = 2An, \alpha_1 = 0$$
$$y_{20} = -A, \dot{y}_{20} = -\sqrt{3}An, z_{20} = \sqrt{3}A, \dot{z}_{20} = -An, \alpha_2 = 120 \deg \tag{7}$$
$$y_{30} = -A, \dot{y}_{30} = \sqrt{3}An, z_{30} = -\sqrt{3}A, \dot{z}_{30} = -An, \alpha_2 = -120 \deg$$

The change in inclination and right ascension to achieve this desired motion when the chief satellite is on the equator is

$$\delta i = \dot{z}_0 / Rn, \delta\Omega = -z_0 R \qquad (8)$$

where R is the radius of the chief's orbit. Thus, the inclination and right ascension changes for the three satellites are

$$\begin{aligned}\delta i_1 &= 2A/R, \delta\Omega_1 = 0 \\ \delta i_2 &= -A/R, \delta\Omega_2 = -\sqrt{3}A/R \\ \delta i_3 &= -A/R, \delta\Omega_3 = \sqrt{3}A/R\end{aligned} \qquad (9)$$

Gravitational Perturbation Effects

The primary gravitational perturbation effect is due to the equatorial bulge term, J_2. The J_2 term changes the orbit period, a drift in perigee, a nodal precession rate and periodic variations in all the elements. Let's consider the right ascension rate which is

$$\dot{\Omega} = -\frac{3}{2} J_2 \left(\frac{R_e}{p}\right)^2 n \cos i \qquad (10)$$

If a change in inclination is used to create out-of-plane motion a differential nodal precession rate occurs which causes the planes to slowly separate. The differential rate is

$$\delta\dot{\Omega} = -\dot{\Omega}_c \tan i \, \delta i \qquad (11)$$

Consider the case when the out-of-plane motion is caused by only an inclination change. This means that at equator crossings there is no out-of-plane separation. Letting $\rho = 2A$ be the radius of the relative motion orbit the growth in the out-of-plane separation at the equator is

$$\frac{\delta\rho_i}{\rho}(per\ day) = 0.118 \qquad (12)$$

Thus, the circle begins to distort. The rate of distortion is a function of the changes in inclination and right ascension used to create the out-of-plane motion, thus it would be different for each satellite. The question is what is the effect on fuel consumption and system performance of using Hill's equations for the control system model and continuing to let the control system correct this growth. Including these effects in the design of the relative motion orbits can minimize this growth[7] and including them in the control system model may improve fuel consumption and reduce the frequency of control. In addition to these secular out-of-plane effects there are in-plane secular effects and the effects of the short period variations due to J_2 and the orbit eccentricity.

Developed in this paper are two analytic methods that can be used to evaluate the effects of neglecting these terms in the orbit design and control.

PROBLEM FORMULATION

To evaluate the effect of neglecting the chief satellite eccentricity, the gravitational perturbations and nonlinearities the control needs to be evaluated with and without these effects included in the model. Thus, the objective is to obtain a state transition matrix with these effects included. In this paper it will be assumed that the eccentricity is small, basically it will be assumed that $e = O(J_2)$. A state transition matrix for small eccentricity has been obtained by Melton [10]. There are two approaches for obtaining the state transition matrix. One approach is to use the geometry of the problem and realize that the deputy relative motion is a result of small changes in the chief satellite orbital elements. This method will be called the geometric method and is developed in this paper. A second approach is to write the equations of motion in the rotating reference frame, but include the gravitational perturbations and not assume a circular orbit for the chief satellite. In addition, the non-linear terms in the relative motion variables can be included. A solution using a perturbation method can then be used to obtain a solution. There are several perturbation methods that can be used to obtain the problem. A perturbation solution of this problem using Hamiltonian mechanics and Lie Series is currently underway and will be reported on in a later publication. The primary reasons for using this approach are:

- The large amount of algebraic manipulations required are easily implemented on the computer.
- The solution will be in a form to easily evaluate what effects need to be included in the state transition matrix. For example, do the periodic effects due to J_2 need to be included or is it sufficient to just include the secular effects.

This method will be referred to as the Hamiltonian method.

<u>Geometric Method</u>

Let

$$\mathbf{x}^T = (x, \dot{x}, y, \dot{y}, z, \dot{z})$$
$$\mathbf{e}^T = (a, \theta, i, q_1, q_2, \Omega) \tag{13}$$

where θ is the argument of latitude and $q_1 = e\cos\omega, q_2 = e\sin\omega$. These variables are used because the true anomaly and argument of perigee are undefined for zero eccentricity. Since the relative motion is small the approach will be to express the orbital elements of the deputy as a Taylor series about the chief satellite elements. Thus,

$$\mathbf{e}_d = \mathbf{e}_c + \delta\mathbf{e} \tag{14}$$

W now want to relate the $\delta \mathbf{e}$ to the relative motion state $\mathbf{x}$. The deputy's position is

$$\vec{R}_d = \vec{R}_c + \vec{\rho} = (R+x)\vec{e}_{xc} + y\vec{e}_{yc} + z\vec{e}_{zc} \tag{15}$$

The deputy's position in the chief reference frame is also given by

$$\vec{R}_d^C = T^{CE}T^{ED}\vec{R}_d^D \tag{16}$$

The transformation from the orbit frame O (C or D) to the inertial frame E is given by

$$T^{OE} = \begin{pmatrix} \cos\theta & \sin\theta & 0 \\ -\sin\theta & \cos\theta & 0 \\ 0 & 0 & 1 \end{pmatrix}\begin{pmatrix} 1 & 0 & 0 \\ 0 & \cos i & \sin i \\ 0 & -\sin i & \cos i \end{pmatrix}\begin{pmatrix} \cos\Omega & \sin\Omega & 0 \\ -\sin\Omega & \cos\omega & 0 \\ 0 & 0 & 1 \end{pmatrix}$$

$$T^{OE} = \begin{pmatrix} c\theta c\Omega - s\theta ci s\Omega & c\theta s\Omega + s\theta ci c\Omega & s\theta si \\ -s\theta c\Omega - c\theta ci s\Omega & -s\theta s\Omega + c\theta ci c\Omega & c\theta si \\ si s\Omega & -si c\Omega & ci \end{pmatrix} \tag{17}$$

Some identities that will be used are

$$e\cos f = e\cos(\theta - \omega) = q_1 \cos\theta + q_2 \sin\theta$$
$$e\sin f = e\sin(\theta - \omega) = q_1 \sin\theta - q_2 \cos\theta$$
$$V_t = R\dot{\theta}$$
$$V_r = \dot{R} = \frac{R^2\dot{\theta}}{p}e\sin f = \frac{h}{p}[q_1 \sin\theta - q_2 \cos\theta], \; p = a(1-e^2) \tag{18}$$
$$\dot{\theta} = \frac{h}{R^2} = \sqrt{\frac{\mu}{p}}(1+e\cos f)^2$$
$$\dot{\theta} = an\eta^{-1}(1+q_1 \cos\theta + q_2 \sin\theta)^2$$

Now expand eq. (16) about the chief satellite motion.

$$\vec{R}_d^C = T^{CE}\left(T^{EC} + \delta T^{EC}\right)\begin{pmatrix} R_c + \delta R_c \\ 0 \\ 0 \end{pmatrix} = \left(I + T^{CE}\delta T^{EC}\right)\begin{pmatrix} R_c + \delta R_c \\ 0 \\ 0 \end{pmatrix}$$

$$\vec{R}_d^C = \begin{pmatrix} R_c + \delta R_c \\ 0 \\ 0 \end{pmatrix} + R_c T^{CE}\begin{pmatrix} \delta T_{11} \\ \delta T_{12} \\ \delta T_{13} \end{pmatrix} \tag{19}$$

The δT_{ij} are

$$\begin{aligned}
\delta T_{11} &= T_{21c}\delta\theta - T_{12c}\delta\Omega + (T_{13c}\sin\Omega_c)\delta i \\
\delta T_{12} &= T_{22c}\delta\theta + T_{11c}\delta\Omega - (T_{13c}\cos\Omega_c)\delta i \\
\delta T_{13} &= T_{23c}\delta\theta + (\sin\theta_c \cos i_c)\delta i
\end{aligned} \quad (20)$$

Substitute eq. (20) into eq. (19).

$$\begin{aligned}
\vec{R}_d^C &= \begin{pmatrix} R_c + \delta R_c \\ 0 \\ 0 \end{pmatrix} + R_c T_{C2E} \begin{pmatrix} T_{21c}\delta\theta - T_{12c}\delta\Omega + (T_{13c}\sin\Omega_c)\delta i \\ T_{22c}\delta\theta + T_{11c}\delta\Omega - (T_{13c}\cos\Omega_c)\delta i \\ T_{23c}\delta\theta + (\sin\theta_c \cos i_c)\delta i \end{pmatrix} \\
\vec{R}_d^C &= \begin{pmatrix} R_c + \delta R_c \\ 0 \\ 0 \end{pmatrix} + R_c \begin{pmatrix} 0 \\ \delta\theta + \delta\Omega\cos i_c \\ -\cos\theta_c \sin i_c \delta\Omega + \sin\theta_c \delta i \end{pmatrix}
\end{aligned} \quad (21)$$

Thus,

$$\begin{aligned}
x &= \delta R_c \\
y &= R_c(\delta\theta + \delta\Omega \cos i_c) \\
z &= R_c(-\cos\theta_c \sin i_c \delta\Omega + \sin i_c \delta i)
\end{aligned} \quad (22)$$

where

$$R = \frac{a(1-e^2)}{1+e\cos f} = \frac{a(1-q_1^2-q_2^2)}{1+q_1\cos\theta+q_2\sin\theta}$$

$$\delta R = \frac{R_c}{a_c}\delta a - 2R_c\left(\frac{a_c}{p_c}\right)(q_{1c}\delta q_1) - \frac{R_c^2}{p_c}\left[(-q_{1c}\sin\theta + q_{2c}\cos\theta_c)\delta\theta + \delta q_1 \cos\theta_c + \delta q_{2c}\sin\theta_c\right] \quad (23)$$

The deputy's velocity in the chief reference frame is

$$\vec{V}_d^C = (V_{rc} + \dot{x} - \dot{\theta}_c y)\vec{e}_x + (V_{tc} + \dot{y} + \dot{\theta}_c x)\vec{e}_y + \dot{z}\vec{e}_z \quad (24)$$

Also,

$$\vec{V}_d^C = T^{CE}T^{ED}\begin{pmatrix}V_{rd}\\V_{td}\\0\end{pmatrix} = T^{CE}\left(T^{EC}+\delta T^{CE}\right)\begin{pmatrix}V_{rc}+\delta V_r\\V_{tc}+\delta V_t\\0\end{pmatrix}$$

$$\vec{V}_d^C = \begin{pmatrix}V_{rc}+\delta V_{rc}\\V_{tc}+\delta V_{tc}\\0\end{pmatrix} + T^{CE}\begin{pmatrix}V_{rc}\delta T_{11}+V_{tc}\delta T_{21}\\V_{rc}\delta T_{12}+V_{tc}\delta T_{22}\\V_{rc}\delta T_{13}+V_{tc}\delta T_{23}\end{pmatrix}$$

(25)

$$\delta T_{21} = -T_{11c}\delta\theta - T_{22c}\delta\Omega + T_{23c}\sin\Omega_c\delta i$$
$$\delta T_{22} = -T_{12c}\delta\theta + T_{21c}\delta\Omega - T_{23c}\cos\Omega_c\delta i$$
$$\delta T_{23} = -T_{13c}\delta\theta + \cos\theta_c\cos i_c\delta i$$

(26)

Substituting gives

$$\vec{V}_d^C = \begin{pmatrix}V_{rc}+\delta V_{rc}\\V_{tc}+\delta V_{tc}\\0\end{pmatrix} + V_{rc}\begin{pmatrix}0\\\delta\theta+\delta\Omega\cos i_c\\-\cos\theta_c\sin i_c\delta\Omega+\sin\theta_c\delta i\end{pmatrix} + V_{tc}\begin{pmatrix}-\delta\theta-\cos i_c\delta\Omega\\0\\\sin\theta_c\sin i_c\delta\Omega+\cos\theta_c\delta i\end{pmatrix}$$

(27)

$$V_{tc} = R_c\dot{\theta}_c = \frac{h_c}{R_c} = \frac{\sqrt{\mu p_c}}{R_c} = \sqrt{\frac{\mu}{p_c}}(1+e_c\cos f_c) = \sqrt{\frac{\mu}{p_c}}(1+q_{1c}\cos\theta_c+q_{2c}\sin\theta_c)$$

$$\delta V_t = -\frac{V_{tc}}{2p_c}\delta p + \sqrt{\frac{\mu}{p_c}}\left[\delta q_1\cos\theta_c+\delta q_2\sin\theta_c+(-q_{1c}\sin\theta_c+q_{2c}\cos\theta_c)\delta\theta\right]$$

(28)

The variation in the radial velocity is

$$V_{rc} = \dot{R}_c = \sqrt{\frac{\mu}{p_c}}(q_{1c}\sin\theta_c - q_{2c}\cos\theta_c)$$

$$\delta V_r = -\frac{1}{2p_c}V_{rc}\delta p + \sqrt{\frac{\mu}{p_c}}\left[(\delta q_1\sin\theta_c - \delta q_2\cos\theta_c) + (q_{1c}\cos\theta_c+q_{2c}\sin\theta_c)\delta\theta\right]$$

(29)

Substituting for δp gives

$$\delta V_t = -\frac{V_{tc}}{2a_c}\delta a + \sqrt{\frac{\mu}{p_c}}(-q_{1c}\sin\theta_c + q_{2c}\cos\theta_c)\delta\theta$$
$$+\left(\frac{V_{tc}a_c q_{1c}}{p_c} + \sqrt{\frac{\mu}{p_c}}\cos\theta_c\right)\delta q_1 + \left(\frac{V_{tc}a_c q_{2c}}{p_c} + \sqrt{\frac{\mu}{p_c}}\sin\theta_c\right)\delta q_2 \quad (29)$$

$$\delta V_r = -\frac{1}{2p_c}V_{rc}\left[\left(\frac{p_c}{a_c}\right)\delta a - a_c(2q_{1c}\delta q_1 + 2q_{2c}\delta q_2)\right]$$
$$+\sqrt{\frac{\mu}{p_c}}\left[(\delta q_1 \sin\theta_c - \delta q_2 \cos\theta_c) + (q_{1c}\cos\theta_c + q_{2c}\sin\theta_c)\delta\theta\right] \quad (30)$$

The velocity development is now complete. Using eq. (22b) we get

$$\dot{x} = \delta V_r$$
$$\dot{y} + \dot{\theta}_c x - (V_{rc}/R_c)y = \delta V_t \quad (31)$$
$$\dot{z} = (V_{tc}\cos\theta_c + V_{rc}\sin\theta_c)\delta i + (V_{tc}\sin\theta_c - V_{rc}\cos\theta_c)\sin i_c \delta\Omega$$

This completes the development. We now have

$$\mathbf{x} = A\delta\mathbf{e} \quad (32)$$

The elements of A and its inverse A^{-1} are given in the Appendix. Please note that we have made no restrictions on the orbital elements. They can be two body elements or they can be the solution from an analytic theory such as Brouwer's theory [11]. Whatever theory is used we can develop

$$\delta\mathbf{e}(t) = \Phi_e(t)\delta\mathbf{e}(t_0)$$
$$\delta\mathbf{e}_0 = A^{-1}(t_0)\mathbf{x}(t_0) \quad (33)$$

giving

$$\mathbf{x}(t) = \Phi_x(t)\mathbf{x}_0$$
$$\Phi_x(t) = A(t)\Phi_e(t)A^{-1}(t_0) \quad (34)$$

The development is now shown for a circular chief orbit. Since there are no perturbations the only time varying element is the argument of latitude. Using the angular momentum integral

$$r^2 d\theta = \sqrt{\mu p}\, dt$$

$$\sqrt{\frac{p^3}{\mu}} \frac{d\theta}{\left(1+q_1 \cos\theta + q_2 \sin\theta\right)^2} = dt \tag{35}$$

Now integrate

$$\sqrt{\frac{p^3}{\mu}} \int_{\theta_0}^{\theta} \frac{d\theta}{\left(1+q_1 \cos\theta + q_2 \sin\theta\right)^2} = t \tag{36}$$

Expand this equation in a Taylor series about the chief orbit.

$$0 = 1.5 \frac{\delta p}{p_c} \sqrt{\frac{p_c^3}{\mu}} \int_{\theta_{0c}}^{\theta_c} \frac{d\tau}{\left(1+q_{1c} \cos\tau + q_{2c} \sin\tau\right)^2} - 2\sqrt{\frac{p_c^3}{\mu}} \int_{\theta_0}^{\theta} \frac{(\delta q_1 \cos\tau + \delta q_2 \sin\tau) d\tau}{\left(1+q_{1c} \cos\tau + q_{2c} \sin\tau\right)^3}$$

$$+ \delta\theta \sqrt{\frac{p_c^3}{\mu}} \frac{1}{\left(1+q_{1c} \cos\theta_c + q_{2c} \sin\theta_c\right)^2} - \delta\theta_0 \sqrt{\frac{p_c^3}{\mu}} \frac{1}{\left(1+q_{1c} \cos\theta_{0c} + q_{2c} \sin\theta_{0c}\right)^2} \tag{37}$$

Now set $e=0$ and use eq. (42).

$$\delta\theta = \delta\theta_0 - 1.5 n_c \frac{\delta a}{a_c} + 2\delta q_1 \sin\theta_c + \delta q_2 (1 - \cos\theta_c) \tag{38}$$

Therefore,

$$\Phi_e(t) = \begin{pmatrix} 1 & 0 & 0 & 0 & 0 & 0 \\ -1.5n_c t/a_c & 1 & 0 & 2\sin\theta_c & 2(1-\cos\theta_c) & 0 \\ 0 & 0 & 1 & 0 & 0 & 0 \\ 0 & 0 & 0 & 1 & 0 & 0 \\ 0 & 0 & 0 & 0 & 1 & 0 \\ 0 & 0 & 0 & 0 & 0 & 1 \end{pmatrix} \tag{39}$$

RESULTS

To evaluate how accurately the method developed here estimates the relative motion and to determine the errors resulting from using Hill's equations the following example was used. The initial conditions used for the deputy result in the Circular Horizontal plane orbit when the Earth is spherically symmetric and the chief satellite orbit is circular.

Chief orbital elements

$a_c = 7100\,km$
$e_c = 0.005$
$i_c = 70\,\deg$
$\Omega_c = \omega = f = 0$
$\theta_c = 0$
$q_{1c} = 0.005 \quad q_{2c} = 0$
$n = \sqrt{\dfrac{\mu}{a^3}} = 1.05531 \times 10^{-3}\,r/s$

Deputy initial conditions and elements

$x = 0$ $\delta a = 0$
$\dot{x} = n_c/4$ $\delta\theta = 0.004055\,\deg$
$y = 0.5\,km$ $\delta i = 0.0040148\,\deg$
$\dot{y} = 0$ $\delta q_1 = 0$
$z = 0$ $\delta q_2 = -3.556 \times 10^{-5}$
$\dot{z} = n/2$ $\delta\Omega = 0$

The method is first compared to Hill's equations and the exact solution for a spherically symmetric Earth. Figure 3 shows the error in the new method for one day. The errors are only several centimeters. Figure 4 shows the errors resulting from the use of Hill's equations. The errors in Hill's equations are considerably larger than the geometric method. For these results to obtain the mean motion in Hill's equation we used the mean semi- The periodic variations result from the circular orbit assumption and the secular growth in the in-track direction is mostly due an incorrect mean motion.

Figure 5a shows the relative motion trajectory with the gravitational perturbations included (J_2-J_4) and Figure 5b show the same trajectory using the Geometric Method. Obviously the Geometric Method is incorporating the primary eccentricity and gravitational perturbation effects.

Figures 6 and 7 show the errors that occur when estimating the motion with the Geometric Method and Hill's equations, respectively. The Geometric results in much smaller errors. The in-track error of 20 m after one day is the result of approximately a 2m error in semi-major axis. In

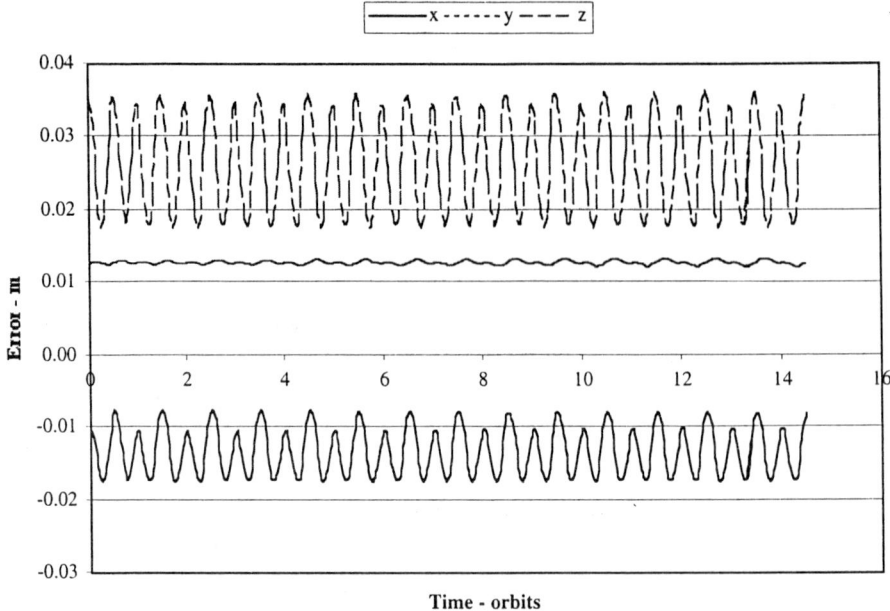

Figure 3 Geometric Method Errors for Spherically Symmetric Earth

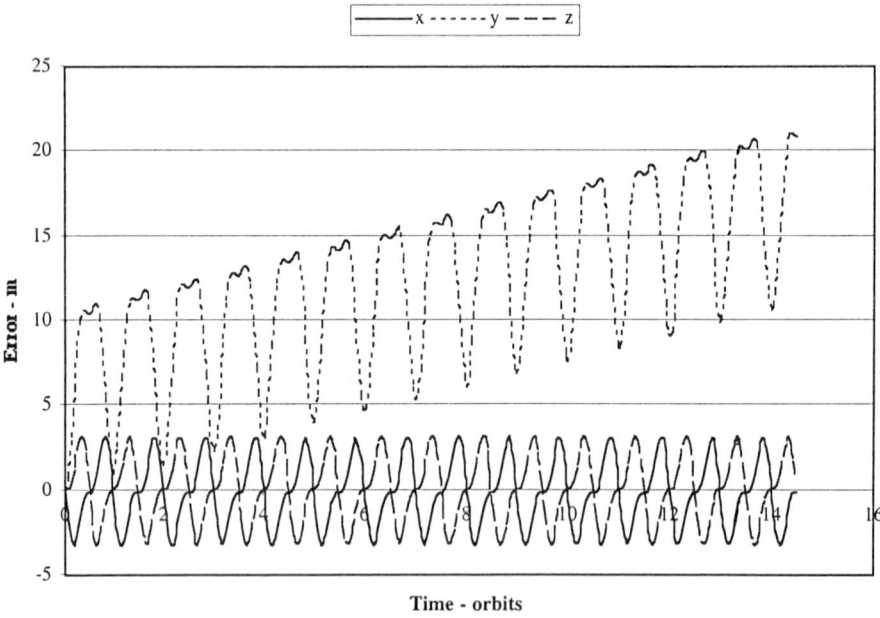

Figure 4 Hill's Equations Errors for Spherically Symmetric Earth

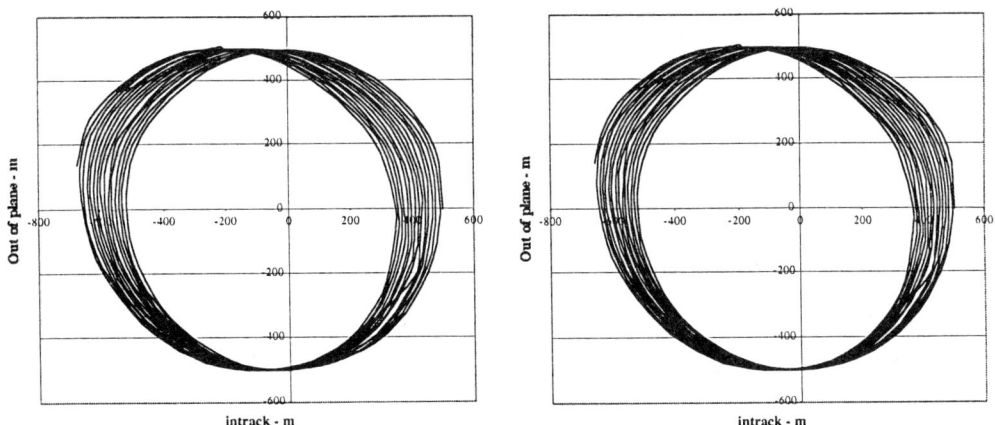

Figure 5 Horizontal Plane Trajectory

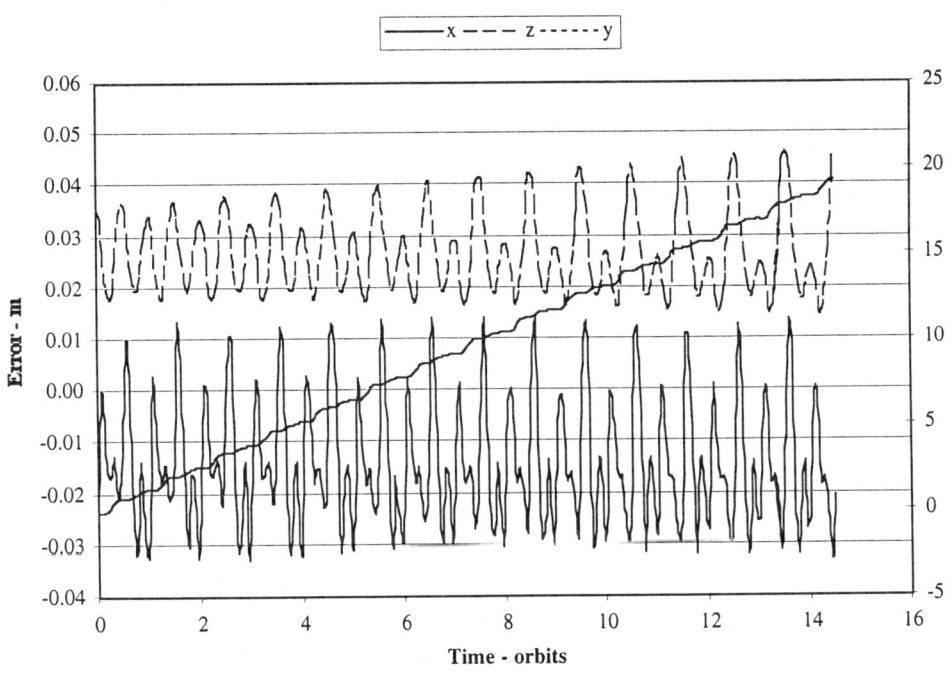

Figure 6 Geometric Method Errors With Gravitational Perturbations

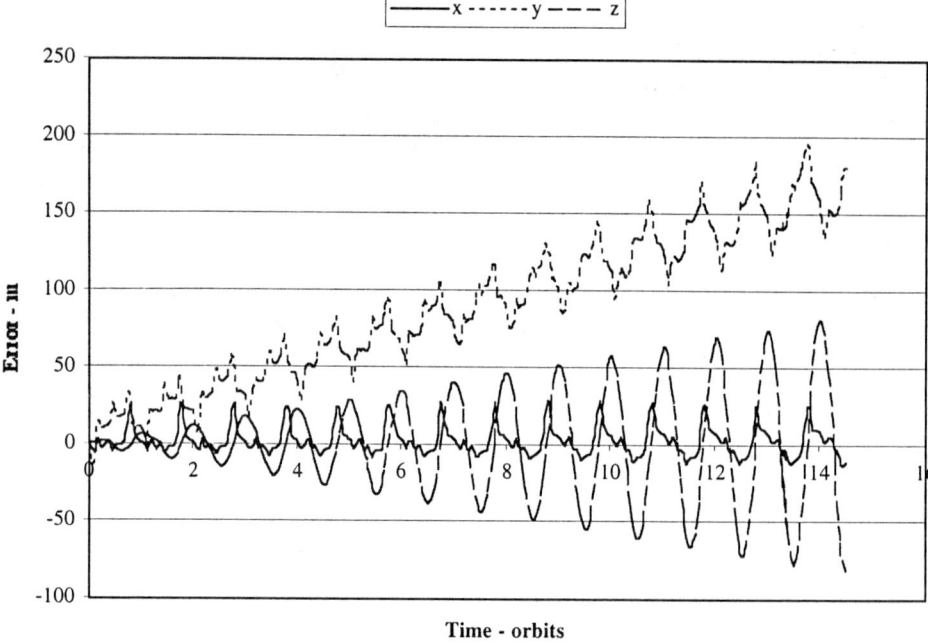

Figure 7 Hill's Equations Errors With Gravitational Perturbations

Hill's equations the mean semi-major axis was used to compute the mean motion, otherwise, the errors would be much larger.

CONCLUSIONS

An algorithm for relating the orbital element changes to the relative motion variables has been developed. This algorithm is used in the development of a state transition matrix that includes the effects of the chief satellite orbit eccentricity and the gravitational perturbations. This state transition matrix was developed by considering the geometry of the problem, not by solving the differential equations. The errors in estimating the relative motion are much less than with using Hill's equations.

Evaluation of the method is continuing. Research is also underway for solving the relative equations of motion using Hamiltonian mechanics and Lie Series.

REFERENCES

1. Folta, D., Newman, L. and Gardner, T., "Foundations of Formation Flying for Mission to Planet Earth and the New Millennium," *AIAA/AAS Astrodynamics Specialists Conference*, July 1996.
2. How, J., Twiggs, R. Weidow, D., Hartman, K. and Bauer, F., "Orion: A Low Cost Demonstration Flying in Space Using GPS," *AIAA/AAS Astrodynamics Specialists Conference*, Boston, MA, August 1998,
3. Campbell, M., Fullmer, R. and Hall C., "The ION-F Formation Flying Experiments", *AAS/AIAA Space Flight Mechanics Conference*, Clearwater, FL, January 23-26, 2000, Paper No. AAS 00-108.
4. Kapila, V., Sparks, A., Buffington, J.M., and Yan, Q.: Spacecraft Formation Flying: Dynamics and Control," *Proceedings of the American Control Conference*, San Diego, CA, June 1999.
5. Kong, E.M., Miller, D.W. and Sedgwick, R.J., "Optimal Trajectories and Orbit Design for Separated Spacecraft Interferometry," TR SERC #13-98, MIT, November 1998.
6. Inalham, G., Busse, F.D. and How, J.P., "Precise Formation Flying Control of Multiple Spacecraft Using Differential Carrier-Phase Differential GPS," *AAS/AIAA Space Flight Mechanics Conference*, Clearwater, FL, January 23-26, 2000, Paper No. AAS 00-109.
7. Schaub, H, and Alfriend, K.T., "J_2 Invariant Reference Orbits for Spacecraft Formations," *Flight Mechanics Symposium,* NASA Goddard Space Flight Center, May 18-20, 1999, Paper No. 11.
8. Schaub, H., Vadali, S.R. and Alfriend, K.T., "Spacecraft Formation Fling Using Mean Orbit Elements," *AAS/AIAA Astrodynamics Specialists Conference,* Girdwood, Alaska, Aug. 16-19, 1999, Paper No. AAS 99-310.
9. Vadali, S.R., Schaub, H. and Alfriend, K.T., "Initial Conditions and Fuel-Optimal Control for Formation Flying of Satellites," *AIAA GN&C Conference*, Portland, OR, Aug. 9-12,1999, Paper No. AIAA 99-4265.
10. Melton, R.G., " Time-Explicit representation of relative Motion Between Elliptical Orbits," ," *AAS/AIAA Astrodynamics Specialists Conference,* Sun valley, ID, Aug. 1997.
11. Brouwer, D., "Solution of the Problem of Artificial Satellite Theory*,"* *The Astronautical Journal*, Vol. 64, No. 1274, 1959, pp. 378-397.

Appendix A - The A Matrix

The non-zero elements of A and A^{-1} are

$$A_{11} = \frac{R_c}{a_c}$$
$$A_{12} = V_r / \dot{\theta}_c$$
$$A_{14} = -\frac{R_c}{p_c}(2a_c q_{1c} + R_c \cos\theta_c) \tag{A1}$$
$$A_{15} = -\frac{R_c}{p_c}(2a_c q_{2c} + R_c \sin\theta_c)$$

$$A_{21} = -05 V_{rc} a_c$$
$$A_{22} = \sqrt{\frac{\mu}{p_c}\left(\frac{p_c}{R_c} - 1\right)}$$
$$A_{24} = \frac{V_{rc} a_c q_{1c}}{p_c} + \sqrt{\frac{\mu}{p_c}} \sin\theta_c \tag{A2}$$
$$A_{25} = \frac{V_{rc} a_c q_{2c}}{p_c} - \sqrt{\frac{\mu}{p_c}} \cos\theta_c$$

$$A_{32} = R_c$$
$$A_{36} = R_c \cos i_c \tag{A3}$$

$$A_{41} = -1.5 V_{tc} / a_c$$
$$A_{42} = -V_{rc}$$
$$A_{44} = 2\sqrt{\frac{\mu}{p_c}} \cos\theta_c + \frac{3 V_{tc} a_c q_{1c}}{p_c} \tag{A4}$$
$$A_{45} = 2\sqrt{\frac{\mu}{p_c}} \sin\theta_c + \frac{3 V_{tc} a_c q_{2c}}{p_c}$$
$$A_{46} = V_{rc} \cos i_c$$

$$A_{53} = R_c \sin\theta_c$$
$$A_{56} = -R_c \sin i_c \cos\theta_c$$
$$A_{63} = (V_{tc} \cos\theta_c + V_{rc} \sin\theta_c) \tag{A5}$$
$$A_{66} = (V_{tc} \sin\theta_c - V_{rc} \cos\theta_c) \sin i_c$$

$$A_{11}^{-1} = \left(\frac{a_c}{R_c}\right)\left[4 + 6\left(\frac{a_c}{R_c}\right)(q_{1c}\cos\theta_c + q_{2c}\sin\theta_c) + 4\left(\frac{V_{rc}}{V_{tc}}\right)\left(\frac{a_c}{R_c}\right)(q_{1c}\sin\theta_c - q_{2c}\cos\theta_c)\right]$$

$$A_{12}^{-1} = 2\left(\frac{a_c}{R_c}\right)^2 (q_{1c}\sin\theta_c - q_{2c}\cos\theta_c)/\dot{\theta}_c$$

$$A_{13}^{-1} = -2\left(\frac{a_c}{R_c}\right)\left(\frac{V_{rc}}{V_{tc}}\right)\left[1 + 2\left(\frac{a_c}{R_c}\right)(q_{1c}\cos\theta_c + q_{2c}\sin\theta_c) + \left(\frac{V_{rc}}{V_{tc}}\right)\left(\frac{a_c}{R_c}\right)(q_{1c}\sin\theta_c - q_{2c}\cos\theta_c)\right]$$

$$A_{14}^{-1} = \left(\frac{2a_c}{R_c\dot{\theta}_c}\right)\left[1 + 2\left(\frac{a_c}{R_c}\right)(q_{1c}\cos\theta_c + q_{2c}\sin\theta_c) + \left(\frac{V_{rc}}{V_{tc}}\right)\left(\frac{a_c}{R_c}\right)(q_{1c}\sin\theta_c - q_{2c}\cos\theta_c)\right]$$

(A6)

$$A_{23}^{-1} = 1/R_c$$
$$A_{25}^{-1} = -A_{65}^{-1}\cos i_c \quad (A7)$$
$$A_{26}^{-1} = -A_{66}^{-1}\cos i_c$$

$$A_{35}^{-1} = \left(\sin\theta_c - (V_{rc}/V_{tc})\cos\theta_c\right)/R_c \quad (A8)$$
$$A_{36}^{-1} = \cos\theta_c/V_{tc}$$

$$A_{41}^{-1} = \sqrt{\frac{p_c}{\mu}}\left[3\cos\theta_c + 2(V_{rc}/V_{tc})\sin\theta_c\right]\dot{\theta}_c$$

$$A_{42}^{-1} = \sqrt{\frac{p_c}{\mu}}\sin\theta_c$$

$$A_{43}^{-1} = -\frac{1}{R_c}\left[\frac{R_c}{p_c}(q_{1c}\sin\theta_c - q_{2c}\cos\theta_c)^2\sin\theta_c + q_{1c}\sin 2\theta_c - q_{2c}\cos 2\theta_c\right] \quad (A9)$$

$$A_{44}^{-1} = \sqrt{\frac{p_c}{\mu}}\left[2\cos\theta_c + (V_{rc}/V_{tc})\sin\theta_c\right]$$

$$A_{45}^{-1} = -q_{2c}\cot i_c\left[\cos\theta_c + (V_{rc}/V_{tc})\sin\theta_c\right]/R_c$$

$$A_{46}^{-1} = q_{2c}\cot i_c \sin\theta_c/V_{tc}$$

$$A_{51}^{-1} = \sqrt{\frac{p_c}{\mu}}\left[3\sin\theta_c - 2(V_{rc}/V_{tc})\cos\theta_c\right]\dot{\theta}_c$$

$$A_{52}^{-1} = -\sqrt{\frac{p_c}{\mu}}\cos\theta_c$$

$$A_{53}^{-1} = \frac{1}{R_c}\left[\frac{R_c}{P_c}(q_{1c}\sin\theta_c - q_{2c}\cos\theta_c)^2\cos\theta_c + q_{2c}\sin 2\theta_c + q_{1c}\cos 2\theta_c\right] \quad\text{(A10)}$$

$$A_{54}^{-1} = \sqrt{\frac{p_c}{\mu}}\left[2\sin\theta_c - (V_{rc}/V_{tc})\cos\theta_c\right]$$

$$A_{55}^{-1} = q_{1c}\cot i_c\left[\cos\theta_c + (V_{rc}/V_{tc})\sin\theta_c\right]/R_c$$

$$A_{56}^{-1} = -q_{1c}\cot i_c\sin\theta_c/V_{tc}$$

$$A_{65}^{-1} = -\left[\cos\theta_c + (V_{rc}/V_{tc})\sin\theta_c\right]/R_c\sin i_c$$
$$A_{66}^{-1} = \sin\theta_c/V_{tc}\sin i_c \quad\text{(A12)}$$

AAS 00-013

NONLINEAR DYNAMICS, TRAJECTORY GENERATION, AND ADAPTIVE CONTROL OF MULTIPLE SPACECRAFT IN PERIODIC RELATIVE ORBITS

Qiguo Yan, Guang Yang, Vikram Kapila and Marcio S. de Queiroz[*]

This paper considers the problem of relative position control for multiple spacecraft formation flying. Specifically, nonlinear dynamics describing the motion of a follower spacecraft relative to a leader spacecraft are developed for the case where the leader spacecraft is in an elliptical orbit. Next, a Lyapunov-based, nonlinear, adaptive control law is designed which guarantees global asymptotic convergence of the position tracking error in the presence of unknown, constant or slow-varying spacecraft masses and exogenous disturbance forces. In addition, a formation initialization constraint is developed for the ideal, unperturbed, periodic relative motion of the spacecraft formation which serves as a desired, relative motion trajectory. Simulation results are provided to illustrate the efficacy of the formation initialization methodology and the adaptive controller performance.

1. Introduction

Multiple spacecraft formation flying (MSFF) has been identified as an enabling technology by the U.S. Air Force and NASA for future space missions [1,3,10–12,16]. For example, U.S. Air Force's TechSat–21 program [1,12,16] and NASA's Earth Orbiter–I (EO–I) and the New Millennium Interferometer (NMI) programs [3,10,11,14], among others, rely on successful development and deployment of MSFF technologies. In particular, MSFF is expected to enable the Earth and space science missions, such as the distributed aperture radar, enhanced stellar optical interferometer, virtual co-observing and stereo-imaging platforms for space science and Earth observing, etc. For a recent collection of related research, the reader is referred to the proceedings of the Air Force–MIT workshop on spacecraft formation flying and micropropulsion [2].

---

[*] Department of Mechanical, Aerospace, and Manufacturing Engineering, Polytechnic University, Brooklyn, New York 11201.

For ideal MSFF, the initial conditions for the formation flying spacecraft must be chosen to enable the spacecraft to undergo periodic motions such that a relative spatial pattern persists for several orbits with minimal propellant expenditure [4]. Using the linearized dynamics of relative motion between a leader-follower spacecraft pair, *viz.*, the Hill's equations, recently, [16] identified the set of feasible initial conditions that annihilate the secular growth in time in the solution of Hill's equations; thus, yielding periodic relative motion between the leader-follower spacecraft pair. Based on the work in [16], spatial patterns for formation design have been proposed in [15]. Unfortunately, however, a caveat of [16] is that it is based on the linearization of nonlinear dynamics of spacecraft relative motion. In fact, it can be shown that the ideal, no-thrust, formation initialization methodology of [16] fails to hold the *designed* formations [15] for the nonlinear dynamics of spacecraft relative motion.

A majority of the MSFF control designs also utilize the aforementioned Hill's linearized relative motion equations to aid in the control synthesis [13, 18]. This linearized spacecraft relative motion dynamics model (also known as the Clohessy-Wiltshire model [6]) was originally developed for spacecraft rendezvous problem in circular orbits and is valid for short period maneuvers. Thus, to reiterate, formation initialization, formation deign, and formation control schemes based on Hill's equations are unlikely to yield good precision for MSFF in general elliptical orbits for long durations.

It is clear from the preceding discussion that there exists an urgent need for developing MSFF nonlinear dynamic modeling, formation initialization, and control schemes. Reviewing the current state of MSFF control research, [19] developed an exact model knowledge relative position controller with local asymptotic position tracking errors. More recently, [8] proposed a nonlinear, adaptive controller which ensures global asymptotic position tracking errors. However, the framework of [8] is based on the assumption that the leader spacecraft remains in a circular orbit.

In this paper, we develop the nonlinear dynamic model describing the relative positioning of MSFF for the case where the leader spacecraft is in an elliptical orbit. Next, a Lyapunov-based, nonlinear adaptive control law is designed which guarantees global asymptotic convergence of the position tracking error in the presence of unknown, constant or slow-varying spacecraft masses and exogenous disturbance forces. In addition, we propose a formation initialization scheme, which in the ideal case yields a no-thrust, periodic relative motion between the leader-follower spacecraft pair, and serves as a desired, relative motion trajectory. In comparison to the previous work [8, 15, 16], the proposed formation initialization, adaptive nonlinear control, and desired trajectory generation schemes exploit the orbital mechanics (barring the Earth's oblateness effects) to aid in formation maintenance and thus minimize propellant consumption. In fact, as demonstrated by the illustrative numerical simulation, the proposed controller utilizes the natural orbits of the leader-follower spacecraft pair for formation keeping and exactly cancels the unknown exogenous perturbations.

2. System Model

We begin with the consideration of a MSFF system composed of two spacecraft, i.e., a *leader* spacecraft and a *follower* spacecraft. The leader spacecraft provides the basic reference motion trajectory and is considered to be in an ideal, elliptical orbit around the Earth. The follower spacecraft navigates in proximity of the leader spacecraft. Here, we will focus on MSFF such that the relative trajectory of the follower spacecraft with respect to the leader spacecraft is periodic, while the two spacecraft orbit a perfectly spherical Earth according to Keplarian Laws.

A schematic drawing of the MSFF system is given in Figure 1 where we make the following considerations *i*) the inertial coordinate frame $\{X, Y, Z\}$ is attached to the center of the Earth, *ii*) $R(t) \in \mathbb{R}^3$ denotes the position vector from the origin of the inertial coordinate frame to the leader spacecraft, *iii*) a right-hand coordinate frame $\{x_l, y_l, z_l\}$ is attached to the leader spacecraft with the x_l-axis perpendicular to the instantaneous vector R and contained in the orbital plane of the leader spacecraft, the y_l-axis pointing along the direction of the vector R, and the z_l-axis pointing along the orbital angular momentum of the leader spacecraft, and *iv*) $\rho(t) \in \mathbb{R}^3$ denotes the position vector from the origin of the moving coordinate frame $\{x_l, y_l, z_l\}$ to the follower spacecraft.

The nonlinear position dynamics of the leader and follower spacecraft in the inertial coordinate frame $\{X, Y, Z\}$ are given by [8, 18]

$$m_l \ddot{R} + m_l (M + m_l) G \frac{R}{\|R\|^3} + F_{dl} = u_l \qquad (2.1)$$

and
$$m_f(\ddot{R}+\ddot{\rho}) + m_f(M+m_f)G\frac{R+\rho}{\|R+\rho\|^3} + F_{df} = u_f, \qquad (2.2)$$

respectively, where m_l, m_f are the masses, $F_{dl}, F_{df} \in \mathbb{R}^3$ are disturbance force vectors, and $u_l(t), u_f(t) \in \mathbb{R}^3$ are the actual control input vectors of the leader and follower spacecraft, respectively, M is the Earth's mass, and G is the universal gravity constant. Since $M \gg m_l, m_f$, (2.1) and (2.2) can be simplified to

$$m_l \ddot{R} + m_l \mu \frac{R}{\|R\|^3} + F_{dl} = u_l \qquad (2.3)$$

and

$$m_f(\ddot{R}+\ddot{\rho}) + m_f \mu \frac{R+\rho}{\|R+\rho\|^3} + F_{df} = u_f, \qquad (2.4)$$

respectively, where $\mu \triangleq MG$. After some simple algebraic manipulations on (2.3) and (2.4), the dynamic equation describing the position of the follower spacecraft relative to the leader spacecraft in the coordinate frame $\{X, Y, Z\}$ can be written as

$$m_f \ddot{\rho} + m_f \mu \left(\frac{R+\rho}{\|R+\rho\|^3} - \frac{R}{\|R\|^3} \right) + \frac{m_f}{m_l} u_l + F_{df} - \frac{m_f}{m_l} F_{dl} = u_f. \qquad (2.5)$$

In the above model, we consider that the spacecraft masses vary slowly in time due to fuel consumption and payload variations. Furthermore, we consider that the disturbance forces result from solar radiation, aerodynamics, and magnetic fields, and hence, are also slow time-varying quantities [8,19]. As such, we can assume m_l, m_f are constant parameters and F_{dl}, F_{df} are constant vectors.

To write the dynamics of (2.5) in terms of the moving coordinate frame $\{x_l, y_l, z_l\}$, we must obtain an expression for $\ddot{\rho}$ in the moving coordinate frame $\{x_l, y_l, z_l\}$. Hence, we consider the homogeneous form of (2.3) given by

$$\ddot{R} + \mu \frac{R}{\|R\|^3} = 0. \qquad (2.6)$$

Using a polar coordinate frame fixed at the center of the Earth (see the R-θ coordinate frame in Figure 1), the motion of the leader spacecraft given by the vector differential equation (2.6) can be alternatively characterized by the planar dynamics

$$\ddot{r}_l - r_l \dot{\theta}^2 + \mu \frac{r_l}{r_l^3} = 0, \qquad (2.7)$$

$$r_l \ddot{\theta} + 2\dot{r}_l \dot{\theta} = 0, \qquad (2.8)$$

where $r_l \triangleq \|R\|$ and θ represents the true anomaly of the leader spacecraft. Employing standard orbital mechanics techniques, simple manipulations of (2.7), (2.8) yield

$$r_l(t) = \frac{a_l(1 - e_l^2)}{1 + e_l \cos \theta(t)}, \tag{2.9}$$

$$\dot{\theta}(t) = \frac{n(1 + e_l \cos \theta(t))^2}{(1 - e_l^2)^{3/2}}, \tag{2.10}$$

where a_l is the semi-major axis of the elliptical orbit of the leader spacecraft, e_l is the orbital eccentricity of the leader spacecraft, and n is the average orbital angular velocity defined by

$$n \triangleq \frac{2\pi}{T}, \tag{2.11}$$

with T as the orbital period. Finally, differentiating (2.10), we obtain

$$\ddot{\theta}(t) = \frac{-2n^2 e_l (1 + e_l \cos \theta(t))^3 \sin \theta(t)}{(1 - e_l^2)^3}. \tag{2.12}$$

Next, note that the relative position vector $\rho(t)$ expressed in $\{x_l, y_l, z_l\}$ is given by

$$\rho = x\hat{i}_l + y\hat{j}_l + z\hat{k}_l, \tag{2.13}$$

where $\hat{i}_l, \hat{j}_l, \hat{k}_l$ denote the unit vectors, while the angular velocity of the moving coordinate frame $\{x_l, y_l, z_l\}$ is given by $\dot{\theta}\hat{k}_l$. Hence, the relative acceleration vector $\ddot{\rho}(t)$ is given by

$$\ddot{\rho} = (\ddot{x} - 2\dot{\theta}\dot{y} - \dot{\theta}^2 x - \ddot{\theta}y)\hat{i}_l + (\ddot{y} + 2\dot{\theta}\dot{x} - \dot{\theta}^2 y + \ddot{\theta}x)\hat{j}_l + \ddot{z}\hat{k}_l. \tag{2.14}$$

After substituting the right-hand side of (2.14) into (2.5), the nonlinear position dynamics of the follower spacecraft relative to the leader spacecraft can be arranged into the following advantageous form

$$m_f \ddot{q} + C(\dot{\theta})\dot{q} + N(q, \dot{\theta}, \ddot{\theta}, R, u_l) + F_d = u_f. \tag{2.15}$$

Note that, in (2.15), $q(t) \in \mathbb{R}^3$ is the relative position vector $q(t) \in \mathbb{R}^3$

$$q(t) \triangleq [x(t) \quad y(t) \quad z(t)]^T, \tag{2.16}$$

$C : \mathbb{R} \to \mathbb{R}^{3 \times 3}$ is the Coriolis-like matrix given by

$$C(\dot{\theta}) \triangleq 2m_f \dot{\theta} \begin{bmatrix} 0 & -1 & 0 \\ 1 & 0 & 0 \\ 0 & 0 & 0 \end{bmatrix}, \tag{2.17}$$

and $N : \mathbb{R}^3 \times \mathbb{R} \times \mathbb{R} \times \mathbb{R}^3 \times \mathbb{R}^3 \to \mathbb{R}^3$ is a nonlinear term defined as

$$N(q, \dot{\theta}, \ddot{\theta}, R, u_l) \triangleq \begin{bmatrix} m_f \mu \dfrac{x}{\|R+q\|^3} - m_f(\dot{\theta}^2 x + \ddot{\theta} y) + \dfrac{m_f}{m_l} u_{lx} \\ m_f \mu \left(\dfrac{y + \|R\|}{\|R+q\|^3} - \dfrac{1}{\|R\|^2} \right) - m_f(\dot{\theta}^2 y - \ddot{\theta} x) + \dfrac{m_f}{m_l} u_{ly} \\ m_f \mu \dfrac{z}{\|R+q\|^3} + \dfrac{m_f}{m_l} u_{lz} \end{bmatrix}, \quad (2.18)$$

with u_{lx}, u_{ly}, u_{lz} being the components of the leader control input vector u_l. In addition, in (2.15), $F_d \in \mathbb{R}^3$ is the composite, constant, disturbance force given by

$$F_d \triangleq F_{df} - \frac{m_f}{m_l} F_{dl}, \quad (2.19)$$

and R in the moving coordinate frame $\{x_l, y_l, z_l\}$ is given by

$$R = \begin{bmatrix} 0 & r_l & 0 \end{bmatrix}^T. \quad (2.20)$$

Finally, note that r_l, $\dot{\theta}$, and $\ddot{\theta}$ in (2.15)–(2.20) are computed from (2.9), (2.10), and (2.12), respectively.

The dynamic model of (2.15)–(2.19) has the following property which will be exploited in the subsequent adaptive control design. The left-hand side of the dynamic equation (2.15) can be linearly parameterized as

$$m_f \xi + C(\dot{\theta}) \dot{q} + N(q, \dot{\theta}, \ddot{\theta}, R, u_l) + F_d = W(\xi, \dot{q}, q, \dot{\theta}, \ddot{\theta}, R, u_l) \phi, \quad \forall \xi \in \mathbb{R}^3, \quad (2.21)$$

where $W : \mathbb{R}^3 \times \mathbb{R}^3 \times \mathbb{R}^3 \times \mathbb{R} \times \mathbb{R} \times \mathbb{R}^3 \times \mathbb{R}^3 \to \mathbb{R}^{3 \times 5}$ is the regression matrix which is composed of known functions and $\phi \in \mathbb{R}^5$ is the system's constant parameter vector. From the form of (2.15)–(2.19), we can define $W(\cdot)$ and ϕ as

$$W(\xi, \dot{q}, q, \dot{\theta}, \ddot{\theta}, R, u_l) \triangleq \begin{bmatrix} \xi_x - 2\dot{\theta}\dot{y} - \dot{\theta}^2 x - \ddot{\theta} y + \mu \dfrac{x}{\|R+q\|^3} & u_{lx} & 1 & 0 & 0 \\ \xi_y + 2\dot{\theta}\dot{x} - \dot{\theta}^2 y + \ddot{\theta} x + \mu \left(\dfrac{y + \|R\|}{\|R+q\|^3} - \dfrac{1}{\|R\|^2} \right) & u_{ly} & 0 & 1 & 0 \\ \xi_z + \mu \dfrac{z}{\|R+q\|^3} & u_{lz} & 0 & 0 & 1 \end{bmatrix}, \quad (2.22)$$

$$\phi \triangleq \begin{bmatrix} m_f & \dfrac{m_f}{m_l} & F_{dx} & F_{dy} & F_{dz} \end{bmatrix}^T, \quad (2.23)$$

where ξ_x, ξ_y, ξ_z and F_{dx}, F_{dy}, F_{dz} are the components of the vectors ξ and F_d, respectively.

3. Adaptive Control Law Design

In this section, we develop an adaptive feedback control law that asymptotically tracks a prespecified spacecraft relative position trajectory. In particular, we consider that a desired position trajectory $q_d(t) \in \mathbb{R}^3$ for the follower spacecraft relative to the leader spacecraft is given. Furthermore, we assume that $q_d(t)$ and its first two time derivatives are bounded functions of time. Now, an adaptive control law is to be designed for u_f such that $\lim_{t \to \infty} q(t) \to q_d(t)$.

In order to state the main result of this section, we define the following notation. The position tracking error $e(t) \in \mathbb{R}^3$ is defined as

$$e(t) \triangleq q_d(t) - q(t) \quad (3.1)$$

and filtered tracking error $\zeta(t) \in \mathbb{R}^3$ is defined as

$$\zeta(t) \triangleq \dot{e}(t) + \Lambda e(t), \quad (3.2)$$

where $\Lambda \in \mathbb{R}^{3 \times 3}$ is a constant, diagonal, positive-definite matrix. Finally, a new regression matrix $W_d(\cdot)$ is defined as

$$W_d(\cdot) \triangleq W(\ddot{q}_d + \Lambda \dot{e}, \dot{q}, q, \dot{\theta}, \ddot{\theta}, R, u_l), \quad (3.3)$$

where $W(\cdot)$ was defined in (2.22).

Theorem 1. Let $K \in \mathbb{R}^{3 \times 3}$ and $\Gamma \in \mathbb{R}^{5 \times 5}$ be constant, diagonal, positive-definite matrices. Then, the adaptive control law

$$u_f = W_d(\cdot)\hat{\phi} + K\zeta, \quad (3.4)$$

$$\dot{\hat{\phi}} = \Gamma W_d^T(\cdot)\zeta, \quad (3.5)$$

where $\hat{\phi} \in \mathbb{R}^5$ denotes the dynamic estimate of the unknown parameter vector ϕ defined in (2.23), ensures the global asymptotic convergence of the relative position and relative velocity tracking errors as illustrated by

$$\lim_{t \to \infty} e(t), \dot{e}(t) = 0. \quad (3.6)$$

Proof. We begin by rewriting the spacecraft relative position dynamics (2.15) in terms of the filtered tracking error variable (3.2). To this end, differentiating (3.2) with respect to time, multiplying both sides of the resulting equation by m_f, using $\ddot{e} = \ddot{q}_d - \ddot{q}$ from (3.1), and rearranging terms yields

$$m_f \dot{\zeta} = m_f (\ddot{q}_d + \Lambda \dot{e}) - m_f \ddot{q}. \tag{3.7}$$

Next, we substitute for $m_f \ddot{q}$ from (2.15) in (3.7), to obtain

$$m_f \dot{\zeta} = m_f(\ddot{q}_d + \Lambda \dot{e}) + C(\dot{\theta})\dot{q} + N(q, \dot{\theta}, \ddot{\theta}, R, u_l) + F_d - u_f$$
$$= W_d(\cdot)\phi - u_f, \tag{3.8}$$

where we used (2.21) and (3.3). The above first-order, nonlinear, differential equation represents the open-loop dynamics of $\zeta(t)$.

Next, define the parameter estimation error vector $\tilde{\phi}(t) \in \mathbb{R}^5$ as

$$\tilde{\phi}(t) \triangleq \phi - \hat{\phi}(t). \tag{3.9}$$

Now, substituting (3.4) into (3.8) and using (3.9) yields the closed-loop dynamics for $\zeta(t)$ given by

$$m_f \dot{\zeta} = -K\zeta + W_d(\cdot)\tilde{\phi}. \tag{3.10}$$

Finally, note that differentiating (3.9) with respect to time and using (3.5), produces the closed-loop dynamics for the parameter estimation error

$$\dot{\tilde{\phi}} = -\Gamma W_d^T(\cdot)\zeta. \tag{3.11}$$

Now, we utilize the error systems of (3.10) and (3.11) along with the positive-definite, candidate Lyapunov function $V : \mathbb{R}^3 \times \mathbb{R}^5 \to \mathbb{R}$ defined by

$$V(\zeta, \tilde{\theta}) = \frac{1}{2}\zeta^T m_f \zeta + \frac{1}{2}\tilde{\phi}^T \Gamma^{-1} \tilde{\phi} \tag{3.12}$$

to prove the above stability result for the position and velocity tracking errors. Specifically, differentiating (3.12) with respect to time yields

$$\dot{V}(\zeta, \tilde{\phi}) = \zeta^T m_f \dot{\zeta} + \tilde{\phi}^T \Gamma^{-1} \dot{\tilde{\phi}}. \tag{3.13}$$

Now, substitution of the closed-loop dynamics (3.10) into (3.13), yields

$$\dot{V}(\zeta, \tilde{\phi}) = -\zeta^T K \zeta + \tilde{\phi}^T \left(W^T(\cdot)\zeta + \Gamma^{-1} \dot{\tilde{\phi}} \right). \tag{3.14}$$

Finally, substituting (3.11) into the parenthetical term of (3.14), we obtain

$$\dot{V}(\zeta, \tilde{\phi}) = -\zeta^T K \zeta \leq -\lambda_{\min}\{K\} \|\zeta\|^2 \leq 0, \qquad (3.15)$$

where $\lambda_{\min}\{\cdot\}$ denotes the minimum eigenvalue of a matrix.

Due to the form of (3.15), we know that $V(\zeta, \tilde{\phi})$ is either decreasing or constant. Since $V(\zeta, \tilde{\phi})$ of (3.12) is a non-negative function, we can conclude that $V(\zeta, \tilde{\phi}) \in \mathcal{L}_\infty$; hence, $\zeta(t) \in \mathcal{L}_\infty$ and $\tilde{\phi}(t) \in \mathcal{L}_\infty$. Since $\zeta(t) \in \mathcal{L}_\infty$, we can utilize Lemma 1.4 of [7] to show that $e(t), \dot{e}(t) \in \mathcal{L}_\infty$; hence, due to the boundedness of $q_d(t)$ and $\dot{q}_d(t)$, we can use (3.1) to conclude that $q(t), \dot{q}(t) \in \mathcal{L}_\infty$. Since $\tilde{\phi}(t) \in \mathcal{L}_\infty$ and ϕ is a constant vector, (3.9) can be used to show that $\hat{\phi}(t) \in \mathcal{L}_\infty$. From the above boundedness statements and the fact that $\ddot{q}_d(t)$ is assumed bounded, the definitions of (2.22) and (3.3) can be used to show that $W_d(\cdot) \in \mathcal{L}_\infty$. It is now easy to see from (3.4) that the control input $u_f(t) \in \mathcal{L}_\infty$. The above information can be applied to (2.15) and (3.10) to illustrate that $\ddot{q}(t), \dot{\zeta}(t) \in \mathcal{L}_\infty$. Thus, we have explicitly illustrated that all signals in the adaptive controller and system remain bounded during the closed-loop operation.

From (3.15), it is now easy to show that $\zeta(t) \in \mathcal{L}_2$. Since we have already proved that $\zeta(t), \dot{\zeta}(t) \in \mathcal{L}_\infty$, we can apply Barbalat's Lemma [7, 17] to conclude that

$$\lim_{t \to \infty} \zeta(t) = 0. \qquad (3.16)$$

Finally, Lemma 1.6 of [7] can be applied to (3.16) and (3.2) to obtain the result of (3.6). □

Remark 1. In the case that the parameter vector ϕ of (2.23) is perfectly known, the proposed control law (3.4) with $\hat{\phi} = \phi$ (i.e., an exact model knowledge controller) would ensure global exponential convergence of the position and velocity tracking errors in the sense that

$$\|\zeta(t)\| \leq \|\zeta(0)\| \exp\left(-\frac{\lambda_{\min}\{K\}}{m_f} t\right). \qquad (3.17)$$

This result can be proven by selecting a candidate Lyapunov function $V = \frac{1}{2}\zeta^T M_f \zeta$ and utilizing similar arguments as in the proof of Theorem 1 along with Lemma 1.5 of [7].

Note that the above stability result ensures global asymptotic tracking for arbitrary, sufficiently smooth, desired relative position trajectories $q_d(t)$. However, tracking an arbitrary spatial formation geometry for formation maintenance will require prohibitive fuel expenditure. Thus, for spacecraft formation maintenance, $q_d(t)$ must be generated to yield an ideal, naturally attractive, no-thrust, spatial formation geometry (see e.g., the following section). As a result, control effort will *only* be required to compensate for non-ideal conditions, e.g., disturbance forces. In contrast, for formation configuration maneuvers, $q_d(t)$ may represent a more aggressive trajectory; hence demanding significant control effort. The efficacy of our nonlinear control design for formation configuration maneuvers has recently been illustrated in [9].

4. Initialization Constraint for Formation Design

As discussed in the introduction, it has recently been shown [15, 16] that the solutions of the unperturbed Hill's equations of spacecraft relative motion yield periodic relative motion for initial conditions that annihilate secular growth in time. Inspired by the results of [15, 16], in this paper we analyze the feasibility of periodic relative motion resulting via the unperturbed relative motion dynamics

$$m_f \ddot{q} + C(\dot{\theta})\dot{q} + N(q, \dot{\theta}, \ddot{\theta}, R, u_l) = 0, \qquad (4.1)$$

with $u_l \equiv 0$.

We begin by considering the homogeneous form of (2.3) given by

$$\ddot{R}(t) + \mu \frac{R(t)}{\|R(t)\|^3} = 0, \qquad \forall R \in \mathbb{R}^3. \qquad (4.2)$$

It follows from [5] that (4.2) can be manipulated to yield

$$\frac{v^2(t)}{2} - \frac{\mu}{\|R(t)\|} = -\frac{\mu}{2a}, \qquad (4.3)$$

where $v \triangleq \|\dot{R}\|$ and a is the semi-major axis of the spacecraft elliptical orbit.

Now, consider a pair of leader-follower spacecraft in an ideal, central gravitational force system. Furthermore, let the leader and follower spacecraft be in natural, elliptical orbits with orbital periods T_l and T_f, respectively. Then, the relative motion between the two spacecraft will exhibit periodic characteristic if and only if $T_l = kT_f$ or $T_f = kT_l$, $k = 1, 2, \ldots$. In this paper, we will focus on the case where $T_l = T_f$, i.e., $k = 1$. For the cases where $k \geq 2$, the relative distance between the leader and follower spacecraft will be quite large, and are not of particular interest for spacecraft formation flying.

Using Keplar's third law, we have that

$$\frac{a^3}{T^2} = \frac{\mu}{4\pi^2}, \tag{4.4}$$

which, with the constraint $T_l = T_f = T$, yields $a_l = a_f = a$. Using (4.3), we obtain a constraint on the initial condition of the leader-follower spacecraft pair which yields periodic, desired, relative motion between the two spacecraft. Specifically, the aforementioned initial condition constraint is given by

$$v_l^2(0) - \frac{2\mu}{\|R(0)\|} = v_f^2(0) - \frac{2\mu}{\|R(0) + q(0)\|}, \tag{4.5}$$

where $v_f \triangleq \|\dot{R} + \dot{q}\|$.

Next, let $\mathbf{v}_l$ and $\mathbf{v}_f$ denote the absolute velocities of the leader and follower spacecraft, respectively, expressed in the moving coordinate frame $\{x_l, y_l, z_l\}$. In addition, let $\mathbf{v}_{rel}$ denote the velocity of the follower spacecraft relative to (as measured in) the moving coordinate frame $\{x_l, y_l, z_l\}$. Finally, let $\mathbf{v}_{l0}$, $\mathbf{v}_{f0}$, $\mathbf{v}_{rel0}$, θ_0, and q_0 denote $\mathbf{v}_l(0)$, $\mathbf{v}_f(0)$, $\mathbf{v}_{rel}(0)$, $\theta(0)$, and $q(0)$, respectively. Then, it follows, that

$$\mathbf{v}_{f0} = \mathbf{v}_{l0} + \mathbf{v}_{rel0} + \dot{\theta}_0 \hat{k}_l \times q_0, \tag{4.6}$$

where

$$\mathbf{v}_{l0} = -r_{l0}\dot{\theta}_0 \hat{\imath}_l + \dot{r}_{l0}\hat{\jmath}_l, \tag{4.7}$$

$$\mathbf{v}_{rel0} = \dot{x}_0 \hat{\imath}_l + \dot{y}_0 \hat{\jmath}_l + \dot{z}_0 \hat{k}_l, \tag{4.8}$$

$$q_0 = x_0 \hat{\imath}_l + y_0 \hat{\jmath}_l + z_0 \hat{k}_l. \tag{4.9}$$

Substituting (4.6)–(4.9) into (4.5) and rearranging terms, yields the following initial condition constraint for (4.1) to exhibit periodic solutions

$$2\mu \left[\frac{1}{(x_0^2 + (y_0 + r_{l0})^2 + z_0^2)^{\frac{1}{2}}} - \frac{1}{r_{l0}} \right] = \dot{x}_0^2 + \dot{y}_0^2 + \dot{z}_0^2 + \dot{\theta}_0^2(x_0^2 + y_0^2 + 2y_0 r_{l0})$$
$$+ 2\dot{\theta}_0(-\dot{x}_0(r_{l0} + y_0) + x_0(\dot{r}_{l0} + \dot{y}_0)) + 2\dot{y}_0 \dot{r}_{l0}, \tag{4.10}$$

where r_{l0} and $\dot{r}_{l0}$ are determined from (2.9) and $\dot{\theta}_0$ is determined from (2.10).

5. Simulation Results

The adaptive control law described in Theorem 1 was simulated for a leader-follower MSFF problem having the following parameters [8, 18]

$$M = 5.974 \times 10^{24} \text{ kg}, \quad m_f = 410 \text{ kg}, \quad m_l = 1550 \text{ kg}, \quad G = 6.673 \times 10^{-11} \tfrac{\text{m}^3}{\text{kg} \cdot \text{s}^2},$$
$$u_l = F_{dl} = 0 \text{ N}, \quad F_d = [1.9106, -1.9106, -1.517] \times 10^{-5} \text{ N}. \tag{5.1}$$

The orbital elements and initial position for the leader spacecraft were selected as

$$T_l = 24 \text{ hours}, \quad 'e_l = 1.425 \times 10^{-5}, \quad \theta_0 = 0 \text{ rad}. \tag{5.2}$$

The initial condition for the desired, no-thrust, periodic trajectory $q_d(t)$ is selected using (4.10). Note that, given an initial relative position, we can choose different sets of initial relative velocity vectors for the follower spacecraft satisfying (4.10). Next, by integrating (4.1) for the sets of valid initial conditions, we can obtain different periodic, relative motion geometries. In this paper, the following initial conditions satisfying (4.10) were chosen to compute $q_d(t)$

$$q_d(0) = [-100 - 50 - 0.2] \text{ m}, \quad \dot{q}_d(0) = [-26.1805, 10.6574, 0.0398]^T \tfrac{\text{m}}{\text{hr}}. \tag{5.3}$$

Next, by integrating (4.1) with the initial condition (5.3), we obtained the desired, no-thrust, periodic trajectory $q_d(t)$ in the moving frame $\{x_l, y_l, z_l\}$.

In this simulation, the actual initial conditions for the follower spacecraft were chosen the same as the desired initial conditions while the parameter estimates were initialized to zero (i.e., $\hat{\phi}(0) = 0$). The control and adaptation gains were tuned by trial and error until a good tracking performance was achieved, and are as follows

$$\Gamma = \text{diag}\,(22.549, 0, 16.85, 14.10, 20.20)\,, \quad K = \text{diag}\,(30.345, 25.475, 31.361)\,,$$
$$\Lambda = \text{diag}\,(24.798, 17.50, 33.4)\,, \tag{5.4}$$

where the $(2, 2)$ element of Γ is zero since $u_l = 0$ in this simulation.

The phase portrait of the actual trajectory $q(t)$ of the follower spacecraft relative to the leader spacecraft is illustrated in Figure 2 where "*" represents the leader spacecraft at the origin of the moving frame. Figure 3 depicts the position tracking error $e(t)$ which approaches zero as $t \to \infty$. Four components of the parameter estimate vector $\hat{\phi}(t)$ are shown in Figure 4. It is interesting to note that the controller estimates the unknown, constant disturbance force F_d accurately and exactly cancels it as shown in Figure 4 and 5. In addition, note that the follower spacecraft tracks the ideal, desired, periodic trajectory exactly in steady-state, despite the presence of the unknown disturbance.

6. Conclusion

In this paper, we developed the dynamic equation of relative motion for MSFF when the leading spacecraft is in an elliptical orbit. This dynamic model has wider application and yields greater precision compared with the linearized Hill's equation model which is restricted to circular spacecraft orbits. Next, using a Lyapunov-based design and analysis framework, we developed an adaptive controller which was shown to guarantee global asymptotic position tracking errors in the presence of unknown spacecraft masses and exogenous perturbing forces. In addition, we derived an initial condition constraint which enables ideal, no-thrust, periodic relative motion trajectory. Simulation results were provided to demonstrate the efficacy of the formation initialization methodology and the adaptive controller performance.

Acknowledgment. This work was supported in part by the Air Force Office of Scientific Research under grant F49620-93-C-0063, the Air Force Research Lab/VAAD, WPAFB, OH, under IPA: Visiting Faculty Grant, and the NASA/New York Space Grant Consortium under grant 32310-5891.

References

1. http://www.vs.afrl.af.mil/factsheets/Tech-Sat21.html.

2. *Air Force Research Laboratory - Formation Flying and Micro-Propulsion Workshop*, Lancaster, CA, 1998.

3. Bauer, F. et al., "Satellite Formation Flying using An Innovative Autonomous Control System (AUTOCON) Environment," *Proceedings of the AIAA Guidance, Navigation, and Control Conference*, New Orleans, LA, 1997, pp. 657-666.

4. Chao, C. C., Pollard, J. E., and Janson, S. W., "Dynamics and Control of Cluster Orbits for Distributed Space Missions," *Proceedings of the AAS/AIAA Space Flight Mechanics Meeting*, 1999, Paper No. AAS99-126.

5. Chobotov, V.A. (Ed.), *Orbital Mechanics*, AIAA, Washington DC, 1996, pp. 31-33.

6. Clohessy, W.H. and Wiltshire, R.S., "Terminal Guidance System for Satellite Rendezvous," *Journal of Aerospace Science*, Vol. 27, No. 9, 1960, pp. 653–658.

7. Dawson, D.M., Hu, J., and Burg, T.C., *Nonlinear Control of Electric Machinery*, Marcel Dekker, New York, NY, 1998, pp. 1-19.

8. de Queiroz, M. S., Kapila, V., and Yan, Q., "Adaptive Nonlinear Control of Satellite Formation Flying," *Proceedings of the AIAA Guidance, Navigation, and Control Conf.*, Portland, OR, 1999, Paper No. 99-4270.

9. de Queiroz, M. S., Yan, Q. , Yang, G., and Kapila, V., "Global Output Feedback Tracking Control of Spacecraft Formation Flying with Parametric Uncertainty," *Proceedings of the IEEE Conference on Decision and Control*, Phoenix, AZ, December 1999, pp. 584–589.

10. Guinn, J.R., "Autonomous Navigation for the New Millennium Program Earth Orbiter 1 Mission," *Proceedings of the AIAA Guidance, Navigation, and Control Conference*, New Orleans, LA, 1997, pp. 612–617.

11. Lau, K. et al., "The New Millennium Formation Flying Optical Interferometer," *Proceedings of the AIAA Guidance, Navigation, and Control Conference*, New Orleans, LA, 1997, pp. 650–656.

12. Leitner, J., Beck, J., and Bell, K., "Advanced Guidance, Navigation, and Control for Remote Sensing," 1997, AIAA Paper.

13. Redding, D.C., Adams, N.J., and Kubiak, E.T., "Linear-Quadratic Stationkeeping for the STS Orbiter," *Journal of Guidance, Control, and Dynamics*, Vol. 12, No. 2, 1989, pp. 248–255.

14. Robertson, A., Corazzini, T., and How, J.P., "Formation Sensing and Control Technologies for a Separated Spacecraft Interferometer," *Proceedings of the American Control Conference*, Philadelphia, PA, 1998, pp.1574–1579.

15. Sabol, C., Burns, R., and McLaughlin, C., "Formation Flying Design and Evolution," *Proceedings of the AAS/AIAA Space Flight Mechanics Meeting*, 1999.

16. Sedwick, R. J., Wong, E. M. C., and Miller, D. W., "Exploiting Orbital Dynamics and Micropropulsion for Aperture Synthesis using Distributed Satellite Systems: Applications to TechSat 21," 1998, AIAA Paper No. 98-5289.

17. Slotine J.J. and Li, W., *Applied Nonlinear Control* , Prentice Hall, Englewood Cliff, NJ, 1991, pp. 122-126.

18. Vassar, R.H. and Sherwood, R.B., "Formationkeeping for a Pair of Satellites in a Circular Orbit," *Journal of Guidance, Control, and Dynamics*, Vol. 8, No. 2, 1985, pp. 235–242.

19. Wang, P.K.C. and Hadaegh, F.Y., "Coordination and Control of Multiple Microspacecraft Moving in Formation," *Journal of Astronautical Sciences*, Vol. 44, No. 3, 1996, pp. 315–355.

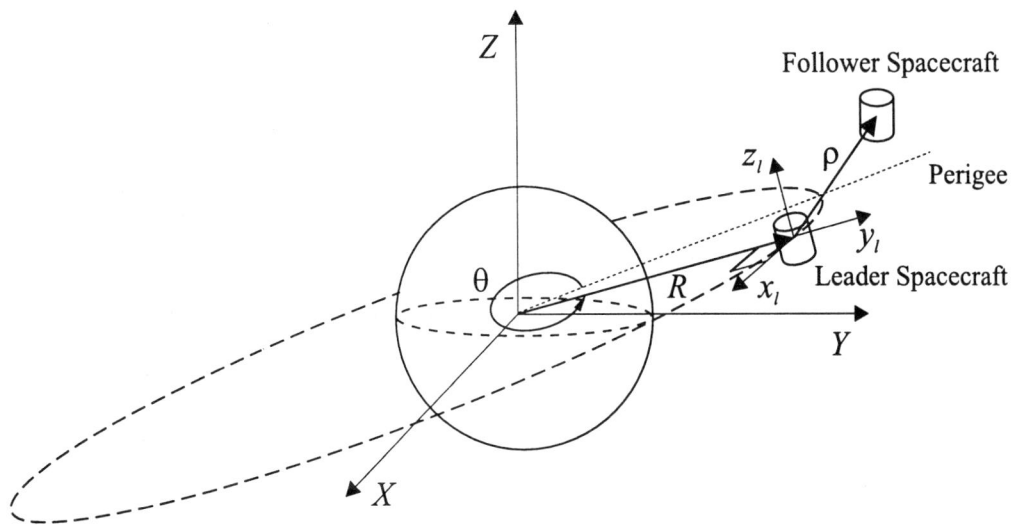

Figure 1: Schematic representation of the MSFF system

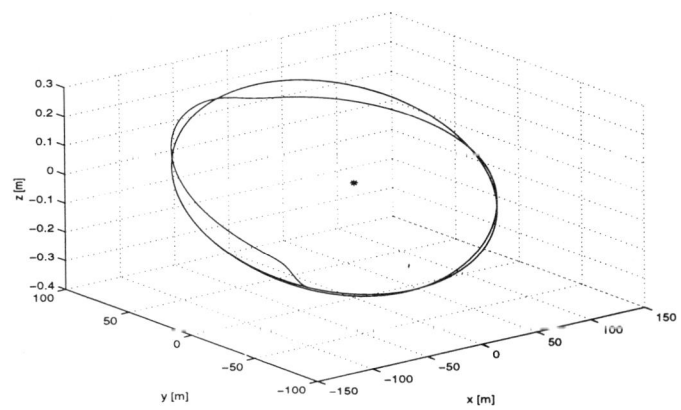

Figure 2: Actual trajectory of follower spacecraft relative to leader spacecraft ('*' denotes the leader spacecraft)

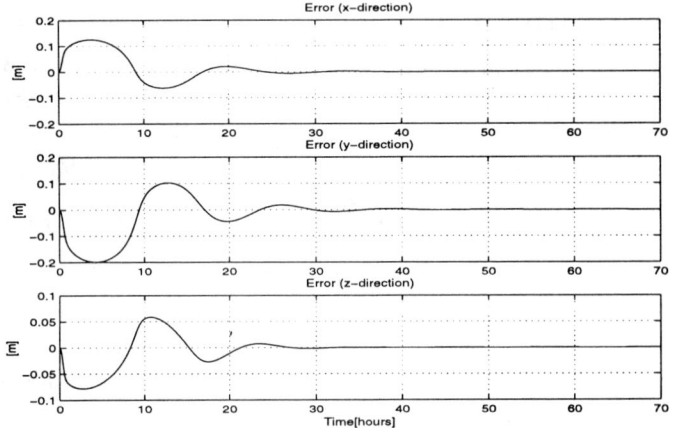

Figure 3: Position tracking error

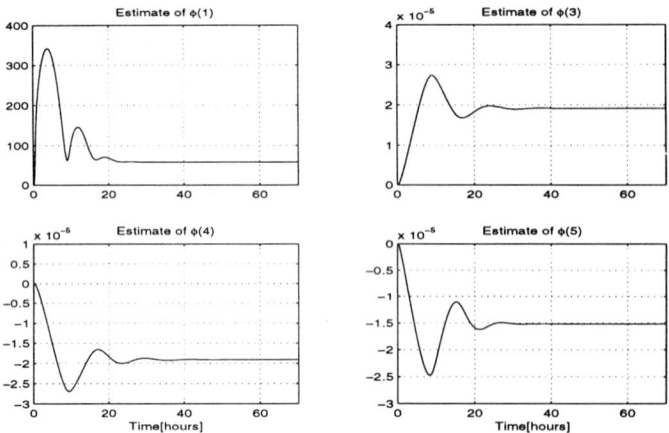

Figure 4: Sample of parameter estimates

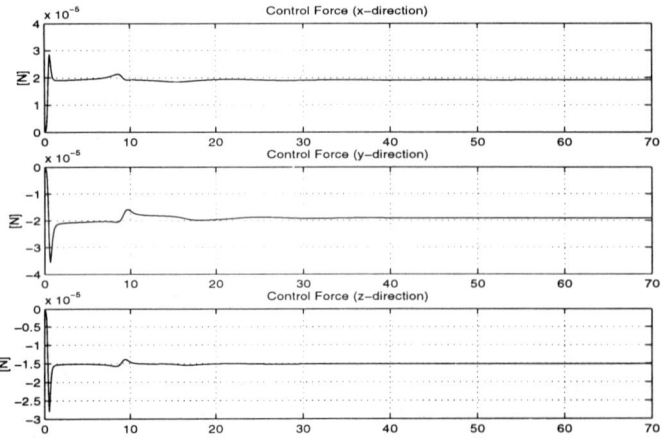

Figure 5: Control Input in $x, y,$ and z directions

AAS 00-014

VALIDATING A FORMATION FLYING CONTROL SYSTEM DESIGN: THE GRACE PROJECT EXPERIENCE

H. D. Stevens,[*] Jack Rodden,[*] Phil Morton[†] and Matthias Fehrenbach[‡]

The Gravity Recovery and Climate Experiment (GRACE) mission will produce a new model of the Earth's gravity field with unprecedented accuracy every 15-30 days for a period of 5 years. The GRACE mission involves flying two satellites in a tandem formation spaced 100 to 500 km apart. The mission uses two co-orbiting satellites which are themselves the instrument. Variations in the Earth's gravity field are manifested in differential orbit perturbations on the two co-orbiting satellites. Inter-satellite range variations are measured using a high frequency inter-satellite cross-link with micron-level precision. Precision 3 axes accelerometer measurements are processed to remove non-gravitational contributions to the range change. The GRACE Attitude and Orbit Control System uses star tracker attitude measurements and GPS based orbit determination to precisely point the two spacecraft at each other using cold gas thrusters and magnetic torque rods. The program is cosponsored by NASA and the German Space Agency, DLR, and is being developed by an international U.S.-German design team. It is scheduled for launch in mid 2001.

Formation flying missions couple not only the attitude and orbit dynamics of a single spacecraft, but of all spacecraft in the formation, complicating the attitude control verification and testing process. This paper provides an overview of the GRACE mission and Attitude & Orbit Control System (AOCS) design, and discusses the tools and processes used by the multi-national, multi-company design team to demonstrate that the control system and flight software meet the performance and robustness requirements. Special emphasis is placed on the effects of different orbit models required by the formation-flying mission.

Introduction

In 1997, the Gravity Recovery and Climate Experiment (GRACE) mission was selected by NASA for development as part of the Earth System Science Pathfinders (ESSP) program under a new Office of Mission to Planet Earth. GRACE will produce Earth gravimetric measurements with an unprecedented accuracy by using the two satellites, following each other in the same orbital plane, as the primary instrument. Direct measurement of non-gravitational forces by an advanced accelerometer further enables this unprecedented accuracy to be achieved. A microwave RF link connects the two satellites enabling measurement of both the exact separation distance, and it's rate of change, to an accuracy of better than 1 μm/sec. In this manner, the satellites themselves become the experiment, allowing a precise 'snapshot' of the gravity field to be measured about every two weeks of the 5 year mission during which the orbit altitude decreases from approximately 500 km to 300 km. This precision allows scientists to use the GRACE mission to 'weigh' various parts of the Earth system, learning about the distribution and changes in ocean mass, the growth or shrinking of the polar ice sheets, the amount of water in underground aquifers and other issues profoundly affecting climate change. It also will provide a never-before-available perspective on global ocean circulation and the time variability of Earth's overall shape.

In addition to the science requirements for the proper functioning of both satellites, mission cost constraints, which dictate ground operations from two German ground stations, place significant emphasis

---

[*] Space Systems/Loral, 3825 Fabian Way, Palo Alto, California 94303.
E-mail: stevensh@ssd.loral.com; roddenj@ssd.loral.com.

[†] Jet Propulsion Laboratory, 4800 Oak Grove Drive, Pasadena, California 91109-8099.

[‡] Dornier Satellitensysteme, D-88039 Friedrichshafen, Germany.

Figure 1: GRACE One Way Formation Flying Mission

on highly reliable and robust Attitude & Orbit Control System (AOCS) design and software implementation. The AOCS design and verification process is both complicated and strengthened by the multi-national, multi-company team developing the GRACE spacecraft. This international team is led by JPL, who is responsible for the systems integration and development of the instrument subsystems. Space Systems/Loral is responsible for the AOCS design and Dornier Satellitensystem (DSS) is responsible for spacecraft design and the flight software, including implementation of the AOCS design.

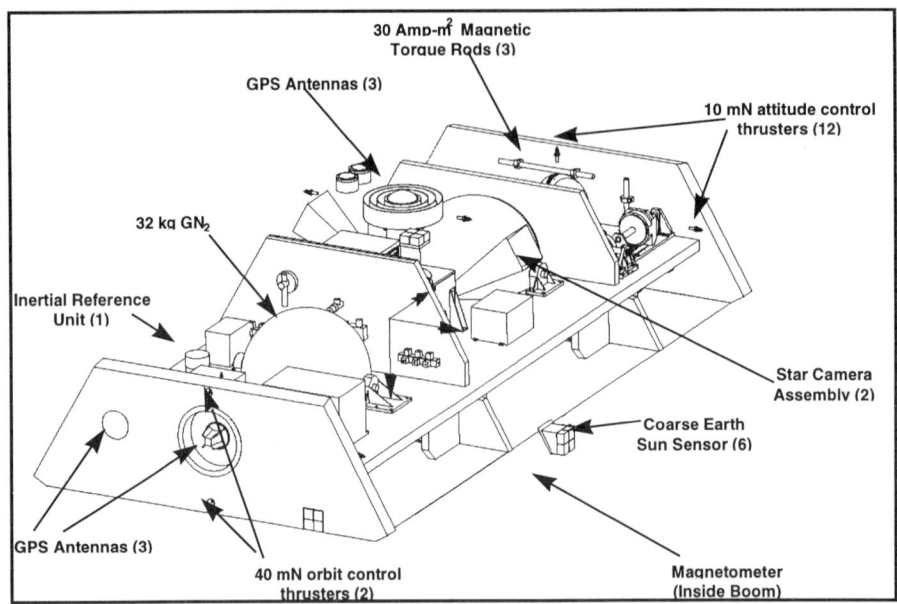

Figure 2: GRACE Satellite Configuration

The validation efforts are subdivided into four categories: design validation, mode and component software verification, software integration tests, and subsystem interface tests. Design validation tasks use systematic simulations of the AOCS modes and mode transitions to demonstrate that the design meets the mission requirements. The design validation results are then used as the basis for the software implementation verification, which demonstrates proper implementation of the design and illuminates any timing issues not accommodated in the initial design. Once the control algorithms are correctly implemented, the subsystem software is fully integrated in the system software on a flight-like processor.

Test of this integration process is completed with mission scenario testing on an integration test bed. Finally, subsystem interface testing is used to demonstrate the proper configuration of the flight system. The entire validation, verification, and testing effort as completed for the two-spacecraft formation is required to demonstrate that the AOCS meets the mission requirements.

Satellite Configuration

The GRACE satellites are tetrahedral in shape, approximately 2 meters in length, 1 meter in height, and 1.5 meters in width. The cost and mass constraints dictate that a simple thruster based control system be used for precision attitude control. However, lifetime constraints dictate that the mass expulsion system be augmented with magnetic torque rods, to the greatest extent possible. As shown in figure 2, the GRACE attitude and orbit control system actuators consist of 12 10 milli-newton GN_2 thrusters, 2 40 milli-newton GN_2 orbit control thrusters, and 3 30 amp-m^2 magnetic torque rods. The GRACE sensor suite consists of a three axis inertial measurement unit, a Coarse Earth Sun Sensor for use in acquisition and contingency modes, 2 star camera assemblies and associated processing to function as a star tracker, and a geodetic grade GPS receiver. At launch the satellite has a mass of 425 kg, including 32 kg of GN_2 propellant.

GRACE Attitude Control

The attitude control approach for GRACE is an implementation of a "one way" formation-flying concept [1]. Fundamentally, the attitude for each spacecraft is controlled to point the science payload at the "expected" position of the "other" spacecraft. Because there is no direct feedback between the two spacecraft, this approach is also known as "following the dot." The ground controllers are responsible for calculating the required orbit parameters, and uploading those values to the respective spacecraft each day. This approach reduces the on-board computations and sensors, thereby enabling a low-cost mission.

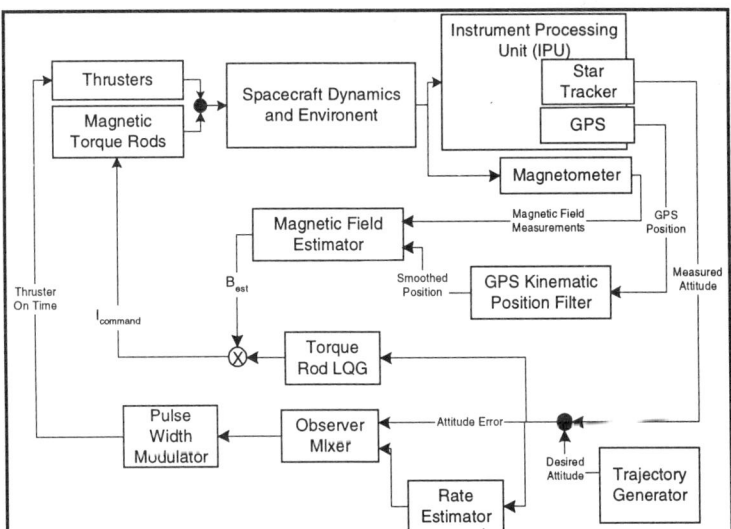

Figure 3: Overview of Precision Pointing Attitude Control System for GRACE Science

Figure 3 shows a block diagram of the attitude control system required to meet the science objectives of the GRACE mission. Cost, power, and mass constraints dictate that a thruster based attitude control system is required. However, GRACE uses magnetic torque rods to improve the propellant efficiency by off loading the thrusters when possible. Clearly, the magnetic torque rods always have one direction that is uncontrollable. However, the combination of the two systems yields a simple, robust design that enables the fine pointing required for the GRACE mission. A complete discussion of the control system design can be found in Reference [2].

The GRACE implementation of a "one-way" formation flying concept uses orbit information for both spacecraft to generate the required attitude profile. In this manner, the attitude control for each spacecraft is coupled with the orbit dynamics of the formation. This coupling is accomplished using the Trajectory Generator, which computes the required attitude to point "THIS" spacecraft at the expected position of the "OTHER" spacecraft.

The GRACE pointing budget has been developed with a requirement on the knowledge of the relative position between the two satellites of 50 meters or better. Since both spacecraft are equipped with geodetic grade GPS receivers, the obvious approach is to use GPS as a real-time sensor for this purpose. However, examination of the position accuracy statistics, and the cost/complexity of incorporating communication on the cross link, lead the system designer to quickly discount this "obvious" approach. Trade studies completed for the GRACE mission show that a simple, well-known orbit model (SGP4), applied to both satellites, provides a system that meets the relative position requirement of less than 50 meters. The SGP4 model generates the expected position for the center of mass of each satellite. As shown in Figure 4, the controller for "THIS" satellite must not only target the expected position of the "OTHER" satellite but in addition follow a superimposed bias and sinusoidal profile for calibration purposes. The target position bias and calibration sinusoid are computed in the OTHER orbit frame, rotated to the inertial frame, and added to the position of the OTHER satellite to create the Target Position. This target position is then used, in combination with the position of THIS satellite, to compute the desired attitude quaternion.

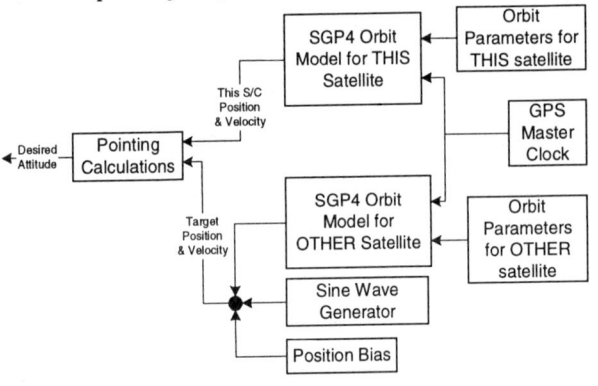

Figure 4: Trajectory Generator Block Diagram

This coupling of the attitude dynamics and control with orbit knowledge from BOTH spacecraft complicates the validation and verification efforts. Multiple orbit models must be used, with the appropriate uncertainties applied to the data. For example, the "true" orbit information must be available for both spacecraft. In addition, the corresponding orbit parameters as derived from the day-old GPS measurements must also be available for the Trajectory Generator. Thus, for one-way formation flying missions, orbit information complicates the AOCS design, validation, and testing efforts.

Validation and Verification Environments

To enable the validation and verification process, several environments have been developed. This section introduces the objectives and configurations for each of the validation and verification environments.

Mode simulations: Clearly the cornerstone for the development of any attitude control system are the simulations for the individual control modes. These mode specific simulations, developed using Matlab and Simulink, are used for mode development, verification of stability analysis results, and algorithm documentation. These simulations contain the most detailed models and information for each mode.

Multi-mode simulation: Because control algorithms are being designed in one country and implemented in flight code in another, the AOCS design team is developing a multi-mode simulation to assist in the "over seas" software verification process. The simulator demonstrates the performance of the attitude control system through both mode transitions and the fault detection logic flow. This simulation, developed using Matlab, Simulink, and Stateflow, enables the design team to model and simulate

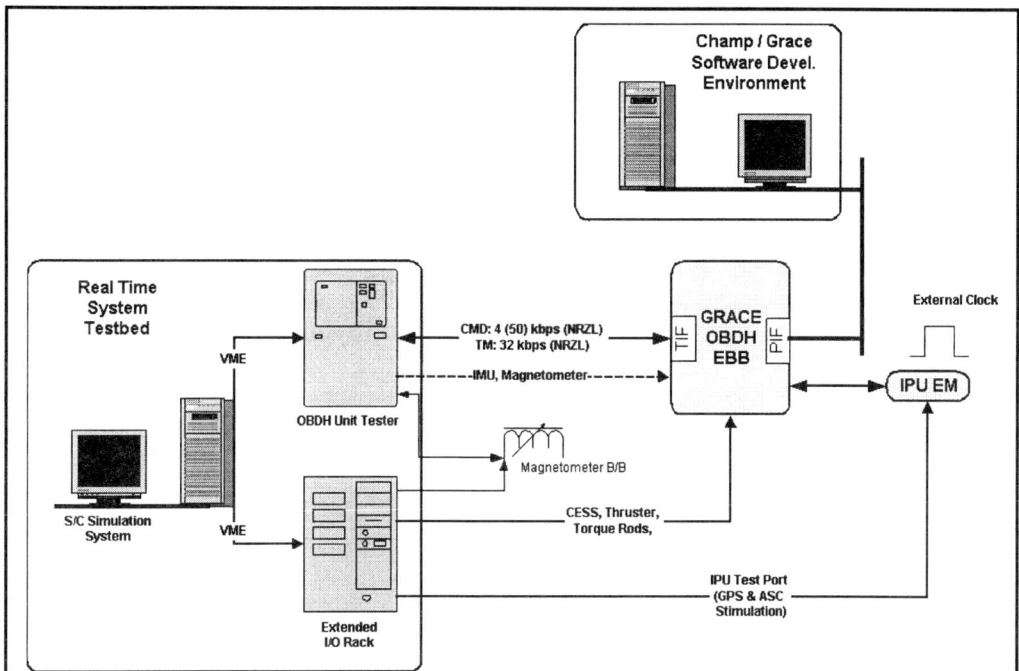

Figure 5: Flat-Sat block diagram

spacecraft performance without all of the flight software. This simulation is not only used in the verification process, but also to support anomaly resolution in the Launch and Early Operations Phase.

Software Development Environment: A second cornerstone in the software development process is the software development environment. The flight software, written in ADA, is tested for algorithmic errors by comparison with the data produced by the multi-mode simulator. This data is referred to as a "foot print" of the input and output characteristics of each component. This development environment consists of a PC workstation and the TLD ADA compiler. No effort is made to simulate the performance of the flight processor at this level; therefore no data is available on the timing, memory, or absolute accuracy of the algorithms.

Real-time system test bed (Flat-Sat): This test bed includes an engineering model of the flight electronics, referred to as the On-Board Data Handling electronics or OBDH, and the OBDH unit tester (i.e. the electronics required to interact with the OBDH). Additional ground support equipment is provided to simulate all of the OBDH interfaces. At later stages of development, an engineering model of the Instrument Processing Unit (IPU) is incorporated, which provides GPS and Star Tracker data for the AOCS. This testbed is used to verify the AOCS ADA code data exchange with the OBDH application software and to verify memory and timing requirements on the OBDH software. Once the IPU is included, end-to-end and closed-loop testing of the entire software system can be completed. Figure 5 shows a block diagram of the real-time system test bed.

Validation, Verification, and Testing Process

The simulation and test environments are used to complete a battery of testing on both the AOCS design and the flight software in order to ensure the performance requirements are achieved and that the software will maintain the health and safety of the spacecraft. Figure 6 provides an overview of the process. The process is divided into six inter-related subprocesses. This section introduces the subprocesses and discusses the objectives, methods, and outputs for each step.

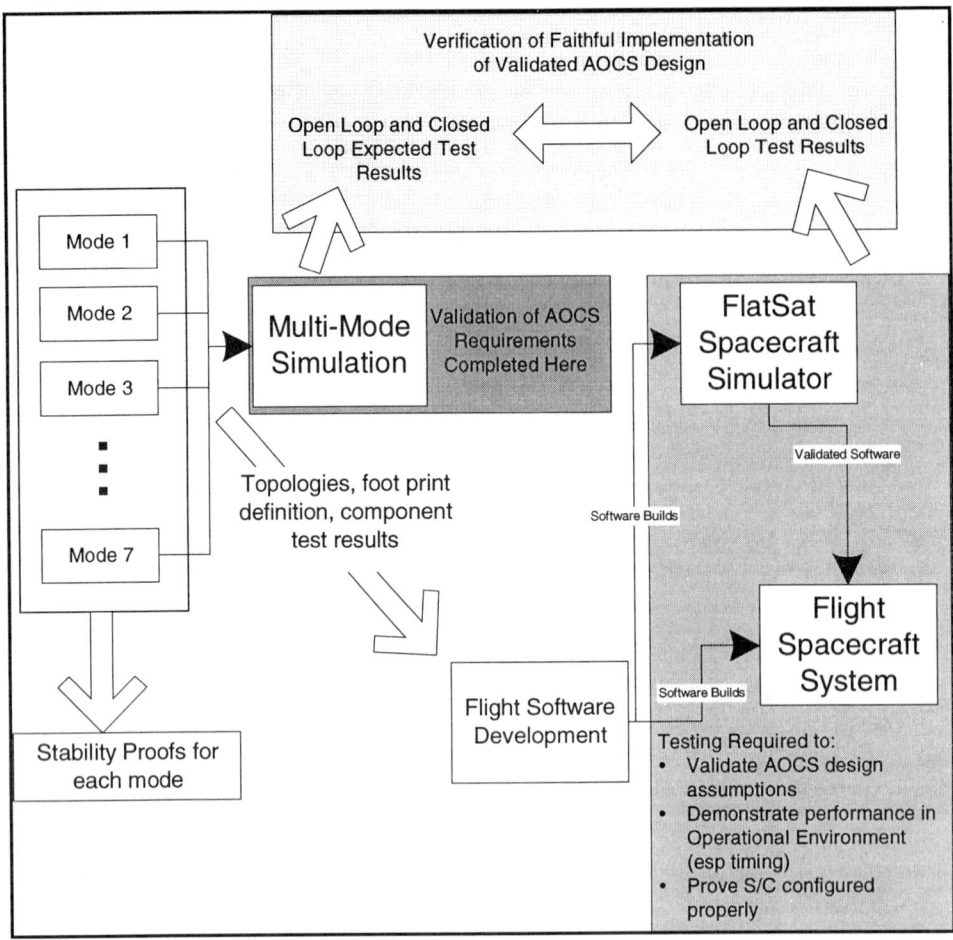

Figure 6: Overview of the GRACE Validation, Verification, and Testing Process

Design Validation: The objective of this step is to demonstrate that the AOCS design fully meets the science requirements and that the AOCS mode switching logic has been designed properly. Validation is accomplished by executing AOCS algorithm models, implemented in Matlab and Simulink, in the multi-mode simulation environment. Performance characteristics are measured and critical parameters (eg. gains) are fine-tuned. Key visibility data are identified and categorized according to availability (via downlink) during each test and operations phase. These data parameters form a performance "footprint" which can be compared as testing progresses. Once validated, this design is used as the reference for the remaining steps in the validation, verification, and testing process.

Component Verification: To effectively document the AOCS design and facilitate the design transfer from SS/L to DSS, the algorithms have been broken into small elements or components. As these elements are coded by DSS, they are tested against the reference design to ensure the algorithm has been properly understood and coded in the flight software. Testing takes place in the software development environment and components are tested at the unit level. This step involves significant interaction between SSL and DSS to ensure that all aspects of the algorithm have been demonstrated using the flight software. To complete verification of software mode and component implementation open- and closed-loop mode testing are performed as discussed next.

Open-loop mode testing: As all of the elements of specific modes are assembled, open loop mode testing is conducted. The objective of this mode testing is to demonstrate the proper algorithmic implementation from input to output using a reference "footprint", and also to demonstrate the timing and phase delay characteristics of the full flight software load. To accomplish the timing and phase delay measurements, open loop testing must be completed on the 1750 target processor (i.e. the real-time system test bed) using the full flight software load.

Closed-loop mode testing: The objective of this step is to demonstrate the performance of the closed loop spacecraft system and relate that performance back to the performance of the design simulations. This set of tests demonstrates the expected performance of the flight system, including all aspects of the SGP4 orbit model that is incorporated into the trajectory generator. This testing is completed using the real-time system test bed with integrated IPU, GPS, and Star Tracker simulation capabilities.

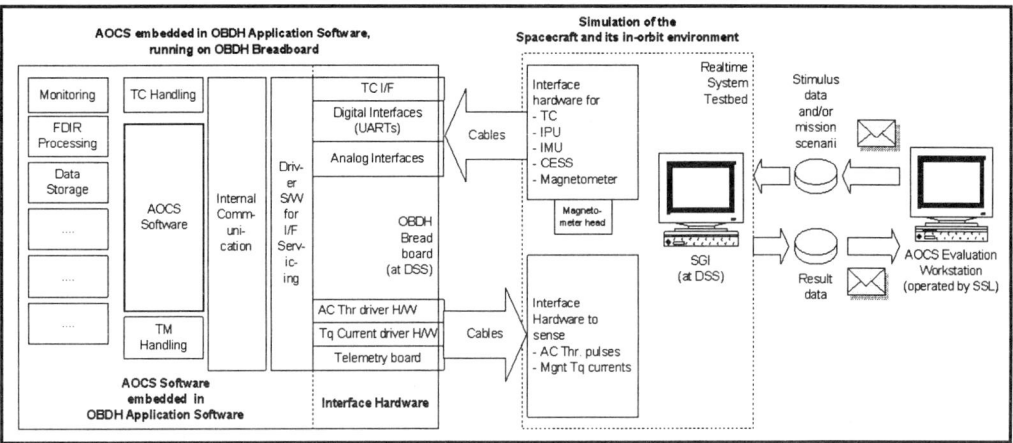

Figure 7: Real-Time System Test Bed for Closed Loop Testing

Mode handler testing: The objective of this step is to confirm that the AOCS subsystem code integrates properly into the system flight software. This step consists of hosting the full flight software on the Flat Sat OBDH. Specified mission scenarios are then initialized and, at the specified times/conditions, commands, fault conditions, and other stimuli are introduced into the system. This allows the design team to demonstrate the full-up performance of the flight system, and allows the operations team to understand how the spacecraft will be operated. The nominal operations procedures are used for this testing.

End-to-End system testing: The objective of this step is to demonstrate that the spacecraft is configured properly and is ready for launch. This step is completed at SSL after the spacecraft has been fully integrated and is in the final test process. Sensor polarity and functionality are demonstrated, as well as proper actuation for a specified stimulus. These tests are open loop tests only, there is no attempt to close the loop and demonstrate performance on the spacecraft, as the test interfaces available on the Flat-Sat are not available on the flight spacecraft and the closed loop testing has been completed using Flat-Sat.

It is the combination of this entire verification process, anchored by the validated design simulations, that enables the development team to confirm that the spacecraft AOCS and flight software are both designed and implemented properly for flight.

Conclusion

This paper has provided an overview of the GRACE mission and design of the Attitude and Orbit Control System (AOCS) required to meet the mission requirements. The validation, verification, and testing process for GRACE is complicated both by the coupling between orbit information and attitude control and by the international, multi-company program structure. This paper has discussed, at a high level, how the AOCS design is validated, how the software implementation is verified, and how the international development team demonstrates the proper performance of the control system and flight software. The design, verification, and validation effort has required the development of several simulations and test beds. Through the use of these tools, in conjunction with the entire verification process, anchored by the validated design simulations, enables the development team to confirm that the spacecraft AOCS and flight software are both designed and implemented properly for flight.

Acknowledgements

Spacecraft development for the GRACE program is funded by the National Aeronautics and Space Administration's Earth System Science Pathfinders (ESSP) program, and managed by the Jet Propulsion Laboratory, California Institute of Technology. The Principal Investigators for GRACE are Professor Byron Tapley, from the Center for Space Research at the University of Texas at Austin, and Professor Chris Reigber, from GeoForschungsZentrum-Potsdam (Germany). The authors are grateful to all of these people and organizations for the opportunity to participate in this exciting program and the funding to complete this work.

References

[1] Bauer, Hartman, and Lightsey. Spaceborne GPS Current Status and Future Visions. Institute of Navigation ION-GPS conference, September 1998.

[2] H. Stevens, J. Rodden, S. Carrou. Mass Expulsion Control for Precision Pointing Spacecraft. AAS 99-318. AIAA/AAS Astrodynamics Specialists Conference, August 1999.

A TETHERED FORMATION FLYING CONCEPT FOR THE SPECS MISSION

David A. Quinn[*] and David C. Folta[†]

The Sub-millimeter Probe of the Evolution of Cosmic Structure (SPECS) is a bold new mission concept designed to address fundamental questions about the Universe, including how the first stars formed from primordial material, and the first galaxies from pre-galactic structures, how the galaxies evolve over time, and what the cosmic history of energy release, heavy element synthesis, and dust formation is. Half of the luminosity and 98% of the post Big-Bang photons exit in the sub-millimeter range. The spectrum of our own Milky Way Galaxy shows this, and many galaxies have even more pronounced long-wavelength emissions. There can be no doubt that revolutionary science will be enabled when we have tools to study the sub-millimeter sky with Hubble-Space-Telescope-class resolution and sensitivity. Ideally, a very large telescope with an effective aperture approaching one kilometer in diameter would be needed to obtain such high quality angular resolution at these long wavelengths. However, a single aperture one kilometer in diameter would not only be very difficult to build and maintain at the cryogenic temperatures required for good seeing, but could actually turn out to be serious overkill. Because cosmic sub-millimeter photons are plentiful and the new detectors will be sensitive, the observations needed to address the questions posed above can be made with an interferometer using well established aperture synthesis techniques. Possibly as few as three 3-4 meter diameter mirrors flying in precision formation could be used to collect the light. To mitigate the need for a great deal of propellant, tethers may be needed as well. A spin-stabilized, tethered formation is a possible configuration requiring a more advanced form of formation flying controller, where dynamics are coupled due to the existence of the tethers between nodes in the formation network. The paper presents one such concept, a proposed configuration for a mission concept which combines the best features of structure, tethers and formation flying to meet the ambitious requirements necessary to make a future SPECS mission a success.

---

[*] Aerospace Engineer, Guidance-Navigation and Control Center, Code 572, NASA Goddard Space Flight Center, Greenbelt, Maryland 20771.

[†] Aerospace Engineer, Guidance-Navigation and Control Center, Code 571, NASA Goddard Space Flight Center, Greenbelt, Maryland 20771.

INTRODUCTION

Our current picture of the Universe at large is at best, incomplete. The answers to many of the most basic questions in cosmology remain elusive. What is the history of element formation and energy release in the Universe? When could life have first formed in the history of the Universe? It has been said that we are literally stardust, so when we study the origins of the elements, we are in a very real sense studying our own origins.

Interstellar dust has two effects on the light that reaches us from the cosmos. First, thermal emission from dust grains bathed in starlight produces the bulk of the far infrared emission that scientists wish to image. Second, through absorption and scattering, the dust attenuates the ultraviolet and visible emissions from galaxies and star and planet forming regions. Conventional optical telescopes provide essential but incomplete information. Nearly half of the luminosity and 98% of the photons in the post-Big Bang Universe are in the far-infrared and submillimeter wavelength range. That is to say, nearly half of the photons in the Universe are not being scrutinized to same fidelity as that of shorter, less prevalent wavelengths. This is not an intentional bias towards shorter wavelengths, merely the result of the fact that ability to study long wavelengths at high resolution has never been considered feasible, until now.

To address these science challenges, a science team led by several COBE (Cosmic Observer of Background Emissions) science team members (Mather, Moseley, Leisawitz) have proposed a 1 Km submillimeter interferometer mission called the Submillimeter Probe of the Evolution of Cosmic Structure (SPECS) [1]. The SPECS concept has been presented to several panels and has been identified as one of the SEU vision missions targeted for launch after 2014. Over this last year, a team of scientists and engineers have been looking at what architectures could accomplish a mission of this nature. With this in mind, a workshop was held in College Park, Maryland in February 1999 to begin looking at the, requirements, issues and problems associated with just such a mission.

PROBLEM DEFINITION

Driving Requirements

The top level scientific requirements of the SPECS mission were provided by the scientists who first envisioned the mission. They are as follows:

- Spectral range from 40-500μm, 15 arc-minute FOV with resolution comparable to that of the Hubble Space Telescope.

- Capable of observing as much of the sky as possible.

- Capable of completing a single observation inside 72 hours.

- Minimum goal of 5 year mission life.

Hubble-Space-Telescope-class resolution in the sub-millimeter range implies a resolution of about 0.05 arc-seconds for the desired wavelength range from 40μm to 500μm. Selecting a mean frequency acceptable to the scientific community of 250μm as the design point, the approximate separation between the collection elements of the interferometer may be computed:

$$resolution = \frac{\lambda}{s} \quad [1]$$

yielding a maximum separation of about 1031.3 meters, in turn becoming the maximum baseline requirement for the interferometer.

Theoretically, two mirror/collectors at the desired separation should be sufficient to accomplish the science goal; however, experience in ground based interferometry demonstrates that the single baseline provided by two light collectors may be subject to the effects of atmospheric or phase distortion of the wave front. To avoid this problem, typically three collectors have been employed resulting in three baselines. This permits phase distortion effects to be mathematically minimized using well established phase closure algorithms. Although there is no atmosphere in space, it is not yet known whether similar effects should be expected for as yet unknown reasons. Preferring to be conservative, this design assumes that three baselines are desirable for reasons similar to those discussed above and provide for redundancy concerns.

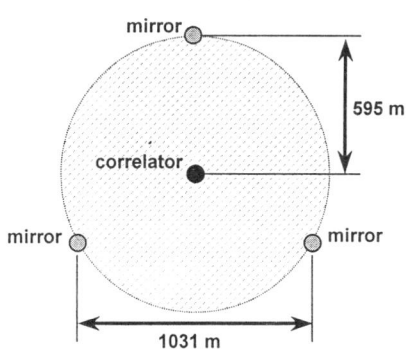

Figure 1. Three small mirrors form a single (595 meter radius) synthetic aperture.

The maximum separation between three mirrors has now been determined. Simple geometry tells us that the most efficient configuration of three such collectors is that of a regular triangle with a beam correlator at the center. Such a configuration yields an interferometer with a 1031.3 meter baseline or a synthetic aperture with a effective radius of 595.4 m (Figure 1). To be fully effective, the collecting mirrors must also be capable of movement within the plane perpendicular to the primary optical axis relative to the central beam correlator in such a way that permits collection of light over the entire area of the synthetic aperture.

Design Challenges

Operating four such vehicles to the control accuracy required in low Earth orbit (LEO) would be an expensive proposition to say the least. The amount of fuel needed to overcome relative orbital dynamics while maintaining the triangular co-planer formation of the separated vehicles, not to mention meeting the requirement to reorient the configuration so as to see as much of the observable sky as possible, would easily forfeit the five year mission life requirement. For similar reasons and others related to the clarity of astronomical 'seeing' in the near Earth environment and the cryogenic temperatures demanded of the science optics, it was decided that the optimal location for such a facility as SPECS would be the intermediate, co-linear Sun-Earth Lagrange Point, L_2. Still, operating a large formation such as this with the relative motions required over an extended period of time would be very fuel intensive, and now that fuel must be carried all the way to L_2. Further, the control requirements could potentially require constant correction thruster firings if some other means of maintaining the formation cannot be found.

Summing up, to meet the SPECS requirements, we need to design a facility which can be deployed at L_2, consisting of at least four spacecraft in a tightly controlled planer formation capable of maintaining an equilateral triangle whose sides can vary in length from virtually 0 to 1031 meters for each observation. An observation should take no more than 72 hours to complete. Finally, it should be capable of repeated observations of any part of the sky over a five year mission life.

A PROPOSED SOLUTION

Although there are literally an infinite number of possible configurations to choose from, we have selected one that satisfies the science requirements and is relatively easy to deploy and operate. We have settled upon a configuration that rotates the entire formation about the primary optical axis while each mirror is free to move along lines radiating out from the central beam collector Figure 2).

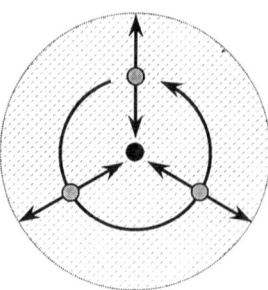

Figure 2. Relative degrees of freedom for the desired configuration.

In this way, conservation of angular momentum lends a hand in maintaining and controlling the spin rate of the central beam correlator, but only if the mirrors are somehow connected to the central beam correlator. Introduction of tethers into the design allows us to tie the formation together in a way that permits angular momentum to do most of the radial station keeping for us. Like a ball on the end of a string, centripetal forces will tend to keep the tethers taught and properly aligned with the central beam correlator. While conservation of angular momentum helps to solve one problem, it introduces another. As the radial length of the tethers change, the inertia tensor of the now connected formation changes and with it, the angular rate of the entire system. With only three outboard masses, the angular rate could be expected to increase by several orders of magnitude unless compensated. Compensation is accomplished by a triad of ballast masses equal to that of the mirrors themselves and positioned at the opposite ends of each tether. As the mirror is retracted, so the ballast mass is extended providing a momentum balance that will cause the angular rates to vary by only a factor of 2, which is manageable.

The proposed configuration, called SPECS-HEX employs three tethers, each of which is 600 meters long (Figure 3). At one end of each tether is a mirror and all the associated spacecraft hardware. At the other end is a ballast mass, a spacecraft in its own right whose primary function is to balance the mass of the mirror at the other end (Earlier designs employed six mirrors operated in two triads. This doubled overall productivity, but also doubled optics requirements on the central vehicle and was therefore deemed impractical). A kind of 'come-along' is employed on the central spacecraft allowing the tethers to be managed without the need to reel them in during every observation. Booms radiate out in six directions along the three axes of the tethers to provide momentum stiffness and a lever arm for performing reorientation maneuvers. It also serves as a framework to support a solar array and sunshield arrangement, as well as a stable platform for the mirrors to run along while performing observations at close quarters.

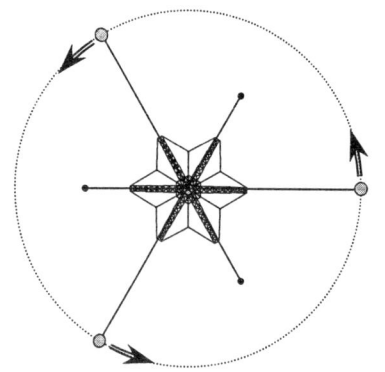

Figure 3. Balanced design of the SPECS-HEX configuration.

This design provides an excellent hybrid of the three major design possibilities. Structure at the center makes the system deployable and stable when the mirrors are closest together. Tethers allow distance to be well known and controllable while helping reduce fuel requirements on the overall system. Precision formation flying techniques may be applied as a means of providing navigation corrections to the positions of the mirrors and ballast masses relative to the central beam collector.

Operational Concept

This configuration of SPECS could be placed into low Earth orbit by the current Space Transportation System (Space Shuttle). Upon release from the shuttle, the spacecraft is spun up to about 90 rpm and a boost engine is fired putting it on a path that will carry it to L_2. Once there, the boost engine is discarded and some 100 days after the initial release from the Shuttle, a pair of 44 Newton ($I_{sp} \sim 300sec$) main engines fire for approximately 3 hours placing SPECS in a Lissajous trajectory about L_2 with a period of approximately six months. Periodic maintenance burns can impart approximately 3 m/s per year to keep SPECS on course over its expected 5 year lifetime.

Final deployment is accomplished in three stages (Figure 4). During each stage of the deployment, the mass of the system is distributed further from the center, slowing down the spin rate from its initial 90 rpm. In stage 1, the six booms fold open, sliding and locking them-selves into position, and the redistribution of mass causes the rotation rate to decrease. During stage 2, the booms are extended locking into their operational configuration, and the rotation rate decreases still further. Finally, during stage 3, the tethered spacecraft are released and the system rotation rate begins to approach the 0.01 rpm which is to be used during operation.

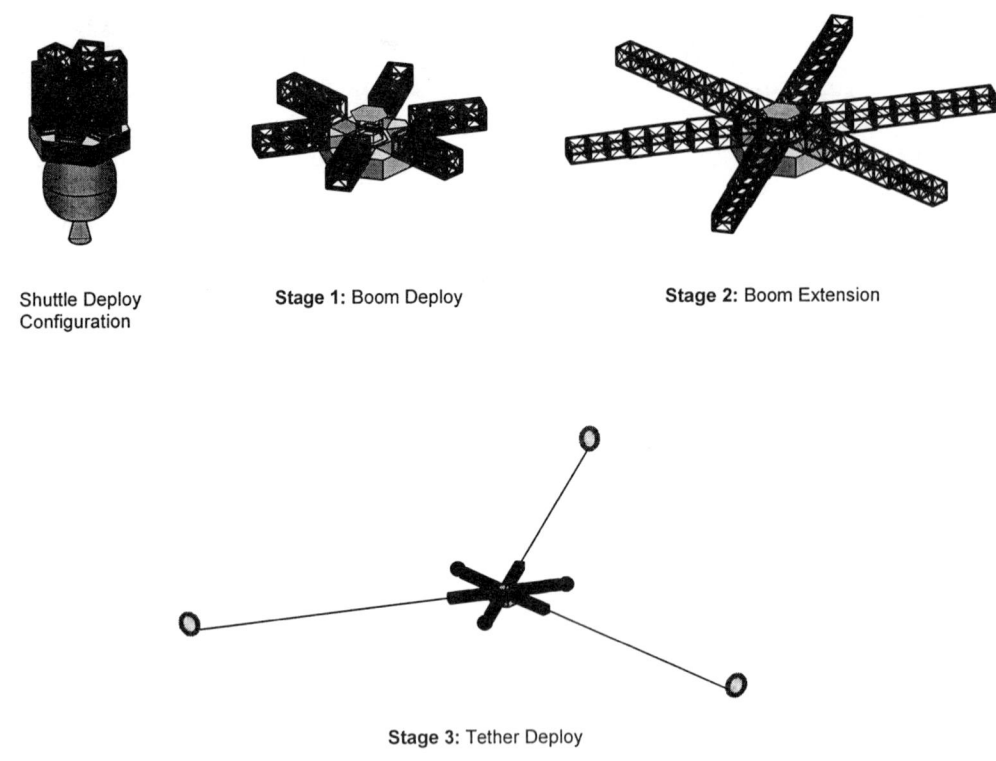

Figure 4. Launch configuration and deployment strategy of the SPECS-HEX concept.

Once on station at L_2, the fully deployed SPECS-HEX may begin its operational life. Outwardly, operations of the SPECS-HEX is a relatively simple matter. An observation is begun with the mirrors at their minimum separation from the central beam correlator and the ballast masses at their maximum distance. As the entire system rotates, the mirror tethers are extended at a rate which causes the mirrors to spiral out, completely filling in the synthetic aperture, as required. For this to work, the radial extension speed of the mirror tethers must be a function of the rotational speed which itself will change so as to maintain a constant angular momentum as the inertia tensor of the total system changes. When the mirrors have reached their maximum extension, the next observation may begin by reversing the process. By using three tethers each with a mirror and a ballast mass at each end, this gradual exchange of mass will contain rotation rates by keeping mass distribution of the entire system as balanced as possible. In this way consecutive observations of a single starfield can be made before re-orienting the system for the next observation.

Reorientation of the SPECS-HEX design may be accomplished in a relatively simple manner as well. This is executed by reeling in all six daughter spacecraft (all three mirrors and all three ballast spacecraft) to the ends of the structural deployment booms. While the spin rate will increase dramatically, science observations are suspended during this time so attention is focussed on the re-orientation. Once in the six daughter spacecraft are in place at the ends of the booms, the entire system may be treated as a single spin-stabilized spacecraft and reoriented accordingly. Consecutive thruster firings by the daughter spacecraft as the system rotates, will serve to retarget the overall system. Although each daughter spacecraft has a spooling mechanism at each end of the three tethers, we see here that operations of SPECS-HEX during observations does not require exercising the tether spools, this is only necessary for reorientation. During observations, the tethers need move only back and forth whereas during re-orientation they must reeled in. Since the platform will be reoriented only between observations, this saves wear and tear on the tether spooling mechanisms.

A SIMPLE MODEL

Intuitively, the SPECS-HEX concept appears to present an elegant solution to the problem of large, space-based imaging interferometry but the question remains can it work in reality? While a full three dimensional non-linear dynamics model will require years to formulate (and is well outside the scope of this conceptual paper) a simple model can be constructed in short order. For starters, we assume zero attitude and position error of the spacecraft mirrors and ballasts relative to the central spacecraft so that the outboard spacecraft formation always maintains themselves along the spokes of a perfect planer hexagon. If we then assume the tethers to always be taut, we can treat the entire idealized system as effectively rigid. Obviously this is not the true situation, but this assumption allows us to calculate the total mass of the system as well as the first order rotational inertia in terms of the distances from the central to outboard spacecraft. Non-linearities and deviations from this assumption (introduced by the tethers) will be dealt with in subsequent analyses. Under this initial idealizing assumption, both the mass and the inertia tensor of the system may estimated as follows:

$$M_{total} = m_{center} + 3(m_{mirror} + m_{ballast}) + 3\left(\frac{m_{tether}}{l_{tether}}\right)[l_{mirror} + l_{ballast}] \quad [2]$$

$$I_{total} = I_{center} + 3\left[(I + ml^2)_{mirror} + (I + ml^2)_{ballast}\right] + \frac{3}{4}\left(\frac{m_{tether}}{l_{tether}}\right)[l^3_{mirror} + l^3_{ballast}] \quad [3]$$

Although the distance from any one mirror or ballast changes as observations are made, the sum of the distance from one mirror to its opposing ballast remains constant. As expected, the total mass of the system remains constant, but the rotational inertia will change due to the redistribution of mass. Maintaining angular momentum of the system as constant therefore requires the angular rate to change as the mirrors are brought closer and further away from the central spacecraft. Now the purpose of the ballast masses become clear. If only three spacecraft-mirrors were employed and the system rotating at an acceptable rate, rotation rates would increase by several orders of magnitude as the mirrors were reeled in towards the center. The ballast masses allow for a mass exchange to occur such that as the mirror is brought in, a ballast mass is extended. This reduces the variations in mass distribution and allows rotation rates to increase only on the order of about 2:1.

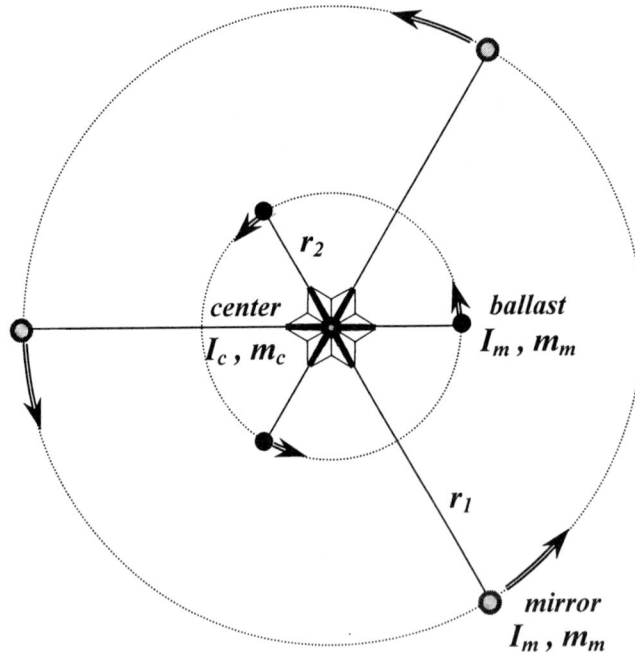

Figure 5. Assumed configuration for first order mathematical model.

As an example, let us assume the total mass of the system to be approximately 5000 kg, using 2000 kg as the mass of the central spacecraft leaving 3000 kg for the total of the six daughter spacecraft at 500 kg each. Allowing for a mass of 1.36×10^{-3} kg/meter for three 600 meter tethers adds only another 2-3 kg to the system and may be considered negligible, even though they will be carried in the mathematical development. The equations above now permit the inertia tensor of the entire system to be computed as a function of the distance of the daughter spacecraft from the central hub (assuming also that the center of the mother spacecraft is both the center of mass of the system and the origin of coordinates). Recognizing that completely filling the synthetic aperture requires that the each mirror follows a spiral path so as to minimize overlap as the system rotates, demands that radial position and angular position of the mirrors be related. To determine how the relationship changes over time, we recognize that a simple relationship between radial and angular position exists which will permit a complete fill of the synthetic aperture:

$$\theta(r) = \left(\frac{d\theta}{dr}\right)r + r_0 \qquad [4]$$

Take the time derivative:

$$\left(\frac{d\theta}{dt}\right) = \cancelto{0}{\frac{d}{dt}\left(\frac{d\theta}{dr}\right)} + \left(\frac{d\theta}{dr}\right)\frac{dr}{dt} + \cancelto{0}{\frac{dr_0}{dt}}$$

By definition $\omega \equiv \left(\frac{d\theta}{dt}\right)$, therefore: $\omega = \left(\frac{d\theta}{dr}\right)\dot{r}$ ...and since: $\omega = \frac{h}{I(r)}$

We may now define $q \equiv \dfrac{1}{\left(\frac{d\theta}{dr}\right)}$ $\Rightarrow$ $qh\,dt = I(r)\,dr$

$$qh\int dt = \int I(r)\,dr$$

$$qh(t-t_0) = \int_{r_0}^{r} I(r)\,dr$$

Recalling that the distance between a mirror and its opposing ballast mass is constant:

$$r_1 + r_2 = l \quad \Rightarrow \quad r_2 = (l - r_1)$$

...where r_1 is the distance between a mirror and the center and r_2 is the distance between the ballast mass and the center (Figure 5). Compute terms for r_2 in terms of r_1:

$$r_2^2 = \left(l^2 - 2lr_1 + r_1^2\right) \qquad [5]$$

$$r_2^3 = \left(l^3 - 3l^2 r_1 + 3lr_1^2 - r_1^3\right) \qquad [6]$$

Substitute Eqs. [5] & [6] into Eq. [3]:

$$I(r_1, r_2) = I_c + 6I_m + \frac{3}{4}\left(\frac{m_t}{l}\right)\left[r_1^3 + r_2^3\right] + 3m_m\left[r_1^2 + r_2^2\right] \qquad [7]$$

Leaving the inertia in term of r_1 only. Rearrange Eq. [7] dropping the subscript on r_1:

$$I(r) = \left[\tfrac{9}{4}m_t + 6m_m\right]r^2 - \left[\tfrac{9}{4}m_t l + 6m_m l\right]r + \left[\tfrac{3}{4}m_t l^2 + 3m_m l^2 + I_c + 6I_m\right] \qquad [8]$$

Integrate Eq. [8]:

$$qh(t-t_0) = \int_{r_0}^{r}\left\{\left[\tfrac{9}{4}m_f + 6m_f\right]r^2 - \left[\tfrac{9}{4}m_f l + 6m_f l\right]r + \left[\tfrac{3}{4}m_f l^2 + 3m_f l^2 + I_s + 6I_f\right]\right\}dr \qquad [9]$$

For convenience, define A_0, A_1, A_2 such that:

$$qh(t-t_0) = \int_{r_0}^{r} \{A_2 r^2 - A_1 r + A_0\} dr \quad [10]$$

$$qht = \tfrac{1}{3} A_2 (r^3 - r_0^3) - \tfrac{1}{2} A_1 (r^2 - r_0^2) + A_0 (r - r_0) \quad [11]$$

$$qht = \left(\tfrac{1}{3} A_2\right) r^3 - \left(\tfrac{1}{2} A_1\right) r^2 + (A_0) r + \left(-\tfrac{1}{3} A_2 r_0^3 + \tfrac{1}{2} A_1 r_0^2 - A_0 r_0\right) \quad [12]$$

using Eqs. [10], [11] and [12] in Eq. [9] yields time as a function of radial position:

$$t = \left[\tfrac{1}{3} \tfrac{A_2}{qh}\right] r^3 - \left[\tfrac{1}{2} \tfrac{A_1}{qh}\right] r^2 + \left[\tfrac{A_0}{qh}\right] r + \left(\tfrac{1}{qh}\right)\left[-\tfrac{1}{3} A_2 r_0^3 + \tfrac{1}{2} A_1 r_0^2 - A_0 r_0\right] \quad [13]$$

If we set the mirror diameters at 4 meters, we find that there are constants of integration which remain undefined. This final ambiguity is solved by noting the scientists' desire for the mirrors to move gradually during observations. While the position of the mirrors will not be held fixed, the linear tangential speed of the mirrors is desired to be capped at approximately 1 m/s. From Eq. [13] above, a sixth order curve fit is employed to find radial position as a function of time. For this specific example, the equations have been set up and the total momentum of the system held constant, but that constant is tuned so as to yield maximum tangential rate of no more than 1 m/s, as requested. Parametric equations of radial and angular position in time may now be set up such that:

$$r(t) = (C_6) t^6 + (C_5) t^5 + (C_4) t^4 + (C_3) t^3 + (C_2) t^2 + (C_1) t + (C_0) \quad [14]$$

$$\theta(t) = \left(\tfrac{C_6}{q}\right) t^6 + \left(\tfrac{C_5}{q}\right) t^5 + \left(\tfrac{C_4}{q}\right) t^4 + \left(\tfrac{C_3}{q}\right) t^3 + \left(\tfrac{C_2}{q}\right) t^2 + \left(\tfrac{C_1}{q}\right) t + \left(\tfrac{C_0}{q}\right) \quad [15]$$

From which angular and radial rates can be easily computed:

$$\dot{r}(t) = 6(C_6) t^5 + 5(C_5) t^4 + 4(C_4) t^3 + 3(C_3) t^4 + 2(C_2) t + (C_1) \quad [16]$$

$$\dot{\theta}(t) = \left(\tfrac{6C_6}{q}\right) t^5 + \left(\tfrac{5C_5}{q}\right) t^4 + \left(\tfrac{4C_4}{q}\right) t^3 + \left(\tfrac{3C_3}{q}\right) t^2 + \left(\tfrac{2C_2}{q}\right) t^1 + \left(\tfrac{C_1}{q}\right) \quad [17]$$

So, for a given (constant) angular momentum we find that these equations describe a spiraling motion of the mirrors over time which will completely fill the synthetic aperture. It is interesting to notice that neither the radial or angular rates are constant. One might desire to set the radial rates at some constant value such that the tethers could reeled in and out at a constant rate. While this approach could be taken, it results in a very inefficient fill of the synthetic aperture taking far too long with a great deal of overlap if the rate is too low or leaving gaps in the fill if the rate is too high.

SPECS Concept Example

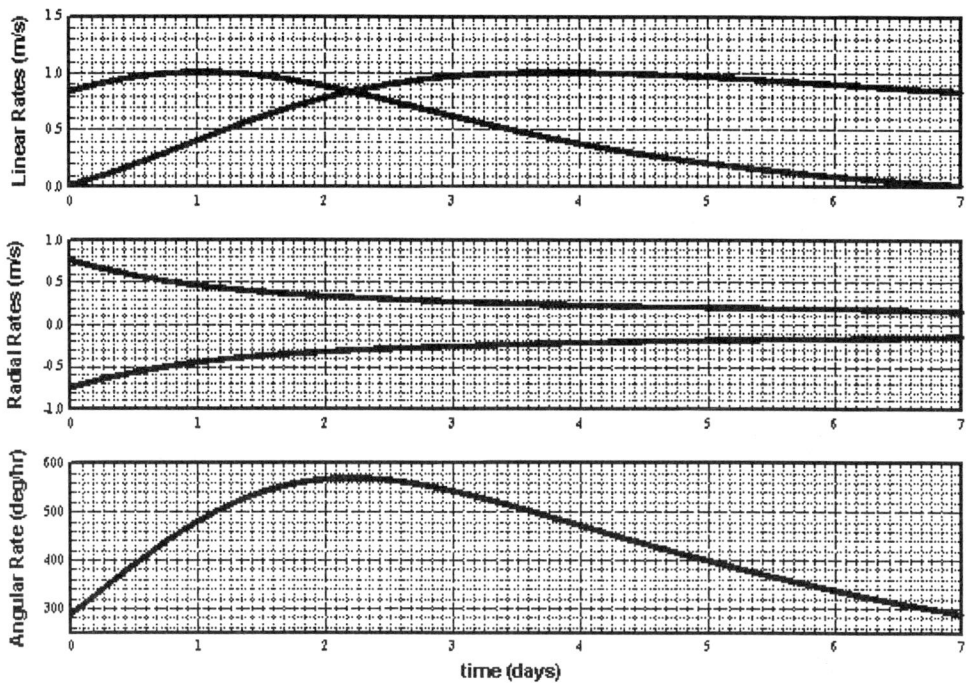

Figure 6. Linear, Radial & Angular Rate of the given mathematical model.

Figure 6 shows the results of these computations. The first graph shows the tangential rates of the mirrors and ballasts as an observation progresses. The total system momentum was tuned so that these curves (one for each end of a given tether) reach a maximum of 1 m/s, consistent with the desires of the scientists. The second graph shows the radial rates of each end of a given tether, or the speed with which the tethers must be pulled through the come-along at the central spacecraft. The last graph shows the angular rate of the system as a whole. Notice that the variation of angular rates ranges from about 300 – 600 deg/hr resulting in an efficient fill of the entire synthetic aperture in about 7 days.

If we wish to reorient the SPECS-HEX system, then all six daughter spacecraft must be reeled to a position at the ends of the central booms where they are held fixed to the central body. If the booms are assumed to be approximately 50 meters in length, doing this increases the angular rate of the system to about 1 rpm. Since observations can be assumed to be suspended during this operation, the 1 m/s tangential speed limit is no longer an issue. Once all the daughter spacecraft are secure at the ends of the booms, the system may now be treated as single spin stabilized spacecraft, for the purposes of retargeting. Each of the daughter spacecraft possesses a full compliment of thrusters, so a retargeting burn may executed by firing thrusters on any of the daughter spacecraft. If we wish to be capable of reorienting the system from any target to any other target in one day, we assume the maximum reorientation rate λ_{max} accordingly. We can then estimate the force required to initiate the reorientation to be:

$$F = \frac{\lambda_{max} I_\perp}{\Delta t \; r} \qquad [18]$$

where $I_\perp$ is the off-axis inertia, Δt the thrust duration and r the length of the boom. For this example a thrust of approximately 0.058 Newtons is required to initiate the reorientation and similar thrust is required to stop once the new target has been attained making a total 0.116 Newtons required per reorientation. After the new target orientation has been attained, the daughter spacecraft may be redeployed on their tethers and observations of the new target may begin. If we allow for one such reorientation (on average) every 10 days, we find that 20 grams of fuel per daughter spacecraft is more than enough to complete a five year mission. This remarkably low fuel requirement is due primarily to the long lever arm provided by the booms and the slow rate (λ_{max}) at which the reorientation is to occur.

It must be noted again that this very simple model was developed to parameterize the problem to first order. The tethers here are assumed to be rigid, effectively massless and always perfectly aligned along the axes of a planer hexagon. Under these highly idealized assumptions a mathematical model was developed in order to determine if it is even possible for the scientific requirements of such an observatory could be met.

IMPLICATIONS OF THE FIRST ORDER MODEL

Several implications come out of our first order model. First is that the science desire to complete an observation in 72 hours is subject to the efficiency of the observation. As we have seen, the 1 m/s tangential rate limit flows from the fact that a complete spiral scan of the synthetic aperture requires a minimum of seven days under the best of circumstances. If the 72 hour limit becomes a driving factor then either the 1 m/s tangential limit or the fully populated scan of the synthetic aperture must be compromised.

Secondly, since the radial speeds of the mirrors are not constant we can be sure that some method of reeling in and out the tethers at variable predetermined rates will be required. An efficient means of pulling in the outer spacecraft while releasing the inner ones at the same rate would need to be developed. Perhaps a 'variable come along' mechanism could be devised which would not require the tethers to be reeled in during observations.

Further, the fuel requirements seen in this simple model are reasonable. L_2 insertion and subsequent maintenance can be met with a minimum of 400 kg of fuel on the central spacecraft. Reorientation burns require only 20 grams per spacecraft for each of the six daughter spacecraft. While the true dynamics of the full tethered system have yet to be determined and the demands on any thruster based control system have yet to be established, there is nothing else preventing the SPECS-HEX configuration from meeting the mission objectives.

Finally, and perhaps most important of all, some means of controlling the outboard spacecraft relative to the desired formation is in order. For first order purposes, we have made simplifying assumptions about the tether being rigid and straight, neither of which is true in reality. The true non-linear nature of tether dynamics in a rotating system such as this will need to be more fully explored before any kind of control scheme can be designed and implemented. Obviously modeling the dynamics and control of such a complex rotating tethered system will be require a considerable amount of time and expertise, but this simple study shows that the investment may be made with a reasonable expectation of sucess.

FUTURE STUDIES

Obviously, much more work must be done to fully explore the possibilities of SPEC-HEX configuration. Primarily, the non-linear dynamics of tethers under circumstances such as these must be more completely understood and a means of controlling the daughter spacecraft developed. Work at the Naval Research Laboratory in modeling tether dynamics [2] as well as efforts at the Goddard Space Flight Center in decentralized control methods [3, 4 & 5] are being examined in the belief that they may be adaptable to address the specifics of this problem. Tether deployment mechanisms will need to advance to permit deployment, retraction and re-deployment of tethers in the manner required of SPECS-HEX. Tether materials will need to endure the rigors of long term exposure to deep space and deployable booms will be required to complete the SPECS-HEX design. These are formidable obstacles, but with the proper expertise brought to bear on this most challenging problem, there can be little doubt that a funded study of a large rotating tethered observatory can be brought to life.

CONCLUSIONS

The key conclusion of this cursory study is to confirm the possibility that a rotating tethered distributed spacecraft system is at least feasible and worthy of further investigation. Once tether dynamics are understood, and a means of control has been developed, we will be able to reap the substantial benefits offered in the ability to create enormous observatories with relatively small spacecraft. In this way, an economically viable means of examining the long wavelength emissions of the Universe to a high precision will finally become available. As new and more detailed information becomes available, unexpected results will lead us to startling conclusions. Our understanding of Nature will increase as we paint a clearer picture of the Universe and our place within it.

REFERENCES

1. Mather, J.C. et al., "The Submillimeter Frontier: A Space Science Imperative," 2000, Rev. Sci. Inst.

2. Barnds, Coffey, Davis, Kelm and Purdy, "TiPS: Results of a Tethered Satellite Experiment", AAS 97-600.

3. Carpenter, J. R., "Feasibility of Decentralized Linear-Quadratic-Gaussian Control of Autonomous Distributed Spacecraft", 1999 Flight Mechanics Symposium, NASA/CP-1999-209235, pp. 345-357.

4. Folta, D. C., and D. A. Quinn, "A Universal 3-D Method For Controlling The Relative Motion Of Multiple Spacecraft In Any Orbit", Paper AIAA-98-4193, AIAA/AAS Astrodynamics Specialist Conference, August 10-12, 1998, Boston, MA.

5. Quinn, D. A., and D. C. Folta, Patent Rights Application and Derivations of Autonomous Closed Loop 3-Axis Navigation Control Of EO-1, 1996.

AAS 00-016

PROJECT ORION: CARRIER PHASE DIFFERENTIAL GPS NAVIGATION FOR FORMATION FLYING[*]

Franz D. Busse,[†] Gokhan Inalhan[‡] and Jonathan P. How[**]

This paper describes the guidance, navigation, and control algorithms that are being developed for the Orion-Emerald Project. The Orion-Emerald Project at Stanford University is a mission designed to demonstrate key technologies required for formation flying, which is a critical technology for many planned space missions. This paper presents a general description of the Orion-Emerald mission and the satellite design. The primary focus of this mission is to demonstrate carrier-phase differential GPS (CDGPS) as the primary relative navigation sensor for controlling the formation. A GPS receiver has been developed to provide absolute state, relative state, and attitude data. Note that this receiver has been extensively tested on ground testbeds with multiple vehicles. Simulation results are presented in this paper to demonstrate the estimation and control for multiple vehicles on-orbit (e.g. the Orion mission). These results show reliable relative position estimation with a 5-10cm accuracy, and relative velocity estimation with 0.5 mm/sec accuracy.

INTRODUCTION

NASA and DOD/USAF both envision formation flying as a key technology for both deep-space and orbital applications [26, 28]. There are several reasons for this shift in space mission paradigms. One motivation is to replace the traditional single monolithic satellite with a virtual spacecraft bus consisting of clusters of autonomously controlled vehicles [27]. This can lead to less expensive, lower risk missions, and also speed deployment of scientific missions.

Formation flying also enables a whole new range of orbital and deep-space missions. For some applications, such as earth mapping, synthetic aperture radar, or interferometry, a long baseline separation (on the order of many kilometers) is required to met the desired signal or image resolution (inversely proportional to baseline) [30, 24, 25]. Furthermore, the ability for individual vehicles within a formation to move and realign their relative positions, provides a much more flexible science instrument.

These missions all require precise alignment between the vehicles in the fleet [17, 20], and there are major technical challenges in achieving these levels of control. In addition, the fleet control design must be done with careful consideration of the onboard computation, inter-vehicle communication, and power requirements for the formation flying spacecraft. Thus a systems-level approach is essential, allowing explicit inclusion of the hardware limitations

---

[*] This work is funded in part under Air Force grant #F49620-99-1-0095 and NASA GSFC grant #NAG5-6233-0005.

[†] Research Assistant, Department of Aeronautics and Astronautics, Stanford University, Stanford, California 94305. E-mail: teancum@stanford.edu.

[‡] Research Assistant, Department of Aeronautics and Astronautics, Stanford University, Stanford, California 94305. E-mail: ginalhan@stanford.edu.

[**] Assistant Professor, Department of Aeronautics and Astronautics, Stanford University, Stanford, California 94305. E-mail: howjo@sun-valley.stanford.edu.

(power, mass, computation) in the theoretical analysis of the various multi-level control and estimation architectures [21, 22, 23]. For this reason, the following system level challenges for such distributed systems have been identified under the overall formation flying problem:

1. Sensing and metrology (relative/absolute sensing, sensor fusion)
2. Aperture optimization (orbit and formation selection)
3. Fleet and vehicle autonomy (control architecture)
4. Control computation (formation planning, maneuvering and data collection)
5. Spacecraft bus design (RCS, crosslink communications and CDH)

This paper describes ongoing research at Stanford University to address these systems-level issues for a future on-orbit demonstration of formation flying called the Orion-Emerald project. The particular emphasis of this work is to demonstrate precise relative sensing and formation flying control of multiple spacecraft via carrier-phase differential GPS.

The Global Positioning System has been instrumental in advancing formation flying [29] since it provides a robust, reliable, low-mass, and relatively inexpensive sensor for formation flying. It provides an array of sensing capabilities, including absolute state, relative vehicle states within the formation, attitude of individual vehicles, and precise timing. For precision attitude and relative state determination, carrier-phase differential GPS techniques are employed, which can provide $\approx$5 cm accuracy levels. These features, and the three-in-one nature of the GPS sensor, make it an attractive option for formation flying.

A key objective of the Orion-Emerald project is to demonstrate the use of GPS as the primary sensor in sensing and controlling a satellite formation. Project Orion-Emerald is being designed, developed, and managed at Stanford University and Santa Clara University.

ORION-EMERALD MISSION

Orion-Emerald is a high-risk, low-cost NASA-funded project dedicated to the research and development of microsatellites that are capable of performing distributed relative sensing and formation flying. The mission will involve three vehicles, two Emerald microsatellites and a single Orion microsatellite. These three vehicles will operate together in an actively controlled formation.

The Emerald spacecraft is being designed and constructed by a student managed team at Stanford University and Santa Clara University [10] (See Fig. 2). Each Emerald vehicle has a mass of 15 kg. It has a hexagonal layout, with a width of 45 cm, and a height of 29 cm. The Emerald vehicles will serve as a bus for several experimental packages. They will each have a two-antenna GPS receiver. They will have downlink capability as well as crosslink capability. The primary position control capability will be achieved using panels that will allow control over the differential drag between vehicles.

The Orion spacecraft is being designed by a Stanford University team [13] (See Fig. 1). The vehicle is 35 kg, and is cubic, with each side 45 cm long. The Orion vehicle features a 6-antenna GPS receiver. This receiver will provide attitude sensing as well as absolute and relative state information. Orion will have crosslink communication capability shared with the Emerald vehicles, a necessary feature to allow relative navigation. The communications, command/data handling, and spacecraft housekeeping processor are common to both Emerald and Orion. The crosslink will operate at 437 MHz using SpaceQuest modems,

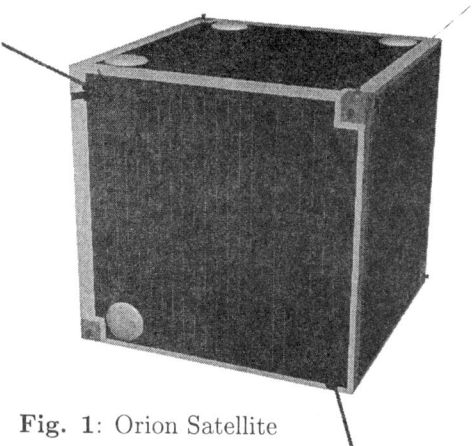

Fig. 1: Orion Satellite

Fig. 2: Emerald Satellites

and provide 9600bps baud rate. The uplink/downlink for Emerald-Orion will also be at the 437 MHz band, which is an amateur radio frequency. Command and Data Handling, which will run the low-level house keeping of the spacecraft will be performed using a SpaceQuest FCV53 microprocessor and will be running BekTek Spacecraft Operating System. Power will come from Gallium-Arsenic solar cells, which will be used to charge Lithium-Ion batteries.

Perhaps the most distinguishing feature of Orion, however, is the attitude and position determination and control system. Using a cold gas propulsion system, with pressurized Nitrogen, Orion will have 12 thrusters, allowing full translational and attitude control. In particular, the Orion vehicle will be able to perform much higher bandwidth station-keeping control than that allowed with drag control panels. Orion will also have magnetometers and magnetic coils, which can provide backup/coarse attitude sensing and control. The other distinguishing characteristic of the Orion vehicle is that it will have much greater computational power than the Emerald vehicles. Linked to the GPS receiver will be a Pentium-equivalent processor, which will handle the high-level fleet coordination, run a higher fidelity orbit propagator, and manage the autonomous formation control commands for all three vehicles within the formation.

These vehicles will provide the testbed for demonstrating formation flying concepts. The focus of the work will be on the navigation and control features of the formation. Specifically, the goals are to demonstrate:

- Ability to organize a group of small satellites into a pre-determined formation on-orbit. In particular, this will include the ability to exchange GPS, position and scientific data between the satellites, as well as execute pre-planned, organized maneuvers [31]. The execution of the maneuvers will be governed by on-board real-time autonomous control software.
- Use and operation of a low-power, low-cost, multi-channel GPS receiver for real-time attitude and position determination of the satellites (note the antenna layout in Fig. 1).
- On-orbit autonomy using various control architectures. Demonstrate autonomous mode switching from coarse to fine (and back) formation flying.
- Ability to perform formation flying and station keeping with various baselines and configurations.

AUTONOMOUS FORMATION CONTROL

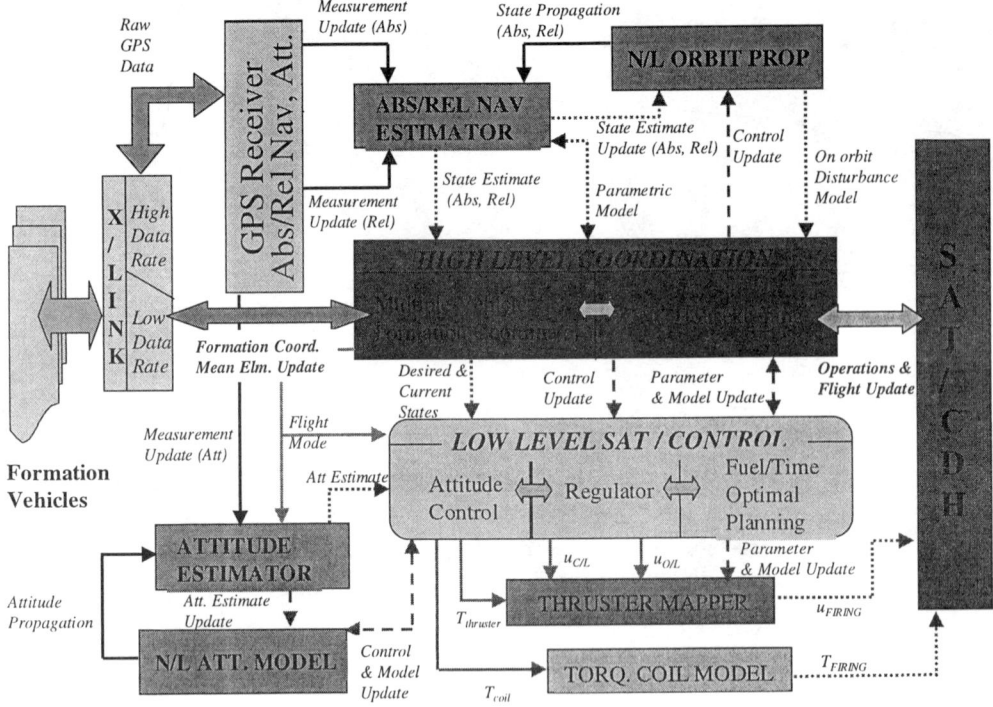

Fig. 3: Autonomous Control Architecture

With these main objectives for the Orion-Emerald project, we have developed, simulated, and experimentally demonstrated basic features of an autonomous guidance, navigation, and control architecture for formation flying. Fig. 3 shows the basic elements and the information flow for this control architecture.

The process begins with the reception of the RF signals by the GPS receiver. The receiver takes the measurements from these signals and passes them to an estimator. The estimator uses these measurements in combination with a dynamic model (provided by a nonlinear model propagator) to arrive at a current estimate of the state. Note that this process explicitly depends on an exchange between all vehicles in the formation of the raw GPS phase measurements using the GN&C cross-link. The current state information is provided to both a high-level formation coordinator as well as the low-level satellite controller. The high-level formation controller takes the state information, as well as information from the other vehicles within the formation, and determines the commands/trajectories for each vehicle. The low-level controller uses the current attitude information along with the desired control commands from the coordinator to determine actual satellite thruster commands or other low-level satellite functions.

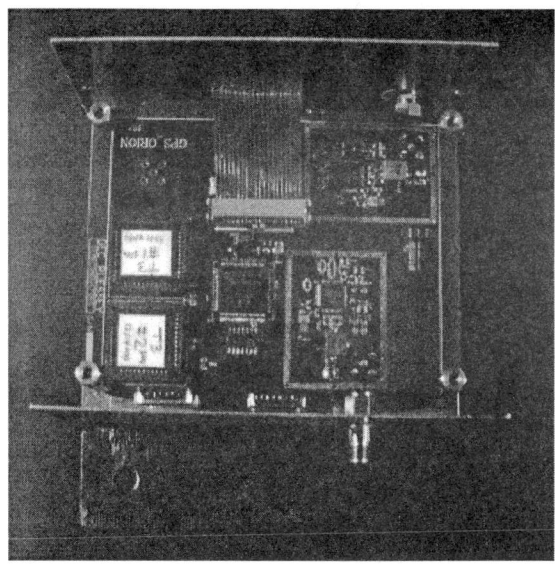

Fig. 4: Modified GPS Receiver – 2 Antenna Configuration for Emerald

GPS RECEIVER

The central element for this navigation and control system is the GPS receiver. Project Orion will use a modified receiver based on the Mitel/Plessey GPS chipset [7]. The receiver shall be referred to the Stanford-Modified Receiver (SMR) within this work. The SMR uses the GP2015 RF front end and the GP2021 12-channel correlator. The SMR only operates using the L_1 frequency, 1575.42 MHz. An ARM60B processor is used to run the software and control the tracking loops on the correlator. There are several hardware modifications of the SMR. The first primary change is the addition of a second RF front end for each correlator. The twelve channels can be allocated to the two antenna inputs as desired. Each upper receiver board (which consists of 2 RF front ends, a correlator, and a processor) also has a port for an external clock signal. This allows multiple boards to be synchronized to the same clock, a necessary requirement for an attitude capable receiver. For the Orion vehicle, three of these upper boards are interfaced together, providing a 6-antenna (36-channel) GPS receiver. The Emerald vehicle will use only a 2-antenna receiver, and cannot determine attitude (see figure 4). The 6-antenna receiver draws approximately 8 Watts of power.

The SMR uses modified carrier and code tracking loops. For search, initial acquisition, and lock, it operates at 1 kHz, using a frequency-lock loop (FLL). Once the signal is locked on, it drops to a 50 Hz carrier and code tracking loop, using a Phase-Lock Loop (PLL). This saves computational effort for the receiver, saves power, and allows it to perform the rigorous estimation calculations. The SMR software has also been modified to slew the clock. Every second, the clock is adjusted so that it is synchronized to GPS time. This insures that the phase measurements are taken on different receivers within 100 nanoseconds of each other. The software also indicates whenever cycle slips occur, in which case the measurement is not used during that period. It should also be noted that the GPS chipset used for this

Data	Size	Total (example)
Header	1 byte	1
PRN for this channel	$1 \times k$ bytes	12
Time Tag	4 bytes	4
Slip Flag	2 bytes	2
Integrated Carrier Phase	$4 \times n \times k$	48
Code Phase	$4 \times n \times k$	48
Doppler	$4 \times n \times k$	48
Total		163

Table 1: Crosslink Data Budget

receiver has space heritage. It is the same set that has recently been employed successfully by the University of Surrey satellite UoSAT-12, in their Space GPS Receiver (SGR) [9].

The GPS receiver needs to interface with other components within the Guidance Navigation and Control system. To perform relative navigation, it must receive data from the other satellites within the formation. At each epoch the integrated carrier phase, the Doppler shift, and the code phase are all recorded for each tracking channel. A flag is also set to indicate whether the satellite is visible and whether a cycle slip was detected. This data is time-stamped and the identifying PRN number for each signal recorded as well. Generally, this data must be received from at least one antenna on each vehicle.[1] Table 1 shows the data cross-linked in real-time for every epoch. The receiver passes this data and performs the estimation routines at 1 Hz. Using this data, and its own measurements, an estimator produces three solutions: the attitude and absolute state of this particular user vehicle, and the relative state between the other user vehicles.

ESTIMATOR

The absolute state is determined by standard GPS pseudo-ranging techniques [2]. The time of transmission for RF signals from at least 4 NAVSTAR satellites is measured. These measurements are used in a weighted least square routine to solve for the absolute position and time. Velocity is measured by differencing the position over consecutive time steps. The current work assumes Selective Availability (S/A) is present. With S/A, there is approximately 25 meters of error in each signal.

The basic GPS pseudo-ranging equation, at antenna i from transmitter k, is

$$\rho_{ik} = |X_k - X_i| + b_i + B_k + E_{ik} + \nu_{ik} \tag{1}$$

where

ρ_{ik} = measured range for vehicle i from transmitter k

$|X_k - X_u|$ = distance from vehicle i to transmitter k expressed in Cartesian frame

---

[1] Improved accuracy can be achieved using an "all-baseline-in-view" approach. Here, data from all antennas on a vehicle tracking any satellites is shared, so that any two antennas within the formation could potentially provide differential measurements. This improves the accuracy and robustness of the estimation, but increases the required data rate.

b_i = clock bias for receiver i
B_k = clock bias for transmitter k
E_{ik} = other error sources (i.e. Ionosphere, Ephemeris, etc.)
ν_{ik} = receiver noise

This is a nonlinear relationship between position and range measurement. This can be linearized by taking the Jacobian of the position matrix, which is the line-of-sight matrix

$$G = \begin{bmatrix} los_1 & 1 \\ los_2 & 1 \\ \vdots & \vdots \\ los_n & 1 \end{bmatrix} \qquad (2)$$

where los_k is the line of sight vector from the vehicle to the kth NAVSTAR satellite (which will also be referred to as the geometry or observation matrix). The additional column of ones is for the clock error term.

Given the pseudo-range measurements, ρ, the position and clock error of vehicle i is estimated by

$$\delta \hat{X}_i = (G_i^T G_i)^{-1} G_i^T \delta \rho \qquad (3)$$

where

$\delta \hat{X}_i$ = increment to estimated position of vehicle i in Earth Centered Inertial Cartesian frame from expectation
G_i = geometry matrix for the ith user
$\delta \rho$ = difference between actual measurements and expected measurements

Clearly G is a function of the absolute position of the user vehicle, so an iterative process must be used to estimate the absolute position. This problem must be performed first, because the G matrix is also used for the relative and attitude solutions. Estimation error for absolute position in simulation matches actual terrestrial experience, and is on the order of 50 m.

The relative state solution uses Carrier-phase Differential GPS techniques [1]. Here the phase of the carrier wave of the signal is measured, giving accuracy to a fraction of the wavelength (a fraction of 19.2 cm, depending on phase resolution). The carrier phase at any two antennas can be measured, and the difference in phase will indicate the distance between the two antennas. This approach requires synchronous measurements be taken on both receivers, which is achieved by the slewing of the receiver clock to synchronize with GPS time [18]. When measurements are received via the crosslink from other spacecraft, the time-stamps on the data are compared to insure synchronous measurements.

The measurement equation for the carrier observable, at antenna i from transmitter k, is very similar to the general pseudo-range equation

$$\phi_{ik} = |D_{ik}| + b_{vi} + B_k + \beta_{ik} + \nu_{ik} + E_{ik} \qquad (4)$$

where

ϕ_{ik} = measured carrier phase vehicle i from transmitter k

$|D_{ik}|$ = distance from vehicle i to transmitter k
b_{vi} = clock bias for receiver i
B_k = clock bias for transmitter k
β_{ik} = carrier-phase bias for measured phase at vehicle i from transmitter k (including line bias)
ν_{ik} = receiver noise
E_{ik} = other error sources (i.e. Ionosphere, Ephemeris, etc.)

Taking the difference between measurements at two different antennas, i and j, the common terms are eliminated, leaving:

$$\Delta \phi_{ij} = G_i \begin{bmatrix} x_{ij} \\ b_{ij} \end{bmatrix} + \beta_{ij} + \nu_{\text{carr}} \tag{5}$$

where

x_{ij} = the relative position between vehicles i and j
b_{ij} = the difference between clock errors for receiver i and receiver j
ν_{carr} = all differenced noise and error terms on carrier phase measurement

Of course, the main challenge with using carrier-differential GPS, is that upon initial acquisition of a signal, there exists an unknown number of integer cycles (β) between the two antennas. There are many approaches to solving this integer ambiguity. In terrestrial applications, there are generally two approaches to determining the integer biases [14]. Recall that the observation matrix is based on the lines-of-sight between the receiver and the transmitter. To determine the constant bias, this line-of-sight matrix must change sufficiently. One approach is to wait for the NAVSTAR satellites to move (which can be too long for many terrestrial applications), the other is to move the user vehicle with respect to a local transmitter. However, the orbital dynamics helps solve this problem. Due to the high velocity of the user with respect to the NAVSTAR satellites, it is possible and practical to depend on just the relative motion to provide the needed observability to determine the biases. The disadvantage of LEO is that the NAVSTAR satellites are being acquired and lost at a much higher rate than Terrestrial users experience, therefore requiring the re-initialization of the bias each time one is gained.

The first task is to decouple the bias estimation from the relative position problem (see Lawrence [6]). Let L be a matrix of the orthonormal basis for the left null space of the geometry matrix, G. Premultiply equation 5 by L, and by definition, the G term is eliminated, leaving a new measurement vector equal to the biases,

$$L\Delta\phi_{ij} = LG_i \begin{bmatrix} x_{ij} \\ b_{ij} \end{bmatrix} + L\beta_{ij} + L\nu_{\text{carr}} \tag{6}$$

leaving,

$$L\Delta\phi_{ij} \equiv z_k = L\beta_{ij} + L\nu_{\text{carr}} \tag{7}$$

The new measurement set, z_k, is then used within an Extended Kalman Filter [18]. Note this formulation depends on the receivers being close enough together to assume the same line-of-sight vectors. But this assumption only gives an error on the order of 10 cm when there is a difference of 5 kilometers between vehicles.

For the relative velocity, the exact same approach is followed as for the differential carrier phase measurement, using the difference in measured Doppler shifts in the received RF signals. There is no bias ambiguity issue for the Doppler measurement, making it a much more accurate estimate (which will be seen in the results)

$$\Delta\dot{\phi}_{ij} = G_i \begin{bmatrix} \dot{x}_{ij} \\ \dot{b}_{ij} \end{bmatrix} + \nu_{\text{Dopp}} \tag{8}$$

where

$\Delta\dot{\phi}_{ij}$ = measured Doppler on carrier signal at vehicle i from transmitter k
$\dot{x}_{ij}$ = relative velocity of vehicle i with respect to vehicle j
$\dot{b}_{ij}$ = relative clock drift between receiver i and receiver j
ν_{Dopp} = differential noise and error on Doppler signal

With the complete set of relative position and velocity measurements, a Kalman filter can be employed to estimate the relative state [3]. Two Kalman filters are actually used, one for the relative state, and one for the biases. For the bias estimation, the filter equations are:

$$K_\beta = P_{\beta,k}^- L^T \left[L P_{\beta,k}^- L^T + R_\beta \right]^{-1} \tag{9}$$

$$\hat{\beta}_k^+ = \hat{\beta}_k^- + K_\beta (z_k - L\hat{\beta}_k^-) \tag{10}$$

$$P_{\beta,k}^+ = [I - K_\beta L] P_{\beta,k}^- \tag{11}$$

where

K_β = Kalman gain matrix for bias estimates
$P_{\beta,k}^-$ = Bias covariance matrix at time-step k before the measurement update
$P_{\beta,k}^+$ = Bias covariance matrix at time-step k after the measurement update
L = Bias observation matrix (in this case, the left null space matrix)
$\hat{\beta}_k^-$ = Bias estimate before the measurement
$\hat{\beta}_k^+$ = Bias estimate after the measurement

and

$$R_\beta = E\left[L\nu_{carr}\nu_{carr}^T L^T\right] = R_{carr} \tag{12}$$

And since the biases should be constant as long as the transmitter is in view, the propagation is direct,

$$P_{\beta,k+1}^- = P_{\beta,k}^+ \quad , \quad \hat{\beta}_{k+1}^- = \hat{\beta}_k^+ \tag{13}$$

Note that the number and elements of β, and its corresponding covariance matrix, will change frequently as NAVSTAR satellites come into and exit from view. At each time step the matrix must be remade. The variance for the new measurement is set to the variance of the differential code noise.

The Kalman filter for the relative state is very similar. The state propagation depends on position and velocity together, so the measurements are combined,

$$y = \begin{bmatrix} \Delta\phi \\ \Delta\dot{\phi} \end{bmatrix} \tag{14}$$

and the observation matrix is

$$H = \begin{bmatrix} G & 0 \\ 0 & G \end{bmatrix} \tag{15}$$

The covariance matrices are also combined,

$$R_{\delta x} = \begin{bmatrix} R_{\text{carr}} & 0 \\ 0 & R_{\text{Dopp}} \end{bmatrix}, \quad R_{\text{carr}} = E\left[\nu_{\text{carr}}\nu_{\text{carr}}^T\right], \quad R_{\text{Dopp}} = E\left[\nu_{\text{Dopp}}\nu_{\text{Dopp}}^T\right] \tag{16}$$

Using this formulation, the measurement update for the kth timestep with the filter is

$$K_x = P_{x,k}^- H^T \left[H P_{x,k}^- H^T + R_{\delta x} \right]^{-1} \tag{17}$$

$$\hat{x}_k^+ = \hat{x}_k^- + K_x(y_k - H\hat{x}^-) \tag{18}$$

$$P_{x,k}^+ = [I - K_x H] P_{x,k}^- \tag{19}$$

where K_x is the Kalman gain matrix for the relative state, and $P_{x,k}$ is the relative state covariance matrix at time k.

Here, however, the propagation to the next measurement step is more involved. For the covariance matrix, the state dynamics can be linearized for the given absolute position. Note that this is similar to Hill's equation, but remains in the inertial frame, rather than performing the transformation into the local-vertical local-horizontal frame. This requires the dynamics matrix, A, be re-derived (linearized) at each time-step using the current estimated absolute position.

For two vehicles, the approximate equation governing the relative motion in the inertial frame is [4]

$$\ddot{\mathbf{x}} = n^2 \left[-\mathbf{x} + 3 \left(\frac{\mathbf{X}}{|\mathbf{X}|} \cdot \mathbf{x} \right) \frac{\mathbf{X}}{|\mathbf{X}|} \right] + u_n \tag{20}$$

where

n^2 = mean orbital motion
$\mathbf{x}$ = relative position vector (ECI frame)
$\mathbf{X}$ = absolute position vector (ECI frame)
u_n = normalized accelerations from control input

The state is propagated forward from time k to time $k+1$ by integrating this equation of motion using a single-step, fourth-order Runge Kutta integration routine. The initial conditions used are the current absolute and relative state information, $\mathbf{X}_k^+$ and $\mathbf{x}_k^+$, as well as the control inputs, u_n. The covariance is propagated over time by:

$$P_{x,k+1}^- = A_k P_{x,k}^+ A_k^T + Q_x \tag{21}$$

where
$$A_k = \begin{bmatrix} 0 & I \\ \left(-I + 3\frac{XX^T}{|X|^2}\right) & 0 \end{bmatrix} \qquad (22)$$

with
$$Q_x = E\left[\omega\omega^T\right] \qquad (23)$$

and ω is the expected disturbances or model uncertainty.

The attitude is also determined using carrier differential GPS [1, 19]. At least 6 antennas are mounted in known locations on the satellite body, in an orientation that allows them to receive the common NAVSTAR satellite signals. By determining the position of the antennas in inertial space relative to each other, and knowing their fixed location in the body coordinate frame, a transformation between the body frame and the inertial frame can be resolved. Note that for attitude, because the distance between the antennas is a fixed and a known constant, a different bias resolution technique is used. Although this technique is nonlinear, requiring an iterative solution approach, the final result is typically more accurate, allowing finer attitude resolution. Also note that in general, the attitude accuracy is a function of antenna baselines. Since Orion is a "micro-satellite", the attitude accuracy is limited. For example, following the analysis of Lightsey [5], with a 40 cm separation between antenna phase centers the RMS attitude error should be approximately 1-2 degrees.

SIMULATION SETUP AND RESULTS

A simulation was generated to help assess the expected performance of the estimator. Almanac data was taken for the NAVSTAR constellation, and the motion of the NAVSTAR satellites was simulated. The initial state of two user satellites was also specified, either in Cartesian Earth Centered Inertial or Kepler Orbital elements, and their orbits were propagated forward with time. At each time step, the visibility of each NAVSTAR satellite at the receiver was determined. For those satellites which were visible, measurements were simulated to each one. The measurements taken were pseudo-range (absolute position), differential code phase, differential carrier phase, and differential Doppler measurement. The assumed noise levels are shown in Table 2.

Signal	Noise Level
Range measurement	25 m
Differential Code Phase	2 m
Differential Carrier Phase	.02 m
Differential Doppler	5 mm/sec

Table 2: Signal Noise Levels

These noise levels are consistent with current terrestrial experience. With an improved multi-path environment, and without ionospheric and tropospheric effects, these levels should represent conservative noise models.

The number of channels that the receiver could track was also a changeable parameter. For the results shown, 9 receiver channels were assumed the maximum available. Note that in LEO, there are generally 11-14 NAVSTAR satellites visible at any given time. Cycle slips

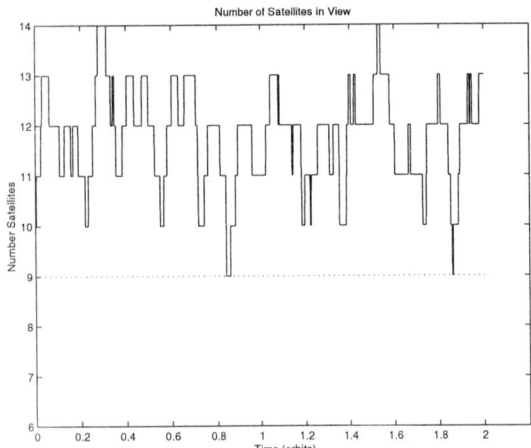

Fig. 5: Number of NAVSTAR satellites visible and number used

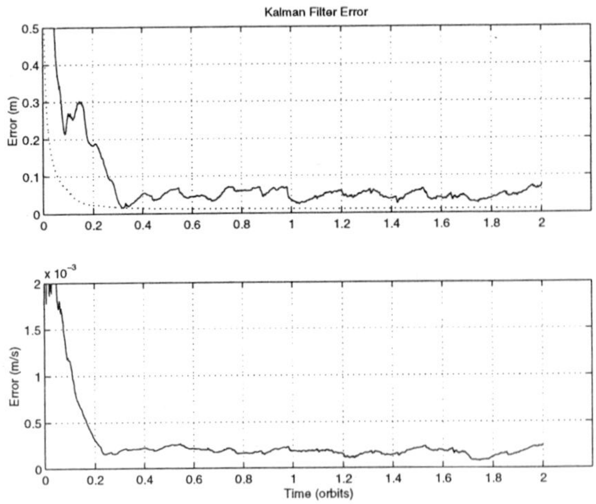

Fig. 6: Velocity and Position Error

were not simulated in these tests. Anti-nadir pointing antennas were assumed. With this simulated setup, figure 5 shows the number of satellites tracked during the simulation, and the number actually used (which is generally all available channels).

The performance of the estimator in the simulated LEO environment is shown in figures 7 through 10. These plots are from a single simulation, which are a characteristic representation of the estimator performance. After settling, the position estimate can be seen to be on the order of 5 cm. Figure 7 shows that the radial position error is worse, as would be expected due to the geometry of the NAVSTAR satellites. The velocity estimate is well below 0.5 mm/sec, and again, figure 8 shows the radial component to have the largest error. Figure 10 shows how the filtered solution compares with a simple weighted least squares solution. The velocity performs orders of magnitude better. However, though the position estimate smooths some of the spikes of the WLS solution, it does not provide a significant increase in accuracy. This is partly due to the inherent error in the bias estimate. This error "envelope" is shown by figure 9, which shows the maximum and minimum bias errors in the set of measurements used. The large number of spikes demonstrate the difficulty of the LEO environment; each jump is the addition of a new signal, with an un-initialized bias. This bias error sets the relative position error floor.

Multiple simulations were run with random noise included, and the final RMS results from these are in Table 3. Note that the units for absolute position is meters, relative position is centimeters, and velocity is millimeters per second.

The GPS constellation and estimator simulator was written in modular form, to allow interface with other simulation packages. After initial estimator capability was demonstrated, it was then incorporated in a simulation with a High-Fidelity Orbit Simulator (Free Flyer [11]) and an active control formation flying operation [12]. The performance of this closed loop control scheme is shown in figure 11 and figure 12. These figures demonstrate that in a closed loop control routine (where control thrusts are present) and a very accurate orbit environment simulation is used, the estimator still performs nominally.

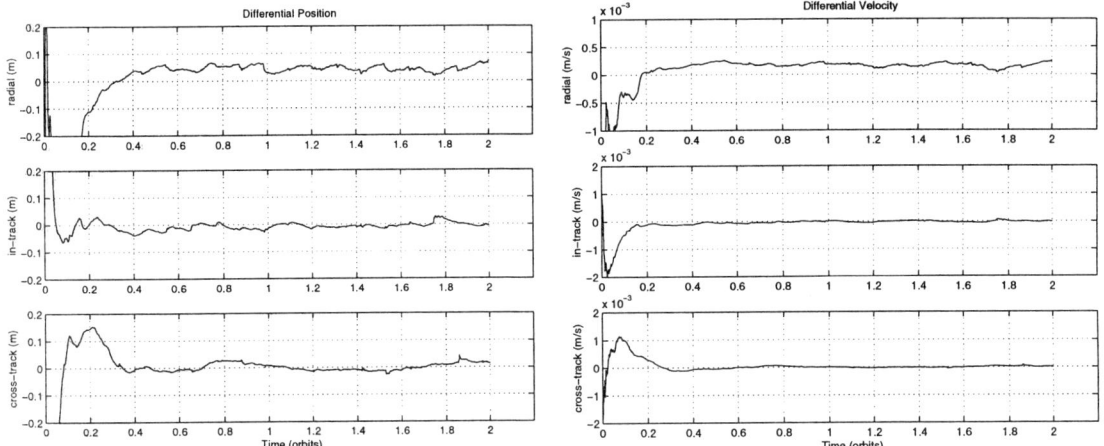

Fig. 7: Position Error **Fig. 8**: Velocity Error

Error	Mean	Variance
Absolute Position (m)	12.07	36.49
Radial Position (cm)	4.30	1.37
In-track Position (cm)	0.49	1.31
Cross-track Position (cm)	0.37	0.81
Radial Velocity (mm/s)	0.160	0.038
In-track Velocity (mm/s)	0.029	0.031
Cross-track Velocity (mm/s)	0.003	0.019

Table 3: Error for Estimation

CONCLUSION

Precise knowledge of relative position and velocity plays a crucial role in minimizing fuel usage in formation flight /cite[gkijph]. Potential algorithms such as formation keeping and formation initialization depend on exact knowledge of current states to generate thrusting sequences(open loop) to obtain desired relative locations. If there are measurement errors, motion replaning or relative error correction(closed loop) is required to keep the vehicles within desired relative accuracy levels. However this is done in expense of the most abundant resource, fuel.

For LEO missions, inaccurate knowledge of relative velocity can be a dominant source of error when compared to position errors. Essentially to have comparable effects on the drift of the spacecraft, the effective resolution of the relative velocity error must be 3 orders of magnitude below the radial position error.

It should be noted that others have simulated better performance on relative position estimation than has been presented in this work [8]. However, these other simulations have depended on widelaning techniques (using L_1 and L_2 frequencies), more computationally intensive integer estimation routines, significantly more expensive receivers, or post-processing techniques. This work provides a relatively simple, low-cost, single frequency, real-time relative position sensor.

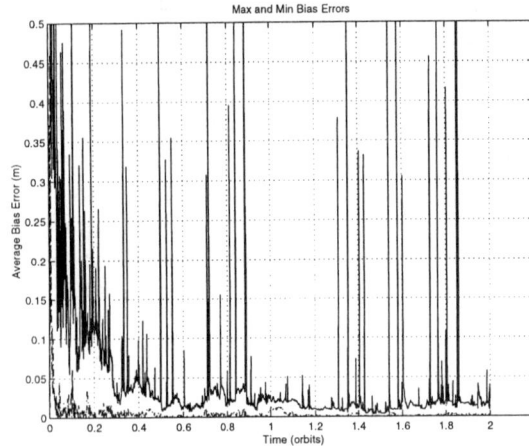

Fig. 9: Maximum and Minimum Bias Error

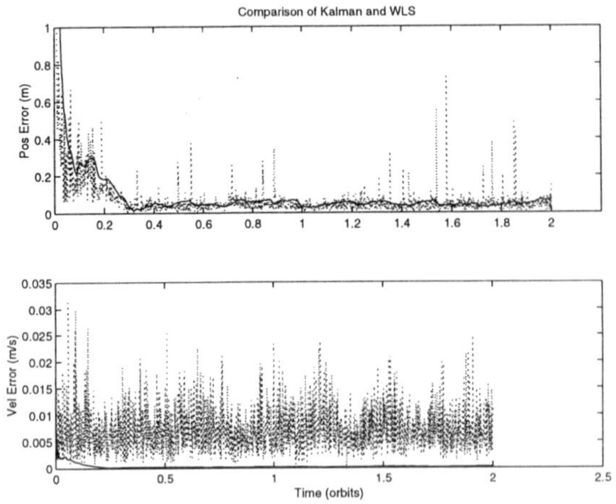

Fig. 10: Kalman Filter compared to Weighted Least Squares

In closing, this work has presented a general overview of the Orion-Emerald mission. It has focused on the control and navigation subsystems within that project. The design of a space-capable carrier-differential attitude GPS receiver has been presented. The estimator for the relative navigation and formation control has also been presented. Performance for the estimator has been simulated, in a high fidelity orbit simulator, and results show that estimation error for relative position ot be on the order of ≈5cm, and more importantly, the estimation error on velocity to be less than 0.5 mm/sec. The receiver/estimator has successfully met all design requirements, and development and integration of the GN&C subsystem with the Orion-Emerald vehicles continues.

References

[1] E.G. Lightsey, C.E. Cohen, "Development of a GPS Receiver for Reliable Real-Time Attitude Determination in Space", *Proc. of the Institute of Navigation GPS-94 Conf.*, Sep 1994, Salt Lake City, UT, vol.2,pg. 1677

[2] P. Axelrad, R.G. Brown, "GPS Navigation Algorithms," Global Positioning System: Theory and Applications, Vol. I, ed. B.W. Parkinson, J.J. Spilker Jr., AIAA, 1996.

[3] A. Gelb, "Applied Optimal Estimation," 1996,pp.102-132.

[4] M.H. Kaplan, "Modern Spacecraft Dynamics and Control," p.110, 1976.

[5] E.G. Lightsey, "Development and Flight Demonstration of a GPS Receiver for Space", Ph.D. Thesis, Stanford University, Dept. of Aeronautical and Astronautical Engineering, Feb. 1997.

[6] D.G. Lawrence, "Aircraft Landing Using GPS, Development and Evaluation of a RealTime System for Kinematic Positioning Using the Global Positioning System",Ph.D. Thesis, Stanford University, Dept. of Aeronautical and Astronautical Engineering, Sep. 1996.

[7] E. Olsen, "GPS Sensing for Formation Flying Vehicles," Ph.D. Thesis, Stanford University, Dept. of Aeronautical and Astronautical Engineering, Dec 1999.

[8] P.W. Binning, "Absolute and Relative Satellite to Satellite Navigation Using GPS," Ph.D. Thesis, University of Colorado, Dept. of Aerospace Sciences, 1997.

[9] M.J. Unwin, M.K. Oldfield, S. Purivigraipon, Y. Hashida, P.L. Palmer, "Preliminary Orbital Results from the SGR Space GPS Receiver", in *Proc. of ION GPS, Nashville*, Sep. 1996.

[10] Available at **http://ssdl.stanford.edu/Emerald/home.html**

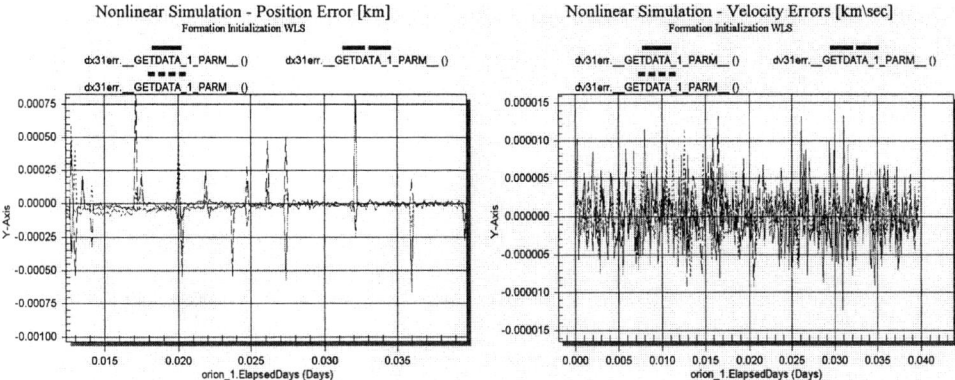

Fig. 11: HighFidelity Position Estimate Error

Fig. 12: High Fidelity Velocity Estimate Error

[11] A.I.Solutions, "FreeFlyer User's Guide," Version 4.0, Mar. 1999.
[12] G. Inalhan, F. Busse, J.P. How, "Precise Formation Flying of Multiple Spacecraft Using Carrier-Phase Differential GPS," *Proc. of AAS Space Flight Mechanics Winter Meeting, Clearwater, FL.*, AAS 00-109, Jan. 2000.
[13] Z. Kiraly, B. Engberg, F. Busse, R. Twiggs, J.P. How, "The Orion Microsatellite: A Demonstration of Formation Flying on Orbit," *Conference Proc. of AIAA Space Technology 99*, SSC99-VI-8, Sep. 99.
[14] B.W. Parkinson, P.K. Enge, "Differential GPS," Global Positioning System: Theory and Applications, Vol. II, ed. B.W. Parkinson, J.J. Spilker Jr., AIAA, 1996.
[15] D. Folta, L. Newman, T. Gardner, "Foundations of Formation Flying for Mission to Planet Earth and New Millennium," *AIAA/AAS Astrodynamics Specialists Conference*, July 1996.
[16] M. Colavita, C. Chu, E. Mettler, M. Milman, D. Royer, S. Shakian, and J. West, "Multiple spacecraft interferometer constellation (MUSIC)," Tech. Rep. JPL D-13369, JPL - Advanced Concepts Program, February 1996.
[17] Timothy H. Dixon, "SAR Interferometry and Surface Detection," Workshop Report, Boulder-Colorado, Feb. 1994.
[18] C.W. Park, E.A. Olsen, J.P. How, "Sensing Technologies for Spacecraft Formation Flying," presented at *Institute of Navigation 2000 Technical Meeting*, Anaheim CA, Jan 2000.
[19] J.C. Adams, "Robust GPS Attitude Determination for Spacecraft," Ph.D. Thesis, Stanford University, Dept. of Aeronautical and Astronautical Engineering, Dec 1999.
[20] R. S. Wimokur, "Operational Use of Civil Space-Based Synthetic Aperture Radar," *NOAA JPL Publication* 96-16, Aug 1996.
[21] D.C. Folta and D.A. Quinn, "A Universal 3-D Method for Controlling the Relative Motion of Multiple Spacecraft in any Orbit," in *Proc. AIAA/AAS Astrodynamics Specialists Conf.*, Aug 1998.
[22] M. Meshabi and F. Y. Hadaegh, "Formation Flying Control of Multiple Spacecraft: Graph Theoretic Properties and Switching Schemes," in *Proc. of the AIAA GNC*, Aug. 1999.
[23] P.K.C. Wang and F.Y. Hadaegh, "Coordination and Control of Multiple Microspacecraft Moving in Formation," *Journal of Astronautical Sciences*, Vol 44, No. 3, 1996, pp. 315-355.
[24] T. Alfery, "Request for proposal for deep space 3 (DS-3)spacecraft system industry partner," Tech. Rep. RFP No. N01-4-9048-213, Jet Propulsion Laboratory, November 1998.
[25] H. Hirosawa, H. Hirabayashi, T. Orii and E. Nakagawa, "Design of the Space-VLBI Satellite MUSES-B," *19th International Symposium on Space Technology and Science*, Yokohama, May 1994.
[26] Available at **http://gnctech.gsfc.nasa.gov/gto/tech/formation/roadmap.html**.

[27] J.C. Adams, A. Robertson, K. Zimmerman, and J.P. How, "Technologies for spacecraft formation flying," *Proc. of the ION GPS-96 Conference*, (Kansas City, MO), Sep. 1996.

[28] F.H.Bauer, K. Hartman, J.P. How, J. Bristow, D. Weidow, F. Busse, "Enabling Spacecraft Formation Flying through Spaceborne GPS and EnhancedAutomation Technologies," *Proc. of Institute of Navigation*, Jul 1999?

[29] F.H. Bauer, K. Hartman, E.G. Lghtsey, "Spaceborne GPS Current Status and Future Vision," *1998 IEEE Aerospace Conference. Proc. Snowmass at Aspen CO*, Mar 1998.

[30] Available at **http://www.vs.afrl.af.mil/factsheets/TechSat21.html**

[31] G.Q. Xing, S.A Parvez and D.C.Folta, "Implementation of Autonomous GPS Guidance and Control for Spacecraft Formation," in *Proc. of the ACC,* June 1999.

[32] G. Inalhan, J. P. How, "Relative Dynamics & Control of Spacecraft Formations in Eccentric Orbits," submitted to *AIAA/GNC Conference 2000* , January 2000.

AAS 00-017

MODE AND LOGIC-BASED SWITCHING FOR THE FORMATION FLYING CONTROL OF MULTIPLE SPACECRAFT

Mehran Mesbahi[*] and Fred Y. Hadaegh[†]

Mode and logic-based switching for formation flying control is motivated and addressed in the context of separated spacecraft optical interferometry. It is shown that logic-based switching is an attractive approach for satisfying multiple performance criteria during the different phases of a representative formation flying mission. Simulation results are included to demonstrate the need for introducing logic-based switching, its mechanism, and its performance; the relevant theoretical results are also presented.

Introduction

Traditionally, mode switching has played a central role in the operation of space systems. The Cassini attitude control system (ACS), for example, operated around six distinguished control modes [1]. Generally, there is a set of actuators, sensors, and control laws which are specifically selected to satisfy the requirement for each mode.

The principles employed for identifying these functional/control modes are often quite intuitive. This *mode approach* also has very attractive features as compared to an alternative, which attempts to design a *global* control law, capable of satisfying multiple mission requirements during the *entire* mission life time (for the Cassini mission, this would be a period of seven years!). The latter approach is clearly prone to many difficulties and its success, highly improbable.

The classification of the functional modes for a space system is often a by-product of a careful mission planning and system engineering; if done correctly, it is also a blessing for the control system analyst. The underlying basis for introducing these modes of operation is essentially the golden rule of *divide and conquer*: partition the mission into different modes, design a control laws for each- then "glue" these controlled modes together. The designer's concern is not only the control design for each mode, but of course the "putting together" of them: given n modes and the corresponding set of candidate control laws, how should the control law for each mode be selected and switched in and out of the spacecraft control system?

---

[*] Assistant Professor; Department of Aerospace Engineering and Mechanics, University of Minnesota. E-mail: mesbahi@aem.umn.edu.

[†] Senior Research Scientist and Group Supervisor, Jet Propulsion Laboratory, Califronia Institute of Technology, 4800 Oak Grove Drive, Pasadena, California 91109-8099. fred.y.hadaegh@jpl.nasa.gov.

The concern regarding the judicious orchestrating of the control switching is quite relevant- the state of the spacecraft is generally not in unique correspondence with the modes of operation of the space system and its control laws. Consequently, there is a need to introduce a *logical mechanism* which decides, in real time, which control law has to be placed in the feedback loop. The criteria for these logical decisions are based on a set of performance requirements for each mode, as well as the available set of candidate control laws.

This paper is concerned with logic-based control switching for multiple spacecraft formation flying in an optical interferometry mission. Formation flying is among the most relevant applications for adopting a *hybrid* control approach, incorporating mode and control logic-based switching as its core algorithmic component.

The contribution of the paper is twofold. First, by providing an overview of a representative optical interferometry mission, the paper identifies the basic modes of operation of the multiple spacecraft formation flying. For each mode of operation, a class of control laws is then presented which account for the particular requirements of that mode. Second, the paper provides a framework for the "putting together" of these control modes. For each scenario, simulation results supplement the discussion. The associated analysis for the switching control laws conclude the paper.

Setting the Stage

Let us begin by briefly describing one of the most important science applications for multiple spacecraft formation flying.

Optical interferometry

Space-borne optical interferometry holds the promise of revolutionizing our understanding of the origins and evolution of the planetary systems. By synthesizing the image of a star, or possibly its orbiting planets, through interfering light gathered by two or more collectors, optical interferometry provides a unique window into characterizing size, temperature, and orbital parameters of heavenly bodies- possibly identifying their habitable zones [2], [4].

Imaging a star can be thought as the evaluation of the intensity function $\mathcal{I}$, in a frame which is attached to the star. The van Cittert-Zernike formula [5],

$$\underbrace{\mathcal{I}(X,Y)}_{\text{intensity}} = \int_u \int_v \underbrace{\mu(u,v)}_{\text{mutual coherence}} e^{i2\pi(uX+vY)} dv\, du \tag{1}$$

relates the evaluation of $\mathcal{I}$ to the inverse Fourier transform of the mutual coherence function μ, which can be measured by optical instruments [3]. In the van Cittert-Zernike formula (1), one has

$$u = \frac{1}{\lambda}(x_1 - x_2) \quad \text{and} \quad v = \frac{1}{\lambda}(y_1 - y_2), \tag{2}$$

where λ is the wavelength of the light emitted by the star- (x_1, y_1), (x_2, y_2) are the positions of the aperture in the physical space; see Fig. 1.

Imaging a star thus amounts to the evaluation of the mutual coherence function μ over a disk of infinite radius in the uv-plane. By choosing a disk of finite radius and sampling (evaluating the

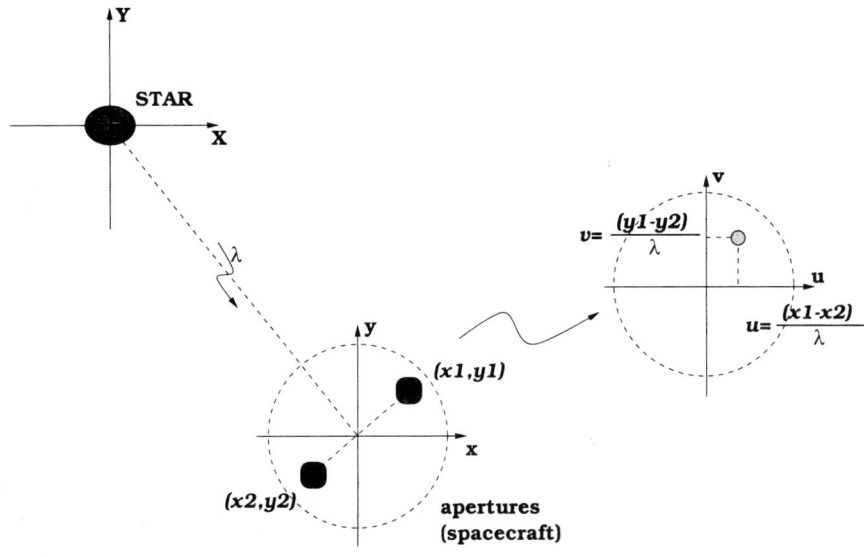

Figure 1: Imaging a star via optical interferometry

mutual coherence function) at a finite number of uv-points, $\mathcal{I}$ is, in a sense, approximated by keeping a range of its frequency content.

Suppose each of the aperture is placed on a separate spacecraft (Fig. 1). The process of moving the two spacecraft (aperture) in a formation [6, 7, ?], in order to sample a specified points in the uv-plane, is called the uv-plane coverage. Selecting the uv-points based on some apriori assumptions about the source (star) is an interesting problem in its own right. For the purpose of the present discussion, we assume that for each star, the uv-points have already been specified. For N stars, this amounts to providing a list of the form

$$\underbrace{(u_1,v_1),\ldots,(u_{n_1},v_{n_1})}_{\text{star 1}},\ldots \underbrace{(u_1,v_1),\ldots,(u_{n_2},v_{n_2})}_{\text{star 2}},\ldots \underbrace{(u_1,v_1),\ldots,(u_{n_N},v_{n_N})}_{\text{star }N},$$

where n_j is the number of uv points for star j.

The geometry of every optical interferometry mission is dictated by the path lengths that the starlight travels to the location where the image is finally synthesized. The Space Technology 3 (ST3) mission for example, is a two spacecraft optical interferometry mission [9]. In ST3, one spacecraft is at the focus of, and one moves on the perimeter of a paraboloid, in an earth-trailing orbit around the sun; see Figure 2. The spacecraft at the focus of the paraboloid, the combiner, carries the instruments necessary for collecting and combining light beams. The other spacecraft, the collector, carries the instruments for collecting light only. The light gathered at the collector is redirected to the combiner. Interference of light beam occurs at the combiner; see Figure 3. Note that other geometrical configurations can be adopted for an optical interferometry mission based on the number of spacecraft and the science objectives. For example, for planet finding, and ultimately, for planet

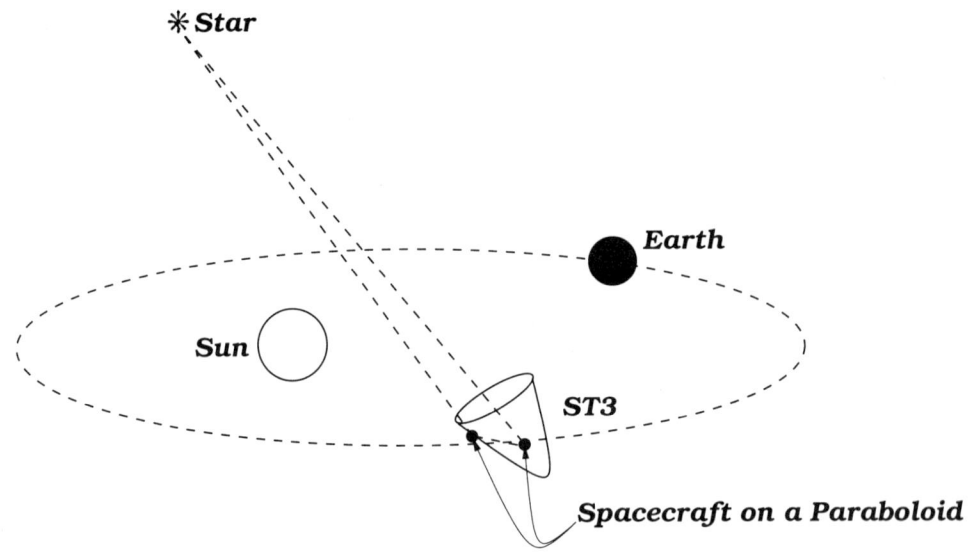

Figure 2: ST3 in a helio-centric orbit

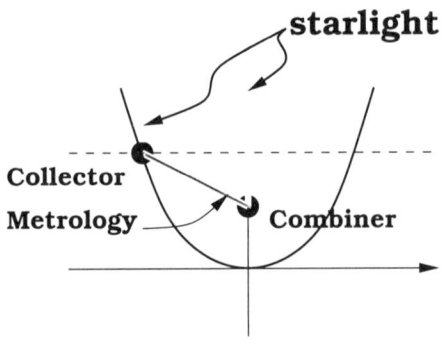

Figure 3: Light path and metrology in the ST3 configuration

imaging, a *nulling* capability is also required, necessitating a formation flying of five spacecraft [4][1]. Nevertheless, the *uv*-plane coverage still provides the central theme for the required spacecraft moves during all optical interferometry formation flying missions.

Mapping the *uv*-plane to the physical space

Given a set of *uv* points for a particular star, it is necessary to obtain the corresponding set of spacecraft positions; this correspondence is not unique. To make this observation more concrete, we briefly digress from our main discussion, to point out how in general, *uv*-point moves translate to the spacecraft moves in the physical space. For this purpose, the paraboloidal shape of the ST3 mission is used to demonstrate the procedure.

Let $\mathcal{F}^I$ denote the inertial reference frame. Let $\mathcal{F}^O$ denote the reference frame attached to the ST3 paraboloid, with its z-axis parallel to the axis of observation- the **xy**-plane of $\mathcal{F}^O$ is denoted by $\mathcal{F}^O_{\mathbf{xy}}$. Let M_I^O denote the direction cosine matrix that associates vectors represented in $\mathcal{F}^I$ to those that are represented in $\mathcal{F}^O$; similarly $(M_I^O)_{\mathbf{xy}}$ denotes the direction cosine matrix associated with the respective **xy**-planes.

Denote by **f** and **c**, the locations of the combiner and the collector spacecraft in the inertial reference frame. Let

$$\mathbf{r} := \mathbf{c} - \mathbf{f}.$$

Denote the projection of **r** onto the **xy**-plane of $\mathcal{F}^O$ by $\tilde{\mathbf{r}}$. Using (2) one has

$$\begin{bmatrix} x_2 \\ y_2 \end{bmatrix} = \begin{bmatrix} x_1 \\ y_1 \end{bmatrix} - \lambda \begin{bmatrix} u \\ v \end{bmatrix}.$$

Motivated by the required geometry of the ST3 and the associated *uv*-plane, we proceed to let

$$(\tilde{\mathbf{r}})_{\mathcal{F}^O_{\mathbf{xy}}} = \lambda \begin{bmatrix} u \\ v \end{bmatrix} = \begin{bmatrix} \tilde{r}_1 \\ \tilde{r}_2 \end{bmatrix}, \ (\tilde{\mathbf{f}})_{\mathcal{F}^O_{\mathbf{xy}}} = \begin{bmatrix} x_1 \\ y_1 \end{bmatrix}, \quad \text{and} \quad (\tilde{\mathbf{c}})_{\mathcal{F}^O_{\mathbf{xy}}} = \begin{bmatrix} x_2 \\ y_2 \end{bmatrix};$$

refer to Fig. 4. Then

$$(\tilde{\mathbf{r}})_{\mathcal{F}^O} = \begin{bmatrix} \tilde{r}_1 \\ \tilde{r}_2 \\ 0 \end{bmatrix}, \quad (\mathbf{f})_{\mathcal{F}^O} = \begin{bmatrix} 0 \\ 0 \\ f \end{bmatrix}, \quad \text{and} \quad (\mathbf{c})_{\mathcal{F}^O} = \begin{bmatrix} \tilde{r}_1 \\ \tilde{r}_2 \\ \frac{1}{4f}(\tilde{r}_1^2 + \tilde{r}_2^2) \end{bmatrix}.$$

The relative position of the collector with respect to the combiner is thus

$$(\mathbf{r})_{\mathcal{F}^I} = (M_I^O)^T (\mathbf{c} - \mathbf{f})_{\mathcal{F}^O}$$
$$= (M_I^O)^T \begin{bmatrix} \tilde{r}_1 \\ \tilde{r}_2 \\ \frac{1}{4}(\tilde{r}_1^2 + \tilde{r}_2^2) - f \end{bmatrix}.$$

Specifying a set of (u,v) points for imaging a given star thus translates to a sequence of vectors **r** in the Euclidean space. In other words, *uv*-plane coverage for N stars is simply a collection of the

---

[1] The five spacecraft TPF mission is the current baseline- the number of the spacecraft and its geometry is subject to change.

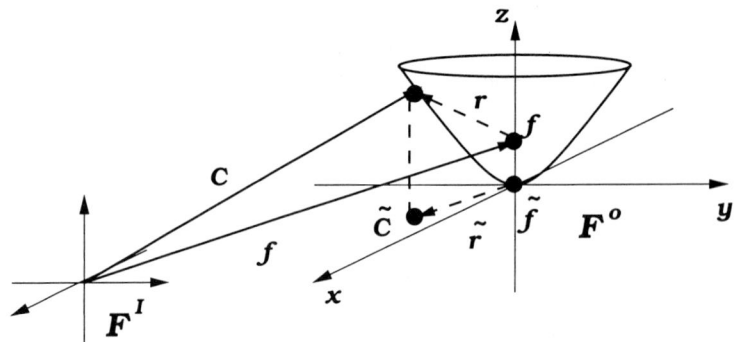

Figure 4: Projections from the ST3 paraboloid to the uv-plane

form

$$\underbrace{\mathbf{r}_1,\ldots,\mathbf{r}_{n_1}}_{\text{star } 1},\ldots\underbrace{\mathbf{r}_1,\ldots,\mathbf{r}_{n_2}}_{\text{star } 2},\ldots\underbrace{\mathbf{r}_1,\ldots,\mathbf{r}_{n_N}}_{\text{star } N},$$

where n_j is the number of uv points scheduled to be sampled for star j.

Mission modes

The main objective of an interferometry mission is to synthesize the images of distant objects, e.g., stars and planets. Achieving this objective via formation flying translates, in principle, to traversing a sequence of vectors in the Euclidean space. In reality however, image synthesis requires more than the re-positioning of the distributed spacecraft. The interaction between the operation of the science (optical) instruments and the control of the spacecraft bus on one hand, and the presence of different classes of uncertainties and constraints, necessitates the introduction of the mission modes.

In order to motivate the introduction of these modes of operation, consider again the ST3 mission. To image a given star, first it is required that a uv-pattern is chosen. This is then followed by choosing the sequence by which the formation shall traverse the uv points (these issues are not considered in this paper); see Fig. 5. Given a sequenced uv-pattern for each star, the ST3 mission is envisioned to proceed as follows [10] (also refer to Fig. 3):

1. Let $\text{star}_{\text{index}} = 1$.

2. <u>Mode 0-</u> Re-position/re-orient the spacecraft for imaging star $i = \text{star}_{\text{index}}$.

   (a) Let $\text{uv}_{\text{index}} = 1$.
   (b) <u>Mode 1-</u> Re-position/re-orient the spacecraft for sampling uv-point $j = \text{uv}_{\text{index}}$ using minimum amount of fuel and within the allocated amount of time.
      i. <u>Mode 2A-</u> **combiner starlight acquisition:** acquire starlight into the combiner instruments bore-sight. There is a requirement on the attitude control and a period of no thruster firing.

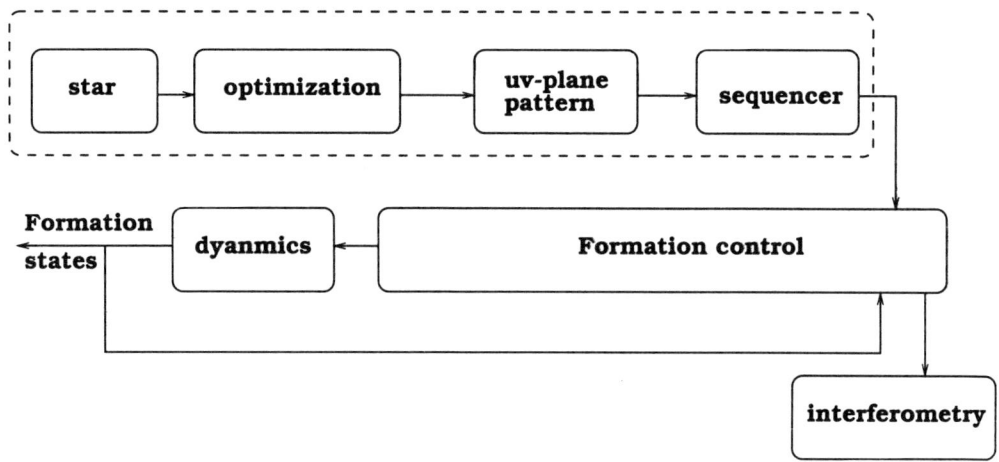

Figure 5: The general architecture

  ii. <u>Mode 2B-</u> **measure and reduce combiner rotation rate:** slow the rotation of the combiner spacecraft such that the instrument coarse stage can be locked down. There is a requirement on the spacecraft attitude. Limited thruster firing is allowed, provided that the instrument does not loose lock on the target star.
 iii. <u>Mode 2C-</u> **acquire angular metrology:** point the instrument laser metrology towards the collector spacecraft's intensity gradient detector. Limited thruster firing is allowed, provided that it does not lead to loosing the metrology lock.
  iv. <u>Mode 2D-</u> **acquire linear metrology:** acquire the linear measurements obtained via the use of collector retroreflectors using the laser metrology. Limited thruster firing is allowed, provided that it does not lead to loosing the metrology lock.
   v. <u>Mode 2E-</u> **acquire starlight from the collector:** obtain starlight gathered at the collector spacecraft. There is a required window of no-thruster firing.
  vi. <u>Mode 2F-</u> **align metrology and starlight:** during this step, metrology and the left combiner starlight alignment is performed. The formation flying control is expected to satisfy the requirement of the previous mode.
 vii. <u>Mode 2G-</u> **estimate formation attitude and separation and measure and reduce formation attitude rotation rates:** these two steps are intended to allow the instrument to precisely determine and correct for the formation attitude and separation, i.e., instrument delay, along with formation attitude rate (instrument delay rates). There is a requirement on the window with no-thruster firing.
viii. <u>Mode 2H-</u> **find fringe and stabilize and track fringe:** these two steps allow the instruments to search for and track the starlight fringe. The requirements are as in the previous mode.
  ix. <u>Mode 2I-</u> **measure fringe:** this step allows the instrument to measure the starlight fringe. Thruster firing is allowed as long as the instrument fringe tracking loop does

not loose lock. The requirements are as in the previous mode.

x. Let $uv_{index} = uv_{index} + 1$; if uv_{index} is less or equal to the allocated maximum number of uv-points for star j, go to 2b; otherwise, exit the loop.

(c) Let $star_{index} = star_{index} + 1$; if $star_{index}$ is less or equal to the allocated maximum number of stars to be imaged, go to 2; otherwise, exit the loop.

3. End of mission.[2]

Generally there is a specific sensor suite available for each mode of operation.

A question that naturally arises at this point is the following "Can a single control law satisfy the requirements of the ST3 operational modes?"

Control Modes

Reflecting on the operational modes of a typical optical interferometry formation flying mission, such as ST3, we are inevitably led to consider using a dedicated control law during each mode. For example, since the main requirement during the uv-plane coverage is minimizing the formation total fuel usage, an optimal control law appears to be a natural choice for this phase of the mission. During the transition (hand-off) to the imaging phase (Mode 2) however, fuel considerations are replaced by the accuracy requirement on the relative position, position rate, attitude, and angular velocity between the spacecraft in the formation. See Figs. 6,7. In the subsequent sections, we confine ourselves to the following two distinct formation modes (Fig. 8):

1. <u>Mode 1:</u> moving between the uv points, and

2. <u>Mode 2:</u> hand-off to the imaging phase.

To make the results specific, our presentation will be in the context of a two spacecraft formation flying; the reader shall not have a difficulty seeing how the approach can be generalized to n spacecraft scenarios.

For each of the two formation modes, two distinct control laws, suitably designed to satisfy the requirement of that mode, are first presented. We then proceed to *piece* these control modes together by introducing a switching mechanism. The resulting control architecture, by the virtue of its logical components (switches), is called a hybrid control architecture [11, 12, 13]; see also [14, 15, 16, 17].

Optimal control mode

The optimal control laws for the uv plane coverage serve two important roles in the analysis of the required formation maneuvers. First, as it is shown shortly, the optimal control laws provide an attractive scheme for formation reconfiguration, given an *apriori* fuel and life time allocation for the mission. Second, these control laws can be used during the early mission studies to determine the relationship between the fuel budget, the mission life time, and the number of science sources which can be imaged.

For this purpose, let us initially consider moving one spacecraft in the formation. The proposed optimal maneuvers are then subsequently generalized for the two or more spacecraft moves.

---

[2] For brevity, few other modes, such as the safing mode and the earth downlink, have been omitted.

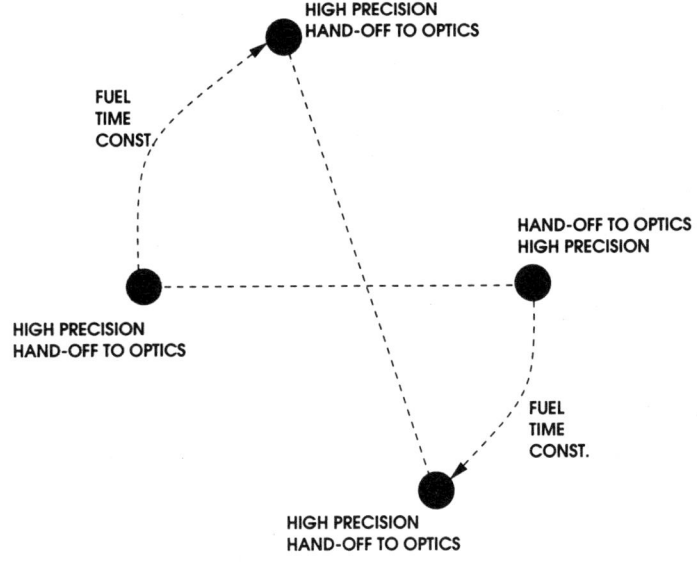

Figure 6: Optimal maneuvers and the hand-off regime

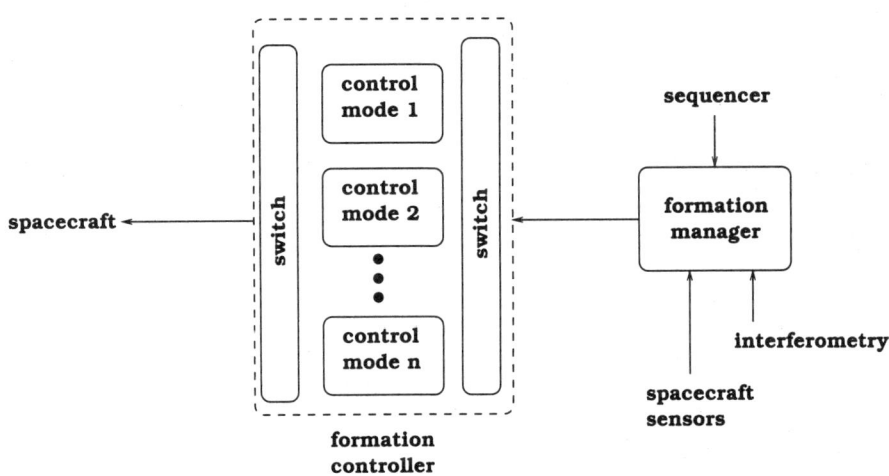

Figure 7: Switching between the control modes

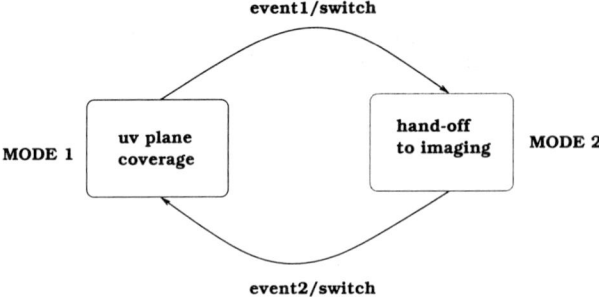

Figure 8: The two main control modes

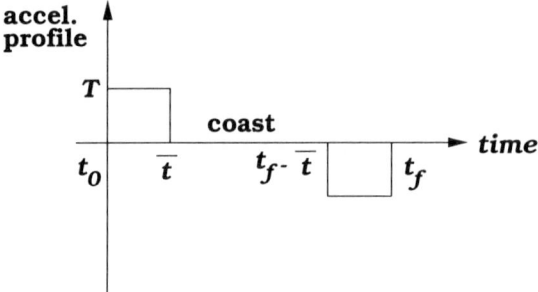

Figure 9: The acceleration profile

Let the fuel depletion rate for each spacecraft at time t be represented by

$$\frac{dF(t)}{dt} = \begin{cases} -\gamma T & \text{when} \quad F(t) > 0 \text{ and } \mathbf{u}(t) \neq 0 \\ 0 & \text{otherwise,} \end{cases}$$

where $\gamma = 1/(gI)$, T is the maximum available thrust, I is the thrusters' Isp, and $\mathbf{u}$ is the control force applied on the spacecraft. Assume that the spacecraft has already been oriented such that the thrust direction is parallel to the desired direction of the move, $\mathbf{d}$. Consider the scenario where full thrust T is applied on the spacecraft in the direction of $\mathbf{d}$ during the interval $[t_0, \bar{t}]$, the spacecraft coasts during $[\bar{t}, t_f - \bar{t}]$, followed by a full thrust T in the direction of $-\mathbf{d}$ during $[t_f - \bar{t}, t_f]$, where t_0 and t_f are the initiation and termination time of the maneuver (Fig. 9). This control law has the form of a bang-coast-bang control, a class which is known to contain an optimal fuel maneuver [18].

Integrating the spacecraft equation of motion twice and plugging in the boundary conditions we obtain

$$\bar{t}(d) = \frac{t_f(d)}{2} - \sqrt{\frac{t_f(d)^2}{4} - \frac{Md}{T}}, \tag{3}$$

where M denotes the mass of the spacecraft, and $d = \|\mathbf{d}\| := \mathbf{d}^T\mathbf{d}$, is the distance that the spacecraft has moved during the interval $[t_0, t_f]$ [19, 20].

The fuel usage for a move of length d is therefore

$$F(d) = 2\gamma T \bar{t}(d)$$

and thus,

$$t_f(d) = \frac{F(d)}{2\gamma T} + \frac{2\gamma M d}{F(d)}; \qquad (4)$$

Consider the following minimization problem

$$\min_{t_f(d), \bar{t}(d)} J[F(d), t_f(d)] := F(d) + \Lambda t_f(d). \qquad (5)$$

The parameter Λ in equation (5) provides the means of specifying the relative importance of fuel or time in the planning for the spacecraft movement (we say more about the selection of Λ shortly).

Setting the first derivative of the expression for $J[F(d), t_f(d)]$ (5) with respect to $F(d)$ to zero, after replacing $t_f(d)$ with its equivalent expression (4), one obtains

$$F^*(d) = 2\gamma \sqrt{\frac{\Lambda T}{2\gamma T + \Lambda}} \sqrt{Md} = 2\gamma \sqrt{MTd}\beta(\Lambda) \qquad (6)$$

and

$$t_f^*(d) = \frac{F^*(d)}{2\gamma T} + \frac{2\gamma M d}{F^*(d)} = \sqrt{\frac{Md}{T}}[\beta(\Lambda) + \frac{1}{\beta(\Lambda)}], \qquad (7)$$

where

$$\beta(\Lambda) = \sqrt{\frac{\Lambda}{2\gamma T + \Lambda}}; \qquad (8)$$

see Fig. 10. Recall that the uv-plane coverage translates directly to traversing a sequence of vectors in the Euclidean space. Suppose that the (u, v) points for a given star have already been sequenced so that the sequence

$$\mathbf{r}_1, \mathbf{r}_2, \ldots, \mathbf{r}_n$$

corresponding to the desired uv-plane coverage is given.

Let $d_{ij} = \|\mathbf{r}_i - \mathbf{r}_j\|$ $(i = 1, \ldots, n-1, j = i+1)$. Define

$$D := \frac{1}{N} \sum_{j=1}^{N} \sum_{i=1}^{n_j} \sqrt{d_{ij}},$$

where N is the number of stars scheduled to be imaged. The number D is the average square root distance traveled by the formation per star.

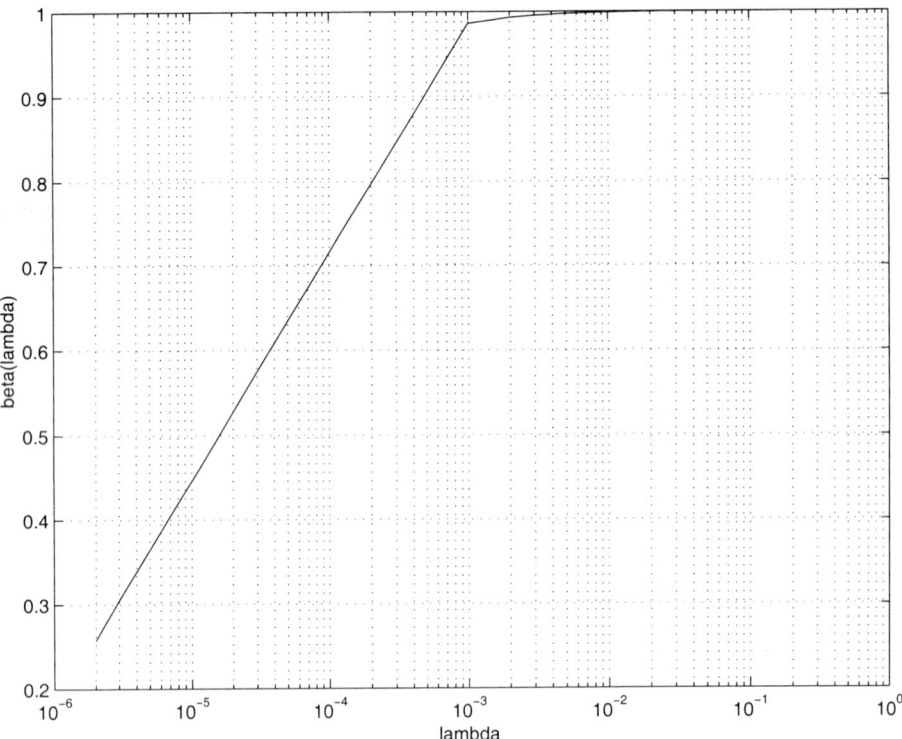

Figure 10: $\beta(\Lambda)$ ($T = 9$ mN, Isp = 65 sec)

Using equations (6)-(7), the total fuel and time required for the uv-plane coverage of N stars, using a spacecraft of mass M, is thus

$$f_{\text{total}} = 2\gamma DN\sqrt{MT}\beta(\Lambda)$$

and

$$t_{\text{total}} = DN\sqrt{\frac{M}{T}}[\beta(\Lambda) + \frac{1}{\beta(\Lambda)}].$$

Given that the maximum available thrust T, the Isp I, and the mass of the spacecraft M is fixed, Λ is the only parameter which can chosen to satisfy a given mission requirement. The choice of Λ effects:

- the acceleration/deceleration time, determined by $\bar{t}(d)$,
- the fuel consumption for a given desired repositioning of the spacecraft, and
- the time required to complete the maneuver for a given repositioning of the spacecraft.

How Λ effects these parameters is through the multiplier $\beta(\Lambda)$. We note that when $\Lambda \gg 2\gamma T$, $\beta(\Lambda) \approx 1$.

Example: Let $\Lambda = 2.572 \times 10^{-6}$, $T = 9$ mN, $m = 250$ kg, Isp $= 65$ sec, and $D = 152\sqrt{m}$. Then $\beta(\Lambda) = 0.2888$, and the total fuel usage for uv-plane coverage for N sources would be $f_{\text{total}} = 0.207 \times N$ kg, and the total time of the coverage would be $t_{\text{total}} = 1.1 \times N$ days. Given 20 kg fuel allocation and 100 days for the mission life time, the number of sources that we can be observed with this choice of Λ is 90. For $\Lambda = 1.0$, the same analysis shows that $f_{\text{total}} = 0.716 N$ kg, and $t_{\text{total}} = 0.567 \times N$ days; only 27 sources can be observed in this case.

Recall that one of the reasons in using optimal control laws for the uv-plane coverage is the convenience of being able to estimate the relationship between the mission life time, mission fuel usage, and the science return of the mission. For example, given f_{total} and t_{total}, the number of sources which can be observed by the formation while moving only one of the spacecraft in a two spacecraft formation, is

$$N = \lfloor \min\{\frac{f_{\text{total}}}{2\gamma D\sqrt{MT}\beta(\Lambda)}, \frac{t_{\text{total}}}{D\sqrt{\frac{M}{T}}[\beta(\Lambda) + \frac{1}{\beta(\Lambda)}]}\} \rfloor. \quad (9)$$

The mission runs out of fuel and time at about the same time if

$$\frac{f_{\text{total}}}{t_{\text{total}}} = \frac{2\gamma\sqrt{T}\beta(\Lambda)}{\frac{1}{\sqrt{T}}(\beta(\Lambda) + \frac{1}{\beta(\Lambda)})},$$

or, after some simplification

$$\alpha := \frac{f_{\text{total}}}{t_{\text{total}}} = \frac{\gamma T \Lambda}{\gamma T + \Lambda};$$

this expression can be solved for Λ; denote the solution by $\bar{\Lambda}$.

Example: Suppose that $M = 250$ kg, $T = 9$ mN, Isp $= 65$ sec, $D = 152\sqrt{m}$, $f_{total} = 20$ kg and $t_{total} = 2$ months. Then $\alpha = 3.858 \times 10^{-6}$, $\bar{\Lambda} = 5.307 \times 10^{-6}$, and $\beta(\bar{\Lambda}) = 0.398$. Using (9), we conclude that for this choice of $\bar{\Lambda}$, 70 sources can be observed.

More generally, given a fuel and time allocation, one can determine the number of sources which can be observed by the formation as follows:

1. Let $\alpha = f_{total}/t_{total}$ and $\bar{\Lambda} = (\alpha\gamma T)/(\gamma T - \alpha)$.

2. Set $\beta(\bar{\Lambda}) = \sqrt{\frac{\bar{\Lambda}}{(2\gamma T + \bar{\Lambda})}}$.

3. The number of sources which can be observed is then,

$$N = \frac{f_{total}}{2\gamma D \sqrt{MT} \beta(\bar{\Lambda})}.$$

The corresponding control law for a move of length d can also be determined as follows:

1. Find $t_f^*(d)$ using (7), i.e.,

$$t_f^*(d) = \sqrt{\frac{Md}{T}} (\beta(\bar{\Lambda}) + \frac{1}{\beta(\bar{\Lambda})}).$$

2. The acceleration time parameter is then found using (10),

$$\bar{t}(d) = \frac{t_f^*(d)}{2} - \sqrt{\frac{(t_f^*(d))^2}{4} - \frac{Md}{T}}.$$

The parameter $\bar{t}$ determines the form of the control law used for repositioning of the spacecraft for a move of length d. For example, given a predetermined set of moves, one would proceed to calculate the corresponding $\bar{t}$'s and have them organized in a lookup table as part of the control algorithm- see Figure 11. An example of the lookup table used by the control algorithm is shown in the following table, where the data of previous example has been used to generate the numerical values.

distance (meters)	control parameters (seconds)	
d	$t_f(d)$	$\bar{t}(d)$
39	3.0294×10^3	414.2515
68	4.0002×10^3	546.9987
87	4.5247×10^3	618.7161
136	5.6571×10^3	773.573
168	6.2875×10^3	859.7783
199	6.8431×10^3	935.7468
279	8.1027×10^3	1.108×10^3

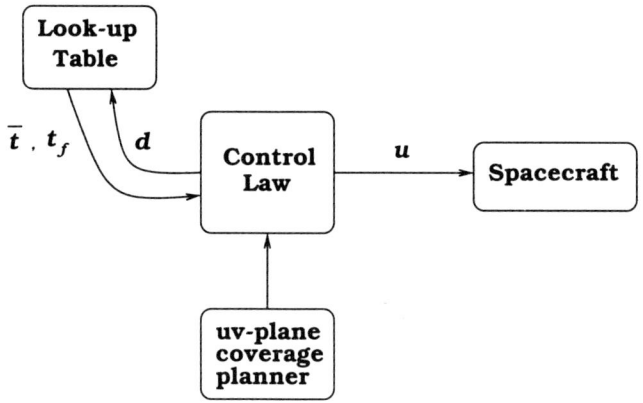

Figure 11: uv-plane coverage with one spacecraft

Consider now deriving an optimal control strategy for moving two (or more) spacecraft in the formation during the uv-plane coverage. This case is particularly relevant, when due to some mission considerations, it is desired that the area to mass ratio of both spacecraft remain close during the mission life time.

Denote the spacecraft in the formation as i and j such that $m^i = M$, and $m^j = \theta M$, where $\theta \geq 1$. Suppose that in order to move between two uv points, the *formation* is required to travel a distance d, i.e., the total movement of the formation adds up to d. Let i move δd and j move $(1-\delta)d$, where, $0 \leq \delta \leq 1$. According to equations (6)-(7) the corresponding fuel usage for spacecraft i and j would be $2\gamma\sqrt{m^i \delta T d}\beta(\Lambda^i)$ and $2\gamma\sqrt{m^j(1-\delta)Td}\beta(\Lambda^j)$, respectively, where Λ^i and Λ^j are the optimization parameters in the optimal control laws for spacecraft i and j.

The total fuel usage for the uv-plane coverage of N sources when moving both spacecraft is thus

$$f_{\text{total}} = 2\gamma DN\sqrt{MT}[\sqrt{\delta}\beta(\Lambda^i) + \sqrt{(1-\delta)\theta}\beta(\Lambda^j)].$$

Similarly, the total mission time would be

$$t_{\text{total}} = \max\{DN\sqrt{\frac{M\delta}{T}}[\beta(\Lambda^i) + \frac{1}{\beta(\Lambda^i)}], DN\sqrt{\frac{M\theta(1-\delta)}{T}}[\beta(\Lambda^j) + \frac{1}{\beta(\Lambda^j)}]\}.$$

The two spacecraft in the formation use the same amount of fuel during their respective moves and finish their maneuvers at the same time if we let

$$\Lambda^i = \Lambda^j \quad \text{and} \quad \delta = \frac{\theta}{1+\theta}.$$

Comparing the total fuel usage for the single spacecraft moves and the two spacecraft moves, we note that in the first case, the fuel usage is determined by $\beta(\Lambda)$ and in the second by

$$[\sqrt{\delta} + \sqrt{(1-\delta)\theta}]\beta(\Lambda).$$

Of course the Λ's in the two cases (moving spacecraft i only and moving both spacecraft i and j) do not have to be equal. If the two Λ's are *chosen* to be equal, since $(\sqrt{\delta} + \sqrt{(1-\delta)\theta} \geq 1$ when $\theta \geq 1$, moving both spacecraft always results in higher fuel usage (the mission life time shortens). However, it is best to choose $\bar{\bar{\Lambda}} < \bar{\Lambda}$ as we proceed to show ($\bar{\Lambda}$ was defined in the previous section).

Setting $\theta(1-\delta) = \delta$, the time required for the uv-plane coverage when moving both spacecraft is

$$t_{total} = DN\sqrt{(M\delta/T)}[\beta(\Lambda) + \frac{1}{\beta(\Lambda)}].$$

The corresponding fuel usage is thus $f_{total} = 4DN\gamma\sqrt{mT\delta}\beta(\Lambda)$. Similar to the single spacecraft case, the mission runs out of fuel and time at the same time by letting

$$\alpha := \frac{f_{total}}{t_{total}} = \frac{2\gamma T\Lambda}{\Lambda T + \gamma}. \tag{10}$$

Given α, γ, and T in equation (10) one can solve for Λ; denote the solution by $\bar{\bar{\Lambda}}$, i.e.,

$$\bar{\bar{\Lambda}} = \frac{\alpha\gamma T}{2\gamma T - \alpha}. \tag{11}$$

Note that $\bar{\bar{\Lambda}} < \bar{\Lambda}$; in this case the ratio between the fuel usage for one spacecraft moves and the two spacecraft moves is

$$\frac{\beta(\bar{\Lambda})}{2\sqrt{\delta}\beta(\bar{\bar{\Lambda}})}.$$

In view of equation (11) the number of sources which the formation can observe with two spacecraft moves is

$$N = \lfloor \frac{f_{total}}{4\gamma D\sqrt{MT\delta}\beta(\bar{\bar{\Lambda}})} \rfloor.$$

Example: Again, let $f_{total} = 20$ kg, $t_{total} = 2$ months, $m_1 = 250$ kg, $m_2 = 350$ kg, $T = 9$ mN, $Isp = 65$ sec, and $D = 152\sqrt{m}$. Then $\alpha = 3.858 \times 10^{-6}$, $\theta = 1.4$, $\delta = 0.583$, $\bar{\bar{\Lambda}} = 2.234 \times 10^{-6}$, and $\beta(\bar{\bar{\Lambda}}) = 0.271$. The number of sources that can be observed in this case is 67.

The control architecture for the uv-plane coverage while moving both (or multiple) spacecraft in the formation based on an optimal control law is shown in Fig. 12. As it has been shown, these control laws highlight the relationship between the number of stars which can be imaged using the formation and the fuel and time allocation required for the mission.

Coordination and disturbance rejection mode

Precision is the main requirement during the hand-off to the imaging phase of the mission. This requirement necessitates a coordination among the multiple spacecraft to the extend that the formation is "quiet" enough for initialization of the imaging process. An optimal control strategy

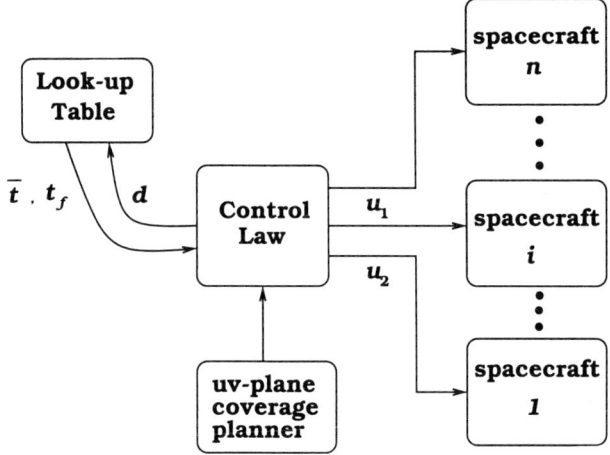

Figure 12: uv-plane coverage with two (or more) spacecraft moves

however, as described in the previous section, is not suitable for the level of precise coordination that is required during the hand-off regime. This is mainly due to the presence of different classes of uncertainty, including, but not limited to, the differential misalignment in the thrust direction, timing issues, environmental disturbances (e.g., solar pressure), and uncertainties in the modeling of the formation.

Consider for example, using the optimal control law of the previous section for moving one of the spacecraft (the collector) in order to sample *three* consecutive uv points. Furthermore, assume that the imaging phase is initialized if the sum of the position and velocity errors is within a ball of radius 10^{-4} (units for position and velocity are in meters and meters per second, respectively). Fig. 13 shows the performance of this optimal control maneuvers. The simulation was developed so that it provides means of capturing the interaction between the uv-plane coverage mode and the imaging phase. The parameters used for the simulation were as follows

$$M = 0.25 \text{ kg}, T = 9 \times 10^{-3} \text{ Newtons}, I = 65 \text{ sec},$$
$$g = 9.8 \ m/s^2, f_{\text{total}} = 20 \text{ kg, and } t_{\text{total}} = \text{ two months.}$$

Note that contrary to the requested set of maneuvers, only *two* consecutive uv maneuvers actually occur! In fact, since the timing of the optimal control algorithm is off by 0.02%, right after the sampling of the first uv-point, the state error never enters the region required by the imaging phase; see Fig. 14,15.

In order to effectively address the limitations of the optimal control methodology during the hand-off mode, two additional control strategies are introduced. The first strategy builds around a proportional-derivative-integral (PID) controller [7]- the second, adopts a robust control strategy formulated in terms of a system of linear matrix inequalities. In order to retain most of the optimality properties of the optimal control laws, however, we only resort to the latter control laws when the formation is in the "vicinity" of the desired uv point. This translates to designating a capturing

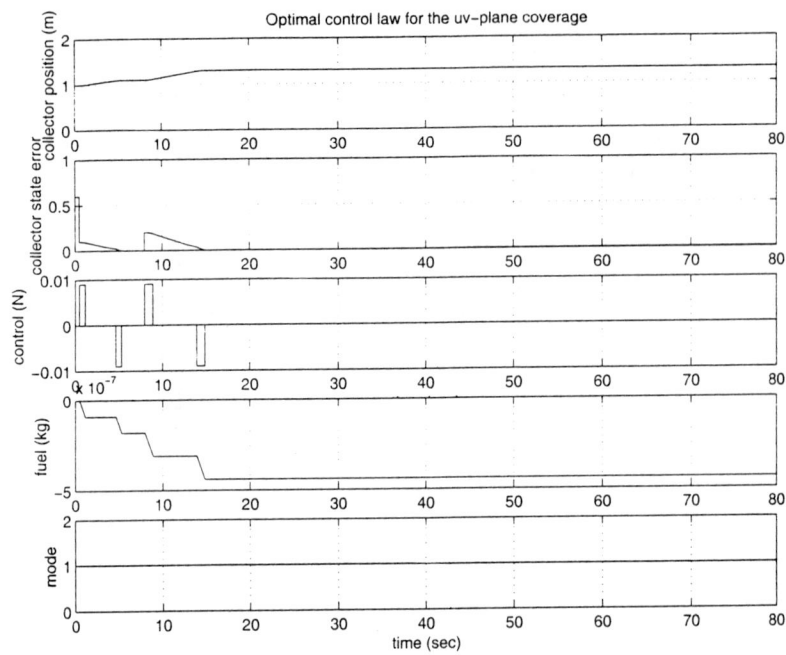

Figure 13: Performance of the optimal control for covering three uv points

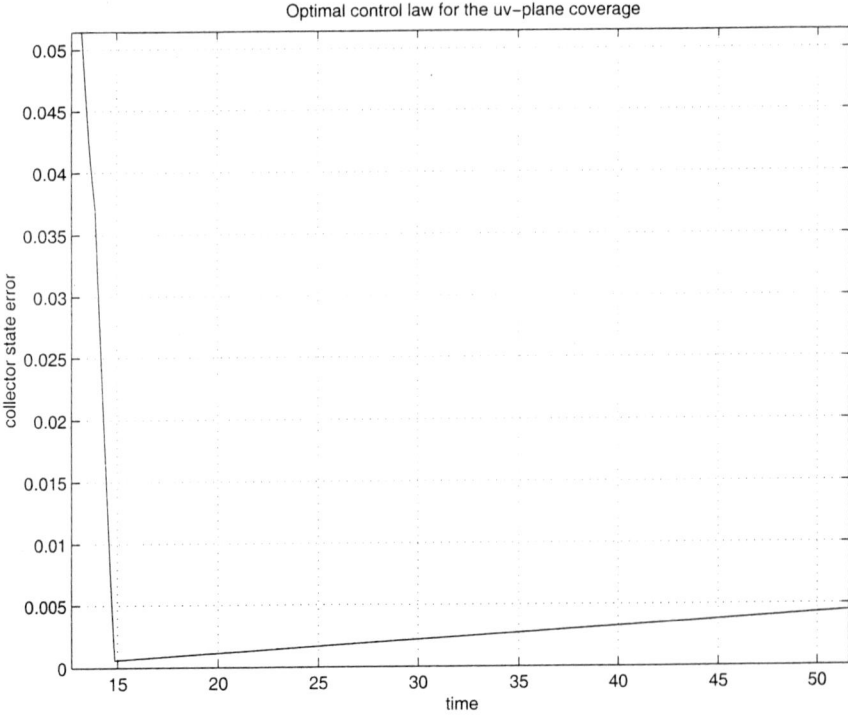

Figure 14: Optimal control performance- handoff never occurs

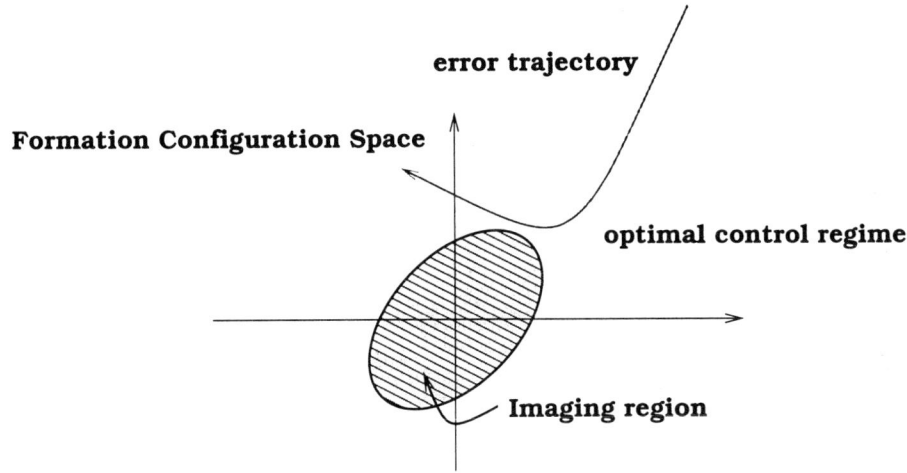

Figure 15: Optimal control performance- handoff never occurs

region in the formation state space (positions/velocities) and switching between multiple controllers depending on where the formation is in this state space at a given time; see Fig.16. The approach leads us directly to the hybrid control architecture as shown in Fig. 17. First, let us derive the candidate controllers for the hand-off regime.

Suppose that spacecraft j is required to keep a relative position $\mathbf{h}^{ji}(t)$ with respect to spacecraft i for all $t \in [t_0, t_f]$- since our interest is mainly in the interferometry applications, we consider the case where $\ddot{\mathbf{h}}^{ji}(t) = 0$, for all t.

The position error for spacecraft i is now,

$$\mathbf{e}^j(t) = \mathbf{r}^j(t) - \mathbf{r}_d^i(t) = \mathbf{r}^j(t) - \mathbf{r}^i(t) - \mathbf{h}^{ji}(t) \ .$$

Let m^i, m^j, $\mathbf{f}_c^i$, $\mathbf{f}_d^i$, $\mathbf{f}_c^j$, $\mathbf{f}_d^j$, be the mass of spacecraft i and j, and the forces acting on them due to control action and disturbances, respectively. Let

$$\mathbf{u}_c^k(t) = \frac{\mathbf{f}_c^k(t)}{m^k(t)}, \quad \text{and} \quad \mathbf{u}_d^k(t) = \frac{\mathbf{f}_d^k(t)}{m^k(t)} \quad (k = i, j).$$

Thereby

$$\ddot{\mathbf{e}}^j(t) = [\mathbf{u}_c^j(t) - \mathbf{u}_c^i(t)] + [\mathbf{u}_d^j(t) - \mathbf{u}_d^i(t)].$$

Let $\mathbf{z}(t) = \begin{bmatrix} \mathbf{e}^j(t) & \dot{\mathbf{e}}^j(t) \end{bmatrix}^T$. Suppose that

$$\mathbf{u}_c^j(t) = \mathbf{u}_c^i(t) - Kz(t).$$

The error dynamics for spacecraft j can thus be represented via

$$\dot{\mathbf{z}}(t) = (A + B_u K)\mathbf{z}(t) + B_w \mathbf{w}(t) \tag{12}$$

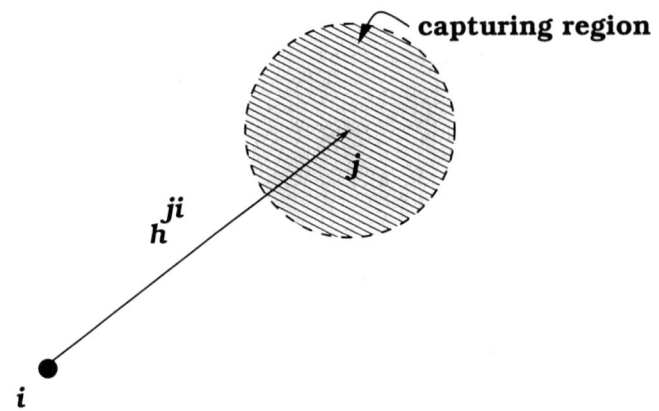

Figure 16: Capturing region

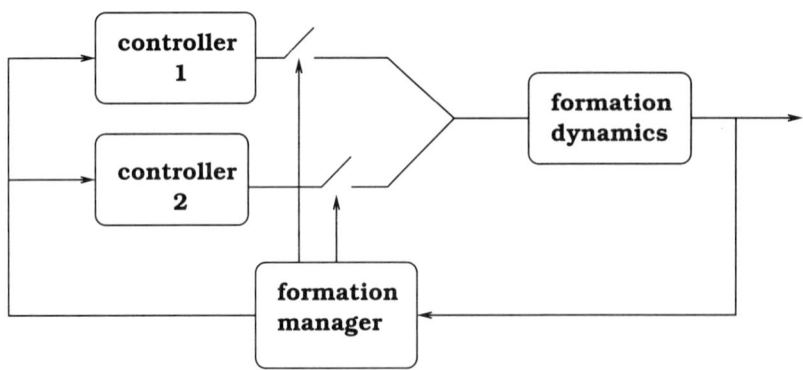

Figure 17: Hybrid control architecture

where

$$A = \begin{bmatrix} 0 & I \\ 0 & 0 \end{bmatrix}, \quad B_u = B_w = \begin{bmatrix} 0 \\ I \end{bmatrix}, \quad \text{and} \quad \mathbf{w}(t) = \mathbf{u}_d^j(t) - \mathbf{u}_d^i(t).$$

Control law

The plant model described by equation (12) is clearly in a form which has been considered extensively in the control literature. For example, it is well known that in the case where the misalignment in the thrust direction is negligible and $\mathbf{w}$ is an unknown constant disturbance, a PID controller of the form

$$\mathbf{u}(t) = K\mathbf{z}(t) + \mu \int_0^t \mathbf{e}(\tau)d\tau, \tag{13}$$

provides an effective means of assuring that $\mathbf{z}(t) \to 0$ as $t \to \infty$ (internal model principle [21]). Of course, one still has to choose the state feedback gain $K \in \Re^{3 \times 6}$ and the constant μ in some judicious manner. Moreover, the performance of the resulting control system has to be studied in light of a bound on the maximum available thrust. It should be noted that the matrix gain K in equation (13) can be found by solving the linear matrix inequality (LMI)

$$AQ + QA + BY + Y^T B^T < 0, \quad Q > 0$$

and then letting $K = YQ^{-1}$ [22]. In fact, due to the simple structure of the matrices A and B in equation (12), it can be verified that for any $\alpha > 0$, a constant feedback gain of the form $K = -\alpha[I\ I]$, provides sufficient means of guaranteeing the global asymptotic stability of the origin for the state error governed by equation (12).

In the case where the disturbance can not be adequately modeled as a constant vector and there is an uncertainty in the thrust direction, a more elaborate control synthesis approach can be adopted. For this purpose, assume that a polytopic region is used to capture the uncertainty in the thrust direction. Consider the state feedback of the form $\mathbf{u}(t) = K\mathbf{z}(t)$. The resulting closed loop system is therefore

$$\Sigma: \quad \dot{\mathbf{z}}(t) = (A + B_u K)\mathbf{z}(t) + B_w \mathbf{w}(t), \quad B_u \in \text{Co}\,\mathcal{B} := \{B_1, \ldots, B_n\}, \tag{14}$$

where Co A denotes the convex hull of the set A. The problem of disturbance rejection can now be formulated in terms of minimizing the RMS gain of the closed loop system (Fig. 18). The RMS gain of the system shown in Fig. 18 is defined as

$$\sup_{\text{RMS}[\mathbf{w}(t)] \neq 0} \frac{\text{RMS}[\mathbf{z}(t)]}{\text{RMS}[\mathbf{w}(t)]}$$

where

$$\text{RMS}[\zeta(t)] := \left[\lim_{T \to \infty} \sup \frac{1}{T} \int_0^T \zeta(t)^T \zeta(t)\, dt\right]^{\frac{1}{2}}.$$

For this purpose, consider the quadratic Lyapunov function $V[\mathbf{z}(t)] := \mathbf{z}(t)^T P\mathbf{z}(t)$, for some positive definite matrix P. Suppose that the matrix P is chosen in such a way that there exists a nonnegative number γ satisfying

$$\frac{d}{dt}V[\mathbf{z}(t)] + \mathbf{z}(t)^T \mathbf{z}(t) - \gamma^2 \mathbf{w}(t)^T \mathbf{w}(t) \leq 0. \tag{15}$$

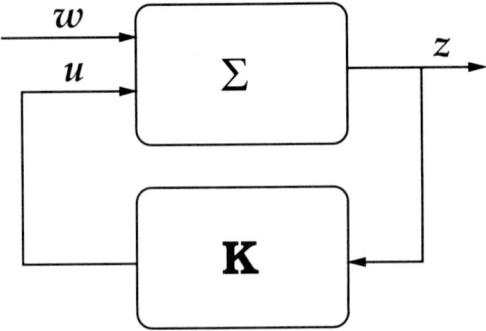

Figure 18: The closed loop system and the disturbance rejection problem

Provided that the inequality (15) holds, one can conclude that

$$V[\mathbf{z}(T)] + \lim_{T\to\infty} \frac{1}{T} \int_0^T \mathbf{z}(t)^T \mathbf{z}(t)\, dt - \lim_{T\to\infty} \frac{1}{T} \int_0^T \gamma^2 \mathbf{w}(t)^T \mathbf{w}(t)\, dt \leq 0. \tag{16}$$

Since $V[\mathbf{z}(T)] \geq 0$, the inequality (16) on the other hand implies that

$$\frac{\text{RMS}[\mathbf{z}(t)]}{\text{RMS}[\mathbf{w}(t)]} \leq \gamma,$$

i.e., ensuring that inequality (15) holds is equivalent to guaranteeing that the RMS gain of the closed loop system is bounded by γ.

We now express the inequality (15) is terms of a family of LMIs. Expanding the inequality (15) for the *family* of plants described by the differential equation (12), one obtains

$$[(A + B_u K)\mathbf{z}(t) + B_w \mathbf{w}(t)]^T P \mathbf{z}(t) + \mathbf{z}(t)^T P[(A + B_u K)\mathbf{z}(t) + B_w \mathbf{w}(t)]$$
$$+ [\mathbf{z}(t)]^T [\mathbf{z}] - \gamma^2 \mathbf{w}(t)^T \mathbf{w}(t) \leq 0, \tag{17}$$

for all $B_u \in \mathcal{B}$, and all time signals $\mathbf{z}(t)$ and $\mathbf{w}(t)$. This expression can be represented in terms of the inequality

$$\begin{bmatrix} (A + B_i K)Q + Q(A + B_i K)^T + B_w B_w^T & Q \\ Q & -\gamma^2 I \end{bmatrix} \leq 0 \quad (i = 1, \ldots, n),$$

where the inequality for matrices is interpreted in the sense of Löwner, i.e., $A \leq 0$ if and only if A is negative semi-definite. By letting $Y = KQ$, the matrix inequality (18) can be written as an LMI

$$\begin{bmatrix} AQ + QA^T + B_i Y + Y^T B_i^T + B_w B_w^T & Q \\ Q & -\gamma^2 I \end{bmatrix} \leq 0 \quad (i = 1, \ldots, n),$$

where $Q > 0$ [23]. In order to find a synthesis a controller which aims at minimizing the RMS gain of the family of closed loop systems, representing the follower spacecraft dynamics, one can proceed to solve the semi-definite program (SDP),

$$\min_{\gamma>0, Q>0, Y} \gamma \qquad (18)$$

$$\begin{bmatrix} AQ + QA^T + B_i Y + Y^T B_i^T + B_w B_w^T & Q \\ Q & -\gamma^2 I \end{bmatrix} \leq 0 \quad (i = 1, \ldots, n); \qquad (19)$$

then let $K = YQ^{-1}$. Since these matrix inequalities can be efficiently solved via the interior point methods [24], the above approach provides an effective procedure for addressing the problem of thruster misalignment and disturbance rejection during the precision coordination required during the hand-off regime.

Putting the modes together

Up to now, we have proposed different classes of controllers for two distinguished modes of operation of the formation. In the first phase, an optimal control law is used to reconfigure the formation in order to bring it to the vicinity of final desired configuration. Subsequently, a PID-based or a robust control control mode takes over and coordinates the spacecraft movements, guaranteeing a disturbance rejection performance measure- taking into account the possible uncertainties in the thrust direction. The performance of this switching control scheme is shown in Fig. 19. The capturing region suggested in Fig. 16, can in principle, be any closed geometric figure which encompasses the origin. However, due to computational considerations, it is customary to choose these regions to be ellipsoidal. Note that from a fuel optimality point of view, the ellipsoidal region should have a small volume. From the robustness point of view however, large volumes are preferred for the capturing ellipsoid. Consequently, there is a trade-off in sizing the capturing region. For the present discussion the imaging requirements essentially dictated the size of this capturing region. A representative set of simulation results for the proposed switching control laws for uv-plane coverage is shown in Figs. 19-20. The simulation was developed in the STATEFLOW environment [25]. In the case where the capturing region has an infinite diameter, one essentially recovers the PID-based/robust controller. Fig. 21- 22 demonstrates the simulation results for the case where a PID-based controller is used during the complete uv-plane coverage. As can be seen from these simulations results, the PID-based controller, in general, leads to higher fuel usage for the uv plane coverage as compared with the switching control approach. A formation control architecture, which takes into account the requirements for each operational modes of the formation and *autonomously* chooses the appropriate control law, thus emerges as shown in Fig. 23.

The proposed control switching strategy for putting the optimal and coordinated control in the formation control loop is essentially a *proximity* control switching. The proximity switching strategy is motivated by the desire to maximize the benefit of using the optimal control while guaranteeing a certain degree of robust performance in the control design.

Consider the ellipsoid defined by the positive definite matrix Q,

$$\mathcal{E}_Q := \{\zeta : \zeta^T Q^{-1} \zeta \leq 1\}.$$

The proximity switching strategy is essentially the selection of the positive definite Q; in the case of the uv-plane coverage using a dual mode controller, this translates to switching to the second control mode if the state of the formation enters $\mathcal{E}_Q$. Note that if $Q = P^{-1}$ where P is the solution of the LMIs (19), then, in the absence of disturbances, the ellipsoid $\mathcal{E}_Q$ is *invariant* with respect to (12),

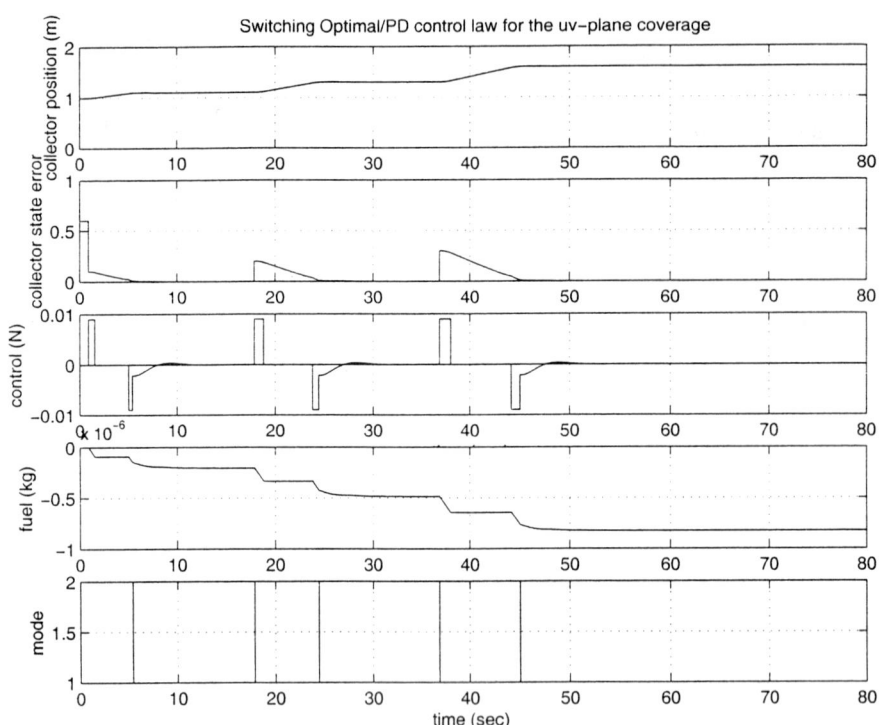

Figure 19: Switching optimal/PD controller for the uv-plane coverage

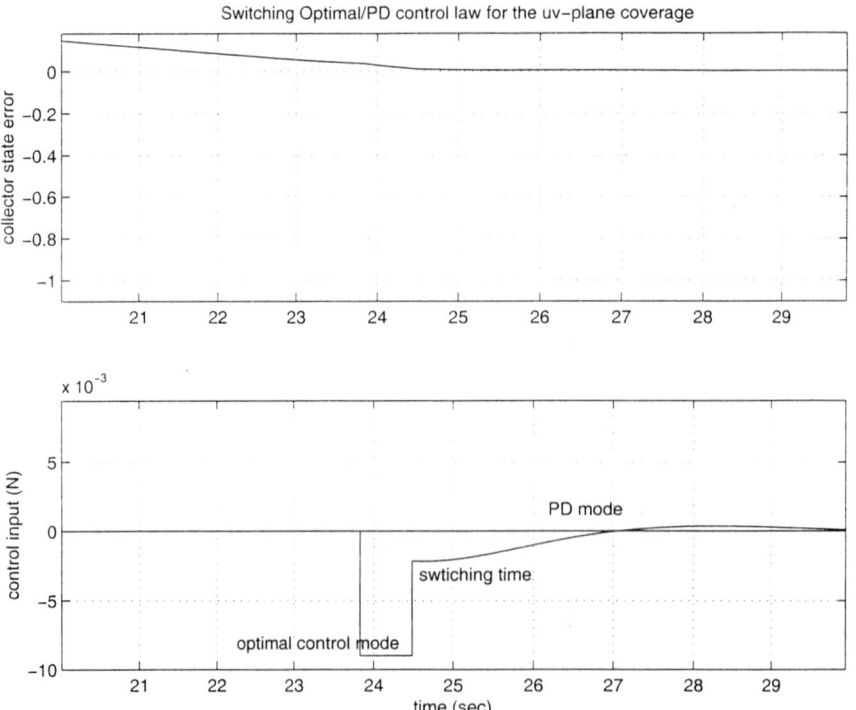

Figure 20: The state error and control for switching optimal/PD controller

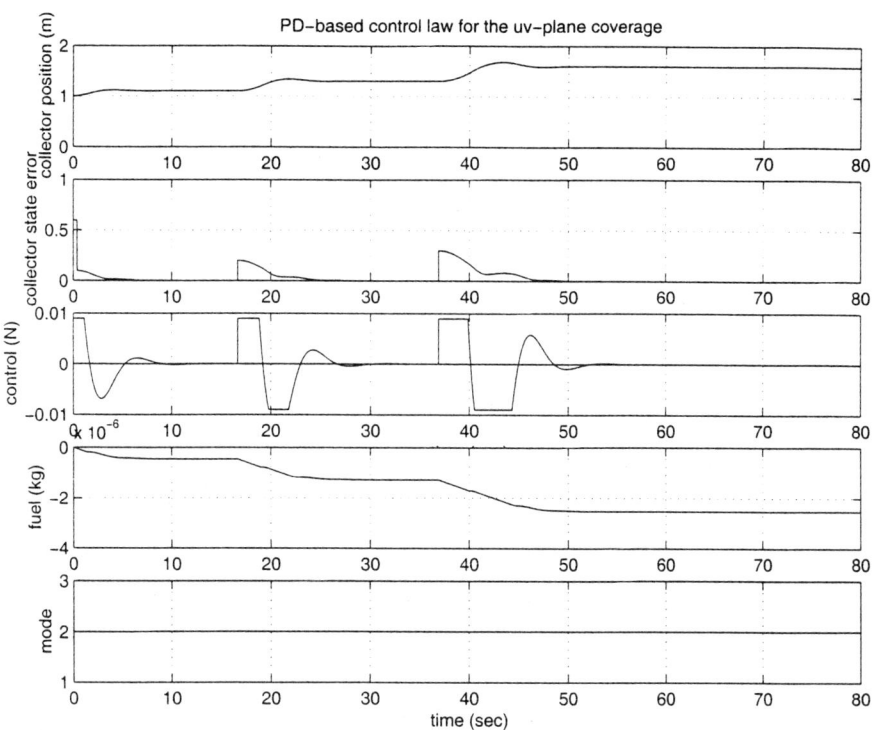

Figure 21: PD-based controller for the uv-plane coverage

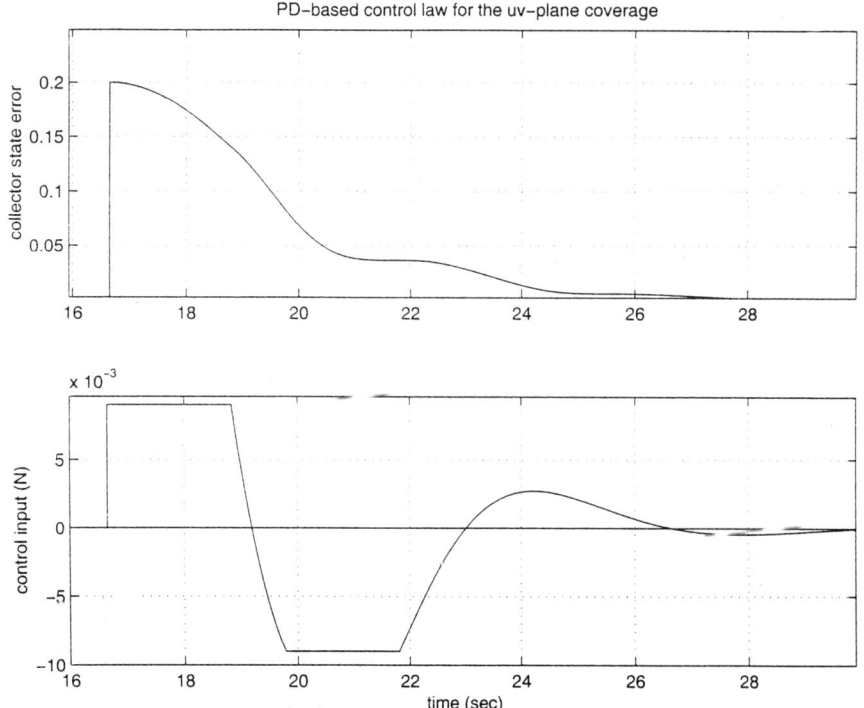

Figure 22: The state error and control for PD controller

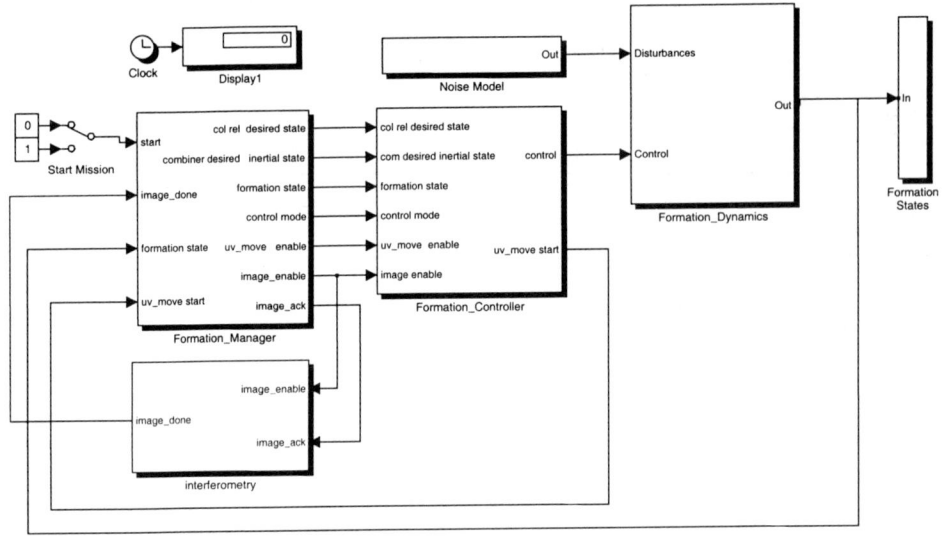

Figure 23: Formation Control Simulink Diagram

i.e., the trajectories of the differential equation (12) remains inside the ellipsoid once it enters it [22]. This invariance property guarantees that, in the absence of disturbances, *chattering*, the phenomena of going back and forth between the modes, does not occur. In the presence of disturbances however, it is judicious to allow switching back to the first mode only if the trajectory leaves an ϵ-*neighborhood* of $\mathcal{E}_Q$; see Fig. 24.

Concluding Remarks

Starting with a reflection on the operational modes of a separated spacecraft optical interferometry mission, the paper proceeded to show that multiple controller switching provides an attractive way of satisfying multiple performance criteria during the different phases of the formation flying mission. Simulation results and theoretical analysis supplemented the case for the viability of the proposed hybrid control architecture for the uv-plane coverage.

Acknowledgment

The research of the authors was carried out at the Jet Propulsion Laboratory, California Institute of Technology, under a contract with the National Aeronautic and Space Administration.

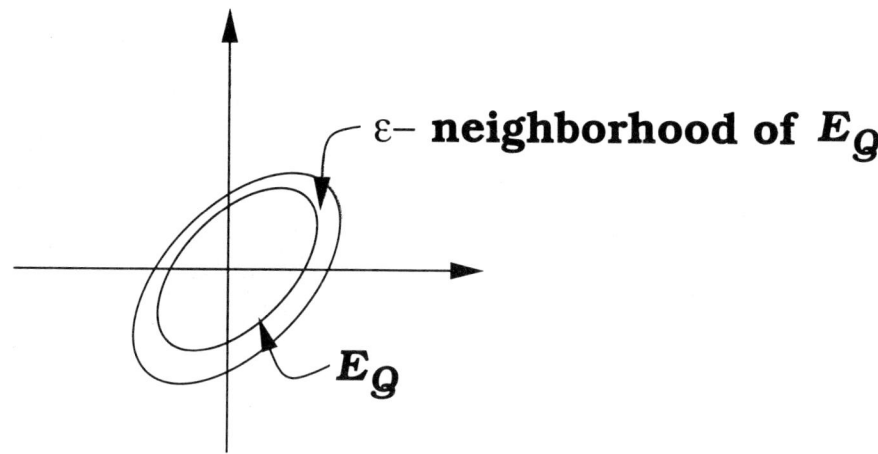

Figure 24: $\mathcal{E}_Q$ and its ϵ-neighborhood

References

[1] E. Wong (Editor), *Cassini Project Control Analysis Book*, Jet Propulsion Laboratory, California Institute of Technology, 1997.

[2] J. Davis, "Measuring stars with high angular resolution: current status and future prospects," in *Calibration of Fundamental Stellar Quantities*, D. S. Hayes, L. E. Pasinetti, and A. G. D. Philips (Editors), Reidel Dordrecht, 1985.

[3] R. Stachnik, K. Ashlin, and K. Hamilton, Space-Station-SAMSI: A spacecraft array for Michelson spatial interferometry, in *Bulletin of the American Astronomical Society*, 16:818–827, 1984.

[4] *Terrestrial Planet Finder: Origins of Stars, Planets, and Life*, Jet Propulsion Laboratory, California Institute of Technology, May 1999.

[5] M. Born and E. Wolf, *Principles of Optics* (sixth edition), Cambridge, 1997.

[6] A. B. Decou. Multiple spacecraft optical interferometry- Preliminary feasibility assessment. Technical report, Jet Propulsion Laboratory, 1991.

[7] P. K. C. Wang and F. Y. Hadaegh, Coordination and control of multiple microspacecraft moving in formation, *Journal of the Astronautical Sciences*, 44(3):315–355, 1996.

[8] P. K. C. Wang, F. Y. Hadaegh, and K. Lau, Synchronized formation rotation and attitude control of multiple free-flying spacecraft, *Journal of Guidance, Control, and Dynamics*, 22: 28–35, 1999.

[9] *The Space Technology 3 Mission*, Jet Propulsion Laboratory, California Institute of Technology, 1998.

[10] K. Lau, *The ST3 Requirements Document*, Jet Propulsion Laboratory, California Institute of Technology, 1999.

[11] P. Antsaklis, W. Kohn, A. Nerode, and S. Sastry (Editors), *Hybrid Systems 4*, Lecture Notes in Computer Science. Springer-Verlag, 1997.

[12] M. Mesbahi and F. Y. Hadaegh, Graphs, Matrix Inequalities, and Switching for the Formation Flying Control of Multiple Spacecraft, *Journal of Guidance, Navigation, and Control* (submitted).

[13] A. S. Morse. Control using logic based switching, In A. Isidori, editor, *Trends in Control: A European Perspective*, Springer-Verlag, 1995.

[14] M. S. Branicky. Multiple Lyapunov functions and other analysis tools for switched and hybrid systems, *IEEE Transactions on Automatic Control*, 43:475–482, 1998.

[15] R. W. Brockett. Hybrid models for motion control systems, In H. L. Trentelman and J. C. Willems, editors, *Essays in Control*, pages 29–53. Birkhauser, 1993.

[16] J. Malmborg, B Bernhardson, and K. J. Astrom, A stabilizing switching scheme for multi-controller systems, In *Proceedings of the IFAC World Congress*, San Francisco, CA, 1996.

[17] T. Pavlidis. Stability of systems described by differential equations containing impulses, *IEEE Transactions on Automatic Control*, 12:43–45, 1967.

[18] R. F. Stengel, *Optimal Control and Estimation*, Dover, 1994.

[19] R. W. Beard and F. Y. Hadaegh, Fuel Optimization for Unconstrained Rotation of Spacecraft Formations, *Journal of Astronautical Sciences* (submitted).

[20] R. W. Beard and F. Y. Hadaegh, Finite thrust control for satellite formation flying with state constraints, In *Proceedings of the American Control Conference*, 1998.

[21] T. Kailath. *Linear Systems*, Prentice-Hall, Englewood Cliffs, New Jersey, 1980.

[22] S. P. Boyd, L. EL Ghaoui, E. Feron, and V. Balakrishnan, *Linear Matrix Inequalities in System and Control Theory*, SIAM, Philadelphia, 1994.

[23] M. Mesbahi, M. G. Safonov, and G. P. Papavassilopoulos, Bilinearity and complementarity in robust control, In *Recent Advances on LMI Approach in Control*, SIAM, Philadelphia, 1999.

[24] Y. Nesterov and A. Nemirovskii, *Interior-Point Polynomial Algorithms in Convex Programming*, SIAM, Philadelphia, 1994.

[25] *StateFlow user manual*, Mathworks, 1998.

Section III
GUIDANCE AND CONTROL ISSUES FOR THE INTERNATIONAL SPACE STATION

SESSION III

Joint Session Chairperson: Karen Frank
NASA Johnson Space Center

Joint Session Charperson: Thomas Russell
Boeing Space and Defense

Local Session Chairperson: Dr. Steve Jolly
Lockheed Martin Astronautics

The following papers were not available for publication:

AAS 00-023   "Flight Control Algorithms for International Space Station Propulsion Module," by C. Rui, M. Rivera, G. Cortes, The Boeing Company

AAS 00-024   "Operational Constraints Associated with the Interim Control Module," by J. Bendle, United Space Alliance

The following paper numbers were not assigned:
AAS 00-028 to -030

AAS 00-021

EVOLUTION OF INTERNATIONAL SPACE STATION GN&C SYSTEM ACROSS ISS ASSEMBLY STAGES

Roscoe Lee[*]

INTRODUCTION

The GN&C system for the International Space Station (ISS) will evolve with the deployment of a sequence of elements: (1) Functional Cargo Block (FGB); (2) Service Module (SM); and (3) U.S. Laboratory. The GN&C capabilities and components of each of these elements are described below.

FUNCTIONAL CARGO BLOCK (FGB)

The Functional Cargo Block (FGB) was built by the M. V. Khrunichev State Scientific Research Space Center under contract to the Boeing Defense and Space Group Co., was launched on 20 November 1998 and is currently operating on-orbit. Table 1 presents general technical characteristics of the FGB. Figure 1A provides a general view of the FGB and Figure 1B provides the relationship of the FGB to other ISS elements through Stage 1R (Reference 1).

TABLE 1
BASIC FGB TECHNICAL SPECIFICATIONS

Launch vehicle	Proton
Launch mass	24,100 kg
Orbital mass	20,040 kg
Length	12,991 mm
Maximum diameter	4,100 mm

The FGB is the exclusive ISS GN&C system until after the Service Module is docked to the ISS. The FGB GN&C system performs the following functions:

- Determines attitude and attitude rate and propagates state vector based on ground updates
- Performs orbital correction maneuvers (reboosts)
- Performs attitude maneuvers (to conserve propellant, the FGB performs X-nadir spin the majority of the on-orbit time)
- Controls FGB motion during acquisition, rendezvous, and docking of SM
- Performs on-orbit monitoring of GN&C equipment and operations and automatically aborts current active control operations, if off-nominal conditions are detected.

---

[*] TRW Systems and Information Technology Group, 1110 NASA Road 1, Suite 303, Houston, Texas 77058.

Figure 2 provides an overview of the FGB GN&C system. The sensing components of the FGB GN&C system are:

- Four infrared horizon sensors (operate in pairs) for determining attitude
- One set of single-axis rate gyros for determining attitude rate
- One X-axis accelerometer to support translational maneuvers (not used during ISS operations)

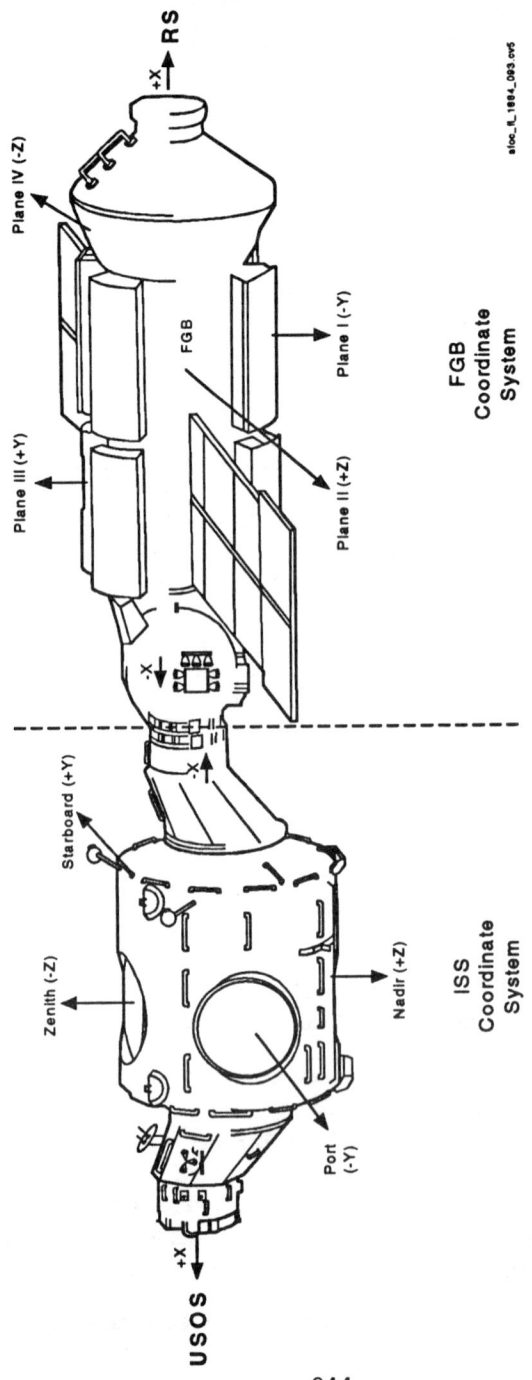

Figure 1A View of FGB and Node

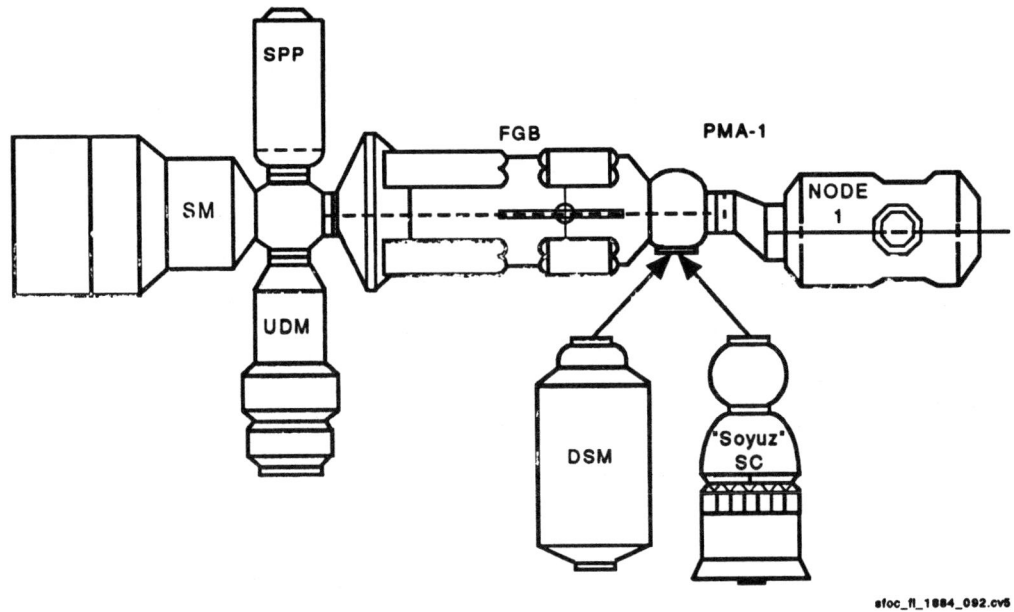

Figure 1B FGB in Relation to Other ISS Elements

The key computational components of the FGB GN&C system are:
- A digital computer which is internally redundant and performs two-of-three voting on a continuous basis
- Two Read-only Memories, which store basic programs and constants of the digital computer algorithms
- Three Special Memory Devices, which are non-volatile memory components and store flight plans and data uplinked from the Russian Ground Stations
- Input/output devices to relay commands to FGB thrusters and engines.

The FGB GN&C system control effectors are:
- Two main engines (417 kg_f) for reboosts
- Twenty-four 40 kg_f attitude control thrusters (two redundant sets controlled by manifold valves)
- Sixteen 1.3 kg_f vernier attitude control thrusters (two redundant sets controlled by manifold valves)

The locations and orientations of these engines and thrusters are depicted in Figures 3A – 3E (Reference 1).

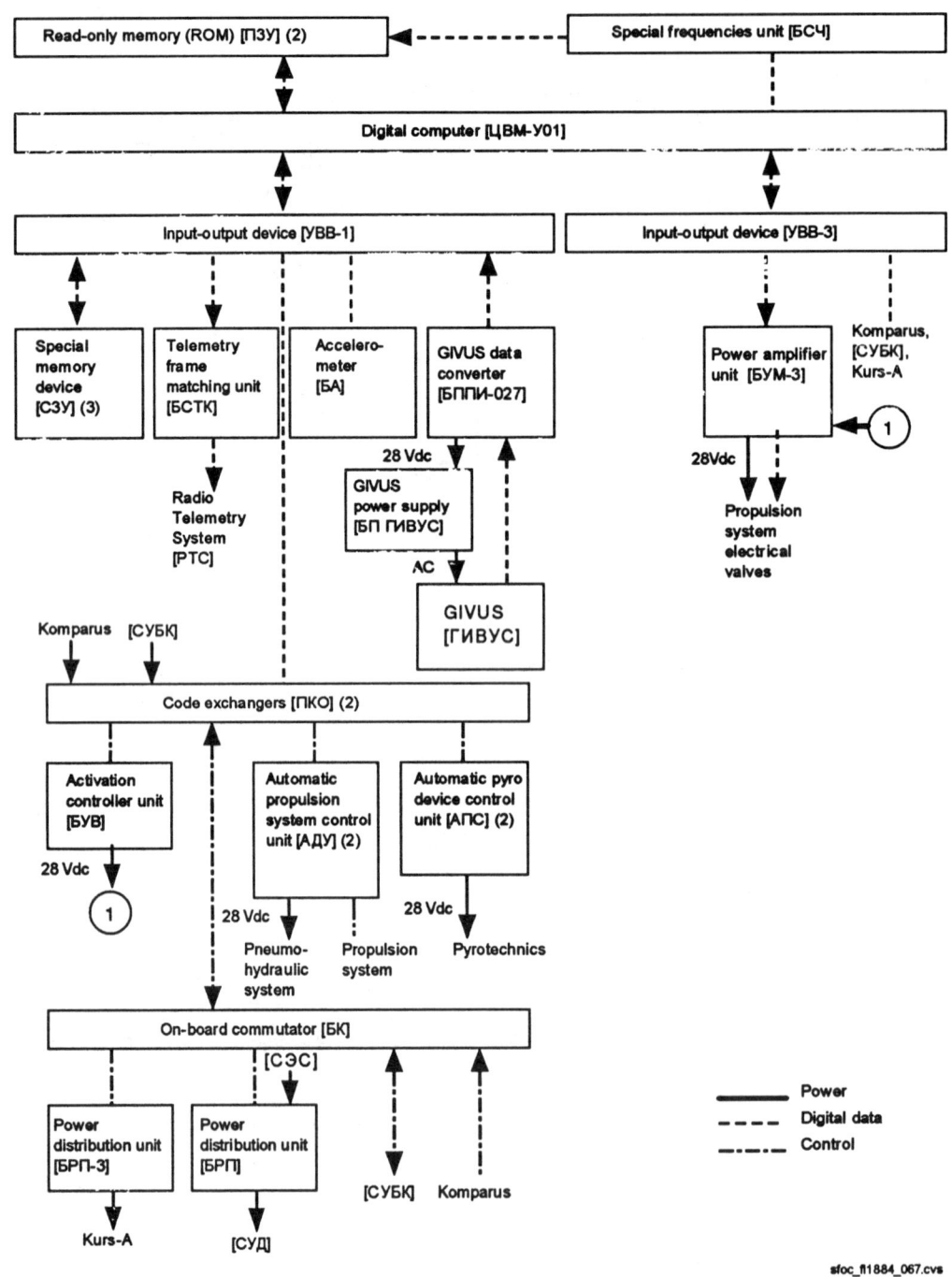

Figure 2 Overview of FGB GN&C System

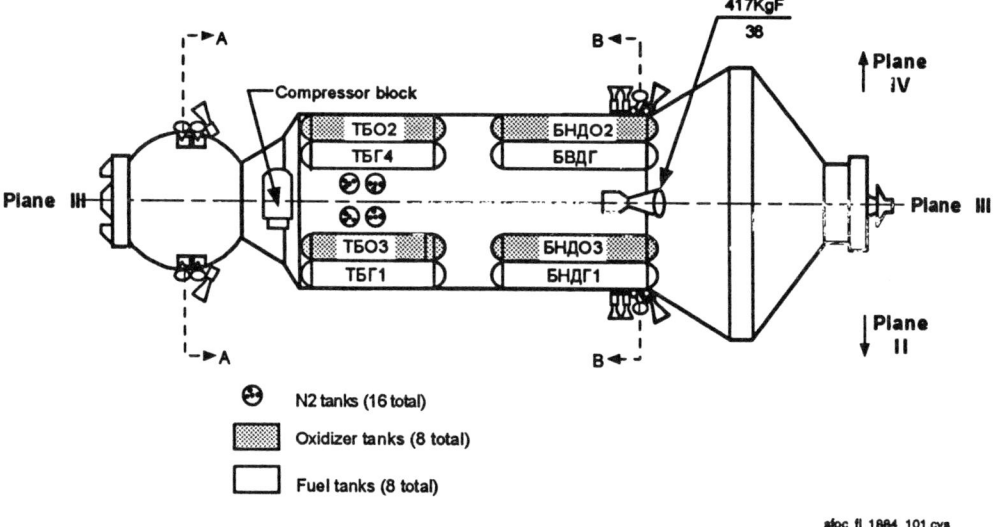

Figure 3A Plane I View of FGB Engines and Thrusters

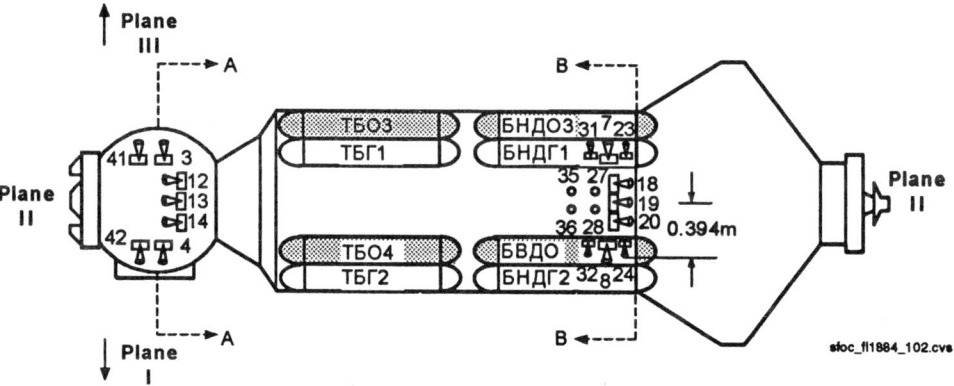

Figure 3B Plane II View of FGB Engines and Thrusters

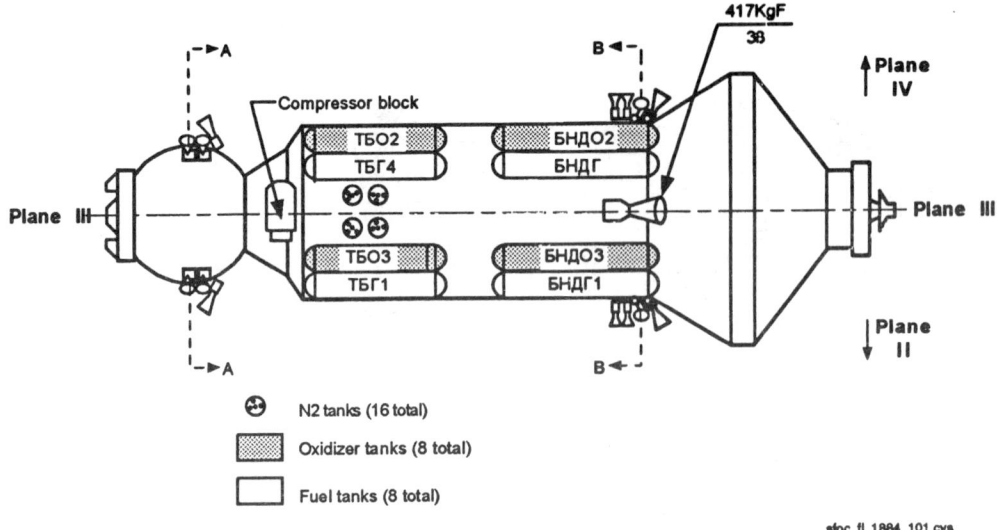

Figure 3C Plane III View of FGB Engines and Thrusters

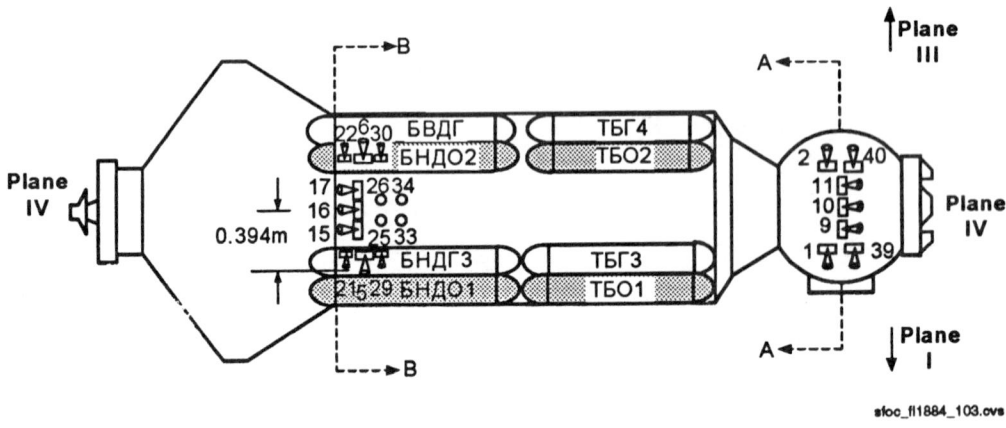

Figure 3D Plane IV View of FGB Engines and Thrusters

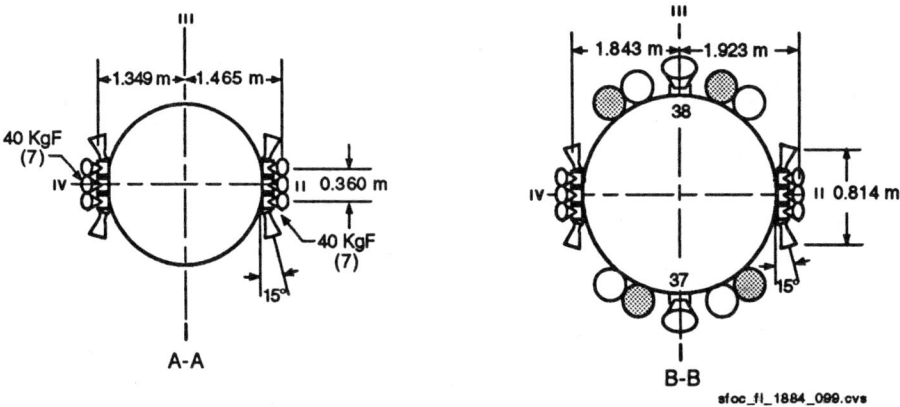

Figure 3E Cross-sectional View of FGB Engines and Thrusters

The FGB performs the active rendezvous and docking with the SM, when it is launched (TBD). After completion of docking and activation of the SM, the FGB GN&C will no longer be used on the ISS.

SERVICE MODULE (SM)

The Service Module is built by the Rocket Space Corporation-Energia and is one of Russia's major contributions to the ISS. It provides the exclusive GN&C capability from Stage 1R until arrival and activation of the U.S. Laboratory. Table 2 provides an overview of the SM technical specifications. Figure 1B provides the relationship of the SM to other ISS elements through Stage 1R. Figure 4 provides an isometric view of the SM and FGB (Reference 2).

TABLE 2

BASIC SM TECHNICAL SPECIRICATIONS

Launch vehicle	Proton
Launch mass	
Orbital mass	
Altitude of working orbit	410–460 km
Inclination of working orbit	51.64°
Length	13.0 m
Maximum diameter	4.2 m

Figure 4 – Exterior View of the SM

Following activation of the SM, the SM GN&C performs the following functions for all subsequent stages:

- Determines the attitude and attitude rate of the ISS
- Determines/maintains the ISS state vector (orbital position and velocity)
- Provides thruster only attitude hold and attitude maneuvers
- Provides orbital correction maneuvers (reboosts). When a Progress vehicle is docked to the SM aft end, the SM will use the Progress thrusters/engines for translational thrust.
- Control ISS motion during docking of Russian vehicles (e.g., Progress, Soyuz)
- Performs on-orbit monitoring of GN&C equipment and operations and automatically aborts current active control operations, if off-nominal conditions are detected.
- Fires thrusters to desaturate Russian gyrodynes

Following activation of the U.S. Laboratory, the SM GN&C performs the following additional functions.

- Fires thrusters to desaturate U.S. CMGs based on commands from the U.S. GN&C
- Accepts unconditional and conditional "Take Control" commands from the U.S. On-Orbit Segment (USOS).
- Provides attitude, attitude rate, and state vector data to the U.S. GN&C (only data source from Stage 5A, until activation of Stage 8A).
- Following activation of Stage 8A, accept attitude, attitude rate, and state vector data from the U.S. GN&C

Figure 5 provides an overview of the SM GN&C components. The sensing components of the SM GN&C system are:

- Three Star Trackers [••••]
- Three Infrared Horizon Sensors [256K]
- Two Magnetometers [CM-8M]
- Four Sun Sensors [251K2]
- High-Accuracy Rate Sensor with four single DOF rate gyros [GIVUS]
- Moderate-Accuracy [ORT]
- Navigational Satellite Instrumentation [ASN], which is similar to U.S. GPS system [two receiver/processors and four antennas]
- Radar system to support Russian vehicle docking operations [Kurs]
- Television system to support Russian vehicle docking operations

The key computational components of the SM GN&C system include:

- Three Terminal Computers [TBM]
- Matching Units (Digital-to-Analog and Analog-to-Digital), which allow interfacing between the Terminal Computers and the GN&C sensors and effectors.
- Central Post Computers and Russian Lap Top Computer

The key crew command input components of the SM GN&C system include:

- Russian Lap Top Computer which allows flight crew interaction with the Terminal Computer
- Rotational and Translational Hand Controllers, which allow SM flight crew to manually fly Russian incoming vehicles for docking operations. Note that these hand controllers are <u>not</u> used for control of the ISS.

The SM GN&C system control effectors are (see Figure 6, Reference 3):

- Thirty-two 13.3 kg_f thrusters separated symmetrically across two propulsion manifolds
- Two 315 kg_f gimbaled engines. SM engines can be gimballed +/- 5 degrees in each of two axes.
- Six gyrodynes (Momentum storage of 2,400 n-m-sec each) [implementation is TBD]

Other Russian Segment thrusters/engines which can be commanded by the SM GN&C system:

- The roll-axis (ISS X-axis) control authority will be augmented by a set of thrusters mounted on the Solar Power Platform (see Figure 6).
- When the Progress is docked to the ISS and a matching unit is connected between the SM TC and the Progress, the SM GN&C can issue commands to Progress thrusters (see Figure 7) for attitude control and CMG desaturation.

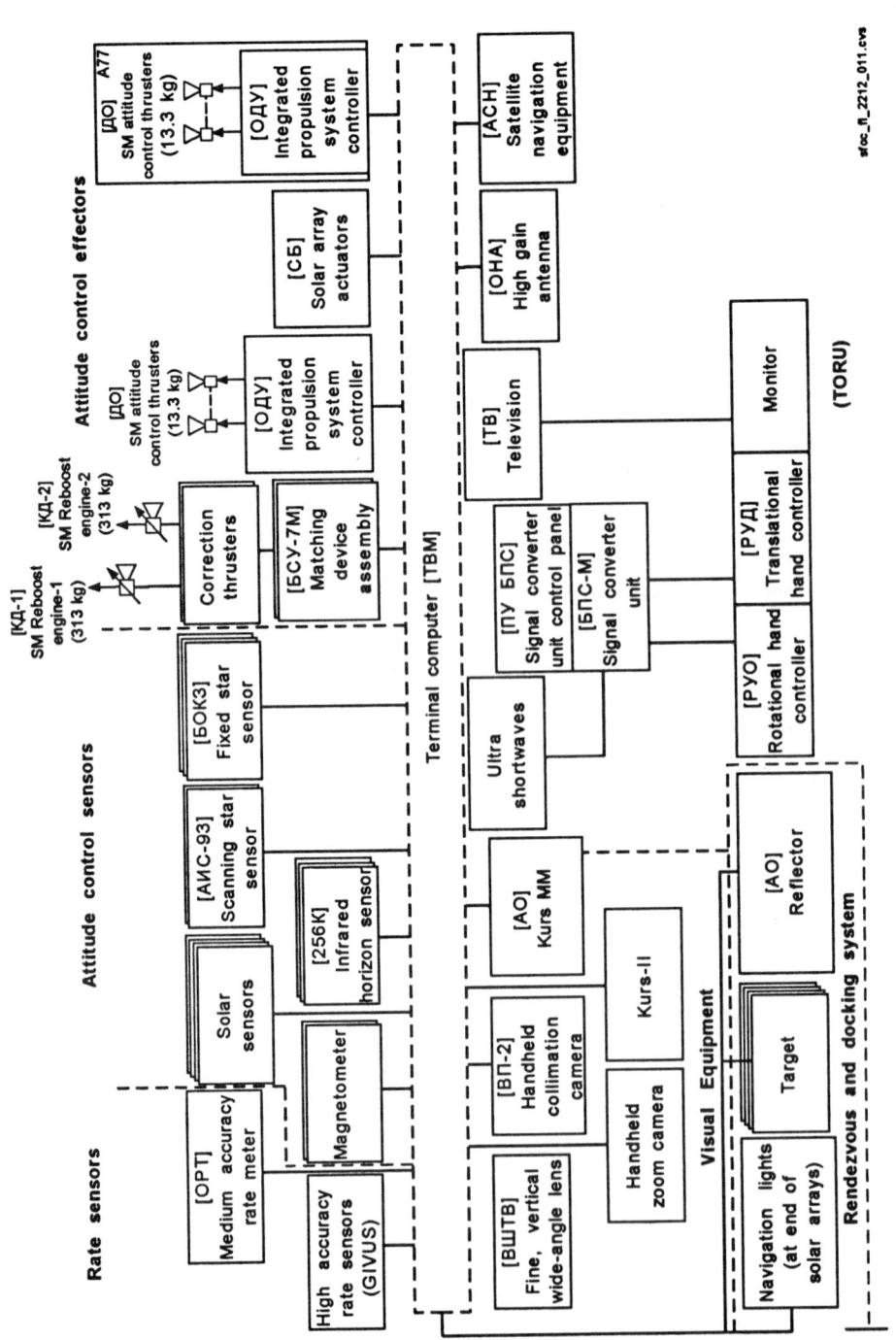

Figure 5 Overview of SM GN&C System

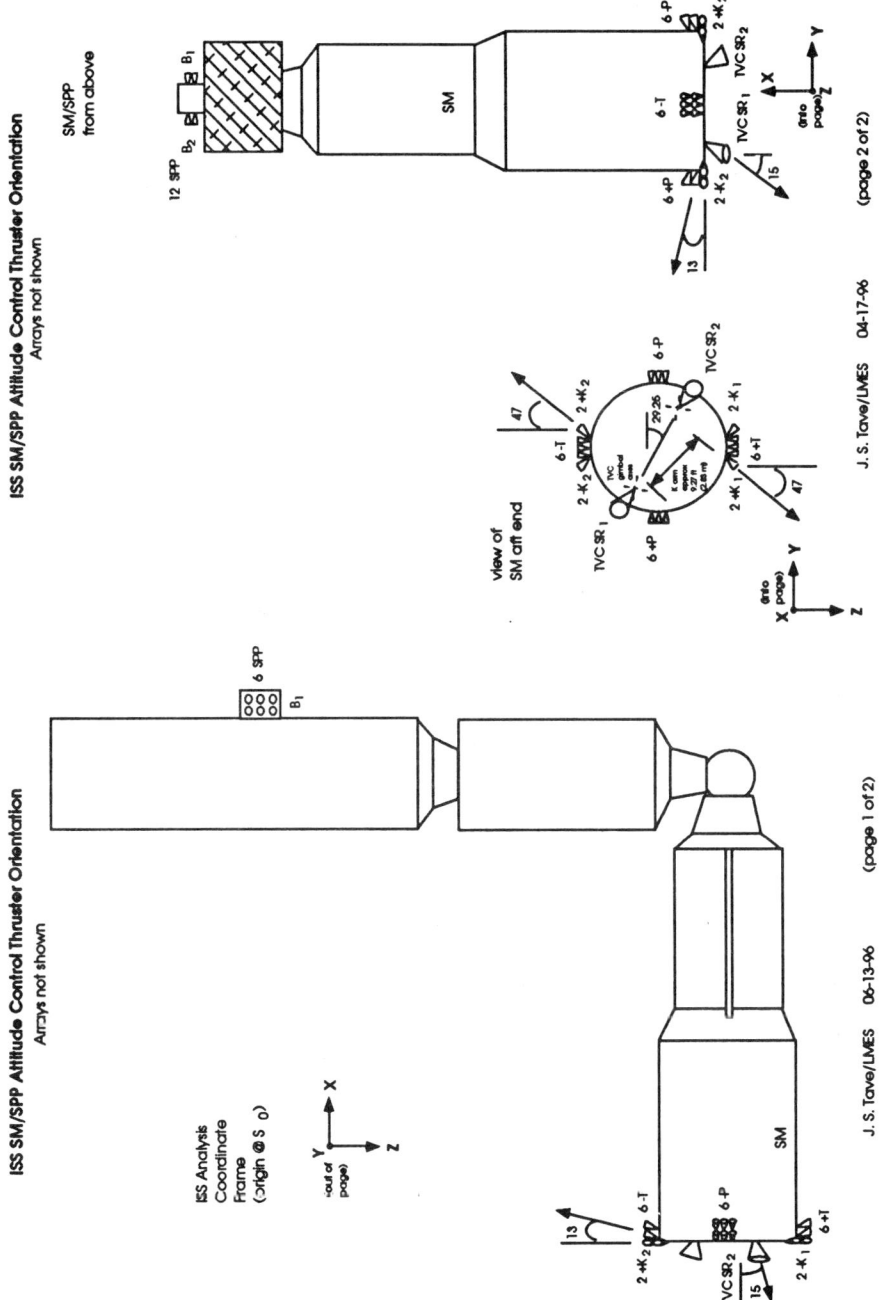

Figure 6 SM and Solar Power Platform (SPP) Engine/Thruster Configurations

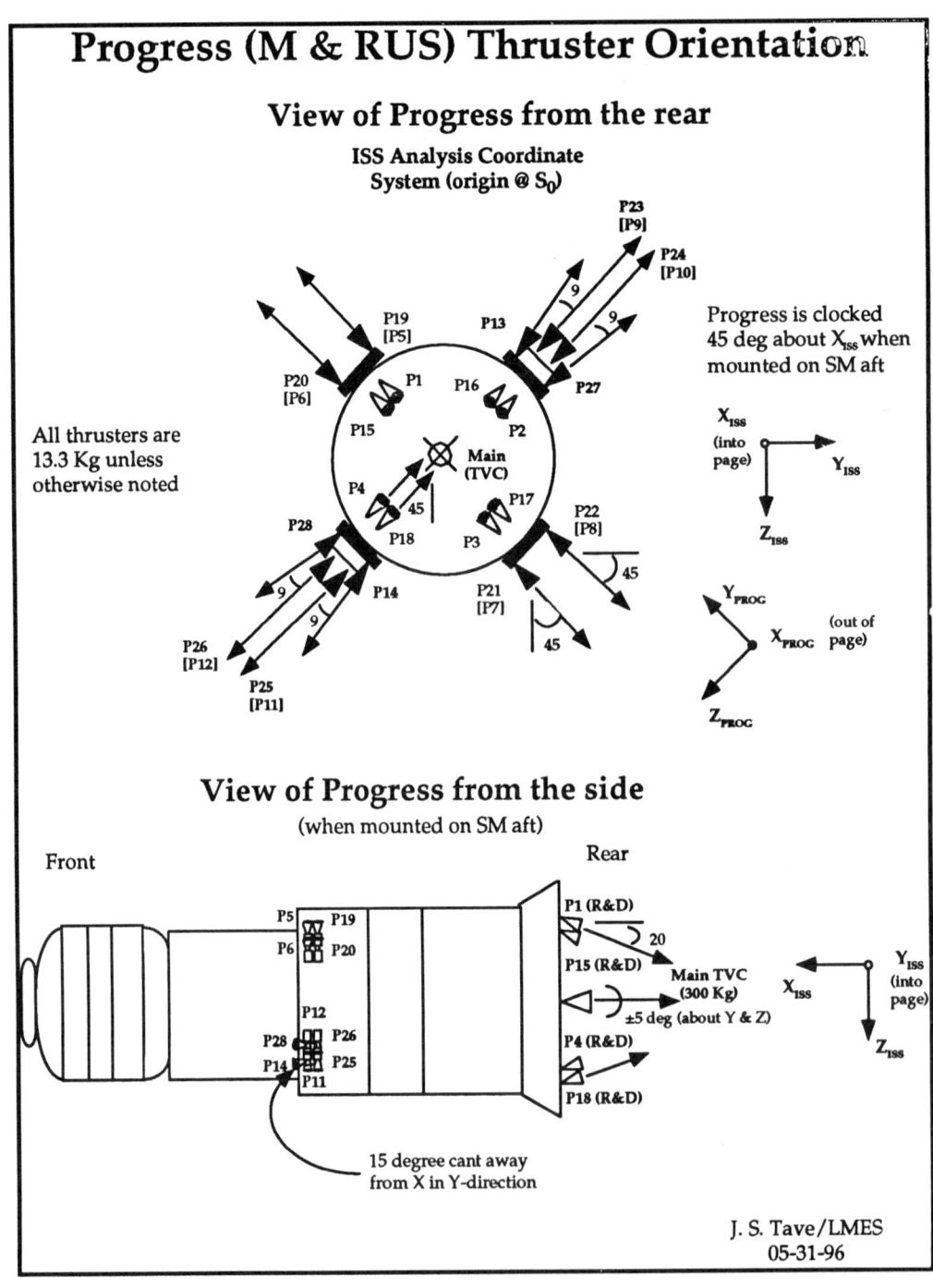

Figure 7 Progress Thrusters

UNITED STATES ON-ORBIT SEGMENT (USOS)

The U.S. GN&C components are delivered incrementally and initial U.S. GN&C functions begin after activation of the U.S. Laboratory on Stage 5A. Following its activation, the U.S. GN&C performs the following functions from Stage 5A until activation of Stage 8A:

- Non-propulsive attitude control using CMGs
- Provides pointing data to U.S. systems including solar arrays, thermal radiators, communications antennas, based on attitude, attitude rate, and state vector data provided by the SM GN&C system.
- Performs on-orbit monitoring and redundancy management of the CMGs and GN&C computers
- Issues commands for CMG desaturation to the SM GN&C
- Supports hand over of control between the USOS and RS.

After activation of Stage 8A, the U.S. GN&C performs the additional functions:

- Determines the attitude and attitude rate of the ISS and provides these data to the SM GN&C
- Determines/maintains the ISS state vector and provides these data to the SM GN&C.
- Performs on-orbit monitoring and redundancy management of the U.S. navigation components.

An overview of the U.S. GN&C functionality is presented in Figure 8

The sensing components of the U.S. GN&C are:
- Two rate gyro assemblies (RGA) are attached on the S0 element of the ISS (Stage 8A)
- Two Global Positioning System (GPS) Receiver/Processors are installed in the U.S. Laboratory (Stage 5A), but are not functional until the arrival of the four GPS antennas on Stage 8A.
- Four GPS antennas on Stage 8A

The key computational components of the U.S. GN&C system include:

- Two Multiplexer-Demultiplexers (MDM's), which are connected via 1553 data busses to the Russian Terminal Computers.

The U.S. GN&C control effector components consist of:

- Four dual-gimbal, Control Moment Gyros (CMG's) are installed on the Z1 truss, which is delivered on Stage 3A. However, these control effectors do not become operational until Stage 5A, when the U.S. Laboratory is delivered.

Figure 9 shows the interfacing between the U.S. and Russian GN&C subsystems beginning on Stage 5A to prior to activation of Stage 8A. During this operational period, the U.S. GN&C system is totally reliant on the SM GN&C for attitude, attitude rate, and state vector (orbital position and velocity). After activation of Stage 5A (U.S. Laboratory), attitude control using CMG's will be the primary mode of operations to minimize the use of propellant and to provide a degree of low-gravity operations. The Russian Segment thrusters will be used to "desaturate" the U.S. CMG's.

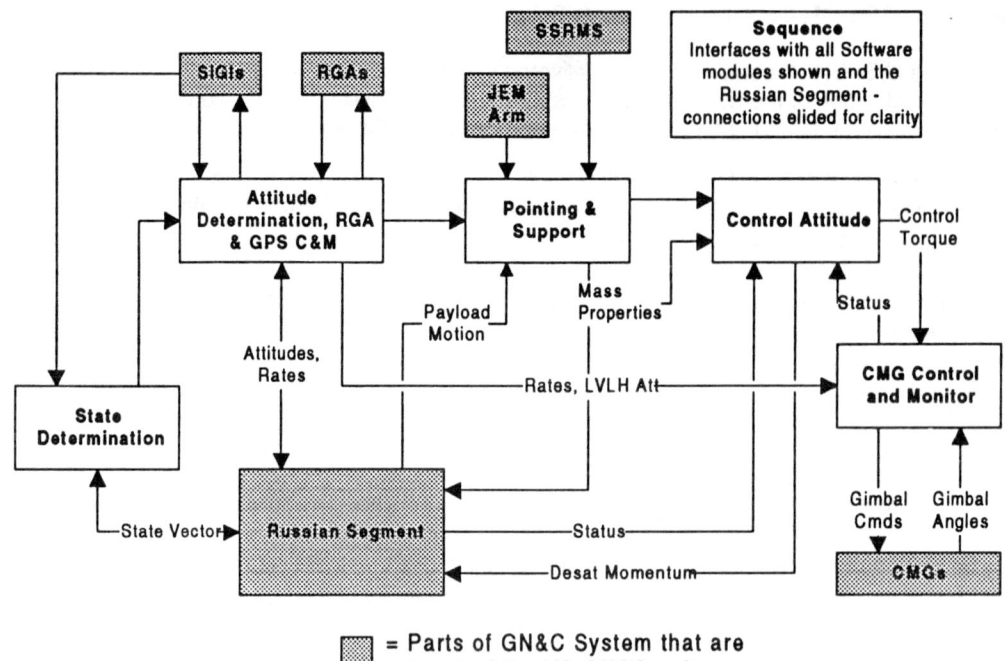

Figure 8 Overview U.S. GN&C Functionality (Reference 4)

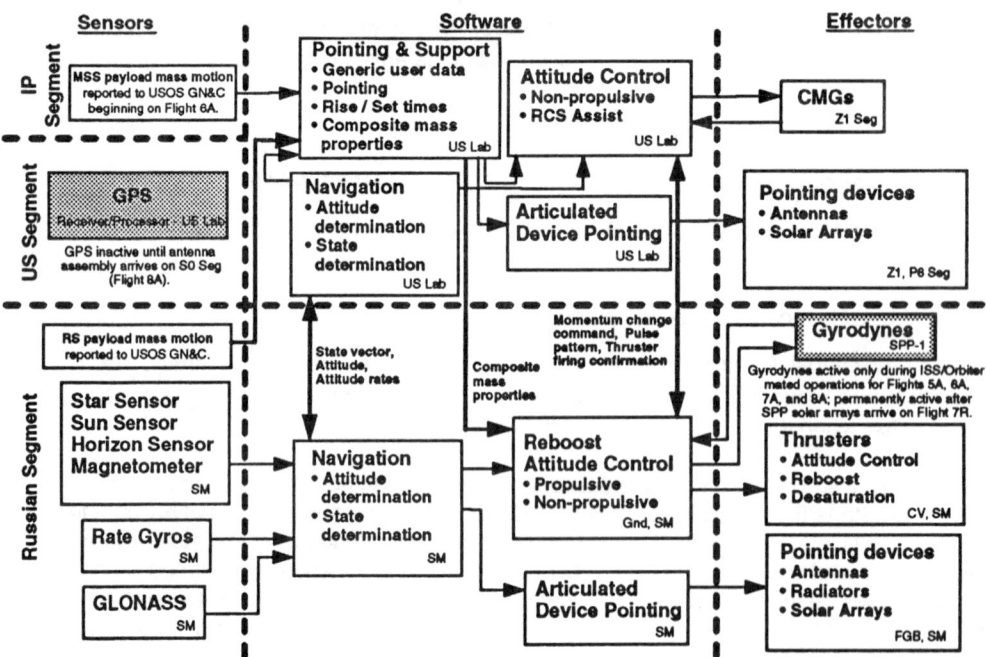

Figure 9 U.S. and SM GN&C Interactions on Stages 5A – pre-activation of Stage 8A

Figure 10 provides an overview of the interfacing between the U.S. and Russian GN&C subsystems after activation of Stage 8A, when all U.S. GN&C system assets are available and active. Beginning on Stage 8A, a U.S. and a Russian set of attitude, attitude rate, and state vector data are available to either GN&C system.

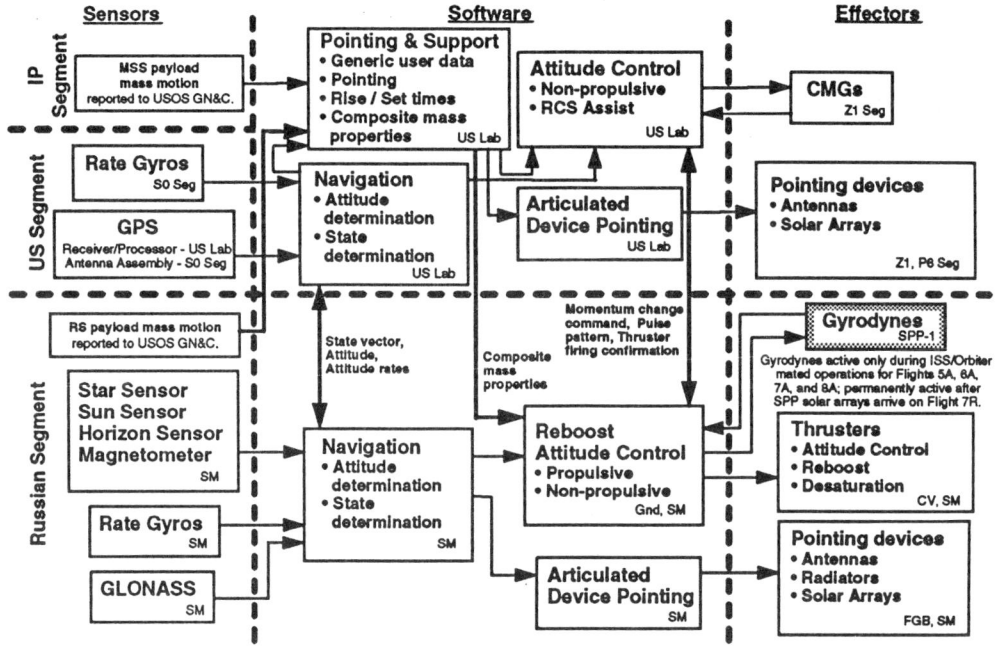

Figure 10 Overview of U.S. and SM GN&C Interaction, Post-Stage 8A Activation

REFERENCES:

1. SFOC FL1884, "Data Book, Functional Cargo Block, Rev A, dated 1 May 1999.

2. SFOC FL2212, "Data Book: Service Module," dated 15 October 1999

3. "Thruster Diagrams," J. Tave, Lockheed/Martin, dated May – June 1996.

4. D684-10506-02-01, "Guidance, Navigation & Control Architecture Description Document, Volume 2. GN&C Subsystem Architecture Notebook, Revision B/Version 1, dated 1 November 1999.

AAS 00-022

INTERNATIONAL SPACE STATION ASSEMBLY AND OPERATION CONTROL CHALLENGES

Nazareth Bedrossian[*]

For the International Space Station program to be successful mission requirements must be robustly met in the presence of uncertainties. Changing Station characteristics during assembly such as variability in structural flexibility and mass properties pose unique control challenges. Three examples of control challenges are reviewed in this paper, the first two arising out of variability in flex structure and the last one due to mass property variation. Controller/flex structure interaction issues and their origins are reviewed as well as representative examples. CMG momentum desaturation issues during robotic operations are addressed and an operational solution is reviewed. Finally, CMG attitude control issues during payload robotic operations are presented as well as an issue resolution technique.

Introduction

The assembly and operation of the Space Station poses unique control challenges due to its complex, variable flexible structure and mass properties during assembly stage as well as variety of operational modes and nonlinear control systems. Operational modes include attitude control for mated and unmated configurations, reboost, and robotic payload operations. For the program to be successful requires that mission requirements be met using a variety of attitude and other control systems from various International Partners. These include the Russian propulsive systems on the FGB cargo block, Service Module (SM) and Progress resupply vehicle, the US Control Moment Gyroscope (CMG) non-propulsive systems and the US propulsive options using the Interim Control Module (ICM) and Propulsive Module (PM), the Canadian Space Station Remote Manipulator System (SSRMS) as well as the Mobile Transporter (MT) etc. An overview of the various Space Station elements and participating International Partners is shown in Figure 1. Further, these attitude control systems must be robust to system uncertainties and variations in flex structure, mass properties, failures and initial conditions. Hence, it is critically important that an integrated system assessment of the GN&C system robust stability and performance be carried out in order to certify the vehicle for flight readiness.

Of paramount importance is avoiding control system/flex structure interaction, or dynamic interaction specially when the Station is under propulsive attitude control. These

---

[*] Principal Member of Technical Staff, The Charles Stark Draper Laboratory, Inc., 2200 Space Park Drive, Suite 210, Houston, Texas 77058.

feedback instabilities can compromise the structural integrity of the vehicle due to their large induced loads, the lifetime of the vehicle by using excessive propellant and result in planned operations having to be postponed or cancelled. The integrated control system assessment and flight readiness certification process must identify potential for dynamic interaction and if necessary develop operational workarounds. For the early stages, where flex dynamics are dominant, flex filter architecture and coefficients have to be carefully tailored to preclude the possibility of the attitude control system responding in phase with these resonances.

Attitude control during robotic payload operations poses a particularly challenging issue. In general, robotic arm flexibility is large and low enough in frequency to be inside the attitude controller bandwidth. This is a particular issue especially during attitude maneuvers, which can and usually do excite robot arm flexibility. Another robotic flex issue is the impact of open–loop thruster firings during Station momentum desaturation. In this case the issue is peak induced structural loads which impact the operational lifetime of the arm. Hence, the desaturation firings must be performed in a manner that do not compromise the structural integrity of the arm. Another issue with robotic payload operations is the resulting variability in mass properties and their impact on CMG attitude control. The issue in this case is CMG momentum and torque saturation. This situation is undesirable because it impacts the operational timeline and also requires the use of attitude control thrusters to desaturate the CMGs.

This paper addresses control challenge issues related to assembly and operation of the International Space Station. Changing Station characteristics during assembly such as variability in structural flexibility and mass properties create control challenges that must be addressed for a successful Station program. Three examples of control challenges are reviewed in this paper, the first two arising out of variability in flex structure and the last one due to mass property variation during robotic assembly operations. Controller/flex structure interaction issues and their origins are reviewed as well as representative examples for a particular assembly stage. CMG momentum desaturation issues during robotic operations are addressed and an operational solution to induced structural loads is reviewed. Finally, CMG attitude control issues during payload robotic operations are presented as well as an issue resolution technique utilizing a momentum optimal CMG maneuver logic.

Control System Flex Structure Interaction (CSI)

A basic requirement for any space vehicle is that its control systems are stable and meet performance specifications in the presence of uncertainties. For the Space Station attitude control system to be stable, avoiding control system/flex structure interaction, or dynamic interaction is necessary. Dynamic interaction is most likely when the Station is under propulsive attitude control due to the large control authority of the Reaction Control System (RCS). The standard definition of dynamic interaction is that of feedback control instability. For phase plane controlled systems such as the Station, this implies a growing limit cycle. These feedback instabilities can compromise the structural integrity of the vehicle due to their large induced loads, the lifetime of the vehicle by using excessive propellant and result in planned operations having to be postponed or cancelled. The cause is prolonged flex induced firings which reinforce structural flex, i.e. the thruster firings don't damp flex but rather add energy to the system. The cause for this phenomenon is in phase, with respect to flex, thruster firing. However, even persistent high frequency limit cycles are classified as instabilities as they lead to unacceptably high rate of propellant use. In general, for phase plane based attitude control systems inadequate control system attenuation and excessive lag are the cause for dynamic interaction.

The integrated control system assessment and flight readiness certification process must identify potential for dynamic interaction and if necessary develop operational workarounds. For the early Space Station stages, where flex dynamics are dominant, flex filter architecture and coefficients have to be carefully tailored to preclude the possibility of the attitude control system responding in phase with these resonances.

Space Station Structural Flexibility Overview

During assembly and operation of the Space Station, the complex and varying flexible structure provide one of the necessary ingredients for potentially undesirable or even unstable CSI. The main culprit is low frequency and large amplitude flexibility primarily due to articulating solar arrays but also can be caused by robotic arm flexibility. Further as the assembly progresses from early nearly symmetrical stages to late stage complex configurations the structural frequency spectrum exhibits progressively more complex characteristics such as densely packed modes, lower frequency modes, dominant mode directions changing etc.

A brief overview of structural flexibility during assembly is shown in Figure 2, Figure 3, Figure 4, and Figure 5. Each of these figures shows a particular stage in the Station assembly process, a frequency response plot (Singular Value Decomposition) from thruster commands in Roll, Pitch and Yaw to the Rate Gyro sensor (in deg/sec), and the input/output directional decomposition (singular directions) for the three highest amplitude flex modes. Figure 2 shows the characteristics for Stage 1R, which is currently the next stage in the assembly sequence. Stage 1R is formed when the Russian Service Module vehicle docks to the FGB module already in orbit. Figure 3 depicts Stage 4A during robotic assembly using the Space Shuttle Remote Manipulator System (SRMS) with payload the first US solar arrays. Figure 4 shows the completed assembly for Stage 4A, and Figure 5 provides detail on Stage 13A. From the figures, it is evident that 1R has two dominant Roll and largest amplitude modes, 4A+Robotics has very low frequency dominant modes which are lower than the lowest Station structural mode, while 4A and 13A show signs of modes in lower frequencies with higher density.

A further complication in the Station structural flexibility is the issue of uncertainty. The flex models and their characteristics discussed in the previous paragraph are based on finite element models. Based on test results and estimates made, flex model uncertainty levels have been established for flex body robustness analyses and flight parameter verification [1]; modal damping ratio of 0.5%, uncertainty in modal frequency ±20%, uncertainty in mode shape ±50%. The uncertainty in the mode shape results in a peak amplitude uncertainty of 125%. Further, given the uncertain nature of integrating

US and Russian hardware and models and taking into consideration the uncertainties used by the Shuttle program office for Shuttle mated structural models [2], the uncertainty in frequency can be prudently expanded to ±30%.

Space Station Propulsive Flight Control System Overview

The International Space Station propulsive attitude control system provides rotational and orbital velocity control. The propulsion capabilities of the ISS reside entirely within the Russian segment on three separate vehicles; the Functional Cargo Block (FGB), the Service Module (SM), and the Progress resupply vehicle. However, for the majority of the Space Station's lifetime, the Service Module will provide the propulsive flight control function and will be used in this paper to illustrate the CSI problem. The SM Motion Control System (MCS) [3] provides all the algorithms for control of the Station including flex rate filters, maneuver logic, phase plane control law, reboost control law, thruster restriction and jet select. The MCS algorithms assume only control of the rigid-body attitude and rate, thus requiring attenuation of any flexible dynamics

The flex rate filters are used to provide an estimate of the rigid-body rate for use in the phase plane control law. An overview of the filter architecture is shown in Figure 6. A second-order low-pass ~0.05Hz bandwidth filter in each axis is used to attenuate low-energy high-frequency bending modes and to minimize the effects of sensor noise and quantization. In estimating the rigid body vehicle rate, the flex filters rely on feedforward prediction of the jet firing rate change to account for the slow, 5Hz, sensor update rate and to allow use of a low bandwidth low-pass filter, i.e. the filter is only required to filter the error in the prediction of the jet firings and not the entire rate change. Additionally, a disturbance filter is also used to estimate slowly varying unmodeled disturbances. To attenuate low frequency bending modes that may not be adequately attenuated by the low-pass filter, a series of three second-order notch filters per axis are also used. The algorithms also provide for an adaptive tuning of the notch frequencies during flight to account for uncertainties in the structural model. The frequency response of the flex filter from measured rate input to estimated rate output is shown in Figure 7. Note that the notch locations in Figure 7 have been purposely selected in order to illustrate the CSI example that follows.

CSI Example – Stage 1R

In this section, examples of unstable Controller/flex Structure Interaction (CSI) will be presented for assembly stage 1R and the causes of the interaction identified. Given the nonlinear nature of phase plane control laws in general, standard stability techniques are difficult if not impossible to apply. A linear phase plane equivalent cannot be developed that adequately represents the nonlinear controller, and standard nonlinear analysis techniques, e.g., describing function analysis, have been shown to be so conservative as to return unusable results [4].

In general, for switching control systems such as phase planes the typical CSI occurs when filtered flex oscillations are large enough to exceed both positive or negative rate deadbands. In an idealized situation, with the rigid rate estimate centered between the switch curves (assuming symmetric rate deadbands), rate-to-rate deadband exceedence results in bipolar firings. Some of the causes of this phenomenon are insufficient attenuation of flex dynamics in the rate estimate, very small rate deadbands, very high jet thrust, high frequency jet firings, and excessive phase lag. Therefore, if the filtered rate estimate does not exceed the rate deadband, a dynamic interaction cannot be induced, hence, an unstable dynamic interaction cannot occur. Under appropriate assumptions, filtered flex amplitude less than the rate deadband provides a sufficient condition for flex stability. However, if the rate deadband is exceeded it is not necessary that an unstable dynamic interaction develop. One condition that can lead to instability is sufficient phase lag such that the deadband exceedence induced firings are in-phase with the structural (unfiltered) flex dynamics. However, even then a rate-to-rate deadband oscillation may not result unless jet thrust is large or frequent. Note that the phase lag applies to the transfer function from the actuator node to the phase plane input. An example for Stage 1R flex structure is shown in Figure 8, which represent the transfer function from three-axis jet command to rate sensor. The phase lag for the dominant mode @0.12Hz is nearly zero, 0.4deg. For the transfer function just described, and using appropriate assumptions, 90deg is sufficient phase lag to cause in-phase thruster firings. Hence, one of the necessary conditions for instability is that phase lag exceeds 90deg.

Due to the nonlinear nature of the phase plane control system, predicting instances of unstable dynamic interaction can be challenging. An example of the type of radically different responses that can be obtained due to small changes in maneuver angle is shown in Figure 9, a CSI instability, and Figure 10, which shows a stable CSI example. The example configuration is Stage 1R where the flex frequencies have not been perturbed while the peak flex amplitude has been increased by 100%. The commanded maneuver rate in both cases is 0.2deg/sec. Figure 9 depicts the results for an 18.65deg Roll maneuver, while Figure 10 shows the results for a 19.2deg maneuver. The flex filter notches, for both examples, are initialized as shown in Figure 7, i.e. the first Roll notch is purposely initialized at 0.2Hz while the Roll dominant mode is at ~0.12Hz. In a real world situation the dominant flex modes will not be known precisely, hence without in-flight system ID or adaptation the discrepancy between notch and dominant mode could be large. This example shows the severe repercussions that may result.

From Figure 9, it is evident from the phase plane plot that rate deadband to rate deadband flex oscillations are observed. The jet firings are in-phase with the measured structural modes as can be seen from the measured vs jet on-time plots. This is an example of an unstable dynamic interaction. It is caused by the secondary rigid body rate damping firings that occur when the attitude error is 15deg, as indicated on the phase plane. This behavior is a feature of the SM MCS phase plane. Rate damping is used to decelerate the vehicle during the terminal phase of the maneuver. This firing which occurs at ~20sec is sufficiently large and in-phase with the structural mode at 0.12Hz, thus exciting the mode sufficiently that rate-to-rate deadband flex oscillations result. It may be noted that the rate damping firings are a function of the 15deg transition zone rate estimate. From the phase plane plot, the rate error is positive, indicating that the actual maneuver rate at the instant of transition is actually greater than the commanded rate. This is the cause for the large rate damping firing.

Comparing the results in Figure 10 with those in Figure 9, it is clear that the results in Figure 10 indicate a stable dynamic interaction. In this case the rigid-body rate damping firing at 15deg or ~25sec is out-of-phase with the structural flex, thus attenuating flexure. Even though the flex filter phase lag exceeds 90deg @ 0.12Hz, the CSI is stable because the flex induced firings are not large enough to excite rate-to-rate deadband flex oscillations. Further, the rate damping firings are substantially smaller in this case. From the phase plane plot, the rate error is negative indicating that the actual maneuver rate at the instant of transition is actually smaller than the commanded rate. This is the cause for the smaller rate damping firing.

CMG Momentum Desaturation

During assembly of the International Space Station (ISS), robotic manipulation of various payloads will be required. The Shuttle Remote Manipulator System (SRMS) and the Space Station Remote Manipulator System (SSRMS) will manipulate the payloads. During these assembly operations, attitude control is provided primarily by the US Control Moment Gyroscope (CMG) system. Since the angular momentum of the CMG system is limited, manipulation of relatively heavy payloads may cause large momentum transients resulting in momentum and torque saturation. Hence, momentum desaturation requires the use of thrusters.

Thruster activity during robotic payload operations poses certain sub-system specific issues. The primary issue is excessive robotic structural loads either due to large or periodic thruster firing. The desaturation induced firings should avoid exciting the robotic arm to its load limits, and must also be robust to variable arm dynamics. Further, frequent desaturations are undesirable as they increase operational complexities and time required completing the robotic operation. The implementation should also be compact with respect to on-board computer memory utilization. Also, limited desaturation times are desirable in order to minimize impact to operational timeline. To reduce program costs both in design and verification phases, a single approach is to be used for all Station assembly stages and robotic operations. Finally, the desaturation must also be flexible so that the technique used to satisfy robotic structural loads does not hinder the ability to remove momentum during docking operations. In the following sections the CMG momentum desaturation method developed by Draper Laboratory for the International Space Station for use during robotic payload operations is reviewed.

Thruster momentum desaturation can be accomplished in a variety of ways. It can be classified as either closed loop (feedback) or open loop (feedforward). For closed loop methods, individual firings are commanded based on current CMG momentum states until momentum reaches the desired levels. The pulse pattern can range from a single large pulse to multiple high frequency small pulses. An issue with such an approach is pulse phasing which could coincide and excite a robotic structural mode and violate load limits. Also this approach may take an unacceptably long time to complete. Another approach is to use non-periodic, i.e. variable delay time, fixed pulse-width firings, which also terminate when momentum reaches a pre-specified value. A concern with such an approach is that loading frequency content varies with firing termination on different pulse of pre-defined pattern. This makes loads verification difficult. For these reasons a feedback strategy was not considered.

For open loop momentum desaturation, a pre-defined number of individual thruster firings are commanded at pre-determined start times while the CMGs are commanded in a feedforward manner to counter the thruster induced momentum. The thruster pulsetrains and CMG feedforward are designed to desaturate a specific amount of angular momentum. The pulse pattern can range from a single large pulse to multiple pulses with either periodic or non-periodic delay times. A single large pulse is quick, efficient and simple to implement. However, this method results in large induced loads [5]. An alternative is to consider using a form of "command preshaping" [6][7][8] which effectively notch out the dominant resonant frequencies. However, since preshaping is a model-based approach it requires knowledge of the structural modal frequencies. For flexible structures like the Station, the uncertainty and variability in mode frequencies makes it difficult to use such an approach.

Using multiple fixed pulse-width firings at periodic delay times between firings does substantially reduce loads [5], however, the phasing of the multiple firings remains an issue as detailed in the previous section. This is the preferred strategy when structural characteristics are known. This type of pulsing strategy is currently being used by the Space Shuttle Alternate Primary Reaction Control System (PRCS) mode [9]. Using multiple fixed pulse-width firings at non-periodic delay times between firings resolves the phasing issue with periodic firings while at the same time reducing loads. This approach can be extended further by also considering non-periodic thruster pulse-width firings. Hence, this desaturation strategy requires choosing delay times to robustly minimize loads. The advantages of such an approach are that the pulse pattern can be chosen early in the Space Station design phase and verified for all vehicles and operations. It also requires minimal ground support during flight as well as satisfies the issues associated with desaturation. Due to the desirable features of an open-loop approach to momentum dumping, multiple fixed pulse-width firings at non-periodic delay times was chosen as the design architecture to robustly meet the design goals mentioned previously [9].

The design objective for the firing strategy is to minimize structural loads in the presence of plant uncertainty, such that a new pulse pattern is not required for every possible Space Station configuration. One approach is to implement a non-periodic delay pulsing pattern where the delays are randomly chosen in order to avoid interaction with shifting and unknown resonances. This approach, however, is robust with respect to on-time but not with respect to frequency. Our approach is to consider the frequency content or Power Spectral Density (PSD) of the pulsing pattern and to choose the delay time such that the flexible body mode excitation is minimized over a certain frequency range [9]. This results in a pulsing strategy, which is robust with respect to frequency. Hence, the optimization problem for a specified frequency range $[\omega_1, \omega_2]$ can be stated as [9]:

$$\min_{\tau_i} \max_{\omega \in [\omega_1, \omega_2]} \Phi_{uu}(\omega) \qquad (1)$$

subject to
  a) Fixed thruster on-time/pulse
  b) Fixed total delay time
  c) Fixed delay time range

In the above, $\Phi_{uu}(\omega)$ is the PSD of the thruster pulsetrain. Note that since the PSD of the input signal is a function of the number and magnitude of delay time parameters, τ_i, it can be shaped by a judicious selection of the τ_i. Such a selection has the effect of maximizing the uniformity of the power in the chosen band and reducing the concentration of power at any one particular frequency.

For the International Space Station CMG desaturation during robotic payload activity, a threshold of 10,000ft-lb-sec is used with a CMG feedforward torque of approximately 170ft-lb resulting in total desaturation time of approximately 1min. Further, the thruster pulse widths were assumed to be 1.5sec based on a 5 pulse/4 delay pattern to reflect a worst case scenario of a minimum of 1333ft-lb thruster torque in a worst axis during assembly of the Station. This results in a constraint of 50sec on the total allowable delay time. With regard to frequency, a range of 0.01Hz-1Hz was used as a representative sample of expected Station/Robotics structural modes. Solving for the optimal delay pattern using these constraints results in the delay and start times listed in Table 1.

CMG Attitude Control

Since the angular momentum of the CMG system is limited, manipulation of relatively heavy payloads may cause large momentum transients resulting in momentum and torque saturation. This situation is undesirable because it impacts the operational timeline of the assembly operation by requiring the use of Reaction Control System (RCS) to desaturate the CMGs, thus interrupting the payload operation. It is also undesirable because of the fuel use by the RCS and because of the structural load issues that arise by using the RCS while a massive payload is attached to one of the robotic arms.

CMG momentum use during the robotic operation depends on several factors. Clearly it depends on the motion of the payload. The momentum use also depends on the commanded attitude of the Station during the payload operation. Since external torques are being applied to the Station, the system angular momentum is not conserved. This means that the final CMG momentum depends on the path between the initial and final attitudes and not simply on the attitudes themselves. The initial and final attitudes are usually chosen to balance the external torques acting on the Station. Since the payload operation changes the mass properties of the Station, these equilibrium attitudes will also change. This requires an attitude maneuver to be performed during the payload operation. Some results on attitude maneuvers in the absence of external torques using momentum wheels and CMGs can be found in [11][12]. In [11] both open-loop and closed-loop optimal control strategies are considered. However, since in general angular momentum is not conserved on Station due to presence of gravity gradient and aerodynamic torques and the fact that payload motion complicates the vehicle dynamics motivates the analysis approach presented in the following.

Payload Dynamics Formulation

The payload dynamics formulation relies on efficiently modeling the dominant dynamics of a connected system of rigid bodies moving relative to each other by using appropriate simplifying assumptions. The standard rigid body formulation of Station robotic operations include 6 Degree Of Freedom (DOF) Shuttle/Station base body dynamics coupled to either a 6 Degree Of Freedom (DOF) Shuttle Remote Manipulator System (SRMS) or 7 DOF Space Station manipulator (SSRMS) with attached payload. To reduce the dynamical complexity of the standard formulation, first the robot arm dynamics are neglected resulting in 6 DOF base body and 6 DOF payload dynamics. This is a reasonable assumption since the arm mass properties are much smaller than either the payload or the base body. To further simplify the dynamics, the payload motion kinematics are assumed to be prescribed, i.e. the robot joint controllers track the commanded arm trajectory with minimal error. This assumption eliminates the need to model the payload dynamics and hence reduces the problem to 6 DOF for the base or core body dynamics. Finally, by using relative (to base body) payload position and orientation parameters and a standard center of mass identity, the system angular momentum is decoupled from its linear momentum and is expressed compactly as,

$$H_{sys} = D_0 \omega_{ss} + h_{pld} + h_{cmg}$$

where:

$$h_{pld} = D_{1a} \omega_{pld/ss} + D_{1l} v_{pld/ss}$$

$$D_0 = I_{ss} + \Re_{ss}^{pld} I_{pld} \Re_{ss}^{pld\,T} - \frac{m_{ss} m_{pld}}{m_{ss} + m_{pld}} \left[p_{pld/ss}^\times \right] \left[p_{pld/ss}^\times \right]$$

$$D_{1a} = \Re_{ss}^{pld} I_{pld} \Re_{ss}^{pld\,T}$$

$$D_{1l} = \frac{m_{ss} m_{pld}}{m_{ss} + m_{pld}} \left[p_{pld/ss}^\times \right]$$

In the above all parameters are expressed in the core body or Space Station reference frame unless otherwise noted, h_{pld} represents the payload angular momentum relative to the core body, h_{cmg} represents the CMG angular momentum, $p_{pld/ss}$, $v_{pld/ss}$ represent the payload translational position and velocity relative to the Space Station, $\Re_{ss}^{pld}$, $\omega_{pld/ss}$ represent the payload angular orientation (rotation matrix) and velocity relative to the core body, and I_{pld} is the payload inertia expressed in the payload axis system.

With these simplifications, the system dynamics are reduced to a 3 DOF Newton-Euler type formulation [13],

$$D_0\dot{\omega} = -\dot{D}_0\omega - \omega \times [D_0\omega + h_{pld} + h_{cmg}] - \dot{h}_{pld} - \dot{h}_{cmg} + \tau_{gg} + \tau_{aero} \qquad (1)$$

In the above, τ_{gg} and τ_{aero} represent the gravity gradient and aerodynamic torques on the Space Station. In this paper aerodynamic torques are neglected for convenience. The corresponding attitude kinematics are obtained from standard quaternion formulation for angular velocity [14],

$$\dot{\varepsilon} = 0.5(\varepsilon^\times + \sqrt{1-\varepsilon^T\varepsilon}\, 1_3)(\omega - \omega_{orb}(\varepsilon)) \qquad (2)$$

where ε is the vector part of the quaternion representing the attitude of the Station body fixed frame with respect to the rotating local vertical local horizontal (LVLH) reference frame, and ω_{orb} is the angular velocity inherent in maintaining attitude in the rotating LVLH reference frame. Augmenting the core body dynamics with the CMG momentum controller then forms the complete set of governing equations,

$$\dot{h}_{cmg} = k_1 D_0 (\omega - \omega_{orb}(\varepsilon_c)) + k_2 D_0(\varepsilon - \varepsilon_c) \qquad (3)$$

where k_1 and k_2 are the proportional and derivative gains of the CMG controller, and ε_c is the commanded attitude of the Station.

Attitude Control During Payload Operations

Maintaining CMG attitude control during robotic payload operations without exceeding the momentum limits requires a commanded attitude control strategy during the robotic operations. A standard approach is to command an instantaneous Torque Equilibrium Attitude (TEA). A torque equilibrium attitude is any constant spacecraft attitude with respect to a rotating reference frame that results in the sum of all external and gyroscopic torques being equal to zero [15]. The benefits of such a strategy are that they are (locally) optimal solutions. Commanding an instantaneous TEA minimizes the external disturbance torques on the vehicle. However, in cases where the TEA is not constant due to payload motion, following such a strategy does not account for the cost (in momentum) of following the TEA trajectory. It also does not anticipate the change in TEA due to payload motion. Since the payload trajectory is assumed known in advance, this information could be used in generating momentum minimizing attitude command schedules.

From an operational standpoint it would be desirable to avoid saturating the CMGs during payload motion. In general, to minimize CMG momentum use for attitude hold prior to and after the payload operation it is envisioned that the local TEA attitude would be commanded. Hence, an optimization problem can be defined to minimize peak momentum magnitude by appropriate choice of attitude command during the payload operation, i.e.

$$\min_{\varepsilon_c} \left\| h_{cmg}^T(t) h_{cmg}(t) \right\|_\infty \qquad (4)$$

subject to the Space Station dynamic equations (1) and (2), the CMG control law, (3), and the following boundary conditions:

$$\varepsilon(0) = \varepsilon_0 \qquad \varepsilon(t_f) = \varepsilon_{t_f}$$
$$\omega(0) = \omega_{orb}(\varepsilon_0) \qquad \omega(t_f) = \omega_{orb}(\varepsilon_{t_f})$$
(4.a)

where ε_0 and ε_{t_f} are the initial and final TEA attitudes.

To solve this problem, an approach based on converting this problem to a nonlinear constrained function optimization approach was considered [13]. Briefly stated, this approach first removes the dependency on commanded attitude as it simplifies the constraint dynamics and results in the optimization being a function of the Station attitude. Once the optimum (or any) attitude trajectory, $\varepsilon(t)$, and its corresponding CMG momentum, $h_{cmg}(t)$, have been obtained, the corresponding attitude command $\varepsilon_c(t)$ can be obtained by solving the CMG control law equation (3) for ε_c. The next step was to convert the optimal control problem to a function optimization problem, which in many cases is much simpler to solve. The conversion was accomplished by removing the derivative dependence from the constraints, which results in a nonlinear algebraic dependence of h_{cmg} on ε using finite difference approximations. At this stage standard function optimization results can be applied. However, the mini-max criterion (4) is not amenable to analytic solutions due to the lack of differentiability. However, equation (4) can be upper bounded by [13]:

$$\min_{\varepsilon} \quad \left\| h_{cmg}^T(t) h_{cmg}(t) \right\|_2 \tag{5}$$

Using the reformulated cost function (5), the optimization problem can be solved either using standard Lagrange multiplier techniques for the constraints or alternatively the boundary conditions can be treated as known constants and eliminated from the optimization problem.

Simulation Results

The methodology developed in [13] was applied to International Space Station robotic assembly for Stage 12A [16]. The payload consists of the inboard port integrated truss segment P3/4 which is initially unberthed from the Shuttle payload bay by the SRMS, then handed off to the SSRMS for installation on the Station as shown in Figure 14. The vehicle principal axes corresponding to the pre-planned payload trajectory are shown in Figure 15.

The optimal attitude command trajectories were compared to three other alternative strategies [13]. In the following the notation TEA is equivalent to principal axes:

- **Follow Principal Axes**: Command instantaneous TEA
- **Mid Course Correction**: Command initial TEA for the first half of the maneuver then command final TEA for the second half of the maneuver

- **Linear Principle Axes**: Command a linear interpolation between the initial TEA and final TEA
- **Optimal Attitude Trajectory**: Command momentum optimal attitude

The time step for all testcases is 10 sec. The peak and final CMG momentum magnitudes for all four testcases are summarized in Table 2. It is evident that the optimal attitude trajectory results in the best CMG momentum performance and theoretically does not require momentum desaturation. The optimal attitude command trajectory results are shown in Figure 17. Using this attitude history as the CMG attitude command resulted in reducing the peak CMG momentum to 8,400 ft-lb-sec, which is below the momentum desaturation limit. One interesting characteristic of this trajectory is the jump in CMG momentum at the beginning and end of the maneuver. This is similar in nature to the bang-coast-bang fuel optimal solution for a maneuver.

Conclusions

This paper reviewed control challenge issues related to assembly and operation of the International Space Station. Changing Station characteristics during assembly such as variability in structural flexibility and mass properties were identified as issues that must be addressed for a successful Station program. Three examples of control challenges were reviewed in this paper, the first two arising out of variability in flex structure and the last one due to mass property variation. Controller/flex structure interaction issues and their origins were reviewed as well as representative examples for a particular assembly stage. CMG momentum desaturation issues during robotic operations were addressed and an operational solution was reviewed. Finally, CMG attitude control issues during payload robotic operations were presented as well as an issue resolution technique.

Acknowledgments

The author would like to acknowledge the assistance, Edward McCants, and Dr. Jian-Woei Jang, who contributed to the work documented here and to the preparation of this paper. The work documented in this paper was completed under NASA contract NAS9-1556.

References

[1]. NASA/Boeing, "International Space Station GN&C Design Guidelines and Assumptions", Revision B, 1997.
[2]. Hall, R., Kirchwey, K., Martin, M., Rosch, G., Zimpfer, D., "Flight Control Overview of STS-88, The First Space Station Assembly Flight", *Procc 1999 AAS/AIAA Astrodynamics Conference*, AAS Paper No. 99-371, Aug. 1999
[3]. RSC-Energia, "ISS Item 17Km N128 Terminal Computer Software Guidance, Navigation and Control System Information Memorandum", Software Version 4.0 Document, 1999.

[4]. Zimpfer, D., Kirchwey, K., Hanson, D., Jackson, M., and Smith, N., "Shuttle Stability And Control Of the STS-71 Shuttle/MIR Mated Configuration", *Proc. 1996 AAS/AIAA Spaceflight Mechanics Conference*, AAS Paper No. 96-131, Feb. 1996.

[5]. Bedrossian, N., "SSRMS Loads During CMG Desaturation", Draper Memo SSDI-94-011, Jul. 1994

[6]. Meckl, P. H., and Seering, W. P., "Minimizing Residual Vibration for Point-to-Point Motion", *Journal of Vibration, Acoustics, Stress and Reliability in Design*, Vol. 107, Oct. 1985, pp. 378-382.

[7]. Seering, W. P., and Singer, N. C., "Using Acausal Shaping Techniques to Reduce Robot Vibration", *Proc. 1988 IEEE International Conference on Robotics and Automation*, Apr. 1988, pp. 1434-1437.

[8]. Wie, B., *Space Vehicle Dynamics and Control*, AIAA Education Series, 1998.

[9]. Kubiak, E. T., and Sargent, D., "CR 89921C: Alternate Primary Jet Mode", Presentation to the Shuttle Avionics Software Control Board, Johnson Space Center, Houston TX, Jun. 1989.

[10]. Bedrossian, N, Lepanto, J., Adams, N., Sunkel, J., and Hua, T., "Jet Firing Strategy To Minimize Structural Loads ", *Proc. 1995 American Control Conference*, pp. 3622-3626.

[11]. Vadali, S. R., and Junkins, J. L., "Optimal Open-Loop and Stable Feedback Control of Rigid Spacecraft Attitude Maneuvers," *Journal of the Astonautical Sciences*, Vol. 32, No. 2, 1984, pp. 105-122.

[12]. Carrington, C. K., and Junkins, J. L., "Spacecraft Attitude Control Using Generalized Angular Momenta", *Proc. AIAA/AAS Astrodynamics Conference*, AAS Paper No. 85-361, Aug. 1985.

[13]. Bedrossian, N., and McCants, E., "Space Station Attitude Control During Payload Operations", *Proc. 1999 AAS/AIAA Astrodynamics Conference*, AAS Paper No. 99-372, Aug. 1999.

[14]. Hughes, P. C., *Spacecraft Attitude Dynamics*, J. Wiley & Sons, 1986.

[15]. Elgersma, M. R., and Chang, D. S., "Determination of Torque Equilibrium Attitude For Orbiting Space Station", AIAA, 1992.

[16]. "International Space Station Robotic Assembly Analysis For Design Analysis Cycle#5, Flights UF1 Through 12A" DAC 5 Document, NASA JSC-38409.

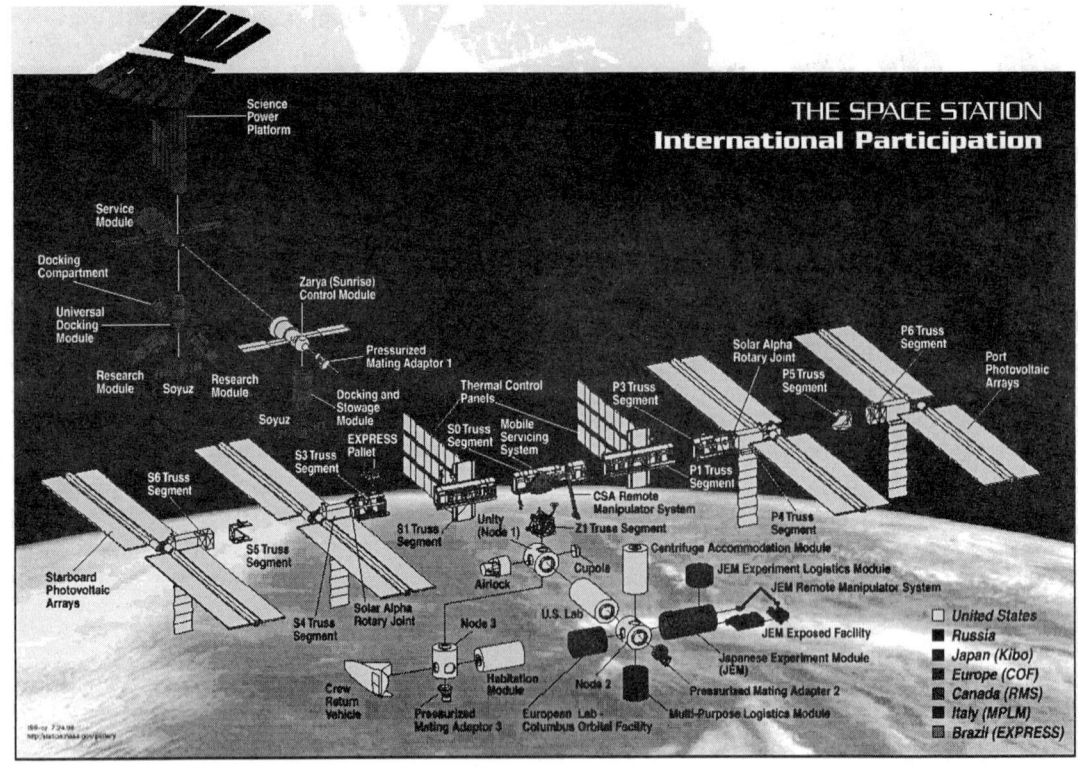

Figure 1. International Space Station – Partner Nations & Elements

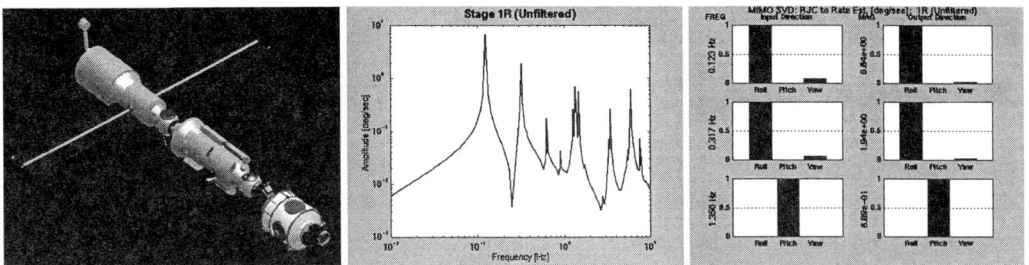

Figure 2. Stage 1R: Configuration, frequency & directional response

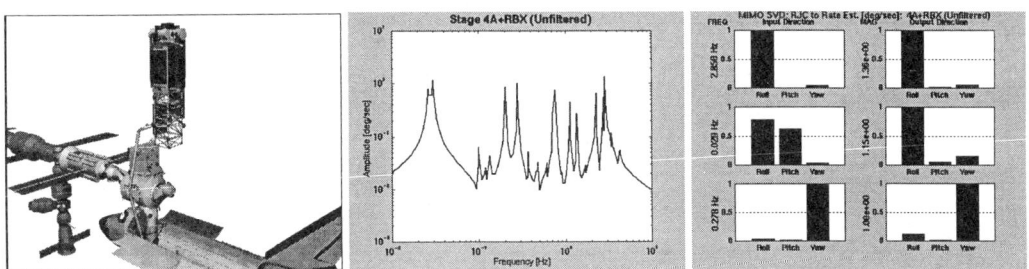

Figure 3. Stage 4A+Robotics: Configuration, frequency & directional response

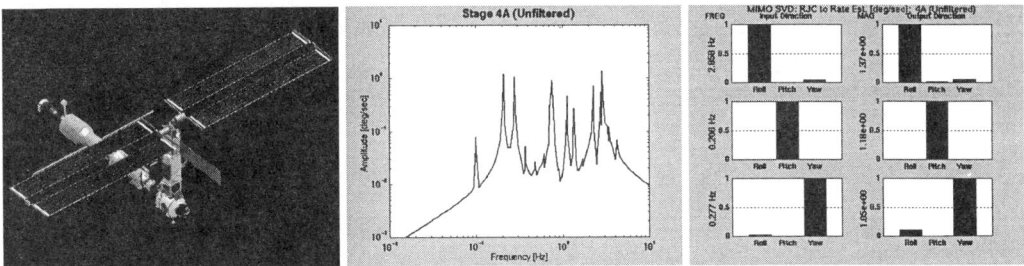

Figure 4. Stage 4A: Configuration, frequency & directional response

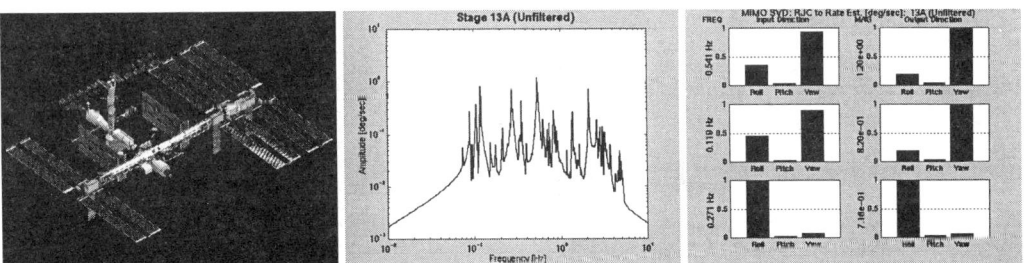

Figure 5. Stage 13A: Configuration, frequency & directional response

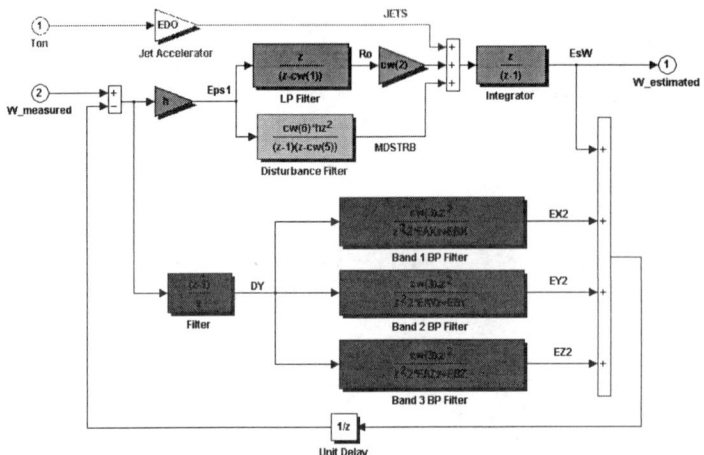

Figure 6. Russian SM MCS – Flex Filter Architecture

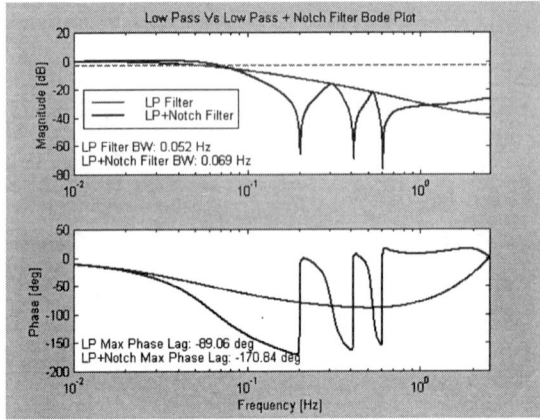

Figure 7. Russian SM MCS – Flex Filter Frequency Response

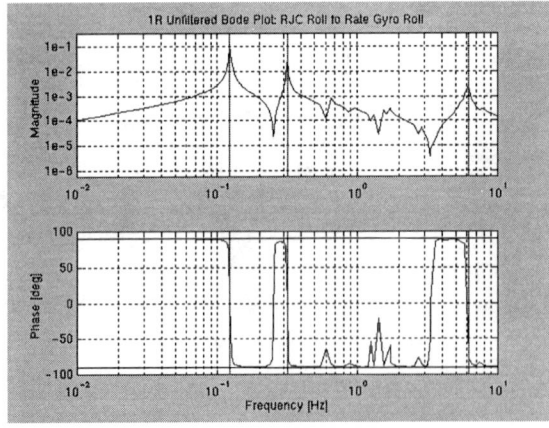

Figure 8. Stage 1R flex model Roll axis magnitude & phase response

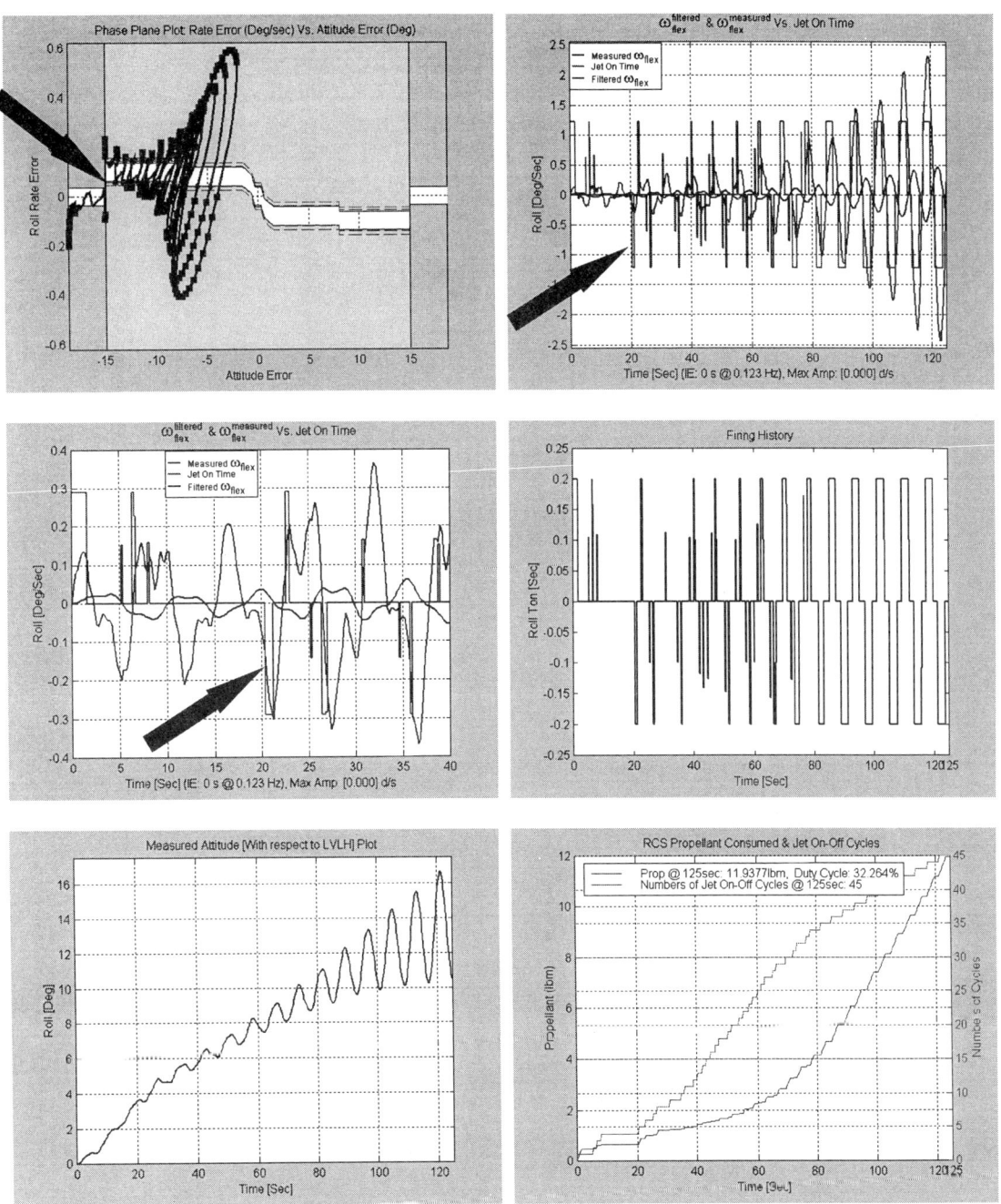

Figure 9. CSI Example: 18.65deg Maneuver Angle - Unstable ACS: Phase Plane, Measured & Filtered Flex Vs Jets (125sec & close-up), Jet On-Time, Attitude, and Fuel Use

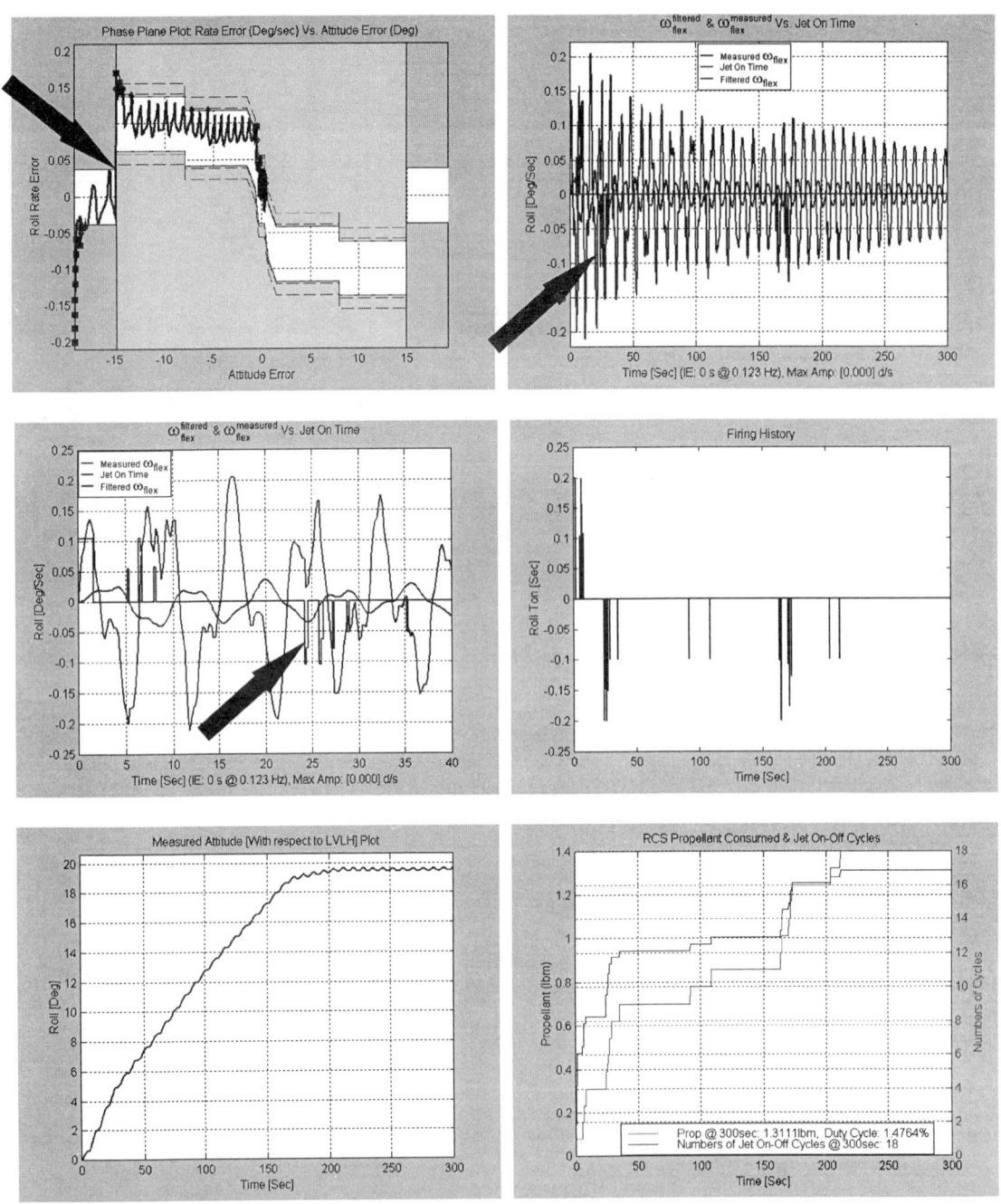

Figure 10. CSI Example: 19.2deg Maneuver Angle - Stable ACS: Phase Plane, Measured & Filtered Flex Vs Jets (300sec & close-up), Jet On-Time, Attitude, and Fuel Use

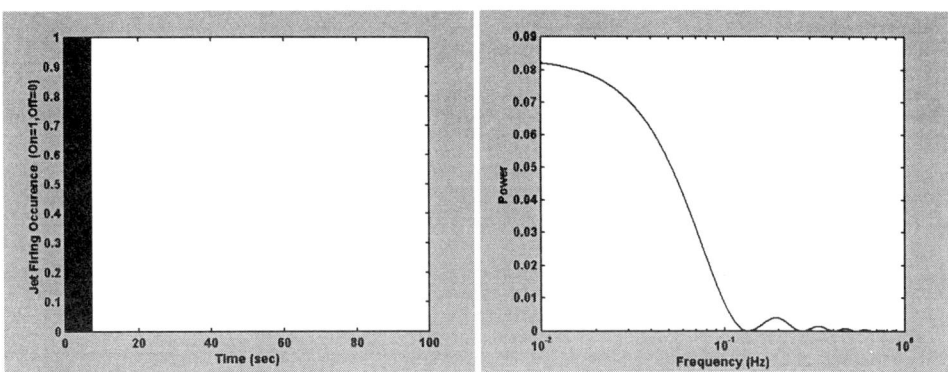

Figure 11. Single pulse in time & frequency (PSD) domain

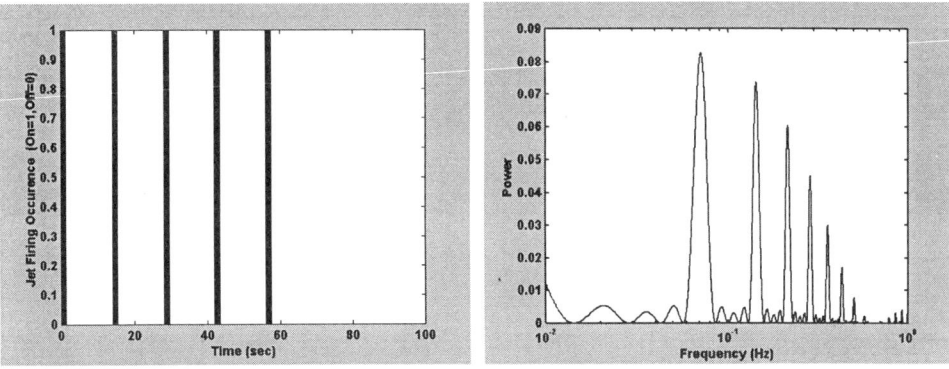

Figure 12. Fixed delay pulse in time & frequency (PSD) domain

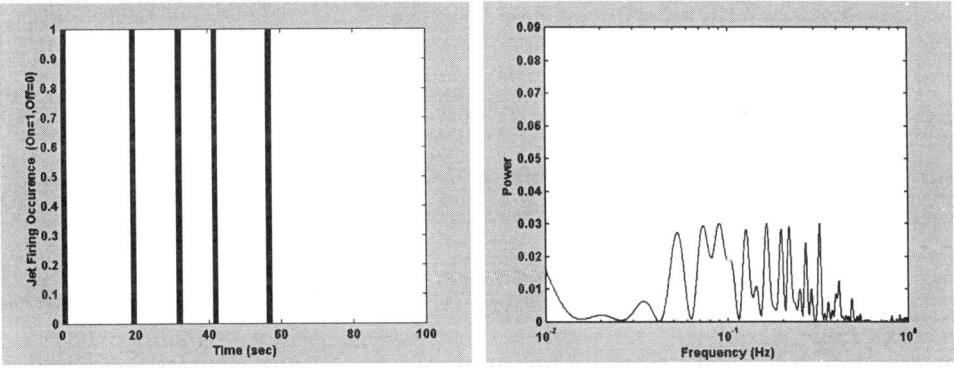

Figure 13. Power optimal pulse pattern in time & frequency (PSD) domain

Delay Number	1	2	3	4	
Optimal Delay (sec)	17.3	10.9	8.4	13.4	
Pulse Number	1	2	3	4	5
Optimal Pulse Start Times (sec)	0	18.8	31.2	41.1	56

Table 1. ISS CMG momentum desaturation optimal delays and pulse start times

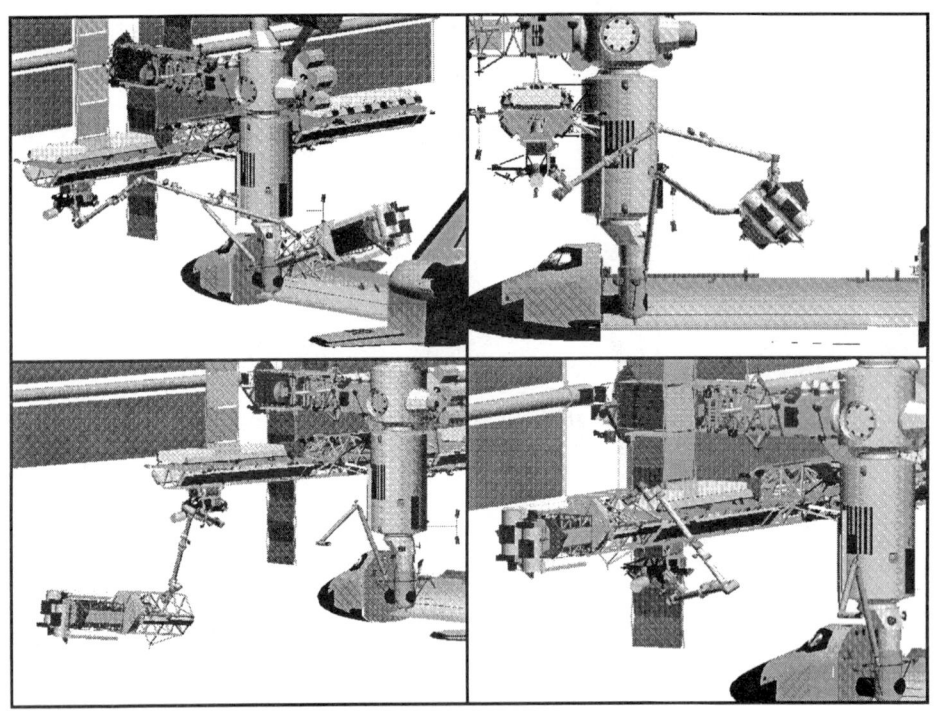

Figure 14. ISS Stage 12A robotic payload assembly operation

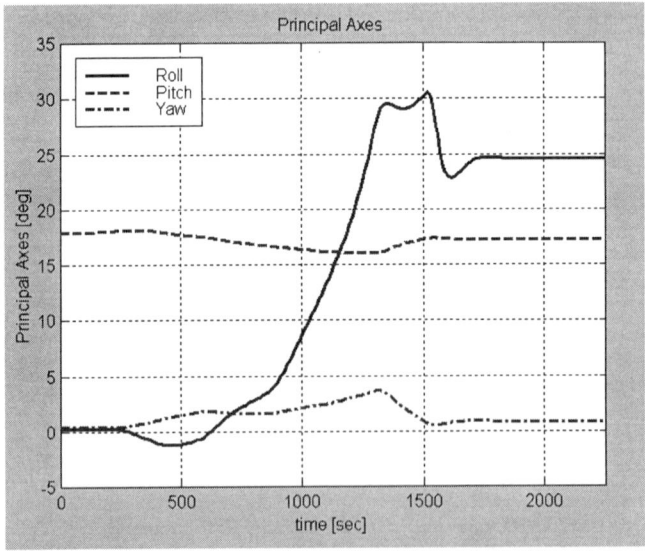

Figure 15. ISS Stage 12A principal axes during robotic payload assembly operation

Table 2. ISS Stage 12A CMG peak momentum use for various strategies

TESTCASES	CMG MOMENTUM [ft-lb-sec]	
	Peak 2-norm	Final 2-norm
Follow Principle Axes	47,696	3,251
Mid Course Correction	73,513	2,565
Linear Principle Axes	20,256	2,376
Optimal Attitude Trajectory	8,396	5,194

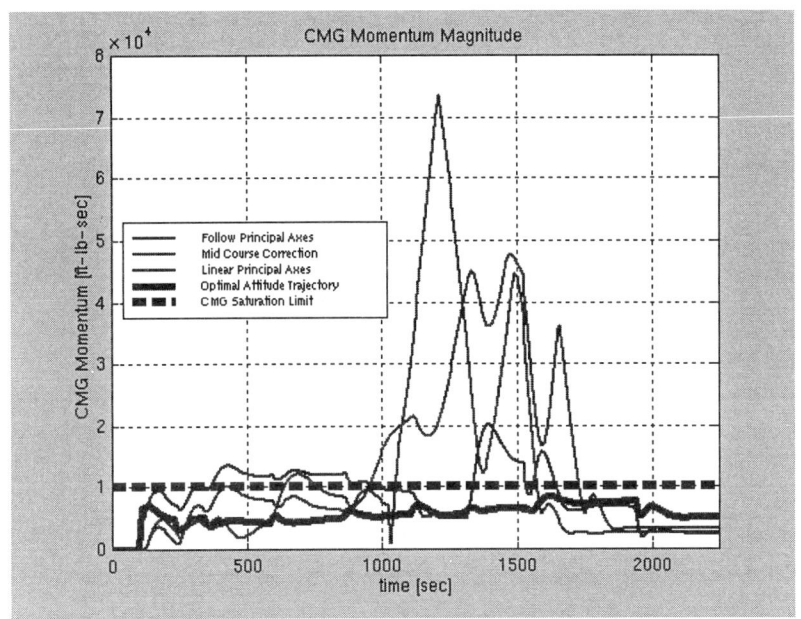

Figure 16. ISS Stage 12A CMG momentum use for various attitude command strategies during robotic payload assembly operation

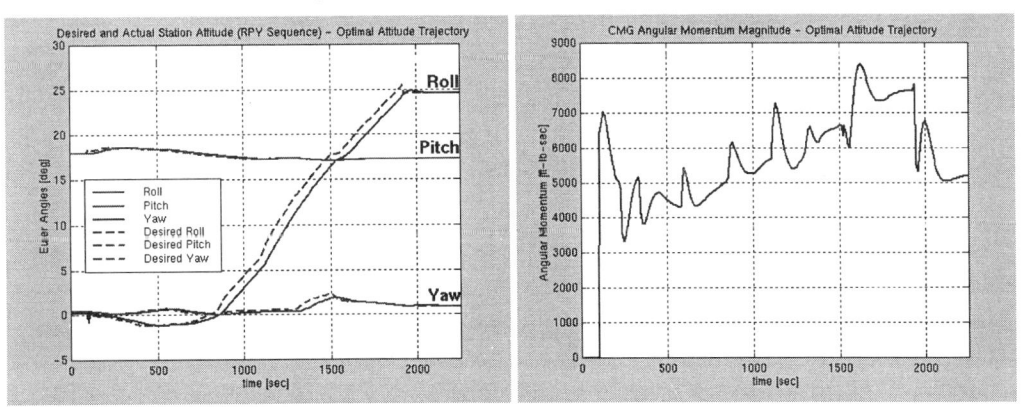

Figure 17: Optimal Attitude Trajectory results: Commanded and Actual Euler Angles (RPY sequence) and CMG Momentum Magnitude

AAS 00-025

STUDIES ON THE ATTITUDE CONTROL SYSTEM DESIGN FOR THE CREW RETURN VEHICLE (X38)

K. Abdel-Motagaly,[*] O. Rombout,[*] R. Gonzalez,[†] K. Berrier,[*] D. Hasan,[*] D. Strack[*] and B. Rishikof[*]

Work is under way by NASA for development of the new emergency Crew Return Vehicle (CRV) for the International Space Station (ISS). The CRV will be used as a "lifeboat" for the International Space Station. This paper presents an overview of studies for the attitude control system design of the on-orbit test vehicle (X38) for the CRV. Attitude control for X38 is achieved via 2 different sets of thrusters, the attitude control subsystem (ACS) thrusters and the deorbit propulsion stage (DPS) reaction control subsystem (RCS) thrusters. The attitude control system can be divided into two main components, attitude controller and jet selection algorithm. The controller used in this study is the phase plane controller. Three different versions of the phase plane controller were used and compared from propellant usage and jet duty cycle points of view during on-orbit coasting and deorbit bum. A jet-selection tool was developed to facilitate the generation of jet-mappings using different jet selection methods, namely dot-product, pseudo-inverse and optimal-type algorithms. Simulation results are given for the study of X38 attitude control system performance under different scenarios including deorbit burn, using the DPS axial jets, and translation maneuvers, using the ACS jets. Studies were performed to examine the effects of jet failures, center of mass variations, thrust blow-down, and phase plane controller settings on the vehicle stability and performance.

1. INTRODUCTION

The International Space Station (ISS) is an earth-orbiting vehicle providing science, engineering, and microgravity research facilities. The ISS is expected to provide a habitable environment for up to seven crewmembers. The Crew Return Vehicle (CRV) is a vehicle being designed by NASA to ensure the safe return of the crewmembers under various planned and unplanned scenarios (e.g., medical emergencies and loss of station control, etc.). The CRV will to be attached to the ISS on the starboard side of the Node 3 element as shown in Fig. 1. In emergency cases, the CRV will separate from the ISS and autonomously fly back to Earth.

X38 is a NASA test vehicle designed to demonstrate on-orbit, deorbit burn, entry, and atmospheric flight phases of the CRV. X38 consists of a lifting body entry vehicle (EV) and deorbit propulsion stage (DPS), Fig. 2. The DPS module will separate from the EV before atmospheric entry. X38 is equipped with 2 separate propulsion systems: cold gas attitude control subsystem jets (ACS) and DPS jets. Two sets of ACS jets, 8 jets each, are

---

*   LinCom Corporation, 1020 Bay Area Blvd., Suite #200, Houston, Texas 77058.
†   NASA Johnson Space Center, Houston, Texas 77058.

mounted exclusively on the EV body and used for attitude control during on-orbit coasting and before entry. The DPS jets consist of 8 axial jets and 8 reaction control system (RCS) jets. The axial jets are used for deorbit burn while the RCS jets are used for attitude control during on-orbit coasting and deorbit burns.

Fig. 1 Crew Return Vehicle (CRV) attached to the ISS

Fig. 2 Crew return testing vehicle (X38) with the DPS attached

The focus of this paper is the design and performance of the X38 attitude control system. Control systems for space vehicles equipped with on/off thrusters can be divided into two principal elements: controller and jet-selection (Fig. 3). The controller generates the acceleration vector required to minimize the error between guidance and navigation inputs. The jet-selection module determines which jets need to be fired to yield the required acceleration vector. Different control and jet-selection algorithms have been used in the literature for thruster-controlled space vehicles. One of the most popular control techniques is independent phase plane control [11]. This technique assumes that the equations of motion are uncoupled for small rotation rates, and employs phase-plane analysis techniques for each channel independently (roll, pitch, and yaw). Another class of controllers is based on the use of thruster pulse-width or pulse-frequency modulation along with linear feedback controllers [2, 8]. A third class of controllers is the phase space control law, which is a nonlinear, coupled, 6 degree-of-freedom control law [3]. Some of the jet-selection techniques used in the literature are fixed lookup-table [13, 14], maximum dot-product [14], pseudo-inverse [7] and linear programming [3, 5].

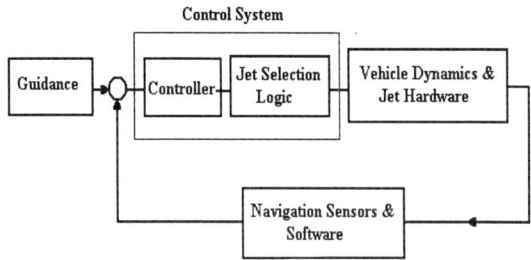

Fig. 3 Overview of control loop for thrusters equipped space vehicles

In this paper, simulation studies for the attitude control system design of X38 during on-orbit coasting and deorbit burn are presented. One of the main goals is to develop the vehicle with particular emphasis on efficiency, taking advantage of existing technologies. Based on this criterion, a control system similar to the shuttle attitude control system is being used for X38. Specifically, three different versions of the phase plane controller are considered in this study, namely the shuttle on-orbit [14], shuttle transition [13], and modified phase plane controller with rate change output [6]. For the jet-mapping, a fixed lookup-table similar to the Shuttle's primary jets [14], and an on-line dot-product jet-selection similar to that used for the Shuttle's vernier jets [14], were implemented in the simulations. A tool was developed to generate lookup-table jet-mappings using different methods. The methods considered were dot-product, pseudo-inverse [7], and optimal type jet-mapping based on linear programming [5]. For the dot-product method, a new cost function is used to minimize the normal component while maximizing the dot-product between the commanded acceleration and the resultant jet acceleration. This tool also evaluates the theoretical controllability of the vehicle for a specified set of jets [5]. Different 6 DOF (degrees of freedom) simulations are presented to study the performance of X38 attitude control system. Effects of using different versions of the phase plane controller, different phase plane controller settings, various jet selection schemes, center of mass variations, jet failures, and thrust blow-down are considered. Sample results are presented for the X38 attitude control using DPS-RCS during on-orbit coasting and deorbit burn, and using ACS jets during EV/DPS separation.

2. PHASE PLANE CONTROLLER

Phase plane control is based on a minimum time linear optimization problem for bang-bang control of a second order linear system [4]. The solution results in a set of switching curves in the phase plane, attitude error vs. rate error, domain. The switching curves separate the regions of positive and negative control action as shown in Fig. 4. The resulting switching lines are quadratic in the form, $x_1 = k\, x_2^2$, where k is a constant, x_1 is the attitude error, and x_2 is the rate error.

To use this idea for spacecraft attitude control, both angular rates and displacements must be small. With these assumptions, the coupled nonlinear equations of motion become a linear uncoupled set of second order differential equations for roll, pitch, and yaw respectively. In practical applications, the attitude and rate errors never hit zero exactly due to measurement errors, disturbances, noise, and jet quantization firing time.

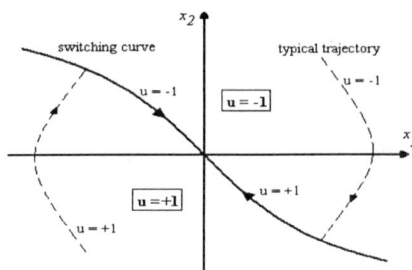

Fig. 4 Phase plane trajectories and switching curves for 2nd order linear system.

SHUTTLE ON-ORBIT PHASE PLANE CONTROLLER

This phase plane controller is used for the shuttle on-orbit digital autopilot (on-orbit DAP) which is used by the shuttle during all on-orbit operations. It determines whether a rotation acceleration command should be issued to the jet selection module for roll, pitch and yaw axis. A complete description of this controller, including switching lines and firing commands, can be found in [10] and [14]. Attitude and rate errors are compared against a set of phase plane regions to determine the required command, Fig. 5. Note that the switching lines for this controller are different from those for a simple linear second order system, Fig. 4. The differences are mainly due to practical considerations. Some of the features of this controller include a dead band (DB) to save propellant, a dead band shelf of 20% due to measurement quantization and digital system delay, and a disturbance hysteresis region to prevent excessive jet firings due to undesired disturbances.

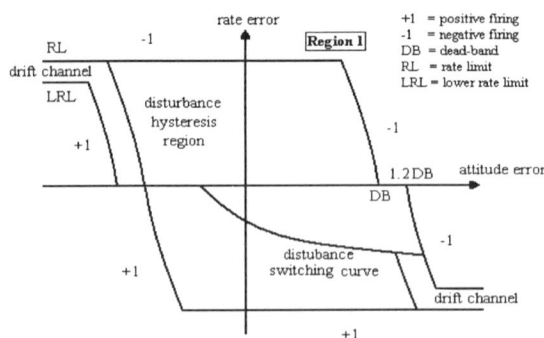

Fig. 5 Shuttle on-orbit phase plane controller.

The switching curves are based on the predicted jet acceleration to produce efficient limit cycle behavior. The purpose of the disturbance switching line is to reduce jet duty cycles by extending jet firings in the hysteresis region to counteract some of the anticipated effects of the undesired disturbance. Undesired acceleration estimate is fed to the controller from the state estimator. The output of this controller is the sign of the desired rate change when shuttle primary jets are used. The output can also contain off-axis preference when shuttle vernier jets are used (this is because the vernier jets use an on-line dot-product jet selection algorithm). The off-axis preference can never be weighted more than 80 % of a firing command (see [14] for more details).

SHUTTLE TRANSITION PHASE PLANE CONTROLLER

The transition phase plane controller [13] is used by the shuttle transition digital autopilot (Trans DAP) which is active during orbital insertion and deorbit burns. This controller only operates the shuttle primary jets and hence the output is only the sign of the desired rate change. Since no payload operations are performed during transition phases, the attitude pointing accuracy is reduced; hence no state estimator is used. And since attitude measurements are obtained directly from the IMU (inertial measurement unit), the controller is subjected to substantial measurement noise. One feature of the transition phase plane controller is to move the dead-band switching curve further from the origin when the attitude error magnitude causes jet firings. This is done to avoid excessive jet chatter due to measurement noise. Another difference between the transition and on-orbit phase plane controllers is the elimination of the disturbance hysteresis region. As it will be seen in the results section, the disturbance switch curve is required for X38 during deorbit burn to compensate for disturbance due to center of mass variations.

Fig. 6 Shuttle transition phase plane controller.

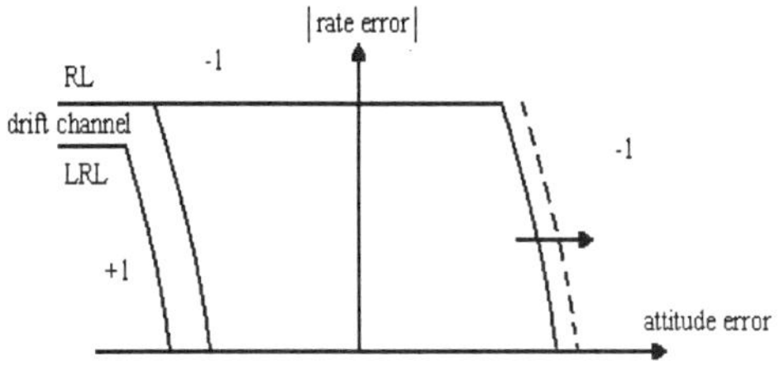

MODIFIED PHASE PLANE CONTROLLER

The modified phase plane controller was developed by Kubiak and Davison [6], to interface with optimal type jet-mapping algorithms. This phase plane controller logic is very similar to the shuttle on-orbit phase plane controller but with changes to output commanded rate change. As shown in Fig. 7, it can be seen that there are additional switching curves and phase plane regions. For example, the on-orbit phase plane controller in region 1, Fig. 5, is "-1", which means fire the negative torque jets. For the modified phase plane controller, the control action in region 1D, Fig. 7, is given by:

$$X_C = -[X_2 + 0.5 (RL+LRL)] \qquad (1)$$

Where, $X2 = |\dot{\theta}_E|$, $\dot{\theta}_E$ is the attitude rate error, RL is the rate limit, and LRL is lower rate limit. Refer to [6] for complete description of this controller logic. The control action in different regions is determined based on minimum impulse limit cycle to minimize propellant consumption.

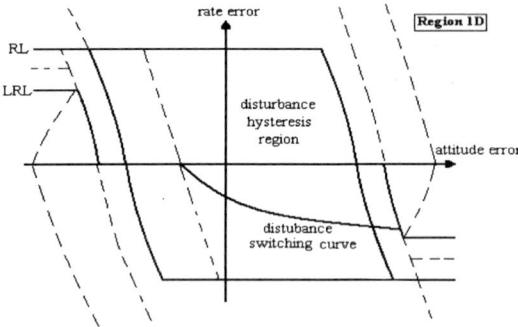

Fig. 7 Modified phase plane controller with rate change output.

3. JET SELECTION AND CONTROLLABILITY

This section presents a description of the jet selection methods used in this study. As mentioned before, a look-up table has been selected as the main jet-mapping method. Look-up tables are simple, easy to test, and work efficiently for vehicles with simple symmetry and jet configurations and with no large variations in vehicle mass properties. Similar to the shuttle system, the selected jets from both the ACS and the DPS are fired for the same amount of time. Another assumption used in this study is that the minimum jet on time equals the closed loop sampling time.

To facilitate the generation of look-up table jet-mappings and to determine the vehicle controllability for a specific jet configuration, a jet selection tool was developed using MATLAB. The main method used to develop look-up tables is the maximum dot-product algorithm. Generally speaking, this method determines the set of jets, which maximize the dot-product between the commanded acceleration (from phase plane controller) and the resultant jet acceleration. A cost function which, maximizes dot-product and minimizes the off-axis component is used. This cost function is more efficient, especially for DPS-RCS jets where jet firings result in coupled commands. The cost function used is of the form,

$$J = \frac{c_1(r \bullet r_c) - c_2|r \times r_c|}{n_j} \quad (2)$$

Where, r is the resultant jet acceleration, r_c is the commanded acceleration from the phase plane controller, n_j is the number of jets fired and c_1, c_2 are weighting constants. The term n_j is the denominator is used to minimize the number of jets fired and consequently propellant used. For the purpose of look-up table generation, the commanded rotational acceleration is discretized to 0, -1 and +1 corresponding to no jets fired, negative acceleration jets fired and positive acceleration jets fired respectively. Using a simple search method, through all possible combinations of jets, The set of jets which maximizes the cost function J for each command is selected.

Two other methods are implemented in the jet-selection tool, namely pseudo-inverse and optimized jet selection logic. These two methods are optimal because they determine the jet on-time such that the propellant consumption is minimized. For these two methods the jet-selection problem is in the form of solving the set of linear equations given by:

$$[A]\{t_{jets}\} = \{r_c\} \quad (3)$$

Where [A], is an n×3 rotational acceleration matrix, where n represents the number of jets available and the columns represent the roll, pitch and yaw channels respectively; and $\{t_{jets}\}$, is the on-time vector for the available jets.

Using the pseudo-inverse method, the solution for jet firings is given by $\{t_{jets}\}$ = $[A]^*\{r_c\}$, where $[A]^*$ is the pseudo-inverse of the acceleration matrix, $A^* = A^T(AA^T)^{-1}$. Use of the pseudo-inverse results in the minimum length for $\{t_{jets}\}$. If a negative firing time is obtained for any jet, then this jet is removed from the available jets and the solution process is repeated.

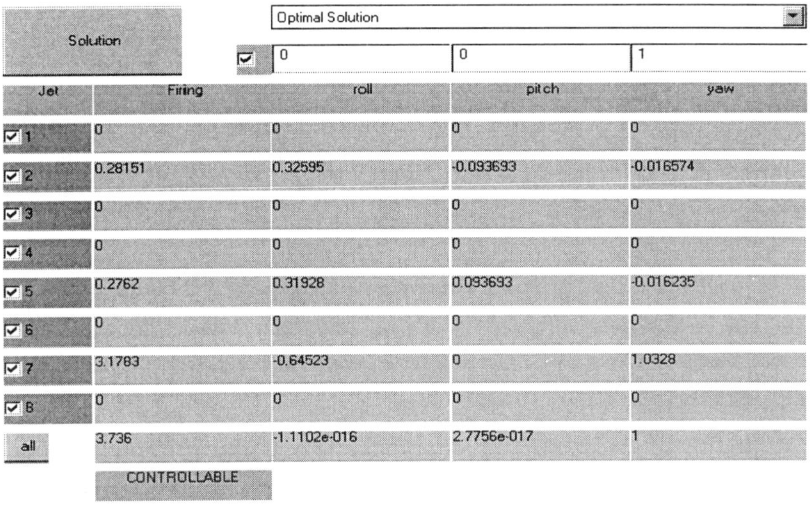

Solution	Optimal Solution			
	☑ 0		0	1
Jet	Firing	roll	pitch	yaw
☑ 1	0	0	0	0
☑ 2	0.28151	0.32595	-0.093693	-0.016574
☑ 3	0	0	0	0
☑ 4	0	0	0	0
☑ 5	0.2762	0.31928	0.093693	-0.016235
☑ 6	0	0	0	0
☑ 7	3.1783	-0.64523	0	1.0328
☑ 8	0	0	0	0
all	3.736	-1.1102e-016	2.7756e-017	1
	CONTROLLABLE			

Fig. 8 Sample output for ACS jets using the jet-selection and controllability tool.

The other method used is the optimized jet selection logic [5], which is an analytic, non-iterative approach. The basic idea is to find the jet firing times, $\{t_{jets}\}$, which minimize propellant usage subject to the constraint of positive jet firing times. This algorithm has additional criteria to test the vehicle controllability for a given set of mass properties and jet configuration. This criteria essentially tests whether or not the available jet accelerations span the entire command space. In other words, is any required acceleration command achievable using the available jets? This controllability condition is necessary and sufficient. However, it must be noted that a "controllable" result may still be impractical if the system is very close to the uncontrollability limit and requires long duration jet firings. This becomes evident when dynamic simulation is performed.

Fig. 8 shows a sample output from the jet-selection tool for ACS jets. In this tool, the vehicle configuration is specified, including jet locations, mass properties and thrust levels; then the jet selection is determined for a specific acceleration command using dot-product, pseudo-inverse or optimized jet selection logic. This tool provides very useful insight into the jet selection and it allows for straightforward comparisons between the various jet selection algorithms. It also provides a necessary and sufficient test to check if the vehicle is uncontrollable under different jet failure conditions.

4. ANALYSIS AND RESULTS

This section presents some of the analysis performed for the X38 attitude control system. The following cases were considered in this study:
- Attitude control using DPS jets
  - Comparison between different phase plane controllers.
  - Effect of center of gravity (CG) variations and jet failures.
  - Comparison between coupled vs. pure command jet-selection.
- Attitude control using ACS jets.

4.1 Simulation Environment and Analysis Tool
The simulation environment used in this analysis was Trick [12]. Trick is an in-house, NASA-Johnson Space Center simulation development environment. It allows users to rapidly develop and operate simulations for different applications while it automatically generates most of the simulation specific input/output source code and database files. Another important part of the X38 attitude control simulation is the "RPOC Models" suite. The RPOC (Rendezvous, Proximity, and Capture) models are a collection of functions and math models developed to assess vehicles operating in the vicinity of another spacecraft (e.g., ISS, Shuttle). The RPOC models suite contains functions for dynamics, orbital mechanics, environment, guidance, navigation and control, etc. The different versions of the phase plane controllers and the jet-mapping and firing algorithms were developed as part of the RPOC models.

4.2 Assumptions
The following are the assumptions used for the analysis performed:
- The modified on-orbit phase plane controller was used (unless mentioned otherwise).
- The attitude dead band used is ± 1 deg. and rate dead band of 0.15 deg/s for Roll and between 0.1 and 0.3 deg/s for Pitch and Yaw.
- Attitude hold is in the Inertial frame.
- The configuration of DPS axial and RCS jets is as shown in Fig. 9. The vehicle CG is to be located within a specified CG box, as shown in the figure.

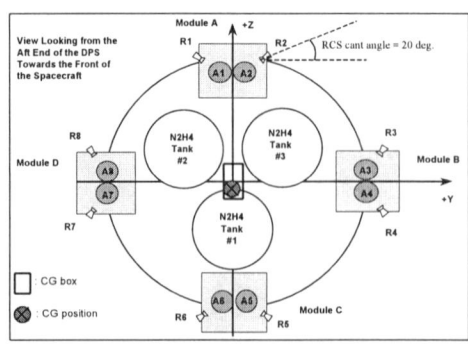

Fig. 9 Configuration of the Deorbit Propulsion Stage (DPS)

- The deorbit burn was performed using 6 axial jets only (e.g. jets A1, A3, A4, A5, A7 & A8) and lasts for 14 minutes. The reason for using 6 instead of 8 jets in the nominal

case is to prevent any large guidance perturbation for an axial jet failure. In such a case, the failed jet and its opposite counterpart are disabled (to minimize the disturbance) and the on command is switched to the two previously disabled axial jets. Therefore, the overall thrust stays the same. However, if two axial jets fail on the same pod, then the opposite pod is disabled and deorbit burn is accomplished using only 4 axial jets.
- The maximum number of jets that can be activated simultaneously is 3 for the ACS and 4 for the DPS-RCS systems.
- Thrust build-up characteristics of the DPS-RCS jets are shown in Fig. 10. No thrust build-up is used for the DPS axial jets due to their low activation frequency.

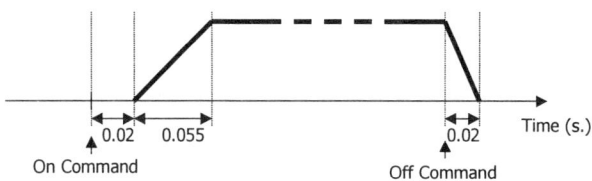

Fig. 10 Thrust build-up characteristics for DPS-RCS jets

- DPS jets are subject to a blow-down effect, i.e. to a loss of thrust due to tank pressure decrease as propellant is consumed. Fig. 11 shows the thrust ratio of DPS-RCS jets to axial jets over the range of operating pressures.

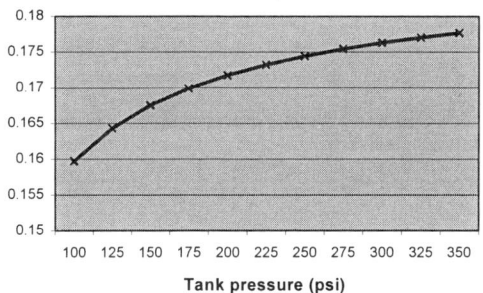

Fig. 11 Variation of thrust ratio of RCS/axial jets due to blow-down effect

- Failure of RCS jets R1 and R3 is used in this analysis as the worst 2-failure case.
- Nominal ACS thrust is 25 lbf, DPS-RCS thrust is 23-8 lbf, and DPS axial thrust is 115-45 lbf.
- No thrust force, jet position and cant angle dispersions taken into account.
- No navigation errors are considered.

4.3 Comparison between Different Phase Plane Controllers

Different phase plane controllers were compared during simulation of on-orbit coasting and deorbit burn flight segments. The purpose of this analysis was to determine which phase plane controller is most suitable for this application. The comparison was performed in terms of attitude hold performance, propellant usage, and jet duty cycle. In this comparison, no jet failure is assumed and the CG is assumed to be at the lower end of the CG box.

Case 1: During on-orbit coasting

The DPS-RCS jets were used for inertial attitude hold with no axial jet burns. The two controllers compared for this case were the shuttle on-orbit and the modified phase plane controllers considered. The transition phase plane controller was not considered since it was originally developed for use during deorbit burns. Fig. 12 and Fig. 13 show the attitude and attitude rate error performance using both controllers considered. The settings used were the same for both controllers.

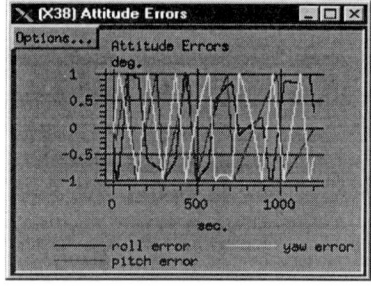 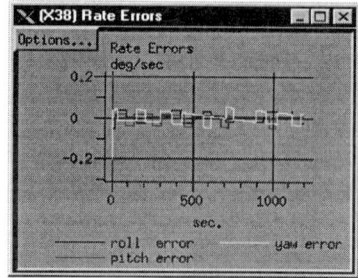

Fig. 12 On-orbit coasting using modified phase plane controller

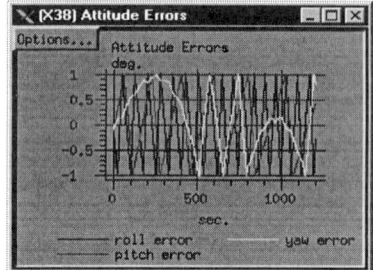 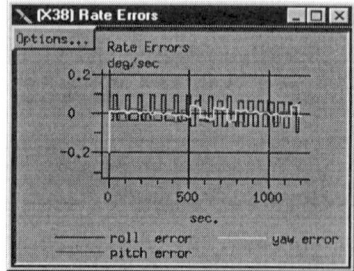

Fig. 13 On-orbit coasting using shuttle on-orbit phase plane controller

The RCS jet propellant consumption during a 20 minutes inertial hold was 0.22 kg for the modified phase plane controller, and 0.35 kg for the shuttle on-orbit controller. These plots show that the two controllers perform very well for attitude and attitude rate errors within the specified dead-bands and rate limits. A slight advantage for the modified phase plane controller is that it makes the correction more smoothly (smaller rate error), resulting in a decrease in the number of jet firings and consequently propellant consumption.

Case 2: During deorbit burn

Transition and modified phase plane controllers were compared for this flight segment. The CG is again placed at the bottom of the CG box and the resulting undesired rotational acceleration is passed as input to the modified phase plane controller. Fig. 14 and Fig 15 show the attitude error and the R8 jet duty cycle using both controllers. The R8 jet behavior is typical for the given initial conditions and configuration. It can be seen that the jet is activated very frequently when the transition phase plane controller is used. This is due to the CG offset disturbance. Unlike the transition phase plane controller, the modified phase plane controller uses some knowledge of the undesired acceleration to reduce jet activity. The RCS propellant consumed when using the shuttle on-orbit controller is 9 kg, while 12 kg is consumed when the modified phase plane controller is used. These quantities are still small compared to 720 kg of propellant consumed over the 14 minutes deorbit burn.

From these comparisons it can be seen that the modified phase plane controller is a better choice for both the on-orbit coasting and the deorbit burn. Another finding is that the use of undesired acceleration as a disturbance estimate is essential in order to reduce the jet duty cycle during deorbit burn.

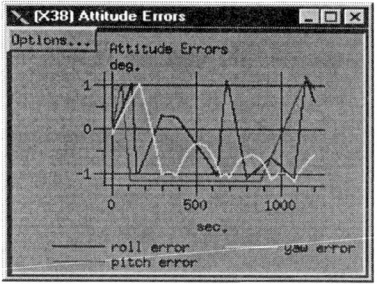 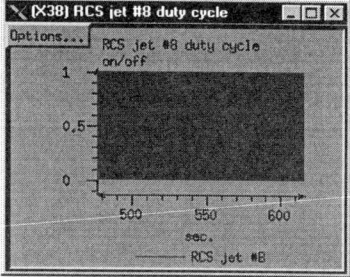

Fig. 14 Control during deorbit coasting using transition phase plane controller

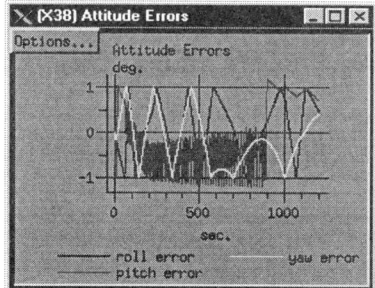 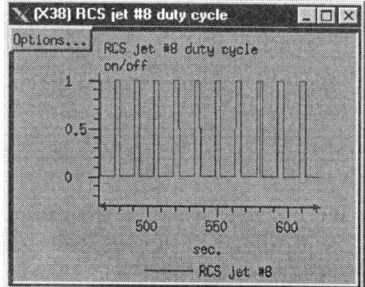

Fig. 15 Control during deorbit burn using modified phase plane controller

4.3 Effect of CG variations and jet failures

The controllability assessment was based on the position of the CG, jet failures, and DPS jets thrust blow-down characteristics. The worst case scenario when considering attitude control occurs during the deorbit burn when the CG is at its largest offset. This is because the disturbance generated by the axial jets is principally due to this offset. Therefore, this analysis was performed only for the deorbit burn segment, with and without jet failures. Fig. 11, which depicts the DPS jet blow-down characteristics, shows that when the tank pressure decreases over time, the thrust capabilities of the RCS jets decrease more rapidly, in a relative sense than the axial jet thrust capabilities. This exacerbates the effect of the CG offset over time and can result in an uncontrollable situation when jet failures occur. This is due to the fact that the RCS jet control torque becomes less effective compared to the axial jet disturbance torque as propellant is used.

Effect of CG variations

Fig. 16 and Fig. 15 (from the pervious subsection) show the attitude control system performance for the case with no jet failures when the CG is at the center and at the bottom of the CG box respectively. From these two figures it can be seen that the CG offset during the deorbit burn results in significantly more jet firings and consequently more propellant is consumed. The RCS propellant used when the CG was at the center was 0.15 kg and for the CG at the bottom of the CG box, it was 12 kg.

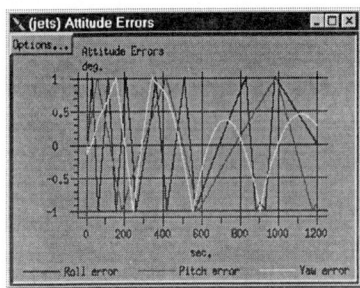 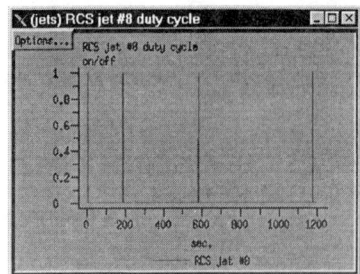

Fig. 16 Control during deorbit burn with CG at the center and no jet failure

Effect of Jet failures
Fig. 17 shows the attitude control system performance for the case of the CG at the center of DPS axial jet configuration and with two RCS jets failures (R1 & R3). It can readily be seen that the effect of two jet failures is more frequent jet firing and therefore, more propellant usage (compare to Fig. 16). The RCS propellant used for this case was 0.4 kg, compared to 0.15 kg for the case without failures.

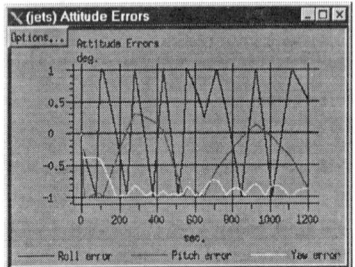 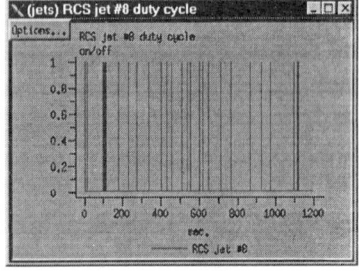

Fig. 17 Control during deorbit burn with CG at the center and 2 RCS jet failures

Effect of CG offset and Jet failures
Now, for the deorbit burn once again, we consider the worst case scenario, which has been defined as two pitch jet failures when the CG is at the bottom of the CG box. For this case, it was determined that the X38 will be uncontrollable (loss of attitude control). Fig. 18 shows the unstable pitch and yaw phase plane trajectories for this case. So, for the CG at the center or close to the center of the CG box, the vehicle remains controllable, but when the CG is moved to the bottom of the box while RCS jets R1 and R3 are failed, the vehicle becomes uncontrollable during the deorbit burn. In other words, the 6 remaining RCS jets cannot overcome the torque generated by the axial jets and maintain attitude.

 A first cut solution to regain controllability for this case was to off-pulse one axial jet (jet A1) at regular intervals to counteract the axial jet disturbance pitch acceleration. In this configuration, the axial jet A1 is off pulsed for between 5 and 8 seconds every 15 seconds. The axial jet propellant consumption stays approximately the same as for the nominal 6 engine deorbit, but the deorbit time increases from 815 sec to 870 sec. For the off-pulsing case the propellant consumption for the RCS jets is 20 kg. Fig 19 shows the attitude error and axial jet duty cycle.

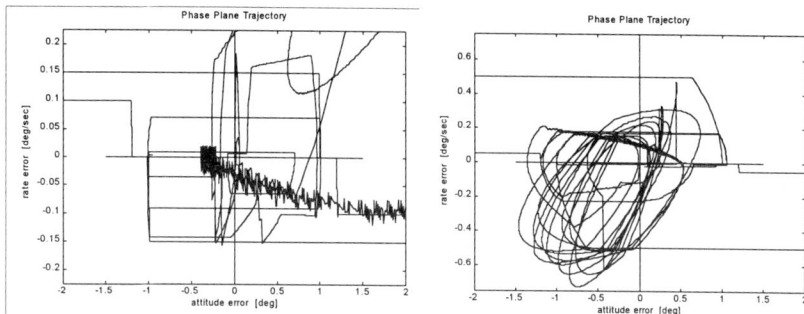

Fig. 18 Pitch and Yaw phase plane trajectory for the case of 2 jet failures and CG offset

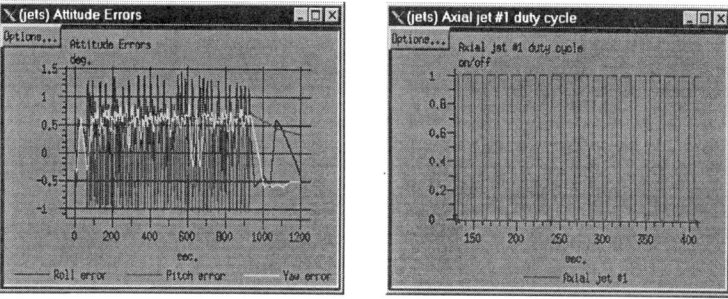

Fig. 19 Control for the case of R1 & R3 failures, CG at bottom and off pulsing A1

4.4 Comparison between Coupled vs. Pure Command Jet-Selection

The purpose of this analysis was to find the optimal set of RCS jets to fire for any given command generated by the phase plane controller. Two options were considered: Pure commands (partial jet-mapping) and coupled commands (full jet-mapping). The option of selecting coupled vs. pure command occurs when the controller sends a command (rate change) on more than one axis (roll, pitch, and yaw) at the same time. In this case, there are two possibilities: either the jet-mapping algorithm selects the best set of jets to execute the command on all axes simultaneously, or it selects the axis with the largest required rate change and finds the best set of jets to perform the maneuver. In the first case, the command is coupled, while in the second case, the command is pure. The coupled command yields better results in the nominal case, when there are no jet failures. This is because it is always possible to find a set of jets that yield the required acceleration commands simultaneously. However, this is not true for the worst case with two RCS jet failures. Here, pure commands yield better results.

Fig. 20 shows the attitude control performance for the case with the CG moved to the worst location (bottom right corner of the CG box) and with RCS jets R1 and R3 failed during the deorbit burn. In the first case (coupled jet command), the controller asks for a "–roll and +pitch" command for the first 400 sec. However, the best set of jets (R5 & R8) for this command does not provide any torque on the roll axis. Thus, the roll attitude is uncontrollable as long as the controller requests this command. At 400 sec., the pitch attitude error is brought back inside the dead-band and the controller switches to a "–roll" command only. The best set of jets for this command (R5, R7 & R8) provides a roll torque, and the torque attitude error finally starts to decrease. In the second case (pure jet command), a pure roll and pure pitch command are alternately requested, thus avoiding loss of control in one axis over a long period of time.

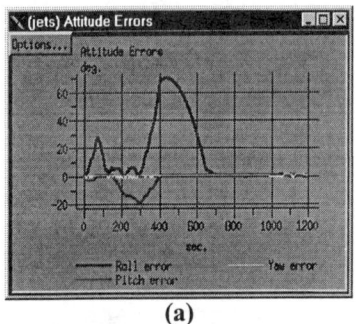 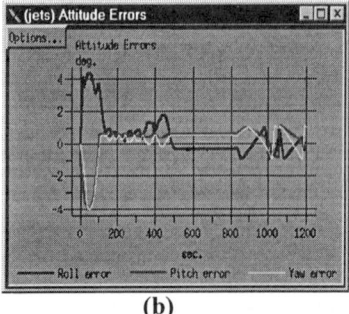

(a) (b)

Fig. 20 Control using (a) coupled and (b) pure jet commands

4.5 Attitude Control Using ACS Jets

All the analysis performed in the previous subsection was for attitude control using the DPS-RCS jets. In this section a sample of the results obtained for attitude control using the ACS jets is presented. The case selected is for attitude control of the EV during separation from the DPS module. Besides a spring mechanism that provides a push-off force at separation, an additional velocity change (ΔV) is applied using the 4 ACS axial jets mounted on the EV.

Fig. 21 shows the attitude control performance of the EV when separating from the DPS. The ΔV commanded to the ACS axial jets is 3 ft/s (~1m/s). It can be seen that the limit cycle of the pitch channel is single sided for about 50 seconds. This is due to a small pitch disturbance from the axial ACS jets. Once the axial burn is complete, the double-sided limit cycle response returns.

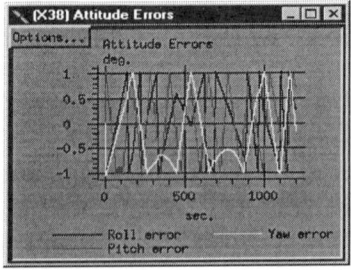 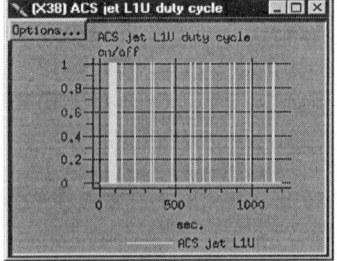

Fig. 21 Attitude control using ACS jets during EV and DPS separation

Another observation is the more frequent firing for the ACS case compared to the DPS-RCS in the nominal case (CG at center, no failure). This is mainly due to the higher level of jet acceleration when the ACS is used on EV only, especially for the roll channel.

For this case, using a wider dead-band may result in considerable propellant savings. This is an important issue for the ACS system since the propellant budget is very limited. Fig. 22 shows the effect of increasing the dead-band on the ACS propellant consumption. It can be seen that about 50% less propellant is used for inertial hold when the dead-band was increased from +/-1 deg. to +/-5 deg. The current recommendation is to use a +/-2 deg. dead-band, but more investigation is needed to find the optimal controller settings.

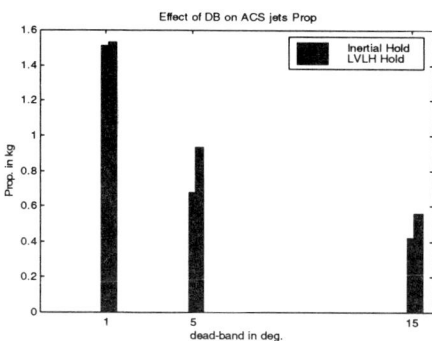

Fig. 22 Effect of phase plane controller dead-band on ACS propellant consumption

5. SUMMARY AND CONCLUSIONS

The attitude control system design for the X38 test vehicle on-orbit flight segments was considered in this paper. Simulation studies were presented for attitude control using both the DPS-RCS jets and the ACS jets. The main objective was to recommend controller settings and jet-mappings for both jet systems that optimize attitude performance, propellant consumption, and jet duty cycle. Three versions of the phase plane controller were used (shuttle on-orbit, shuttle transition, and modified phase plane) along with look-up table jet-maps. The jet selection tool used to facilitate the generation of the jet-maps was discussed. The main method applied in the development of these look-up table jet-maps was the dot-product with a modified cost function used to minimize the number of jets fired and undesired couplings (normal and off-axis components). The pseudo-inverse and optimized jet selection methods were also implemented in the jet selection tool to compare the outcome with the dot product method. And a controllability check was implemented to evaluate this characteristic for specific mass properties and jet configurations.

Various 6 DOF simulations were performed for X38 using the Trick simulation environment and the RPOC models suite to test the attitude control system of the vehicle. Comparisons between different phase plane controllers were performed and it was recommended to use the modified phase plane controller during on-orbit coasting and deorbit burn. It was also seen that the undesired acceleration switching curve is required during deorbit burn to reduce jet duty cycle. The effects of CG offset and jet failures in the presence of blow-down effects were also considered. The CG variation and jet failures were seen to result in more jet activity and consequently more propellant usage. For two RCS jet failures and the CG at the worst location, the X38 is uncontrollable. A proposed solution to regain controllability was the off-pulsing of an axial jet to counteract the disturbance torque due to the CG offset. Comparison between coupled and pure command jet selection was performed. A pure command jet-map was seen to outperform a full command jet-map for the 2 RCS failure case. Simulations were also flown to demonstrate the attitude control of the EV using the ACS jets during separation from the DPS.

REFERENCES

1. B. Rishikof et al. "Crew return vehicle (CRV) escape trajectories", AIAA ISS Service Vehicles Conference, April 1999.
2. D. Zimpfer, L. S. Shieh, and J. W. Sunkel, "Digitally redesigned pulse width modulation spacecraft control", Journal of Guidance, Control and Dynamics, Vol. 21, No. 4, July 1998, pp.529-534.
3. E. Bergmann, et al. "An advanced spacecraft autopilot concept", Journal of Guidance and Control, Vol. 2, No. 3, 1979.
4. E. Bryson, and Y. Ho, Applied optimal control, Hemisphere Publishing Corp., 1975.
5. E. T. Kubiak, and D. Johnson " Optimized jet selection logic", Johnson Space Center, Houston, TX, March 1987.
6. E. T. Kubiak, and R. Davison, "A phase plane logic design to interface with optimal type jet selection algorithms", Johnson Space Center Report EH2-86L-001, Houston, TX, July 1986.
7. L. Kellogg, " Precise nulling of attitude and motion errors of a spacecraft using a phase space autopilot", Master thesis, MIT, Sept. 1978.
8. M. Elgersam, et al., "Space station attitude control using reaction control jets", Proceedings of the 31st conference on Decision and Control, Tucson, AZ, Dec. 1992.
9. N. Penchuk, P. D. Hattis, and E. T. Kubiak, "A frequency domain stability analysis of a phase plane control system", Journal of Guidance and Control, Vol.8, No.1, 1983, pp. 50-55.
10. P. D. Hattis, "Qualitative differences between on-orbit and transition RCS control", Proceedings of the AIAA Guidance and Control conference, San Diego, CA, Aug. 1982, pp. 437-436.
11. P. D. Hattis et al., "Shuttle on-orbit flight control characterization, Simplified digital autopilot", NASA JSA 18511, Aug. 1982.
12. R. W. Bailey, and E. J. Paddock, Trick User's Guide, Johnson Space Center/LinCom Corp., Houston, TX, 1998.
13. STS83-0008V2-28, Space shuttle orbiter operational level C, functional subsystem software requirements, Guidance, Navigation, and Control, Part C, Flight Control Ascent, Volume 2, United Space Alliance, April 1998.
14. STS83-0009-26, Space shuttle orbiter operational level C, functional subsystem software requirements, Guidance, Navigation, and Control, Part C, Flight Control ORBIT DAP, Rockwell Aerospace, May 1996.

AAS 00-026

THERMAL RADIATOR POINTING FOR INTERNATIONAL SPACE STATION

Scott A. Green[*]

In order to provide thermal radiation environments that result in adequate heat rejection, the single-phase, liquid ammonia heat rejection system on the International Space Station (ISS) requires that its two thermal radiator wings be dynamically rotated as the ISS travels through its orbit. This paper discusses the closed-loop, thermal radiator pointing system that is used on ISS to ensure adequate heat rejection by the radiators, while preventing freezing of the ammonia under low heat loads and cold environmental conditions.

Although initial designs used an open-loop approach for radiator pointing, concerns about performance robustness, algorithm complexity, memory requirements, and sustaining support drove the development of a more robust, simpler, closed-loop system. Hence, the challenge of the closed-loop system was to utilize existing sensors, actuators and computers to fit into the existing hardware and software architecture of the ISS. Using a proportional-integral (PI) control architecture with limited output and an anti-windup integrator, the temperature of the ammonia coming out of the radiator is measured and controlled by adjusting the radiator wing orientation. The radiator wing orientation for the local minimum environment is fed forward to the control system, and the closed-loop controller is used to generate a bias off of that local minimum environment in order to heat up the ammonia when necessary to avoid freezing. In the earth's shadow, the controller is suspended and the radiator wing is oriented to face the earth, the local maximum thermal environment which further prevents freezing of the ammonia. This control architecture is shown to provide adequate heat rejection and avoid freezing of the ammonia, even though the physical system consists of large transport delays and time-varying, nonlinear dynamics which change dramatically due to orbit motion and variable heat loads.

INTRODUCTION

The External Active Thermal Control System (EATCS) on the International Space Station (ISS) utilizes a single-phase, liquid ammonia system to collect waste heat from the ISS and radiate it out into space using two radiator wings. In order to provide thermal radiation environments that result in adequate heat rejection, the EATCS requires that its two thermal radiator wings be dynamically rotated as the ISS travels through its orbit. Although adequate heat rejection can be achieved using simple pointing algorithms that orient the radiator wings to a local minimum thermal environment throughout the orbit, such orientations can cause the ammonia in the radiators to freeze, thus potentially rupturing the pressurized system when thawed. A modified approach that orients the radiators to a local maximum

---

[*] Dr. Green is a Principal Engineer with The Boeing Company, 5301 Bolsa Avenue, MC H020-F508, Huntington Beach, California 92647-2099. E-mail: scott.a.green@boeing.com.

thermal environment when in the earth's shadow will also achieve adequate heat rejection, but can still result in freezing of the ammonia. Hence, a more sophisticated pointing approach is necessary.

Initial designs for the EATCS achieved a balance between adequate heat rejection and freezing avoidance by pointing the thermal radiators using an on-orbit estimation of the thermal radiation environment.[1] This "open-loop" approach estimated the incident heat flux from the sun, earth and surrounding ISS structure, and derived an equivalent thermal radiation environment which was used to point the thermal radiators. However, as this approach was developed, concerns emerged regarding the algorithm's performance robustness to modeling errors, its complexity, and its requirements for significant on-board computer memory and considerable sustaining support from operators on the ground. These concerns resulted in a reevaluation of the pointing algorithm design, and the eventual transition to a closed-loop control approach.

Due to cost and schedule limitations, alternative control approaches for thermal radiator pointing are constrained to utilize existing sensors, actuators and computers in order to fit within the existing hardware and software architecture of the ISS. Thus, the straightforward approach of feedback control using thermal environment sensors on the radiator panels can not be considered, even though this design could ensure both adequate heat rejection and prevention of ammonia freezing. The constraints limit all candidate closed-loop approaches to the use of ammonia fluid-temperature feedback, even though the physical separation between sensors and actuator result in long, time-varying transport delays. The closed-loop control design is further complicated by nonlinearities from the thermal radiation dynamics, coupled with the difficulty that the thermal dynamics, which are related to the fluid mass flow rate through the radiator, also change as a function of time.

Such nonlinear, time-varying systems with significant transport delays are difficult to control. In such cases, control designers often rely upon physical insight into the system dynamics in order to develop effective control solutions. Common insights for these problems have resulted in linear control systems that use gain scheduling, or nonlinear "feedback-linearization" techniques, both of which can accommodate the variable operating points associated with nonlinear, time-varying dynamics.[2,3] Other techniques, targeted specifically to time delays, include Smith predictors or Internal Model Control (IMC), which help compensate for the significant phase lag and bandwidth limitations associated with large transport delays.[3] Further investigation of the literature for this type of system leads to still more complex control approaches, including model following control, model-reference adaptive control (MRAC), and other adaptive control techniques.[2,3] The challenge to the control designer, then, is to find the simplest, robust control approach that can be easily implemented and will still satisfy the performance and stability requirements of the system.

For the ISS EATCS, insights into the physical dynamics of the thermal system led to the development of a proportional-integral (PI) control architecture that feeds forward the local minimum environment and adds a bias in order to heat up the ammonia when necessary to avoid freezing. More sophisticated control techniques such as Smith predictors, Internal Model Control, adaptive control, or nonlinear control were not necessary in order for the system to meet its performance objectives.

This paper presents the closed-loop, thermal radiator pointing system that is used on ISS to ensure adequate heat rejection by the radiators, while preventing freezing of the ammonia under low heat loads and cold environmental conditions. The details of the EATCS physical system are first presented. The challenges associated with the control design for this system are then discussed, followed by a description of the closed-loop control architecture that was developed. Finally, simulation results demonstrating the nonlinear performance of the system are presented and discussed.

PHYSICAL SYSTEM

This section describes the details of the EATCS physical system, from heat acquisition through heat rejection, including the pump and bypass control used to maintain a system mass flow rate and a set-point temperature for the ammonia leaving the pump. The EATCS loop used for heat acquisition and heat rejection on the ISS is shown schematically in Figure 1. Each radiator wing on the ISS provides heat rejection for a separate and independent EATCS loop, though the system hardware and functional operation is identical for both loops. Hence, discussion can be limited to a single EATCS loop and thermal radiator without loss of generality.

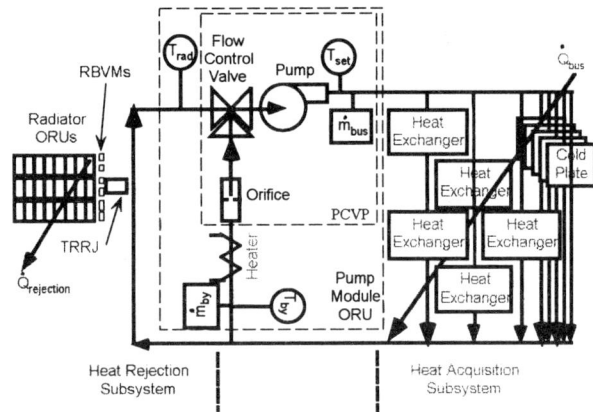

Figure 1 EATCS Schematic

Heat is acquired from the ISS laboratory and habitation modules through heat exchangers and cold plates in the heat acquisition subsystem shown schematically on the right side of Figure 1. The total heat load acquired is labeled "Q_{bus}" and can vary from 0 to 35 kW. The ammonia fluid temperature entering the heat acquisition subsystem, at the top of Figure 1, is controlled to a set-point of 37°F as it comes out of the pump. The ammonia fluid temperature leaving the heat acquisition subsystem, at the bottom of Figure 1, is elevated, depending upon the heat load acquired.

Ammonia exiting the heat acquisition subsystem travels either through a bypass leg or through the heat rejection subsystem. A flow control valve, located just before the pump, controls the relative pressure between the two legs and, thus, controls the mass flow rate split between the two paths. Ammonia traveling through the heat rejection subsystem, entering at the bottom left side of Figure 1, is transported through the Thermal Radiator Rotary Joint (TRRJ), through the Radiator Beam Valve Modules (RBVMs), and into one of three radiator Orbit Replacement Units (ORUs) where the waste heat is radiated into space. The total heat rejected is labeled "$Q_{rejection}$" in Figure 1. The cooled ammonia leaves the radiator ORUs, returns through the RBVMs and the TRRJ and then travels to the flow control valve and pump module.

The flow control valve maintains a 37°F set-point temperature at the pump outlet by adjusting the mix between the relatively cold ammonia coming from the thermal radiators with the relatively hot ammonia coming through the bypass leg directly from the heat acquisition subsystem. The pump head accommodates pressure drops through the system and maintains a mass flow of ammonia through the EATCS loop.

Several temperature and mass flow rate sensors are located throughout the system, and are available for closed-loop control. Starting from the outlet of the pump in Figure 1 and working around the loop, the system mass flow rate is measured and labeled as "$\dot{m}_{bus}$" in lb_m/hr. The pump outlet temperature is measured and labeled as "T_{set}" in °F, which is used by the flow control valve for closed-loop control in order to maintain the loop set-point temperature of 37°F. The temperature of the fluid leaving the heat acquisition system is measured along the bypass leg, denoted "T_{by}" in °F. The mass flow rate of ammonia traveling through the bypass is measured and labeled as "$\dot{m}_{by}$" in lb_m/hr. The mass flow rate through the heat rejection subsystem can then be approximated as the difference between the total system mass flow rate and the mass flow rate through the bypass, $\dot{m}_{rad}=\dot{m}_{bus}-\dot{m}_{by}$. The temperature of the fluid coming from the heat rejection subsystem as it enters the pump module ORU is measured and labeled as "T_{rad}" in °F.

Ammonia fluid temperature sensors also exist on the thermal radiator wings themselves, labeled "T_{rad}" as shown in the thermal radiator wing schematic in Figure 2. There are six fluid temperature sensors for each thermal radiator wing, one sensor mounted on each RBVM assembly. There are two RBVMs and two RBVM manifolds per radiator ORU in order to allow for redundant fluid flow paths. Each radiator ORU consists of eight panels, and each panel has 11 tubes per manifold, for a total of 22 tubes per panel.

The ammonia travels through the TRRJ, is split into six paths, travels up an RBVM manifold, through a panel, and then back down to return to the RBVM. The fluid temperature sensors are mounted on this return path, measuring the ammonia temperature after the fluid from each of the panels has mixed together.

Compared with the radiator outlet fluid temperature measured at the pump module ORU, as discussed above, measurements of the radiator outlet fluid temperature at the RBVM are much closer to the heat rejection source. This results in a shorter transport delay between the sensor and actuator, as the cooled ammonia traveling from the radiator panels will reach the RBVM fluid temperature sensors long before it will reach the pump module ORU. Shorter transport delays are of particular benefit to a closed-loop design due to the reduction in the associated phase lag in the loop and the corresponding ability to increase the control system bandwidth. Additionally, the six RBVM sensors can be polled to provide fault detection capabilities. Hence, feedback using the RBVM sensors is the preferred choice for fluid temperature feedback control, with fluid temperature feedback at the pump module ORU available as a backup.

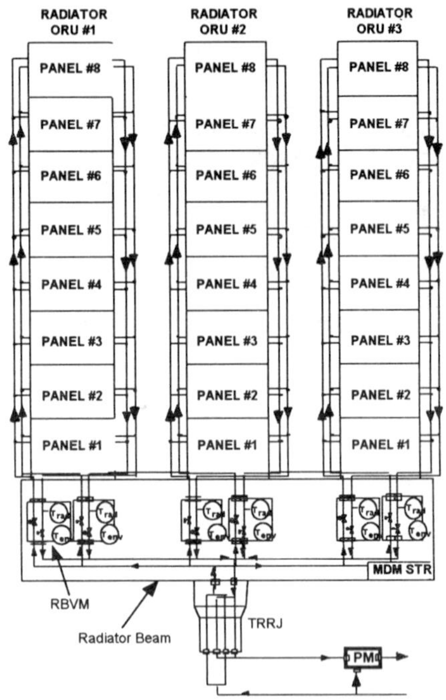

Figure 2 Radiator Wing Schematic

The TRRJ allows for one-degree-of-freedom articulation of the thermal radiator wing. Its design enables the radiator to be rotated such that the unit normal of the radiator wing is kept perpendicular to the vector between the ISS and the sun, thus providing a minimum thermal environment for maximum heat rejection when exposed to the sun. By off-pointing the thermal radiators from this local minimum thermal environment orientation, the TRRJ can then be used as an actuator that effectively decreases the amount of heat rejected by the thermal radiators by increasing the equivalent thermal environment. Hence, the radiator outlet fluid temperature, as measured at the RBVMs or at the pump module ORU, can be roughly manipulated by rotation of the TRRJ. This use of the TRRJ to effectively heat up the radiator fluid is exploited in a closed-loop control system designed to prevent freezing of the ammonia, though the transport delays and nonlinearities of the radiation dynamics make it a challenging system to control, as discussed in the next section.

CONTROL OBJECTIVE AND CHALLENGES

The system-level objective of the EATCS is to collect and reject waste heat from the ISS while maintaining the 37°F set-point temperature of the fluid entering the heat acquisition subsystem. The supporting objective of the thermal radiator pointing system is to orient the radiator wings so as to ensure adequate heat rejection, up to 35 kW of waste heat per loop, while simultaneously preventing the fluid in the radiators from freezing. The freezing temperature of ammonia is -108°F. This section discusses the complexities associated with the system operation and system dynamics of the EATCS, which ultimately challenge the design and development of the thermal radiator pointing system as it works to meet its control objective.

The simplest pointing approach to ensure adequate heat rejection is to always orient the thermal radiators to their locally minimum thermal environment so as to achieve maximum heat rejection. However, when the ISS moves into the earth's shadow, known as the umbra, where the space station has no exposure to the sun, the minimum thermal environment can get cold enough to freeze the ammonia in the radiators. Thus, analysis shows that this approach can not achieve the complete control objective of adequate heat rejection while ensuring the ammonia in the radiators will not freeze.

A modified pointing approach is to orient the radiators to their locally maximum thermal environment in the umbra, which is maximum exposure of the radiator panel surface to the earth, while maintaining the local minimum, edge-to-sun, orientation of the radiators when not in the umbra. This modified pointing approach is still relatively simple to implement, and analysis shows it still provides for adequate heat rejection by the thermal radiators. However, again, under some orbital circumstances this radiator pointing profile can result in freezing of the ammonia in the radiator panels. Nonetheless, because this radiator pointing profile satisfies the control objective in most cases, it is used as the default pointing profile for the thermal radiator system. For the purposes of this discussion, this profile of edge-to-sun on the sun-side, earth-face in the umbra is named the "gamma theoretical" orientation, γ_t, a name derived from the TRRJ rotation angle, which is called the gamma angle, γ. A more complete discussion of the gamma theoretical orientation, γ_t, and its derivation is presented in Reference 1.

Using the gamma theoretical profile, γ_t, as the nominal pointing command, adequate heat rejection by the thermal radiators is assured, but a more sophisticated pointing approach is necessary to prevent freezing of the radiators under all conditions. As discussed before, any such control modifications designed to further prevent freezing of the thermal radiators are constrained to use existing sensors, actuators and computers in order to fit within the hardware and software architecture of the ISS. Thus, closed-loop control systems are restricted to using ammonia fluid temperature feedback for sensing and TRRJ rotation for thermal environment control. This has many important implications to the physical system that must be controlled. The three most challenging implications are associate with using the TRRJ as an effective actuator, the inherent nonlinearities of the radiation dynamics, and the time-varying transport delays of the system. These challenges are discussed next.

Time-Varying Thermal Environments

The TRRJ angular position, γ, can be used to manipulate the equivalent thermal environment seen by the radiators, thus changing the radiator heat rejection capability and ultimately adjusting the ammonia fluid temperature coming out of the radiator panels. Computational models of the space station and the heat sources and heat sinks of the space environment are used to compute the equivalent thermal environmental temperature, T_{env}, to which energy from the surface of the radiator wings is effectively emitted. The computational models for ISS analysis use the Thermal Radiation Analyzer System (TRASYS) and the Systems Improved Numerical Differencing Analyzer (SINDA).[1] Thermal environment temperatures are computed for various positions of the ISS relative to the earth and the sun. They are also determined for various rotation angles of the TRRJ.

Figure 3 shows a typical curve of T_{env} as a function of the TRRJ gamma angle, γ. It is plotted for a point in the ISS orbit when the sun appears at its maximum height above the space station with the earth directly below. For the purposes of this discussion, this point is called "orbit noon" and is the initialization point from which the orbit position is set equal to zero. Figure 3 is also plotted for a solar beta angle of 25°. The solar beta angle is the angle between the ISS orbit plane and the solar line-of-sight vector. This angle changes as the ISS orbit regresses around the earth, and as the earth

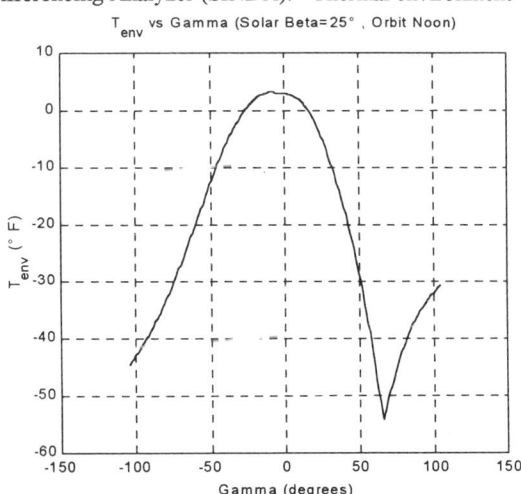

Figure 3 Typical T_{env} vs. γ Curve

orbits around the sun.[4] Due to the 51.5° inclination of the ISS orbit and the 23.5° tilt of the earth, the solar beta angle can vary between ±75° over the course of a year, though it remains approximately constant on the time scale of a single orbit.

The local minimum T_{env} shown in Figure 3 occurs when the TRRJ is in the edge-to-sun orientation, an orientation in which the unit normal of the radiator wing surface is perpendicular to the vector between the ISS and the sun. For Figure 3, edge-to-sun occurs at the gamma theoretical orientation of $\gamma_t=65°$, earth facing is at $\gamma=0°$, and full sun facing is at $\gamma=-25°$. The local maximum thermal environment is a blend between the earth-facing and full sun-facing orientations. In general, the view factors, or form factors of the earth and sun dictate the characteristic shape of the T_{env} vs. γ curve, and define the local minima and maxima. These view factors characterize the percent of the sphere surrounding the radiators that is covered by the earth or sun.

A local minimum thermal environment on the sun-side of the orbit occurs at γ_t, when the radiators are oriented edge-to-sun. By biasing off of this γ_t orientation, the thermal environment can be increased up to a point. If the bias gets too large, the gamma angle can exceed the local maximum thermal environment and an additional bias off of γ_t would cause the thermal environment to decrease. It should be noted that the T_{env} vs. γ curve shown in Figure 3 is only plotted within the physical angular rotation limits of the TRRJ, between ±105°.

Figure 4 plots a family of T_{env} vs. γ curves for various orbit positions, ranging from the exit of the umbra (250°) up until just before orbit noon (320°). All curves in Figure 4 are plotted for a solar beta angle of 25°. The edge-to-sun orientation, γ_t, is identified by an "X" on each curve. The local maximum T_{env} is identified buy an "O" on each curve. Notice that the T_{env} curves vary considerably as a function of orbit position, changing both in shape and in the relative locations of minima and maxima. This variation of the T_{env} vs. γ family of curves demonstrates one of the most significant challenges associated with using the TRRJ as an actuator to adjust T_{env}, and ultimately to control the radiator outlet fluid temperature.

Just after leaving the umbra, at an orbit position of 250° in Figure 4, the local minimum is around $\gamma_t=-40°$ and the local maximum is around $\gamma=+25°$. For this case, a *positive* $\Delta\gamma$ bias added to γ_t would result in the largest increase in T_{env}, since the slope of $\partial T_{env}/\partial\gamma$ is largest in that direction. For the orbit position of 320° in Figure 4, a *negative* $\Delta\gamma$ bias added to γ_t would result in the largest increase in T_{env}. Thus, the sign of the desired $\Delta\gamma$ bias changes as a function of orbit position. In general, the desired $\Delta\gamma$ bias off of γ_t should be added or subtracted such that the radiators are turned away from the nearest position limit of the TRRJ, and toward the local maximum of the T_{env} curve. Following this rule ensures that the $\Delta\gamma$ bias will travel along the largest $\partial T_{env}/\partial\gamma$ slope.

From Figure 4, it is clear that full sun facing, 90° away from γ_t, is not necessarily the maximum thermal environment. Again, this is because the true maximum T_{env} comes from a relative blend of the view factors to both the sun and the earth, which change as the ISS moves through its orbit. It is also clear that the edge-to-sun orientation, γ_t, must switch relative locations with the local maximum T_{env}, as occurs between 270° and 280° in Figure 4. During this transition, the $\Delta\gamma$ bias off of γ_t that achieves a maximum thermal environment must transition through zero. Because the $\Delta\gamma$ bias should always be chosen to move the radiators towards the local maximum, and because that local maximum is a blend between full sun-

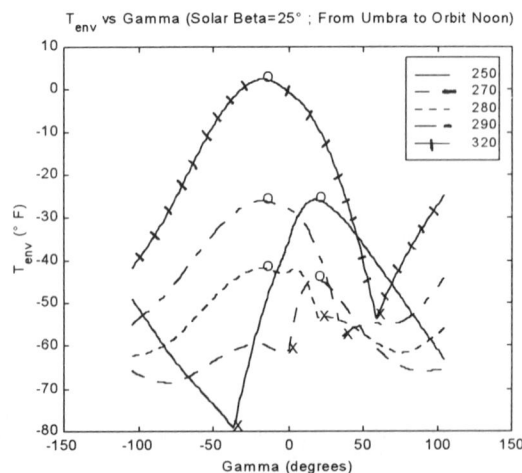

Figure 4 T_{env} vs γ (at Various Orbit Positions)

facing and full earth-facing, a conservative estimate of the Δγ bias limit is the absolute difference between full earth-facing and edge-to-sun. Though this is not always the local maximum T_{env}, it is a conservative estimate that ensures the Δγ bias off of γ_t will always maintain the same sign of its $\partial T_{env}/\partial \gamma$ slope. This is necessary to ensure the stability of a closed-loop control system that actuates with a Δγ bias off of γ_t. Thus, within limits, the thermal environment can be manipulated by biasing the TRRJ gamma angle off of γ_t.

Radiator Dynamics & Transport Delays

Also of interest to the control system development is an understanding of the radiator dynamics and transport delays. The nonlinear radiation dynamics of the radiator wings are based upon a well-known dependency of temperature to the fourth power. Hence, the nonlinear radiation dynamics are left to discussions in the literature, such as Reference 5. It is sufficient to say for this application that these radiator dynamics are nonlinear (with T^4), and that they are time varying due to their dependency on the radiator mass flow rate. These nonlinear, time-varying dynamics must be accommodated in the development of the thermal pointing closed-loop control system.

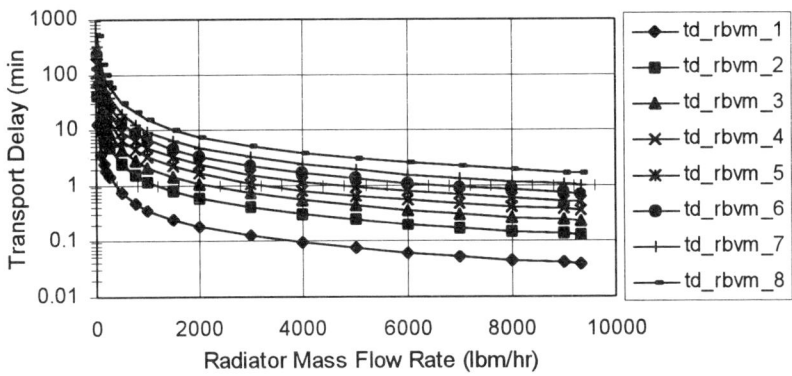

Figure 5 Transport Delay vs. Radiator Mass Flow Rate

The transport delays of the system are also dependent on the time-varying, radiator mass flow rate. The transport delay from an individual panel to its corresponding RBVM sensor is proportional to the relative length of the tubing separating the two, and inversely proportional to the velocity of the fluid traveling through that tubing. Figure 5 plots the transport delays from each radiator panel to its RBVM assembly. As shown in Figure 5, the transport delay is inversely proportional to the radiator mass flow rate. In fact, as the mass flow rate drops to zero, the transport delays go to infinity. As a result, Figure 5 must be plotted on a log-linear scale.

Due to the large transport delays for low radiator mass flow rates, the system becomes more challenging to control. Fortunately, however, the panels close to the RBVM have relatively short transport delays, and so they provide a relatively fast reaction to changes in the thermal environment or changes in the acquired heat load. This allows the thermal radiator pointing system to have sufficient bandwidth so as to respond relatively quickly to system disturbances, as will be discussed with the nonlinear simulation results presented in a later section.

Summary of General Observations of the Dynamic System

In summary, there are several useful insights into the dynamic system that can be exploited in the development of a closed-loop control design that will meet the control objective. To ensure adequate heat rejection of the radiators, the gamma theoretical profile, γ_t, should be taken as the nominal pointing command. On the sun side of the orbit, biasing off of this γ_t orientation will increase the thermal environment. This allows for indirect control of the radiator outlet fluid temperature and provides the capability to prevent freezing of the ammonia in the radiators. Thus, the total control objective can be achieved.

Note, however, because γ_t is the local minimum thermal environment, a $\Delta\gamma$ bias off of γ_t can only be used to warm up the radiator outlet fluid temperature. Additionally, there are limits in both the direction in which the $\Delta\gamma$ bias should be added, as well as the maximum magnitude allowed to ensure stability. To ensure maximum impact of the $\Delta\gamma$ bias, it should always be added or subtracted such that the radiators are turned away from the nearest TRRJ position limit and toward the local maximum T_{env}. For stability, it is necessary to limit the maximum $\Delta\gamma$ bias to ensure it does not exceed the maximum T_{env}. A conservative limit is the absolute difference between the earth-facing γ-angle and the edge-to-sun γ_t-angle. These rules, which are based upon evaluation of the T_{env} curves for many solar beta angles and orbit positions, are helpful guidelines offering valuable insights that can aid in development of the closed-loop control system architecture. The control architecture for the system designed is discussed in the next section.

FLUID TEMPERATURE CONTROL DESIGN

This section presents the fluid temperature control system architecture designed to ensure adequate heat rejection, while simultaneously preventing freezing of the ammonia in the radiators. Figure 6 shows a block diagram schematic of the system using fluid temperature feedback from the RBVM sensors. This same architecture can be used for fluid temperature feedback at the pump module sensor, though the control system gains must be reduced in order to accommodate the increased phase lag from the additional transport delay.

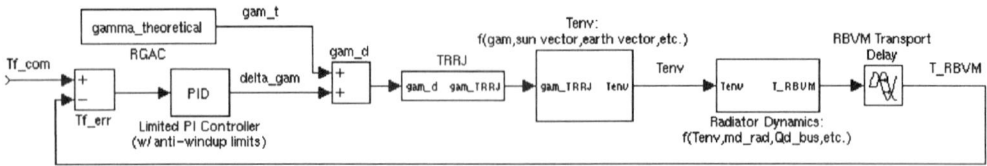

Figure 6 Block Diagram Schematic of Fluid Temperature Control System

To ensure adequate heat rejection by the thermal radiators, the γ_t orientation is fed forward to the control system. This defines the local minimum thermal environment on the sun-side of the orbit and the local maximum environment in the umbra. The radiator outlet fluid temperature, as measured at the RBVM, is fed back and compared with a fluid temperature command, labeled "Tf_com" in Figure 6, to generate the temperature error, "Tf_err".

On the sun side of the orbit, a PI control law is used to convert the temperature error into a desired $\Delta\gamma$ bias off of the γ_t orientation. Because the γ_t orientation defines a local minimum thermal environment, the PI control law can only be used to increase the fluid temperature. Thus, the $\Delta\gamma$ bias must be limited to a minimum value of 0°. The maximum limit of the $\Delta\gamma$ bias can be set to a constant value, though this could result in a temporary instability during periods in the orbit when the bias angle drives the TRRJ command beyond the maximum T_{env}, so that an increasing $\Delta\gamma$ bias results in a decreasing T_{env} temperature.

To avoid this potential instability, a "smart limiter" is employed. For the smart limiter, the maximum $\Delta\gamma$ bias limit is set as the absolute difference between the earth-facing γ-angle and the edge-to-sun γ_t-angle. This ensures the slope of the T_{env} vs. γ curve always maintains the same sign so that stability can be maintained. It also has the effect of smoothing out the motion of the TRRJ by eliminating large TRRJ rotations, as will be demonstrated in the next section.

Because the PI controller must use output limits, it is necessary to implement an anti-windup integrator to stop the integration of the temperature error when the output is on the upper or lower limit.

This anti-windup integrator is also used in the umbra, when the PI controller is temporarily suspended as the TRRJ points the radiators to an earth-facing orientation.

As discussed in the previous section, the $\Delta\gamma$ bias computed from the limited PI control law should be added or subtracted from γ_t such that the radiators are turned away from the nearest TRRJ position limit and towards the local maximum T_{env}. For the case of the constant limiter, this means that the $\Delta\gamma$ bias is added when $\gamma_t<0$ and subtracted when $\gamma_t>0$. Unfortunately, this logic can result in large rotations of the TRRJ when passing through $\gamma_t=0$ if the bias value is large, as will be demonstrated in the next section. For the case of the smart limiter, the $\Delta\gamma$ bias is always combined with γ_t to ensure it is moving towards the local maximum T_{env}.

The combined γ_t and $\Delta\gamma$ bias are called the "gamma desired" command and labeled as "gam_d" in Figure 6. This "gamma desired" is sent to the TRRJ, which tracks the command within its own position, velocity, and acceleration constraints.[6] The actual angle achieved by the TRRJ, combined with the relative position of the sun and earth, dictate the T_{env} to which the radiators are exposed. The thermal radiator dynamics can then be used to determine the heat rejection, and ultimately the radiator outlet temperature that is measured at the RBVM. Thus, the feedback loop in Figure 6 is closed.

Linear analysis is used to compute the PI control system gains and ensure adequate stability margins to accommodate the nonlinear radiation dynamics and fluid transport delays. These control gains are used in a nonlinear simulation to analyze and verify system-level performance. Nonlinear simulation results for three performance scenarios are presented and discussed in the next section.

NONLINEAR SIMULATION RESULTS

Three scenarios are evaluated in order to demonstrate the nonlinear performance and robustness of the thermal radiator pointing system. These scenarios include both variations in the heat acquisition load, $\dot{Q}_{bus}$, as well as ΔT_{env} biases off of the nominal thermal radiation environment. All simulation results presented are for a solar beta angle of 25° and a time-varying orbit position. The simulations use fluid temperature feedback from the RBVM sensors, and apply a command temperature of -40°F to the fluid temperature control system. The first scenario uses a constant limit of 65° for the $\Delta\gamma$ bias. The other two scenarios use the smart limiter bias.

Unlike results presented in previous sections, the simulation results discussed here include shadowing effects of the local space station structure on the thermal radiator wings. These shadowing effects result in decreased thermal environments due to the blockage of sun light on local areas of the radiator wings when on the sun-side of the orbit.

Additionally, the effects of the ISS torque equilibrium attitude (TEA) are also included in these results. The TEA is the ISS vehicle attitude flown in order to balance gravity gradient torques with aerodynamic torques. The TEA used for these simulations is based upon a yaw-pitch-roll rotation off of the local vertical, local horizontal reference frame; [yaw, pitch, roll]=[-4°, -5.2°, -0.2°] for this case. The local vertical, local horizontal reference frame is defined with the x-axis along the velocity vector in the orbit plane, the y-axis normal to the orbit plane, and the z-axis in the orbit plane pointed towards the center of the earth.

Scenario 1: Nominal Performance with Constant Limiter

The first scenario is a nominal performance case, using a constant limit of 65° for the $\Delta\gamma$ bias. This scenario uses a nominal thermal environment, so $\Delta T_{env}=0$. The run begins at orbit noon, with a 35 kW heat load coming into the heat acquisition subsystem. After one complete orbit (one orbit is 360°), the heat load is stepped down to 2 kW. It remains at the 2 kW level for three-and-a-half orbits, then, just after exiting the umbra on the fifth orbit, the heat load is stepped back up to 35 kW. The nonlinear simulation results for the

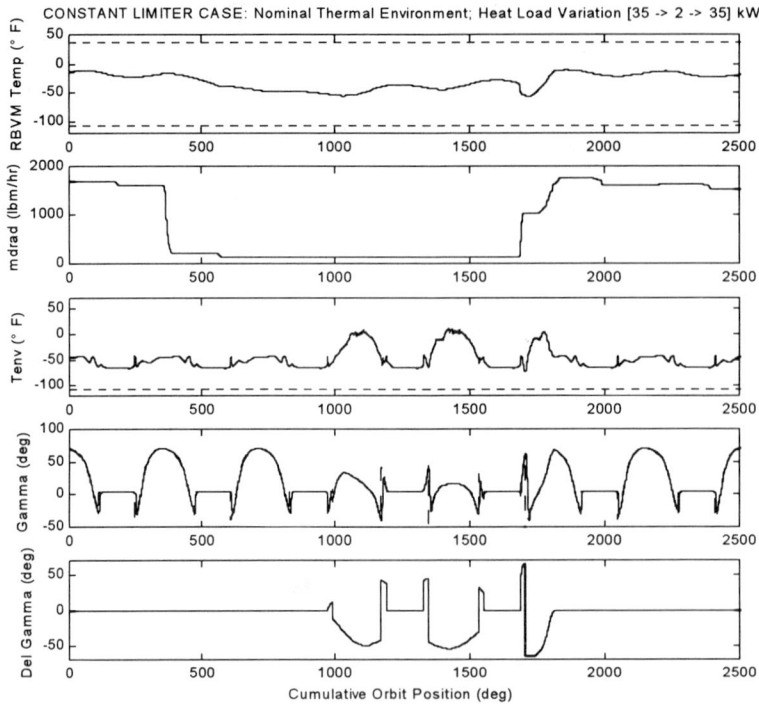

Figure 7: Nominal Performance, Constant Limiter Scenario

RBVM fluid temperature, radiator mass flow rate, thermal environment temperature, TRRJ angular position, and Δγ bias are all shown in Figure 7. The results are plotted for eight complete orbits of the ISS around the earth.

The fourth strip in Figure 7 shows the TRRJ angle, γ, as a function of orbit position. It begins at orbit noon, as the radiator tracks edge-to-sun with a zero bias up until the space station enters the umbra. When in the umbra, the TRRJ angle transitions to approximately 0° where it remains at a constant angle until the space station exits the umbra. The TRRJ then shows a step change in the gamma angle, as it goes back to tracking edge to sun. During this first orbit, the RBVM fluid temperature is hovering around -20°F, the radiator mass flow rate is stable around 1700 lb_m/hr, the environment temperature is around -50°F, and the Δγ bias is at zero, as expected since the fluid temperature is above the -40°F temperature command.

At orbit noon of the second orbit, when the TRRJ gamma angle is again at its maximum point, the heat load into the system is stepped down to 2 kW. The system responds immediately by dropping the mass flow rate through the radiator down to approximately 120 lb_m/hr, as shown in the second strip plot in Figure 7. This has the effect of increasing the transport delay of the fluid traveling from the radiator panels down to the RBVM temperature sensors. As a result of the lower heat load and the correspondingly lower mass flow rate through the radiator, the fluid temperature measured at the RBVM slowly begins dropping.

By the third orbit, as the station again exits the umbra for the third time, the fluid temperature at the RBVM has dropped below the -40°F temperature command, and so the Δγ bias begins to build up as the PI control system starts to work. As a result of the bias, the thermal environments get hotter, achieving a peak of 11°F at orbit noon of the fourth and fifth orbit.

Notice in the fifth strip plot in Figure 7, that the Δγ bias is temporarily set to zero during the umbra event, as the PI control system is suspended until the station again enters the sun-side of the orbit. Notice also that the Δγ bias has a large flip just before and after the umbra event. This is due to the change in sign of the Δγ bias, which occurs when the γ_t passes through zero. One of the purposes of the smart limiter, discussed in the next scenario, is to eliminate this unnecessary flip of the TRRJ.

As the station exits the umbra for the fifth time, the heat load to the system is again stepped back up to 35 kW. The EATCS responds immediately with an increase in mass flow rate through the radiators, as shown in the second strip plot in Figure 7. As the mass flow rate increases, the cold fluid in the radiators is flushed out, resulting in a relatively abrupt temperature drop as measured at the RBVM sensors. This temperature drop initially causes the Δγ bias to increase, and actually hit its position limit. However, as the fluid temperature rises above the -40°F temperature command, the Δγ bias quickly drops back down to zero by orbit noon at the beginning of the sixth orbit.

For the sixth and seventh orbit, the remainder of the run, the system returns to its state of quasi-equilibrium, back to where it was at the beginning of the scenario. Hence, this case demonstrates the nominal operation of the EATCS loop as it receives step changes in its heat load input. As expected, for high heat loads, the pointing control system has the TRRJ track the γ_t profile in order to achieve maximum heat rejection. When the heat load drops, the pointing control system builds up a sun-soaking bias in order to elevate the temperature of the fluid coming out of the radiators, and thus prevent freezing of the ammonia in the radiators.

Scenario 2: Nominal Performance with Smart Limiter

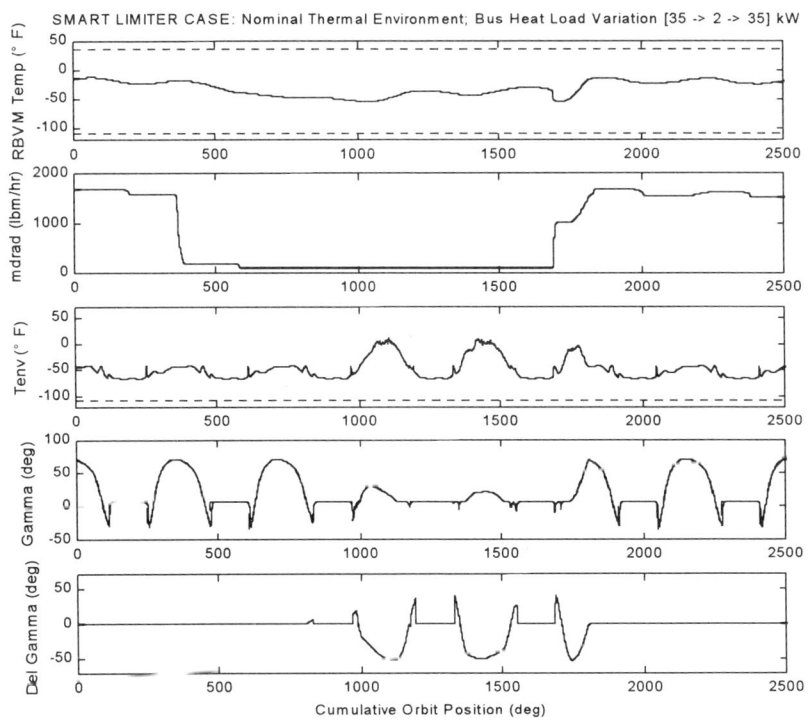

Figure 8: Nominal Performance, Smart Limiter Scenario

The second scenario is exactly like the first, except it uses the smart limiter instead of the constant limiter for the $\Delta\gamma$ bias. For the smart limiter, the maximum $\Delta\gamma$ bias is set as the absolute difference between the earth-facing orientation and edge-to-sun. This allows the $\Delta\gamma$ bias to have a smooth transition through the points on the T_{env} vs. γ curve when the local minimum and local maximum critical points change relative location, as discussed in previous sections. This allows for a much smoother motion of the TRRJ, as shown when comparing Figure 7 and Figure 8.

The results of the second nonlinear simulation are plotted in Figure 8. Note that the results for the first three strips in Figure 8 are almost identical to the same parameter strip plots in Figure 7. This is because the smart limiter bias and the constant limiter bias achieve approximately the same thermal environments, thus maintaining approximately the same RBVM fluid temperatures and radiator mass flow rates.

The primary difference between the results for the constant limiter, shown in Figure 7, and the results for the smart limiter, shown in Figure 8, can be seen in the TRRJ gamma angle motion, plotted in the fourth strip plot of both figures. The smart limiter effectively eliminates the large flips of the TRRJ near the umbra event, letting the TRRJ remain at a nearly constant angular position when a large bias is required. The resulting thermal environments are effectively the same, which demonstrates the superiority of the smart limiter approach. The same thermal performance can be achieved with significantly less motion of the TRRJ. This helps reduce the wear on the TRRJ mechanism, while also limiting momentum disturbances associated with large flips of the thermal radiators.

Scenario 3: Robustness Stress Test with Smart Limiter

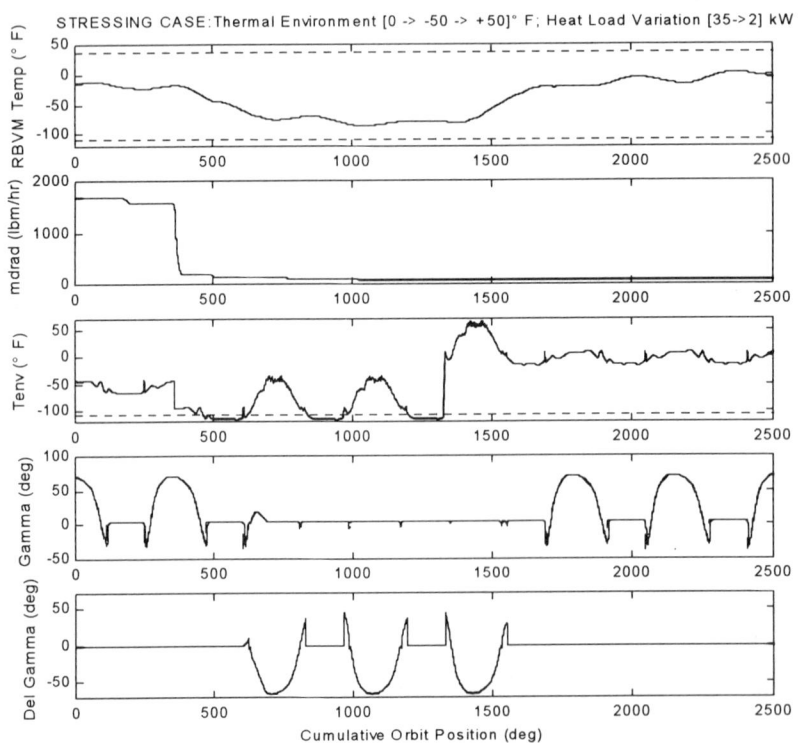

Figure 9: Robustness Stress Test, Smart Limiter Scenario

The third scenario is a stressing case of the fluid temperature control system, using the smart limiter for the $\Delta\gamma$ bias. This scenario is designed to test the control system robustness to potentially large variations in the thermal environment when the transport delays are at their worst. The nonlinear simulation results for this scenario are plotted in Figure 9.

The run begins at orbit noon, with a 35 kW heat load coming into the heat acquisition subsystem, and a nominal thermal environment with $\Delta T_{env}=0$. After one complete orbit, the heat load is stepped down to 2 kW, while the thermal environment is simultaneously stepped down by $\Delta T_{env}=-50°F$. This causes the transport delays to increase, as the radiator mass flow rate drops to approximately 70 lb_m/hr. It also causes the thermal environments to drop below the freezing point of ammonia when in the umbra. This combination of low flow rates and cold thermal environments creates a situation in which the ammonia can easily freeze in the radiator panels unless corrective action is taken by the fluid temperature control system.

Although the transport delays are large, particularly for the outer panels, the fluid temperature measured at the RBVM begins to drop almost immediately after the two steps. By the time the station leaves the umbra for the second time, the fluid temperature controller is biasing the TRRJ to its maximum allowable thermal environment. This biasing on the sun-side of the orbit is sufficient to keep the RBVM temperature from dropping below the freezing temperature of ammonia, -108°F, even though the thermal environment temperature gets below the ammonia freezing temperature every time the station is in the umbra.

With the system settling into a quasi-equilibrium, the thermal environment temperature is then stepped up by 100°F, so $\Delta T_{env}=+50°F$. This environment temperature step occurs just after the station exits the umbra for the fourth time. The resulting step change in thermal environment, combined with the maximum bias being commanded from the control system, causes the thermal environment to peak at 66°F, well above the allowable radiator return temperature, which must always be kept less than the set-point temperature of 37°F. Meanwhile, the heat load into the system is kept low so as to keep the thermal radiator mass flow rates low, and the transport delays from the panels to the RBVM sensors high. This combination of high heat loads and low mass flow rates creates a situation in which the fluid temperature coming out of the radiator panels could exceed the set-point temperature, meaning that the system is not rejecting as much heat as it is acquiring.

Regardless of the large transport delays, though, the panels close to the RBVM sensors provide a sufficiently timely feedback signal so that within 1/4 of an orbit, the RBVM fluid temperature begins to rise in response to the step change in environment temperature. Within 2/3 of an orbit after the step change in environment temperature, the $\Delta\gamma$ bias has been removed, and the system once again starts tracking the γ_t orientation profile in order to reject as much heat as possible when in the sun-side of the orbit. Although the radiator outlet fluid temperature, as measured at the RBVM, rises as high as 0°F, it never gets large enough to exceed the set-point temperature of 37°F. Thus, the fluid temperature controller used for the thermal radiator pointing system is shown to be robust to worst case temperature step changes, even when the transport delays are very large.

CONCLUSION

For the ISS EATCS, insights into the physical dynamics of the thermal system led to the development of a proportional-integral (PI) control architecture with limited output and an anti-windup integrator. The temperature of the ammonia coming out of the radiator is measured and controlled by adjusting the radiator wing orientation. The radiator wing orientation for the local minimum environment is fed forward to the control system, and the closed-loop controller is used to generate a bias off of that local minimum environment in order to heat up the ammonia when necessary to avoid freezing. In the earth's shadow, the controller is suspended and the radiator wing is oriented to face the earth, the local maximum thermal environment which further prevents freezing of the ammonia. This control architecture is shown to provide adequate heat rejection and avoid freezing of the ammonia, even though the physical system consists of large transport delays and time-varying, nonlinear dynamics which change dramatically due to orbit motion and variable heat loads. More sophisticated control techniques such as Smith predictors, Internal Model Control, adaptive control, or nonlinear control were not necessary in order for the system to meet its performance objectives.

REFERENCES

1. R.L. Broeren, "Radiator Wing Rotation Algorithm Overview for the International Space Station's External Active Thermal Control System," *Proceedings of the 28th International Conference on Environmental Systems*, SAE #981732, July 1998.
2. J.E. Slotine and W. Li, *Applied Nonlinear Control*, Prentice Hall, New Jersey, 1991.
3. W.S. Levine, *The CONTROL Handbook*, CRC Press, Inc., 1996, pp. 206, 215-237.
4. J.R. Wertz, *Spacecraft Attitude Determination and Control*, Kluwer Academic Publishers, 1990.
5. J.P. Holman, *Thermodynamics*, Third Edition, McGraw-Hill Book Company, 1980.
6. D.R. Nowlan, "Rotary Joint Models to Support Microgravity and Multiple Controller Interaction Studies," McDonnell Douglas, A3-J092-DRN-M-9800428R1, 23 October 1998.

AAS 00-027

COMMAND LEVEL MANEUVER OPTIMIZATION FOR THE INTERNATIONAL SPACE STATION

Gregory E. Chamitoff,[*] Adam L. Dershowitz[†] and Amy L. Bryson[‡]

This paper presents a maneuver commanding optimization tool for minimizing fuel requirements during International Space Station (ISS) guidance, navigation, and control (GN&C) maneuvers.

The dynamic behavior of the ISS varies significantly throughout its construction, as mass properties and external aerodynamic shape change with the addition of new components and the attachment and movement of various earth-to-orbit vehicles and payloads. Performance of the GNC system is similarly affected by the vehicle configuration as well as the sequence and timing of commands. In particular, fuel required to achieve and maintain desired attitude depends on the configuration, the environment, the controller used, and the GN&C command sequence. GN&C operations for the ISS are constrained by the ability to issue a specific set of commands, which are limited to certain mode changes, controller parameter updates, attitude commands, and control moment gyro (CMG) momentum commands.

An approach for optimizing the command sequence was developed with the objective of minimizing fuel utilization. A high-speed and variable fidelity simulation for the ISS dynamics, environment, and GNC system, was developed and integrated into an optimization code that includes a complete model for the operator commanding capability. Commands, parameters, and time-tags can all be treated as optimization variables. An initial guess for the optimal solution can be supplied by a simplified model using the differential inclusions method. Optimal solutions for the full nonlinear problem are obtained via 3D visualizations of the search space combined with local gradient techniques. This tool will provide ground-based flight controllers with an automated command sequence optimization capability for any GNC operations scenario. Ultimately, this tool could be extended to provide the crew with an onboard autonomous capability for attitude control planning, optimization, and operation.

---

[*] NASA Johnson Space Center, 2101 NASA Road 1, Mail Code CB, Houston, Texas 77058.
[†] United Space Alliance.
[‡] Lockheed Martin Space Operations, 2400 NASA Road 1, Houston, Texas 77058.

INTRODUCTION

The assembly of the International Space Station (ISS) is a complex endeavor involving the coordination of requirements and constraints from many disciplines. Mission planners design a time-line based on the mission objectives, but rely on inputs from flight controllers, crew, system experts, designers, the payload community, and many others in the effort to refine a mission plan. From a Guidance, Navigation, and Control (GN&C) perspective, it is important to execute this plan in the most efficient manner; this usually means executing vehicle maneuvers in minimum time, or using minimum fuel. Maneuver planning is also subject to numerous factors, including vehicle capability, operational constraints, and environmental effects. For example, attitude time-lines must be acceptable for communications, thermal considerations, solar power generation, and lighting.

Provided that other constraints can be met, propellant consumption is the most important consideration for ISS maneuver planning. At approximately $10,000 per pound of fuel delivered to orbit, efficient attitude control can save substantial cost for the program. More importantly, however, the more fuel available onboard, the less susceptible the station is to problems with propellant resupply capability and launch schedules.

The actual cost of ISS maneuvers (in terms of fuel) depends on when and how the maneuvers are performed. Particularly during the ISS construction phase, there are many variables, such as the configuration of the station, control effectors available, vehicle mass properties, daily atmospheric properties, aerodynamic shape, payload motion, and scenario specific requirements. For example, the time and propellent required to perform a maneuver from an XPOP (solar inertial) attitude to a LVLH (local vertical local horizontal) attitude has been seen to vary drastically depending on when in the orbit you start the maneuver. Likewise, the propellant required to achieve and maintain a docking attitude can vary from nothing to unacceptable levels, depending on the vehicle configuration and time required at that attitude.

How can we optimize all ISS maneuvers in view of the variety of scenarios? Theoretical techniques for spacecraft maneuver optimization only address an idealized version of the actual problem. The actual implementation of the ISS GN&C system is limited in that there are certain command capabilities available to the operator (ground or crew). To achieve an efficient and practical attitude maneuver, it is necessary to optimize over the set of command capabilities.

The tool discussed in this paper provides a fast and powerful means for optimizing the actual command sequence to be used by the ground or the crew. It also combines ideal theoretical maneuvers with actual vehicle capability to determine the best operationally feasible sequence.

ISS ATTITUDE CONTROL

More than any other system on the Space Station, the GN&C system is an integrated US and Russian (RS) capability. At least initially, all thrusters are on the RS side, while the US system uses Control Moment Gyros (CMGs) for attitude control. Although the US system is "prime" for attitude control most of the time after the US Laboratory arrives in orbit (flight 5A), it is always critically dependent on the RS system.

There are two basic types of ISS US GN&C attitude controllers: Attitude Hold (AH) and Momentum Management (MM). Attitude Hold controllers are used for maneuvering or maintaining a fixed attitude. Momentum Management controllers are used for micro-gravity operations, and are required to control attitude without jet firings for up to 30 consecutive days. CMGs are capable of producing torque within a limited angular momentum envelope. Once that limit is reached, a desaturation request is sent to the RS system resulting in a jet firing to "desaturate" the CMGs. While MM controllers can keep the CMG momentum within this envelope for extended periods of time, getting into a micro-g configuration, or transitioning to another one with a new ISS configuration, will require jet firings. Likewise, any large maneuvers using CMGs are likely to require desaturation burns. Large maneuvers may therefore be performed by the RS system, which requires handing over attitude control back and forth between the US and RS systems. An option called RCS Assist allows the US system to remain in control while requesting excess torques (beyond CMG capability) to be achieved by the RS thrusters. As such, US flight controllers may prefer to perform some large maneuvers using RCS Assist.

For the purposes of discussing maneuver optimization, it is also necessary to understand that the US GN&C system is limited to a certain set of commands, options, reference frames, limits, and flags, that are either accessible to the operator or included in controller configuration database files called CCDBs. Another set of parameters that change less frequently belong to PPLs (Pre-Position Loads). The operator on the ground or onboard configures the GN&C system through direct commands, or through the changing of these files. The system's "command flexibility" is essentially limited to attitude commands, controller changes, manual desaturations, and the setting of certain flags. Moreover, the maneuver logic is limited to performing eigenaxis maneuvers. This is a significant limitation with regard to maneuver optimization, but can be overcome by optimizing intermediate attitude commands and time-tags as will be shown later.

From this discussion, it is clear that any practical maneuver optimization scheme for the ISS must fit within the framework of the actual (limited) operational capability.

ISS ATTITUDE MANEUVER PLANNING

The current ISS mission planning approach within Mission Operations at NASA is very inefficient at designing attitude maneuvers. A number of very high fidelity simulation tools are connected into what's called the Integrated Planning System. The purpose of this is to allow the outputs of each discipline to feed into the other disciplines. For example, a new attitude time-line will affect communication opportunities, power management, heating, lighting, and so on. While the integrated analysis capability of these combined tools is necessary, it's not a good place to start with maneuver optimization. The complexity and fidelity of this system means that single simulations take hours to set up and run. The result is that the effort and time available to spend iterating and optimizing the plan is very limited.

A practical ISS attitude maneuver optimization tool needs the following:
- Good models of the vehicle configuration, environment, and control system
- An approach to finding the best theoretical maneuver for an "ideal" vehicle
- An approximation to the ideal solution with actual vehicle commands
- Command level optimization capability using "semi-high-fidelity" models within the framework of actual operator capability
- To be fast, simple, and automatic – (maximizing operational utility)

Once a solution is obtained in this manner, it can be validated using the Integrated Planning System described above.

MANEUVER & TRAJECTORY ANALYSIS TOOL (MTAT)

The concept of MTAT began with the idea that some basic ISS GN&C analysis code could be made into a powerful tool for maneuver planning and analysis in Mission Control. The original analysis tool was developed by the C.S. Draper Laboratory (see Acknowledgements). Some of the original dynamics models are still used, but the program has evolved into a complete design and analysis package with many applications.

First of all, MTAT can be used as a flight simulator for the ISS. In fact, combined with new graphics tools described later, a user can configure and command the simulated GN&C system, and then view the attitude time-line graphically in 3D with a virtual exterior view of the Station. MTAT has enough fidelity to adequately simulate the ISS attitude trajectory including the response of the control system. Any orbit, ISS configuration, controller CCDB, payload, robotics operation, and GN&C commanding sequence can be programmed into MTAT. Aside from maneuver optimization, this simulation capability is very useful for visualizing ISS maneuvers. The user can get a "feel" for how the ISS flies (at different stages), and how long a maneuver will take. As such, flight controllers, trainers, and crew can use MTAT as a training tool for understanding the GN&C system.

Secondly, MTAT can help to answer "What If?" questions about GN&C operations. Planned scenarios can be entered or changed and the result can be quickly analyzed. For example, if attitude control was lost during a critical operation to mate components with the robotic manipulator system (RMS), what would happen to the attitude? How much time would there be before attitude constraints were violated? Another example would be to determine the time and fuel required for a docking operation as a function of the docking attitude chosen. These kinds of questions are easily answered by MTAT.

Finally, MTAT has been designed to optimize command sequences automatically. The only required user input is the arrangement of commands and designation of which parameters are allowed to change. MTAT can run multivariable simulation grids to analyze the effect of parameter variations. In this mode, the output is in the form of 3D performance analysis diagrams showing cost (fuel or time) as a function of specified variables. Alternatively, MTAT can perform a constrained nonlinear optimization to simply determine the optimal command sequence. Doing all of this easily and quickly is an important contribution of MTAT.

MTAT Capabilities

Figure 1 illustrates a summary of MTAT's structure and capabilities. The User Interface provides displays for editing and selecting all values related to the vehicle, control system, payload, orbit, and commands. Simulation options include choices of model fidelity as well as initial conditions and simulation run-time. Model fidelity options allow coarse solutions to be found very quickly, while high fidelity solutions can be fine tuned later. MTAT executes one or multiple simulations of the maneuver/command sequence, depending on optimization options, and generates all flight data required for analyzing maneuver performance. Control system design is also an integrated feature of MTAT. Momentum Management controllers are specific to the vehicle configuration and must be designed for each scenario that requires an MM controller. Analysis tools include 2D plotting of simulation results and 3D performance analysis plots. These plots allow the user to visualize the sensitivity of CMG momentum or fuel consumption to selected parameter variations. Essentially, the optimal maneuver corresponds to the minima of a multidimensional performance plot.

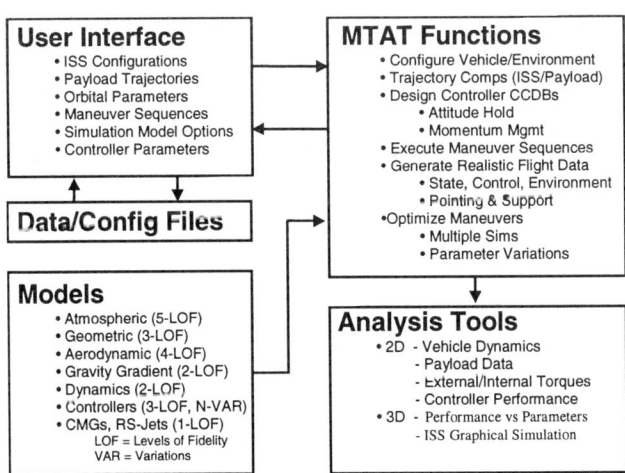

Figure 1: MTAT Structure and Capabilities

MTAT User Interface

The Matlab Graphical User Interface is designed for quick and easy simulation configuration as shown in Figure 2. The **Configuration** button on the main display accesses all data entry panels, which are mostly self-explanatory. Detailed features, such as CCDB controller design, payload trajectory editing, or TEA attitude estimation, are available from these panels. The **Simulation** button on the main display starts a single MTAT run or a maneuver optimization sequence. If selected, it will also drive the 3D graphical ISS simulation display. The **Analysis** button brings up all the tools for plotting and performance analysis.

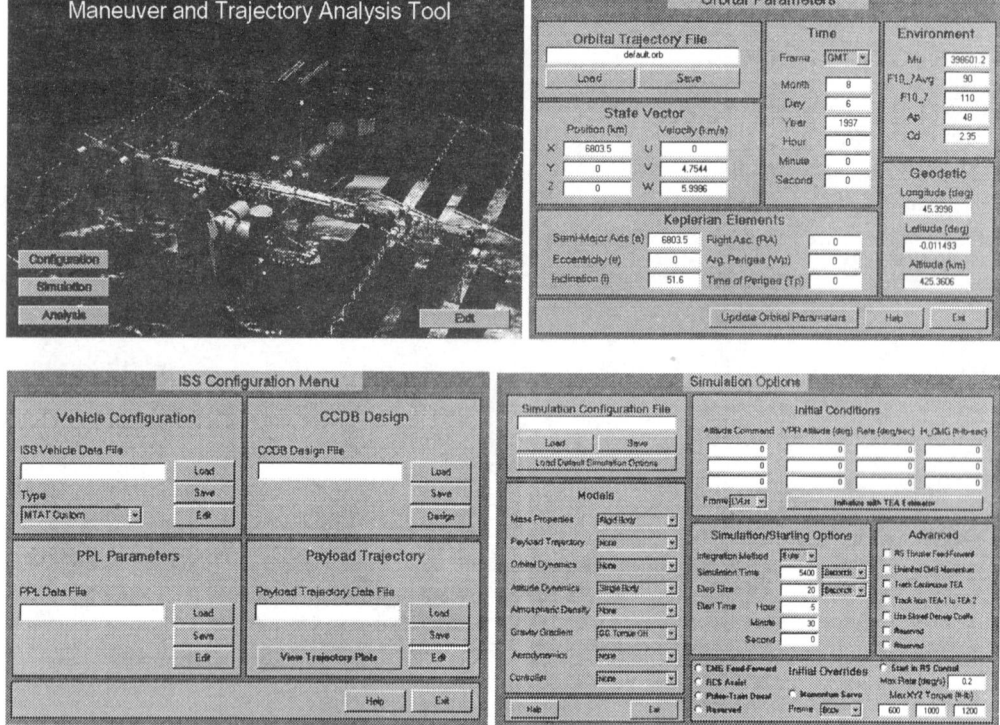

Figure 2: MTAT Graphical User Interface

COMMAND LEVEL OPTIMIZATION

The most important display for accessing MTAT's maneuver and command optimization capabilities is shown in Figure 3. All commands, mode changes, and options available to the system operator are selectable from the **Maneuver Command** pull-down menu. These include control handovers between US and RS systems, attitude commands, going to Drift, manual CMG desaturation, loading new controller CCDBs, changing some controller parameters by command, and options for stopping and starting payload motion on the robotic arm. When selected, each of these actions requires a time-tag indicating when it should occur in the maneuver sequence. The maneuver plan that will be executed by MTAT appears in the **Command Sequence** window. If the user specifies the exact sequence, then MTAT will perform a single simulation and generate appropriate results for analysis. However, if the user designates any of the parameters as variables, MTAT then treats the command sequence as something to be optimized. The time that a command is issued, as well as any parameters it may include, can all be treated as optimization variables. For each variable, a range of allowed values is also specified. For example, when optimizing an attitude command, any, or all, of the roll, pitch, and yaw values can be optimized over a specified range. The Maneuver Command section of this display handles all of the options for each type of command.

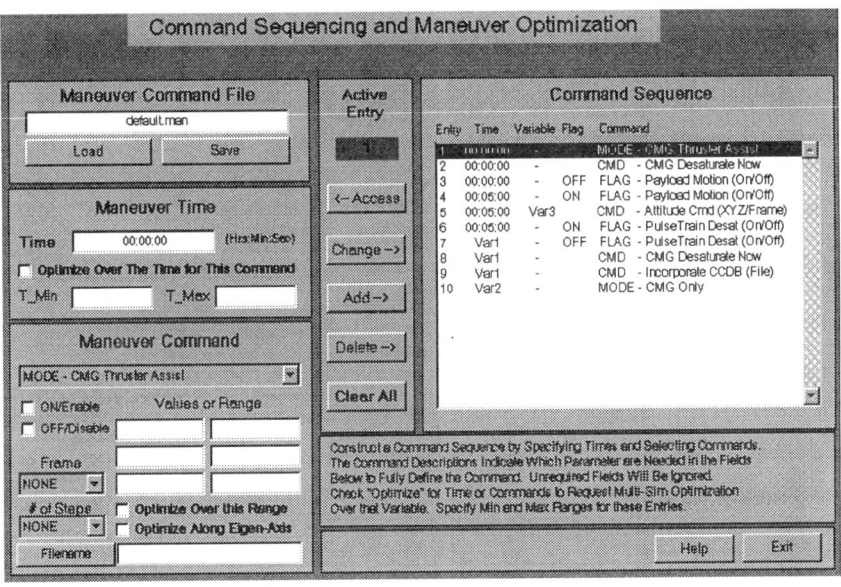

Figure 3: MTAT Command and Maneuver Optimization Control Panel

Sample Results

Some example results from MTAT are shown in Figure 4. The CMG momentum during the start-up of a Momentum Management controller is shown in Figure 4A. Prior to converging into a cyclic attitude, rate, and momentum condition, desaturation firings are seen during the first few orbits. Figure 4B shows the results of an optimization run with pitch command as one of the variables. As pitch varies from the equilibrium attitude, the cost in CMG momentum to maintain attitude climbs sharply. The jagged "bumps" on the plot are desaturation burns keeping CMG momentum within defined limits. Figure 4C shows the results of an optimization in two variables for an attitude command. The minimum momentum (and minimum fuel) solution is attained by the attitude command giving the lowest point on this surface. In Figure 4D, Fuel used is plotted as a function of Yaw command and elapsed time during the maneuver. While propellant is required to achieve and maintain all attitudes examined, the minimum fuel Yaw attitude is easily identified.

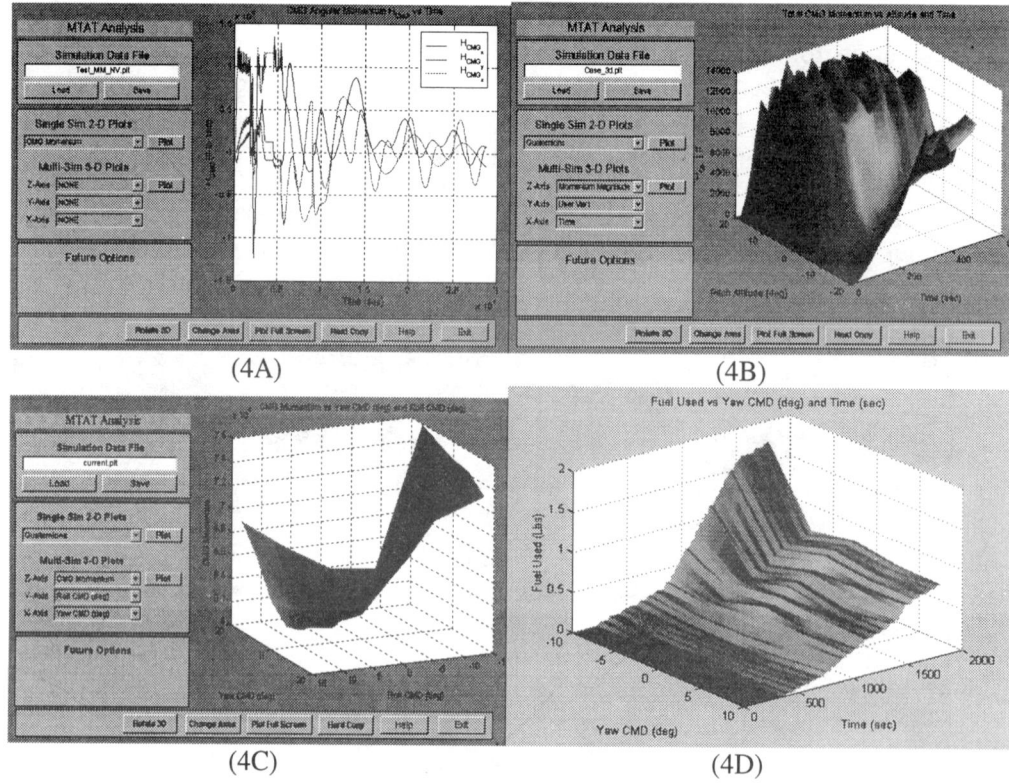

Figure 4: Sample Results from MTAT Simulation & Optimization Runs.
(4A) CMG Angular Momentum during start-up of a Momentum Management Controller. (4B) Sensitivity of Momentum Required to Pitch Attitude Variations. (4C) Optimizing an Attitude Command in Yaw and Pitch. (4D) Fuel required as a function of Yaw Command.

IDEAL MANEUVER: DIFFERENTIAL INCLUSIONS

The achievable performance of an optimized maneuver in MTAT can be dramatically improved through the proper design of the initial guess for the command sequence. This idea is being explored by a current master's thesis [1], in which the differential inclusion (DI) technique is used to first solve an idealized space station maneuver optimization problem. This idealized problem is not limited by the commanding constraints of the ISS GN&C system, and uses a simplified model of the vehicle dynamics and environment. Arbitrary attitude trajectories are allowed, but all other dynamic, operational, or actuator constraints are imposed. In particular, maneuvers are not limited to eigenaxis rotations as in the ISS GN&C system. The DI approach is applied to finding an ideal attitude trajectory, which can then be used as an initial guess for the detailed GN&C command sequence. A series of small attitude commands and associated time-tags can then be optimized within MTAT using higher fidelity models. The combined approach can result in very significant propellant savings, especially for large angle maneuvers.

DI Method for ISS Maneuvers

The DI algorithm is discussed in depth in reference [2]. The approach taken here is outlined in Figure 5. Essentially, a nonlinear constrained optimization problem is solved by searching over discretized values for the states, while using the equations of motion as equality constraints. These constraints are enforced half-way between the nodes, by computing average values for the states and their derivatives. Depending upon the control actuators being used (CMGs, jets, or both), a control equation is used to specify the control necessary to achieve the state trajectory. External torques and desaturations can be accounted for in this step, by simply adding their effect to the required control torque. Inequality constraints, such as maximum rates, or maximum CMG momentum, are also included. The cost function is a weighted combination of fuel consumption and maneuver time, and a standard constrained nonlinear optimization code is use to find the solution.

DI Results

The advantage of combining the DI approach with the full Command Level Optimization in MTAT is clearly illustrated by the results below. Figure 6 shows an MTAT maneuver from the XPOP attitude to an LVLH [0 0 0] attitude (a 90 degree yaw maneuver). As expected, the GN&C system performs an eigenaxis maneuver to the final attitude. The second figure shows that seven desaturation firings were required during the maneuver. Now Figure 7 shows the same maneuver performed by the DI algorithm. Given the flexibility of an unconstrained attitude trajectory, the DI algorithm found a maneuver that requires no desaturation firings at all. Zero fuel was required to perform the DI maneuver! In general, the next step would be to approximate the DI maneuver with a series of smaller attitude commands, and use this as the initial guess for the MTAT optimization.

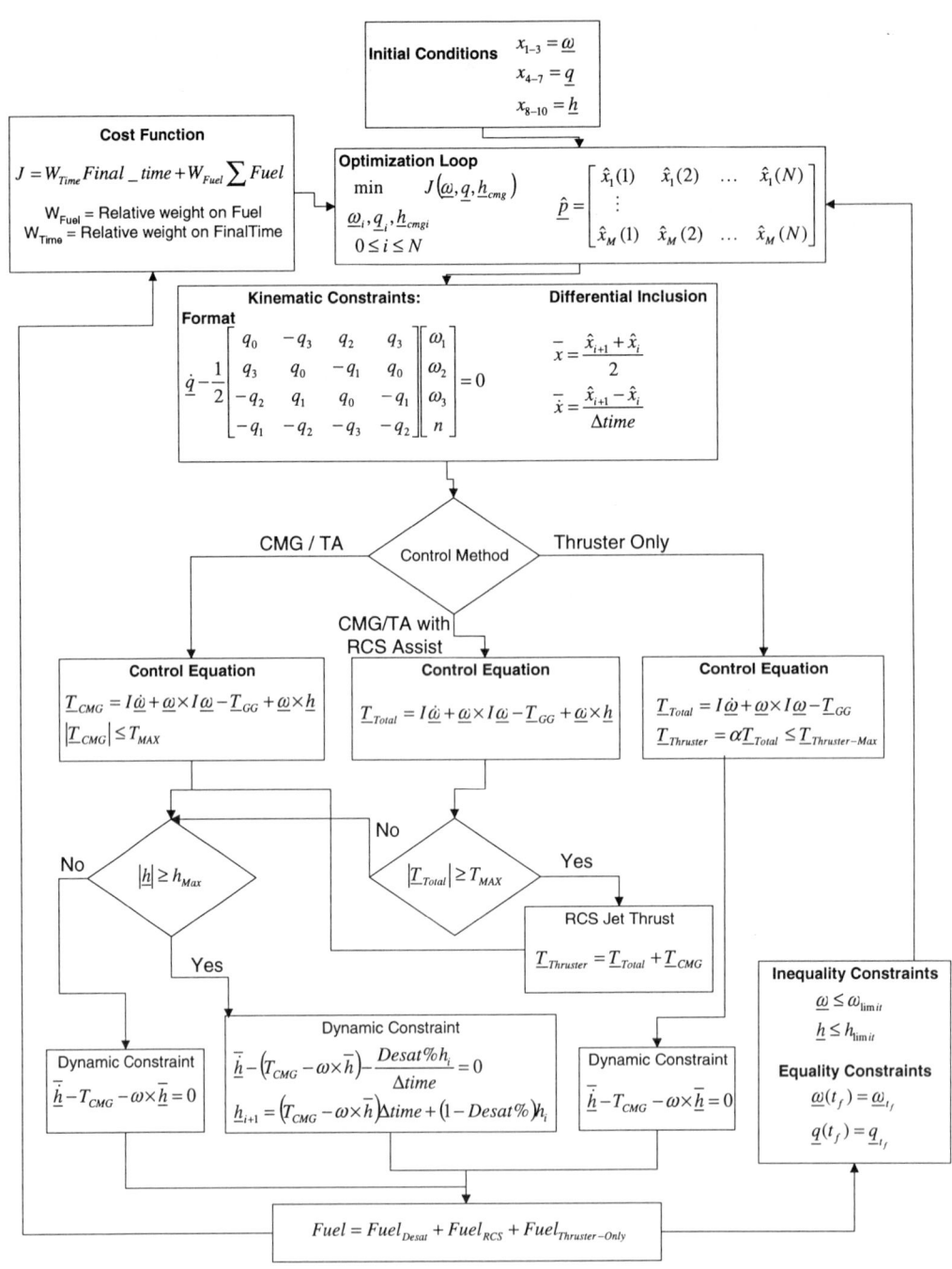

Figure 5: Differential Inclusions Method for ISS Manuever Optimization

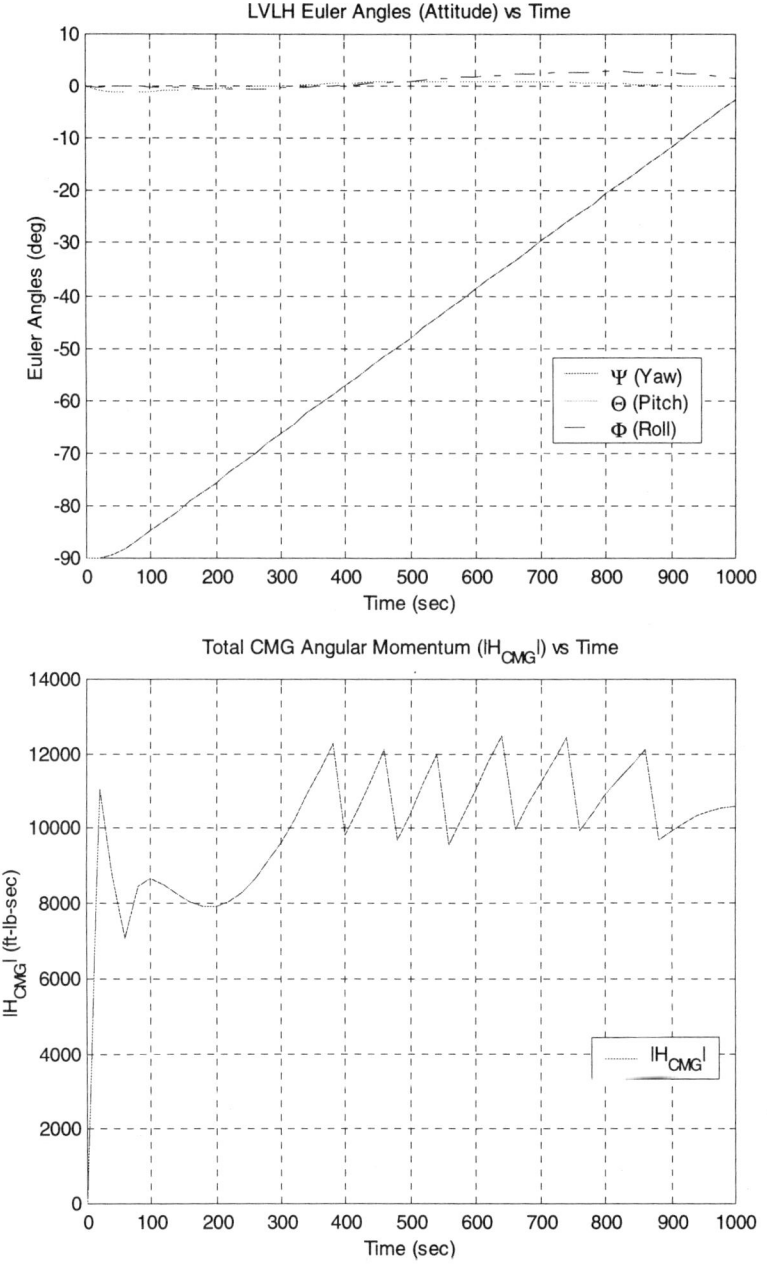

Figure 6: MTAT Eigen-Axis Maneuver - LVLH Mnvr [-90 0 0] to [0 0 0]

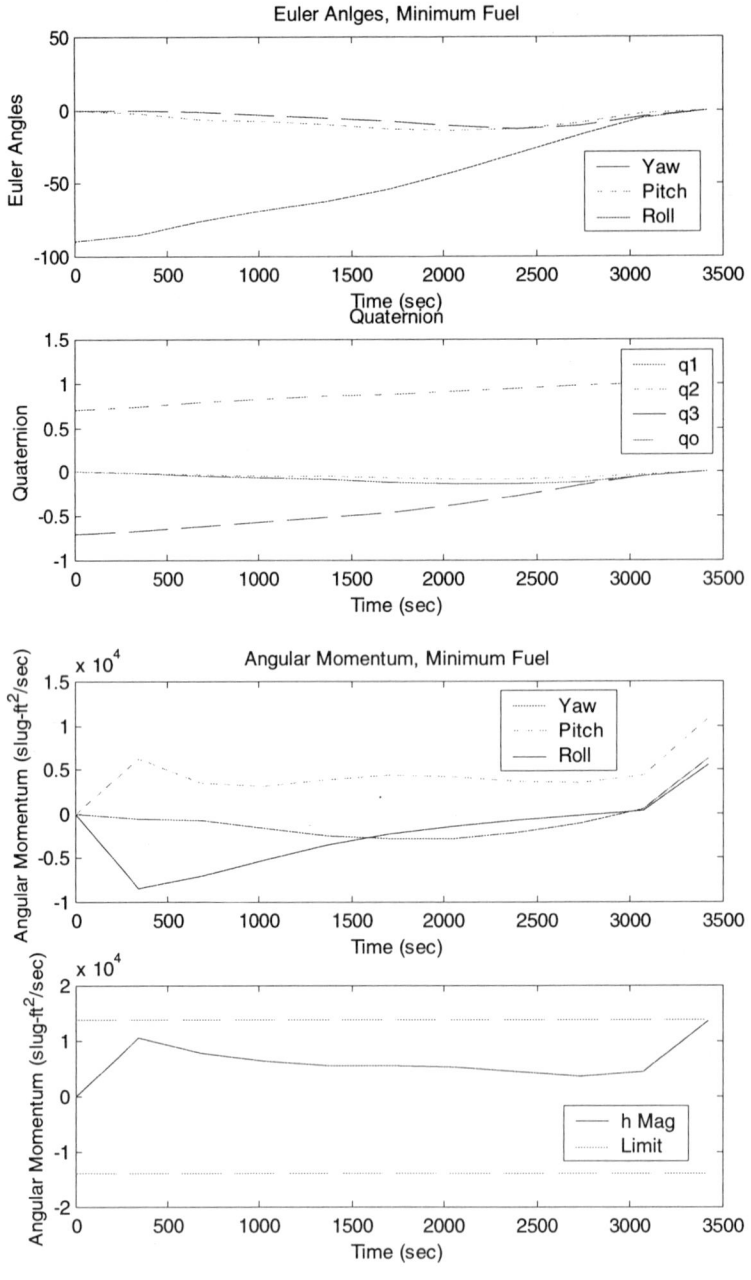

Figure 7: Differential Inclusions - Ideal LVLH Mnvr [-90 0 0] to [0 0 0]

APPLICATIONS

The tools discussed in this paper have numerous applications, some of which are already being developed for the ground as well as the crew onboard. Having a fast simulation tool allows the ground or crew to "look-ahead" and ask "what if" questions. A highly desirable capability is to evaluate contingency scenarios, simulate and analyze the outcome, and then decide on a course of action. This is especially true during off-nominal situations where power, communication, fuel, temperature, and other factors may become critical.

Real-Time Analysis & Planning Tools

A real-time version of MTAT is being developed for use in the Mission Control Center (MCC) as well as for onboard the ISS. Real-time simulation-based tools are a new technology for the MCC, and are now being considered for Shuttle applications as well. The Navigation, Attitude, Pointing & Support (NAPS) tool will be used to monitor the GN&C system in real-time and predict ahead using MTAT models during loss-of-signal (LOS) periods. Communication blockage will occur for a significant fraction of the time for the ISS, so a continuous actual or predicted vehicle state is a very useful tool for MCC flight controllers. As discussed earlier, the predictive capability can also be used to evaluate contingency scenarios. The difference between NAPS and MTAT, however, is that NAPS is connected to actual telemetry and can predict from the last known valid data.

The NAPS display shown in Figure 8 allows the ground or crew to compute desired attitudes for contingency situations. For example, one could determine an ISS attitude, for say one hour from now, such that a stuck antenna on a docked vehicle points directly at the desired communication satellite or ground site. Moreover, since different vehicles use different reference frames, this desired attitude could be converted to an appropriate command for the vehicle that will carry out the maneuver (e.g. Shuttle). Another example might be to figure out the minimum attitude maneuver required to keep some station element out of the Sun. Note that these problems cannot be solved on the ground prior to flight, because the solution is dependent on the current time and vehicle state. Hence, this on-line version of MTAT brings new capabilities to the flight controllers and crew.

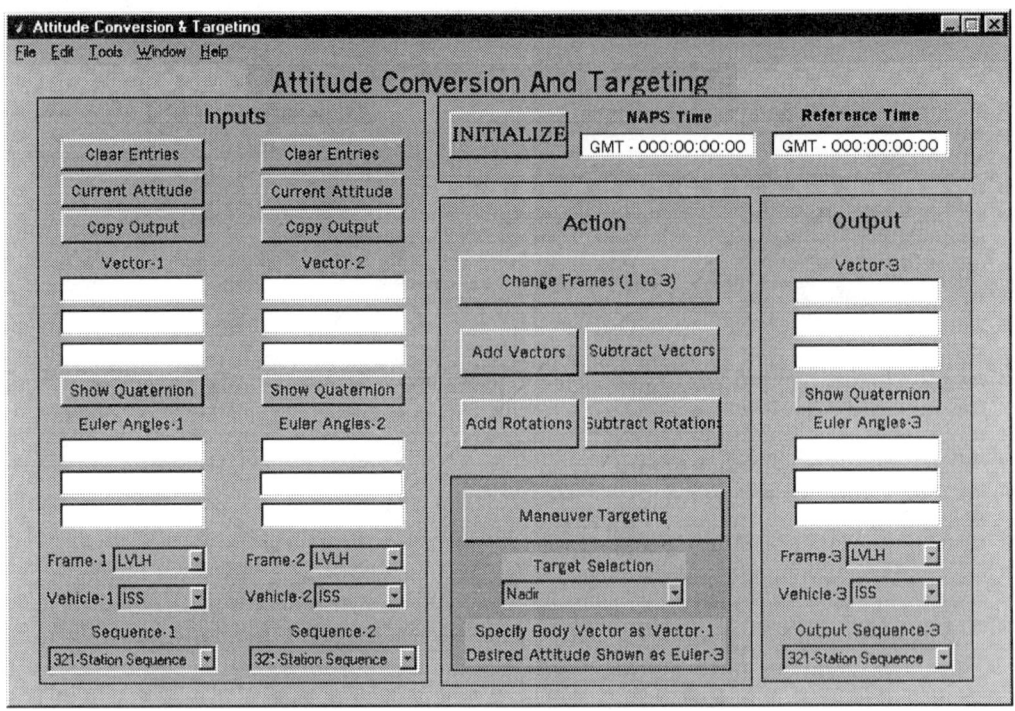

Figure 8: Real-Time Attitude Targeting Display

Bird's Eye View (BEV) - Situational Awareness & Visualization

Another application mentioned earlier is a virtual camera view of the Space Station driven by actual or simulated telemetry (Figure 9). With limited access to windows onboard, this is a powerful situational awareness tool for the crew. It is based on a customized 3D graphics package for displaying important GN&C related information. It can be driven, however, by a variety of sources, such as actual telemetry, prediction models, or even optimized maneuver plans from MTAT. It displays the ISS vehicle and its position and attitude with respect to the Sun, Earth, orbit, and other programmable targets. Features include a visible orbit track, multiple camera views, command reference frames, space and ground target acquisition, and a 2D overlay showing important GNC parameters and commanded vs actual attitude. This Bird's Eye View (BEV) tool will be used onboard the ISS as a situational awareness tool beginning on flight 5A (US Lab activation). It is also a very powerful way to present the output of MTAT or any derivative applications. As a visualization tool, it will be used in conjunction with NAPS to solve and display real-time problems onboard and on the ground.

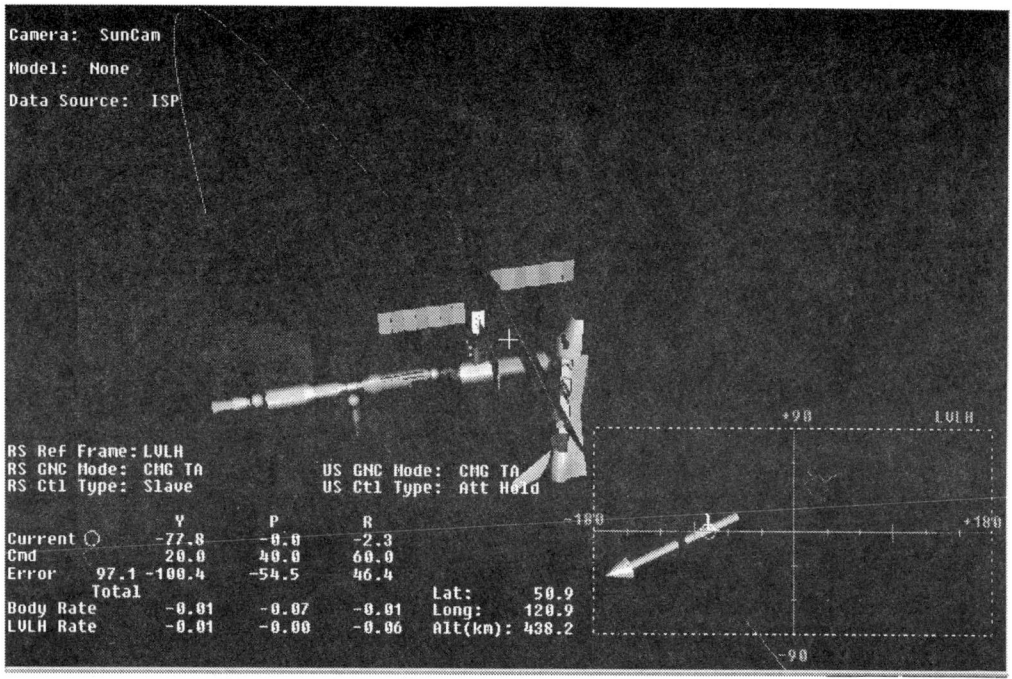

Figure 9: ISS Bird's Eye View Situational Awareness Tool

CONCLUSION

While the International Space Station is not as dynamic a vehicle as the Shuttle, it is required to perform a wide range of attitude maneuvers at the expense of valuable propellant. Performing these maneuvers efficiently would save the program money while providing a measure of robustness against launch schedule uncertainties. This paper has presented a maneuver optimization approach that operates at the command level, meaning that the result is a solution that can be implemented on the real system using real commands. The Maneuver and Trajectory Analysis Tool can optimize over all inputs in a command sequence. A good initial guess, however, can significantly improve performance. An approach using the Differential Inclusions method was demonstrated for generating ideal maneuver trajectories. For the case shown, a simple yaw maneuver, it was demonstrated that this approach could reduce the number of desaturation firings from seven to none. Approximating this with several small maneuvers, this can be incorporated into the full command-level optimization. The tools described in this paper provide flight controllers with powerful maneuver planning and analysis capabilities for all GNC operations. New applications extend many of these capabilities onboard.

ACKNOWLEDGEMENTS

In addition to the authors, several individuals have contributed significantly to the work described in this paper.

>Nazareth Bedrossian [3] – C.S. Draper Laboratory
>David Boyle – Texas A&M Visiting Researcher (from University of Sydney)
>Andres Mur-Dongil – Texas A&M University (Master's Thesis)

REFERENCES

1. A. Bryson, *to be completed*, Space Station Maneuver Optimization Using Differential Inclusions, Master's Thesis, Texas A&M University, 2000.

2. H. Seywald, "Trajectory Optimization Based on Differential Inclusion," *Journal of Guidance, Control, and Dynamics*, Vol. 17, No. 3, May-June 1994.

3. N. Bedrossian and E. MacCants, Space Station Attitude Control During Payload Operations, *AAS/AIAA Astrodynamics Conference*, Alaska, Aug 1999.

4. Technical Description Document for the PG-1 Guidance, Navigation, and Control System, McDonnell-Douglas, MDC-95H0223D, Jan 17, 1997.

5. *International Space Station On-Orbit Assembly, Modeling, and Mass Properties Data Book*, JSC-26557 (LESC-31166), Revision J, August 1999.

Section IV
GUIDANCE AND CONTROL STORYBOARD DISPLAYS

SESSION IV

Joint Session Chairperson: Dr. Ian Gravseth
Ball Aerospace & Technologies Corp.

Joint Session Charperson: Rick Jackson
Lockheed Martin Astronautics

The following papers were not available for publication:

AAS 00-039   "A Portable Artifical Star for in Situ Testing of Ultraprecision Star-Tracker Telescopes," by E. Acworth, Stanford University

AAS 00-040   "Solid State Rate Gyro for Spacecraft Applications," by D. DeBelle, V. Selby, B. Wolfe, BF Goodrich

AAS 00-042   "Attitude Control System Design for the Modern Spacecraft," by L. Schlom, R. McNaughton, SimuLogix

AAS 00-047   "Star Tracker/IRU Attitude Determination Filters," by C. Gray, The Aerospace Corp.

AAS 00-048   "GyroWheel^{TM} & BrITES^{TM} - Innovative New Spacecraft Attitude System Hardware for 3-Axis Earth Pointing Control," by G. Tyc, D. McCabe, Bristol Aerospace Limited; D. Staley, Ancon Space Technology

The following paper numbers were not assigned:
    AAS 00-033, -051 to -060

AAS 00-031

ADAPTATION OF A SPACEBORNE GEOLOCATION SYSTEM TO AIRBORNE EXPERIMENTS

Alan S. Hope,[*] Henry M. Pickard[†] and Jay W. Middour[*]

On August 1, 1998 the Naval Research Laboratory (NRL) conducted an experiment using a High Definition Television Camera (HDTV) and a slightly modified previously space flown HERCULES geolocation system. The experiment was flown on an airship flight originating in Elizabeth City, NC. Although no strict requirements were set for the geolocation accuracy, a goal of 0.5 kilometers was adopted. Based on the space experiments and scaling those results to a low-altitude blimp flight, sub-meter accuracy should be attainable. However, time constraints prevented the level of modification and system integration that would have been required to achieve accurate results at this level. The geolocation accuracy for the examples presented vary from just over 100 meters to 655 meters.

The technique used for geolocating images of Earth requires the following: Precise knowledge of the pointing direction (attitude) of the camera boresight; precise knowledge of the camera position; and an Earth model to use in determining the intersection of the camera boresight with the Earth.

The changes required for collecting the geolocation parameters moving from a space based to an aircraft based system are discussed. Experiment errors and geolocation results are presented.

INTRODUCTION

The airborne experiment was flown on an AIRSHIP 600 blimp. A previously space flown ring laser gyro was co-boresighted with a HDTV camera. A Global Positioning System (GPS) receiver was installed on the airship to collect data during the flight for post processing position information.

The HERCULES system [1] contains a ring laser gyro (RLG) which provides stable and accurate attitude determination. The gyro initialization process was modified for this airship flight. Under space flight conditions, the heavy system (15 lbs.) was weightless and the camera boresight to gyro axis alignment was performed using stars. This Earthbound test required the adaptation of the alignment objects to be the center of the Earth, determined using a bubble level, and the sun, sighted while holding a solar filter over the lens of the alignment camera. The initialized gyro was then bolted to the HDTV gimbal

---

[*] Aerospace Engineer, Naval Research Laboratory, Code 8103, 4555 Overlook Avenue, S.W., Washington, D.C. 20375. E-mail: hope@ssdd.nrl.navy.mil; middour@ssdd.nrl.navy.mil.

[†] Physicist, Naval Research Laboratory, Code 8103, 4555 Overlook Avenue, S.W., Washington, D.C. 20375. E-mail: pickard@ssdd.nrl.navy.mil.

platform without being able to align it, other than by sight, to the expected HDTV boresight. It was thought that any misalignment of gyro axes to the HDTV boresight could be determined post-flight and the bias applied to the measurements. Because equipment was unavailable before the test, the integration took place in the field and a proper alignment could not be performed. The initialization of the gyro and the methods used are discussed.

To determine the position of the camera throughout the flight, a GPS receiver was installed on the airship. The raw GPS measurements were collected and a second receiver was set up at the blimp flight origination location. These two receivers collected simultaneous measurements throughout the blimp flight and the data set was to be post-processed twice. The first process was to be performed using the ground data and a nearby IGS GPS site with known precise coordinates. The result would have been the precise position of the base station set up at the Weeksville site. The second process would have used the ground collected measurements, the precision-determined ground position and the blimp data in a differential carrier phase process to determine the absolute Earth Centered Earth Fixed (ECEF) location of the blimp throughout the flight. Due to the ground receiver memory filling up halfway through the blimp flight, the data was unusable for the precision-position determination needed. A private site in Virginia Beach, VA which logs data at 5 seconds was used as an alternate and provided excellent measurements for our post processed position determination. The number of satellites tracked by the airship and the number of satellites in common view to the airship and ground station are shown. A comparison of corrected blimp positions to the in-flight computed positions is also shown.

The blimp altitude can be referenced to many different altitude models, but a best guess of the altitude above ground level was used to determine the intersection of the camera boresight with the Earth for the geolocation process. Several still frames from the HDTV video were used as reference points to determine the accuracy of the geolocation process. The WGS-84 ellipsoid was used for the earth model, but care had to be taken in dealing with where the actual surface of the earth was in relation to the ellipsoid. The method used and the assumptions made for the determination of this ground level altitude will be discussed as well as the final geolocation results. Also the sensitivity of the geolocation to the assumed ground elevation are presented.

SYSTEM DESCRIPTION

The HERCULES system flown in space was able to instantly compute and display the geolocation of images taken. However, the geolocation determination system for this airship demonstration was assembled with only a postprocessed capability in mind, since there was insufficient time to integrate the new hardware and software components for the flight. Each subsystem needed to run throughout the duration of the airship flight and store all data required for geolocation determination after the flight had ended. The major subsytem components were: the HERCULES attitude determination system which consists of a ring laser gyro, the attitude processor unit, and a Personal Computer (PC) for gyro initialization and data archiving; two ASHTECH Z-XII GPS receivers for position determination on the airship and base station collection for differential post processing.

The gyro was mounted to the camera control gimbal so that the alignment boresite was approximately parallel to the HDTV camera boresite. Because there was very little time to adapt this package, no alignment was performed other than trying to mount the gyro such that the gyro boresight was in the camera boresite direction. No calibration was performed before the test, but it was hoped that the misalignment would be determined in the postprocessing.

The GPS antenna was placed above the airship gondola in the small space between the gondola and the gas bag of the airship. All of the rough and sharp surfaces of the GPS antenna were covered with tape and cellulose sponges to protect the gas bag of the airship. Field testing on the blimp revealed that while the gondola was transparent to the GPS signals, the firewall blanket above the gondola was not. This forced the placement of the GPS antenna above the firewall to collect sufficient GPS satellite data for good post processing differential results.

The gyro was initialized on the ground minutes prior to the flight. The gyro and HERCULES Attitude Processor (HAP) were then kept powered and loaded into the gondola. The gyro can was passed through the hole on the floor of the gondola and was secured to the gimbal frame. The gyro initialization was performed with a laptop PC and then the HAP was transferred to another PC that archived the HAP flight data along with the HDTV time stamp records so that the gyro data could be associated with each frame of the HDTV data. The GPS receivers were turned on prior to take off so that they could acquire GPS satellites and log data from preflight through post-flight without any gaps in data. A data collection interval of 10 seconds was selected because airship motion was expected to be benign. GPS data was collected in memory that is internal to the GPS receivers.

TEST DESCRIPTION

The test, as designed, required that all of our equipment had to be integrated and packaged for easy and secure installation onto the airship. Total weight of the flight equipment was a concern but time did not allow for weight optimization of any components. The airship platform required that all equipment was rack mounted if interior to the airship and firmly secured if exterior to the airship. This proved to be challenging with integration onto the airship taking place in a small town in North Carolina. Any equipment or fixtures that were needed had to be purchased or fabricated in a short amount of time without detailed drawings or a large selection of vendors. All equipment eventually met all the requirements and we were finally ready to fly the TV and the geolocation determination package.

The HERCULES gyro and the HAP system require continuous uninterrupted power after initialization to maintain inertial attitude knowledge. This required that the airship engines be powered and generating the required electricity for the HAP and gyro. With the airship engines powered, the HERCULES alignment was performed by temporarily attaching a Nikon F4 camera to the gyro can and triggering the HAP while pointing the camera at known inertial directions. For this test, software was developed to provide the inertial directions to the Sun, moon and the center of the Earth versus time. One person held the camera and gyro while pointing at the earth or sun and another person sent a trigger command to the HAP to record the attitude data for the target to initialize the gyro solution. After the initialization was performed, the extension cord, which provided power to the HAP and gyro was coiled up and the attitude determination system was handed through the gondola window. The HAP was secured to the top of an equipment rack inside the airship and the gyro was passed through the hole in the gondola floor and was bolted into position on the gimbal frame that would point the HDTV camera during the flight.

After the attitude subsystem was securely attached the airship was detached from the mast car and then took off for its flight. The HAP data was archived along with the TV frame time codes and the GPS data was stored on internal GPS receiver memory.

GEOLOCATION

The geolocation method incorporated for this test has three data requirements: the camera boresight direction; the camera location; and an Earth model. Each of these data are discussed in the following sections. Over 300,000 HDTV video images were collected and geolocated as a result of this test. The analysis in this paper covers only 3 of these images.

CAMERA BORESIGHT DIRECTION

The boresight direction was determined by using gyros to continuously track the intertial pointing direction of the camera. Three Honeywell RLG-1320 ring laser gyros mounted orthogonally, together with supporting electronics built at NRL, formed an attitude reference unit (ARU). This ARU also maintained a temperature compensated real time clock accurate to about 0.25 seconds per month. The clock provided the necessary, accurate time to the HERCULES computer for ephemeris generation. The packaging of the RLG was required to meet space shuttle operations requirements. Once powered on, the gyro package tracked attitude in an inertial, but unknown reference frame. To be useful the reference frame of the gyro had to be related to a known reference frame. An initialization procedure was developed to do this.

Gyro Initialization Procedure

The previously space flown HERCULES system was designed to be initialized by pointing at 2 identified stars and taking 3 triggers [1,2]. The first selected star was placed in the lens crosshairs and a trigger was sent to the attitude processor. Then the camera was rolled about the boresight and a second trigger was taken with the same star in the crosshairs of the lens. Finally a second star was sighted and a third trigger performed with the second star in the lens crosshairs. This procedure provided the necessary transformation matrix between the gyro reference frame at startup and a known reference frame, the Earth Centered Inertial (ECI) frame. The second trigger on the first star was used to determine the boresight of the camera with respect to the gyro axes at every system power up and alignment. Because the system was not engineered to place the camera boresight precisely along one of the gyro axes, the second trigger in the procedure provided a means of calibrating out the misalignment. This recalibration at every power up was important because several lenses were flown and the camera was removed and stowed after each HERCULES session on the Space Shuttle. It was assumed that the boresight misalignment would change slightly with each reassembly of the system. The alignment software accessed a database of bright stars along with their exact inertial locations. Because stars are basically inertially fixed and there is negligible parallax to the stars from Earth orbit, this provided a reliable and accurate inertial alignment process. Once initialized, the ARU tracked the camera boresight direction in ECI with an accuracy of about 0.03 degrees per hour. The initialization itself had an error of 0.01 to 0.05 degrees depending on the ability to hold the system steady on a given star while triggering. The system was able to roughly assess the accuracy of the marks by comparing the included angle between the 2 stars as measured by the gyros with that calculated by knowing the stars' positions from a star catalog. Thus the overall accuracy of the ARU typically started out between 0.01 and 0.05 degrees and degraded after an hour of operation to between 0.03 and 0.06 degrees. The details regarding the algorithms and techniques used in the ARU can be found in [3,4].

For the blimp flight star alignments were impractical, because using stars would have required doing the alignments the night before (or very early morning of) the flight. It was necessary that the previous alignment procedure be modified in some way so as to allow for initialization without having to initialize several hours before the actual flight when stars were still visible. It also was not practical to rely on the weather conditions to provide a clear sky near the dawn of day for star alignment. Also, atmospheric bending would change the apparent position of the stars sighted in a land based alignment. Due to constraints on the airship flight, it was decided to use the local down direction and the Sun and/or moon as inertial targets. Given the short time available to prepare for the flight, it was deemed unwise to attempt to change the HERCULES software to accommodate the new targets automatically. In effect, the sun and earth had to be treated as stars for the purposes of the alignment software. The Sun, Moon, and the Earth were manually entered into the database as "stars" for each new alignment performed. However, unlike real stars, all of these objects are moving at a substantial rate with respect to inertial space. The Sun moves on the order of a degree per day, The moon close to 13 degrees per day and the Earth rotates roughly 15 degrees per hour. A software program was quickly adapted from software written for the Clementine Program at NRL [5] to determine the directions from the blimp hanger field in Weeksville, NC to the Earth center, Sun and Moon as a function of time. A daily table was generated with inertial directions every minute. As the time for launch of the airship approached, the inertial locations for the Earth center and Sun were input to the database for a specific time in the near future. The HERCULES system was then triggered for each target direction and roll at the prescribed time. This technique, while cumbersome, proved to provide a sufficiently accurate initialization of the gyro. It should be noted that as a final check of the alignment, the system was pointed in the general direction of the north star and the readout for declination was verified to be near 90 degrees.

There were 2 main problems with the adaptation of the space alignment technique to the blimp flight. The first was the difficulty in working with the star substitutes. The need to input target coordinates into the database and then trigger on these targets at a precise time was an unwelcome complication at a critical time just prior to flight. As noted earlier, this whole procedure had to take place with aircraft power because any interruption of power to transfer from ground to aircraft power was unacceptable. If the time for triggering on a target was missed, the whole procedure of modifying the database had to be repeated. In addition, the star substitutes adversely affected the accuracy of the alignment. Because stars are point sources, they are excellent targets for precise alignments, and routinely provided alignments of between .01 and .05 (typically .03). The visual determination of the center of the sun was probably accurate only to about 0.1 degree, and that of the center of the earth about 0.2. It is estimated that the alignments using the modified procedure were accurate to 0.3 degrees, an order of magnitude worse than what was achieved in space.

The second problem in accurately initializing the system was the inability to align the HDTV boresight to the gyros. In the space flights, as described above, the boresight of the camera was aligned to the gyros in the initialization procedure. Also, well before the flight, the boresight of the camera was associated with a specific pixel of the images produced by the camera. Thus the space system provided real time geolocation for that specific pixel of the image. For the blimp flight, however, it was not possible to determine the pixel of the image that corresponded to the HDTV boresight. Moreover, the system was not even aligned with the gyros attached to the HDTV camera, but to another (Nikon F4) camera. The HDTV camera had to remain mounted underneath the gondola, hence it was not practical to use the HDTV to perform the gyro alignment. Thus, the Nikon camera was attached to the gyros, the alignment procedure executed, and the gyros then transferred to the HDTV camera. The errors introduced by this procedure were estimated to be about 1 degree. Overall, the alignment procedure for the blimp had errors estimated at over a degree, compared the 0.06 degree maximum for space. Again, however, this error was acceptable for the goal of the flight.

It should be said that the awkwardness of these procedures is due to a lack of time to properly adapt the software to terrestrial and airborne constraints. Work has already been completed to alleviate the problems and will be reported on in the future.

GPS POSITION DATA

The GPS data was processed after the flight using both the real-time measurements and fiducial base station data. A trajectory was generated using only the airship data to coarsely determine the path taken during the test. This set of data is corrupted by selective availability (SA), the errors intentionally introduced to GPS psuedorange measurements to allow only 100 meter position accuracy to non-authorized users. The number of satellites tracked by the airship and the Position Dilution of Precision (PDOP) are shown in Figure 1. The PDOP is determined by the geometry of all tracked satellites and the lower the PDOP, the better the theoretical position solution. Data collected simultaneously at the airship launch site ended prematurely due to the data capacity of the receiver being exceeded during the flight. This made the data set insufficient for reconstructing the entire blimp trajectory and so an alternate source of data was determined. A site that stores raw GPS data at a 5 second interval was used to differentially process the airship data. This data set was obtained via Magellan, Ashtech Products and is from the City of Virginia Beach (CVB1) site. The position of CVB1 is shown in Table 1.

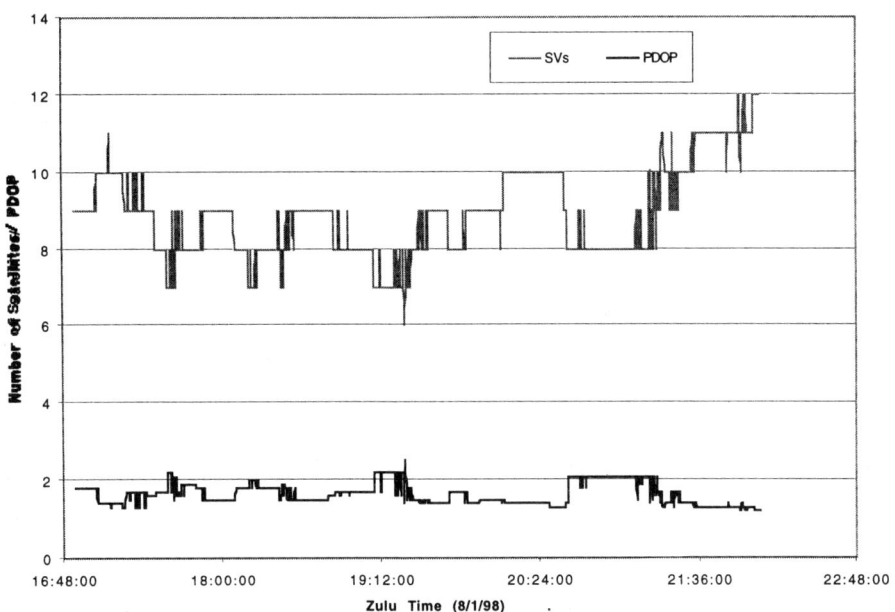

Figure 1. Real-time GPS satellites tracked and PDOP

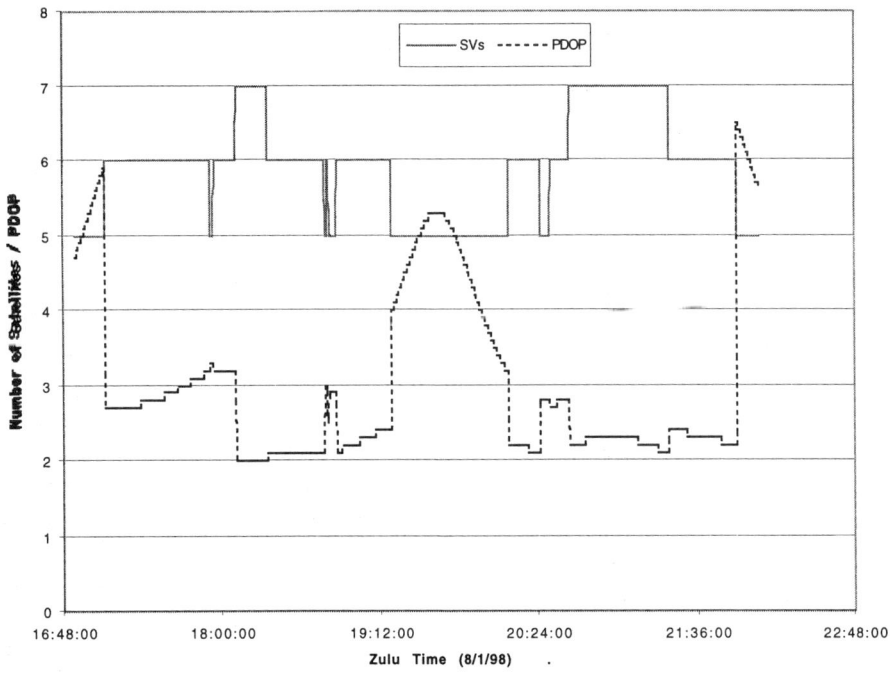

Figure 2: Differentially Corrected GPS Satellites used and PDOP.

WGS 84 Cartesian Coordinates (m)	1232837.347	-4965833.739	3795253.063
WGS 84 Geographical Coordinates	36° 45' 2.57764" N	76° 3' 26.68903" W	-17.053 m

Table 1. Location of fiducial site used for GPs differential processing.

The data from CVB1 combined with the airship data produced a converged fixed ambiguity solution for the entire airship flight. The position error of the antenna during the flight is estimated to be 2-3 cm for all computed positions. The post-processed trajectory require simultaneous measurements to the same satellite at both the base station and the kinematic airship platform. The number of satellites common between the base and the airship and the corresponding PDOPs are shown in figure 2. Note that while the PDOP numbers are higher for the differential case, the measurement errors are greatly reduced producing an overall much more accurate position history. The latitudes and longitudes of the flight path are shown in figure 3 and the altitudes of the airship in figure 4.

A comparison of the kinematic post processed solution and the realtime solutions are shown in figure 5. Due to the nature of GPS position solutions, the altitude errors are typically 2-3 times worse than the horizontal component errors. The figure shows both the 2D (local horizontal) and 3D errors in the real-time uncorrected position solution. Future tests will utilize a real-time differential correction to produce 2-3 cm position accuracy on the airship in real-time.

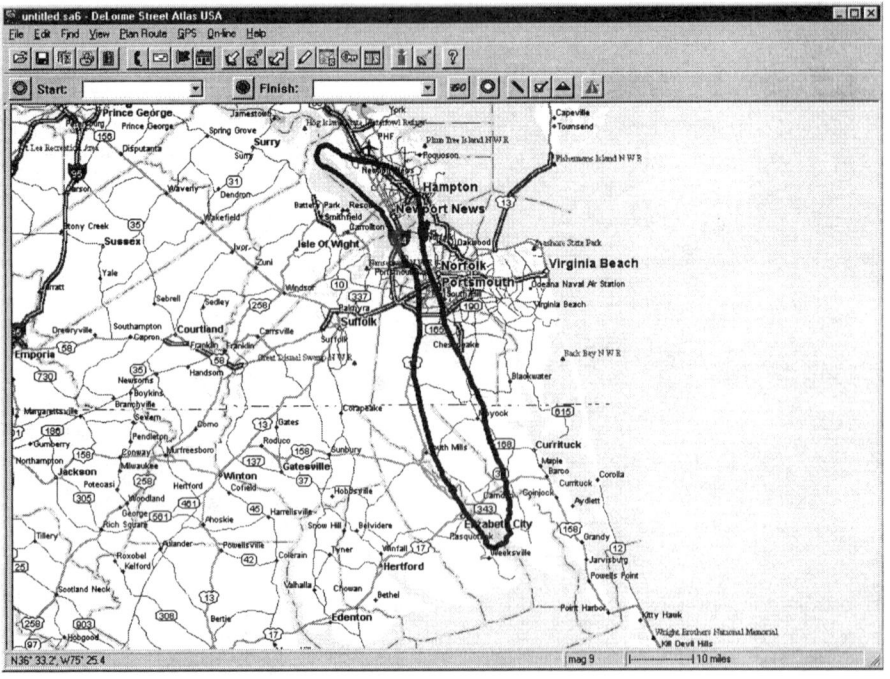

Figure 3: Airship Trajectory for August 1, 1998.

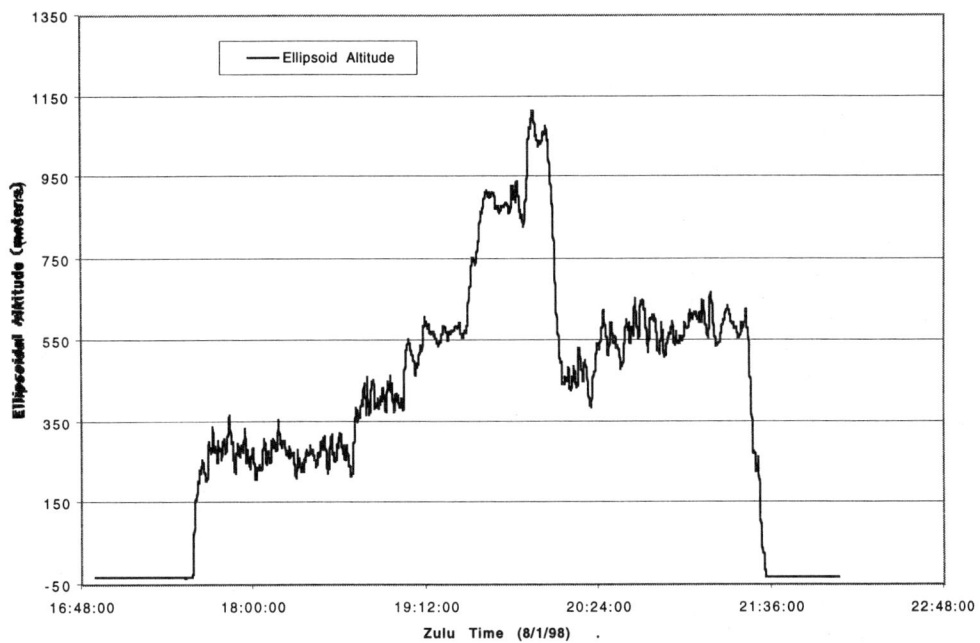

Figure 4: Airship altitude for August 1, 1998.

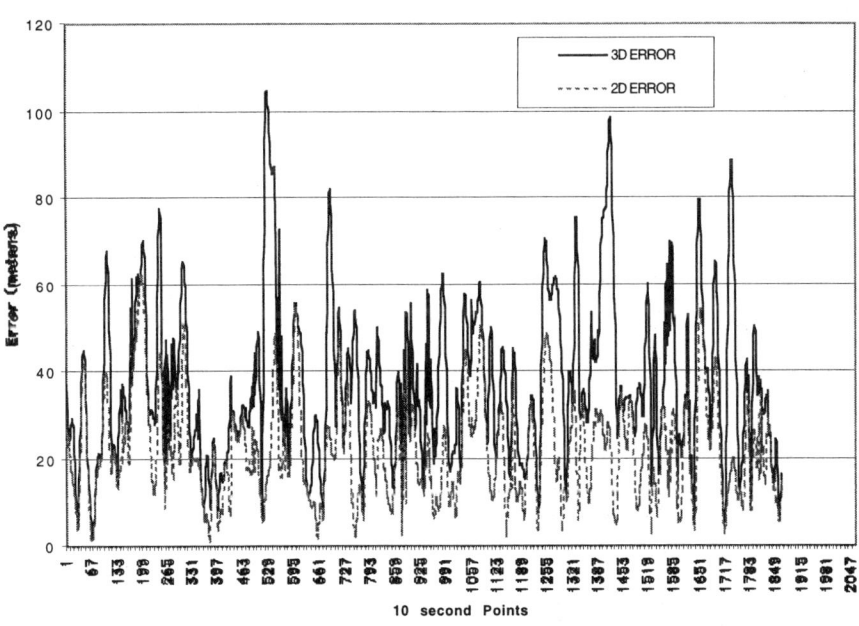

Figure 5: 2D and 3D error for real-time vs. post-processed airship trajectory at 10 second intervals.

EARTH MODEL

The final geolocation solution is obtained by solving for the intersection of the camera boresight vector from a specific location with the earth's surface. A model of the earth's suface is required. The model used for the space flights was the WGS84 ellipsoid. Given that the earth's surface deviates from the ellipsoid by many meters depending on location, errors due to this modeling can be several tens of meters. In space, where the goal was a geolocation accuracy of 2 km., this error is relatively small. On airborne platforms, however, where accuracies better than 1 meter are possible, this error is relatively large and becomes dominant. The_____4_____ geolocation error ε is equal to the geoid/ellipsoid error, b, times the tangent of the angle from nadir of the boresight, α. At 45 degrees this error is equal to b. For long slant ranges with large nadir angles (near 90 deg) this error is very large.

Figure 6 illustrates the method used to fix the problem. Two possible fixes were considered: shrink the earth ellipsoid to make it "fit" at the location of interest; or add a delta height to the GPS measured height to place the blimp as high above the ellipsoid as it actually is above the earth. Although the first approach is the better of the 2 (and is the approach taken in subsequent work), the second was the easier approach in this instance because it required fewer software changes. Both methods produce slight modeling errors. The error in the figure is exaggerated because the curvature of the earth is exaggerated. In practice the area of the earth visible from the altitude of the blimp is essentially flat and the error illustrated becomes negligible.

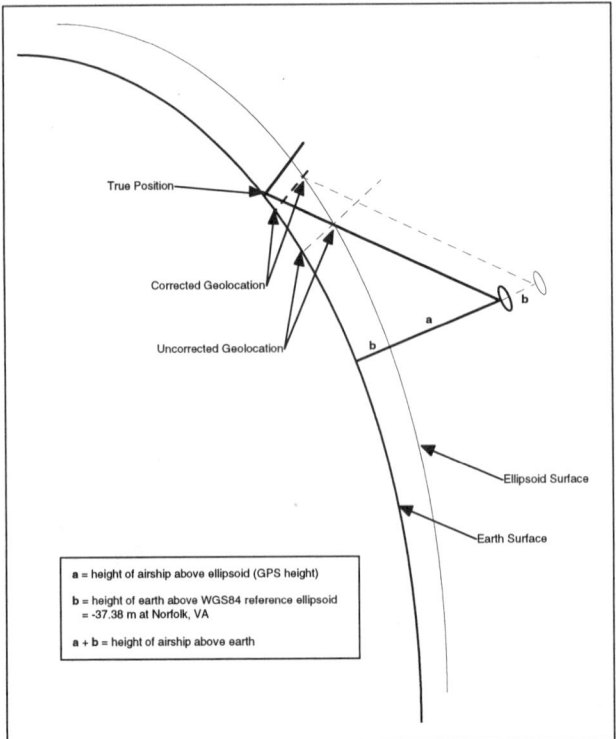

Figure 6: Representation of geolocation error due to geoid modeling.

Once all three requirements for geolocation were addressed, the data was used to produce the desired geolocations. As stated above, the geolocation solution was a post-processed solution in contrast to the real-time solution provided by the space system. The HERCULES software ran on a PC in space and was required to handle a single serial data stream from the gyro electronics. A real-time capability for the airship flight would have required 2 additional serial streams into the PC: one from the GPS receiver and the other from the HDTV recorder. The required changes in both hardware and software interfaces were impractical in the time available. Therefore, although the HERULES software was used to carry out the gyro initialization, a separate data collection program was written to do the actual data acquisition during the flight. It collected the raw data from the gyros as well as the HDTV data in an interleaved fashion. This allowed the precise time available from the gyro subsystem (initially synchronized to GPS time) to be available as a precise time stamp for the HDTV image frames. The GPS data was collected separately. A subset of the HERCULES software was then adapted to post-process the collected data files into the desired geolocations.

POST FLIGHT ANALYSIS

The post flight analysis consisted of processing the collected GPS data to centimeter level accuracy and then entering the accurate GPS data, gyro data, and image time tags into a program to generate the geolocation of each frame. Once completed, several frames were identified to check the geolocation validity and accuracy. The video recordings were viewed and frames selected which had identifiable features. Further the frame selection was such that the feature of interest was in the center of the frame. Providing truth for the geolocation of the features was somewhat circuitous. Unfortunately a direct visit to the site to survey the feature position was not possible. Instead, the position of feature was identified on a map by using the Street Atlas USA software program (version 4). The scale of the frames did not provide sufficient perspective to correlate with the maps directly. It was necessary to find imagery of the scene of sufficiently small scale to place the feature in the same image as landmarks that would show up on the maps. The web site www.terraserver.com provided this capability. Aerial imagery provided images that showed the features in the vicinity of roads and boundaries that show up on maps.

The results for 3 frames are shown in figures 7-9. Figures 7a, 8a, and 9a are the HDTV frames. Figures 7b, 8b, and 9b are Street Atlas USA maps with the feature location marked. The latitude/longitude readout from the maps provided the truth values. Figures 7c, 8c, and 9c show the terraserver images used.

Table 2 gives a brief comparison between the space-based and airship-based geolocation systems.

Knowledge of camera position	Spacecraft Ephemeris Prediction	Differential GPS
Reference frame of position measurements	ECI	ECF
Expected error in position knowledge	5000 feet	1 meter
Knowledge of Camera boresight direction	Ring Laser Gyros	Ring Laser Gyros
Reference frame of Gyro measurements	ECI	ECI
Gyro initialization method	Sight on 2 stars	Sight on sun, center of Earth
Expected error in star sighting measurement	0.03 degrees	0.1 degrees for sun, 0.2 degrees for center of Earth
Earth Model	WGS84 Ellipsoid	WGS84 Ellipsoid adjusted for Norfolk elevation

Table 2: Comparison of Space-based and Airship-based geolocation.

Table 3 shows the geolocation results for the 3 frames shown in the figures. For each scene the following are shown: the position of the blimp when the image was taken; the angle from nadir that the camera was pointing; the slant range from the blimp to the center of the scene; the geolocation of the scene as determined by the system described in this paper; the "truth" location of the scene; and the geolocation error found by differencing the HERCULES geolocation with the "truth" location.

Fig	Description	Blimp Lat °'	Blimp Lon °'	Blimp Alt. (m)	Angle	Slant Range (m)	HERC Lat °'	HERC Lon °'	Map Lat °'	Map Lon °'	Err (m)
1	Water Tower at Fort Monroe	37 1.254	-76 21.174	741	79	4239	37 0.360	-76 18.600	37 0.191	-76 18.595	311.1
2	Intersection in Hampton, VA	37 2.142	-76 21.582	815	41	1137	37 2.532	-76 21.726	37 2.612	-76 21.747	127.4
3	Weeksvillle Hangar	36 20.370	-76 14.374	628	88	15753	36 13.854	-76 7.602	36 13.98	-76 8.00	655.6

Table 3: Geolocation Statistics for analyzed points.

Figure 7a: HDTV Imagery of Fort Monroe, VA

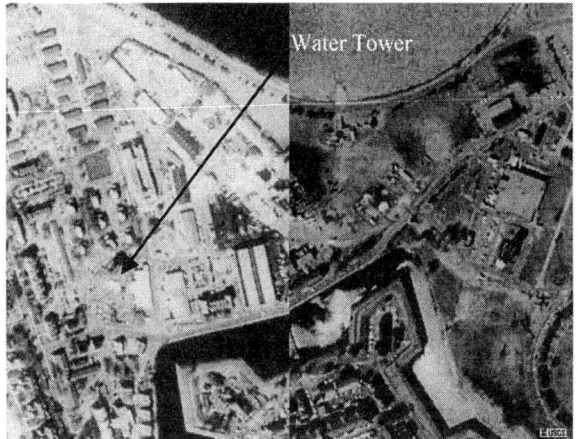

Figure 7b: Aerial Imagery of Fort Monroe, VA

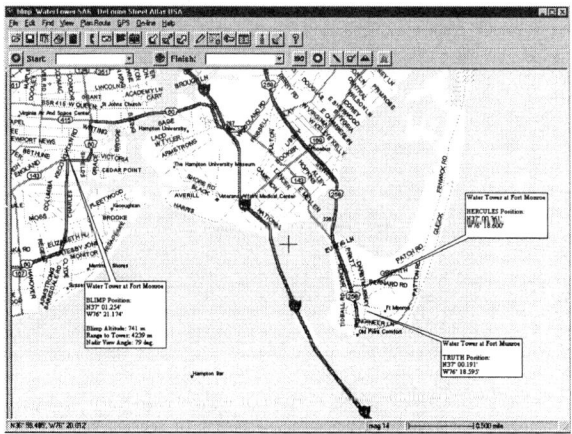

Figure 7c: Vicinity of Fort Monroe, VA

Figure 8a: HDTV Imagery of Hwy 167 crossing US 268 in Hampton, VA

Figure 8b: USGS Aerial Imagery of Hwy 167 crossing US 268 in Hampton, VA

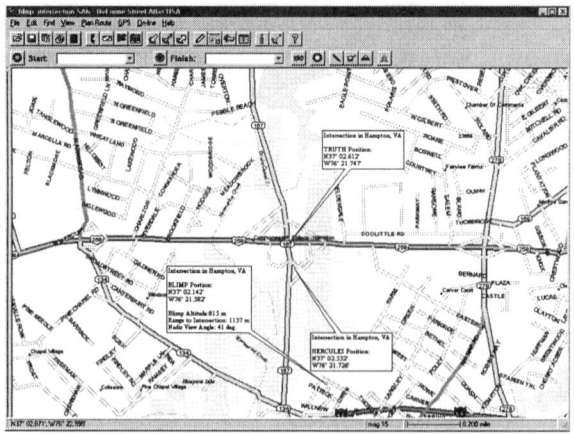

Figure 8c: Hwy 167 crossing US 268 in Hampton, VA

Figure 9a: HDTV Imagery of Hangar in Weeksville, NC

Figure 9b: USGS Aerial Imagery of Hangar in Weeksville, NC

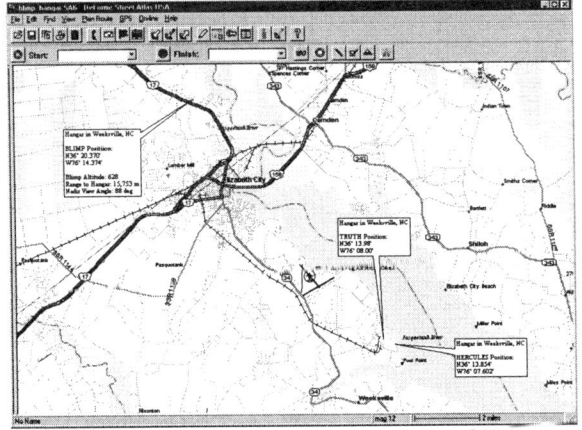

Figure 9c: Vicinity of Weeksville, NC

ACTUAL EVENTS

The airship test was accomplished in less than two weeks. This hurried approach to adapting the HERCULES system for the airship flight was met with many shortcomings, some prior to the test, some during the test, and some discovered after the test was completed. A brief discussion of these shortcomings will now be discussed.

HERCULES was designed as a self-contained geolocation system to be utilized in Low Earth Orbit (LEO). The system operated on a portable PC with input from the user via the keyboard and a single serial interface to the HERCULES Attitude Processor (HAP) which provides the gyro orientation data. This test required two serial interfaces on a single PC to interleave the HDTV time stamp and the gyro data. Only one set of hardware existed and two development groups required access to accomplish test goals. Neither group had sufficient time to identify problems that would later affect the test.

A second shortcoming was the lack of formal procedures. Integration was performed in the field and adaptation to the airship was more difficult than expected. Procedures were modified nearly every day and this caused delays and foul-ups in the execution of the test.

The gyro initialization procedure needed to be performed near the airship so that power could continuously be applied to the HAP during the alignment and uninterrupted through the test flight. The day of the first unsuccessful flight it was discovered that the airship gas bag blocked our view to the Sun with the prevailing wind conditions. An extension cord that allowed us to move the HAP and gyro to the nose of the airship was added so that airship power could be applied continuously. This problem was solved in future tests by the addition of an Uninterruptable Power Source (UPS) to allow for initialization away from the airship and then transfer of a powered system to the airship for flight. The extension cord method proved to be insufficient when an airship ground crew member, not present when the method was incorporated the day before, when asked to roll up the extension cord for loading onto the airship for the flight, untied the knotted extension cord and unplugged the system. This left the current successful alignment invalid and required another alignment, which takes approximately 15 minutes, to be performed prior to airship takeoff. This reduced the amount of flight time due to burning of fuel for an additional 15 minutes.

The first flight attempt on July 31, 1998 was unsuccessful after several circuit breakers were blown due to overloading the circuit. The load was better distributed for the next day's flight. In addition to circuit overloading a serial cable needed to collect the time stamp data was left on the ground. With the power problem identified after the first take off, the serial cable was of no consequence. The first flight also identified the lack of balance of the HDTV and Gyro on the gimbal. The gimbal motors were unable to point the camera near nadir due to this imbalance. The balance was also corrected for the next day's flight.

During the first successful flight on August 1, 1998 it was discovered part way into the flight that data was not coming from the HAP. A reset was initiated during the flight and this necessitated an alignment after the airship landed. While the cause for the data stoppage was not determined until months later, the post flight initialization was successful and the flight data after the reset was preserved. The problem arose from the last version of the HERCULES software. The PC for the space flight would not boot with data streaming from the HAP. Therefore, when the HERCULES software was exited, a command was sent to the HAP to stop the data streaming and when the program was started, a command to start the data stream was sent. The HAP would continue to integrate the rate gyro, however, the 30 Hz data packets to the serial port were stopped. The HDTV group developed a special version of software that captured only the gyro data and the HDTV time code data. Due to time and equipment constraints, this software was never tested with the gyro after the HERCULES software was used with the HAP for an initialization. When first turned on, the HAP sends data out the serial port, and it is only after it has successfully communicated with the HERCULES software that the data stream is turned off. The software was developed on an uninitialized HAP and the flight on August 1 was the first time that this was encountered. The reset of the HAP during the flight reset the data output and corrected the problem.

As mentioned earlier, the ground GPS receiver data memory filled up. This should not have happened if the data from the day before had been offloaded and deleted. A lack of procedures and neglecting to check the remaining memory available could have prevented this. The support group at Ashtech was very helpful and was able to identify another source of base station data with at least a 10 second interval for post processing the GPS flight data. Most reference stations store data at a 30 second interval. Also, the data interval that GPS data was collected was insufficient for high precision geolocation data. While it was sufficient for this test with no clear requirements, all future flights have collected data at a 1 second interval for both real-time and post-processed position information.

ACKNOWLEDGEMENTS

The authors would like to thank the airship ground crew who helped to rapidly install the test equipment onto the airship. Also, thanks to Mark Bryant and Bob Lemoine of Magellan, Ashtech products for their help in finding a base station site in the vicinity of the test with data at the required interval.

REFERENCES

1. M. T. Soyka, and P. J. Melvin, ``HERCULES: Gyro-Based, Real Time Geolocation for an Astronaut Camera," (AAS 91-130), *Spaceflight Mechanics 1991*, Vol. **75**, Part I, *Advances in the Astronautical Sciences*, J. K. Soldner, A. K. Misra, L. L. Sackectt, and R. Holdaway, eds., Proc. AAS/AIAA Spaceflight Mechanics Meeting, Houston, TX, Feb. 1991 (Univelt, San Diego, CA, 1991), 471-490

2. P. J. Melvin, M. T. Soyka, H. M. Pickard, S. N. Lam, K. H. Little, R. R. Dasenbrock, and T. W. Murphy, ``Geolocation Accuracy of HERCULES on STS 53,"

(AAS 93-604), Presented at the AAS/AIAA Astrodynamics Specialist Conference, Victoria, B.C., August 16-19, 1993

3. R. F. Higgins, ``HERCULES Attitude Processor (HAP)," Dissertation, Master of Science in Electrical Engineering, Virginia Polytechnic Institute and State University, Blacksburg VA, 1992

4. R. F. Higgins and J. G. Tront, ``HERCULES Attitude Processor: Gyro Data Processing System for Real-Time Geolocation of Images Captured by Astronauts," 1993 IEEE Aerospace Applications Conference Digest, S. Swift, Conf. Dir., IEEE Document Catalog Ref No 92TH0526-4, Library of Congress Cat.No.88-640065

5. Middour, J.W., "A New Trajectory Propagation Program For DSPSE Mission Support", September 1993, Presented at the 7th Annual Utah State University / AIAA Conference on Small Satellites.

AAS 00-032

NANOSOL – A NEXT GENERATION SUN SENSOR

John Glaberson[*]

A new precision sun sensor has been developed which uses a single optical assembly to sense two axes, and to combine the functions traditionally performed by separate fine and coarse sensors. This sensor, named Nanosol, provides high accuracy, wide field of view, high reliability, radiation tolerance, small size, and low mass. Preliminary test results are presented and error sources are discussed.

INTRODUCTION

The Nanosol sun sensor uses a new method of sensing sun angle which improves accuracy, field of view, size, weight, and ease of use, when compared to conventional approaches.

It is a unitary sensor, combining the functions which are customarily performed by fine and coarse sensors; it also senses both axes (α and β) of sun angle. In addition, the optics and electronics are housed in a single compact housing, using a single optical path.

The new sensor is a high accuracy, wide field sensor, having a 3σ error of .03° over a range of ±64°. In addition, it can function with reduced accuracy over a ±80° cone. A unique feature of the sensor is that it is highly immune to Earth Albedo.

Combining the two axes, the coarse and fine functions, and the optics and electronics in a single package has many advantages, including reduced size and weight, simplified testing and calibration, and simplified integration.

PRINCIPLE OF OPERATION

The Barnes Nanosol Sun Sensor uses the well known principle of Moiré Interferometry, which has been used successfully in fine sun sensor (FSS) designs on many missions[1]. Moiré techniques allow extremely high accuracies to be achieved with simple hardware.

Previous Moiré sensor designs have used one-dimensional Moiré patterns to perform single axis sun angle measurement, requiring the use of two separate sensors for the fine measurement function. This sensor instead uses two-dimensional Moiré patterns, which enables a single sensor to measure two axes simultaneously without an accuracy penalty.

---

[*] Mr. Glaberson is a Systems Engineer with Barnes Engineering. He can be contacted at Barnes Engineering, 88 Long Hill Cross Roads, Shelton, Connecticut 06484.

The output of a Moiré sensor is cyclic, and therefore gives an attitude measurement which is ambiguous. In prior systems, a separate coarse sensor was used to resolve this ambiguity. By contrast, the Nanosol sensor performs the coarse and fine measurement functions in a single sensor. This is accomplished by arranging the Moiré patterns so that multiple signals are available in which the phase changes at differing rates as a function of sun angle. By combining these signals, it is possible to deduce which of many cycles the current measurement represents.

The construction of the Nanosol sensor is extremely simple. A fused silica block is coated on one side with a two-dimensional Moiré "encoder" pattern. The opposite side of the block is coated with an "analyzer" pattern, which is optimized to operate with the encoder pattern. Behind the block is a set of silicon photodiodes, connected to a multiplexer and preamplifier.

The output of the mux goes to an A/D converter, which is connected to a digital ASIC which performs the demultiplexing, formatting, and communications functions.

The digital values are transmitted to the mission computer where they are translated to attitude measurements.

SPECIFICATIONS

The tentative specification for the Nanosol sensor is shown in the following table.

Table 1
NANOSOL TENTATIVE SPECIFICATIONS

Measurements	Both α and β (with single telescope)
Accuracy	$\pm 0.03°$ in α and $\pm 0.03°$ in β (3σ)
Operating Range	$\pm 60°$ in α and $\pm 60°$ in β (with full accuracy)
	80° half angle cone (with reduced accuracy)
Reliability	0.99943 (single redundant telescope - 15 year mission)
Redundancy	Full internal redundancy
Size	Cylinder 2.8 inch diameter (7cm), 3 inches long (7.5cm)
Mass	19.6 ounces (<0.55 kg)
Output Signals	Serial digital (asynchronous or synchronous data stream). Sensor temperature measurement output provided. 15 pin EMI connector
Input Power	± 15 VDC and $+5$ VDC (regulated voltages)
	500 mWatts/channel typical
α/β Data Refresh Rate	10 Hz
Radiation Tolerance	100 Krad tolerant components
	Sensor designed for long life in high radiation orbits

PERFORMANCE

The calculated error sources, and their contributions to total error are shown in the following error budget. As is standard in the industry, these errors refer to either the α or β output, as opposed to the combined error magnitude.

Table 2
NANOSOL ERROR BUDGET

NanoSol Errors (+/- 60 deg), α or β (All Errors 3 Sigma)		
Fixed Errors		
Alignment Uncertainties		0.010
Transfer Function Deviation		0.014
	RSS	0.017
Short Term Errors		
Noise		0.001
Quantization		0.003
	RSS	0.003
Orbital Period Errors		
Earth Albedo		0.000
Long Term Errors		
Thermal Static		0.004
Electronic Drift		0.000
	RSS	0.004
Total Error		**0.024**
(Linear Sum of Fixed, Short, Orbital and Long Term Errors in α or β)		

ERROR SOURCES

Alignment uncertainty refers to the accuracy limitations imposed by the test stand, including the motorized stages, autocollimators, and the ability of the operator to set up the sensor on the stage for testing. It also includes the combined effects of beam quality. The Nanosol sensor has been designed to be immune to many of the known deficiencies of sun simulators, particularly in regard to beam quality.

Transfer Function Deviation is any deviation from a straight line transfer function which remains after the application of calibration constants. The value in the table represents a modest reduction from tested values, which we believe can be achieved by minor design modifications and improved algorithms.

Electronic noise is a very small factor in the sun sensor. The value in the table is based on test results, and includes sensor noise, test setup noise, and atmospheric noise.

Quantization noise can be calculated in a straightforward manner, based on the A/D converter characteristics and the signal amplitude. The signal amplitude has been verified by test.

Earth Albedo is inherently rejected by the Nanosol sensor. It can cause two different problems - first, the Earth may contain spatial intensity variations which are at the same spatial frequency as the Moiré mask. By simulation with actual Earth images, this effect has been demonstrated to be on the order of .0001°. Second, the Earth's radiation adds a constant bias to all of the detectors in the sensor, which is rejected to first order by the algorithms. The value in the table represents the higher order effects of this bias.

Thermal Static errors are caused by the change of index of refraction of the fused silica block with temperature. Uncompensated, this would produce an error of about $\pm.035°$, but by using a thermistor to monitor the temperature of the block, this is reduced to the value in the table.

Electronic Drift has no effect on the sensor because it is a multiplexed system, and any change in the electronic path affects all detectors equally. The algorithms reject any change in gain or offset which is common to all detectors.

PERFORMANCE VERIFICATION

The performance of the Nanosol sensor has been verified both by analysis and by test.

A breadboard of the Nanosol sensor has been built and tested (Figure 1), verifying the fine sensing accuracy of the technology. The performance of the Nanosol sensor, including coarse sensing, has been demonstrated by a software model, and a physical model is currently being built.

Figure 1 Sun Sensor Breadboard

Testing of the breadboard was performed at JPL's Table Mountain heliostat facility with encouraging results – a standard deviation of .01° was achieved over a ±60° range. The α and β errors from this test are shown in the following plot.

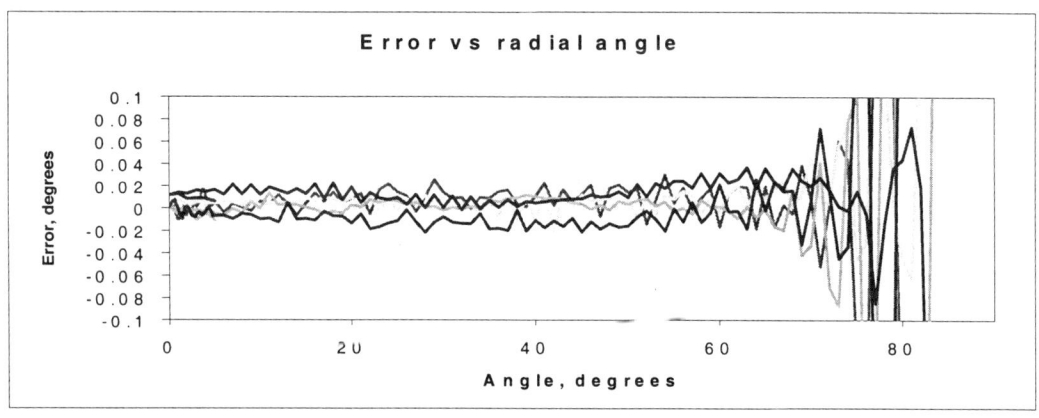

Figure 2 Breadboard α and β errors

While the test results were excellent for a hand built breadboard, certain deficiencies in the design and construction were demonstrated by the test. These can be seen in the small scale "ripple" in the plot, and in the dramatic increase in the errors above 70°. Some of these errors can be attributed to specific errors in materials or assembly, while others can be removed in the algorithms.

Even with these problems, the standard deviation of the errors was .01°. The distribution of the errors is shown in the following histogram. Note that the distribution does not have long tails - this is a product of the underlying error phenomena, which are inherently bounded. The "3σ" error will therefore be smaller than would otherwise be indicated by this standard deviation.

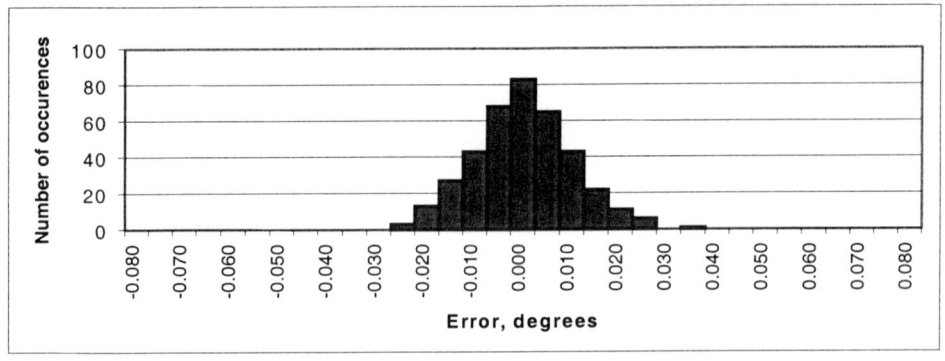

Figure 3   Distribution of errors

USAGE AND INTEGRATION

The Nanosol sensor provides serial digital outputs. 78 bytes are required to transmit the information for one full attitude update from one sensor. With a bit rate of one megabit per second and (N,8,1) asynchronous communications parameters, it would take 780 microseconds of bus time to transmit the data. At four samples per second, this represents a bus utilization of 0.3%.

The algorithms required to support an earlier Nanosol sensor have been coded in C and benchmarked on a 100 MHz 80486 computer. The time required to execute the algorithms for a full attitude update is about 64µsec. Assuming the same update rate as above, this represents a processor utilization of 0.025%.

The required code space is 4 Kbytes (assuming that floating point libraries will be available) and the required data space is less than 2 Kbytes.

DEVELOPMENT STATUS

Nanosol technology has been under development at Barnes since 1997.

In early 1998, a prototype was constructed to prove the principle of measuring two axes with a single optical path. While it did not achieve the accuracy we are currently achieving, it did prove that it was possible to measure two independent axes with a single sensor.

Later in 1998, Barnes won a NASA contract to study the feasibility of producing a minature high accuracy sun sensor using Nanosol technology. In this study, we performed a paper design of an extremely small sensor utilizing a single optoelectronic integrated circuit. The relevant error sources and their effect on the operation of the instrument were analyzed in great detail.

Barnes has since initiated the development of a Nanosol sensor which does not rely on a custom optoelectronic IC. This was done to reduce time-to-market and technical risk, while accepting a modest increase in size and weight.

A breadboard using these techniques was built to demonstrate the feasibility of reaching a goal of .03°, 3σ accuracy. It was tested at Table Mountain, and came close to providing the final performance required. Many of the errors are attributable to specific construction faults of the hand built model. Program risk has been dramatically reduced by this testing.

A software model has been developed to allow the evaluation of various optical geometries, to determine the effects of radiation and ultraviolet exposure, and to test algorithms for calculating sun angle from the detector outputs. As a result of this software, we now have an optical design which we believe will meet all requirements.

As of this writing, we are constructing a brassboard having the correct optical geometry and most of the correct optical materials. This model will allow us to prove out the fine/coarse operation, and to demonstrate the effectiveness of temperature compensation. It will allow us to simulate the effects of radiation and micrometeorite exposure.

Concurrent with the design work, we are also pursuing a major upgrade of our sun simulation facility. This work is addressing issues of electronic noise, beam uniformity, beam drift, and stage accuracy. Since the Nanosol sensor has been designed to be insensitive to many of the known deficiencies of sun simulators, this upgrade will yield an extremely robust test environment.

CONCLUSION

Do we need one?

REFERENCES*

1. J. Wertz, ed.: Spacecraft Attitude Determination and Control, D. Reidel Publishing Co., Dordrecht, Holland, pp. 166, 227-230, 1978.

AAS 00-034

THE TERMA STAR TRACKER FOR THE NEMO SATELLITE

L. Maresi, T. Paulsen, R. Noteborn, O. Mikkelsen and R. Nielsen[*]

TERMA Elektronik AS, Denmark, has developed a Star Tracker for the NEMO (Naval EarthMap Observer) Remote Sensing Spacecraft. NEMO is a joint effort between Space Technology Development Corporation and the Naval Research Laboratory. The NEMO Star Tracker is equipped with a compact low f-number optics, a large format CCD and a powerful microprocessor. It is designed for measuring attitude with accuracy better than 2 arcseconds for pitch and yaw and 14 arcseconds for roll. The Star Tracker is an autonomous system able to determine the attitude from lost in space conditions and is able to perform the operations of attitude acquisition and tracking under stressing conditions of slew rate and acceleration.

The performance assessment of a high accurate, wide field of view Star Tracker is a challenging aspect of the process of system verification. Performance has been measured by means of laboratory equipment and during outdoor tests using the natural night sky. A test campaign was carried out at the Teide Observatory, at Institute of Astrophysics of Canarie, also to explore to which extend the Star Tracker can be operated when the spacecraft is performing on-orbit attitude maneuvers.

The paper presents an overview of the design and illustrates test methods, experiences, and results of the qualification and test campaign. A comparison of the Night Sky Test with the results obtained in the laboratory is also presented.

INTRODUCTION

The Star Tracker presented in this paper is intended for use on the NEMO spacecraft developed by the US Naval Research Laboratory. This spacecraft will provide high resolution images of the Earth and use two TERMA Star Trackers for geo-location of the images.

In this paper, first the NEMO Star Tracker is presented in an overview of the architecture and the performance. In the following sections the performance tests performed on the sensor in the laboratory and using the night sky are described and their results are presented. The results from laboratory and night sky tests are then compared. Following this, the qualification testing performed on the Star Tracker is presented. Finally conclusions are gathered.

---

[*] TERMA Elektronik AS, Bregnerødvej 144, DK-3460 Birkerød, Denmark. Web Site: www.terma.com.

NEMO STAR TRACKER

In this section the NEMO Star Tracker is described with regard to architecture of the sensor, algorithms and performance.

Architecture of the Star Tracker

The Star Tracker consists of a Camera Head Unit (CHU) shown in Figure 1 and a Data Processing Unit (DPU) shown in Figure 2. Key characteristics of the system are presented in Table 1 (ref. 1).

Figure 1 The Camera Head Unit   Figure 2 The Data Processing Unit

Table 1
KEY CHARACTERISTICS OF THE NEMO STAR TRACKER

CHU Mass	0.9 kg
DPU Mass	1.0 kg
Baffle Mass	1.9 kg
CHU Dimensions	135x120x93 mm
DPU Dimensions	260x195x33 mm
Baffle Dimensions	393L x 344Ø mm
Total Power Consumption	11.5 W
Nominal Input Bus Voltage	28 V
CHU Thermal Load to Spacecraft	3 W
DPU Thermal Load to Spacecraft	9 W
Data Interface	RS-422
Field of View	22° x 22°
Number of Stars Tracked	64
Sun Exclusion Angle	40°
Earth/Moon Exclusion Angle	30°
Radiation Tolerance	15 kRads
Design Life Time	5 years

The DPU features a powerful Digital Signal Processor (TEMIC TSC-21020E) for fast 32 bit floating-point computations. It provides memory capacity in the form of EEPROM memory for permanent program and star catalogue storage and static RAM for data and program memory. A DC/DC converter regulates the main bus voltage.

The interfaces consist of:

1. A 28 VDC power interface with the spacecraft;
2. An RS-422 data interface with the spacecraft computer;
3. A Timing Correlation Pulse interface with the spacecraft;
4. An interface with the CHU for control and image readout.

The CHU contains a front side illuminated 1024x1024 EEV-47-20. The optics has an f-number of 1.4 and the field of view is a square of 22° wide. An FPGA is used for controlling the interface with the DPU, driving the CCD, reading out of the A/D Converter and setting gain and offset parameters. The 8-bit A/D Converter digitizes the amplified and noise processed signal from the CCD readout register.

Finally, a baffle is mounted on the CHU for stray-light rejection (see Figure 3). The Star Tracker can therefore operate with nominal performance for Sun exclusion angles larger than 40° and Earth/Moon exclusion angles larger than 30°.

The Star Tracker has four modes of operation in flight (see Figure 4):

1. A boot mode is used for check of memory, loading of software and catalogue. From here the system can proceed in either of the following three modes:
2. Standby Mode is among others used for setting of parameters. No attitudes are delivered in this mode;
3. The Lost In Space problem is solved by the Initial Acquisition Mode, which provides a pair of coarse attitudes that seed the Attitude Update Mode;
4. The Attitude Update Mode is the normal mode of the Star Tracker in which it is tracking stars and regularly providing attitudes.

If the Attitude Update Mode looses track the system transits back into Initial Acquisition Mode and stays there until a valid attitude is found. This can happen for example when the Sun enters the field of view.

The Lost In Space problem is solved by generating triangles from spots visible in the CCD image

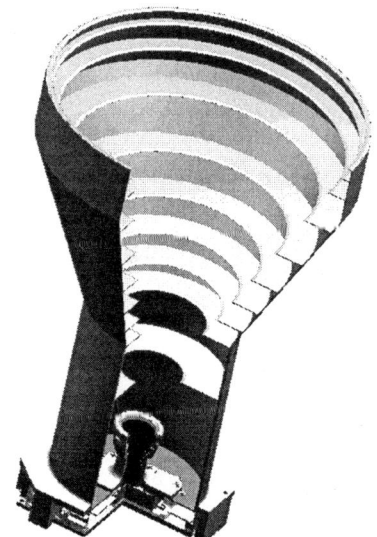 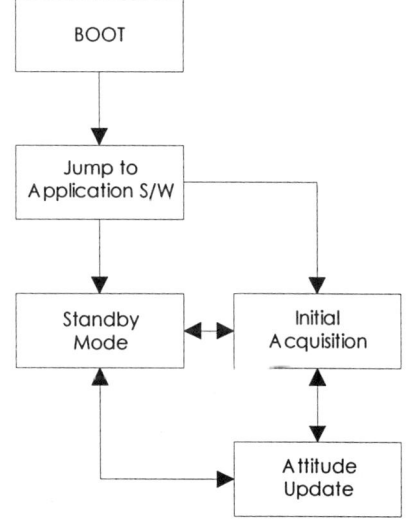

Figure 3 The Camera Head Unit with Baffle   **Figure 4 State Diagram of the Star Tracker**

that are looked up in the onboard catalogue. An attitude is constructed from all the triangles that are identified. The system filters out triangles that give false attitudes. Initial Acquisition provides a coarse attitude plus rate estimation (internally). This provides a means to predict what the attitude will be at the mid integration time of the next image for Attitude Update. The stars that are visible in the field of view at that instant of time are determined and looked up in the image. A least squares estimator constructs an attitude from at most 64 tracking stars. If not enough stars are available in the image, the system concludes it has lost track and will transit back into Initial Acquisition Mode.

Star Tracker Performance

Theoretically, the attitude can be calculated by knowing the line of sight of only two known stars. The NEMO Star Tracker operates calculating the attitude over a large number of stars. This method allows to significantly decrease the effect of random errors affecting the measurement of each single star. The position of a star in the focal plane is measured by calculating the barycenter of its defocused image. Two major errors occur in the measurement: a Temporal Error and Random Spatial Error. The temporal error is due to stochastic process of measuring the CCD pixel signal. The standard deviation of the distribution, translated in angle, is usually defined as the Noise Equivalent Angle (NEA) of the system. Random Spatial Errors are systematic errors varying with small displacements of the star position and with a zero mean value, as the error of the interpolating algorithm, the residual errors after optics calibration, and chromatic aberrations. These errors are systematic, i.e. measurements carried out under the same conditions have the same error. In practice, when considering a large set of measurements with stars randomly distributed in the field of view, we obtain a set of stochastic values with a zero mean value and a nonzero rms value. On the other hand, some errors may generate an offset in the measurement of the attitude, e.g. thermo-mechanical stability or a residual error after the compensation of relativistic effects.

The Performance Analysis (ref. 2) estimated that random errors on a single star measurement are 8.9" (BOL). Offset errors are considered negligible for the specific case of NEMO. In fact, the Star Tracker is mounted on a thermally stabilized Optical Pallet to avoid any thermal drift. Moreover, the Star Tracker software compensates the relativistic effect using velocity information provided by the on-board computer. The total accuracy of the Star Tracker can then be translated to 2" in pitch and yaw and 14" in roll while tracking 25 stars. The dynamical properties for which this accuracy should be reached is a maximum slew rate of 0.5 °/sec with a maximum acceleration of 0.04 °/sec^2 (ref.1). The NEMO Star Tracker is required to solve the Lost in Space problem within 10 seconds and should give attitudes in the tracking mode at 1 Hz.

LABORATORY TESTING

This section describes the laboratory performance tests that have been performed on the Star Tracker Engineering Demonstration Model (EDM). Also the results and a discussion on them are provided. The tests have been performed in a clean room laboratory environment.

Laboratory Test Description

The laboratory at TERMA is completely equipped to do various kinds of testing on the Star Tracker, including thermal vacuum testing. Main tests and operations that are done here include:

1. Functional tests on a Star Pattern Simulator;
2. Random Error Verifications on a Single Star Simulator.

TERMA has developed a Star Pattern Simulator that can be placed in front of the camera. The Simulator provides a 30 degree sky pattern which is imaged by the Star Tracker. It is also possible to move a few degrees in pitch and yaw and there is total freedom of the roll. The setup is shown in Figure 5.

This setup is used for functional tests of the algorithms. Initial Acquisition, Mode Transition and Attitude Update can be effectively tested under controlled circumstances.

The Pointing Accuracy was measured on the Single Star Simulator, a calibrated and collimated beam of light simulating a single star of selectable color temperature and brightness. The Star Tracker is positioned in the beam and can be rotated using a 1" accurate two-axis rotating table (accuracy is for positioning, as the stability is at least one order of magnitude better). This setup is also used for calibration of the optics, measuring focal length, lens distortion, optical center and tilt of the CCD. By measuring the spot centroid on a large series of camera images, the NEA is verified. Interpolation Error can be verified by making images for different angles of the beam, moving the beam from one pixel to the next. By doing this for a larger angle, moving the beam from the center to the edge of the CCD, the residual error after calibration is measured. Chromatic aberration is tested by changing the color temperature of the simulated star and subsequently measuring the average centroid position of a large set of measurements.

Figure 5 The Star Tracker in front of the Star Pattern Simulator

Results from Laboratory Tests

Functionality of the algorithms is tested in the laboratory and it was proved that the sensor is able to do the Initial Acquisition very well within the required time. It takes on average 3 to 4 seconds to do a pair of measurements. Transition into Attitude Update Mode can be achieved and the tracking shows that all of the available stars are detected and identified. Small changes can be tracked as well and sudden changes in attitude cause the system to go out of lock and re-acquire as designed.

The more sophisticated testing of the Random Errors show the results that are summarized in Table 2. In Figure 6 the periodic error due to interpolation inaccuracy can be seen. Very well visible is the period of 1 pixel that this error has. The residual error after calibration is shown in Figure 7. The high frequency component here is the interpolation error while the low frequency signal shows the residual after calibration. This is caused by the imperfection of the model for the lens calibration and to some extent the error in calibration of the model parameters.

Table 2
POINTING ACCURACY (SINGLE STAR)

	Design Budget (")	Verification (")
NEA	5.8	5.9
Interpolation Error	1.5	2.9
Residual Distortion	3.0	3.0
Chromatic Aberration	6.0	<1.0
Total (rms)	8.9	7.3

Chromatic aberration was tested for 4 color temperatures ranging from 3000 to 6500 K. This was done at the edge of the CCD taking 50 measurements for each temperature. The result of this is that the aberration is very small, which shows that the correction of this in the optical system is very successful.

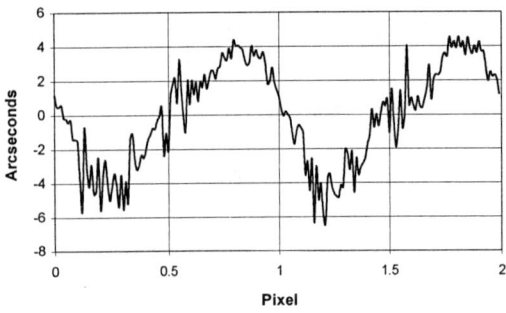
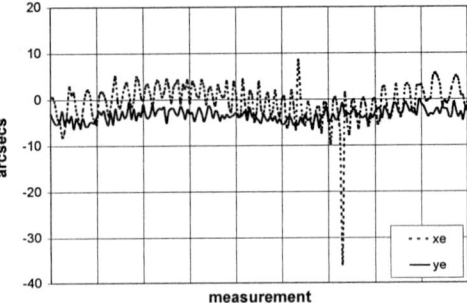

Figure 6 Interpolation Error   **Figure 7 Residual Error after Calibration**

Discussion of Results

The tests on the Star Pattern Simulator show that the Star Tracker functionality is very satisfactory for time performance of Initial Acquisition, which is easily verifiable, as well as for mode transitions. Tests with the Star Pattern Simulator do not give quantitative measurements of accuracy of attitude determination, since the accuracy of the position of simulated stars is one order of magnitude less than the capability of the Star Tracker. The accuracy of attitude measurements is obtained by analysis of the test results with single stars. The test campaign has shown that the Star Tracker is well within the NEMO requirements. In fact, error contributors that are larger than designed are compensated for by the chromatic aberration that is much smaller than designed. Specifically the interpolation error is larger than designed and this could have been covered by a calibration of the centroiding algorithm. As the total accuracy was satisfactory, it was decided not to start this effort. Combining these values result in a total accuracy that is compliant with the requirements when tracking 14 stars or more.

NIGHT SKY TESTING

In this section the Night Sky Test Campaign performed on the Engineering Demonstration Model (EDM) is described. The results, as far as available, are presented with a discussion of them.

Night Sky Test Description

The NEMO Star Tracker was regularly tested on the night sky in Denmark but for a good evaluation of the performance it is necessary to go to a high location to have a better view of the stars with less atmospheric refraction. TERMA chose the island of Tenerife where the El Teide Observatory is located at 2400 meters altitude. A representative of NRL witnessed all tests. The test is intended to confirm the laboratory tests and have the opportunity to functionally test the system at dynamical conditions that are not possible with the Star Pattern Simulator. A re-calibration of the optical system is required here as the atmosphere acts as an added lens. The centroids and attitudes that are provided by the Star Tracker are fed to a calibration program that in turn fits the focal length, the optical center and the distortion parameter.

The functional dynamical tests consist of sweeps over the sky at known velocities and acceleration to see if the system can get in lock and keep tracking at these speeds. Another functional test is to do Initial Acquisitions at different locations to measure the reliability of this mode. The performance of the system is tested in a series of static tests where the sensor is pointed towards zenith and put in tracking for a prolonged period of time. Influence of different settings like integration time, gain and offset, are evaluated by repeating the test with different settings of parameters.

It is not possible to measure the different contributors to the accuracy in a Night Sky Test. The signal can be analyzed to find the total accuracy and what could be called the total Noise Equivalent Angle. As opposed to the NEA measured in the single star test, this time we can only estimate the error of all stars together as formed by the noise, interpolation error, calibration residuals and chromatic aberration. Taking subsequent measurements of the attitude and comparing their differences for a large series of measurements estimates this error. (Ref.4)

Results from Night Sky Test

Although the night sky represents only a part of the total sky, the measurements explored the reliability of the Initial Acquisition Mode on a large part of the celestial vault. About 90 locations were tested and only 4% of these gave a false identification. Most of these false identifications were corrected by the instrument after some time had elapsed.

The dynamic performance of Initial Acquisition was tested up to 6 °/sec and it was found that even at these high speeds a correct identification could be established. The system is compliant with the requirements for NEMO but it is clear that this is not the limit of the envelope. The extremes of the instrument are not yet established. Tracking stars was done for velocities of up to 3 °/sec which was possible by adapting the CCD settings (lower integration time, higher gain). Also here the sensor is fully compliant but a limiting envelope has not been established. The total accuracy of the Star Tracker is evaluated in series of 1-hour measurements. These measurements show an increase in Right Ascension of approx. 15"/sec due to the rotation of the Earth. The expected values for the attitude are computed from estimations of the rotation rate and the nutation of the Earth and Astronomical Aberration. One of these series is presented in Figure 8.

Although at the moment of writing the process of full performance test is still ongoing, it is already visible that the Star Tracker is indeed meeting the requirements of performance. The shown series features standard deviations as presented in Table 3. The figure also shows the dependency on the distribution of stars. This causes a low frequency signal with typical periods of one hour. Stars entering and leaving the Field of View are the cause of this.

The Total Noise Equivalent Angle measurements show results that are very encouraging. Table 4 presents the results. Note that these measurements give the accuracy of all stars together not of just one star.

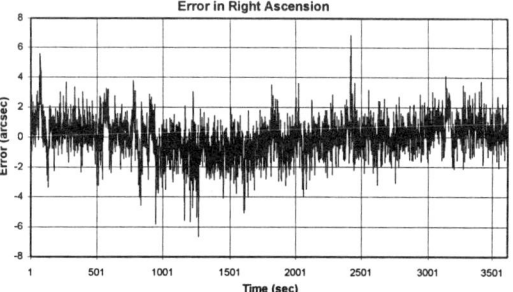
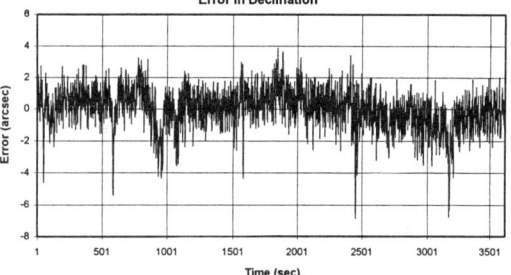
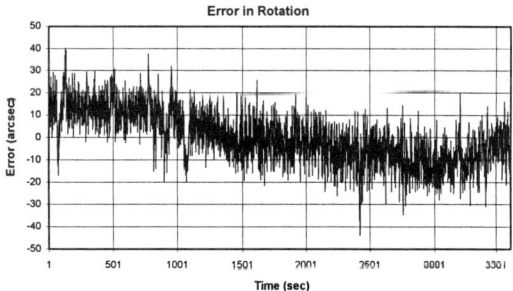

Figure 8 Right Ascension, Declination and Rotation Error Measurements from the Night Sky Test

Table 3
Standard Deviations of Corrected Measurements from Night Sky Test
(Ref. 4, TC01-01B2)

	Stddev. (")
Right Ascension	1.4
Declination	1.2
Rotation	10.9

Table 4
Total Noise Equivalent Angle Test Results
(Ref. 4, TC01-02A)

	Stddev. (")
Right Ascension	1.1
Declination	0.8
Rotation	6.9

Discussion of Results

The results of the Night Sky Test show that the Star Tracker is giving the desired accuracy. It must also be noted that these results are still quite different from what might be expected in space, as the Earth's atmosphere influences the tests. Overall the Star Tracker is expected to perform better in space.
The functional tests give a good feeling of the envelope of the instruments. The ability to acquire and start tracking and coping with the dynamical requirements of NEMO is demonstrated. Higher requirements than those can be met although this can not be totally proven with the present results. Tracking was done at several degrees per second and gave only minor problems, so it is made probable that it is possible to do this but no information on how reliable it is has been obtained.

COMPARISON OF LABORATORY AND NIGHT SKY TESTING

In this section the results from Laboratory and the Night Sky Test are compared. The nature of the two tests is quite different. In the laboratory functional tests are very limited, especially when dynamical requirements are tested. Subsystems of the algorithms have been tested with simulators but the optical part of the system had to be tested at the night sky. The coverage of Initial Acquisition was already tested in simulation using synthetic images but a real test also had to wait until the Night Sky Test. Both methods have their shortcomings. The synthetic images represent only in a limited way what the Star Tracker will see and give therefore a bit too optimistic view. The night sky does not give access to the whole celestial sphere. Both tests show almost full coverage of the sky. It is expected that further enhancement can be done by changes to the catalogue and fine-tuning of the algorithm.
Correlation between the single star results of the laboratory and the Night Sky results has to rely on extrapolation of the laboratory results. It is shown that the total accuracy based on laboratory results is compliant with the requirements and this is confirmed by the Night Sky Test results.

QUALIFICATION CAMPAIGN

Recently (December, 1999), the NEMO Star Tracker went successfully through its qualification campaign. This section describes the various qualification tests performed on the Engineering Qualification Model (EQM) in order to qualify the Star Tracker for use on the NEMO spacecraft. Also, it is documented if performance related calibration parameters turned out to be sensitive to the vibration, the thermal vacuum and to the EMC testing.

Vibration Test

The Camera Head Unit has been designed to withstand the random vibration environment of 9.8 G rms shown in Table 5. The vibration test on the Engineering Qualification Model was performed with a 6 dB margin and twice the duration, namely 19.6 G and 2 minutes for each axis. The Star Tracker passed the test without any problems. Performance related calibration parameters were measured before and after vibration. A decrease of 12.4 μm was observed in the effective focal length of the lens.

The change in effective focal length is not alarming, but it would have a small impact on the instrument performance if left unaccounted for, as it would increase the systematic errors of the system. However the system features a means of performing in-flight calibration which eliminates this problem. Also it has to be pointed out that the shift in effective focal length occurred after vibration tests at very high levels of vibration actually 6 dBs above flight level.

Table 5
RANDOM VIBRATION ENVIRONMENT FOR CHU

Frequency (Hz)	G^2/Hz
20	0.01
80	0.2
200	0.2
2000	0.01

Thermal Vacuum Test

The NEMO Star Tracker has been designed to operate over a temperature range from -15°C to 60°C. However it is not required to deliver its full performance over that entire temperature range. To get the full performance the camera must be kept stable at a moderate temperature. On NEMO the camera head unit will be mounted on an optical pallet which will be stabilized at a temperature of 10°C. The Star Tracker will deliver good performance also at higher temperatures, but not all the way up to 60°C.

The thermal vacuum test consisted of 13 temperature cycles between 65°C and -25°C. At each temperature extreme a functional test was performed on the instrument. Performance related calibration parameters were measured before and after the thermal vacuum test. No significant change in parameters was observed.

EMC Test

The last major part of the qualification consisted of an EMC test. The EMC test included measurements of Conducted Emission, Conducted Susceptibility, Radiated Emission and Radiated Susceptibility. Also this test was passed successfully. For details on the EMC requirements see the NEMO Star Tracker Environmental Requirements (Ref. 5). Performance related calibration parameters were measured before and after the EMC test. As expected, no change in parameters was observed.

CONCLUSION

The test campaign consisting of the laboratory testing and the Night Sky Testing on Tenerife shows that the Star Tracker meets the design requirements. It also shows that the two different test setups (Lab and Night Sky) are adding value to each other as both have a limited scope and distinct characteristics.

The qualification campaign has shown that the NEMO Star Tracker fully meets not only the performance requirements but also its environmental design requirements.

ACKNOWLEDGMENTS

The Authors would like to thank Paul DeLaHunt and Marv Levenson at NRL for revising our work, Paul Setze and Lee Niemela at STDC for their support, and Dr. Miquel Serra Ricart and Mrs. Pilar Rapp at the Teide Observatory for their kind hospitality.

REFERENCES

1. L. Maresi and T.E. Paulsen, NEMO Star Tracker System Requirements Specification, Issue 1, Rev. D, TER/NEMO/SE-SRS.

2. L. Maresi, CRI-15AS Performance Analysis Report, Issue 1, Rev. B, CRI/15AS/PERF/1dB.

3. T.E. Paulsen, Performance Test Results on EDM#1, Handout NEMO Star Tracker Technical Interchange Meeting, Birkerød, June 15-16 , 1999.

4. T.E. Paulsen et al.: Night Sky Test Report at IAC of EDM#2, TER/NEMO/SE-TRP/001, October 5, 1999.

5. T.E. Paulsen, NEMO Star Tracker Environmental Requirements Specification, Issue 1, Rev. B, TER/NEMO/SE-ERS.

AAS 00-035

A COST EFFECTIVE, HIGH RELIABLE RWA SOLUTION TO A COMMERCIAL MARKET DEMAND

Terance Marshall, Mitchell Fletcher and Joseph Zuckerbrow[*]

A changing satellite market, with an ever-increasing commercial presence, has driven industry to respond with minimal cost and high-volume production. This emergence fuels an increasing pressure on space component engineers to meet the high-volume, low-cost, short-cycle demand without compromising quality or performance. Honeywell is creatively meeting current commercial Reaction Wheel Assembly (RWA) component challenges with an evolutionary RWA design, the HR14, as part of its Constellation Class family of RWAs. This class of RWAs advances the state-of-the-art for large-volume, low-cost, high-producibility designs while preserving legendary Honeywell value in performance and expected on-orbit longevity.

---

[*] Honeywell, Satellite Systems Operations, P.O. Box 52199, Phoenix, Arizona 85072-2199.
E-mail: terrence.marshall@honeywell.com; mitch.fletcher@honeywell.com; joe.zuckerbrow@honeywell.com.

INTRODUCTION

The successful advancement of technology for space applications, in general, demands careful execution of design principals that have been time proven with on-orbit success. A lesser approach ultimately realizes technical risk in meeting an application's performance targets, with a severe penalty for the program and its objectives. A changing satellite market, with an ever-increasing commercial presence, has driven industry to respond with minimal cost and high-volume production. This emergence fuels an increasing pressure on space component engineers to meet the high-volume, low-cost, short-cycle demand without compromising quality of performance. The door is open for dramatic and revolutionary technical responses that offer an apparently economical technical response, only to fall short in fulfilling a space program's objectives.

Honeywell recognizes due diligence in space application designs and is meeting current commercial Reaction Wheel Assembly (RWA) component challenges with an evolutionary (as opposed to revolutionary) design, the HR14 RWA, of the Constellation Class series, as illustrated in Figure 1. The HR14, a variable momentum (20 to 75 Nms), variable torque output (up to 0.2 Nm) RWA, reflects an advancement of high production RWA design, surpassing the design goals demonstrated by its predecessor, the HR0610 RWA.

Figure 1   HR14 Reaction Wheel Assembly

The HR0610 RWA design successfully supported a build, test, and delivery of 250+ wheels within a 20-month period and ahead of schedule for the Globalstar Satellite Communication System constellation. Like the HR0610, the new HR14 RWA advances the performance level for large-volume, low-cost, and high-producibility, while preserving heritage design principals to ensure legendary Honeywell value in performance and expected on-orbit longevity.

PRESERVING A SUCCESSFUL RECIPE

The commercial market provides the potential for compromising successful heritage design approaches in meeting the demand for cheaper and quicker to market products. However, products that are cheaper and quicker to market can be successfully achieved in design advancements through smart RWA design architecture involving producibility, modularity, fabrication techniques, process control, and forethought to maximizing design re-usage. This preserves the recipe of basic design principals for RWA elements involving successful application and design of mechanical, electrical, and bearing systems, and their components.

The HR14 RWA, illustrated in Figure 2, by design, advances producibility through integration of design function and a reduction of subassemblies and overall parts count. Modularity of design provides the capability for efficient parallel manufacturing of subassemblies, and variability in output performance sizing and electrical interface interchangeability. Implementation of standardized design modules promotes maximum design re-usage of key components and materials, manufacturing processes and Special Test Equipment (STE), and data management systems. Smart design architecture enables effective up or down class sizing of the HR14 to support wide momentum range applications and continuation for reuse of key design modules.

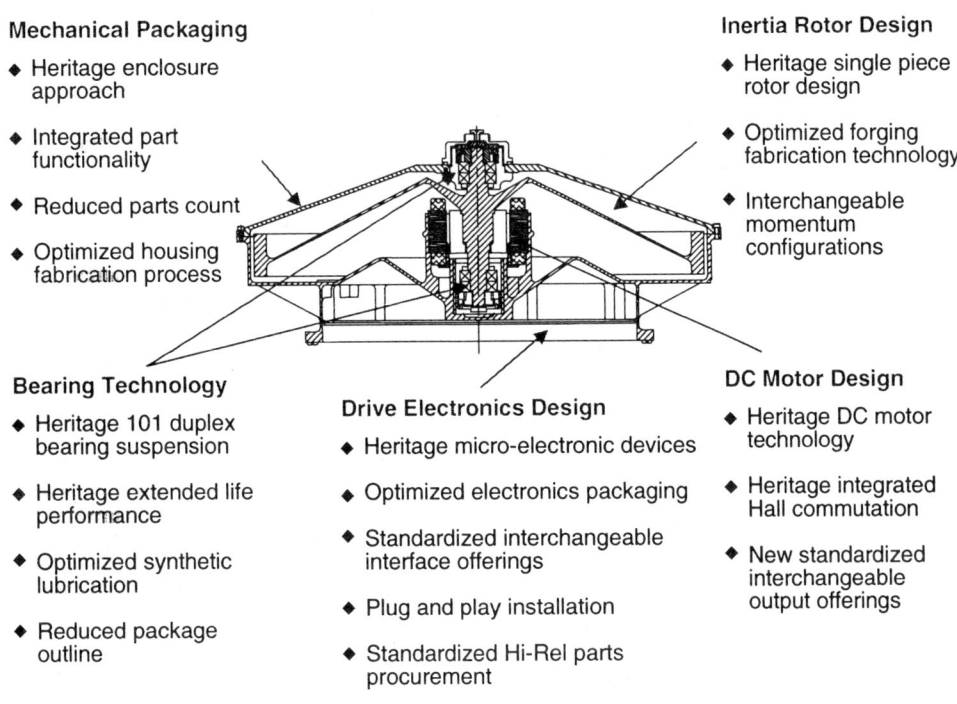

Mechanical Packaging
- Heritage enclosure approach
- Integrated part functionality
- Reduced parts count
- Optimized housing fabrication process

Inertia Rotor Design
- Heritage single piece rotor design
- Optimized forging fabrication technology
- Interchangeable momentum configurations

Bearing Technology
- Heritage 101 duplex bearing suspension
- Heritage extended life performance
- Optimized synthetic lubrication
- Reduced package outline

Drive Electronics Design
- Heritage micro-electronic devices
- Optimized electronics packaging
- Standardized interchangeable interface offerings
- Plug and play installation
- Standardized Hi-Rel parts procurement

DC Motor Design
- Heritage DC motor technology
- Heritage integrated Hall commutation
- New standardized interchangeable output offerings

Figure 2   HR14 RWA Basic Design Elements

RWA DESIGN ARCHITECTURE

As with all successful architectures, the Constellation Class Series, which encompasses the HR14 RWA, is based upon a strong system engineering approach. A responsive RWA design architecture was developed as a result of market surveys indicating that product performance and price are major drivers. Bus applications were found to drive RWA design variations primarily involving torque, momentum, and electrical-interface flexibility to accommodate satellite sizing and a large range of bus voltages architectures. These elements drove the Constellation Class Series to respond with a design architecture flexible in performance and interface that allows for interchangeable momentum (inertia rotor), torque (motor), and electrical interface (drive electronics.) The RWA architectural design response addressed variability by providing a standard mechanical platform and matching single-piece modules of interchangeable design options offering variability in configurations of rotors, motors, and drive electronics. This approach provided the means for an expedient technical solution secured in basic design principals for various spacecraft bus requirements, which maintain heritage on-orbit performance and reliability.

Mechanical Design

The HR14 RWA mechanical architecture, similar to its predecessor, the HR0610, consists of a dual-chamber construction. The top chamber provides an evacuated enclosure containing the inertia rotor, motor assembly, and bearing assemblies. The bottom enclosure provides an unevacuated, isolated chamber that contains the drive control electronics separated from the spinning precision mechanical assembly. The chamber locates the spacecraft interface for electrical connections and physical mounting, as well as providing for direct electronics access without disrupting the mechanical spin assembly located above. Advances in producibility are addressed by this unique approach in mechanical architecture combined with a synergistic relationship involving design modularity, integrated part functionality, component fabrication techniques, and standardization of mechanical component interfaces.

The dual-enclosure architecture supports parallel manufacturing paths for both the mechanical and electrical subassemblies. This allows independent manufacturing paths without procurement or assembly schedule dependencies to support unconstrained, production-inventory accumulation. Advances in producibility were realized in the manufacturing of major design elements, housing, and inertia rotor. Housing elements consist of a top cover and bottom chassis. The housing design architecture was revised to locate the complexity of manufacture within the bottom chassis. Typically machined from aluminum-alloy raw material blank, the bottom chassis fabrication was revised to an aluminum-alloy casting; thereby, reducing extensive machining and leaving only precision-machined, interface surfaces. Therefore, the simplified aluminum-alloy top cover was relieved to support more economical manufacturing techniques. The inertia rotor, typically manufactured from a raw, stainless, forged-material blank, was revised to a near net forging. The near net forging process allowed significant material savings, and provided improved distribution and control of material properties.

Design modularity is supported through design architecture that allows performance sizing for both the inertia rotor and motor, as illustrated in Figure 3. The inertia rotor design allows variability in the outboard rim element. This simple approach, made possible by single-piece rotor construction, allows variability in rotor configurations that support momentum applications for the HR14 from 20 to 75 Nms, all interchangeable within the same housing design. The bottom chassis and rotor designs were configured to accept variability in motor sizing. Both the chassis and rotor provide a common mounting interface that supports interchangeable, low-to-high torque motor configurations.

Standardization was re-addressed through mechanical architecture. The housing design is configured as a common baseline mounting interface to support both electrical and mechanical subassembly configurations, now and in the foreseeable future. Similar to the housing, the inertia rotor was designed for a common mounting interface within the chassis, as well as common rotor cross-section configuration designed to support variability in its inertia element sizing previously mentioned. These architectural approaches implemented standardization throughout the component life cycle and support design reuse, minimizing repetition in analysis modeling, fabrication fixturization, procurement, subassembly, and end-item, STE.

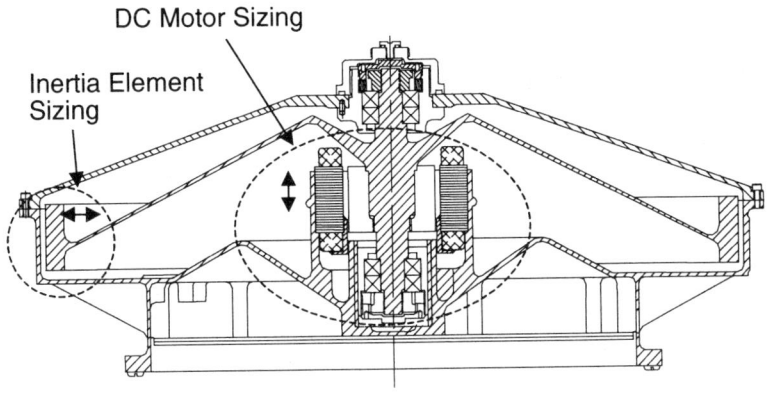

Figure 3   Variable Mechanical Sizing

Bearing System

The HR14 RWA design has implemented Honeywell's heritage bearing design and suspension architecture, previously proven successful, in more than 500 RWAs produced to date. The design principals involved support Honeywell's unblemished record of never failing to meet a program's mission objective.

Though the housing/bearing suspension architecture at first appears dramatically different from previous Honeywell designs, closer inspection reveals that the basic design principals remain intact. The duplex ball bearing system's detailed component designs maintain previous heritage geometric precision control. The bearing suspension system remains consistent in design principals with its heritage fixed and floating cartridge suspension design, which uniquely provides insensitivity to housing deflections imposed by launch and operating temperature environments. The difference, however, rests in two significant advancements, as follows:

1) RWA bearing system lubrication
2) Application of statistical process control in aspects of the bearing manufacturing process

Bearing Lubrication

Honeywell has achieved noteworthy success in the industry concerning RWA on-orbit performance longevity. As important as it is to maintain critical principals of bearing geometrics and suspension design principals, so is the selection of the lubrication system.

In addition to RWAs, Honeywell is a recognized leader in Control Momentum Gyroscope (CMG) technology. CMG technology applies RWA principals supplemented by attaching a second and/or third axis of control via gimbal joint mechanisms and torque motor systems. This allows CMG systems to provide maneuverable satellite performance. Therefore, a CMG system presents more aggravated operating loads upon the momentum spin bearings resulting from gyroscopic force couples or moments. This cyclic-operating load condition, which surpasses typical RWA on orbit bearing loads, demands a robust lubrication system to ensure extended life capability.

The heritage system previously used in both RWAs and CMGs, though extremely successful, required extensive attention to operating temperature due to the highly evaporative characteristics of this mineral, oil-based system. This, in turn, required supplemental bearing lubrication reservoir systems to support the losses realized during life operation. The reservoir systems drove complexity in design, as well as the need for additional parts and space allocation.

Stringent CMG operating conditions required development of a unique formulation of synthetic oil and grease lubrication. This lubrication system has since proven, through extensive characteristic evaluations and CMG life testing, to extend operating life three to five times longer than that of its heritage predecessor.

The evaporative lubrication-loss concerns are now mitigated by this synthetic counterpart, which provides more than two magnitudes of reduced evaporative loss performance. Adaptation of this proven lubrication system from the CMG into the Constellation Class RWA product supported dispensing of the lubrication reservoir systems, a significant contribution in RWA simplification and downsizing, as illustrated in Figure 4. Savings were realized in overall mass, parts count, and supplemental manufacturing processes.

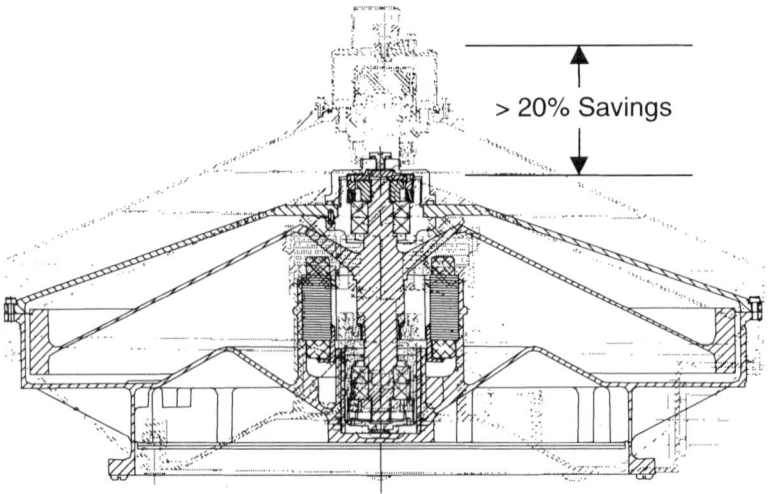

Figure 4   RWA Size Reduction

Bearing Manufacturing Processes

Additional savings opportunities were realized in the refinement of certain bearing manufacturing processes. These highly involved and detailed processes drove extensive component performance, data-recording practices. The substantial quantity of collected data was statistically analyzed and revealed certain controlled precision in performance, indicating good process repeatability. This insight allowed smart reduction in laborious measurement practices to be replaced by random process monitoring that ensured ongoing process control. This process monitoring approach resulted in notable cost-reduction improvement, while maintaining heritage precision and performance capability.

Electrical Design

Similar to the mechanical architecture, the HR14 electrical architecture is based upon heritage design concepts. The implementation was simplified and modularized to allow customization with low, nonrecurring effort. The HR14 electronics design is implemented via current loop control. A torque command is converted into a current command and presented to the forward control path. This current command is converted to average current, through pulse-width modulation, and presented to the motor.

The motor current is continuously sampled and used as the feedback to the control loop. Other elements of the electronics include the Input/Output (I/O) interface, Electromagnetic Interference (EMI) filter, and secondary power supply.

In order to architect a low-cost electronics implementation, the basic loop function was studied and categorized into functional blocks. The optimized architecture was established for each of the blocks along with a target hardware implementation. Completed trade studies, at both the block and system level, resulted in the control architecture. (Refer to Figure 5.) The physical architecture is implemented using three custom hybrids, one gate array, optional interface drive components, and an assortment of standard resistors and capacitors.

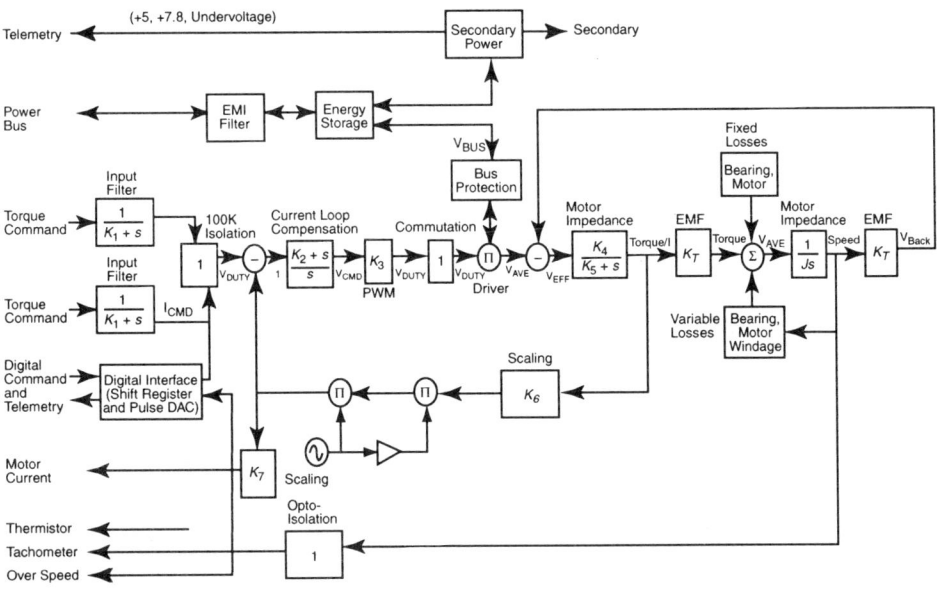

Figure 5   HR14 Control Block Diagram

Trade studies completed within the system flowdown revealed that a hybrid electronics implementation costs approximately one-half of an equivalent discrete implementation. This is primarily driven by the procurement costs of high-reliability parts, their lot charges, and procurement labor for the many components required in a discrete design. A microelectronics design based upon hybrid and gate array implementation requires far fewer parts than a discrete design. The discrete RWA electronics designs completed in the early 1970s required seven Printed Wiring Boards (PWB) populated with 850 parts (of 250 part types.) The first introduction of microelectronics reduced the design to 2 PWBs and 300 components (97 types.) The HR14 takes this reduction one step further by utilizing only 1 PWB populated with 250 components (55 types.) The HR14 electronics layout is illustrated in Figure 6. In addition to parts reductions, the HR14 utilizes smart functional partitioning, which requires reduced nonrecurring effort to implement customer requested scaling changes.

The smart partitioning of the HR14 results in the flexible functional blocks, as follows:

- EMI filter
- Secondary power system
- Control system
- Motor drive
- Bus protection
- Command interface

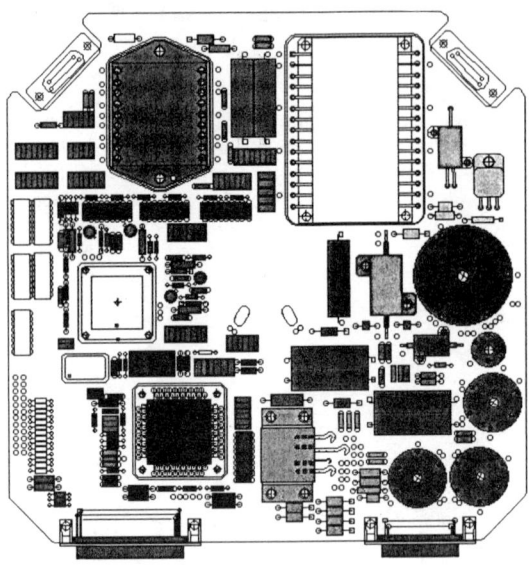

Figure 6   HR14 Electronics Layout

Each block is grouped in an area on the single board PWB that allows a minimum amount of unwanted interaction between functional blocks. This physical implementation prevents reconfiguration in one area from affecting another functional block. Figure 6 and Figure 7 illustrate the interrelationship of the functional blocks.

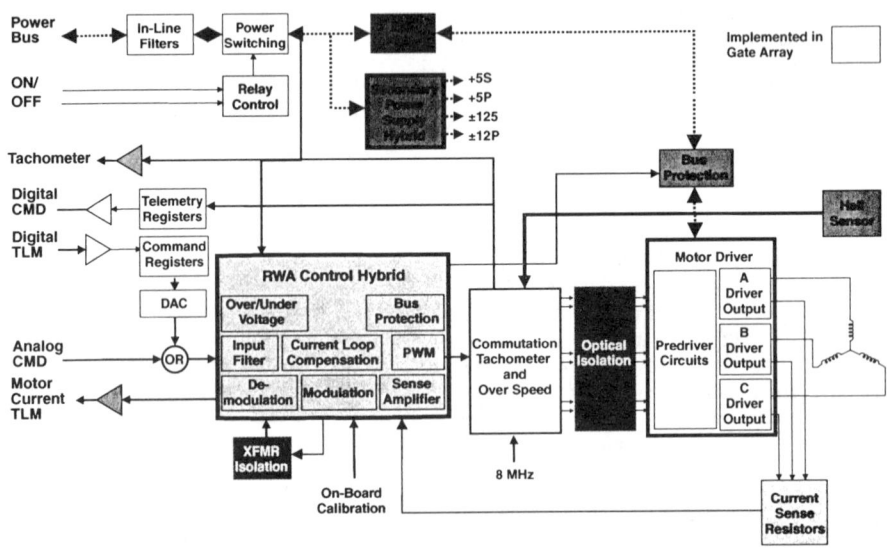

Figure 7   HR14 Electronics Physical Block Diagram

The EMI architecture is based upon a three-stage LC filter. It is designed so that a reduction in overall RWA power results in a reduction in the number of capacitors required to support electronics noise rejection to the spacecraft. The secondary power system is based around a DC-to-DC converter that supports all potential bus voltages and eliminates the need for reconfiguration.

The control law implementation is completed utilizing a custom hybrid device and a gate array. These devices were designed to be flexible with respect to I/O and scaling to allow ease of reconfiguration. The control system provides for simple reconfiguration of the motor torque and voltage range. The hybrid and gate array combination supports both unipolar and bipolar analog commands, in addition to a 32-bit digital interface. Both digital and analog telemetry is provided. Motor current is supplied to the motor through a custom driver hybrid capable of switching up to 10 A.

Custom hybrids were used due to reduced procurement cost, compared to Off-The-Shelf (OTS) common devices. Equally important, they meet the rigorous analysis and radiation requirements needed for space flight applications. The size reduction afforded by the microelectronics implementation allowed for integration into a single PWB. The PWB design incorporates power and signal spacecraft interface connectors as an integral part of the board. This combination eliminates internal harnessing and provides for interchangeable electronics assembly configurations that interface directly with the mechanical assembly, supporting plug and play operation.

These concepts provide for a low-cost, single-board electronics that is flexible in the ability to migrate from one design to another. The low number of components results in both lower procurement costs and high reliability implementation. The simplicity of implementation promotes reduced assembly time at board level and end-item integration, and standardization in tooling and test support.

MAXIMIZING DESIGN REUSE

The Constellation Class, HR14 RWA design was configured with forethought in satellite application sizing, which typically drives RWA momentum sizing. Needed was an RWA platform that could simply provide momentum variability, while avoiding the nonrecurring expense of complex redesign and verification. The HR14 architecture was, therefore, configured to provide independence from momentum sizing. Figure 8 illustrates the HR14 RWA with momentum class sizing options, the HR12, and HR16. Each class represents an optimized momentum to mass configuration.

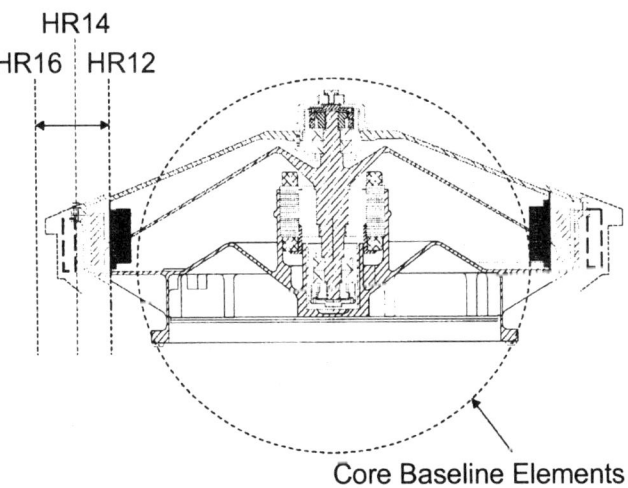

Figure 8   Constellation Series RWA Momentum Sizing

Core design elements involving bearing suspension, motor interface, and drive electronics and its packaging are preserved. Affected design elements were restricted to the case and cover housing components. Within these elements, the form factor and interface approach was maintained. Core baseline design elements and their complexity are preserved. Drive electronics remains interchangeable with each class of RWA. Design reusage was maximized throughout, allowing opportunity for economical high quantity purchasing of the core elements and significantly reducing the potential for nonrecurring design activity, given new satellite applications. Figures 9 and 10 present the resulting individual momentum range applications and a performance overview for the three classes of RWAs: the HR12, HR14, and HR16.

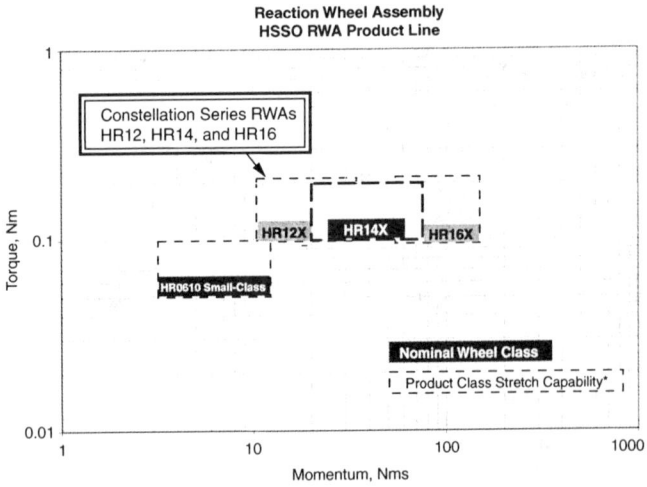

Figure 9   Constellation Series Class RWA Momentum Ranges

Constellation Class Series Reaction/Momentum Wheel Performance Overview				
Performance Item	Units	Capabilities		
Product Number		HR12X	HR14X	HR16X
H, momentum @ 6000 rpm, Mass optimized	Nms	12 to 50 (25)	20 to 75 (50)	75 to 150 (100)
Reaction torque @ 6000 rpm	Nm	Up to 0.2		
Power consumption				
Peak torque @ 3000, 6000 rpm:	W	< 105, 170 @ 0.2 Nm (50 V Bus)		
Steady state @ 3000, 6000 rpm:	W	< 15, 22		
Quiescent	W	< 7		
Bus voltage	Vdc	23 to 57		
Interface	–	Analog or Serial Digital		
Wheel speed range	rpm	±6000		
Mass, (H optimized), bus dependent	kg	6 to 9.5 (7.0)	7 to 10.6 (8.5)	10.6 to 14 (12)
Life requirement:				
On ground and Storage	yrs	> 5		
On-orbit	yrs	> 15		
Integrated electronics	–	Yes		
Radiation hardness (Si)	kRad	> 100		
Motor type	–	DC Brushless		

Figure 10   Constellation Class RWA Performance Overview

PRODUCTIZATION

As detailed as the development of a new product offering can be, just as detailed is the establishment of a means for manufacturing and testing that product. Productization is the disciplined process of concurrent engineering coordination between design engineering and the manufacturing and test groups. It develops the required product's producibility; defines controlled processes involving unit fabrication, assembly, test, and delivery; and appropriately sizes factory capacity, ultimately minimizing Unit Production Cost (UPC).

Productization for the Constellation Series of RWAs was an evolutionary process. Based on lessons learned from our recent successful high-volume, commercial programs, productization execution was further optimized beyond standard concurrent engineering efforts by developing a product towards specific design to cost UPC goals. Extensive cost analyses benchmarks were established for current products resulting in major areas of design improvements to reduce material, build, and test costs. Figure 11 illustrates a producibility summary indicating marked overall improvements in factory build, integration, test, and support labor for the Constellation Series of RWAs vs previous products.

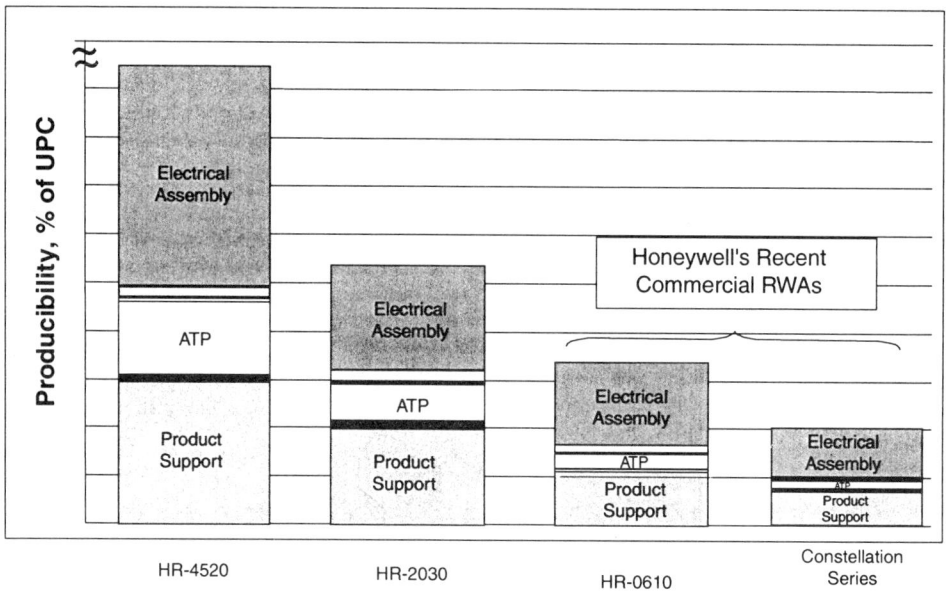

Figure 11 Producibility Comparisons. Using proven techniques for concurrent engineering, Honeywell has significantly improved RWA producibility for the Constellation Series.

Produciblity

Producibility was accomplished by a team consisting of experienced assembly operators, technicians, process/product engineers, design engineers, and design layout professionals. The team conducted producibility efforts in coordinated timing with respect to product development efforts as presented in Figure 12. This discipline consisted of building numerous engineering and production prototype hardware, and conducting manufacturing process evaluations resulting in a product design that meets rigorous spacecraft hardware quality requirements and pre-established UPC goals.

ACTIVITY	PROGRAM START	PDR	CDR	START QUAL BUILD	START PROD BUILD
PRRODUCIBILITY PLAN	PHASE 1				
PRELIMINARY PRODUCIBILITY REVIEW		PHASE 2			
CRITICAL PRODUCIBILITY REVIEW			PHASE 3		
MANUFACTURING REVIEW				PHASE 4	
FINAL PRODUCIBILITY REVIEW					PHASE 5
PRODUCTION SUPPORT CONTINUOUS IMPROVEMENT					PHASE 6

Figure 12 Producibility and Product Design Coordination

Manufacturing Capacity

Honeywell has made significant advancements in manufacturing techniques for commercial RWAs. By developing automated capability for production, inspection, and test processes, dramatic increases in manufacturing capacity (Figure 13) have been realized that are ready to support the commercial reaction wheel market. This capacity expansion, which features built-in quality inspection gates, supports Honeywell's long term goal of maintaining its heritage reputation for delivering spacecraft hardware that has never caused a mission to end prematurely due to hardware anomalies.

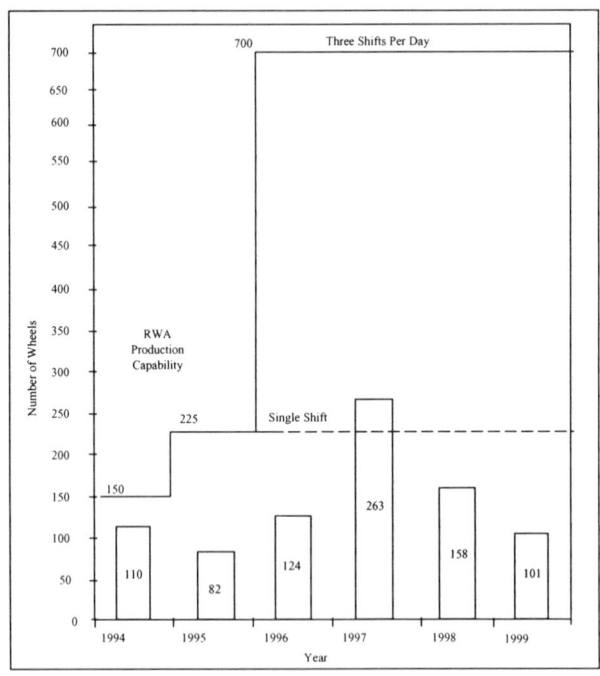

Figure 13 Honeywell Commercial RWA Manufacturing Capacity

High Production Performance Measurement and Test Data Handling Capability

When considering commercial approaches in product verification, it is often the temptation to reduce cost by introducing shortcuts involving product verification. Honeywell's experience, however, has demonstrated that the opposite is the preferred approach. This approach entails development of highly controlled tools and processes which increases a consistent handling and therefore knowledge of the product performance, thus providing early detection of issues preventing logarithmic cost increases for problem resolution later in the life cycle. Cost savings were realized through the result of investment in automation, thereby providing reduced cost of execution and early problem recognition.

As part of the RWA productization effort, automatic performance testing and data collection capabilities are installed to enable operator management of a high-rate of production units using minimal support personnel. Figure 14 illustrates a simplified schematic of the basic automatic testing architecture involved.

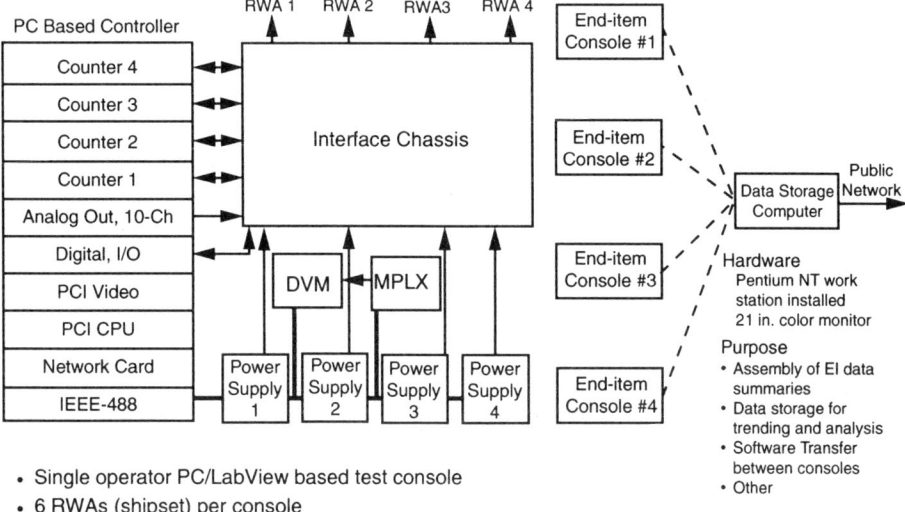

- Single operator PC/LabView based test console
- 6 RWAs (shipset) per console
- Fully automatic execution and data generation
- Automatic unit evaluation vs "family"
- Data stored, networked, and trended

Figure 14 Automatic End-item Testing. Automatic Testing Allowing Detailed Unit Performance Awareness in a Mass Production Environment

The developed data management tools allow the operator direct graphical insight to multi-wheel performances during test with available direct on-line reference, enabling the ability to compare a unit's performance against an extensive collected family of data for that product type's performance parameters. As part of this data management system, graphically supported automated data summaries are generated allowing direct electronic data package submittal. In addition, the data management system provides statistical analysis and performance trend awareness to monitor unit performance throughout the production phase and throughout a product's history, as illustrated in Figure 15. This capability provides added assurance that the performance of all delivered wheel products not only meets all specification requirements, but also maintains a consistent level of expected performance margin.

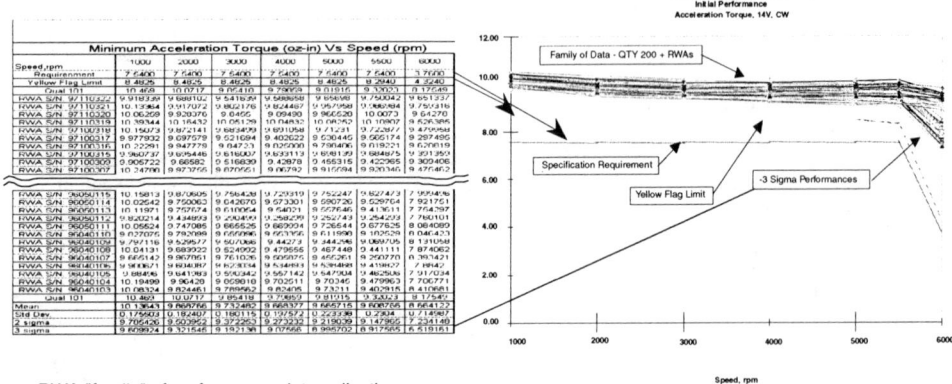

- RWA "family" of performance data collection
- Data automatically amassed and and reviewed
- Operator utilized during RWA unit end-item processing
- Statistically analyzed and graphically presented

Figure 15 Automatic Data Trend Analysis. Automatic trend analysis provides assurance of consistent product performance

SUMMARY

The Constellation Class RWA has introduced an evolutionary, rather than a revolutionary, solution to the high production, minimal cost, commercial market. Through innovative RWA design involving architecture, modularity, interchangeability, maximum design reuse and manufacturing productization, Honeywell's Constellation Class RWA provides an effective, high-quality method of meeting the market demand. By maintaining proven heritage design principals, the Constellation Class RWA supports the customer's expectation and investment for continued and legendary on-orbit performance in total support of the program's mission objective.

AAS 00-036

REDUNDANT LAUNCH VEHICLE GUIDANCE

R. Joe Wright[*]

Honeywell Space Systems Division is developing the next generation Redundant Launch Vehicle Guidance Unit to serve as replacement for existing non-redundant systems. The RLVG features a Honeywell programmed inertial processor providing sensor compensation and fault tolerance and a user programmed flight computer providing vehicle control over standard 1553 interfaces. Engineering Development Units are available for software development. Development Evaluation Units are in fabrication to support an extensive development test program that ensures an expedient and successful program qualification. A systems overview of RLVG, detailed description of the fault tolerant architecture, development unit descriptions, and details of the development test program are included.

INTRODUCTION

Honeywell Space Systems is developing a launch vehicle inertial navigation system that provides fault tolerance in all functional areas for high probability of mission success and high device availability.

The RLVG is designed as a primary avionics component for a launch vehicle guidance, navigation, and control subsystem (GN&C). It performs the inertial measurement functions, as well as provides the computer processing capability for ground and flight software and all the input/output interfaces necessary for the ground and flight software. The RLVG, with its inertial measurement flight software and a customer's mission flight control software, performs incremental angle and velocity measurements, inertial navigation, guidance, flight control, vehicle management, and redundancy management of the avionics subsystem. An exploded view is shown in Figure 1.

The RLVG consists of two principal functional subsystems, a strapped-down inertial measurement subsystem (IMS), and a flight control subsystem (FCS). Figure 2 shows

---

[*] Honewell, Inc., 13350 U.S. Highway 19 North, Clearwater, Florida 33764-7290.

the block diagram of the RLVG. The design process for the RLVG emphasizes the importance of providing single fault tolerance for all functions, and extreme care is taken to ensure that the internal redundancy and the ability to operate following any single failure is never compromised, nor is performance degraded.

RLVG PERFORMANCE BENEFITS

The RLVG implements a strapped-down sensor block accommodating a pentad of inertial sensors: five Honeywell GG1320AF ring laser gyroscopes for angular rate sensing and five Honeywell ISO-COIL2000 pendulous gas filled linear accelerometers for acceleration sensing. Their mounting geometry, shown in Figure 3, is a cone that is arranged symmetrically. The inertial sensors require yearly calibrations to maintain stated performance. No launch day calibrations or optical alignments are required, reducing life cycle costs and launch processing timelines. The GG1320AF and ISO-COIL2000 sensors are high performance, space-borne inertial sensing devices

Figure 1. RLVG Exploded View.

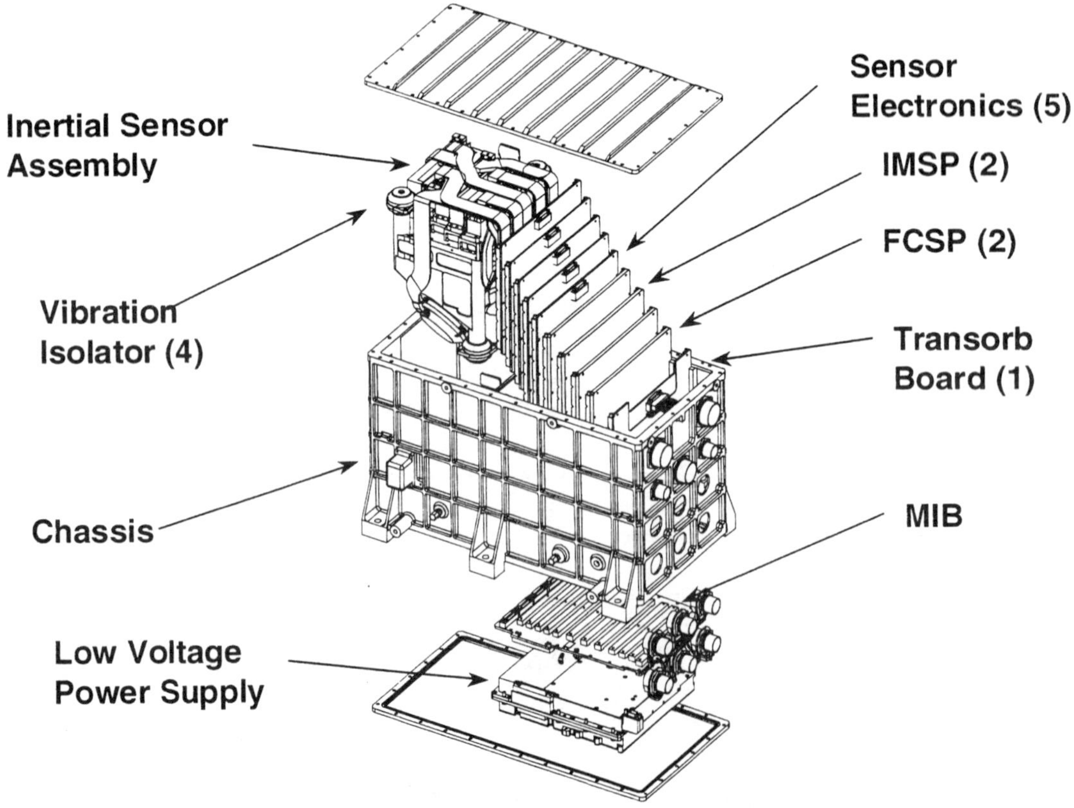

Figure 2. RLVG System Block Diagram.

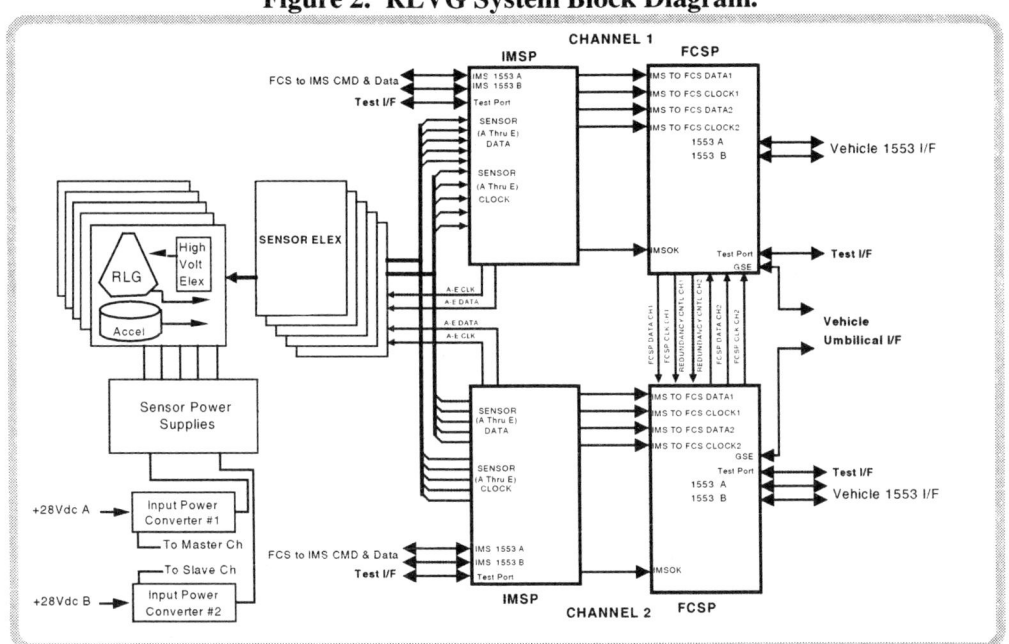

RLVG HERITAGE

The Honeywell RLVG makes use of circuits, software and design techniques developed on high accuracy, high reliable space programs such as: Transfer Orbit Stage, Atlas Centaur, Titan IV, Inertial Upper Stage, Honeywell Space Computer (HSC), Advanced Space-borne Computer Module, and the International Space Station.

Both the IMS and FCS subsystems are controlled by the GVSC processor, a MIL-STD-1750A compliant, radiation hardened chipset. This chipset, developed and fabricated by Honeywell, has been the subject of extensive development testing, characterization, and verification testing. The test results provide great confidence in its ability to meet the throughput and radiation tolerance requirements of the RLVG program. This processor has flown on a number of US military and commercial spacecraft.

The sensor data collection and signal processing circuits make considerable use of existing designs that have been proven on programs mentioned above. These circuits have been extensively characterized over the full operating temperature range as well as under severe physical environments including vibration, shock and acceleration. These characterizations give further confidence in the circuits ability to meet the functional and physical requirements of the RLVG applications and the monitoring capability inherent in the circuits that supports the fault detection and isolation function.

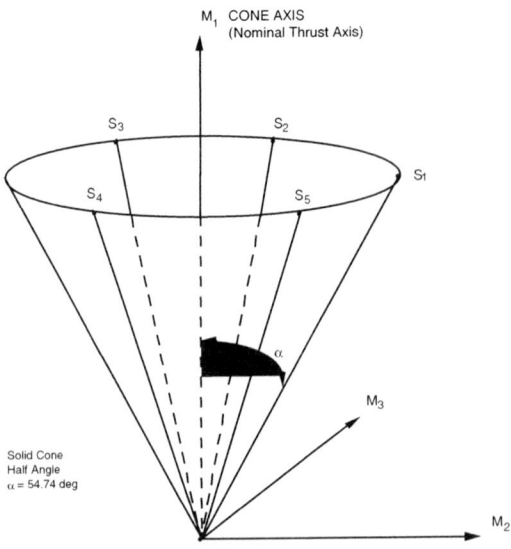

Figure 3. RLVG Sensor Configuration.

The inertial sensors used in the RLVG include the Honeywell developed GG1320 ring laser gyroscope and the Honeywell ISO-COIL2000 Q-Flex Accelerometer. Each of these sensors is in full production on lines certified to provide military quality sensors.

The Honeywell Q-Flex family of accelerometers is used in a large number of commercial and military aircraft and in space launch applications. Honeywell Space Systems has extensive experience using this accelerometer in a variety of applications. The accelerometers provide the desired performance, survivability, and reliability required for space environments.

The GG1320 RLG continues its role as the sensor of choice for a number of applications primarily in the commercial air transport market. The performance requirements for the RLVG inertial navigation system are such that a large portion of the production population of the GG1320 RLG and ISO-COIL2000 accelerometer will meet the desired performance requirements.

Many of the mechanical design and packaging techniques developed on other space programs are used on the RLVG. Techniques developed to withstand high launch vehicle vibration and shock include sensor block isolation, internal molded cables, robust chassis designs, and plug in printed wiring boards. Proven reliable chassis sealing techniques are also used although the RLVG is being designed to withstand a seal leak to ensure no single point failures. Previously developed and utilized surface mount technology is also incorporated to ensure a producible RLVG package.

RLVG SYSTEM ARCHITECTURE

The two RLVG principal subsystems - the Inertial Measurement Subsystem (IMS) and the Flight Control Subsystem (FCS) - are further partitioned into fault containment regions for single fault tolerance.

The IMS consists of:

- 5 sets of sensors and sensor electronics
- 2 IMS Self Checking Pair (SCP) processors
- 2 sets of IMS I/O
- 2 input power converters
- Sensor power supply (provides redundant power to 5 sensors)

The five sets of sensor/sensor electronics are independent from one another to form separate fault containment regions. Each sensor set is powered by independent channels from the sensor power supply. A single fault can disable only one sensor set, gyro and accelerometer, and is detectable by the Fault Detection and Isolation (FDI) algorithms for each IMS processor. Inertial data is distributed to both of the redundant IMS processors over serial data buses. Each processor monitors its own health by operating in a self checking pair configuration. Input power is also redundant, with separate internal power converters supplying power to the redundant IMS processors. Both internal converters interface to the sensor power supply where the five fault tolerant sensor power channels are generated.

The FCS consists of:

- 2 FCS SCP processors
- 2 sets of FCS I/O.

The two FCS processors are identical to the IMS processors and contain MIL-STD-1750A/B computers, Error Detection and Correction (EDAC) protected RAM, and SCP processors. The redundant FCS processors are set up as master/slave channels, each powered by an independent bus. A communication port between the master and slave processors allows information exchange. Each processor has its own FCS I/O distributed in the same independently powered master/slave configuration. The FCS I/O uses MIL-STD-1553B redundant ports with the capability to interface from one through five independent 1553 circuits to each FCS processor. System interfaces are depicted in the system block diagram, shown previously in Figure 2.

RLVG FAULT TOLERANCE DESIGN

The RLVG is a single fault tolerant component and allows continued operation after the first failure without any external interaction. This is achieved by dividing the RLVG design into separate Fault Containment Regions (FCR). If any portion of a FCR fails or malfunctions, the RLVG must continue to operate using the remainder FCRs. Each region has self-contained redundant managers and independent mechanisms for fault detection and isolation. These faults, such as short circuits, cannot propagate into other fault containment regions. The careful selection, isolation, and fault detection capabilities of each region is critical to the RLVG fault tolerance design.

The RLVG design consists of seven primary fault containment regions, five inertial sensor channels and two processing regions. The power system has two inputs designed for two independent 28V sources on the launch vehicle. This bus powers seven independent sets of supply voltages, one for each FDR.

The five sensor FCRs each contain a GG1320 gyroscope, a QA2000 accelerometer, a high volt power supply, and the sensor electronic required to operate and collect data for use by the Inertial Measurement Subsystem. The interface to the IMS is opto-isolated to prevent failures from propagating into the processing FCR.

The launch vehicle is controlled by one of two redundant processing fault containment regions. The two processing regions each contain one IMS processor and one FCS processor. The FCS contains all of the vehicle interfaces. If a failure is detected in one of these regions, vehicle control is diverted to the operational redundant region.

The IMS processor has two main fault detection functions, ensuring the inertial sensors are working correctly and ensuring the IMS data being sent to the FCS has been processed and transferred correctly. The sensor data is protected by a set of gyro and accel parity equations. These equations evaluate the over deterministic data from the five sensors and decide whether all five sensors agree with each other within an acceptable tolerance. If one gyro or accel does not agree well with the data from the other four sensors, the sensor is removed from the solution transferred to the flight control subsystem. The IMS also self evaluates the data it processes by comparing it to a redundant solution. This solution is calculated by a self-checking pair processor that resides on the same board. Any miss-compare is reported to the FCS so that actions can be taken to change over to the backup processing channel.

The FCS processor also has a self-checking pair processor set and can self evaluate its health. Once a failure is detected by the IMS or the FCS self-checking processors, a command is sent to the redundant processor fault containment region and control is transferred. This FCS to FCS interface is opto-isolated and is fault tolerant to ensure required change of control through any failure mode.

RLVG Development Units

The RLVG design will be validated and qualified through the use of development hardware. Qualification and early flight unit deliveries can be greatly expedited by a well-planned development test program. The RLVG evaluation program utilizes Engineering Development Units (EDU), Development and Evaluation Test (DET) units, and Qualification units to validated designs and processes early in development.

The EDUs are presently available for customer use. Primarily used for software development, each EDU contains two IMS processors and two FCS processors installed in a chassis. Each chassis contains a commercial low voltage power supply, a master interconnect board, and input/output connectors. The EDU also contains a software Pulse Accumulator Module (PAM) board which, used along with a PC, can simulate the five channels of sensor data. The EDU contains the vehicle interfaces and the FCS-to-FCS communication so that vehicle simulation can be performed under failed conditions.

The first DET unit is scheduled to complete fabrication in fourth quarter 2000. The DET units are the same as flight except for parts quality. The DET parts will utilize the same technology and have the same physical requirements, although screening and pedigree requirements are reduced to accelerate delivery. Each DET is fully functional and meets the performance requirements of the flight units. The performance of the DET units will be evaluated over all environments to validate the RLVG design.

RLVG Development Test Program

To ensure a smooth qualification and first flight delivery, an extensive development test program is planned through the year 2001. Early development tests exercise the IMS flight software and the FCS test software, as well as the test set software developed to collect and analyze RLVG data. Subassembly tests, such as EMI/EMC tests on the low voltage power supply, are designed to surface issues early so that any modifications can be made easily. Development tests performed on the DET units are planned for all environments.

Development testing using the EDU is underway. Inertial Measurement Subsystem software modules are presently being tested. These modules of the flight IMS software will be combined and validated once all modules are complete. Test software that resides in the FCS processor transfers data from the IMS to the vehicle interfaces. Test interfaces are connected to the vehicle busses so that RLVG data can be collected and analyzed. The EDU is used to develop and test the test set software, as well. Once the IMS flight software, the FCS test software, and the test set software are functioning, an end to end processing test is performed. Inertial simulation data is fed to the IMS processor where it is processed and transferred through the FCS to the test set. This way, all software functionality can be tested before a DET unit is available.

Once the first DET unit is fabricated, the initial calibration is performed. This calibration assures all performance parameters are within specification. Once this calibration is successful, the calibration will be repeated over temperature cycles. This full-temperature calibration ensures inertial sensor performance over the complete temperature range. The RLVG DET then performs a navigation test during temperature cycles to demonstrate both sensor performance and proper system compensation. The calibration and navigation tests are periodically performed throughout the test program. This verifies that performance integrity is maintained as the DET is exposed to environmental extremes.

Before the DET is exposed to qualification level environments, special instrumentation is applied to the unit to monitor device sensitivities to environments. Special thermal tests are performed using thermal couples mounted throughout the unit. Special care is taken to select "hot spots", large temperature gradients, and thermal characteristics of components that effect performance. This specially monitored unit is then exposed to qualification level thermal cycles so that thermal analyses can be verified. A similar test is performed using small accelerometers mounted on subassemblies. These accels monitor resonant frequencies and G levels to verify the structural analyses. Once these analyses are verified, the DET continues to be exposed to the qualification extremes.

The RLVG DET will be exposed to qualification level thermal cycles, vibration, pyrotechnic shock, vacuum tests, linear acceleration, EMI/EMC, and any other environment expected in the qualification and acceptance test programs. In some cases, qualification levels will be exceeded and/or combined to demonstrate additional margin. Once confidence in DET performance to qualification environments is achieved, the qualification program will begin. The qualification unit is exposed first to an acceptance test, then the rigors of the qualification test. The RLVG design becomes flightworthy at the completion of qualification.

AAS 00-037

RADIATION HARDENED POWER PC 603e™ SPACE PROCESSOR

Robert D. Campbell, Richard F. Elmhurst and Gary R. Brown[*]

Honeywell's next generation processor is radiation hardened to withstand the destructive effects of space and is compatible with Motorola's PowerPC 603e™ implementation. This allows 100 percent software compatibility with the commercially available PowerPC 603e™ processor. The RHPPC is expected to perform at 210 MIPS which is ten times the performance of currently available space processors while consuming less power (2.6W typ). RHPPC is designed for military, civil and commercial spacecraft and payload processing applications. These characteristics make it possible to build more capable satellites, while reducing the mass of solar cells, batteries and the total spacecraft.

Honeywell's Silicon On Insulation (SOI) technology provides radiation immunity and high performance, while requiring less power.

---

[*] Honeywell Space Systems, 13350 U.S. Highway 19 North, Clearwater, Florida 33764-7290.

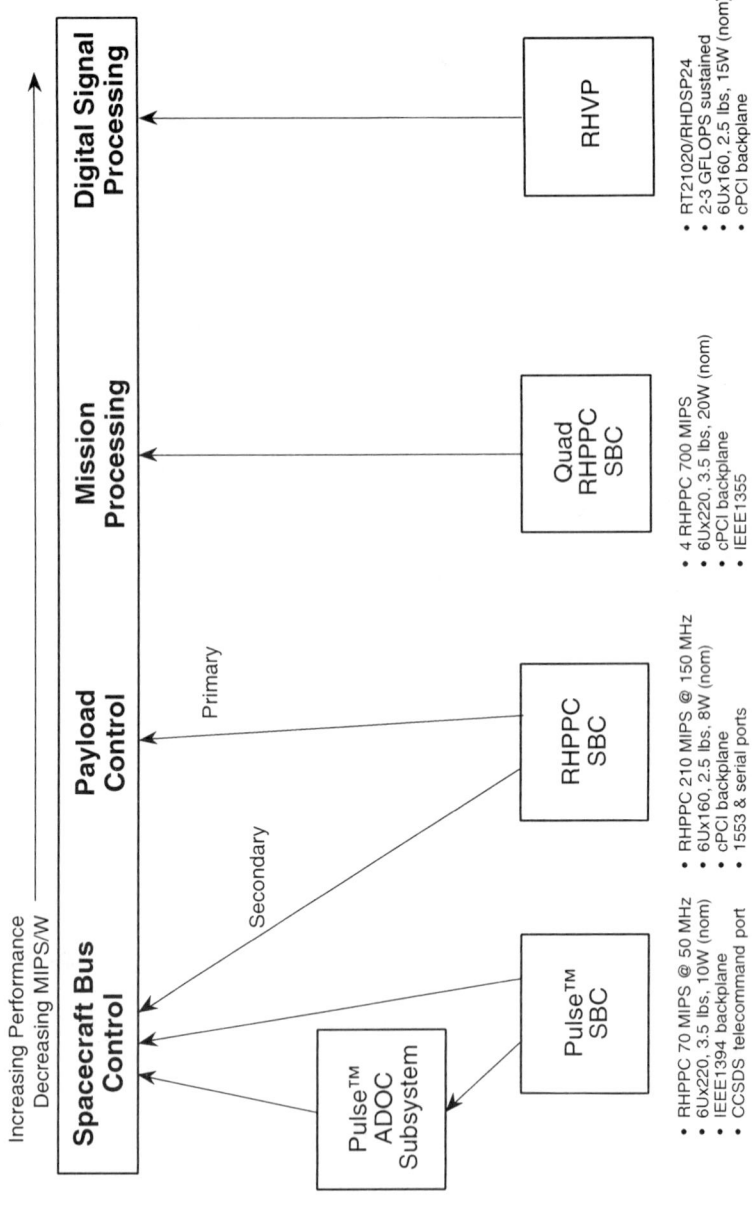

Notional Architecture for Payloads

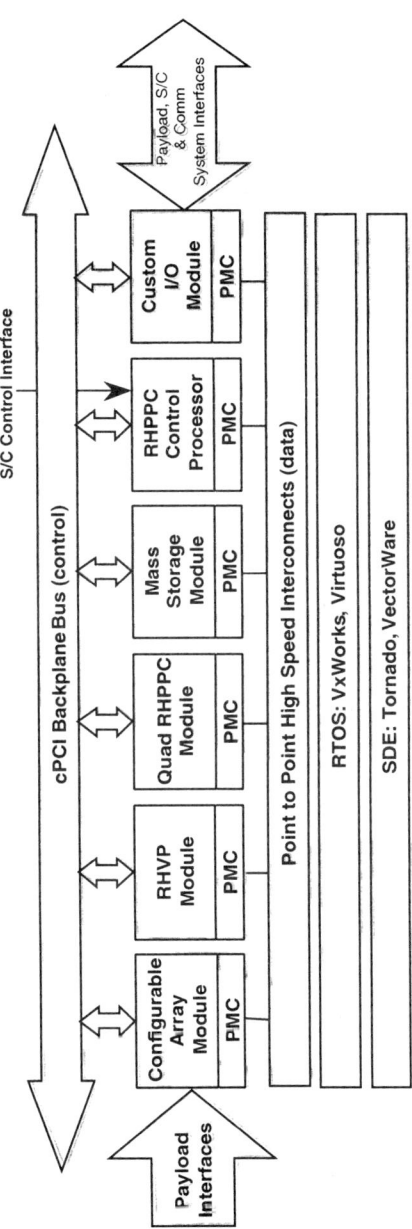

- Highly flexible, scaleable, open architecture based on COTS based engines (instruction set), software tools, RTOS, backplanes, I/O, and form factors
  - Engines: PowerPC603e™ general purpose, ADSP-21020 DSP, and DSP-24/MMU-24 Vector Processor
  - Software Tools: Wind River Tornado™ 2.0, VectorWare
  - Operating Systems: Wind River Systems VxWorks™ 5.4 & Eonic Systems Virtuoso™
  - I/O: IEEE-1355 Link Port Serial Interfaces & MIL STD-1553/1773 Command & Control
  - Backplane: cPCI (rev 2.2)
  - Form Factor: Compact PCI (6Ux160) & PMC
- These building blocks can be mixed/matched to create payload processing solutions
  - Allows efficient subdivision of a task using data flow model
  - Open scalable architecture easily accepts any mix of modules and new module types
- Commercial equivalent hardware modules available for software development

Payload Controller RHPPC SBC Block Diagram

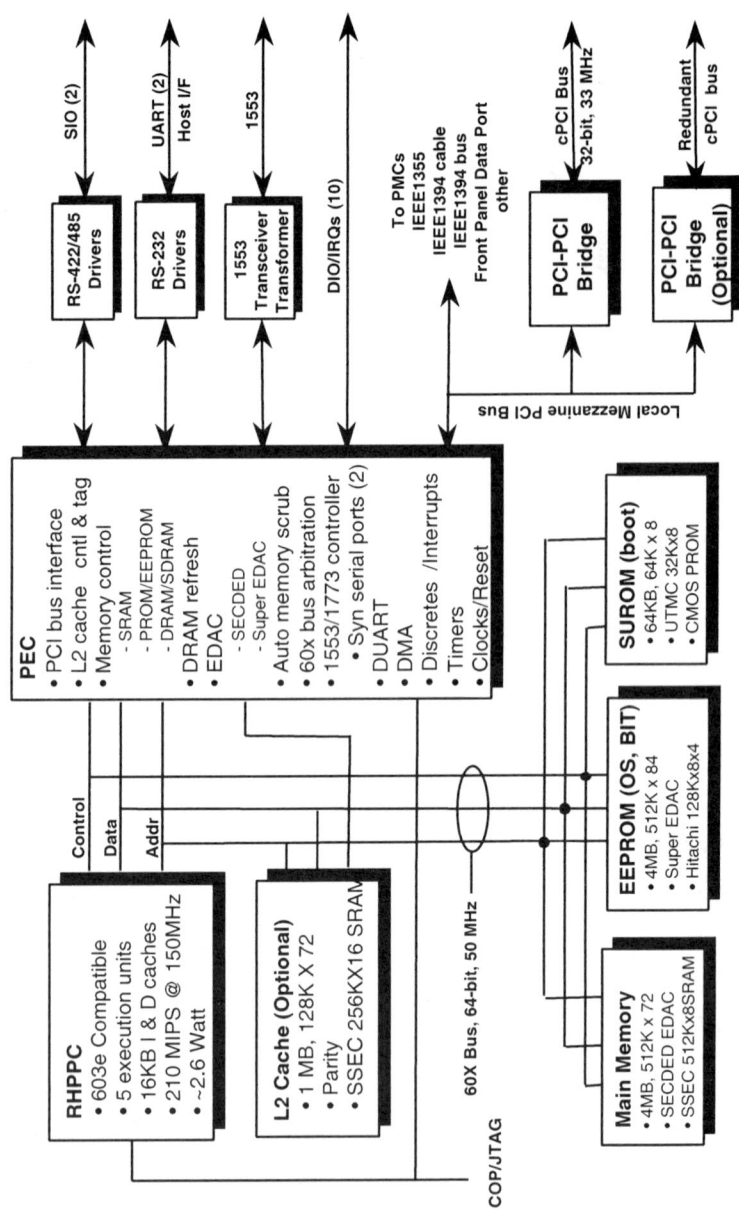

RHPPC SBC Form Factor

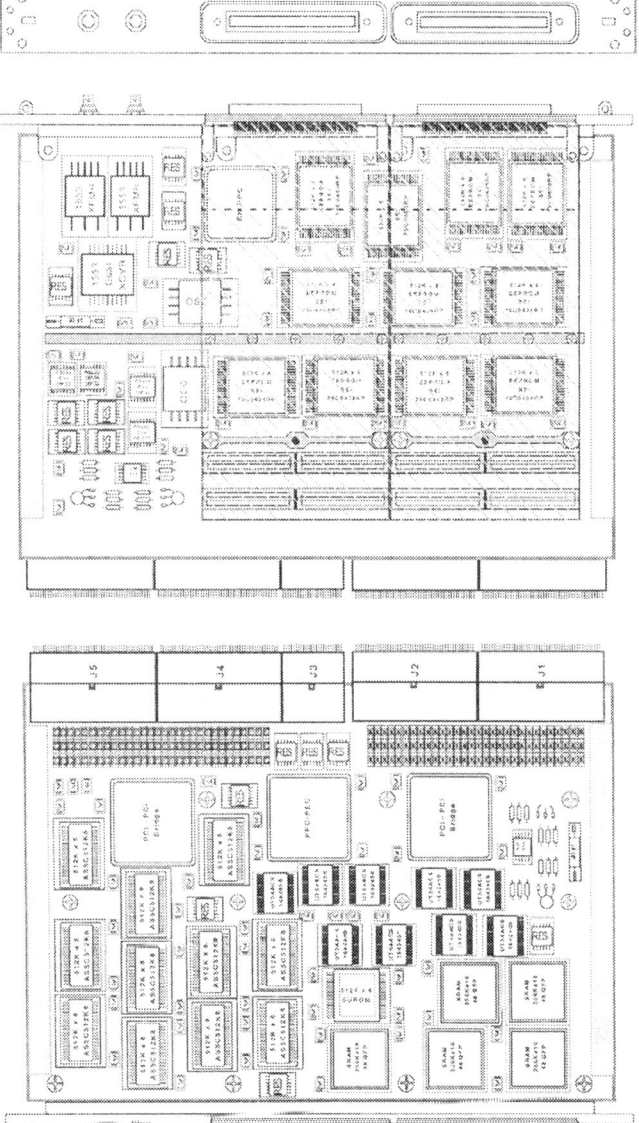

- 6U x 160 (9.187" x 6.299"), 2.5 lb (1120 gr), 8W (nom)
- 2 standard 74 x 149 mm PMCs slots with local mezzanine PCI bus
  - Each PMC slot has 4 standard 64 pin connectors
  - Can be 1 double PMC slot
- Same cPCI backplane connectors as used by X2000 (JPL)
- Standard face panel with 2 1553 connectors & 2 PMC connectors

RHPPC SBC Features

- 6U x 160mm (9.187" x 6.299"), 2.5 lb, 8W (nom)
  - 2 PMC board slots
- 210 DMIPS @ 150 MHz RHPPC processor (cache: 32KB L1, 1MB L2)
  - PowerPC 603e™ technology licensed from Motorola
  - 50 MHz 60x bus, 64 bit data + EDAC, 32 bit address + parity
- Memory
  - 4Mbyte SRAM, SECDED EDAC
  - 4Mbyte EEPROM, super EDAC
  - 64Kbyte PROM (SUROM)
- Backplane: cPCI, 33MHz, 32-bit, redundant
  - Utility bus on J2 connector
- I/O
  - 1553B with 8Kx16 buffer memory, transformer coupled (upgradable to dual rate 1773 with PWB mod)
  - 2 full duplex UART ports, 8250 & 16450 compatible, 9.6K to 1M baud (16 bit rate register)
  - 2 full duplex synchronous serial ports, date/clock/sync, 1.25Mbps or 12.5Mbps (@50MHz)
  - 4 DMA controllers for serial I/O and another DMA for mem-to-mem or mem-to- PCI
  - 10 lines programmable as discrete in, discrete out, or interrupt in

  JTAG/COP debug port
- Timers: 5 general purpose, 2 stage watchdog, mission (20 year)
- Commercial and Military space rad hard
  - TID > 1E5 rad
  - SEU rate < 4.6E-5 e/d (Adams 90% WC, GEO) (without L2), proton immune
  - Dose rate upset > 1E8 rad/s
  - No latchup
- -40°C to 80°C rail temp, 20 Grms
- Ps > 0.99, 15 year, 35°C (with cold spare)

Software Development Environment

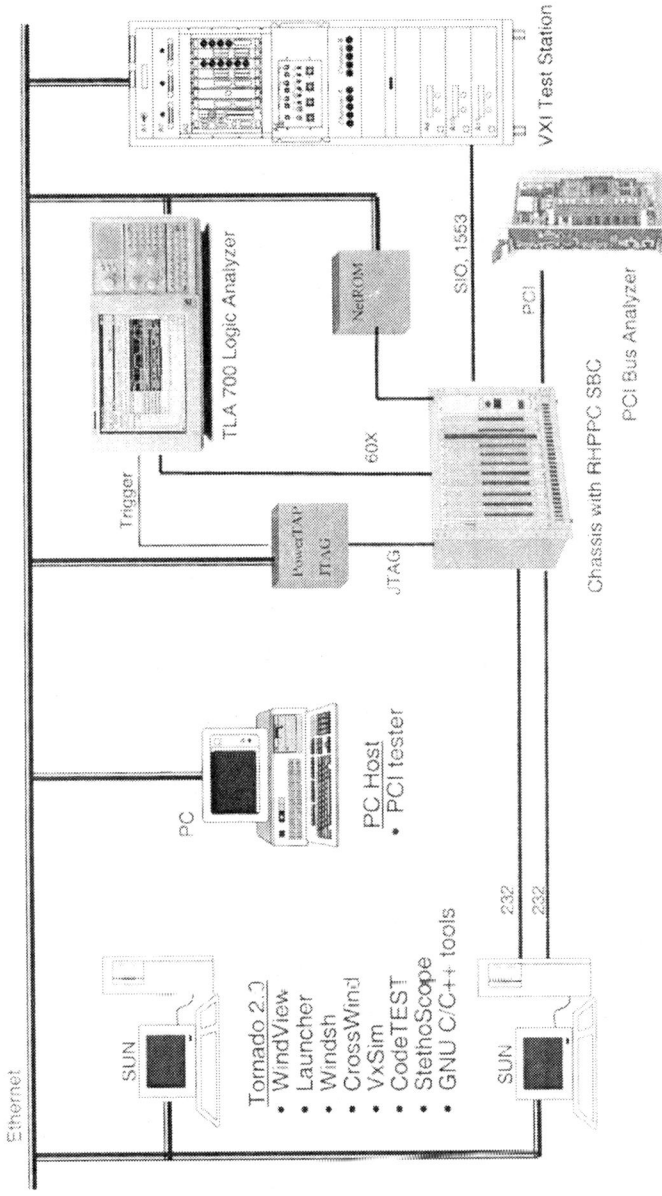

- Honeywell uses the Wind River Tornado 2.0 environment
  - GNU C/C++ (gcc, gdb, xgdb)
  - SUN workstations running Solaris 2.6

Tornado/VxWorks

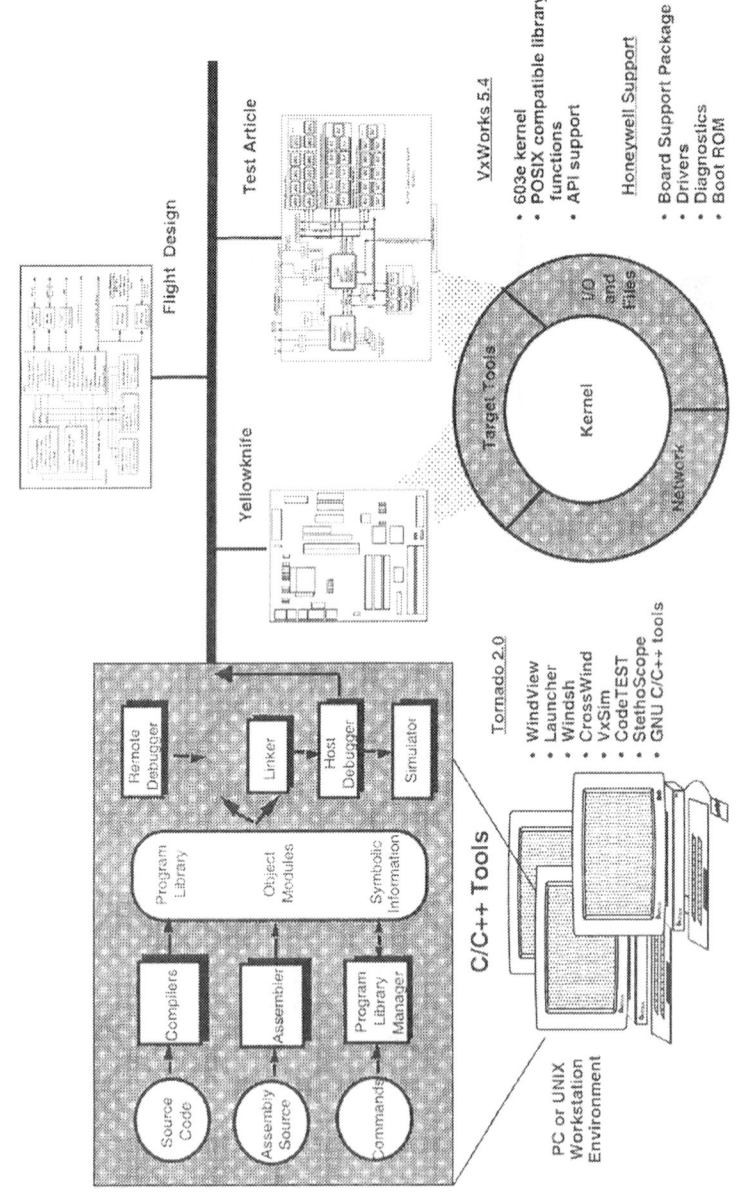

PowerPC™ 60x/7xx Tool Vendors

	Compiler	Debugger	Emulator	OS
Accelerated Technology				X
Applied Microsystems		X	X	
Corelis		X	X	
Cygnus Solutions	X	X		
Diab Data	X			
Enea		X		X
EST		X	X	
Green Hills Software	X	X		
Hewlett-Packard			X	
Integrated Systems Inc. (pSOS)		X		X
Lynx Real-Time Systems	X	X		X
MetaWare	X			
Metrowerks	X	X		
Mentor Graphics (Microtec)	X	X		X
Microware	X	X		X
QNX Software Systems				X
SDS		X		
Sun Microsystems (Chorus)				X
Tektronix			LA	
Wind River Systems	X	X		X

RHPPC Processor

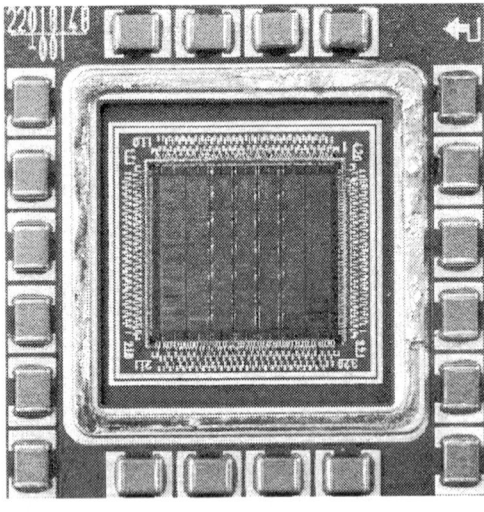

Technology

- 0.35μm, 4 level metal, SOI-V
- HX3000 standard cell with custom drop-ins
- Single chip with CPU, FPU, Cache
- Rad Hard for military space applications

TD	>5E5	Rad(Si)
DRU	>1E9	Rad(Si)/s
DRS	>1E11	Rad(Si)/s
SEU	<4.4E-5	Upset/Chip-Day

  Will not latch up

- Package: 255 lead grid array
  - 21 x 21 mm, 1.27 mm pitch
- Testability coverage > 99%

Features

- Identical to commercial PowerPC 603e™
  - Architecture
  - Programmer's interface
- 210 MIPS (Dhrystone) @ 150 MHz (1.4 IPC)
  - 25 to 150 MHz core frequencies supported
- Icache/Dcache 16Kbyte each
- IEEE 754 floating point
- Supports multiprocessing (Cache Snooping)
- 25, 33.3, 40 or 50 MHz 60x bus
- 3.3V core & I/O
- 2.6W (nom), 3.5W (WC) - 81 MIPS/Watt
- Power management (dynamic, doze, nap, sleep)
- QML Q+ procurable (class S equivalent)
- Footprint compatible with commercial part
- 100% software compatible with commercial part

Application Support

- All commercial PowerPC 603e™ tools and OS
  - C/C++ & Ada 95 tools
  - VxWorks real time OS
- Prototype Development Unit (PDU) for quick prototyping and hardware/software integration ahead of flight hardware

Motorola PowerPC 603e™ G2 Groucho Core

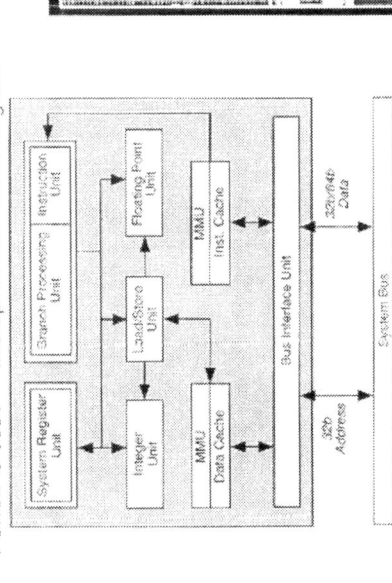

PowerPC 603e™ Microprocessor Layout

Data path (OTS), Control (RLM), Custom, Analog

PowerPC 603e™ Microprocessor Block Diagram

Embeddable core version (G2)
2.6M Transistors
34mm²
0.28µm, 5 LM
1.5W (nom) @ 150 MHz (no I/O)
2.5V operation
Testability > 99%

RHPPC Processor Block Diagram

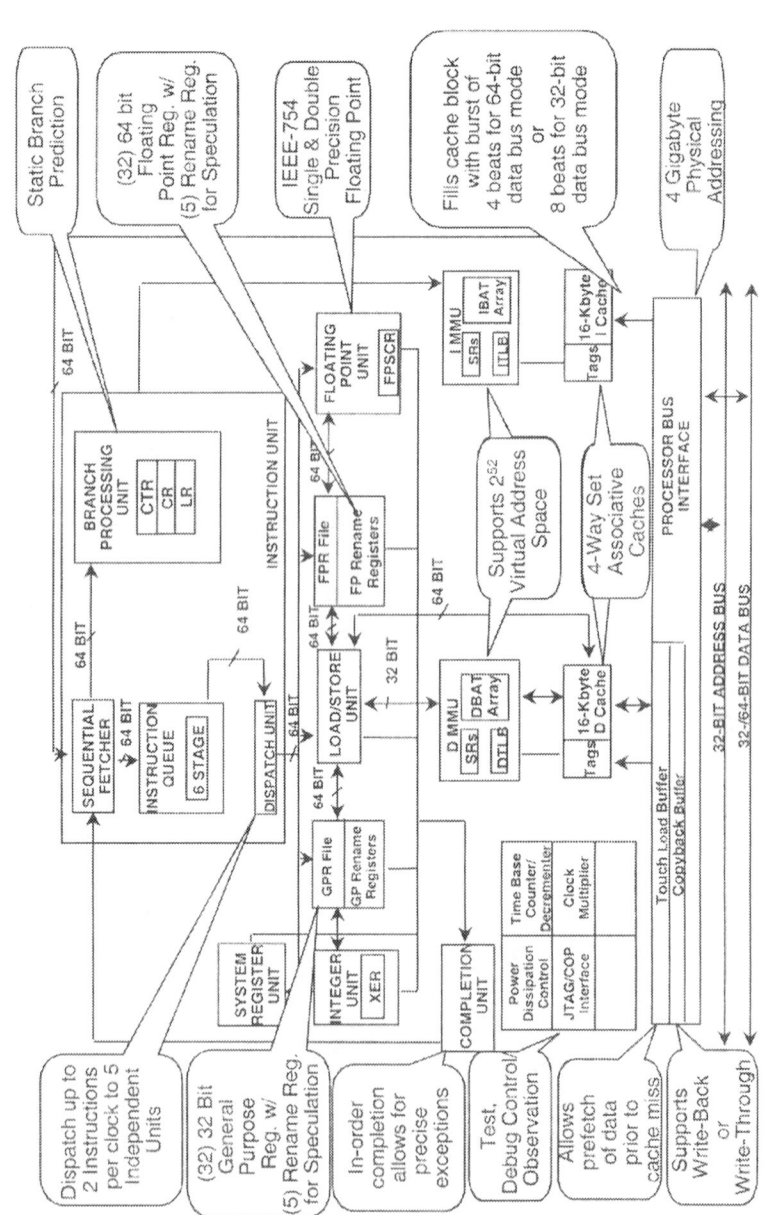

Processor Enhancement Chip (PEC)

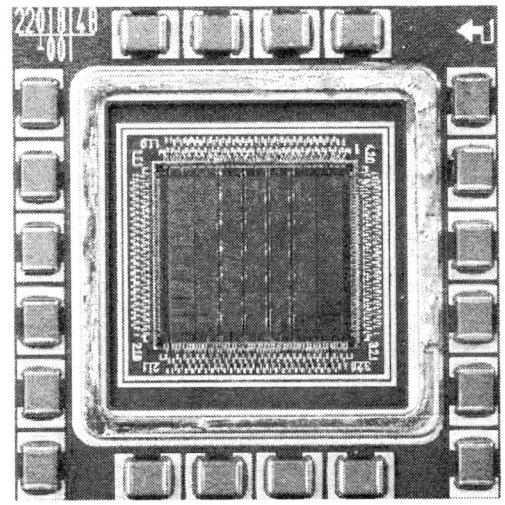

Technology

- 0.35μm, 4 level metal, SOI-V
- HX3000 standard cell with custom drop-ins
- Rad Hard for military space applications

TD >	5E5	Rad(Si)
DRU	>1E9	Rad(Si)/s
DRS	>1E11	Rad(Si)/s
SEU (w/o L2)	<1E-7	Upset/Chip-Day
SEU (w L2)	<6E-5	Upset/Chip-Day

    Will not latch up

- Package: 472 lead grid array
    - 29 x 29 mm, 1.27 mm pitch
- Testability coverage > 95%

Features

- 25, 33.3, 40, 50 & 66 MHz, 64-bit 60x bus
- 512KB/1MB level 2 cache controller
    - On-chip tag RAM, 1 GB cachable space
    - Direct mapped, write through
- Configurable memory controller interfaces to PROM, EEPROM, SRAM, SDRAM and DRAM
    - Contiguous memory space
    - DRAM refresh controller
    - Parity, SECDED EDAC, super EDAC
    - Auto memory scrub
    - Address map B (CHRP)
- Supports up to 4 processors, 3 with L2 cache
- Generates resets & clocks for RHPPC
- 4 General Purpose, Mission, and Two Stage Watchdog Timers
- 16.7, 20, 25 or 33 MHz, 32 bit PCI bus (rev 2.2)
- 1553/1773 protocol controller, internal data buffer
- 2 full duplex programmable synchronous serial ports
    - Baud rate fourth or fortieth SYSCLK (1.25/12.5 Mbps @ 50MHz)
    - Data/Clock/Sync, odd parity
- 2 full duplex 8250 compatible serial UART ports
    - 16-bit rate register (9600 to 1M baud @ 50MHz)
    - Programmable
- 8 programmable interrupts or discretes
- 5 DMA Controllers: 4 for I/O, 1 for memory/PCI
- Software compatible with MPC106 Bridge Chip
    - Implemented functions will remain compatible
- 3.3V, 2.5 W nom, 3.5 W max
    - Supports power management
- QML Q+ procurable (Class S equivalent)

PEC Block Diagram

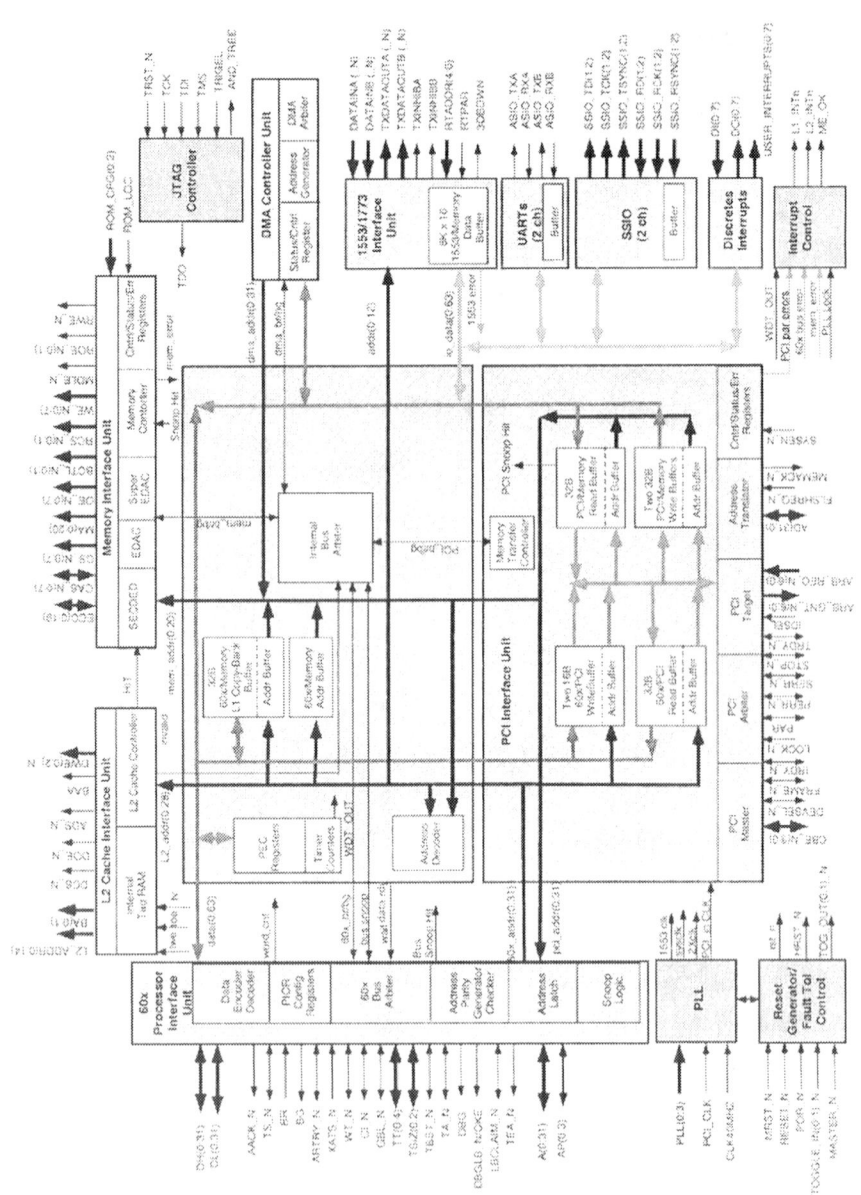

400

RHPPC SBC Benefits

- True commercial compatibility, open architecture
    - PowerPC603e™ technology licensed from Motorola
    - Many mature and supported software development tools available
    - Motorola & IBM plan to support this instruction set for many years
    - VxWorks RTOS
    - cPCI & PMC form factor, cPCI backplane, PCI mezzanine bus
    - MIL-STD-1553, RS422, RS232 I/O
- High performance
    - 210 DMIPS @ 8W => 26 MIPS/W at board level
    - Enables payload functions not possible before
- True space level radiation hardness
- Simple architecture results in:
    - High reliability
    - Low power
    - Low cost

AAS 00-038

ARU ARCHITECTURAL SOLUTIONS FOR LONG LIFE SATELLITE MISSIONS

Edward C. Moulton, Robert H. Fall and Thomas G. Stottlar[*]

INTRODUCTION

This paper discusses and evaluates the redundant architecture of a new product, Evolved Laser Gyro Attitude Reference System (ELARS), being developed by Honeywell for space applications. Comparisons to traditional Box A/Box B redundancy schemes will be used to quantify the improvements.

The Honeywell ELARS design is derived from our Miniature Inertial Measurement Unit (MIMU) and based upon our high volume commercial and military production ring laser GG1320AN Dig-Gyro™. ELARS contains six skewed gyros and I/O electronics that maintains the MIMU serial interface or an optional 1553B interface. The unit is configured with two internal low voltage power supplies (LVPS) and two processors. The ELARS is being developed to satisfy the commercial spacecraft market needs for a cost effective and reliable three axis attitude reference system.

The ELARS is the first element of an emerging product family philosophy. The ELARS architecture is being developed to accommodate the later insertion of a Fiber Optic Gyro (FOG) in place of the Ring Laser Gyro (RLG). This FOG configuration will be referred to as the Evolved Fiber-gyro Attitude Reference System (EFARS).

The ELARS offers a reduced cost by using a highly integrated gyro architecture similar to the Honeywell Dig-Gyro™ presently in high volume production. The Dig-Gyro™, originally developed and qualified in 1991, is a sealed single axis RLG that contains all necessary support electronics including the high voltage power supply to provide gyro output data. The Dig-Gyro™ is being used in many commercial and military aircraft applications. The ELARS integrated product team has designed space hardened electronics to replace the commercial and military electronics while minimizing impact to production operations. Manufacturing experiences from MIMU have been incorporated to improve production flow. For example, the use of the sealed production configuration gyro now allows for a vented chassis design that simplifies final assembly and test. The integrated gyro architecture also reduces cost, improves reliability, and results in lower power by reducing the amount of support electronics necessary for Attitude Reference Unit (ARU) operation. The ELARS exploded view shown in Figure 1 illustrates the simplicity of design. The six radiation-hardened gyros mount to an isolated sensor block assembly. The sensor block mounts into a vented chassis. The dual LVPS are mounted in the front of the chassis. Figure 2 illustrates the small dimensions of the assembled chassis and the system reference axes orientations. Figure 3 shows the placement of the six gyro axes on the sensor block with respect to the system reference axes.

---

[*] Honeywell Space Systems, 13350 U.S. Highway 19 North, Clearwater, Florida 33764-7290.

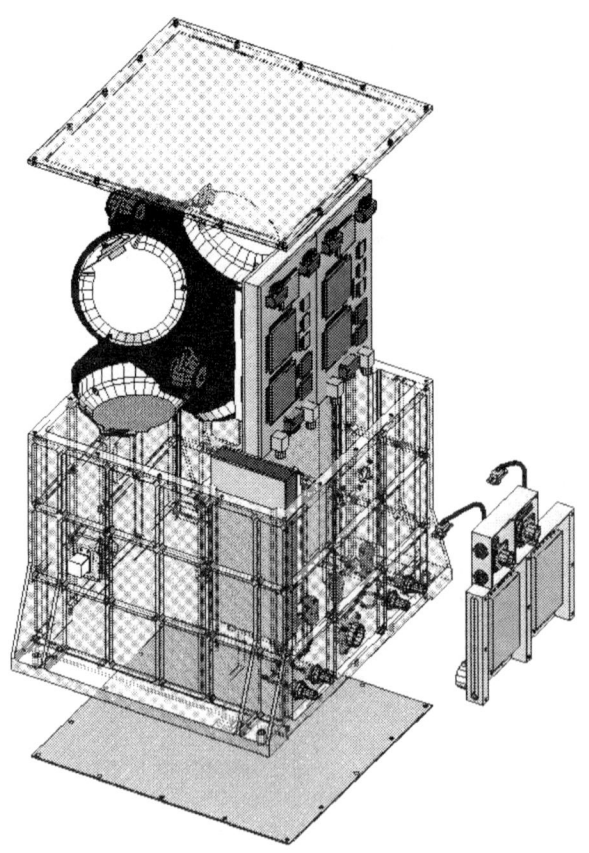

Figure 1 Exploded View and Components

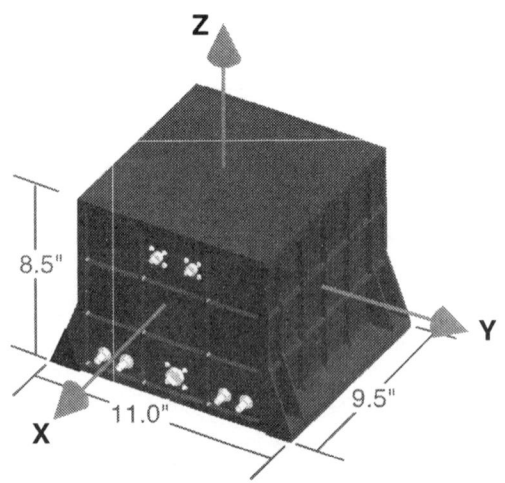

Figure 2 Chassis, Dimensions and ARU Axes

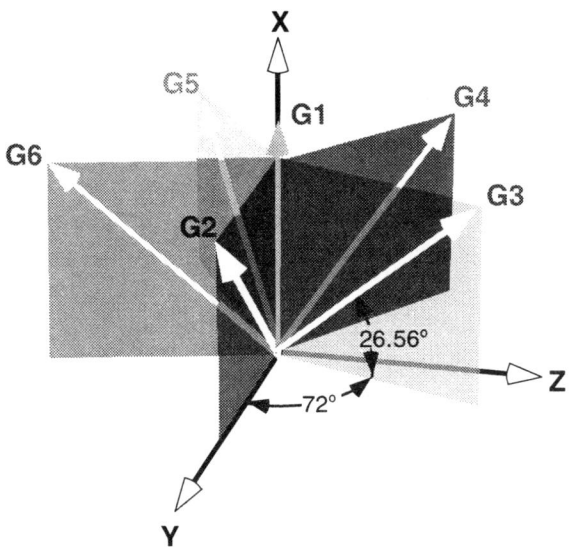

Figure 3 Gyro Axes to ARU Axes Orientations

The size, weight, and power estimates for the ELARS are provided in Table 1.

Table 1
ELARS SIZE, WEIGHT AND POWER

	ELARS
Size	L – 9.7 in. (24.64 cm) W – 11.1 in. (28.19 cm) H – 8.6 in. (21.84 cm)
Weight	21.5 lb (9.77 kg)
Power	22 W, nominal

Customers gain cost and risk benefits from the large production base of GG1320 gyros, presently produced at a rate exceeding 1000 gyros/month with over 45,000 delivered to date.

SYSTEM DESIGN OVERVIEW

The ELARS capitalizes on both the mature GG1320 RLG analog electronics heritage and the Dig-Gyro™ electronics design. The ELARS block diagram shown in Figure 4, illustrates the simple architecture. A block diagram of the ELARS Radiation Hardened Gyro is shown in Figure 5. The increased level of electronics integration internal to the ELARS Radiation Hardened Gyro simplifies the additional electronics necessary to support a six-sensor ARU. The electronics simply provides low voltage power, gyro selection/power control, gyro sampling logic, RS422 or 1553B I/O, and a processor for compensation, alignment transformation, and additional bandwidth filtering if required.

The ELARS architecture combines a cold sparing gyro philosophy with a dual channel processor/LVPS electronics string. Under normal operation, one processor/LVPS string and any three of six gyros will be operational. The other processor/LVPS string and three gyros will be a cold spare. Each LVPS is capable of operating all six gyros if required. Although a number of error detection schemes may be employed, nominally, the process monitors gyro BIT to determine if a gyro channel has failed. Reconfiguration of gyro channels can be initiated autonomously or by satellite command. The satellite will select the processor/LVPS channel similar to customary Box A/Box B methodologies. Both processor/LVPS strings can be operated simultaneously if required.

ELARS Block Diagram

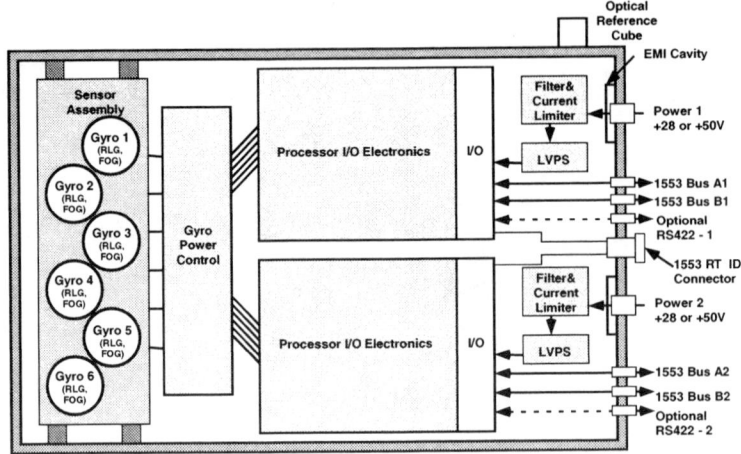

Figure 4 ELARS Block Diagram

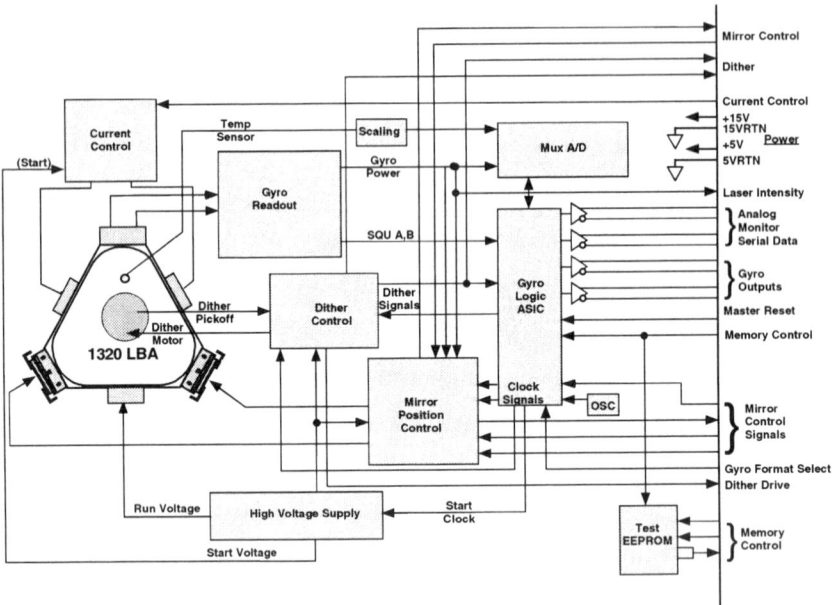

Figure 5 Radiation Hardened ELARS Gyro Block Diagram

PERFORMANCE

The ELARS bias performance is expected to meet or exceed the MIMU performance due to the improved thermal characteristics of individually shielded gyros (0.01 deg/hr, 1 sigma). Angular random walk, gyro scale factor, and alignment performances are expected to be the same as MIMU at the gyro level. At the system level, however, due to the nonorthogonal gyro orientations, the ELARS composite system output rate error about any axis will be higher, but less than 0.05 deg/hour (1 sigma) during any 60-minute period.

The effects of gyro selection geometry on composite axis noise are shown in Figure 6. In the worst case geometries, the RMS noise will increase by a factor of 3 over a single gyro as noise components from three gyros couple in a single axis.

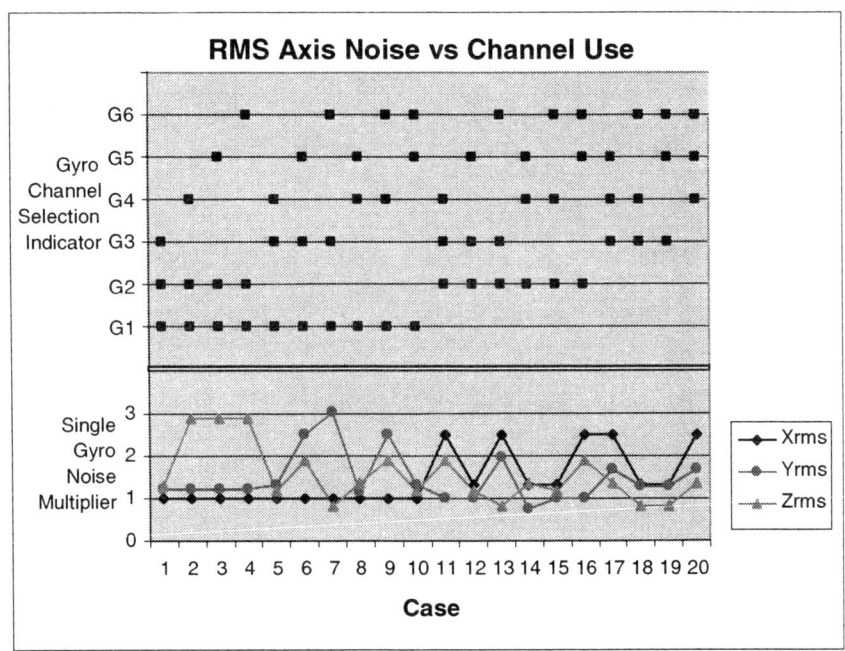

Figure 6 Geometric Effects on System Axes Noise

DATA INTERFACE

The ELARS functions as a 1553B Remote Terminal. The 1553B interface message block contains ten 16- bit data words (Frame Timer, IRU Status Word, X Axis Angle, Y Axis Angle, Z Axis Angle, Multiplexed Gyro 1,4 Angle, Multiplexed Gyro 2,5 Angle, Multiplexed Gyro 3,6 Angle, Multiplexed BIT Status, Checksum) that are updated at a 200 Hz rate. The data format is compatible with existing Honeywell satellite systems using an RS-422 serial interface.

The optional RS-422 interface will be capable of transmitting the same information as the 1553B interface and will provide interface compatibility with existing satellite systems.

POWER INTERFACE

The ELARS is configured with two Low Voltage Power Supplies (LVPS) capable of operating with +28, +50, +70 or +100V inputs.

PACKAGING

A simple aluminum chassis design contains an isolated sensor block assembly and electronics. The sealed ring laser gyros are mounted to the sensor block. The sensor block uses elastomeric isolators to protect the RLGs from launch, transportation, and handling shock and vibration. The sensor block will also accommodate FOG gyros when required. The Isolators also minimize thermal transients to the RLG thus maximizing Ps for Low Earth Orbit (LEO) thermal cycles and attenuate gyro dither to the ARU mounting structure. The ELARS is a vented design.

Radiation

A LEO presents a wide range of possible radiation environments for the ELARS. Additional factors determining the severity of the ELARS radiation exposure include the mission duration, orbital inclination, and the amount of inherent shielding provided by the space vehicle. To provide an optimum radiation hardening solution consistent with the wide range of radiation environments for a 15 year on orbit mission, and with the necessary goal of minimizing cost and weight, Honeywell's ELARS design utilizes radiation tolerant parts. All ELARS semiconductor parts are selected for a radiation tolerance of 100K Rad(Si) ionizing dose and $10E12$ n/cm^2 neutron-equivalent displacement damage (1MeV silicon damage equivalent). These hardness levels are based on allowable degradation of critical performance parameters for each part as derived from a worst-case circuit analysis. The internal components' radiation exposure will be limited to levels below their demonstrated tolerance through the use of adequate radiation shielding. Shielding sources include the ELARS chassis and internal assemblies, semiconductor piece-part packages (and spot shields if needed), and the inherent shielding provided by each spacecraft. The ELARS will use a 100 mil thick aluminum chassis. Applications requiring greater shielding may be accommodated by additional shielding either by bolt on chassis plates, spot shielding, or use of Honeywell's Thermal Flame Spray process to deposit high Z material on component packages or external chassis walls.

The ELARS does not use any digital electronics that are susceptible to proton induced upset to minimize susceptibility to single event effects. The ELARS design also does not contain any parts that are susceptible to single event latchup, and the power metal oxide semiconductor field effect transistor (MOSFETs) used in the design will have adequately derated drain source and gate source voltages to assure immunity to single event gate rupture and burnout.

Table 2 shows the ELARS predicted very low susceptibility to single event upsets for orbits commonly used by customer satellites. These are raw numbers and do not take into account any mitigation by the ELARS BIT or satellite redundancy management schemes.

Table 2

ELARS MEAN TIME BETWEEN SINGLE EVENT UPSETS

Orbit	Mean Time Between SEUs (Years)
Telecommunications LEO (1400km x 50 deg)	3570
Polar LEO	524
Geostationary	147

ELARS RELIABILITY VS COLD SPARING OF CONVENTIONAL SINGLE STRING ARUS

Methodology

A comprehensive study of relative reliability was made via Monte Carlo simulations of the operation of an ELARS system operating three of up to six functional gyros, dual box (A/B) and triple box (A/B/C) single string unchannelized flight sets. The A/B single string ARU flight set uses one active box and one as a cold spare until the first unit fails. The triple box A/B/C single string flight set uses one active box and two cold spares for replacement of the failed unit. To provide an easily understandable comparison, the single string flight set configuration uses three of the ELARS gyros, one ELARS processor and I/O electronics and one ELARS power supply as shown in Figure 7. Currently, there is no Honeywell system using ELARS elements in this fashion. The ELARS configuration was specifically designed to provide a cost effective approach to reliability using low cost commercial parts when feasible, so its element reliability should not be compared on a one to one basis with existing production designs using high reliability parts.

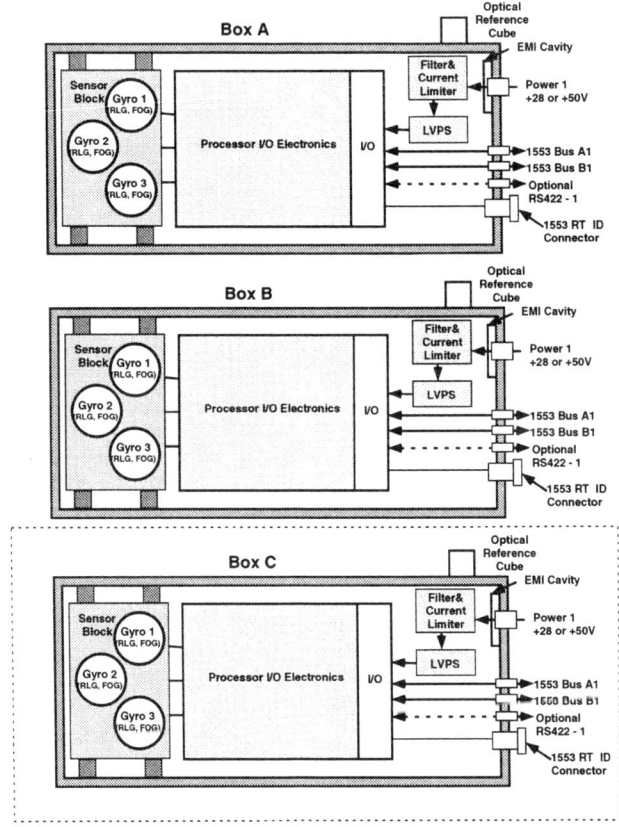

Figure 7 Single String ARU Configuration for Box A/B and A/B/C

Failure rate data is derived using the parts count prediction method within MIL-HDBK-217F Notice 2 assuming a "Space Flight" environment and/or manufacturer life test data. The GG1320 Laser Block Assembly (LBA) failure rate is baselined to the current LBA life determined from life test data and field failures coupled with the most current radiation hardened ELARS gyro electronics design evaluated as above. To protect currently proprietary information, the LBA life is normalized and referred to as 1X. To show the effects of this critical wearout component, evaluations were also made at a gyro LBA life twice as long, referred to as 2X. To approximate the very long life expected of FOGs, an overall gyro life including electronics and optics of six times the current LBA life was used in the evaluations. This is referred to as

6X, although it is a combined optics and electronics life. A fourth case was evaluated to show the architectural comparisons in the ultimate limiting case of infinite gyro life and the same baseline random failure rate.

A Ps was calculated using a Monte Carlo reliability model based on mission parameters. This analysis technique is required to permit the combination of linear failure rates (random failures) with non-linear failure rates (wear-out). The Monte Carlo model simulates the mission profile using a probabilistic approach. Statistics are compiled from multiple iterations to establish reliability. The specific Monte Carlo model permits the use of multiple distributions to characterize the operational and non-operational portions of the mission. These distributions are tracked individually to permit parameters to affect both phases of the mission or only one phase, as required. In addition, multiple distributions of the same type (i.e. exponential, gaussian) may be applied concurrently to each phase of the mission allowing for the direct combination of multiple wearout/random parameters. The statistics generated from these models are based on 100,000 missions (operate to defined failure) per configuration.

The reliability prediction for the ELARS is based upon the following conditions:
- Honeywell ED23055 PMP standard grade with non-hermetic industrial semiconductors with manufacturer failure rate data.
- Power switch circuitry for processor channel selection is external to the ELARS.
- Power switch circuitry for A/B/C box selection is external to the single string systems.
- No single point failures or propagation of failures
- A failure rate factor of 0.1 is applied to unpowered redundant components.

Each configuration assumes operation for 1000 hours at a constant temperature to simulate ground operation and then appropriate low amplitude sinusoidal temperature variations were applied for the remainder of system life to simulate the effects of LEO temperature variations which impact primarily LBA life.

The ELARS configuration shown in the block diagram of Figure 3 was operated with three gyros active with replacement after a failure from the three or less unpowered gyros until there were only two operable gyros and no remaining spares. One processor was operated until failure and then operation was switched to the remaining one until final failure. Only two operable gyros or no functional processor defined the point of mission failure. The failure rates of the channelized Gyro Power Control function were partitioned and assigned to gyro and processor elements. There is no single point failure mechanism in the Gyro Power Control function.

The single string ARUs were operated with all three available gyros and the single processor until at least one gyro or the processor failed. At this point a spare ARU was powered up to replace the failed unit. The failure of the second ARU defined the point of mission failure for the A/B configuration and the failure of the third ARU defined the point of mission failure for the A/B/C configuration.

Results

In all cases, the plots presented below reflect some inevitable scatter caused by the limited number of failures out of the 100,000 samples at short mission times and should not be viewed with the same expectations as plots of analytic solutions. However, sufficient resolution is available to clearly indicate trends and relative differences between configurations.

For ease of understanding, the reliability comparison results are presented as ratios of Probability of Failure (Pf = 1-Ps) rather than the more conventional Probability of Success (Ps). Since the ratios of Ps for short mission times are all very close to 1, differentiating information due to gyro life variations would be lost. The Pf ratio shows these effects more clearly and the differences in failure rates become apparent.

As seen in Figure 8, the ELARS system provides up to a 30 times lower probability of failure than the single string A/B configuration for the baseline (1X) gyro LBA life for shorter missions. The higher ELARS reliability is most pronounced for the shorter mission times and declines in the cases of the very long life (6X) and infinite life gyros to a still respectable factor of >10 for long mission times.

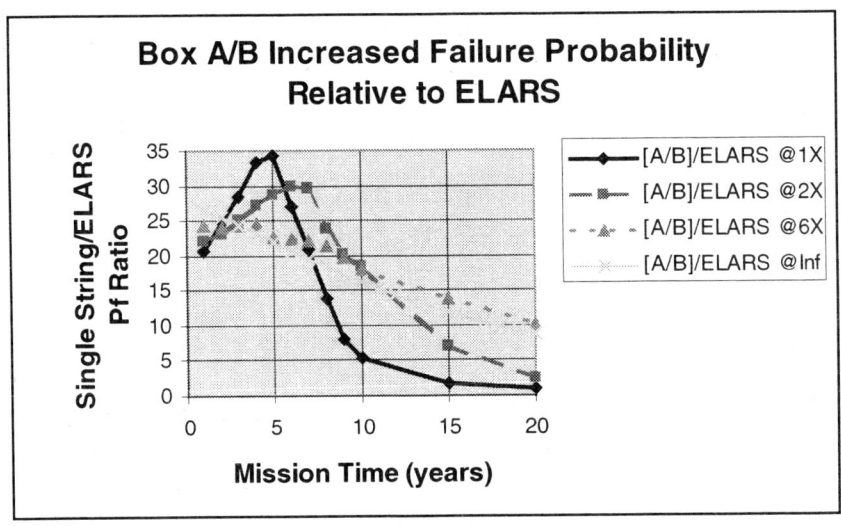

Figure 8 Comparison of Single String A/B Failure Probability to ELARS

The single string A/B/C configuration is much closer to the ELARS in reliability as seen in Figure 9. For short mission times, the reliability difference is much less pronounced, but the long life (6X) and infinite life gyros still maintains an advantage in the ELARS configuration for long mission times.

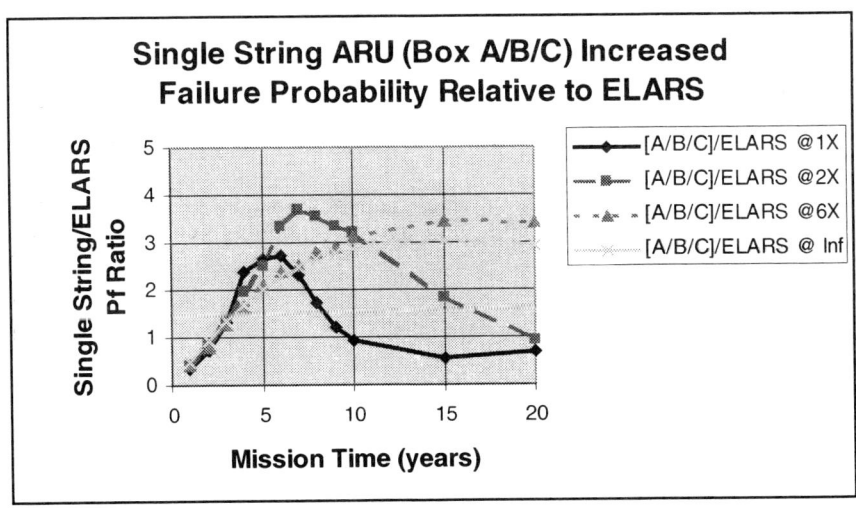

Figure 9 Comparison of Single String A/B/C Failure Probability to ELARS

From this information, we can conclude that the ELARS configuration compared to conventional single string approaches is well suited to both short missions that need to obtain very high reliability from existing gyro technology and to long missions using very long life gyros. For long missions using 1X life gyros, both approaches are equivalent.

In another approach to quantifying reliability differences, mission life comparisons were made by determining the mission life that an ELARS would have for a specific mission duration and resulting Ps for the single string configurations.

Figure 10 shows the additional mission life that can be obtained using an ELARS meeting the same Ps as the conventional single string approach. As a minimum, three to five years can be added to a fixed Ps mission using an ELARS over a single string A/B configuration, and for 6X gyros up to 15 years can be added. Please note that the Monte Carlo simulations were only run out to 20 years so the sum of baseline A/B mission times and ELARS extension cannot exceed this duration – hence the truncated curves.

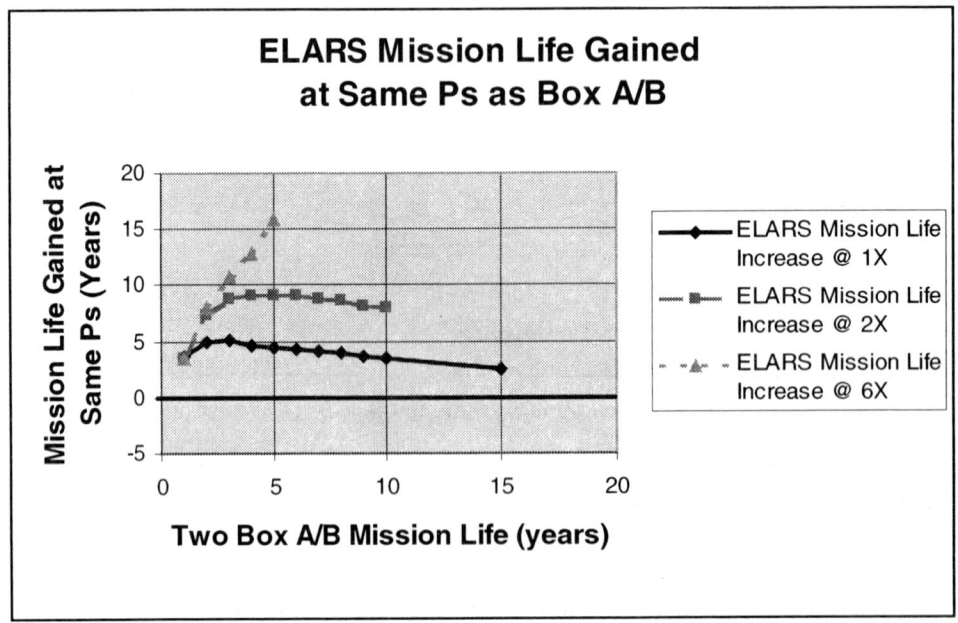

Figure 10 ELARS Increased Mission Life Compared to 2 Traditional ARUs

Figure 11 illustrates the ELARS increased mission life over the single string A/B/C configuration. As expected from the earlier Pf data there is not as a significant gain, but for longer life gyros and longer missions there is still the possibility of significant mission duration extension by switching to an ELARS configuration.

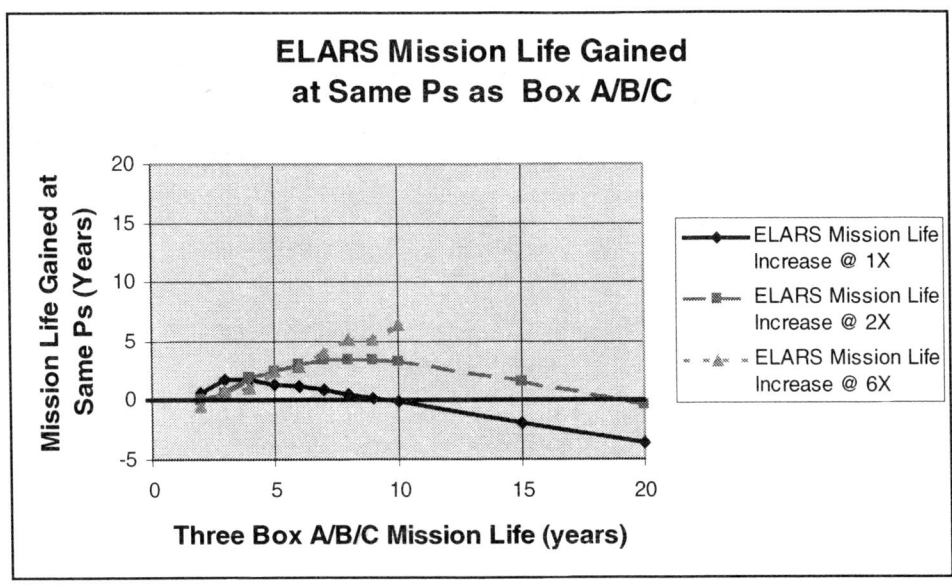

Figure 11 ELARS Increased Mission Life Compared to 3 Traditional ARUs

Finally, to put one more twist on the data, Figure 12 illustrates that an ELARS system using 1X life sensors will produce the equivalent Ps of a traditional A/B approach using sensors with 2X life for missions of up to 10 years.

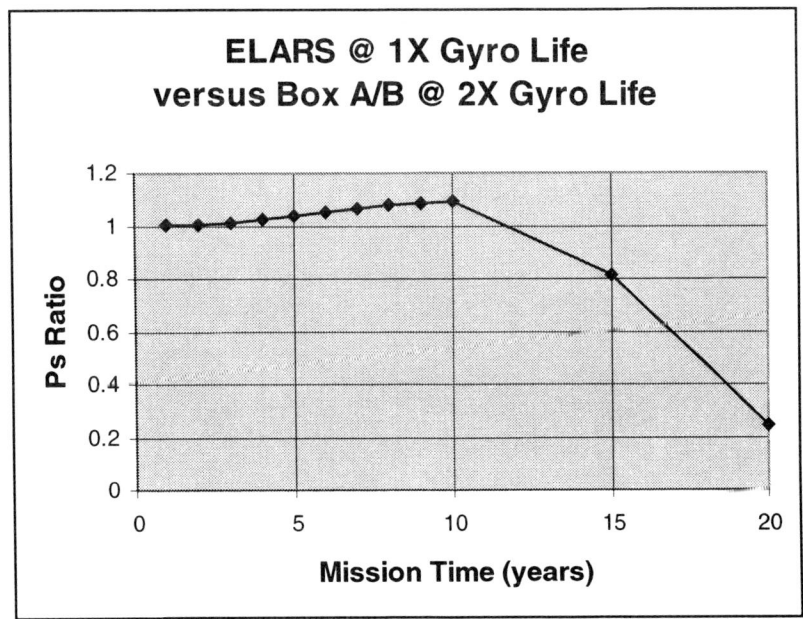

Figure 12 ELARS using 1X Life Gyros Compared to 2 Traditional ARUs with 2X Life Gyros

CONCLUSIONS

A single ELARS significantly increases reliability or mission times over two stand alone conventional single string ARUs using similar gyros and electronics. A single ELARS also achieves the equivalent reliability performance of **three** stand alone conventional single string ARUs that use **1.5 times** the quantity of similar gyros and electronics. These properties allow ELARS to be used for a number of options to enhance reliability of Honeywell systems at a potential cost savings over traditional approaches. Although ELARS development costs exceed the probable costs of modest sensor life enhancements, the architectural solution does not require the lengthy life tests necessary to gather statistically significant data for new sensor technologies.

The Honeywell ELARS offers a robust, low risk, cost effective, high reliability solution using existing sensor technologies to satisfy present and future customer mission requirements. We are presently delivering a broad spectrum of products to meet unique customer requirements for space missions. We work closely with customers to determine the most cost-effective solution. Our 27 years of 100 percent mission success on space programs is an example of the quality and dedication that Honeywell brings to the satellite manufacturing community.

AAS 00-041

TURBO-CHARGED TORQWHEELS™

Bill Bialke[*]

High agility demands imposed on spacecraft designers have increased the typical torque requirements for reaction wheels. Ithaco reaction wheels, or TORQWHEELs, have met this demand by capitalizing on an inherent design feature which allows trading momentum storage for torque, with only a trivial jumper change in the motor termination. This innovative feature affords performance versatility without sacrificing heritage. In addition to these configurations, "Turbo-Wheels" are available with increased current capacity to deliver high torque without giving up momentum storage capacity.

This paper will review the torque versus momentum trade-offs, including power consumption, weight, thermal and reliability issues, and present the performance specifications of high-torque reaction wheel options which are currently available. In addition, summaries of recent micro-vibration test data and lubrication endurance test results applicable to these new requirements are presented.

INTRODUCTION

Higher and higher reaction torque requirements for reaction wheels have been driven by a new generation of scanning and remote sensing spacecraft, where maneuvering time is important for either fiscal or technical reasons. This demand has been addressed with Ithaco TORQWHEELs in several ways. The easiest way of satisfying this hunger for high torque is with a simple jumper change in the motor terminations. An alternative solution pairs a high current motor driver with a small reaction wheel. These high torque options have been successfully implemented on several flight programs.

In addition to high torque, a new generation of remote sensing spacecraft is extremely sensitive to micro-vibrations, due to stringent spacecraft jitter requirements imposed on the attitude control system. Fine balancing capability and high-resolution induced vibration characterization equipment are available to meet these demands.

Most of these high-torque applications are implemented in a maneuverable zero-momentum system with three or four reaction wheels, which invariably must go through zero speed. In some cases, the number of zero speed crossings is significantly higher than would be encountered in previous spacecraft. In response to this and a desire by spacecraft ACS designers to eliminate constraints regarding operating at and near zero-speed, a pair of demanding life tests were recently completed which successfully demonstrate the superior performance of the Ithaco lubricant formulation in this operating

---

[*] ITHACO Space Systems, A Wholly-Owned Subsidiary of B.F. Goodrich, 950 Dandy Road, Suite 100, Ithaca, New York 14850.

environment. The results indicate that previous restrictions imposed on reaction wheels operating at or near zero speed do not necessarily apply to TORQWHEELs.

STOCK DESIGNS

The Ithaco "TW Series" TORQWHEEL product line has evolved into three basic stock classes, identified as Type A, Type B and Type E. The type refers to the chassis size, as shown in Figures 1 and 2. All of these heritage designs feature an ironless armature direct-drive motor, a high-performance synthetic lubrication formulation, and similar chassis construction and suspension systems. Spoked, aluminum-alloy flywheels and a shatterproof window are standard in the stock classes. Side mirrors are optional.

Figure 1: TW-A and TW-B Reaction Wheels

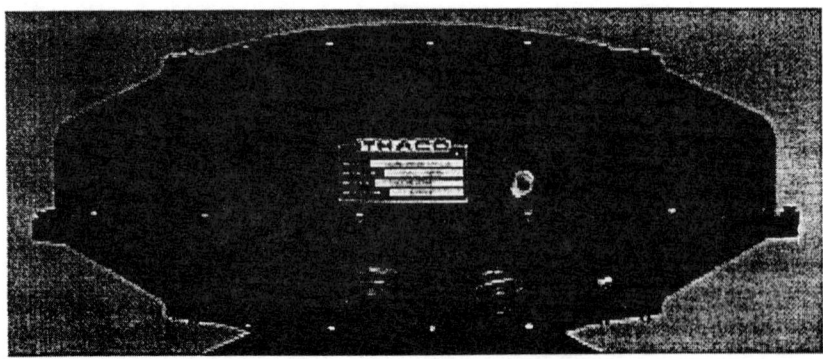

Figure 2: TW-E Reaction Wheel

The Ithaco model number designating the wheel type, reaction torque capability and momentum capacity can be interpreted by the designators shown in Table 1. The model numbers and performance of the three stock designs is shown in Table 2.

Table 1: Model Number Definition

MODEL NUMBER DEFINITION	
TW-8B90 Prefix:	TW = TORQWHEELS
TW-**8**B90 Number:	XXX = Momentum Capacity in N-m-s
TW-8**B**90 Letter:	A, B, E = Reaction Wheel Housing Size
TW-8B**90** Last Number:	XX = Reaction Torque in mN-m

Table 2: Stock RWA Designs

Class		Type A	Type B	Type E
Model Number		TW-4A12	TW-16B32	TW-50E300
Momentum Capacity	N-m-s	4	16.6	50
Reaction Torque	mN-m	12	32	300
Speed Range	rpm	5100	5100	3850
Peak Power	W	25	50	280
Steady State Power at 1000 rpm	W	5	6.5	22
RW Mass	Kg	2.55	5.9	10.6
Motor Driver Mass	Kg	0.91	Included	3.3
Motor Driver Current Capability	A	1.0	2.5	13.0

The stock designs were originally sized with the intent to be able to modify reaction torque capabilities by either increasing motor driver current capacity, or modifying the motor torque constant with a jumper change. In addition, momentum growth was anticipated within each case size by using low-density materials such as aluminum for the flywheel material, which allows for increased inertia with a change to a higher density material such as stainless steel.

HIGH TORQUE MODIFICATIONS

Reaction torque in a TORQWHEEL is generated by an ironless armature brushless DC motor. The unique motor was developed at Ithaco specifically for reaction wheels, since it concentrates all of its mass on the rim of the flywheel, contributing inertia, which results in minimum unit mass. In addition, it has zero cogging and high efficiency, for minimal disturbances and low power consumption.

Increased motor torque capability can be attained by two means. The current driving the motor can be increased, resulting in a proportional increase in motor torque, or the motor torque constant can be increased. An increase in motor current can be accommodated by redesigning the motor driver, or using a higher capacity driver from a larger reaction wheel.

The motor has a torque constant, k_t, and delivers torque proportional to the current driven through it. The Ithaco ironless armature motor is configured such that alternate winding configurations can be selected to easily change the motor torque constant. The windings are brought out to a termination board, and can be connected in series, series-parallel or parallel combinations. The torque constant changes by a factor of two between the intended configurations.

When the torque constant is doubled in a motor, the back-emf is also doubled. These two motor parameters are essentially the same physical constant with different units. It is interesting to note that N-m/amp is exactly equal to V/rad/sec. Go metric! The tradeoff is that the speed capability of the motor is halved when the torque constant is doubled. Trading momentum for torque within a given baseline design is thus a simple jumper change at the motor termination. The change to the motor is transparent to the current-source motor driver, so no other changes are required other than trivial rescaling of the torque command input.

The only other tradeoff with increasing torque in this manner is additional heat dissipation in the motor stator during the application of peak current. Higher conductivity materials are substituted in these cases to reduce thermal impedance from the motor, and maintain acceptable temperatures. These legal modifications to the stock designs have been executed successfully in several TORQWHEEL programs.

The TORQWHEEL configurations which are available with motor connection modifications from the baseline design are listed in Tables 3, 4 and 5 for the three stock TORQWHEEL types.

In addition to the simplicity of the modification, another advantage of trading momentum storage for torque capability is that the peak power consumption is preserved. Since power is the product of torque and speed, doubling the torque while halving the speed results in no net change in peak power consumption. This fundamental characteristic is one of the advantages of using low speed versions of wheels, and is also the reason that the motor configuration change is transparent to the stock motor driver.

In order to increase motor torque without sacrificing momentum storage capacity, the torque scale factor must remain at the stock value, and the current capability of the motor driver must be increased.

The stock Type B TORQWHEEL delivers 2 A of motor current. The capability of the baseline driver was increased slightly to 3 A for the GPS IIF TORQWHEEL, and again to 8 amps for a model in production for ICESAT, affectionately designated the "Killer B" TORQWHEEL. This version will deliver 200 mN-m of reaction torque up to

5100 rpm, thus providing over 16 N-m-s of angular momentum storage while producing the higher torque.

Table 3: Type A RWA Motor Connection Options

Model Number		TW-4A12	TW-2A40
Momentum Capacity	N-m-s	4	2
Reaction Torque	mN-m	12	40
Speed Range	rpm	5100	2500

Table 4: Type B RWA Motor Connection Options

Model Number		TW-16B32	TW-8B90	TW-4B200
Momentum Capacity	N-m-s	16.6	8.1	4
Reaction Torque	mN-m	32	90	200
Speed Range	rpm	5100	2500	1200

Table 5: Type E RWA Motor Connection Options

Model Number		TW-50E300	TW-26E700	TW-13E1000
Momentum Capacity	N-m-s	50	26	13
Reaction Torque	mN-m	300	700	1000
Speed Range	rpm	3850	2000	1000

In order to satisfy the maneuverability demands of impatient commercial remote sensing satellite customers, a Type B TORQWHEEL has been successfully integrated with a 13 amp Type E motor driver, which develops a hefty 300 mN-m (0.22 ft-lb) of reaction torque up to 5100 rpm, producing 220 milli-horsepower, and accelerating from zero to sixty in less than 0.6 seconds. This "Turbo-B" configuration is fully qualified, conduction cooled and designed with space in mind.

The performance specifications for high performance options of the Type B TORQWHEEL are listed in Table 6. The TW-16B200 and TW-16B300 required a change in the motor stator material from fiberglass epoxy to aluminum in order to limit the temperature rise in the motor when full current is applied.

MOMENTUM CAPACITY INCREASE

The stock TORQWHEEL sizes are also capable of higher momentum storage. The original designs utilized low-density materials such as aluminum for flywheel material in order to allow for inertia growth. By changing the material to a higher density material such as stainless steel, the inertia is increased significantly. The structural design of the original hardware anticipated this potential growth. Table 7 shows the potential momentum growth capabilities of the stock TORQWHEEL designs. Higher inertia flywheels can be installed in any of the previously defined TORQWHEEL models, up to the growth percentages in the table.

Table 6: Type B RWA Motor Driver Options

Model Number		TW-16B32	TW-16B50	TW-16B200	TW-16B300
Momentum Capacity	N-m-s	16.6	16.6	16.6	16.6
Reaction Torque	mN-m	32	50	200	300
Speed Range	rpm	5100	5100	5100	5100
Peak Power	W	50	75	250	350
Steady State Power at 1000 rpm	W	6.5	7	7	10
RW Mass	Kg	5.9	7.8	8	5.1
Motor Driver Mass	Kg	Included	Included	Included	3.3
Motor Driver Current Capability	A	2.0	3.0	8.0	13.0
Acceleration (Zero to Sixty rpm)	Sec	6.1	3.9	1.0	0.6
Designation	-	Classic B	GPS IIF B	Killer B	Turbo-B
Sponsor/Driver			Bob/Kevin	Charlie/Alex	Dewey/Scott

Table 7: TORQWHEEL Momentum Growth Capabilities

RWA Size	Baseline Momentum Capacity	Speed Range	Maximum Growth	Maximum Momentum Capacity	Mass Increase at Max Momentum Capacity
	N-m-s	rpm	%	N-m-s	Kg
Type A	4	5100	150	10	2
Type B	16.6	5100	100	30	4
Type E	50	3850	150	125	10

MICROVIBRATION

Higher torque and momentum performance are frequently coupled with low disturbance requirements to minimize spacecraft jitter in remote sensing applications and sensitive astronomic observatories. This is an area in which Ithaco has done considerable research and development over the past several years. In addition to fine tuning the motor and motor driver design to minimize disturbances, a Kistler is being used for characterizing reaction wheel disturbances. This precision state-of-the-art force/torque measurement system has been adapted for performing an optional fine balancing of the flywheel after the bearings are installed, resulting in minimum imbalance. The Kistler table has been used successfully on several programs with jitter sensitive instruments to reduce the imbalance to a minimum and characterize actual disturbance performance.

The stock imbalance is achieved by precision balancing the flywheel relative to the bearing journals on the shaft, and installing the bearings. A small amount of imbalance is introduced due to the runout of the bearing and the bearing fit. For finer balance, the wheel assembly is spun up on the Kistler table, and additional corrections are made after the installation of the bearings. The stock imbalance levels and the optional fine balance levels are presented in Table 8 for the available wheel sizes.

Table 8: Imbalance Specifications

Wheel Class	Static (Bubble Balance)* gm-cm		Dynamic (Spin Balance)* gm-cm^2	
	Stock	Fine Balance Option	Stock	Fine Balance Option
Type A	0.5	0.25	10	5
Type B	1.5	0.5	40	5
Type E	1.8	0.7	60	20

*Note: While all balancing at Ithaco is performed with a rotating wheel, the terms static and dynamic imbalance were coined because static imbalance can be measured statically with a bubble balance, whereas dynamic imbalance can only be measured by spinning a wheel and measuring the resulting force couple.

The characterization of TORQWHEEL disturbances results in a waterfall plot of spectral disturbance versus wheel speed. Sample measurements taken on the TORQWHEELs built for NASA's Space Infrared Telescope Facility, (SIRTF, the fourth and final element in NASA's family of "Great Observatories") are shown in Figures 3 and 4. SIRTF uses a stock TW-16B32 TORQWHEEL with a fine balance option.

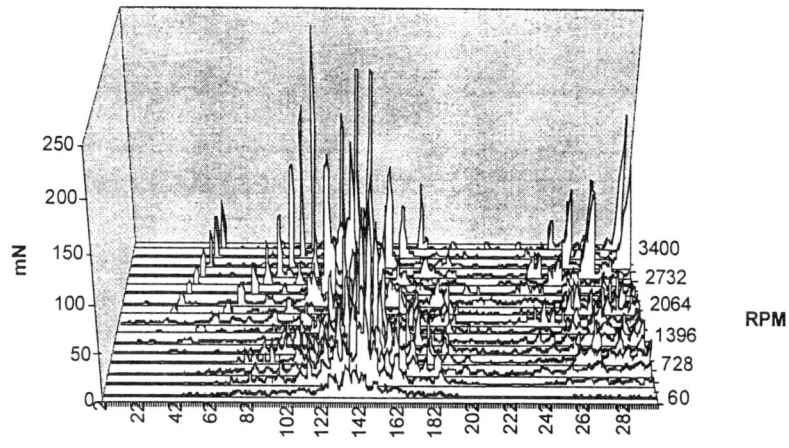

Figure 3: Type B RWA Radial Force Disturbance

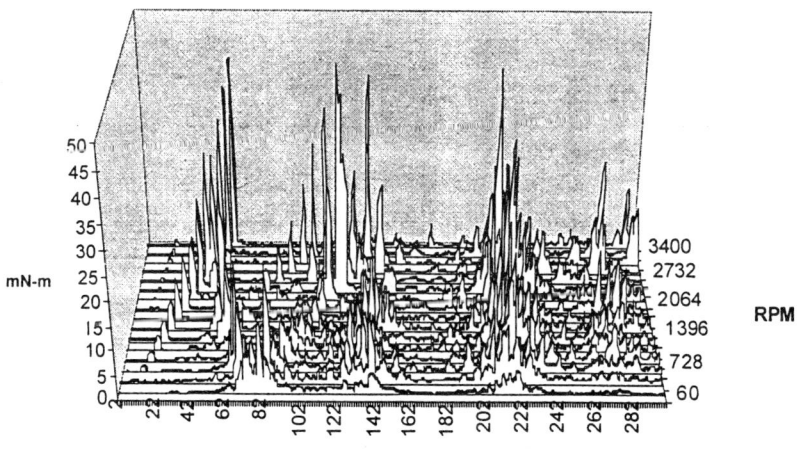

Figure 4: Type B RWA Radial Torque Disturbance

HIGH PERFORMANCE LUBRICATION CONSIDERATIONS

High torque reaction wheel applications typically require a significant amount of zero speed operation and zero speed crossings. Zero crossings can result in wear, since the lubricant is unable to form an elasto-hydrodynamic (EHD) film between the ball and race. In this operating regime, there is metal-to-metal contact between the ball and the race, and the extreme pressure additive provides the wear protection. If an application requires a zero-speed biased wheel, and the error signals driving the torque command are oscillating, the bearings will dither back and forth within the stiction threshold. In this mode, severe wear can occur due to the metal-to-metal asperity contacts, if a suitable extreme pressure additive is not present.

The lubricant formulation used in Ithaco TORQWHEELs includes 5% of a Lead Naphethenate (PbNp) extreme pressure antiwear additive from Bardahl in a synthetic hydrocarbon oil (Pennzane SHF X2000) from Pennzoil. The PbNp adds a protective plating of lead to the metal parts during periods of potential wear. The negligible volatility of the lead ensures its availability for the entire life of the mission, rather than a gradual depletion which can occur with some surface treatments. In order to validate the endurance of the additive, two verification tests have been recently completed on representative bearings in hard vacuum.

The first test was a bearing wear test where the temperature in hard vacuum was elevated to 71C in order to achieve sub-EHD operation at 300 rpm. Over 140 million revolutions were accumulated sub-EHD over ten months. The breakaway torque was trended during the test and showed no signs of degradation, as shown in Figure 5. The visual inspection of the bearings following the test showed no signs of wear, and no measurable lubrication loss.

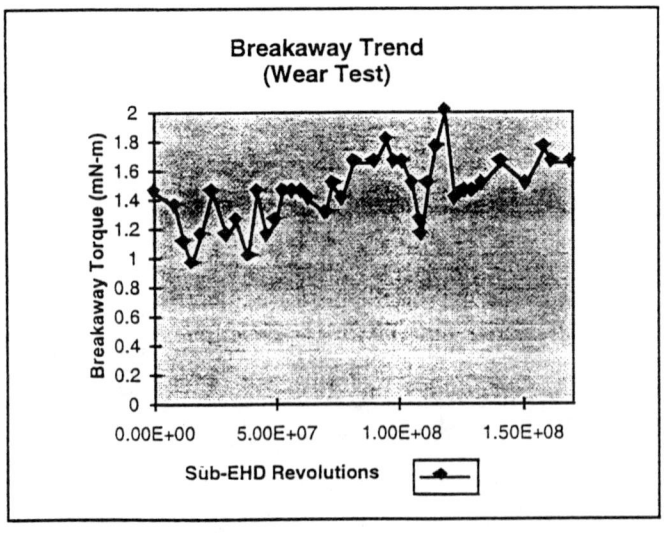

Figure 5

The second test was a dither test, where over 460,000 slow zero crossings were accumulated over eleven months. The zero crossings were accomplished at a rate of once per minute, and utilized a small torque with some AC noise simulating noise at a controller bandwidth to force a small amount of dithering in the bearing stiction deadband. This is the worst case condition for bearing wear in a reaction wheel application. The bearing breakaway torque was trended during the test and showed no signs of degradation, as shown in Figure 6. The visual inspection of the bearings following the test showed no signs of wear, and no measurable lubrication loss.

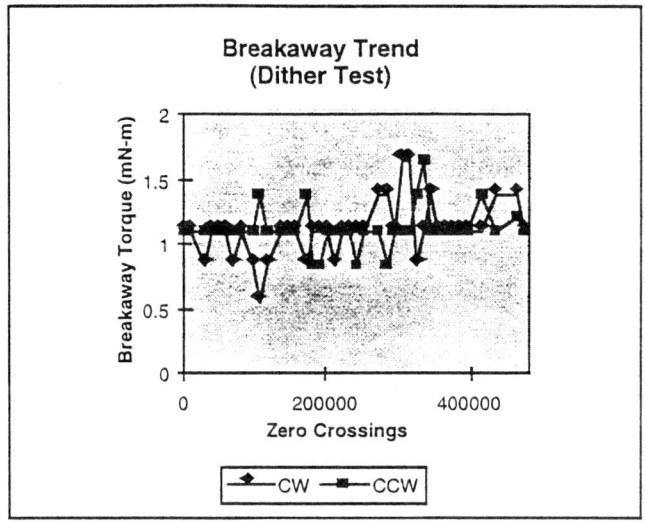

Figure 6

The successful results of these two extremely demanding tests provide critical backup for missions desiring or requiring operation near or through zero speed. Previous limitations imposed on operation below the EHD liftoff speed need not be applied with the PbNp additive.

CONCLUSION

The Ithaco TORQWHEEL product line has realized significant increased performance capabilities by capitalizing on growth provisions planned into the original stock designs. High torque and high momentum are achieved by making minor wire termination changes, incorporating higher current motor drivers, and changing flywheel and stator materials. The result is a broad range of performance capabilities to meet the demands of today's faster, better, cheaper and more agile spacecraft. In addition, significant advancements have been made in the areas of microvibration and low speed endurance, to provide a winning total performance package.

AAS 00-043

PROGRESS OF FIBER OPTIC GYROSCOPE DEVELOPMENT FOR SPACE APPLICATIONS

James Goodwin, Pei-hwa Lo, Mark Mariak and Ming Yu[*]

Users of inertial navigation systems for launch vehicles and spacecraft applications have been demanding, on an increasing basis, systems with better performance, reliability, and life at lower costs. L-3 Communications Space & Navigation has positioned itself to respond to this demand with interferometric fiber optic gyroscope (IFOG) based systems. This paper will present Space & Navigation's current IFOG based inertial reference unit (IRU) products and its activities to improve and expand the technology to meet a wide range of space applications.

INTRODUCTION

Space & Navigation of L-3 Communications has been actively developing and producing inertial navigation systems and attitude control components for launch vehicles and spacecraft for over 35 years. For rate sensors, three principal technologies have been and continue to be utilized; mechanical (spinning wheel), ring laser gyro (RLG), and interferometric fiber optic gyro (IFOG). We have successfully demonstrated these technologies on Apollo and Delta launch vehicle systems, and attitude determination systems on Hubble Space Telescope, classified satellites, P59, P81, Iridium, Ikonos, FUSE, QuickScat, and ICESat. Of these rate sensor types, Space & Navigation has achieved considerable success over the past 15 years in bringing IFOG from an emerging technology to standard product offerings across different performance regimes. By leveraging the knowledge base acquired in developing and producing the high performance ORSA IFOG and the tactical grade SiFORS IFOG, Space & Navigation continues its work to introduce new product lines, improve performance, and reduce costs of IFOGs for space applications.

This paper will present Space & Navigation's background in IFOG development and production and will progress to a discussion of its current activities in three rate sensor performance grades; sensor and appendage control (<25 deg/h), inertial navigation (<0.01 deg/h), and spacecraft pointing (<0.001 deg/h).

---

[*] All authors: Technical Staff, L-3 Communications Space & Navigation, 699 Route 46 East, Teterboro, New Jersey 07608.

BACKGROUND OF IFOG DEVELOPMENT AT SPACE & NAVIGATION

Space & Navigation has viewed IFOG technology as the most promising in the effort to meet the space market's need for a variety of inertial reference unit (IRU) and inertial measurement unit (IMU) grades offering low cost, high reliability, and long life. To this end, it has expanded its internal IFOG development which has resulted in two IFOG IRU/IMU product lines and a number of rate sensors for specialty applications. IFOGs produced by Space & Navigation, and IFOGs in general, offer the following advantages over competing technologies:

- High reliability
- Long life
- Low cost designs
- Low complexity
- No mechanical dither
- Low voltage operation
- High resolution
- Adaptable sensor configuration
- Easily scalable to other performance grades

Due to their scalable nature, IFOGs are re-configurable, at relatively low risk, to meet the full spectrum of performance needs. The two product lines currently offered by Space & Navigation evidence this and are shown in Figure 1:

Performance Parameter	SiFORS *tactical*	ORSA *high performance*
Angle Random Walk (deg/rt-hr)	0.4	0.00025
Bias Stability (deg/h)	25	0.00025
Scale Factor (PPM)	500	1
IRU/IMU volume (sensor) in^3	9	250
Rate range (deg/sec)	±1000	±16
Operational environment	Gun launch	Space

Figure 1. Performance parameters of Space & Navigation IFOGs

The ORSA is a space qualified IRU currently in production. The SiFORS IMU product line is currently in the early stages of high production. The development of these products has been enabled by the advent of highly integrated electro-optical components and the extensive knowledge of systems and functions critical to accuracy, stability, and reliability. Building upon its ORSA and SiFORS IFOG production programs, internal development at Space & Navigation is directed at further reducing costs, improving performance, and expanding its product lines. A discussion of on-going activities across three performance grades for space applications; sensor and appendage control, inertial navigation, and spacecraft pointing follows.

IFOG IMU FOR SPACE BASED SENSOR AND APPENDAGE CONTROL

Development Approach

For applications requiring low cost, miniature IRUs and IMUs for sensor pointing and articulation, Space & Navigation is baselining its SiFORS IMU[1] with additional improvements to lower cost and increase performance. SiFORS, a highly integrated very low cost IFOG based IMU, is currently in production for smart munitions applications. The gyro portion of the IMU is based on silicon integrated optics utilizing Micro-machined Optical-Electronic Systems (MOES), while the accelerometer is based on micro-machined electromechanical systems (MEMS) technologies. Both emerging technologies were selected on the basis of enabling significantly lower sensor costs and sizes while meeting performance requirements.

The rate sensor of the IMU has a unique design employing the best features of both fiber optics and MOES processing. A key component of the SiFORS rotation sensor is the silicon integrated optics Multi-Function Chip (MFC), which provides the optical bench and low noise electronic detection capabilities for this IFOG. With the advancement of the telecommunications industry, electro-optical components have become much less expensive and much more integrated. The fundamental SiFORS design approach is to take full advantage of this industrial trend. Critical electro-optical components in a FOG include the light source, photo-detector, splitter/coupler, polarizer, phase modulator, and fiber coil. Most of these components are also being developed and integrated in the telecommunications industry. However, some significant differences exist between the requirements for the gyroscope and the telecommunication market. For example, a laser diode is a commonly used light source in the telecommunications industry. Laser diodes are inexpensive and can be readily integrated into a silicon-based optical bench. While a laser diode provides high output power and low temperature sensitivity, it presents a discrete narrow band spectrum. It is well known in FOG design that a broadband light source is necessary to maintain desirable in-run bias stability characteristics of the gyroscope. Therefore, the design challenge is to control the operation of the laser diode light source such that bias stability and scale factor meet the gyroscope requirements.

Sensor Configuration

The SiFORS FOG comprises only four parts—the MFC, sensing fiber, sensing spool, and lithium niobate modulator; and the assembly consists of just two operations—winding of the sensing coil and three fiber splices. An open loop architecture is utilized and is operated at the eigen-frequency to minimize electrical and optical error sources. Power control and modulation depth control loops are used to ensure bias and scale factor stability of the gyro. Since a small spool size and unique symmetrical mechanical architecture are used, the SiFORS IMU demonstrates minimal Shupe and other thermal transient effects. This integrated configuration and low part count design results in an IMU weighing less than 0.6 lb and consuming less than 6.5 watts. A photograph of the SiFORS IMU is shown in Figure 2.

Figure 2. SiFORS IMU

The current bias stability performance is under 25 deg/h. An Allan variance analysis depicting bias stability and the angle random walk characteristics for the current configuration is shown in Figure 3.

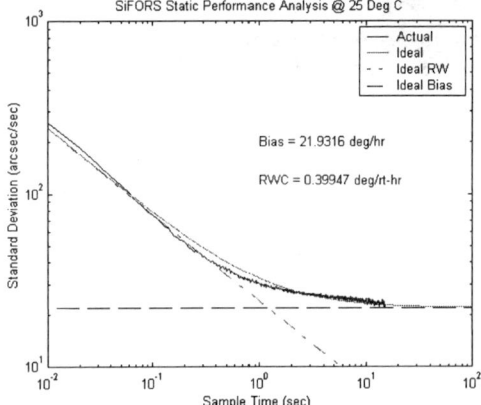

Figure 3. Bias Stability and ARW of SiFORS IFOG

Advanced SiFORS Development

To further reduce costs in material and labor, Space & Navigation initiated a project to further integrate the MFC by adding the functions of the lithium niobate modulator onto the MFC, namely modulation, 50/50 splitting, and polarization. Technical challenges for this task were resolved by a novel design of the polarizer and modulator. Based on analysis, the required polarization extinction ratio for a tactical grade rate sensor must be at least 30 dB. While this may not be a difficult requirement for a discrete polarizer, it becomes increasingly difficult to integrate a reliable polarizer into silicon under tight packaging constraints. Another significant technical challenge is to design an integrated phase modulator that can satisfy SiFORS requirements. Key parameters in phase modulator design include static loss, dynamic loss, modulation phase range and residual amplitude modulation. Successful control or compensation of the phase modulator limitations is critical to the performance characteristics of SiFORS.

Space & Navigation has been successful in overcoming these challenges and has designed and fabricated a MFC with modulation, polarization, and light splitting capability exceeding electro-optical characteristics required by the SiFORS IFOG. A photograph of the advanced MFC integrated to a sensing coil is shown in Figure 4.

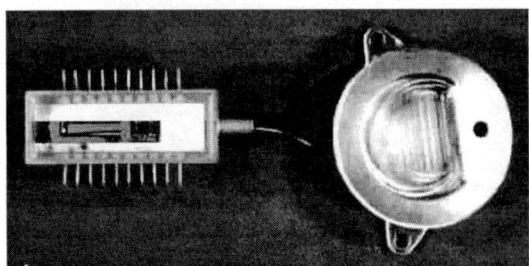

Figure 4. Advanced MFC integrated with Sensing Coil

Future plans include the integration of a upgraded light source onto the MFC and a higher scale factor sensing coil to support bias stability performance of under 1 deg/h for the SiFORS family of rate sensors.

INERTIAL NAVIGATION GRADE IFOG

Development Approach

Space & Navigation has been researching the development of an inertial navigation (IN) grade IFOG for several years. Goals have been to develop an IRU with the characteristics listed in Figure 5:

Performance Parameter	IFOG IN Grade
Angle Random Walk (deg/rt-hr)	<0.005
Bias Stability (deg/hr)	<0.01
Scale Factor (PPM)	<25
Quantization (arcsec/pulse)	<0.05
Weight (lb)	< 16
Power consumption (watts)	< 25
Operational thermal environment	-20 to 60C
Operational life (years)	>15

Figure 5. IN Grade Performance Characteristics

Earlier work centered on the development of a closed loop IFOG featuring a polarization maintaining (PM) sensing coil and a superfluorescent source (SFS). As cost became a key driver for inertial navigation applications the use of expensive PM coils (fiber lengths ~ 1000 meters) and SFS light sources precluded their use. To address this issue Space & Navigation investigated the use of alternate IFOG system designs that would be consistent with the IN grade cost model.

Depolarized IFOG

In order to reduce the sensing coil cost the depolarized IFOG system design[2] was selected which utilizes low cost single mode (SM) fiber for the sensing coil fiber. Space & Navigation has developed a proprietary depolarized (D) IFOG system design which improves upon the performance and cost of previous D-IFOG designs in several areas:

- Reduced magnetic sensitivities
- Reduction of bias drift due to polarization cross coupling
- Compensation for fiber misalignments at fiber-to-fiber and fiber-to-device interfaces
- Utilization of common low cost telecommunication single mode fiber for the sensing coil.

Several D-IFOGs have been built and evaluated including one based on our proprietary design. Configuration and performance requirements consistent with IRU goals have been met with the proprietary design. A photograph of the 3.25 inch diameter sensor head containing the optical circuit (single mode fiber sensing coil, modulator, coupler, passive SFS light source components) is shown in Figure 6. A plot depicting bias stability and angle random walk is shown in Figure 7.

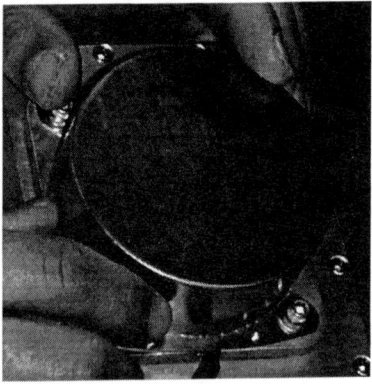

Figure 6. IN Grade IFOG

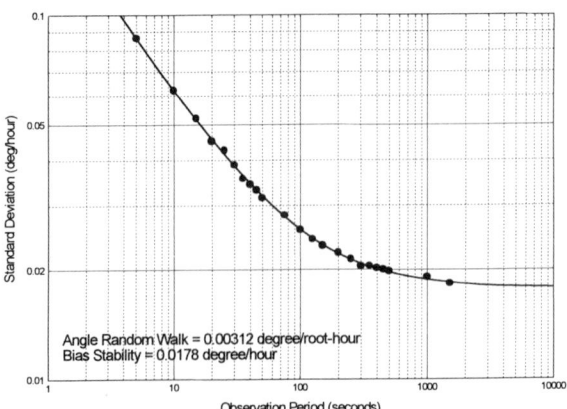

Figure 7. D-FOG Allan variance plot

Light Source Development

To reduce light source costs Space & Navigation is developing lower cost SFS light sources and improved superluminescent diode (SLD) light sources. SFS light sources have been preferred over other light sources due to their broadband spectrum, short coherence length, high wavelength stability, and high power. Historically they have been unattractive due to high cost. But in recent years lower cost SFS components have been developed to support telecommunication industry needs which are consistent with our IN grade cost model. SLD are of lower cost, possess a broad spectrum and short coherence length but lack the wavelength stability required for IN grade performance. Present work is centered on improving this stability through several proprietary techniques.

Other Supporting Development Activities

Other development activities are in progress to further reduce cost and extend IN grade IFOGs into other space based applications. Leveraging the successful application of the silicon integrated electro-optic MFC to the SiFORS line, a new MFC will be developed meeting IN grade requirements. Space & Navigation is also capitalizing on the flexibility offered by extensive utilization of digital processing techniques and modulator system design. With minimum change to hardware and software, changes can be made to baseline configurations to tailor systems to various performance, operating environment, and reliability requirements.

STRATEGIC & POINTING GRADE IFOG

Space & Navigation is recognized in the IFOG market as a leading supplier of pointing grade IFOGs for demanding applications. Space-qualified pointing grade IFOGs used in the ORSA program have been in manufacturing for over 5 years. The ORSA IFOG successfully passed stringent reliability and radiation requirements to meet long life space requirements. Design and development during the ORSA program and internal R&D have given us several key competitive advantages such as IFOG gyro noise source modeling, broadband light source design, detailed FOG scale factor error budgeting, radiation end of life requirement definition & implementation, and IFOG production expertise.

Space & Navigation has committed significant IR&D funding in its efforts to continuously improve several crucial gyro performance parameters. One example is the use of an existing gyro noise model[3] to predict angle random walk performance of less than 100 μ deg/rt-hr using IFOG technology similar to the ORSA hardware and with proven in-house noise reduction techniques.

The IFOG industry has reported IFOG scale factor non-linearities at low angular rates[4], which is commonly referred as a dead band or deadzone. This IFOG deadzone is quite similar to a Ring Laser Gyro (RLG) lock-in error but occurs at a much smaller angular rate range (0.05 deg/h vs. 100 deg/h respectively). Figure 8 illustrates a typical deadzone width for an ORSA IFOG.

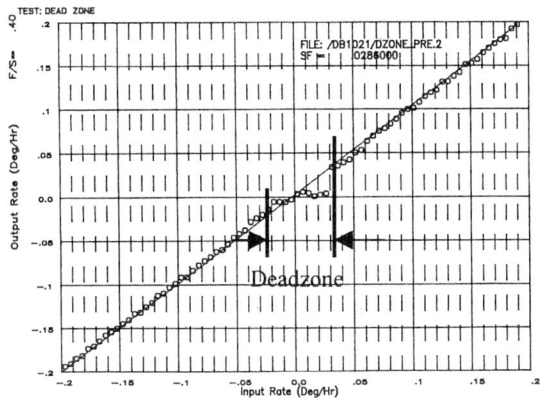

Figure 8. IFOG Output with Deadzone

Recently, we have achieved a critical milestone to eliminate this deadzone with a proprietary electronic dither approach. The result of this approach is showed in Figure 9 demonstrating that the IFOG deadzone was totally eliminated.

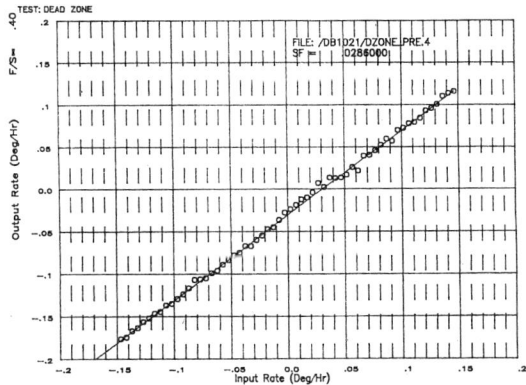

Figure 9. IFOG Output with Deadzone Eliminated

A block diagram of this electronic dither approach is shown in Figure 10 where an external periodic disturbance is periodically added to the gyro closed loop circuit to remove the deadzone.

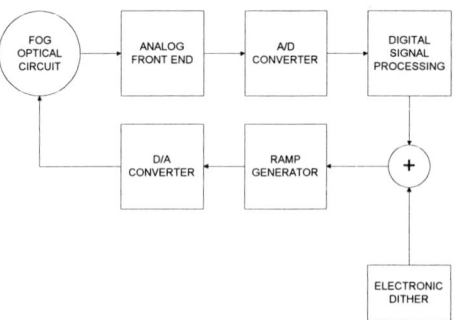

Figure 10. Electronics Dither Approach

Since this is a purely electronic perturbation, there are no technical or cost disadvantages such as the mechanical dither approach used in RLG. The whole approach was achieved in digital processing software, with no additional hardware design or recurring cost to the IFOG.

CONCLUSIONS

L-3 Communications Space & Navigation is building and expanding upon it extensive experience in producing IRUs for demanding space applications. More specifically it is using the knowledge base gained in the development and production of the SiFORS and ORSA IFOG IMU/IRUs in conjunction with recent technological advancements in electo-optics to further reduce cost, improve performance, and increase reliability and life. Production and prototypes units achieving these goals have been delivered and demonstrated respectively.

REFERENCES

1. H. Califano, P. Lo, J. Goodwin, S. Leoni, D. Siebert, "SiFORS Performance: A Low Cost Silicon Integrated Optics Fiber Optic Rotation Sensor", *AIAA 99-4077*, AIAA Guidance, Navigation, and Control Conference 1999.

2. K. Bohm, P. Marten, K. Petermann, E. Weidel, "Low Drift Fibre Gyro using a Superluminescent Diode", Electronics Letters, Vol. 17(10), 352-353, 1981

3. M. Scruggs, P. Wall, M. Yu, "Progress in a Strategic Grade IFOG", *AAS98-051*, 21st annual AAS guidance and control conference 1998.

4. R. Kovacs, " Fiber Optic Gyroscope with Reduced Non-linearity at Low Angular Rates" *AAS98-043*, 21st annual AAS guidance and control conference 1998

AAS 00-044

SOHO: LOSS AND RECOVERY 1998, GYROLESS 1999

A. van Overbeek[*]

The Solar and Heliospheric Observatory (SOHO) is a joint program of the European Space Agency (ESA) and NASA. SOHO finished its commissioning phase and started full scientific observations from its three axis stabilized position in the first Lagrangian point in April 1996.
Control of the spacecraft was lost on June 25, 1998. Based on the last telemetry data received from SOHO we believed that the spacecraft was slowly spinning in such a way that its solar arrays did not receive adequate sunlight to generate power. However due to the yearly orbital movement around the Sun (approximately one degree per day) SOHO's solar panels were exposed to an increasing amount of sunlight each day. On July 23, 1998 the huge antenna of Arecibo confirmed SOHO spacecraft was at its predicted location and moreover that the spin rate was slower than 1 revolution per minute. From August onward, when the first telemetry was collected, a period started which consisted in thawing the frozen hydrazine, while trying to keep the delicate power balance.
In the mean time a number of strategies to escape from the spin were designed and traded off against one another. The big day was September 16, 1998 when the spin was stopped and the spacecraft returned to its normal three axis stabilized position with permanent power from the solar arrays. SOHO was back into Normal Mode on Sept 25.
It turned out that two of the three gyros were gone. Operations continued with only one working gyro. Then just before Christmas 1998 the last gyro died, putting SOHO back into its safe mode. Early 1999 an ad-hoc strategy was designed to get out of the thruster based safe mode to a wheel-based mode in order to stop the excessive fuel use. This was ready and exercised in February 1999.
The spring and summer of 1999 were used to design the 'final and complete' solution in order to increase the safety of the spacecraft and make it as tolerant to equipment failures as it had been before the loss of the gyro. This work resulted in new software for both the AOCS and Data Handling computers, completely new ground procedures and associated changes in the ground system.
In December 1999 we found one problem with the new design which caused an unwanted entry into the safe mode. A temporary fix is now in place and a final fix should be installed in March 2000.

Introduction

SOHO, the Solar and Heliospheric Observatory, was launched in December 1995 in order to observe the sun continuously from the first Lagrangian point (L1) about 1.5 million km from the earth where the gravity forces of the earth and the sun are in equilibrium.

---

[*] European Space Research and Technology Center (ESTEC), Attitude, Navigation and Control Section (TOS-ESC), P.O. Box 299, NL-2200 AG Noordwijk, The Netherlands. E-mail: twovorbe@estec.esa.nl.

Normal scientific observations started in April 1996 and the nominal two-year mission ended in April 1998. On June 25, 1998 SOHO was lost at the beginning of a week with planned maneuvers. Let us first look at the spacecraft (Figure 1).

During normal operations SOHO always looks at the sun (+X-axis). The Y-axis goes through the two solar arrays and the Z-axis is kept aligned with the poles of the sun. The two solar arrays always face the sun and can each provide up to 1500 W of power. There are two 20Ah NiCd batteries for power storage which were foreseen to be only used immediately after launch.

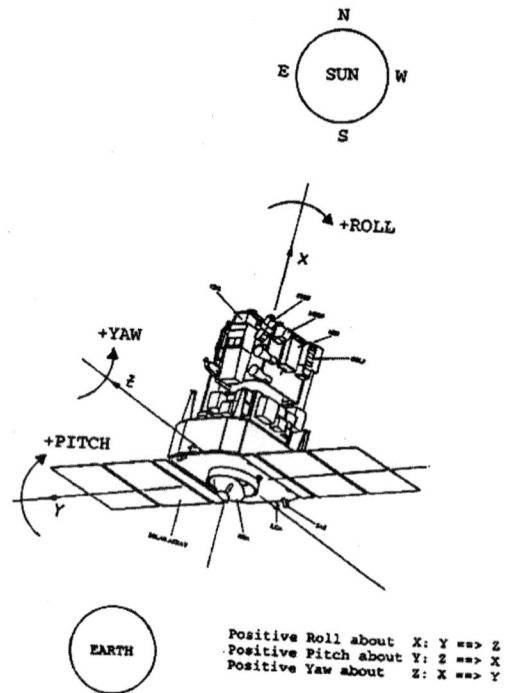

Figure 1: The SOHO Satellite and its Axis Definitions

In case of loss of sun pointing, an Emergency Sun Reacquisition (ESR) is triggered. This is a hardwired thruster based controller. There are eight 4N thrusters per branch. The nominal A branch is used for reaction wheel off loading and orbit correction maneuvers. The ESR controller uses the redundant B branch.

There are three coarse Sun Acquisition Sensors. SAS 1 is located on top of the spacecraft on the +X face. SAS 2 and 3 are located on the bottom on the +Z and −Z face. For nominal control a Fine Pointing Sun Sensor (FPSS) is used for pitch and yaw sensing. Roll sensing is done with a Star Sensor Unit (SSU). Actuation is provided by three out of four reaction wheels. During thruster modes the SSU cannot provide roll information because the rates are too high. Then a single-axis gyro is used. There are a total of three gyros: one for roll control during thruster operations (gyro C), one for failure detection (roll rate/roll angle, gyro B) and one as part of the hardwired ESR controller (gyro A) to avoid build up of excessive roll rate from the pitch and yaw thruster firings in ESR.

The Loss of SOHO, June 25 1998

During the beginning of a set of maneuvers the ESR mode was entered twice due to procedural mistakes. During the recovery of the second ESR the failure detection gyro was switched off because it was (mistakenly) judged faulty. Also gyro A was off. This caused an uncontrolled spin up on the roll axis. The cross products of inertia and thruster misalignments led to an off pointing from the sunline, which triggered a third ESR. Since gyro A was off (no roll-rate control in ESR) the build-up of the roll rate continued which finally led to a total loss of control and power. This is illustrated in one of the computer animations. The animation is based on spacecraft dynamics simulations with our best guess of the initial conditions. Figures 2 and 3 show the attitude angle history and the angles seen by SAS1 on top of the spacecraft. Also in figure 3 you see the actual telemetry points. At simulation t = 0 the spin-rate is already 7.2 degrees/sec and there is already a 5 degrees off-pointing from the sunline.

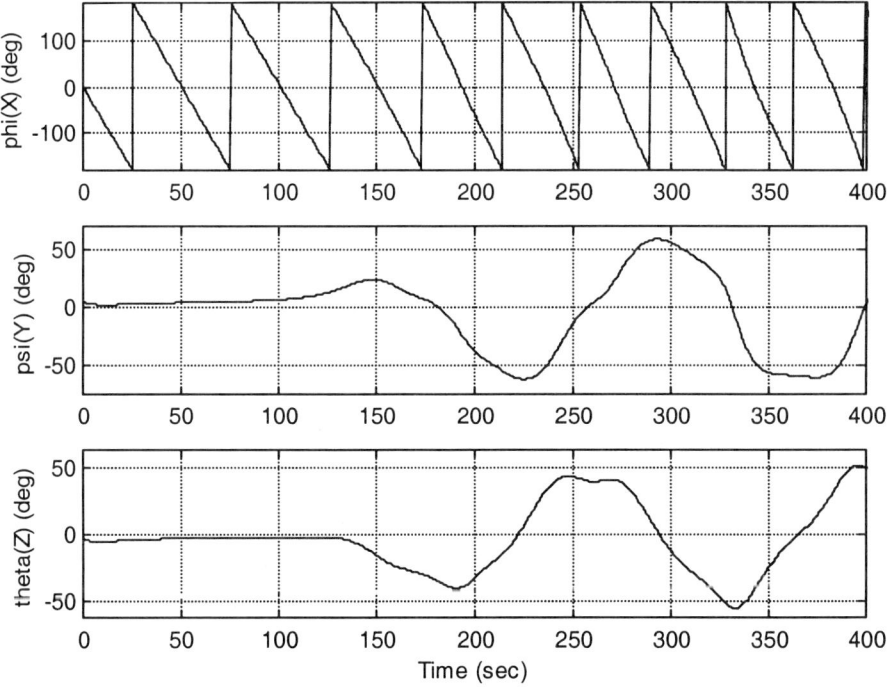

Figure 2: Attitude Euler Angles

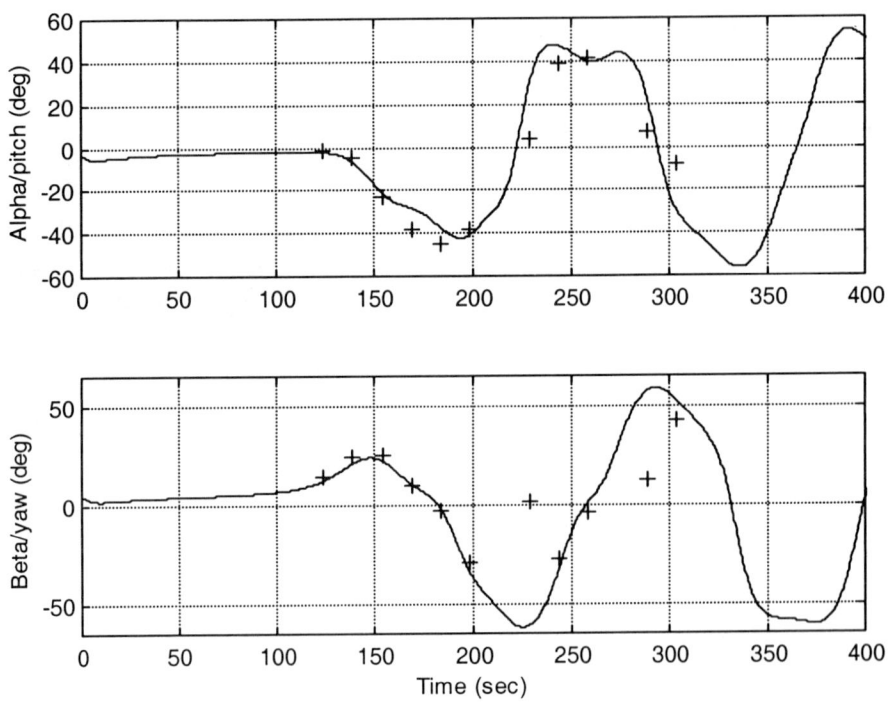

Figure 3: SAS1 Simulated Outputs and SAS1 Telemetry

Getting Telemetry

Flight dynamics analysis and simulation showed that the spacecraft would end up spinning around its major axis of inertia (Z-axis, 3600 kgm^2) after some time. By the end of June, SOHO would have its solar arrays edge on to the sun. The motion around the sun degree by degree, day after day, increases the period with sunlight on the solar arrays. It means that after 3 months the Z-axis would become perpendicular to the sun, which would bring light on the solar arrays for half a rotation period and hence power to the spacecraft. This is also demonstrated in the animation.

Following a proposition by scientists at the US National Astronomy and Ionosphere Center, on July 23 the 305 m antenna of the Arecibo radio telescope (Puerto Rico) was used to perform a bistatic radar test with a power of 580kW and transmitting at 2.380 GHz. The 70-m station at Goldstone was used as the receiver. Strong echoes from SOHO were received and it was still at its predicted location. Moreover the signal width was between 1 to 2 Hz, which is compatible with a spin rate of 52.87 seconds (determined through Fourier analysis by radar experts at Cornell University. The center frequency drifted slowly by a few Hz indicating a non-principal axis of rotation. The analysis of the collected data indicated a radar cross section of 15 to 20 m^2. This implied that SOHO was still in one piece and gave us good hope.

In order to receive telemetry the transmitter has to be turned on (the receiver is on as long as there is power available). However this command has to be sent by the central data-handling computer. So during the (less than) half period, some 20 seconds, when there is solar power, commands have to be sent to switch the computer on, let it boot and send commands to turn the transmitter on and configure the RF system. This sequence of commands was scrutinized and optimized over and over again. The only thing we could expect was to get a spike of the carrier signal. To detect this spike special equipment had to be installed in the DSN ground stations. Also ESA ground stations joined the search for the carrier signal. Finally on August 3, 1998 we got our first carrier spike.

This meant already a lot: the receiver and the tele-command system were working, the central computer, major parts of the data handling subsystem and the power switching were working. Now we needed to charge the battery so we could get power over several revolutions in order to get real telemetry.

It turned out to be necessary to continuously send the Battery Charge Regulator (BCR) on telecommand in order to get the battery charged. The charge regulator switches off automatically when the voltage drops below 20 V. On August 8, after 10 hours of such continuous commanding, battery 2 was charged up and successfully connected to the bus to get telemetry. To avoid discharging the battery, its two Battery Discharge Regulators (BDR) were opened at the end of the test, which switched off the power and hence the transmitter.

These first frames of telemetry showed extreme temperatures in the spacecraft e.g.: batteries at -20 °C, gyros at about -25 °C, some instruments very hot (+80 °C) others very cold (-60 °C).
Analysis of the Sun Acquisition Sensor data confirmed a rotation period of 52.6 seconds and that the +Z axis was facing the sun. This also determined the angle between the spin axis and the sun (approximately 36.7 degrees, as measured on August 11).

Thawing 50 kg of Hydrazine with 30 W

From the first temperature readings and the thermal analyses performed, it was estimated that at least 48 kg of the estimated 200 kg of hydrazine in the tank was frozen. The thrusters on the shadow side of the spacecraft and their associated pipes were also frozen. The thrusters on the +Z sun-lit side showed more favorable temperatures, around 0°C. The propulsion system had to be thawed by starting with the tank and then working outward to the thrusters. In this way any overpressure flows to the tank and pipe rupture is avoided.

The tank thawing operation was started on 12 August, after a cycle of battery charging, but the first signs of a temperature increase were not observed until 25 August. The power of the tank heaters is only 30 W, just a simple light bulb. The tank heating was performed with both batteries on the bus and, after telemetry switch-off, by providing power only to the tank heaters (all other equipment was switched off except for short temperature and battery-voltage telemetry checks every 4 h). The tank thawing process had to be interrupted three times to recharge the batteries. The total power consumption during the heating operation was about 87 W (with telemetry off). The whole story is visible in Figure 4. The helium is used as a pressurizer.

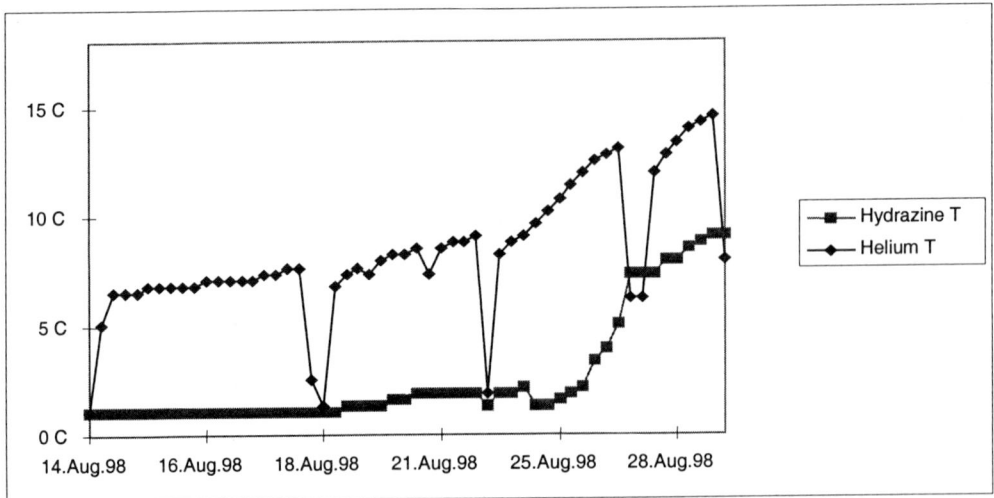

Figure 4: Thawing of the tank

Thawing of the tank was completed on 30 August, after 275 h of heating (more than 11 days, without taking into account the battery charging periods). A very careful balance between the time devoted to battery charging and the power available for thawing the complete propulsion subsystem had to be maintained.

The "Sunheat" mode

For further thawing of the propulsion system and keeping the tank from freezing again more power was needed. The batteries charge with a current of 1 A. When telemetry is on, the power consumption is 105W, which, during each eclipse period, discharges the batteries with a current of 1.25A.

Fortunately it was possible to patch the Central On-Board Software (COBS) to use a current like a "fake thermistor" in order to switch on heaters when power was available from the solar arrays and off if not. The heater was switched on when a solar array current was above a "maximum" and off when below a "minimum". The patch could be done by only changing the values in some data tables. This patch was first tested on August 19. It had to be reloaded after each battery charging period (during charge the on-board computer is off). This became known as the "sunheat" mode. Later on this concept was refined and since it worked so well to heat up the propulsion system without discharging the batteries, it was used for more and more heaters. The total heating power profile had to be within the power provided by the solar arrays during the 26 seconds per revolution with sun on the arrays. At the end (September 1998) 48 heaters were used in "sunheat" mode for a total of 517 W of heating power. See the final "sunheat" profile in figure 5.

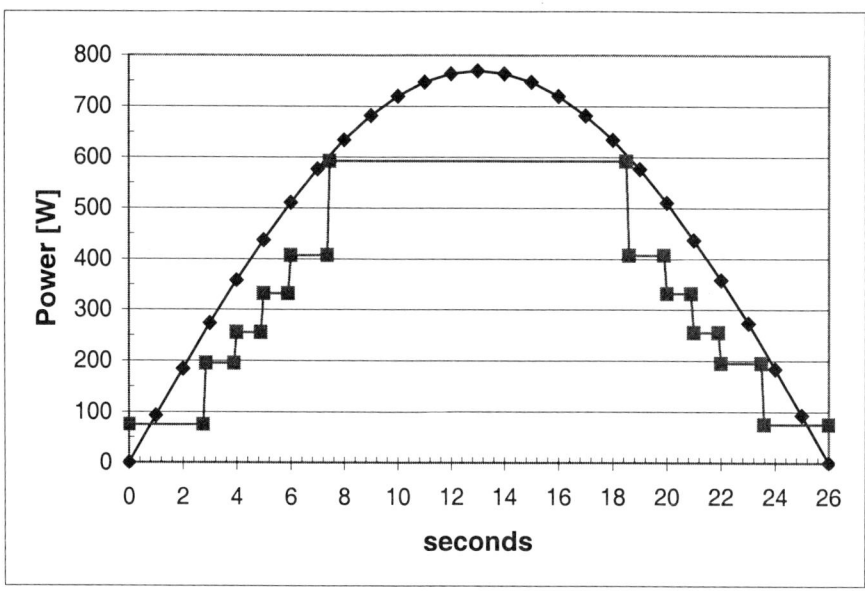

Figure 5: The final "sunheat" profile

Thawing of the Hydrazine Lines

Once the tank had been thawed, the thawing of the pipes and thrusters was started on 3 September. Following a long discharge cycle, it was not possible to recharge the batteries whilst simultaneously maintaining the temperature of the propulsion subsystem. Priority had to be given to recharging the batteries as well as the maintenance of the 28 V bus, which was crucial for communication with the spacecraft. It therefore became evident that the complete thawing of the propulsion subsystem might not be possible. The propulsion subsystem was kept warm by heating/charging cycles. This was done until the final attitude recovery maneuver. Nevertheless it was believed that the thrusters on the spacecraft shadow side were still frozen. Therefore attitude recovery strategies were evaluated which did not need these thrusters.

Possible Recovery Scenarios

Four possible alternatives were identified:

1. Full ESR recovery. A two-step approach based on the assumption that the full B-side of the propulsion system would be available for recovery. The first step would be stepwise spin-down of the existing Z-axis rotation to about 1 deg/s. Full recovery would then be initiated once the Sun was close to the center of SAS-1's field of view. During this ESR recovery, all eight branch-B thrusters would be used with SAS-1 for pitch and yaw control, and with one of the three gyros for roll control.
2. ESR without roll control – this approach took into account the possibility that there might be insufficient power for the above scenario (the gyro is a large power

consumer); ESR recovery without roll control would be pursued, using the Z-axis de-spin, four thrusters of branch-B only, and SAS-1.
3. Dual-spin recovery – this scenario, proposed by NASA, was based on the idea of stabilizing a spinning spacecraft around its axis of minimum moment of inertia. However, this posed nutation problems, and would not achieve a closed-loop coarse Sun-pointing attitude like in ESR.
4. Initial Sun Acquisition (ISA) recovery – similar in concept to ESR recovery, this approach uses the A-side with ISA mode, in case the B-side (ESR mode) would not be available. An important difference is that the ISA mode makes use of the SAS-1 sensor only, leading to less than hemispheric coverage (ESR uses all three SASs, and thus has omni-directional coverage), this would make the timing of the ISA triggering more critical.

The final choice was alternative 2: ESR without roll control.

Back to a normal ESR, 16 September 1998

After a three-week-long period of meticulous preparation the attitude recovery was established on 16 September as follows:

- After a full battery charge, the propulsion subsystem received a six-hour heating boost in order to test the thrusters needed for the recovery. (It was established that all 8 branch-B thrusters were available.)
- A small de-spin was carried out as a calibration, followed by a 2.37 deg/s de-spin (in three steps). A de-spin segment consists of a carefully planned sequence of thruster pulses. The spacing between the pulses is chosen to avoid any nutation build up. In less than an hour, data evaluation confirmed that the thrusters were working as expected and that the target spacecraft spin rate had been achieved.
- A second three-step de-spin maneuver brought the spacecraft rotation rate down to 0.86 deg/s. After a careful check on the success of the maneuver and a 'go' for all subsystems, ESR was triggered on-board to point the spacecraft towards the Sun without roll control. Sun lock was achieved almost immediately at 18:30 UTC. This is demonstrated by a computer animation.

The roll rate was then corrected using thrusters 5 and 6 in open loop from the ground.

From then on it was a busy week to check all the spacecraft subsystems and to perform an orbit correction maneuver. It turned out that all subsystems had survived the ordeal except two of the three gyros. The most probable cause of their failure is the freezing of the liquid, which stressed and broke the very thin wires between the fixed and floating parts (motor and pick-off connections).

On September 25, 19:52:58 UTC SOHO was finally back in its normal operating mode.

Loss of the Last Gyro

During the preparation for a planned momentum management and station-keeping maneuver in December 1998 the last gyro was lost. After gyro switch on it seemed to work OK. When a mode was activated which needed the gyro for control it did not work and SOHO entered ESR again on 21 December 1998, 17:49 UTC.

Now the recovery had to be done without any gyros. Since this required new procedures and validation by simulations it was decided to leave SOHO in ESR over the Christmas holiday period. The thruster firings by the ESR controller disturbed the orbit so a Delta-V maneuver was performed while in ESR using open-loop thruster commands on January 19. The combination of ESR firings and Delta-V meant an average fuel consumption of 7 kg per week. This could not continue (Normal fuel consumption is 2 kg per year). Therefore a scheme was devised to keep the ESR controller from firing by manually firing small (much smaller than the ESR controller) thruster pulses to keep the attitude inside the dead-band. This was mainly performed on the yaw-axis and was baptized "yaw-braking". This scheme could only be applied when there was contact with the spacecraft. Thanks to the good cooperation with the NASA Deep Space Network (DSN) we got extended passes. The flight operations team was very successful in applying this scheme. After it was in operation the net Delta-V caused by the ESR was close to zero.

First Gyroless Operations (February 1999)

In the meantime preparations for the ESR exit and how to operate the spacecraft without gyros continued at a frantic pace. This included the development and validation of AOCS software patches. On January 30 SOHO left ESR again and went straight to a wheel-based mode with roll control provided by the SSU. The roll rate had to be reduced in ESR to a low enough value so the SSU could track a star. This was first done by thruster pulses, but it did not reach the required target. During the maneuver we decided to do the last part of the reduction by changing the wheel speeds: increase the roll momentum of the wheels to decrease the roll momentum of the spacecraft. This was successful and at 17:45 UTC SOHO was out of ESR and in a wheel-based mode.

The main challenge for operating the spacecraft without gyros was how to do thruster based maneuvers (reaction wheel off-loading and orbit corrections) which had required the use of the gyro for roll sensing before. The solution was to fire the thrusters open loop while still in a wheel-based mode with the SSU as roll sensor. Initially this mode used both the SSU (for roll angle) and the gyro (for roll rate) as sensors, since its main purpose was to provide the capability to large roll slews. Now the roll controller was replaced by one that only used the SSU. Of course the pulse lengths have to be chosen in such a way as to stay well within the rate limits of the SSU. Its max allowed rate is 40 arcsec/sec.

In the original design momentum-management was done by commanding the wheels to their new desired speeds. A thruster based control mode took out the disturbances caused by this change of momentum. Now the process is reversed. Precalculated disturbances are introduced by open loop firing of the thrusters, which are taken out by the wheel-based controller and in this way drive the wheels to their desired speeds.

For Delta-V we fire the thrusters in pairs. To take out the disturbances caused by thruster misalignments some of the other thrusters fire small pulses to counteract the disturbance. In this way relatively large Delta-V segments can be done without saturating the wheels.

This was done for the first time on the next day: 1 February 1999.

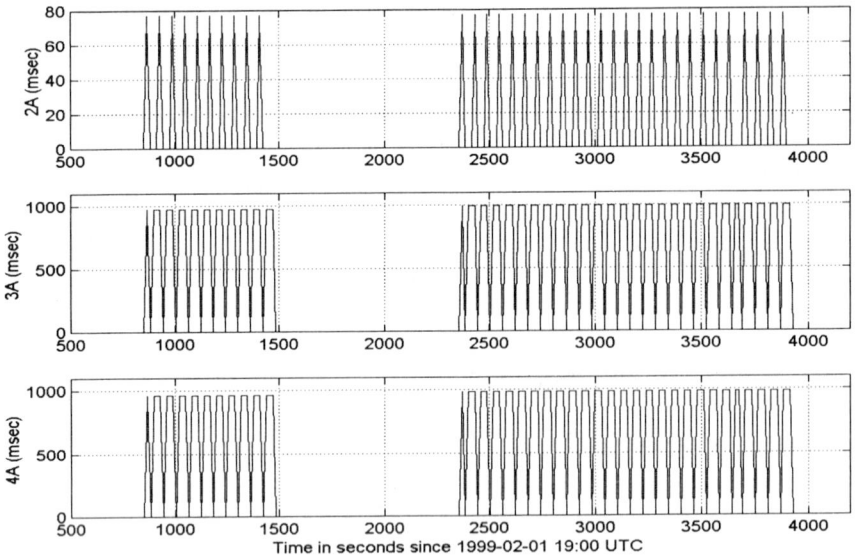

Figure 6: Thruster on-times, Delta-V, 1 February 1999

Thrusters 3 and 4 (See figure 6) were the ones used as Delta-V (1 second pulses). Thruster 2 was used as compensation thruster. The attitude disturbances caused by the thruster firings are shown in figure 7.

The initial disturbance on the yaw axis is caused by one of the delta-v thrusters only firing a partial first pulse. In order to avoid this now before each maneuver 10 short pulses on 4 thrusters simultaneously are fired in order to clear the propulsion lines. When all 4 thrusters fire the net force and torque are zero. If not, the disturbance can easily be taken out by the controller.

This first attempt of compensation was rather good as can be seen in figure 8, which shows the change of wheel speeds during the maneuver.

With this solution in place the spacecraft could be operated without gyros. However this was only a first cut. Now a final solution had to be designed and validated, a complete new user manual with all the associated procedures had to be written and implemented in the ground system. This took most of 1999.

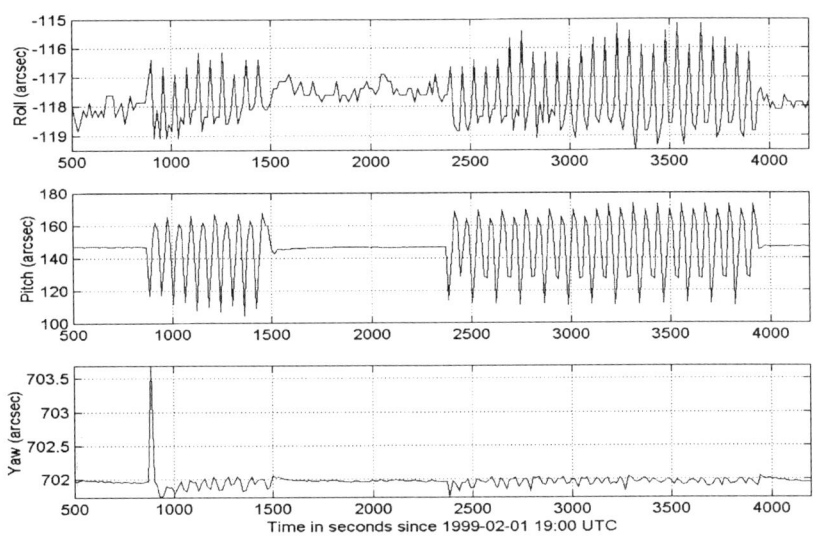

Figure 7: Attitude Disturbances, Delta-V 1 February 1999

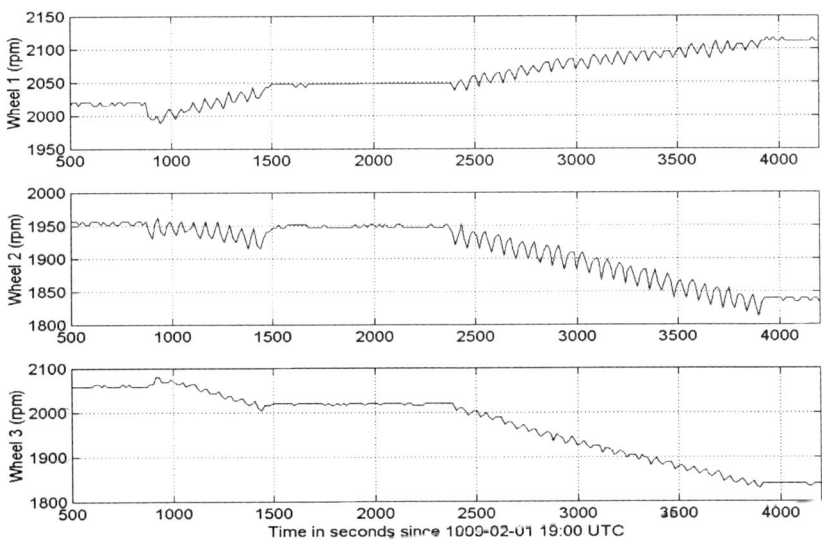

Figure 8: Wheel Speeds, Delta-V 1 February 1999

Development of the Final Solution

The main disadvantage of the ad-hoc solution was that a temporary loss off the SSU due to single event upsets on the CCD would immediately throw us back in ESR. A gyroless/SSU-less wheel based mode was needed. This new mode is now known as CRP (Coarse Roll Pointing). It uses the wheels both as a sensor and actuator. The principle is the conservation of angular momentum. In the short term external torque's can be ignored and hence the direction of the wheel angular momentum remains fixed in inertial space. By trying to keep the direction of the wheel angular momentum fixed in spacecraft coordinates we prevent the spacecraft from rolling.

This led to a new specification of the AOCS software and also to some needed additional functions in the data handling software in order to bring the overall spacecraft tolerance to failures back to its original level. The design was reviewed formally on 19 April 1999. The whole software development was done on a very tight schedule. The architectural principle is shown in figure 9. Since the existing software is being augmented, rather than replaced, it is necessary to place the new software in a previously unused area of memory and for it to be able to refer to existing variables and procedures.

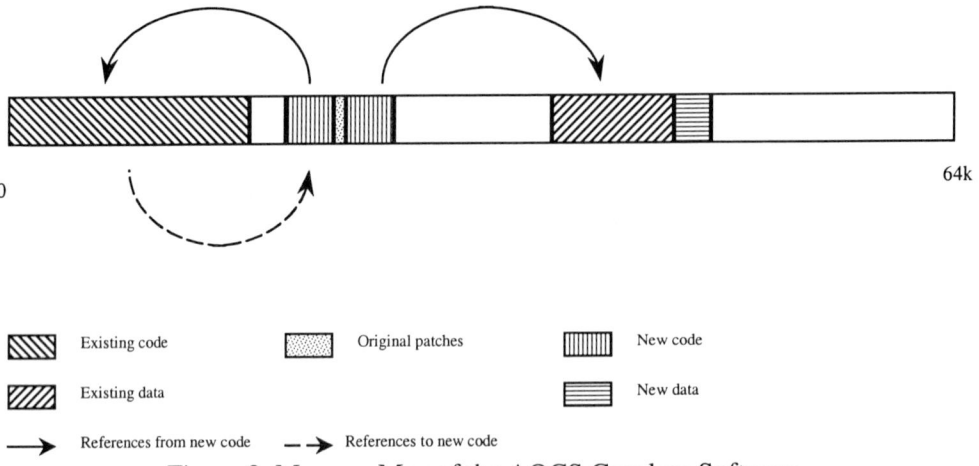

Figure 9: Memory Map of the AOCS Gyroless Software

The total size of the new code is approximately 4 kWord. (Original code size without data is approximately 18 kWord). All this code was ready, tested and delivered in August 1999.

In parallel there was maybe even a larger activity to provide a new user manual and new operational procedures, and implement them in the ground system. This was done in close cooperation with the NASA flight operations team. A co-location in Europe with flight operations personnel at the prime contractor turned out to be very beneficial.

August was spent at GSFC with validation of the ground system, operational procedures and the modified spacecraft simulator. Finally the new software was uploaded to the spacecraft in the period 24 – 27 September 1999. This was followed by a week of successful flight commissioning.

One minor problem was found: one of the new monitoring functions makes use of the wheel speed measurements. These measurements turned out to be noisy (missing bit in read-out) once in a while. A patch has been prepared to make the function robust against a single misreading.

Current Status

On 1 December 1999 the ground triggered an ESR because the CRP mode did not work. The root cause turned out to be a flaw in the AOCS software, which had been there since launch. Under certain circumstances some data memory gets overwritten with zeros. Before the gyroless this data area was not used. Now it contains the data for the new CRP mode. A patch has been prepared to correct this problem.

SOHO is behaving very well with its new gyroless software. After uploading the last few patches, currently planned for March 2000, it will be in such a shape that it can easily fulfill its mission to observe the sun till the solar maximum.

Acknowledgment

This whole effort has been the result of an enormous teamwork of several hundred people, of which the author is only one. The prime organizations involved were the European Space Agency (ESA), NASA (including DSN), Matra Marconi Space, the SOHO prime contractor, and the Allied Signal flight operations team.

More Information

For more information and pointers to other articles look at the SOHO web site:
http://sohowww.nascom.nasa.gov for general info and
http://sohowww.nascom.nasa.gov/operations/Recovery for more information about the subject of this paper.
There is also a European mirror site at http://sohowww.estec.esa.nl

AAS 00-045

A NOVEL NEW GEO EARTH SENSOR PROVIDES HIGH ACCURACY

George Rullman, Richard Burton and Len Anderson[*]

Based on the flight proven MiDES LEO design, a MiDES GEO configuration has been developed. Using the multiple pixel array and digital electronics that has worked so well in the MIDES, a GEO configuration has been developed that provides better than 0.02° accuracy. The design concept and the sensitivity analysis is presented showing the NEA / MRA to be better than 0.006°. The design uses the electronics circuitry and mechanical concept that was qualified and is presently in orbit. The pixel configuration and the optics have been modified to change the FOV to match the GEO requirements. The unit provides an alternate to higher priced systems presently in use and in the applications where higher accuracy control is needed. The unit is capable of life expectancy of greater than 15 years in a GEO environment.

Background

The MiDES Earth Sensor was originally presented as a concept at AAS G&C '94. It was developed for LEO applications to provide high accuracy in a wide FOV, operate while the Sun is in the FOV, reduce radiance error to less than 0.05° 3σ, provide a digital output and perform both pitch and roll sensing in one common assembly. It uses arrays of infrared detectors to provide digitally formatted pitch and roll data to the S/C ACS. It has flown on the ASTRO_SPAS[1] spacecraft launched and recovered by the STS-85 shuttle flight in August of 1997. The comparison to Star-Tracker data showed that the MiDES accuracy was better then the expected 0.06 degrees (3σ) accuracy including radiance errors.

It is presently aboard the UOSAT 12 satellite developed by Surrey Satellite Technology, Ltd. (SSTL) where it is being used as the primary position control device for the last year and providing accuracy of better than 0.06° at an orbital altitude of 600km. In addition a unit has been delivered and is schedule to launch aboard the Orbital Sciences QuickToms satellite in the fall.

The product design is mature and has been fully qualified

---

[*] Servo Corporation of America, 123 Frost Street, Westbury, New York 11590.

MiDES for Geosynchronous Applications (MiDES-G)

A GEO version of the MiDES has been developed that is a direct descendent of the MiDES. A new optical design and a change in the detector pixel pattern were the major changes necessary to turn MiDES-LEO into MiDES GEO. A proprietary pixel pattern design[2] incorporates two orthogonal 16 element arrays, each viewing a 33° FOV. The patterns cross at the center and provide a high and low resolution area within the FOV to accommodate various mission needs in one common unit. The exceptionally wide field of view provides coverage for lower orbits yet provides a high degree of accuracy when at the operational altitudes.

The remainder of the MiDES-G design remains the same as the standard MiDES with design and flight heritage on Astro Spas, UOSAT-12 and QuickToms.

Figure 1 Photograph of MiDES-G

Theory of Operation

The MiDES Earth sensor is designed to provide horizon position information for satellites in two mutually orthogonal axes (X and Y) which form a plane that is normal to the vector passing through nadir. This is accomplished utilizing one assembly that contains two optical heads that view the Earth's lim. These two independent channels share common electronic signal processing circuitry and a common mechanical chopper.

A sixteen element detector array in each of the optical channels is photolithographically patterned on a 40 micron Lithium Tantalate pyroelectric element. The system operates in a quasi-static staring mode, a mechanical chopper providing the radiant contrast needed to produce an AC voltage output from the detectors. The radiance from the chopper (which is essentially the temperature of the MiDES (-30° to +50°C)) is much higher than the scene radiance from space and Earth. This provides a high signal to noise ratio during operation. The S/N is further enhanced by active bandpass filtering located at the output from each channel. A multiplexer then synchronously samples the output from each detector element. The sampling is digitally controlled and is synchronized with the chopper. The samples are processed through a 10-bit A/D converter and then the data is packed and sent out through a serial data port under host control. A supplied algorithm is then used to produce the pitch and roll angle information in the satellite ACS computer or local preprocessor.

The chopper incorporates a flex-pivot operating at a mechanical resonant frequency of twenty(20) Hz. There are no bearings, and in this application the flex pivot has an infinite life expectancy.

The block diagram of the system is shown below. This block diagram is the same as is used in the original MiDES used in LEO applications except that an FPGA has replaced the microcontroller due to radiation requirements.

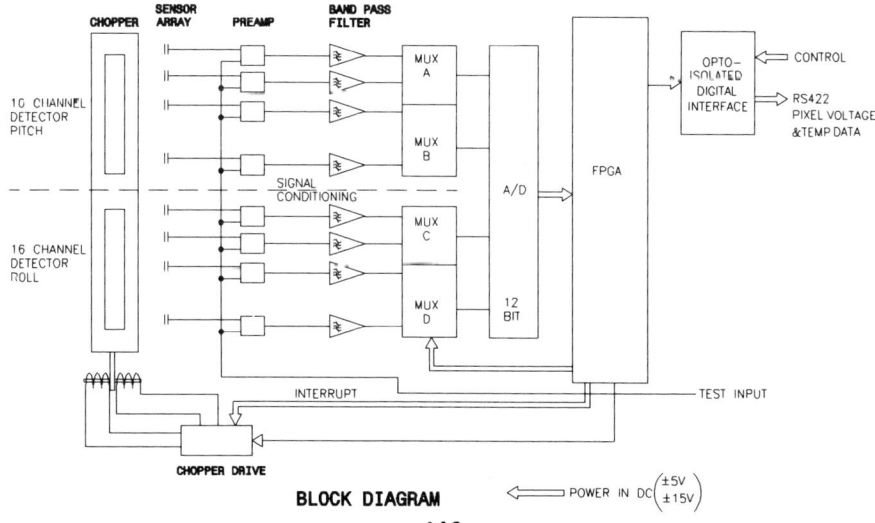

BLOCK DIAGRAM

MiDES-G Technical

The heart of the MiDES-G is in the unique pixel array developed by Servo. The signal processing has been discussed in earlier papers[3]. The modeling of the detector is presented in showing how the extremely high accuracy's that are specified are obtained.

The modeling and the description of the operation is presented. The unit has been designed around general requirements for typical GEO earth sensor requirements and therefore the specifications for performance should be viewed in this light.

Detector Array

Each array is patterned to allow for a high accuracy zone and a moderate accuracy zone. Figure 2 shows how one of these arrays of detector pixels is patterned to provide the accuracy in each of the zones. The high accuracy zone provides coverage from nadir to +/-4° in pitch and roll at an altitude range of from 32000km to 40000km. From +/-4° to +/-8° a decreased accuracy is maintained. These zones can be seen below. The full field of view of the array is 33.6° in both pitch and roll.

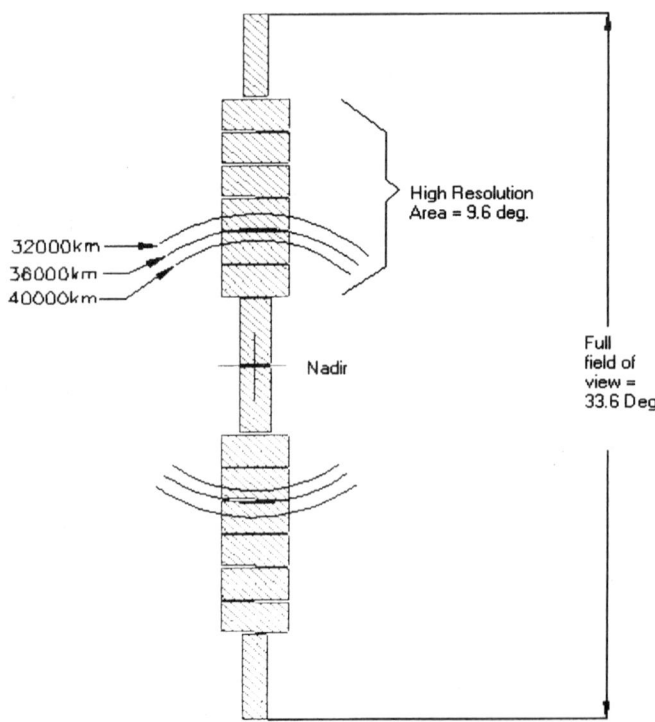

Figure 2- Pixel Pattern

Figure 3 shows a view of the two arrays mapped together with Earth shown imaged on the composite array in various possible positions. For all positions possible while the Earth is within +/- 4° of nadir, the pixels that are shorter in length are the only ones that view the Earth/space transition. We call this the high accuracy region. This would be the region that would be used in normal nadir pointing mode of operation.

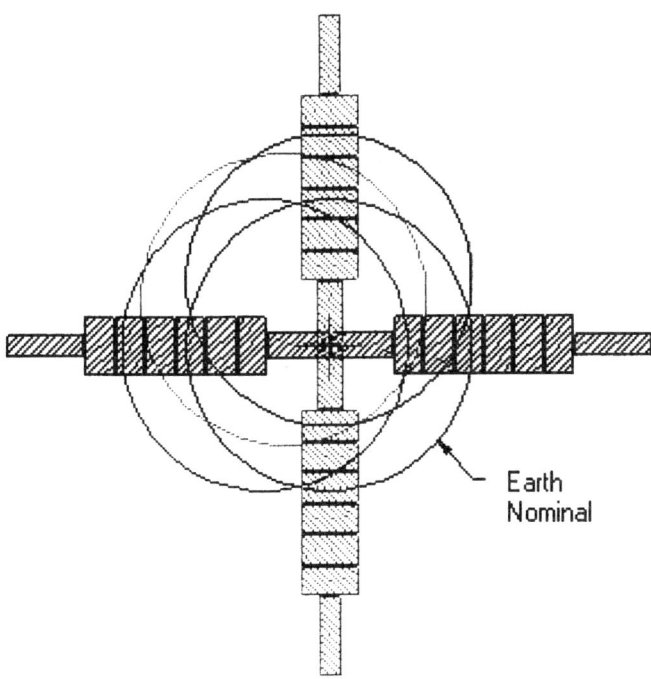

Figure 3 - High accuracy mode-Normal operation

Figure 4 shows the same view of the two arrays mapped together with Earth shown in various positions in the reduced accuracy mode. Note that the horizon positions fall on the long narrow pixels. Further note that there are four pixel crossings at all times, providing one extra pixel than needed to determine nadir. This feature allows the unit to operate with the Sun in the FOV. In addition, the nearest pixel to the crossing is easily determined for radiance mitigation.

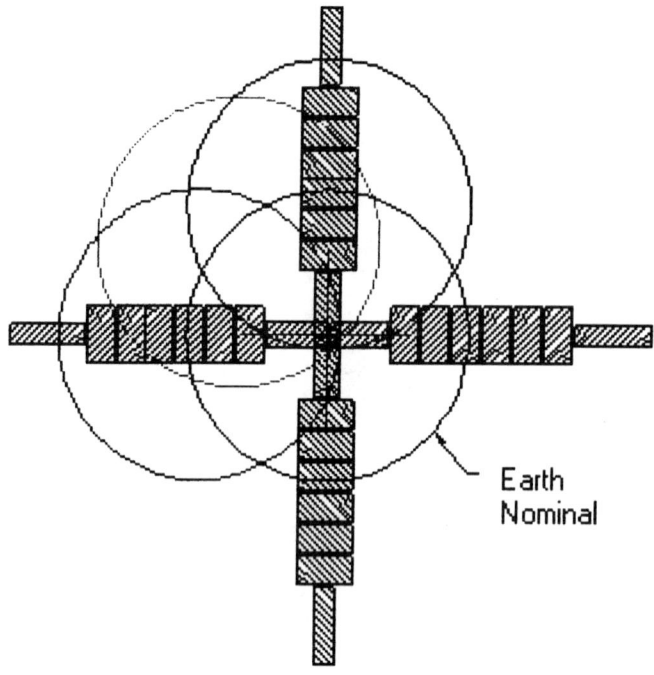

Figure 4 – Reduced accuracy mode

Sensor Accuracy and Output

Total sensor errors including radiance effects are shown in the graphs and curves generated by Mathcad. We have modeled the detector and use the data obtained from the LOWTRAN model for the atmosphere. The model has been found to be an accurate representation of the performance of the sensor. We have correlated the model to the actual space data that we have received from the various flights of the MiDES and continue to check the model against the actual data as additional information becomes available.

For the GEO sensor the model indicates that the accuracy will be better than 0.025° 3σ in the high accuracy zone (including radiance error) and decrease to 0.04° between +/-4° and +/-8°. The noise equivalent angle of the pixels is better than 0.006° (3σ).

Figure 5a shows the voltage output of the pixels as a function of angle, as the horizon moves through the length of the pixel in the high sensitivity zone. In the lower sensitivity zone the slope (volts per degree) is reduced to ½ the value in the high accuracy zone, as the pixels are twice the length.

Note that there are two curves for each of the temperature extremes shown. These curves show the variation in output voltage due to variations in earth radiance caused by seasonal conditions.

Figure 5b shows the normalized curve indicating that the slope does not change as a function of unit temperature.

Figure 5c shows the normalized output when compensated for radiance. This is the final output after the algorithm is applied to the data.

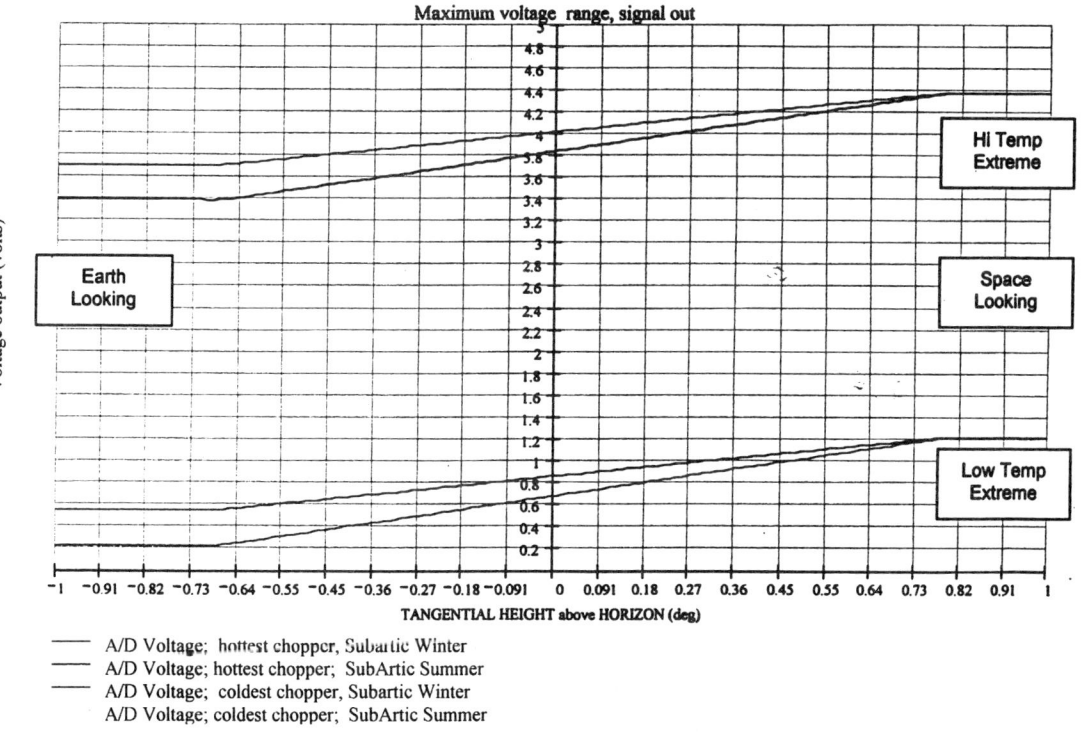

— A/D Voltage; hottest chopper, Subartic Winter
— A/D Voltage; hottest chopper; SubArtic Summer
— A/D Voltage; coldest chopper, Subartic Winter
  A/D Voltage; coldest chopper; SubArtic Summer

Figure 5a Pixel Voltage vs. Horizon position on Pixel
Pixel is 1.6° long

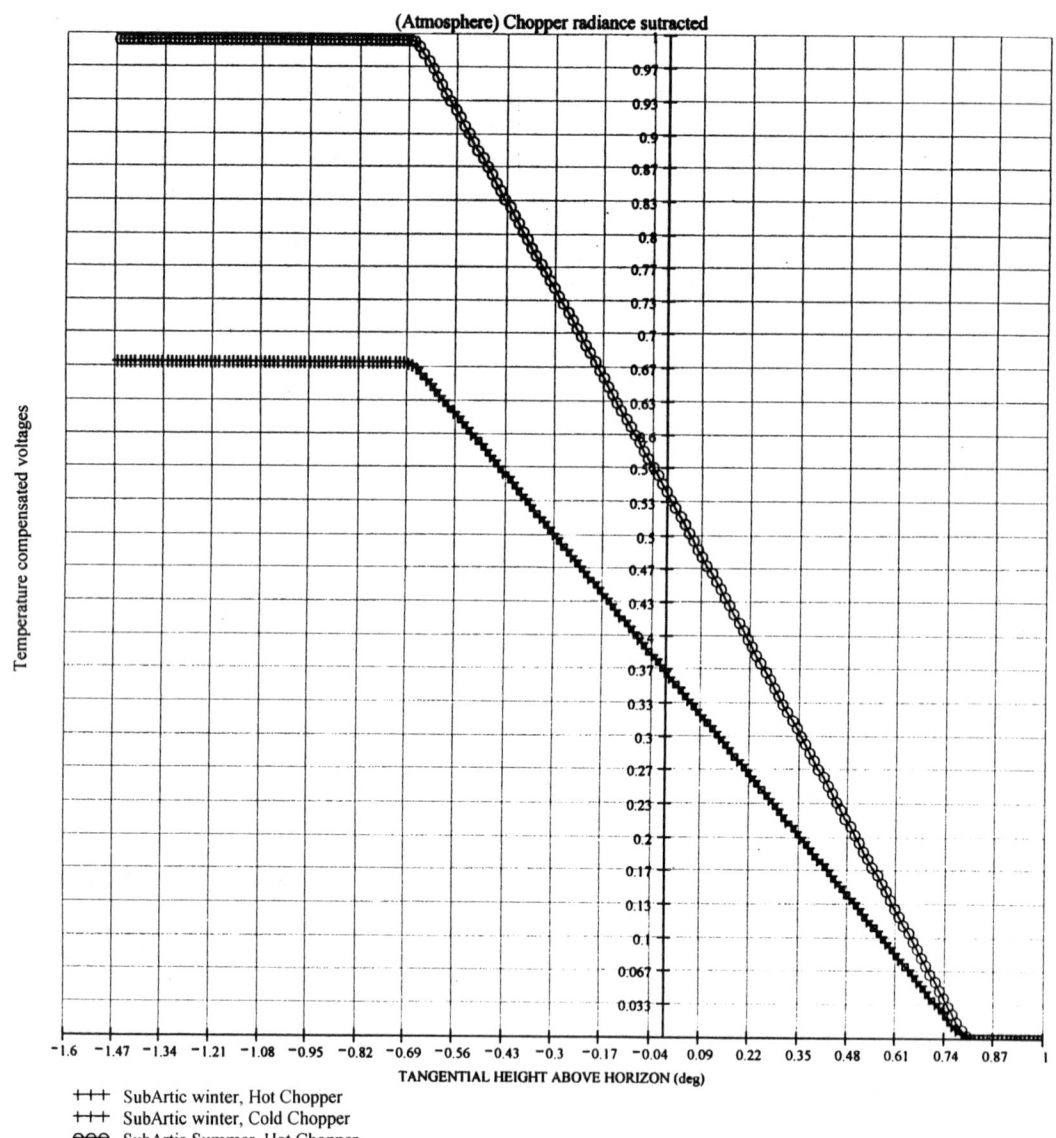

Figure 5b Pixel Voltage Normalized Variation due to Earth Radiance

Figure 5c Composite Output Normalized For Radiance Changes

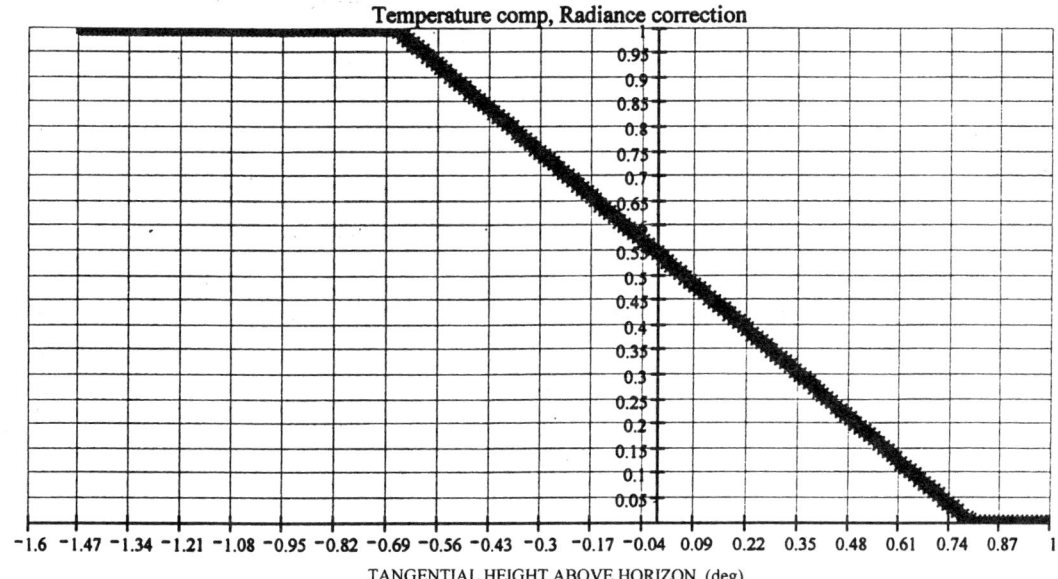

+++ Subartic winter, Hot Chopper
×× Subartic winter, Cold Chopper
⊖⊖⊖ Subartic summer, Hot Chopper
→ Subartic summe, Cold Chopper

Figure 6a shows the NEA of the pixel in the same zone as above. This is based on the modeled and confirmed bu measured output noise voltage of the pixel electronics after passing through the narrow band active filter. The Figure 6b is the system mean resolvable angle(MRA) after A/D conversion assuming a 12 bit converter is used.

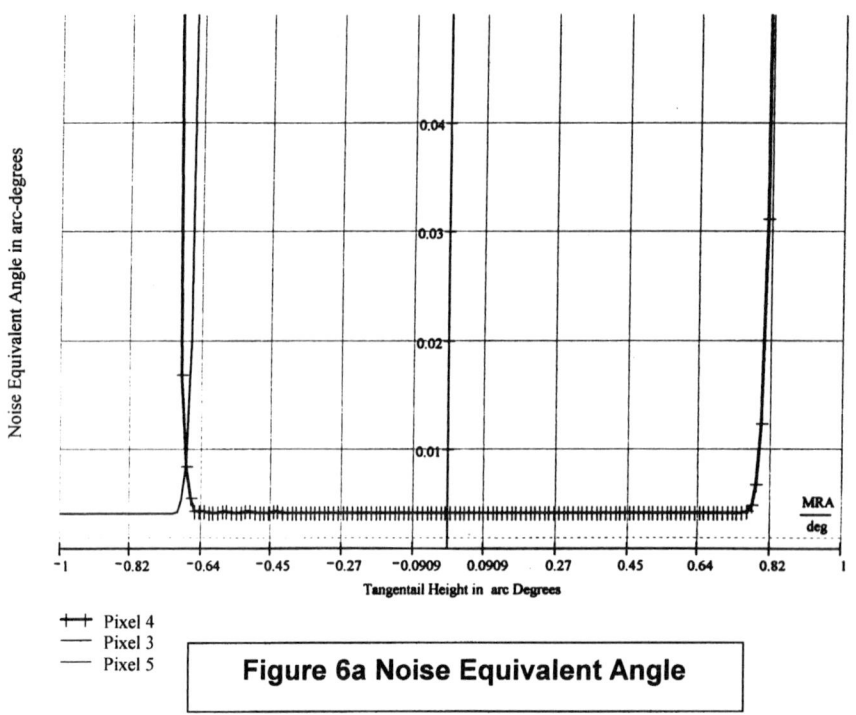

Figure 6a Noise Equivalent Angle

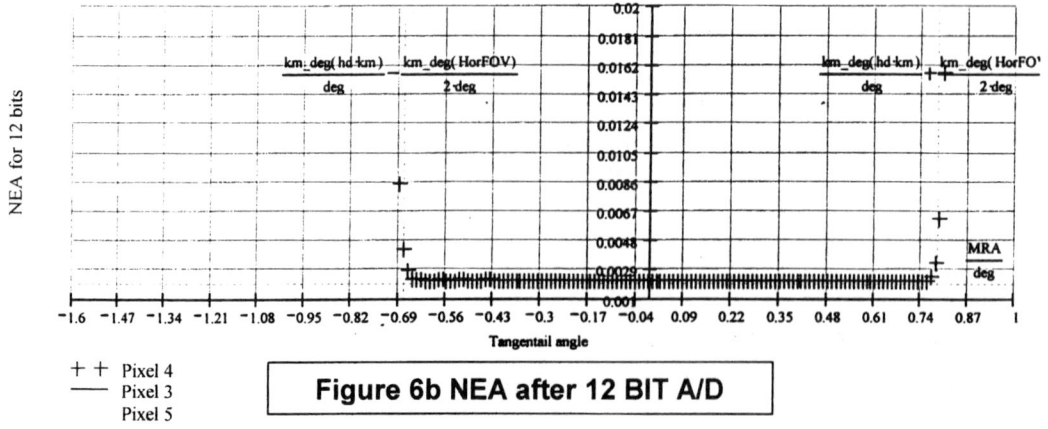

Figure 6b NEA after 12 BIT A/D

Figure 7a shows the slope of the voltage output of the pixel as a function of the horizon position on the pixel. The two curves represent the maximum variation in slope that would be encountered as a function of seasonal extremes.

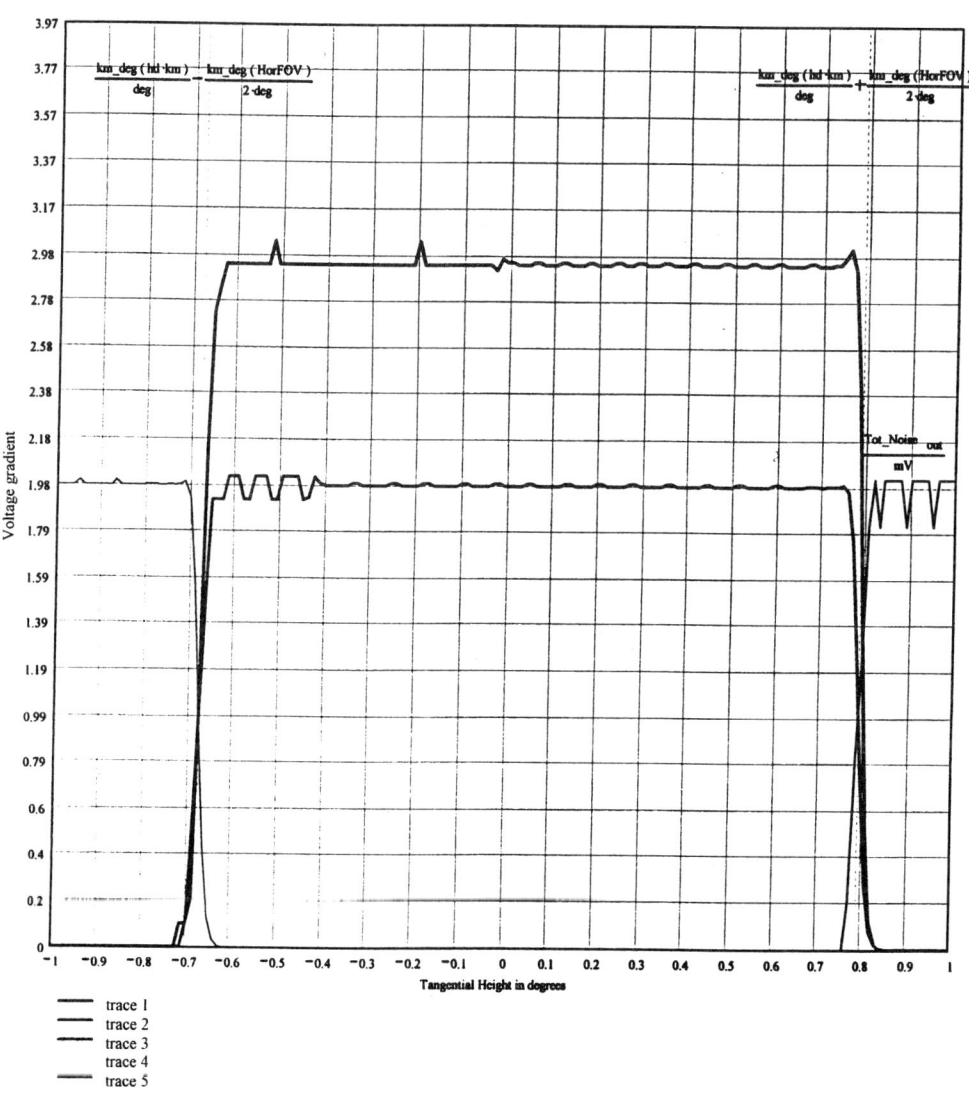

Figure 7a Voltage Gradient of Vout vs. Horizon Position

When the voltages from all the pixels are normalized and assembled, the composite output would look like figure 7b. The first pixel to be fully on Earth or

Space is used to mitigate the radiance effects. This is for one of the two arrays. The other output would be similar.

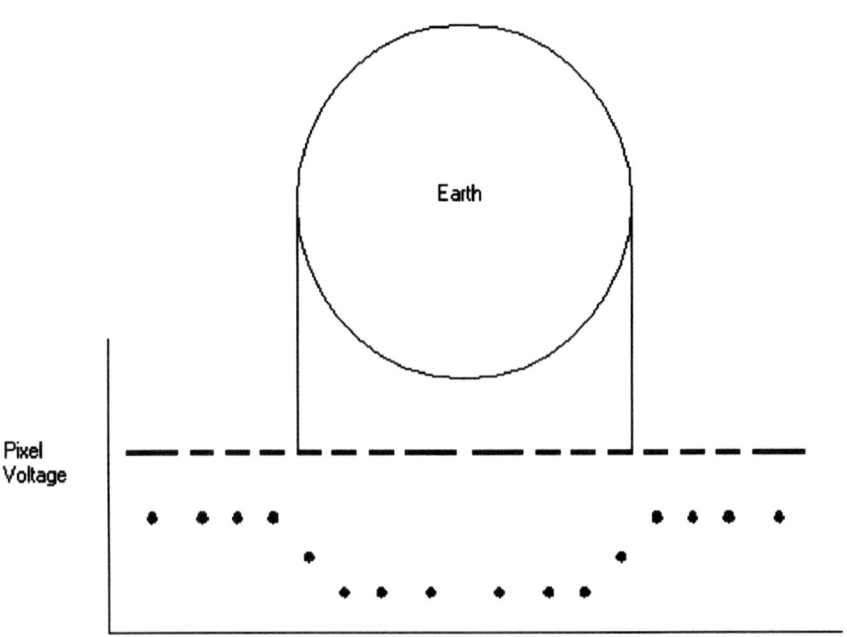

Figure 7b Typical Pixel Output

Optical Description

In order to maintain the uniformity of the image needed for the accuracy over the length of the detector array, a lens system is needed that preserves the scene without introducing additional errors.

The MiDES-G optical system consists of a 3-element germanium aspheric lens system covering a total field of view of 33 degrees. It is designed for the 14 to 16 microns CO_2 absorption wavelength band. We use a f/1.0 lens system as shown in Figure 8. The outer lens is coated with a long-wave pass filter and the band is further restricted as the radiant energy enters the Detector with a narrow band 15 micron filter.

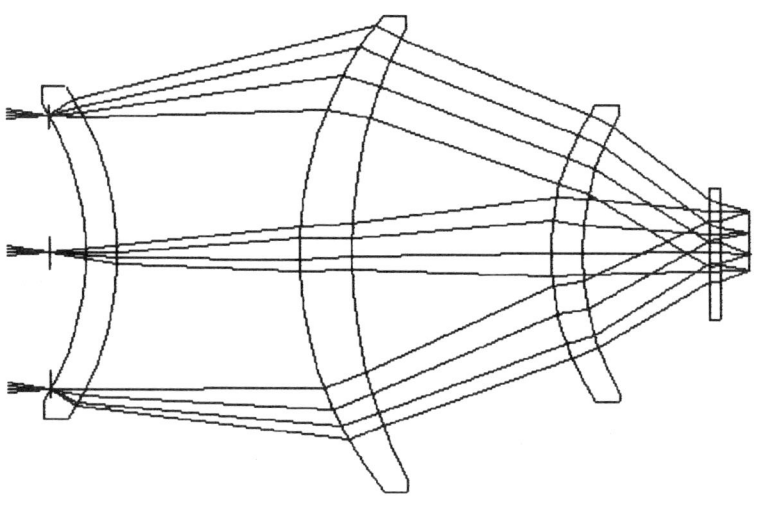

Figure 8 Optical Lens System for 33.6° FOV

Signal Processing

Pitch and roll algorithm uses the ratios of the normalized voltages from the pixels fully in space and fully on Earth to determine position. The horizon-viewing pixel voltage is then compared to Earth/Space to determine position within the pixel that the horizon falls. As can be seen from the graphs, the rate of voltage change is linear. It also has a constant slope for changes in the unit temperature. Each pixel voltage is processed through a 12-bit A/D converter where the noise level of the system is approximately equal to the LSB. The digitally controlled system has developed heritage on various satellites and remains unchanged for this application.

The use of a multi-pixel array inherently minimizes any sun or moon interference problems. The Sun in the FOV will negate data from any one of the four (or-more) pixels that are viewing the Earth's horizon. The others are sufficiently far away so that they are not affected and continue to operate.

Mechanical Description

Shown in figure 9 below is the outline drawing of the MiDES-G package. Note all dimensions are shown in inches and [millimeters].

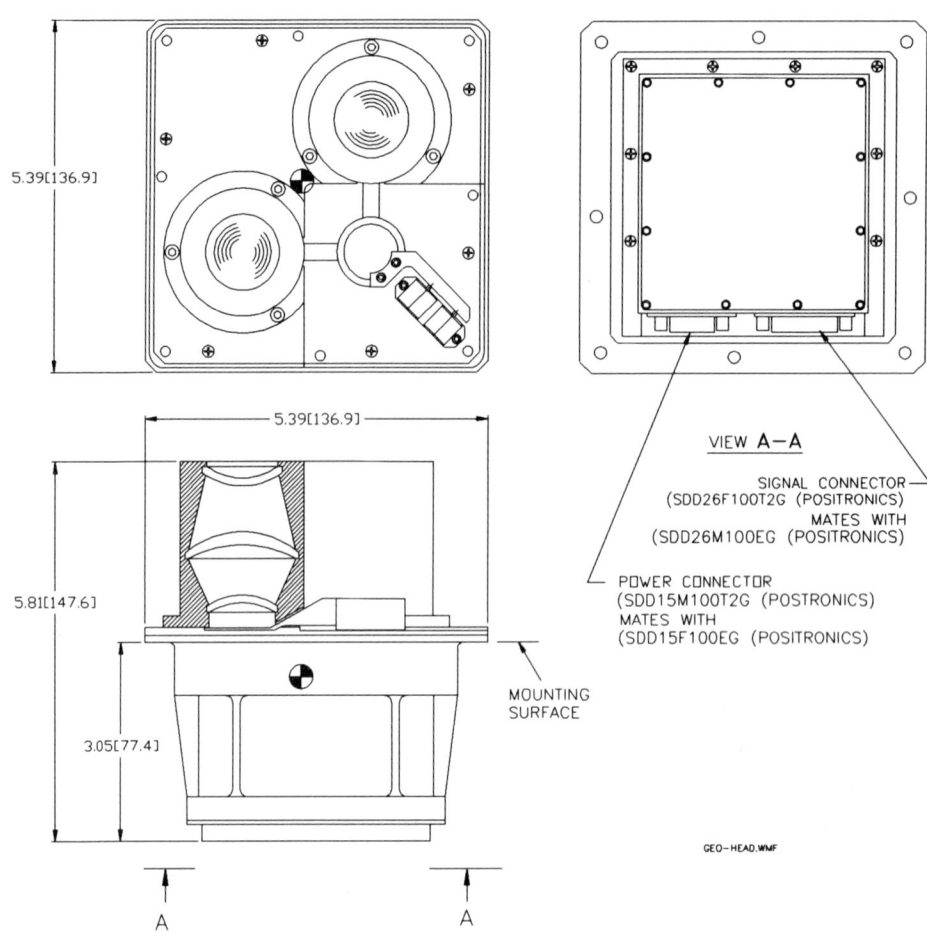

Figure 9 MiDES-G Outline Drawing

Spacecraft Interface

The standard interface supplied with the MiDES-G is RS422. This was a chosen since it makes all testing and alignment of the unit simple with the use of a PC. Other interfaces could be implemented as required by the host spacecraft. The interface is designed to operate synchronously or asynchronously with the host ACS computer. The system can be poled with a "data ready command" from the host. The interface is full duplex.

Radiation

The design of the GEO version of the MiDES Earth sensor includes hardening against the worst case radiation environment in a geosynchronous orbit for a 15 year mission spanning 1.25 solar cycles. We have specified the NASA AE8MAX / AP8MAX trapped particle maps for a parking longitude of 160 degree west, corresponding to L = 6.6 Earth radii, at which the highest electron fluxes are encountered in a geosynchronous orbit. Figure A shows the 15 year integral electron fluence spectrum for this orbit.

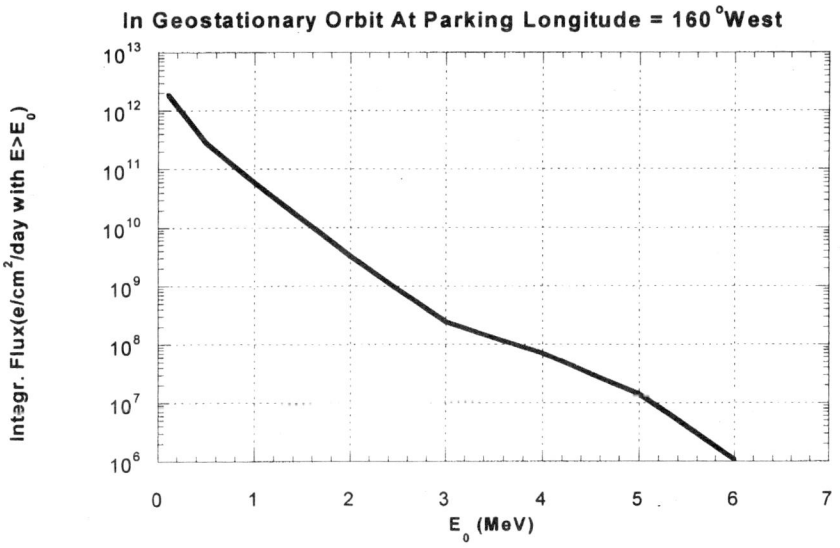

Fig. A. Omnidirectional Trapped Integral Daily Electron Flux (AE8MAX) In Geostationary Orbit At Parking Longitude = 160°West

The trapped protons in GEO, however, are of very low energy (less than a few MeV), are effectively stopped by 0.05 mm of aluminum, They therefore will not penetrate through the sensor housing that has a typical crossection of 7 mm. Solar protons, by contrast, are of high energy and far larger penetration thickness and can be a significant contributor to the mission dose, as well as the principal cause for single event effects (SEE), at least for digital devices with low upset /latchup thresholds. The model ALE integral fluence at the top of the magnitosphere is 1.73E10 p/cm**2/event above 10 MeV (SOLPRO model) with an integral fluence spectrum as shown in Fig. B.

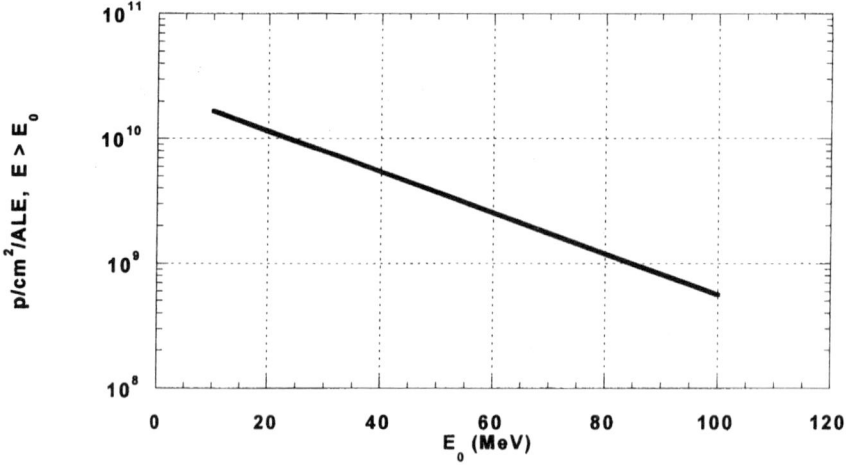

Fig. B. Integral Proton Fluence For Anomalously Solar Large Events (ALE) In Geosynchronous Orbit

Figure B Total Dose Hardening

In addition, the detector sensor housing design represents aluminum equivalent penetration thickness of between ~6.5 and 11.5 mm. Based on the data in Figs. C and D, the expected 15 - year mission dose is no greater than 20 Krad (Si) at the detector location and thus does not represent extreme requirements for component total dose hardness.

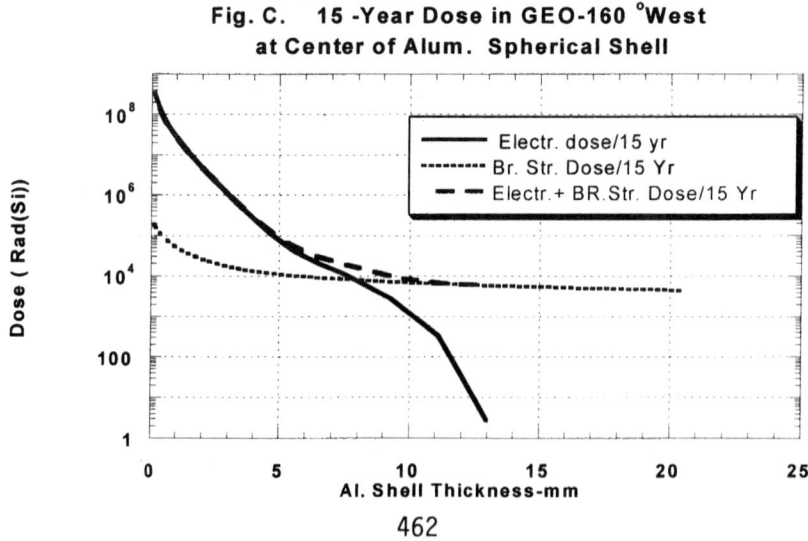

Fig. C. 15 -Year Dose in GEO-160 °West at Center of Alum. Spherical Shell

Single Event Effects (SEE)

The sensor design does not use devices that are susceptible to latchup.

Fig. D. Solar Proton Dose for Six ALE in GEO at Center of Aluminum Spherical Shell

Specification

Performance

Accuracy:	0.025° (+/-4° of nadir)
	0.04° (+/-8° of nadir)
Noise Equiv. Angle (NEA)	0.006 ° 3 sig.
Operational Range:	33° circular (pitch and roll)
Spectral Band:	14 to 16 microns

Mechanical Interface

Weight:	1.5kg.
Mounting:	Mounting flange at base
Alignment:	Integral Alignment Cube

Electrical Interface

Power Input:	50V and 100V DC
Signal Output:	digital word RS-422

Environment

Operating Temperature:	-20 to +60 deg. C
Random Vibration:	28 G rms Qual. level
Radiation:	Able to withstand 15 year GEO environment

Acknowledgement:

The authors wish to thank Dr. Michael Stauber, Radiation Effects Consultant for their contributions to this project.

References:
1. ASTRO-SPAS is a satellite owned by Daimler-Benz Aerospace (DASA) and was launched aboard STS-85 Shuttle flight.
2. Patent Pending George Rullman Servo Corp.
3. AAS_1999 AAS99-041 *MiDES Earth Sensor On Orbit Accuracy Compared to a High Performance Star Tracker* George Rullman[*], Len Anderson[*], Dr. Andreas Langmeier[**], Dr. Folkert Tangerman[***], Glenn VanDerWoude[***]
   ([*]Servo Corporation of America, [**]Dasa/DSS, [***]SUNY Stonybrook)

AAS 00-046

THE DESIGN AND OPERATION OF A COTS SPACE GPS RECEIVER

Martin J. Unwin and Michael K. Oldfield[*]

Spaceborne GPS has proved an invaluable utility for independent orbit determination. This paper describes the SGR Space GPS Receiver, a spacecraft orbit and attitude determination sub-system designed for small satellite applications. Information on the design and characteristics are given and its performance as demonstrated in orbit. Commercial parts are almost exclusively used, and the practical implications are given. Three types of SGR receivers are listed, the SGR-05, -10 and -20, each aimed at a slightly different application. Information is given on how the SGR is integrated into a typical satellite.

INTRODUCTION

The Global Positioning System (GPS) consists of a constellation of 24 satellites at an altitude of 20,000 km which transmit signals at 1.575 GHz to the ground. A GPS receiver decodes these signals, and is able to calculate the user position to an accuracy of 100 metres or better, determine the time and user's velocity, and, furthermore, with an appropriately configured GPS receiver, attitude determination is possible. GPS receivers have become an invaluable piece of equipment for land, air and sea users. The utility of GPS has also been demonstrated in low Earth orbit on numerous satellite missions as the same operating principles apply as for terrestrial users. An on-board GPS receiver provides a low power navigation sensor that can autonomously find the satellite orbit, synchronise the clock and determine the spacecraft attitude.

THE SPACE GPS RECEIVER

The SGR (Space GPS Receiver) is a recent development using state of the art commercial GPS technology applied specifically for space applications. The SGR was developed by Surrey Satellite Technology Ltd in collaboration with ESA for use on small satellites. Surrey specialises in the design of small satellites (15 built and launched to date), and has a long heritage in using GPS in orbit: PoSAT-1 was the first microsatellite to use a GPS receiver in orbit in 1993,[1] and a number of subsequent Surrey missions have since carried receivers. The SGR has been developed by the same team that uses the receivers in orbit, so is designed for practical applications, and has been optimised using real orbital operation.

The functional sub-systems and interfaces of the SGR are shown in Figure 1. The GPS L1 signals at 1.575 GHz are received at the antenna, and amplified by the Low Noise Amplifier (LNA). The RF front-end further amplifies, filters and mixes the signal down to baseband where it is digitally sampled. The sampled data is correlated by the digital section and then formulated

---

[*] Surrey Satellite Technology Ltd., University of Surrey, Guildford, Surrey GU2 5XH, United Kingdom.
E-mail: martin.unwin@ee.surrey.ac.uk.

to make range measurements to the GPS satellites. The SGR has 24 C/A code channels that can be used to independently track up to 24 GPS satellite signals. By using range measurements to four or more GPS satellites, the SGR computer calculates the user position. The accurate time is another by-product of the GPS fix, and the Doppler shift on the received signals is used to recover the user's velocity. Just a single antenna is required for determining position, velocity and time. By using two or more GPS antennas separated by 50 cm to 2 metres, the GPS measurements from two antennas can be differenced. With such a short baseline, the only difference between the two signals (to a first order) is the signal phase difference due to the attitude of the vehicle. Hence with appropriate mathematical algorithms in the receiver, it is possible to use GPS for attitude determination.

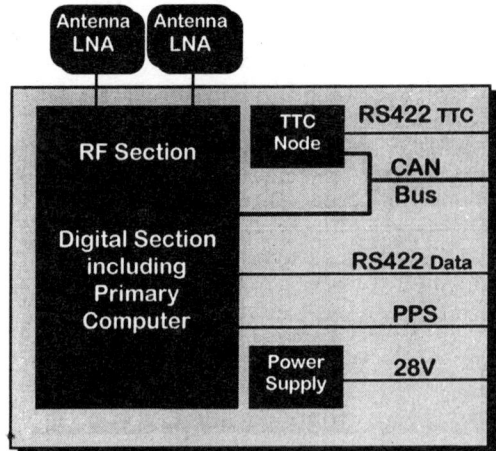

Figure 1 Functional Sub-systems of SGR-10 and 20

The SGR also contains a TTC (Telemetry Telecommand) node, which is a microcontroller specifically for providing distributed telemetry and telecommand capability to the SGR. The TTC node is able to read telemetry and send telecommands (for example to switch in redundancy) even when the primary computer is not operational.

Two electrical communications interfaces are supported, the RS-422 interface and the CAN bus (Controller Area Network). The RS-422 is a point-to-point interface that uses a proprietary binary packet protocol at 19k6 baud. The CAN bus is a communications standard originating from the automotive industry that provides a fault tolerant multipoint network suitable for a distributed telemetry and telecommand architecture. Either the CAN or the RS-422 interfaces can be used for full monitoring and control of the SGR.

Six "pulse-per-second" outputs are provided that can be used to synchronise clocks to an accuracy of 1 microsecond.

The power supply required is nominally 28 volts, although an input voltage range of 18 to 38 V is possible. Each antenna uses a separate LNA and RF front-end and so the more antennas, the more power is consumed. The SGR in 4-antenna mode typically consumes 7 watts, but it will dynamically power down antennas and front-ends according to its mode of operation. In positioning mode using a single antenna, therefore, only 5.5 watts of power is taken. If the customer has a suitable power supply of filtered 5V (e.g. in a centralised power architecture), the SGR DC/DC converter can be bypassed, in which case the power consumption will drop further.

DESIGN OF THE SGR

The core of the GPS receiver is the Mitel Semiconductors GP2000 high performance commercial GPS chipset, comprising of the GP2015 RF front-end and the GP2021 correlator. Each correlator has 12 C/A code channels, so the use of two correlators gives a total of 24 channels (See Figure 2). The primary processor is the ARM60B, a low power 20 MHz 32 bit RISC microcontroller, and this is supported by external EPROM, SRAM and Flash memory. The peripheral components include a CAN adaptor, RS-422 driver and receiver circuitry and a transputer link adaptor. The clock source that serves as a reference to the RF front-ends and primary computer is a 10 MHz TCXO that has a clock stability of 1.5 parts per million. No battery-backed RAM is included as the SGR has a relatively fast time to first fix from cold start due to its large number of channels.

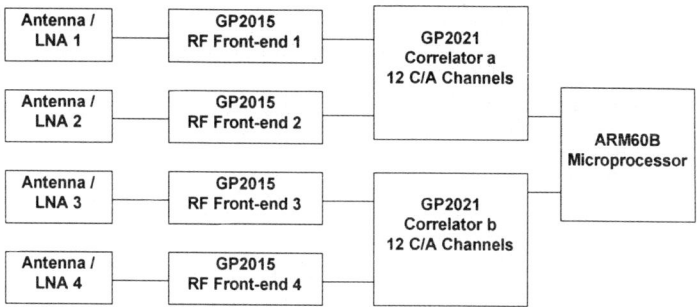

Figure 2 Configuration of Correlators and Front-Ends in SGR

The orbital environment can be very hostile in terms of radiation, and radiation-hardened technology is a fundamental requirement for many space missions. Unfortunately this can lead to expensive sub-systems that are years behind equivalent commercial technology.

The SGR does not use military or radiation-hardened parts but addresses the reliability issue through three approaches:
- A radiation test programme was undertaken by Surrey and ESA to characterise the susceptibility of core components used in the SGR design and to identify weaker components. Several of the components survived after well over 15 kRads, but overall, a figure of approximately 10 kRads (Si) can be assumed. Also where possible, the statistical chance of a Single Event Effect (Upsets and Latch-ups) for each component is determined.[2]
- Specific countermeasures have been included in the design to compensate for Single Event Effects. A bank of SRAM is protected against Single Event Upsets through the use of Error-Detection and Correction (EDAC) circuitry. A fast acting current sensing switch is included to protect against Single Event Latch-ups. In the laboratory, this switch has been repeatedly demonstrated to power down the SGR when a latch-up was induced in the ARM60B apparently preventing any damage. The SGR has a watch-dog timer that will reset and reboot the SGR in the event of code corruption.
- The SGR is being tested and is operating on two research satellites, TMSAT (June 1998) and UoSAT-12 (April 1999). Due to the experimental nature of these satellites, anomalies have been observed which are believed to be due to new versions of software, but no events or over-currents have been observed that can be attributed to radiation effects. Further satellites to be launched in 2000 are carrying the SGR, including ESA's PROBA mission, and as the SGR accumulates more time in orbit, confidence in the reliability of the SGR will grow.

THE SGR FAMILY

Three configurations of the SGR are available.

The **SGR-20** can support up to four GPS antennas. It can provide the user with position, velocity and time. The SGR can take phase differences between the four antennas and determine the satellite attitude if the antennas are configured correctly on the spacecraft using algorithms in the SGR.

The **SGR-10** is the same as the SGR-20 except it can only support two antennas. It can determine position, velocity and time, but not attitude. The SGR can determine phase differences between the dual antennas, and so potentially the measurements could be used to augment the host satellite attitude determination solution.

The approximate specifications for the SGR-10 and SGR-20 are given in the Table below

Table 1 Specifications for SGR-10 and SGR-20

Channels	Up to 24: L1 C/A code
Antennas	SGR-10: 1-2; SGR-20 3-4
Time (UTC)	1 microsecond
3D Position (2σ)	100 m*
3D Velocity (2σ)	2 ms^{-1}*
Attitude Determination	Approx 0.2°
Dynamic Capability	8 kms^{-1}, 2 g
Time to First Fix (TTFF)	Typically 90 s warm start, 7 mins from cold start
Power Consumption	5.5 to 7 W
Dimensions	160x160x50mm (alternatives available)
Mass	1 kg
Random Vibration	15 g
Operating Temperature	-25°C to +55°C
EMC	As per MIL-STD-462D
Cumulative Radiation	> 10 krads (Si)

* These statistics are from a typical 24 hour period of orbital operations but are dependent on Selective Availability imposed on the GPS signals by the US DoD.

The **SGR-05** is a new development and is a miniaturised version of the SGR-10/20, carrying only the basic GPS features. The SGR-05 supports one antenna, and can only give position, velocity and time (using a sub-set of the SGR-10/20 software). The following features are absent: EDAC protection, SEL switch, CAN-bus, RS422 drivers, TTC node or telemetry, power supply, mechanical housing. Instead the user must provide 5V ±5%, and communicate over the TTL UART link, and provide screened housing. As a result of these simplifications, the SGR-05 only consumes **2 watts** and has a mass **50 grams**.

The expected mode of operation of the SGR-05 would be intermittent, so the risk of SEUs and SELs is significantly reduced, but the core components are identical to the SGR-10/20 and so would have the same expected life under radiation. The SGR-05 is particularly suitable for very small satellites, such as nano- or pico-satellites where size and power demands are especially tight.

Flight Heritage

At the time of writing, the SGR is operational on two satellites, and has been selected for several future satellites. Table 2 lists the satellites carrying the SGR. To reach the satellite, each SGR must undergo extensive acceptance tests, including thermal, vibration tests, and often thermal vacuum and EMC tests.

Table 2 Satellites Carrying SGR

Satellite	SGR	Launch	Status
TMSAT	SGR-10	Launched: June 1998	Operational in orbit
UoSAT-12	SGR-20	Launched: April 1999	Operational in orbit
Tiungsat	SGR-10	Mid 2000	Complete
Tsinghua-1	SGR-10	Mid 2000	Complete
PROBA	SGR-20	Mid 2000	Complete
SNAP-1	SGR-05	Mid 2000	Under Test
E-SAT	SGR-10	2001	Awaiting KO
Biltensat-1	SGR-10	2001	Awaiting KO

Figure 3 PROBA SGR-20 Receiver and LNA/Patch antennas

SGR-PC

The control of SGR-PC over the RS-422 port is supported by a PC Windows-based program, SGR-PC. This program can communicate with the SGR over the PC's COM ports (via an RS422 to RS232 converter). It can process files generated by the SGR and store commands in a binary file for transmission to the receiver and can also interpret files containing binary output data from the receiver for off-line control and monitoring. The user can view the packets received from the SGR at byte level and view the interpreted meaningful values from the packets. Command packets can be sent to the SGR via an SGR-PC menu, and the SGR outputs can be programmed. Data received from the SGR can be logged in ASCII format for processing by the user.

Figure 4 shows a display from SGR-PC when processing data from a recent operation of the UoSAT-12 GPS receiver. The windows have been arranged to demonstrate many of the features of SGR. It can be seen that the data in this case is coming from a file that has been downloaded from the satellite. The SGR in this example is operating in a mode using all 24 channels to track 6 satellites on each of its four antennas. The six satellite signals received by antenna 1 are used for positioning, while the remaining antennas are used to track the same satellites in order to take phase differences. The raw data available includes pseudorange, phase, DCO, integrated carrier cycles and signal to noise. The analogue to digital readings give information about the currents taken, the temperature and other useful status information. A Mercator projection plot shows the position of UoSAT-12, which at this moment was flying over the Antarctic.

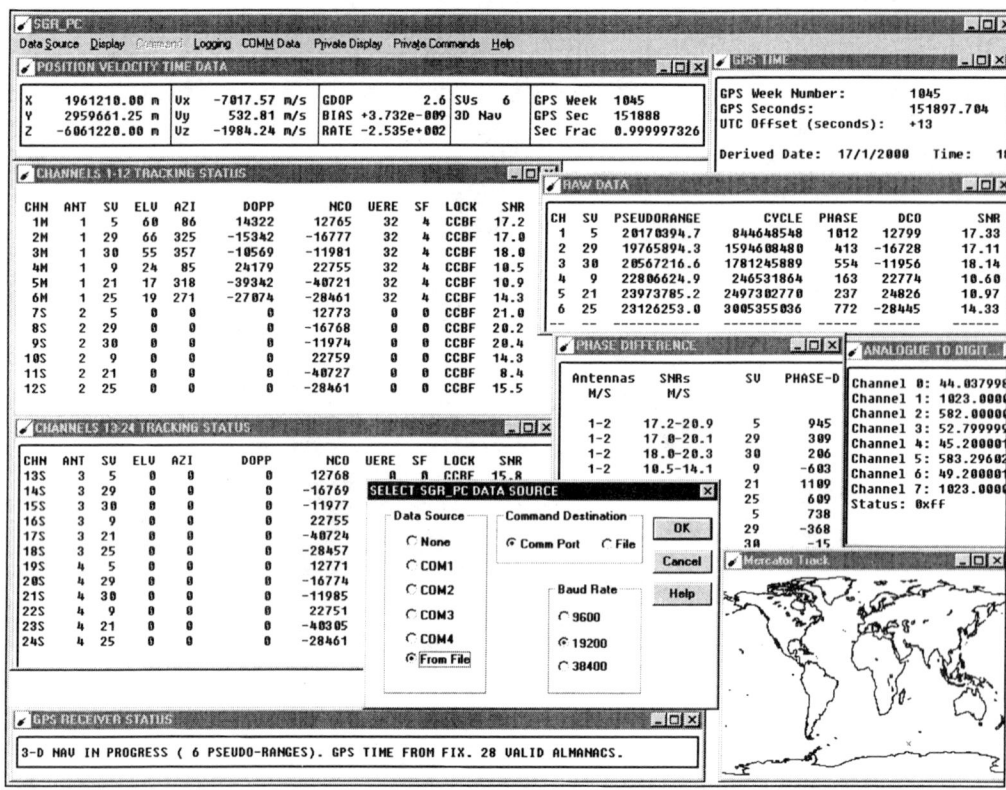

Figure 4 SGR-PC Display from Multi-Antenna operation on UoSAT-12

USING THE SGR IN ORBIT

For positioning, only a single GPS antenna is required, although a second may be included for redundancy's sake. The GPS antenna is mounted externally pointing away from the Earth, and should have a clear view of space such that the maximum number of satellite signals can be received. For attitude determination, three or four antennas are required and there are special mounting constraints. The SGR box is housed inside the satellite, connected to the GPS antennas via RF cables, and is connected to the host satellite system via power and data interfaces.

The host computer communicates over the interface port with the SGR, sending commands and receiving data and responses. The SGR requires no initialisation to successfully acquire and position in orbit and so in the most basic configuration, it is possible to simply record the data output from the RS-422 port. In the other direction, the host may send commands to the receiver to initialise and configure the receiver into different modes. The data format used over the RS-

422 is the SGR Binary Packet Protocol (SBPP), and the protocol used over the CAN-bus is based on the CAN-SU protocol. The SGR User Interface Manual describes these protocols in detail.[3]

If the CAN-bus is used, then only one port is required to access both the primary and the TTC processors, although provision for a redundant CAN-bus is included to increase reliability. If the RS-422 is to be used, then a single port will give full access to the primary computer, and indirect access to the TTC microprocessor via the primary computer and the internal CAN-bus. A second RS-422 port will enable full independent access to the TTC microprocessor.

Acquisition and Tracking

Upon power-up, the SGR enters *cold start* acquisition mode. It has no knowledge of its position, velocity or time, and so begins to search through possible frequencies for GPS satellite signals. Although all GPS satellites transmit at the same frequency, these signals are affected by Doppler shifts caused by relative motions between the GPS satellite and the user. If the user is on the ground, the Doppler shift will be between 0 and ±4 kHz, largely due to the GPS satellites' motion, but if the user is in low Earth orbit (LEO), then the user's velocity may increase the Doppler shift to as much as ±45 kHz. Positioning can only occur when four GPS satellites have been found (3 for 2-D positioning mode) and the latest GPS Ephemeris has been downloaded from each of these GPS satellites. Once this occurs, the SGR changes from *Cold Start* to *Highest Elevation* mode, where satellites are selected based on their elevation above the user's local horizon.

The Time To First Fix (TTFF) from a cold start is typically 7 minutes, and sometimes as fast as 4 minutes. This is because the SGR has 24 channels available for the cold search. However, if the user requires a consistently faster fix, it is possible to initialise the SGR with Almanac, time and user orbital elements to achieve TTFF of around 90 seconds.

The SGR on UoSAT-12 has been proven to be adequately reliable to be used as an integral part of a pioneering experiment in autonomous orbit maintenance over a period of months. To achieve this, there are sanity checks made both within the SGR and externally by the host computer (in this case, UoSAT-12 OBC). The SGR resets itself if there is no signal decoded for 5 minutes, or if no position fix is made for 20 minutes. The OBC keeps a check to ensure the SGR is operating nominally: if the SGR has not communicated for ten minutes or a position fix has not been obtained for 60 minutes, the SGR is powered down for 30 seconds and then powered up again. Checks such as these give a robust system and provide protection against software anomalies, Single Event Upsets and micro-latchups, and give a reset in the event of Single Event Latch-ups.

Positioning and Timing

The SGR takes measurements from the GPS signals latched by its own internal 10 Hz clock. These measurements, or raw data, include the pseudoranges, which contain both range and time information used for calculating the SGR position. Other raw measurements include the DCO, from which the Doppler and hence the user velocity can be found, the signal to noise ratio, carrier phase and integrated carrier cycles, which, while not essential, can be also used for positioning.

As with most GPS receivers, the SGR can output its position in Longitude, Latitude and altitude, which is useful for ground based use. For space-based use, the default output format is WGS-84

Cartesian co-ordinates, i.e. x, y, z in metres, referenced from the centre of the Earth (see Figure 5). Similarly, velocity is defined either along East, North and Up vectors or WGS-84 Cartesian co-ordinates.

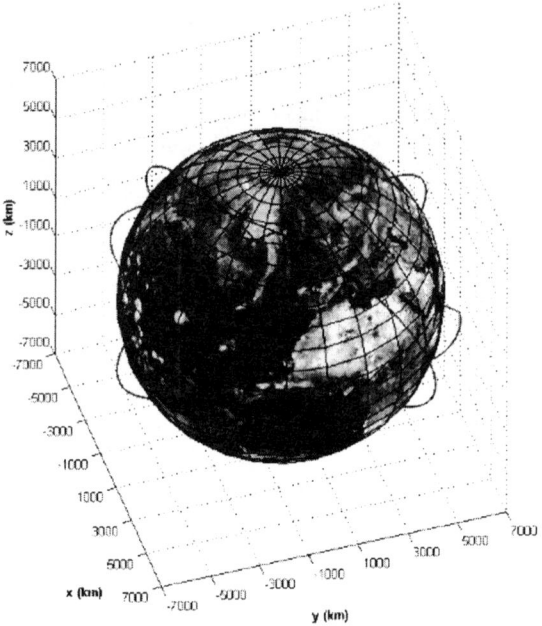

Figure 5 Flight data from UoSAT-12 SGR

The SGR is able to correct its internal clock to UTC extremely accurately as a by-product of the GPS position fix. There are two ways of transferring the time to the host.
- The SGR can inform the user of the time over the serial interface. This will enable a time transfer accuracy of as good as 10 milliseconds, depending on how the serial data is handled by the host.
- If greater accuracy is required, the electrical timing signal from the SGR can be used to synchronise user clocks to an accuracy of approximately 1 microsecond. This consists of a pulse per second signal (PPS) that can be resolved by reading the adjacent serial time message.

Attitude Determination

The implementation of the SGR as an attitude sensor is highly sensitive to the configuration of the antennas on the host satellite. The ideal satellite for this purpose has a flat surface pointing away from the Earth. The four (or three) GPS antennas are located towards the corners of the facet in a square such that the baselines are 1 - 2 metres in length. There are no protrusions on this surface that would cause reflections, and antennas should have 10 cm square ground planes. No satellite conforms to this ideal for GPS attitude determination, and so performance will depend on the configuration.

The phase difference measurements between antennas are the fundamental measurements that are used to determine the vehicle attitude. In the initialisation process prior to attitude solution, two parameters per baseline must found. As each antenna has a separate RF cable and front-end, the reference phase between antennas is not 0° when the GPS signal reaches both antennas at the same time (i.e. GPS satellite line of sight is at right-angles to antenna baseline). This phase offset is referred to as the line bias, and is present on all measurements made from that baseline. Secondly, if the baseline is longer than the GPS signal wavelength, which is 19 cm, then two or more attitudes may give the same phase difference. This integer ambiguity and the line bias must be resolved in the process of initialisation, but then will keep track of these parameters during the process of attitude determination. The SGR may try to resolve these parameters itself, or the host may provide one attitude estimate in order to assist with the attitude resolution. The default attitude output format is Euler angles with respect to the body co-ordinate system.

Programmable Outputs and Software Upgrading

Depending on the application, the user may have quite different requirements from the SGR in orbit. For experimental purposes, it may be necessary to record as much information from the SGR as possible, including position, raw data, position solution residuals, channel tracking status, analogue to digital readings, phase differences, etc. For other applications, the SGR may be required to be silent except for, say, a time message output once every 10 minutes. The SGR data rates can be dynamically programmed at packet level to accommodate these requirements, so that communications and log files can be optimised. The default preset mode on power-up is "quiet", such that many packets are disabled, but position, time and status messages are output every 10-30 seconds. For a particular application, the default data rates and modes can be compiled in the particular SGR such that the requirements for commanding of the SGR upon power-up can be reduced or eliminated.

An important characteristic of the SGR is that the operational GPS software is stored in an internal Flash memory and can be reprogrammed at will. This enables software upgrades to be issued to further enhance the capabilities of the SGR and correct software anomalies after delivery. It is also possible to re-program the flash while the SGR is in orbit, to give further flexibility. Reprogramming a flash memory in the space radiation environment increases the chance of Single Event Effect damage, but the risks may be minimised by selecting a quiet phase of the orbit to do this. Alternatively, new code may be loaded directly into RAM upon every power-up, which is the approach used on UoSAT-12 experiment.

Conclusions

This paper has described the design, specifications and operation of the SGR Space GPS Receiver. A description of the functional sub-systems and the interfaces is given. The SGR-10 is a dual antenna receiver for positioning, the SGR-20 is for attitude determination, and SGR-05 is a positioning receiver that has been streamlined for the smallest applications. Information on the integration into a typical satellite has been presented based on experience from use on the UoSAT-12 mission.

---

[1] Unwin M.J, Sweeting M.N, "A Practical Demonstration of Low-Cost Autonomous Orbit Determination using GPS", Proc. ION GPS-95, Palm Springs September 1995.
[2] Unwin M.J., Oldfield M.K., Underwood C.I.,., Harboe-Sorensen, "The Use of Commercial Technology for Spaceborne GPS Receiver Design", Proc. ION GPS-98, Nashville, Tennessee, Sept 1999.
[3] "SGR GPS Receiver User Interface Manual & Interface Control Document", SPSG-1415-006, January 2000, SSTL.

AAS 00-049

ROCKET SOUNDING BALLOON EXPERIMENTAL SECTION 2000

Kenneth Dalton, Gretchen England, Justin Eisenach, Chris Kilzer, Thanh Tran and Echezona Onwuatuegwu[*]

The RSB XS-00 project was initiated and will be completed by a group of Aerospace Engineering Seniors at the University of Colorado at Boulder. This project design is a single stage rocket carrying atmospheric measuring equipment to be launched from a sounding balloon. The balloon will carry the rocket to near 80,000 feet where the rocket will launch, puncture the balloon and ascend to an altitude of near 150,000 feet. During flight, the telemetry system will be transmitting pressure, temperature, and acceleration data to a ground station for analysis and compilation.

The goal of this project is to learn the complexities of rocketry and ballooning as well as to find a low cost alternative for high altitude measurements. Launch will be conducted during late April or early May and will be documented, whether it is a success or failure. With support from the Department of Aerospace Engineering at CU and the American Astronautical Society, this endeavor has hopes and potential for success.

INTRODUCTION

The RSB XS-00 rocket is being designed to be launched from a sounding balloon at approximately 80,000 feet above sea level and to ascend to its maximum altitude, a projected height of 150,000 feet. Throughout the flight on-board sensors will measure specific atmospheric conditions, namely pressure and temperature. The RSB XS-00 will then descend to the surface of the earth, travelling at survivable speeds due to parachute assistance, where the whole system will be recovered using a Global Positioning System (GPS) transmitter.

To address the more specific requirements of each section, the system and launch will be divided into 5 parts: balloon, launch chassis system, rocket, payload, and recovery system.

Balloon. The balloon portion of the RSB XS-00 system is intended to provide the rocket with a high altitude launch opportunity, significantly decreasing the rocket's requirements for increased size and weight. The balloon can be considered as the first stage of the rocket instead of the conventional propellant system.

Launch Chassis. This system pertains to the apparatus connected to the balloon, which will hold the RSB XS-00 upon balloon ascent and provide a stable platform from which to launch. The launch chassis needs to be capable of supporting the rocket during the ascent of the balloon and during rocket launch. The connections for the launch chassis will provide a vertical launch for the rocket and maintain a vertical orientation.

Rocket. The rocket needs to be light weight, but large enough to carry the intended payload, and strong enough to withstand the pressures and forces during take-off, flight, and recovery.

Payload. The payload will include atmospheric measuring devices, intended for measuring the pressure and temperature, a GPS system for location determination and recovery, and a communications system to

---

[*] Department of Aerospace Engineering, University of Colorado, Campus Box 429, Boulder, Colorado 80309-0429.

transmit the data to a ground system for analysis. It will also include controls for the propulsion and recovery systems.

Recovery System. The recovery system will include an explosive device for deployment of the parachute, and a parachute large and strong enough to allow the RSB XS-00 to have a safe descent and landing. The parachute is constructed of a rip-stop nylon designed for high-speed deployments. This system also includes the GPS system mentioned earlier.

METHODOLOGY

Launch Method

The proposed launch of the RSB XS-00 will be from the balloon at approximately 80,000 feet above sea level. After launch, the rocket is expected to achieve an altitude of near 150,000 feet. The ultimate reason for launching the rocket from the balloon is to minimize the weight requirements of the propulsion system, which is the component of the overall system with the greatest weight. This increased weight would also incur a greater cost due to the high cost of rocket propulsion. A simple rocket motor would no longer be sufficient to achieve the target altitude. If a conventional sounding rocket was used to achieve an altitude of 150,000 feet without the balloon assist, the cost and weight of the sounding rocket would be significantly higher than the projected cost and weight of this project.

Immediately after launch, the rocket will puncture the balloon, through both the base and the top, and travel as near vertical as possible until it reaches apogee. The puncture-balloon launch is illustrated in Figure 1.

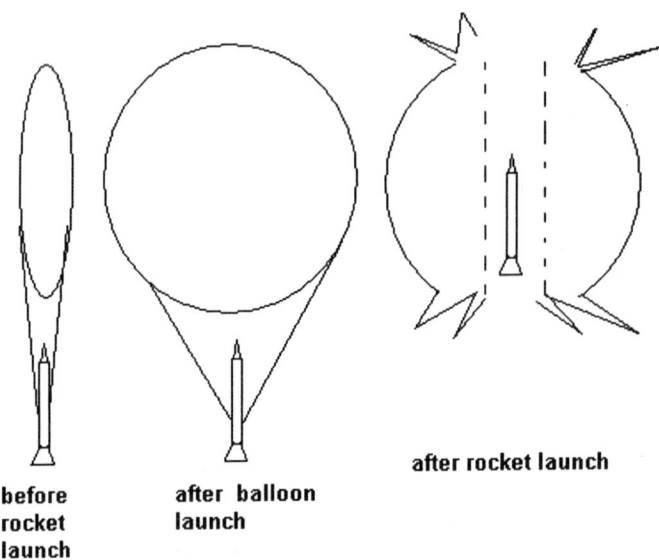

FIGURE 1: Balloon Puncture Rocket Launch

After puncture, the rocket will ascend to a final apogee altitude of near 150,000 feet. During ascent, to ensure vertical flight, the rocket will use fins as lateral stabilizers. The fins will also be positioned to induce a flight-axis spin on the rocket, further increasing flight stability. Once the rocket reaches apogee, the parachute will be ejected and the RSB XS-00 will descend back to earth. Locating the rocket system will be done with an on-board GPS system, a transmitter, and a ground receiver. During the ascent of the balloon, the ascent of the rocket, and the descent of the system, the data acquisition and transmission devices will acquire and transmit flight and atmospheric data to a ground receiver.

Rocket and Balloon Design

The rocket is most easily understood in terms of its six major components: motor, telemetry system, recovery system, instrumentation, GPS, and nose cone section. These components are illustrated in Figure 2.

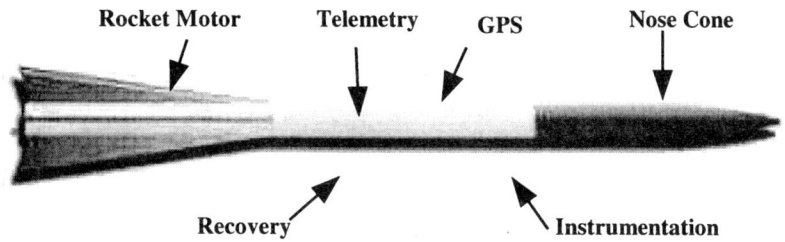

FIGURE 2: The Rocket and its Components

Rocket. The rocket chosen for this project was the Viper L3 rocket. The dimensions of the rocket are 12 feet long and six inches in diameter. In terms of current understanding, the rocket will be constructed of phenolic material with fiber reinforcement for increased structural strength. It has a payload space of two feet for all of the instrumentation on board as well as space for the recovery system and the propulsion system. At the rear of the rocket is a set of fins intended for vertical stabilization during flight. These fins will be attached at a five to ten degree angle to induce spin during flight for further stabilization.

Nose Cone. The nose cone is a hardened plastic piece designed to allow a puncture of the balloon during the launch phase of the mission profile.

Balloon. The balloon will be a standard sounding balloon as used by many atmospheric researchers. The balloon material will be an unstretchable plastic which, when it reaches the desired altitude of 80,000 feet, it will be punctured by the rocket.

Launch Pad and Chassis. The launch pad will be suspended from the balloon and will hold the rocket until ignition. It will be a metal plate, or more likely a grate, which needs to withstand the ignition force. It will be connected perpendicular to both the rocket and the central axis of the balloon. A visual depiction of the launch pad is provided below in Figure 3.

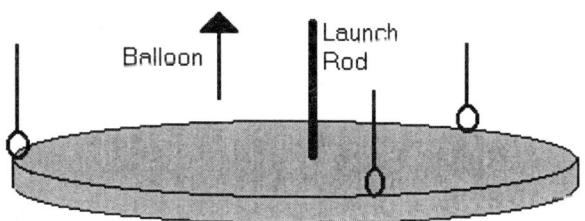

FIGURE 3: Launch Pad and Rod

Rocket Subsystems

Propulsion. The desired propulsion is a single-stage solid rocket fuel system with redundant ignition systems. This propulsion system reduces the complexity, therefore increasing the chance for success. A M-class solid rocket fuel motor will be used because it is less volatile than liquid propellant and significantly less expensive. This will provide a total impulse to the rocket of 2.36 pounds per second or 10,500 Newton-seconds. The ignition system will be a pressure and timer based system, which will be triggered when the balloon/rocket reaches 80,000 feet. A built in redundancy will be included to ignite the rocket if there is any unforeseen problems during balloon ascent.

Recovery Subsystem. The recovery subsystem is a parachute that will deploy when the rocket reaches apogee. A back-charge from the propulsion system will discharge the parachute from its storage inside the rocket and initiate the deployment. The parachute will contain a retaining ring intended to keep the parachute partially open so it does not close or become tangled during descent and rendered useless. The recovery subsystem will also contain a GPS module, which will provide the means for location and recovery of the rocket and its system after descent. Visual representation of the descent can be seen in Figure 4.

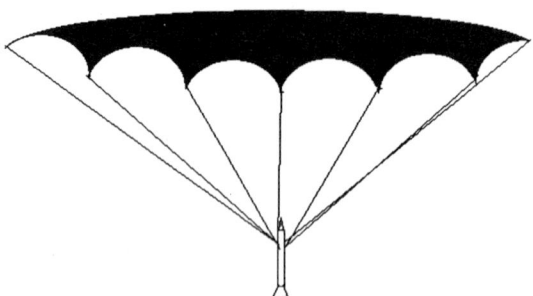

FIGURE 4: Parachute and Rocket Descent

Power Subsystem. In order to power the ignition of the rocket, the data acquisition systems, and the data transmission systems, the rocket will house a battery unit. The battery provides 12 VDC, 1 amp. Research is still being done into the type and the cost of the battery that will be used.

Payload Systems

The payload systems are all contained within the center and front portions of the rocket and include all of the power, the data acquisition, and the data transmission systems. Since the power systems have already been described, this section will focus on the data acquisition and transmission systems.

Telemetry. The telemetry systems housed in the rocket are the transmitter and the antenna. A digital transmitter is desired, but the costs for such devices are too high for the scope of this project, so an analog transmitter will be used. While this will limit the potential transmission distance, the analog system will be sufficient for the distances required for the project.

Instrumentation. The instrumentation systems encompass all of the data acquisition systems, including the atmospheric data, and GPS.
1. *Accelerometer.* This device was fabricated at the University and will be used to measure all of the accelerations experienced by the rocket and the balloon. It is a 50-G accelerometer like the ADXL150 from Analog Devices.
2. *Pressure Gauge.* This device was purchased through SenSym. It is the SCX15 model.
3. *Temperature Gauge.* This device was purchased through National Semiconductor. It is the LM35 model, which measures the temperature in °C. With the pressure and the

temperature data, the other atmospheric properties can be determined, such as density, elevation, etc.
4. *GPS.* This device will be donated or lent to us for the duration of our project. It will be used to find the altitude and the position of the system for information and recovery.

Weight Analysis

Based on best estimation, the gross payload weight of the entire RSB XS-00 structure consisting of the rocket, its payload, and the launch system is approximately 38-45 pounds. The weight distributions of the different systems are as follow:
1. Rocket structure   16 lbs.
2. Communications payload   8-12 lbs.
3. Recovery payload   6-7 lbs.
4. Propulsion system   2 lbs.
5. Battery   6-8 lbs.
6. Launch system   9 lbs.
    a. Pad   4 lbs.
    b. Rail (guide rod)   2 lbs.
    c. Launch tube   2 lbs.
    d. Rope   1 lb.

Testing

Preliminary fabrication and testing will be done on the balloon and the rocket, as well as the rocket payload and the final prototype design of the rocket. At the publishing of this report, three tests of a small-scale, single-stage rocket puncturing the balloon material have been completed. The purpose of these tests was to determine if a rocket could puncture the balloon material, which was successfully demonstrated. Another purpose was to determine if the trajectory or the speed of the ascent would change due to the puncture of the balloon. From the puncture tests, the rocket trajectory and speed of ascent did not change due to the puncture of the balloon. An auxiliary purpose of the tests was to determine the reliability of the ignition system and the recovery system. The ignition of the rockets and deployment of the parachute were successful; however, the deployment of the parachute only occurred in 67 percent of the test launches. It was determined that the packing of the parachute was extremely critical in the successful deployment of the parachute. Further tests will also check the reliability of the parachute deployment. These tests will also include the pressure-based and on-command ignition system for the rocket.

Fabrication and testing has been started and will continue to be done on the payload systems in the rocket. The instrumentation that will collect the high altitude data will include a pressure and temperature gauge, and an accelerometer. The pressure gauge and accelerometer will be tested for reliability at varying altitudes while flying in an aircraft. The temperature gauge will be subjected to a range of temperatures to test the reliability. The instrumentation will be tested in the range of respective areas expected to be experienced in the mission profile.

The tests show that certain aspects of the design are critical. Structural analysis will be required to prevent a failure of the launch pad during the ignition of the solid rocket propellant. Proper wiring of the ignition system is also essential to launch the rocket at the appropriate time. A number of safety measures are included in the circuitry of the ignition system to prevent premature ignition. Several redundant systems have been developed to ensure the ignition of the rocket, which includes a timer and pressure based. The parachute needs to be properly installed so that there is a safe recovery of the rocket.

PROJECT SCHEDULE

The estimated time schedule for the design can be seen in Table 1. The time schedule is reviewed on a biweekly basis to determine progress and efficiency. During the biweekly review, members of the design team are assigned new tasks upon completion of previously finished tasks.

Note: Payload to LEO; Laboratory for Atmospheric and Space Physics, LASP, has offered to launch the payload portion of the RSB XS-00 in a planned launch in June or July of 2000 from White Sands, New Mexico. This is dependent on the recovery of the payload and rocket instrumentation. The project launch altitude of the LASP rocket is 150 km.

TABLE 1
Schedule for RSB-XS-00

Task	Assigned to:	20-27 Sept	4-11 Oct	18-25 Oct	1-8 Nov	15-22 Nov	29 Nov-6 Dec	13-20 Dec	27 Dec-3 Jan	10-17 Jan	24-31 Jan	7-14 Feb	21-28 Feb	6-13 Mar	20-27 Mar	3-10 Apr	17-24 Apr	1-8 May	15-22 May	29 May-26 Jun
Define Problem	Gr	X,C																		
Set Objectives	Gr	X,C																		
Set Guidelines	Gr	X,C																		
Generate Ideas	Gr	X,C																		
Conceptual Design Presentation	Gr, PC	X,C																		
Develop & Write Proposal	Gr		X,C																	
Attend Machining Class	Gr										X									
Cost Analysis	Gr		X,C																	
Set Criteria	Gr		X,C																	
Research of Existing Solutions	Gr		X	X	X	X	X	X	X,C											
Preliminary Design Review	Gr, PC					X,C														
Critical Design Review	Gr, PC						X,C													
Design Revision	Gr, PC							X	X	X	X,C									
Prototype Construction	Gr					X	X	X												
Prototype Testing																				
-Balloon Material Testing	Gr					X,C														
-Hanging & Telemetry Ignition	Gr											X	X,C							
Launch Location/Clearance	Gr						X	X	X	X	X	X	X,C							
Telemetry																				
-Purchase/Fabrication	Gr								X,C											
-Ground Test	Gr														X	X,C				
-Airplane Test	Gr																X,C			
Breckenridge Presentation	Gr									X	X,C									
Final Prototype Construction	Gr														X	X	X,C			
Final Prototype Testing																				
-Ingition/Recovery	Gr																X	X,C		
-Telemetry Integration	Gr																X	X,C		
Final Design Review	Gr, PC									X	X,C									
Launch	Gr, PC									X	X	X	X	X	X	X	X	X	X,C	
Presentation	Gr, PC																X,C			
Payload to LEO (June or July)	Gr																			X,C

Legend
Gr=Group   Sponsor=AAS
PC=Prof Culp
X=Planned Start/Completion
C=Task Completed

BUDGET

With contributions from the Aerospace Engineering Department at the University of Colorado at Boulder and the American Astronautical Society, this project has become a possibility. All of the $0.00 cost estimates in Table 2 are either items that were donated or fabricated. All costs in the table are estimates as of March 25, 2000 and could possibly change.

TABLE 2
Budget for RSB XS-00

Allowable Fund	$4,800.00
Equipment/Materials	Cost (est)
Balloon launch	$300.00
Rocket	$437.00
Ignition System	$650.00
- Launch Pad	$60.00
Recovery System	$100.00
Telemetry	
- Sensors	$100.00
- Transmitter/Receiver	$500.00
- GPS	$0.00
- Telemetry Test	$54.00
Miscellaneous	
- Tools	$20.00
- Materials	$200.00
- Others	$150.00
Launch Cost	
Travel to NM	$240.00
TOTAL	$2,521.00

CONCLUSION

In summary, the RSB XS-00 rocket is being designed to be launched from a balloon at approximately 80,000 feet above sea level and ascend to its maximum elevation, a projected height of 150,000 feet. Throughout the flight sensors will be able to measure specific atmospheric conditions, namely pressure and temperature. The RSB XS-00 will then descend to the surface of the earth, travelling at survivable speeds due to parachute assistance, where the whole system will be recovered using a Global Positioning System (GPS) module and a transmitter. The budget and time schedule are constantly being reviewed for accuracy and efficiency. The projected completion of the RSB XS-00 is set for late April or early May with a launch at that time. Estimated cost of the RSB XS-00 (25 Mar 00) is $2,521 dollars.

ACKNOWLEDGEMENTS

The RSB XS-00 design group would like to recognize the following people for their assistance and encouragement: Dr. Robert Culp, Dr. John Sunkel, Norman Kjome, Larry Germann, Ian Gravseth, Joyce Idler, Pam Marlette, Ron Rausch, Marvin Odefey, Joanne Kitlen, Walter Lund, Bradley Dunkin, Javier Abarca, and Marty Berry.

REFERENCES

1. Frank, R. G., Zimmerman, W. F., Materials for Rockets and Missiles, The Macmillan Company, New York, 1959.

2. Newell, H. E., Jr., Sounding Rockets, McGraw-Hill Book Company, Inc., New York, 1959.

3. Parker, E. R., Materials for Missiles and Spacecraft, McGraw-Hill Book Company, Inc., New York, 1963.

4. HAL5's Project HALO: High Altitude Lift Off, http://advicom.net/~hal5/HALO/.

5. Missile Works Corporation, www.missileworks.com.

6. Paratech Parachutes: Parachute Deployment and Rocket Recovery Systems, www.paratech-parachutes.com.

7. Rocketry Online: Your Global Rocketry Resource, www.rocketryonline.com.

8. Tripoli Rocketry Association, INC., www.tripoli.org.

9. Rocketman Enterprises, Inc., www.the-rocketman.com.

10. Chuck Fosha, University of Colorado, Colorado Springs, experience with balloon launches.

11. Elaine Hanson, Space Grant, experience with sounding balloons.

12. Jack Crabtree, Lockheed Martin, GPS for balloons.

13. Jack Faber, Space Grant, experience with rockets.

14. Shane Brown, ASEN grad, did a rocket launch for his senior design.

15. Viper L3 rocket from GoLive CyberStudio 3

AAS 00-050

IN FLIGHT PERFORMANCE OF THE ZARM MAGNETIC TORQUERS MT80-1/MT140-2 FLOWN ON THE ABRIXAS MISSION

Matthias Wiegand, Oliver Matthews, Peter Offterdinger and H. J. Rath[*]

This paper deals with the in flight performance of the magnetic torquer assembly flown ond the German astronomical satellite ABRIXAS (A BRoad-band Imaging X-ray All-sky Survey). The attitude control system was designed to rely entirely on magnetic actuators from first attitude acquisition to three axis stabilization. A single momentum wheel has been added to provide stability. ABRIXAS has been successfully launched atop a Russian COSMOS launcher from Kapustin Yar April 28, 1999. After release of the satellite, telemetry could be established during the first pass above the ground station Weilheim (GSOC) located near Munich, Germany. At the second contact engineers had noticed a sharp change in temperature in the main battery before the main battery failed. For the next two days the backup batteries (primary cells) have been used to keep the satellite alive. The backup system was designed to provide additional power for initial attitude acquisition and wheel momentum charging. Thus, the attitude control system was mostly not affected by the main battery loss. Housekeeping data have been received every two seconds. The system worked as expected from simulation runs performed prior launch. Shortly after battery discharge telemetry has been lost. The magnetic torquer assembly first flown on ABRIXAS are proofed by the received flight data but it is needless to say that the loss of the mission is a tragedy for the entire scientific team and all the people who worked on the project.

I. Introduction

ABRIXAS (A BRoad-band Imaging X-ray All-sky Survey) is a small astronomical satellite planned by three German institutes[1] and funded by the German Space Agency DLR.[2]

The main scientific objective was a complete deep sky survey with an imaging X-ray telescope at an energy range from 0.5 to 10 keV. ABRIXAS was planned as the follow-up mission to the successful ROSAT mission and as a pathfinder mission for ESA's XMM telescope. The project has been established by the German Space Agency DLR as a pilot system, ZARM was responsible for the attitude determination and control system

---

[1] Astrophysical Institute Potsdam, Max-Planck Institute for Extraterrestrial Physics, Garching, Institute for Astronomy, Tübingen

[2] The satellite was built by OHB System, Bremen; ZEISS, Oberkochem, was responsible for the mirror-

---

[*] Center of Applied Space Technology and Microgravity, ZARM, University of Bremen, Am Fallturm, D-28359 Bremen, Germany. E-mail: wiegand@zarm.uni-bremen.de.

program with a strict cost limitation. Therefore, an innovative and cost effective attitude control system has been selected. The basic approach was to provide flexibility and redundancy while reducing hardware costs by using advanced filters. The momentum biased system relies solely on sun sensor and magnetometer measurements to determine three-axis attitude. Magnetic torquers are used for control only. This paper will not describe the developed algorithms. More information can be found in [1] and especially [4].

II. System Description

Figure 1 shows a rendered CAD drawing [3] of the ABRIXAS satellite. The design is dominated by the tube of the X-Ray telescope. On the bottom the seven 27-fold nested gold-coated mirror system can be seen. The star cameras on the lower right side are used for post-mission attitude determination only. The satellite subsystems like on board computers, batteries, momentum wheel, sun sensors and magnetic torquers are mounted in the bus module. The focal plane instrumentation is located at the top of the satellite. Both magnetometers are mounted on a boom to guarantee a relatively clean magnetic environment.

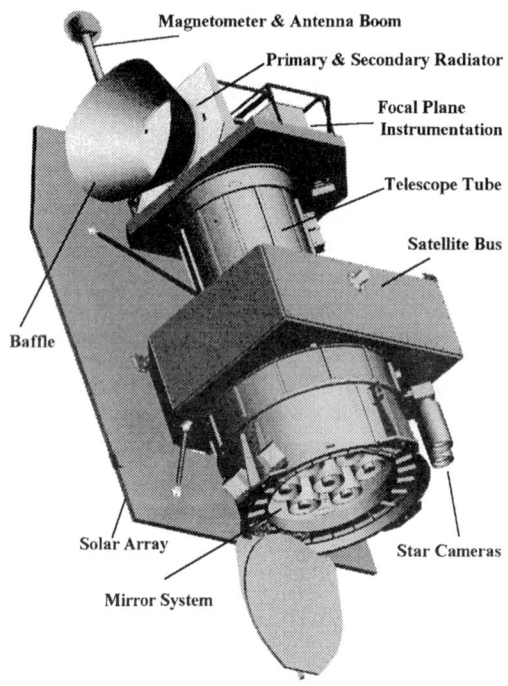

Figure 1: ABRIXAS flight configuration[3]

Total Mass	approx. 500 kg
Payload Mass	approx. 160 kg
Dimensions	height: 2500 mm
	width: 1800 mm
	depth: 1150 mm
Energy Supply	1 battery package
	2 support batteries
	solar generator 4.1 m^2
Average Power	200 W
Onboard Computer	Transputer T800
	80C31
Mass Storage	64 MByte
Software	C, modified RTXC
	real time kernel
TM/TC	S-Band, 1 Mbps/4 kbps
Redundancy	parallel on-board
	electronics and cross
	coupling capability
Attitude Determination	Sun-Sensor
	Magnetometer
	Fibre Optic Gyro (Z-Axis)
	Star-Sensor(postmission)
Attitude Control	Momentum Wheel
	Magnetic torquers

Table 1: System Overview

III. Mission Failure

After only three years of development, integration and testing, the ABRIXAS launch campaign started mid of April. Thanks to the good cooperation of the russian launcher team, integration on the launcher was very smooth (we arrived only one week before launch). Figure 2 shows ABRIXAS during final integration in the cleanroom at OHB-System, Bremen. Figure 3 shows the COSMOS rocket one day before launch at Kapustin Yar launchsite. ABRIXAS has been successfully launched and released into the desired 580 km nearly circular orbit on April 28. Telemetry could be established during the first pass above the groundstation Weilheim located near Munich, Germany. At the second contact engineers had noticed a sharp change in temperature in the main battery before the main battery failed. It is likely that the so called startup battery overcharged the main battery causing it to overheat and fail. The startup battery was on the spacecraft to power the satellite during sun acquisition mode after release from the launcher. For the next two days this battery (primary cell) kept the satellite alive.

IV. Attitude Control System

1. Scanning Motion and Requirements

Figure 4 explains the scanning mode of ABRIXAS. The z-axis is pointing in sun direction and the satellite rotates around this axis once per orbit. Due to the motion of the Earth the sun-pointing z-axis rotates once per year, i.e. the telescope (x-axis) scans the hemisphere twice per year. To avoid the Earth in the telescope's field of view, the nominal scanning law varies with respect to the position of sun and the orbital plane.

The aim of the scan axis control is to maximize the angular distance between the telescope and the horizon for the current orbit. To achieve this goal, the scan direction must be reversed approximately every 30 days. Figure 5 shows the one year simulated angular

Figure 2: ABRIXAS final integration

Figure 3: On the launchpad

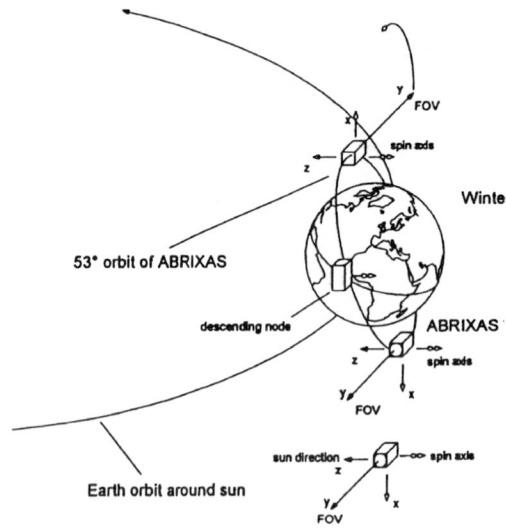

Figure 4: Scanning Motion of ABRIXAS [2]

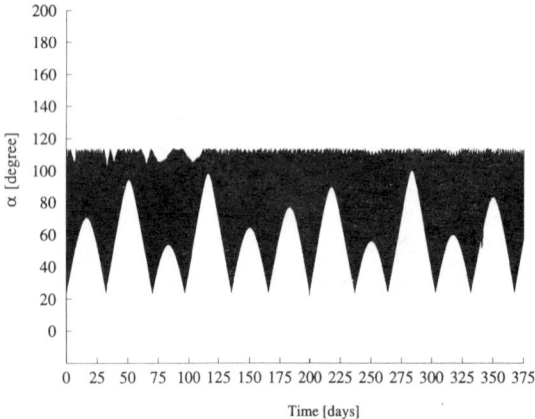

Figure 5: Arc-distance Telescope – Earth-horizon

distance between the telescope and the Earth horizon. Table 2 shows the required attitude accuracy driven by the scientific telescope.

2. Hardware Concept

In case of ABRIXAS the required pointing accuracy is comparable low, however, a regular scanning motion is desired. No X-ray sources shall be missed by random attitude motion, while high accuracy attitude deter-

Accuracy in scan-motion	$\pm 2^o$
Accuracy of z-axis	$\pm 10'$
Arc-distance Telescope - Earth Horizon (mean)	maximum per orbit
Accuracy of post-mission attitude determination	$< 30''$

Table 2: Attitude Requirements

mination using a star-sensor will be done post mission. Therefore, an inherently stable momentum biased system using a single wheel and three magnetic torquers has been selected. The first design included air coils, but

Figure 7: ABRIXAS MT140-2 Magnetic Torquer (Z-axis),ZARM

Figure 8: Bus Modul

these coils are difficult to accommodate. Later on the design was changed to torquers. Unfortunately, there was a length restriction for the X-axis and no suitable torquer was com-

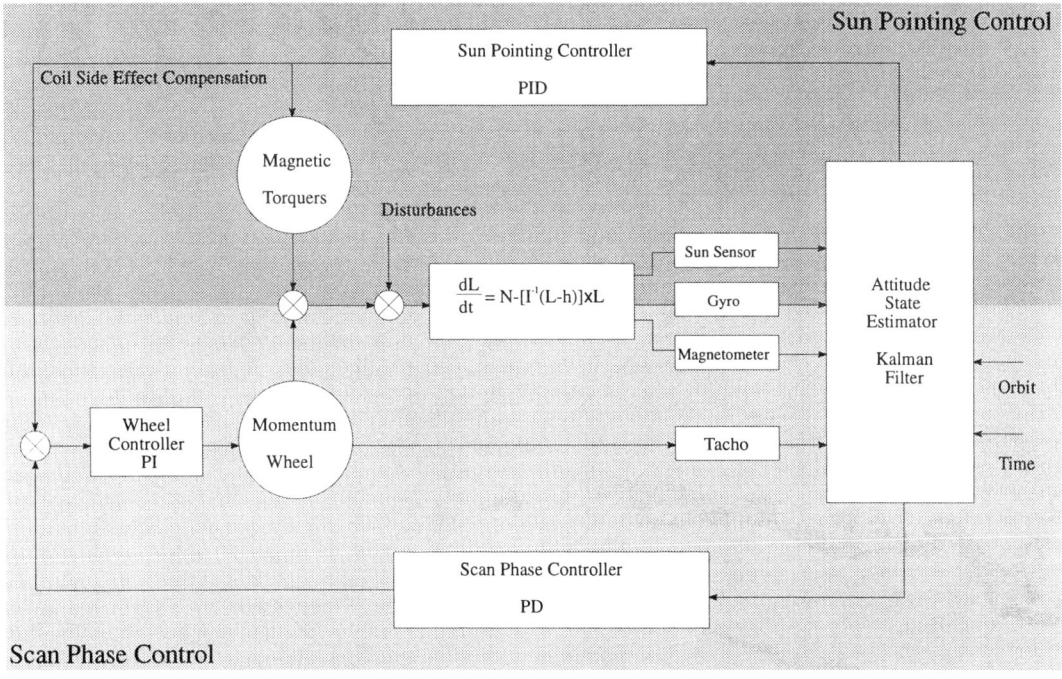

Figure 6: Attitude Control System

mercially available. Due to ZARM's experience with magnetic materials in other fields of science, we decided to develop the magnetic torquers on our own. The first results encouraged us, and by now, different flight models are available. For ABRIXAS we designed two different models. The MT140-2 is 680 mm long and has a dipole moment of 140 Am^2. For the more critical X-axis we designed the MT80-1 with 80 Am^2 at a length of 380 mm.

Figure 7 shows the Z-axis (Y-Axis) magnetic torquer of ABRIXAS. Figure 8 shows the open bus module in the cleanroom. One can see the Z-axis torquer mounted beneath the top plate. The X-axis torquer can be found near to corner strut. The Y-Axis is mounted above the main battery.

Due to nominal sun pointing a sun-senor is used for z-axis determination. In addition magnetometer data are analyzed to deduce three-axis attitude. To derive attitude information from magnetometer data and sun-sensor measurements, an extended Kalman-type filter is used to deduce a minimum error estimate by utilizing the knowledge of system and measurements dynamics, assumed statistics of system noise and measurements errors, and initial condition information. In addition six sun-present sensors are used during initial attitude acquisition. Due to cost efficiency, all grounds test are performed using a PC-based satellite simulator. This approach has been used for the BREM-SAT[3] development and has been shown to be accurate and reliable.

3. ACS Software Application Layer

The ABRIXAS attitude control software has two different operational modes and a safe mode. After deploy, the satellite starts with the initial attitude acquisition. This mode includes three different phases:

---

[3]BREM-SAT was designed by ZARM; launch: February 1994, decay: February 1995

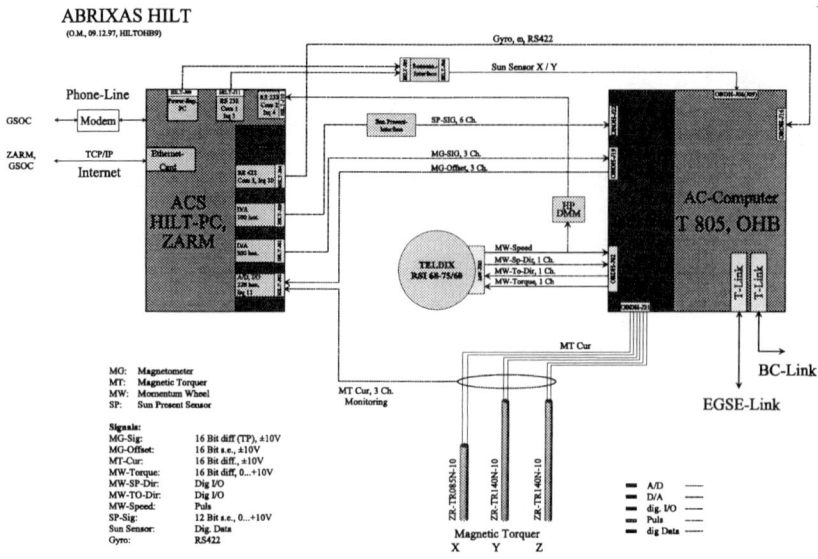

Figure 9: Hardware-In-The-Loop-Test

1. initial rate damping
2. slewing to sun
3. coarse sun pointing

After deploy fairly high angular rates of 1 rpm per axis are expected due to the launcher specification. Thus, ABRIXAS has to get rid of the initial angular momentum utilizing a B-dot procedure. At this phase the torquers operate according to the change of magnetometer readings. The momentum wheel will be spun up to 500 rpm to provide a minimum stability for the slew phase. During the slew phase, a simple algorithm relying solely on the sun-present sensors provides the satellite with coarse information ($\pm 25°$) about the sun direction. When the sun sensors (FOV 64°) detects the sun, the controller will maintain the attitude and spin up the wheel to 5000 rpm utilizing the magnetic torquers to control Z-axis rotation of the spacecraft. Nominal wheel speed will be reached after nearly one day.

The described three phases are part of the so-called PROM software, while the more sophisticated full application software is stored in a (less reliable) EEPROM. The highly autonomous PROM software is designed to slew the satellite to the sun within 4 orbits after deploy to prevent a loss of power. The safety margin is another 4 orbits.

Via telecommand the attitude control system is switched to the nominal scanning mode including

1. fine sunpointing
2. magnetometer based scan motion control

To increase the scan motion stability, a inexpensive fiber optic gyroscope of low accuracy (drift: $6°/h$) has been added to the design. Kalman filter parameters decide, how strong the scan motion control relies on the gyroscope.

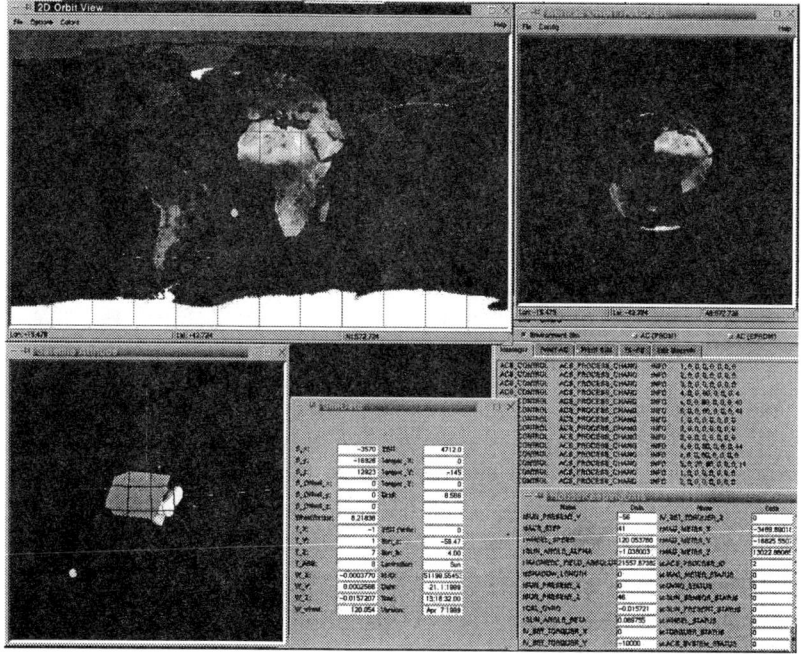

Figure 10: Testbed User Interface

Figure 6 shows a simplified block diagram of the nominal control loop. If one neglects nutation issues, and considers the low scanning motion of 0.001 rad/s, scan phase control and sun pointing control can be treated independently. Due to the momentum bias the sun pointing control is a first order system while the scan phase control is a second order system. As mentioned above, the Kalman filter is used to deduce scan-axis attitude from magnetometer data only. A detailed description can be found in [4]. It can be added that a complete software upload is possible during the mission. Information about the ABRIXAS onboard software design are provided in [6]. Due to the ABRIXAS failure only PROM software (acquisition mode) has been used during the mission. Before the flight data are discussed, the attitude control testbed will be described and simulation are shown. These data are important to evaluate inorbit performance, reliability and, of course, the quality of the testbed itself.

V. Hardware-In-The-Loop-Test

1. Test Equipment

Figure 9 shows the ABRIXAS HILT equipment. It basically consists of an EM or FM of the onboard ACS computer and a simulation PC running the free UNIX clone LINUX. The PC is simulating the space environment and the spacecraft dynamics. The onboard computer can be connected with the PC directly without any changes in hardware or software, to provide most realistic testing. In addition all ACS components (momentum wheel, magnetic torquer, sun sensor, magnetometer, gyroscope, sun-present sensors) can be included in the loop.

2. ACS Tests

Although all ACS components can be operated in closed loop, ACS performance tests under simulated space conditions are only possible using simulated sensor data. The HILT-PC generates these data responding to actuator values from the ACS computer. The

ACS computer receives the data in same format as from the original sensors. All major modes of ABRIXAS attitude control system have been tested this way. The results are discussed in the following section. Since these tests are using the original flight computer and hardware, only real-time simulations can be performed. To use the flight software for fast long-time simulation, a simple wrapper has been written to compile and run most of the flight software directly on the PC. Thus, the simulation can be run three orders of magnitude faster.

VI. ACS Simulation Results

1. Initial Attitude Acquisition

One of the most critical phases for the attitude control system (and for the satellite itself) is the initial attitude acquisition. During this phase the satellite has no ground control and must operate autonomously. Therefore, the software is stored in the radiation hard PROMs. Under worst case conditions four orbits are needed to achieve a stable sun pointing attitude. A special support battery provides power for approximately eight orbits. Figure 11 shows the initial angular rate damping after deploy. Assuming 1 rpm per axis – the worst case scenario – the satellite has to get rid of approx 37 Nms angular momentum. In comparison the momentum wheel can produce a maximum momentum of 68 Nms. The very stable B-dot algorithm needs two orbits for the initial rate damping. Figure 12 depicts the angle between z-axis and sun direction. In addition the momentum wheel loading is shown. Since wheel loading and sun pointing is contrary during coarse mode the emphasize is on stability. Therefore, a certain pointing error is acceptable while the wheel is loading. After one day the wheel reaches the nominal speed of 5000 rpm. At that time ground control can switch to nominal scanning mode.

2. Nominal Mode

The full application software is stored in a less reliable EEPROM due to storage re-

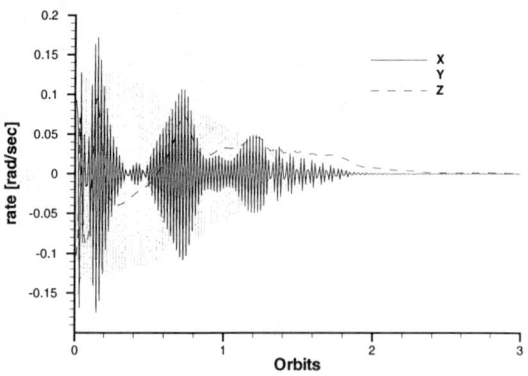

Figure 11: Rate Damping after Deploy, Initial Angular Momentum 37 Nms, Initial Angular Rate per Axis 1 rpm

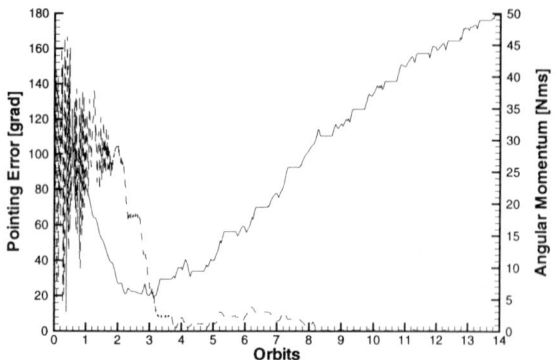

Figure 12: Initial Slew to Sun, Momentum Wheel Loading, Coarse Sun Pointing

striction of the PROMs. The EEPROM is writable and allows a full software upload. To switch to nominal mode requires to reboot the dedicated T800 ACS computer and to load the more complex EEPROM software. The safe mode is still located in the PROM. Thus, in case of an exception the highly reliable PROM software will be loaded.

Figure 13 shows the steady state performance of scan axis determination via Kalman filtering of magnetometer readings. The measurement accuracy is approx 50 nT including analog to digital conversion. The IGRF ac-

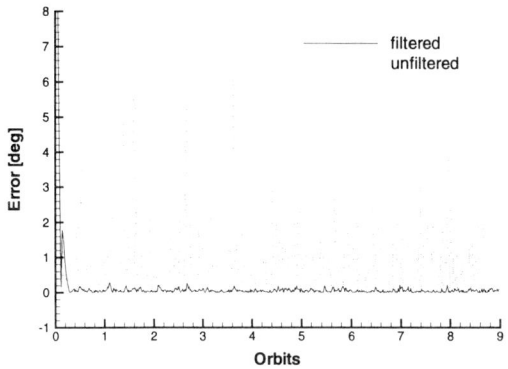

Figure 13: Kalman Filtering of Magnetometer Data to Achieve Scan Axis Attitude, $\sigma = 350\,nT$, Sample Rate 0.5 Hz

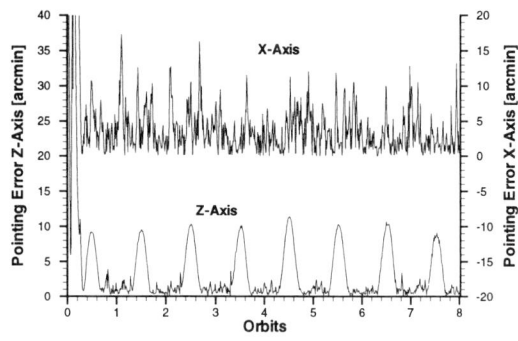

Figure 14: ABRIXAS ACS Steady State Performance, 0.5 Hz Control Loop Update

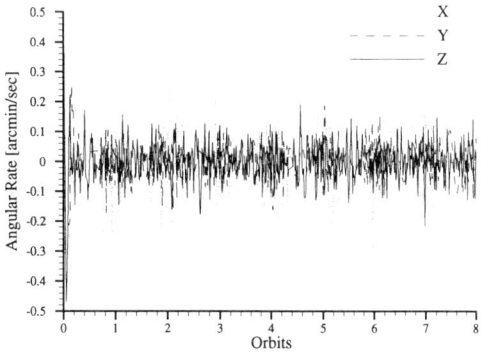

Figure 15: ABRIXAS ACS Steady State Attitude Stability, 0.5 Hz Control Loop Update

curacy is about 300 nT. Detailed information regarding the Kalman filtering are provided in [4]. The magnetometer should be calibrated in orbit via least square batch filtering. The algorithms have been tested and are very robust. As mentioned above, the coupling between sun pointing control loop and scan phase control loop is relatively low. The most important coupling is introduced by the magnetic torquers itself. The control torque is the result of the cross-product with the Earth magnetic field vector. The component in z-direction is directly compensated by the momentum wheel. Thus, loading and unloading the wheel is an issue. Simulation testing and experiences of the BREM-SAT mission have shown that wheel loading or unloading can be performed without a important loss of attitude accuracy during phases where the z-axis is nearly perpendicular to the B-vector [1].

Figure 14 shows the steady state performance of the ABRIXAS ACS. The design fits the requirements of table 2 well. The peaks in the sun angle result from uncontrolled eclipse times, where gravity gradient torque is the primary disturbance. We consider to use the Kalman filter to compensate this disturbance. Figure 15 depicts the angular rates which have an acceptable level with respect to post-mission attitude determination using star images.

VII. Flight Results

As mentioned above more than two days of history data (sample rate 2 sec) have been received at the groundstation. Unfortunately, not only main battery but also some electronics has been damaged. Thus we had some resets of our computers and we lost one of the redundant ACS computers. To discuss all the flight data from the ACS would be much to much for this paper. Therefore, I like to focus on the two most important phases: sun acquisition and momentum charge. Figure 16 shows the acquisition phase from the sunpresent sensors.

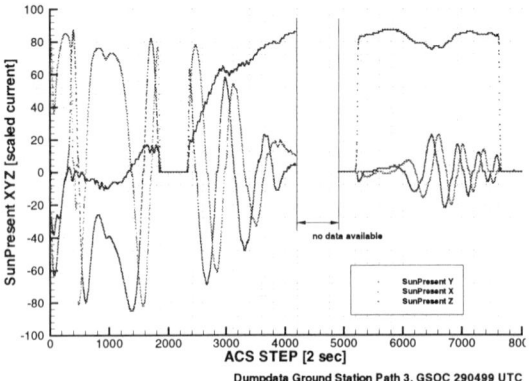

Figure 16: ABRIXAS Acquisition Phase

This is not directly comparable with the simulation results from section VI, Figure 11 and Figure 12. No post mission filtering has been done yet. Instead figure 17 shows the expected unit sunvector from simulation. Full sunlight means 100 and every other light comes from Earth Albedo.

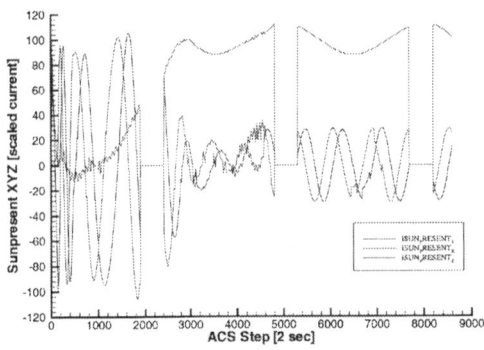

Figure 17: Simulated Acquisition Phase

The release condition used for the simulation are retrieved from the housekeeping data of ACS sensors. The release from the launcher was very calm and so there was no mayor rotation on any axis. The simulated sequels looks very similar to real flight data. The slew worked like in the simulation. Between the shadow time at ACS step 5000 and 8000 one can see the current equivalent from the added +Z and -Z axis. The real data are showing that after the transition from eclipse to sunlight the intensity is slightly increasing before the minimum is reached when the satellite is directly between Sun and Earth. The simulation does not show this because we thought that under sharp angle ($\pm$ 5 deg) the cell current is neglectable. In addition the Earth Albedo is not easy to simulate. But these are minor differences.

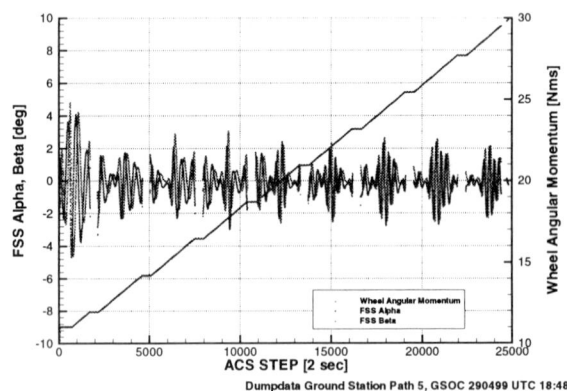

Figure 18: Momentum Wheel Charging

Figure 18 shows the momentum charging of the wheel using the magnetic torquers. It looks much smoother than in the simulation figure 12. Reason for that is not the system momentum but the wheel momentum is shown. The computed system momentum from Z-axis gyro and momentum wheel looks similar to the simulation. The charging rate is also comparable.

From figure 16 and figure 18 we have seen that the attitude control system was fully operational and performed well as long as system power has been provided. The attitude history shows us that the applied torque of the magnetic coils was quite accurate. That was a mayor point during development because of some uncertainties due to disturbances from hard magnetic materials on board the satellite.

Figure 19 shows the applied torque of the magnetic coils needed for an approximately ninety degree slew. The conversion from current to torque bases on ground measurements using a magnetometer. The results has been verified by additional mechanical torque measurements as result of the magnetic dipole moment and the Earth magnetic field vector.

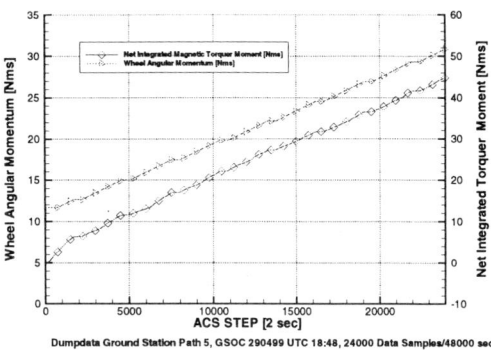

Figure 20: Momentum Wheel Charging

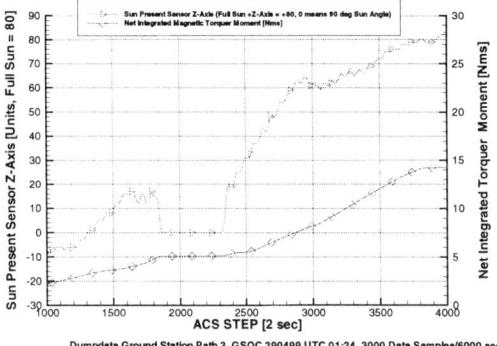

Figure 19: 90 degree slew to sun

Due to the time variant Earth magnetic field vector and the changing direction with respect to the satellite, position and attitude are needed to analyze the torquer performance. During acquisition and coarse mode three axis attitude information are not very well known. The state estimator is not running at this early phases. For that reason figure 19 focus on the applied integrated torque. Due to the wheel momentum of 6 Nms approximately 8.5 Nms are needed to perform a ninety degree slew. The controller is designed to use 50 percent of the applied torque for nutation damping. The applied current and the attitude history matches very well.

Figure 20 shows the applied integrated torque during wheel charging and coarse sun pointing. The controller is designed to use 70 percent for wheel charging and 30 percent for sun pointing and nutation damping. The net applied torque matches very well with the design specification and ground measurements. No disturbances can be seen from the data.

To conclude the ACS worked as good as expected. Unfortunately we lost the mission before the nominal scan phase was reached and the X-ray deep sky survey was started.

VIII. Attitude Estimation after Failure

Until now, the team was not successful to recover the satellite. But there is still a slight hope to get in contact with the satellite during an eclipse free time. But this can only successful if the satellite is stable pointing to Sun. The actual configuration of the system relies on the lost main battery. A reconfiguration is only possible from ground. That means the satellite systems are not running right now. Thus recovering is only possible if the satellite is pointing to sun due to external forces from the Earth's magnetic field or gravity gradient. Computations have been done to find out if there are any stable sun pointing phases and if it is possible to propagate them.

Figure 21 show the result from the computations. The main assumption is a certain damping rate caused by magnetic hysteresis due to motion of the spacecraft w.r.t the Earth's magnetic field. Fortunately, we got some information about the damping rate from FGAN radar measurements to confirm our assumption. From these computations the satellite's Z-axis (largest moment of inertia) will be slewed perpendicular to the orbital plane by gravity gradient forces with magnetic

Figure 21: Stable Areas

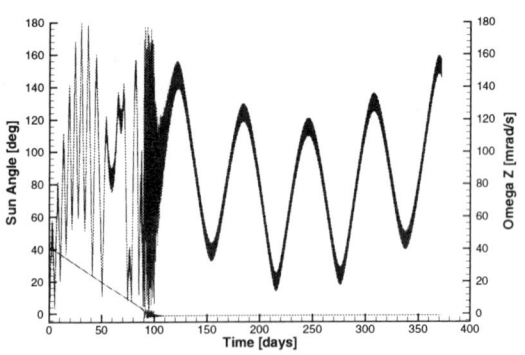

Figure 23: Sun Angle Propagation

damping. Figure 22 shows an example slew. But, of course, the solar panel can pointing north or south of the orbital plane. This problem can be probably solved by optical measurements form ground.

Figure 22: Slewing to a Stable Attitude

Figure 23 shows the angle between solar panel and Sun for about a year. One can easily see that there are some opportunities to get in contact with the satellite. In addition there is a correlation with eclipse free times.

IX. Conclusion

The loss of ABRIXAS is very painful for all the people who worked on the project. It is especially a tragedy for the entire scientific team. Nevertheless the housekeeping data, we received during the first 2.5 days are extremely valuable for our work on attitude control systems. The used magnetic only attitude control system has been proofed as reliable. The system is qualified for most low earth satellites with moderate attitude requirements. It is also suitable as inexpensive backup system for more demanding missions. Our environment simulator and the development testbed has been proofed by the ABRIXAS flight data. Right now we are using that experience for the development of a more generic satellite attitude simulator (SatSim, supported by DLR) for design, development and test of satellite attitude control systems. The magnetic torquers we developed for the project worked excellent and they are now proofed by flight data. We carried on our work on this devices and they are used on many new satellites (Customers: Daimler-Chrysler, Carlo Gavazzi, US-Airforce, Canadian Space Agency, Orbital Science Cooperation). While the scientific mission is lost we still hope that the groundstation team at GSOC will get contact to ABRIXAS during an eclipse free time. The housekeeping data are important for all the subsystems that are flown the first time.

References

[1] KÖNIGSMANN, H.J., WIEGAND, M., "The Magnetic Attitude Control of ABRIXAS ", *USU Conference on Small Satellites*, Logan, USA, 1995

[2] HASINGER, G. ET. AL.,"ABRIXAS - A BRoad-band Imaging X-ray All-sky Survey", Project Report, 1995

[3] GERHARD, I., JUNK, P., LÜBKE-OSSENBECK, B., "ABRIXAS, Satellite Technical Concept", *48th Congress of the International Astronautical Federation*, Turin, Italy, 1997, IAA-97-IAA.11.1.07

[4] WIEGAND, M., "Using Magnetometer and Sun-Sensor to Determine Three-Axis Attitude for the ABRIXAS mission", *IFAC Workshop on Small Satellites*, Breckenridge, USA, Feb 4-8, 1997

[5] WIEGAND, M., "Ground Test of the ABRIXAS Attitude Determination and Control System", *21st AAS Guidance and Control Conference*, Breckenridge, USA, Feb 4-8, 1998

[6] ZAPPEN, F.-P., WINKLER A., KOOPMAN, O., HÜLSING, T., "The ABRIXAS On-board Software Design, Development and Verification Approach", *Conference Proceedings DASIA*, 1997

Section V
GNC TECHNOLOGY FOR MICRO/NANO SPACECRAFT

SESSION V

Joint Session Chairperson: Dr. Frank Redd
Utah State University

Joint Session Charperson: Dr. Darrell Zimbelman
NASA Goddard Space Flight Center

Local Session Chairperson: Dr. Don Mackison
University of Colorado

The following paper numbers were not assigned:
AAS 00-068 to -070

AAS 00-061

FORMATION FLYING AND RELATIVE NAVIGATION – A NANOSATELLITE RESEARCH MISSION

Frank R. Chavez and David K. Schmidt[*]

A student-centered initiative has been undertaken at the University of Colorado at Colorado Springs, which involves the design and development, as well as the planned launch and operation of two nanosatellites. The objectives of the project, entitled NavGold, include the definition of an engineering case study with which to explore open technical issues regarding formation flight, and the possible establishment of an on-orbit test bed for testing, and validation of critical technologies. Some of these technologies include precision relative navigation (both relative position and relative attitude); relative maneuvering and station keeping; command, control and communication architectures; and possibly MEMS devices (e.g., propulsion, sensors). On-orbit experiments are being planned to evaluate concepts and algorithms for formation-flying maneuvers and relative navigation. The "experimental constellation' will consist of two identical nanosats of hexagonal geometry, with targeted spacecraft mass of 14 kg each. Each spacecraft will include maneuvering propulsion plus three-axis active attitude determination, stabilization and control. Absolute and relative navigation will be accomplished with the use of differential and/or augmented GPS. Maximum use of COTS hardware is a goal, to minimize cost and risk, and a 2002-03 launch date is targeted.

Introduction

The concept of a spacecraft constellation has become of interest in the space community, where a constellation is defined a group of two or more spacecraft interacting in some manner to meet common mission objectives. Constellations are viewed as a possible means of achieving greater mission performance, with increased mission robustness, and perhaps less cost, when compared to other mission-design approaches. Many constellation-based mission concepts involve the use of small spacecraft (microsats or nanosatellites), and frequently involve a cluster of spacecraft (2 or more) flying in some specified geometric configuration called a formation. Two potential advantages of using such constellations are reduced mission risk per sensor and reduced development budgets due to smaller spacecraft bus design. One rationale behind this claim is due to the fact that should one spacecraft in a constellation fail, the mission requirement may still be met without any further action, or the failed spacecraft may be replaced.

Examples of missions in which a spacecraft formation may have merit are those involving large baseline interferometry [1-3] (either earth or deep-space observing); stereo, 3-D imaging, or short revisit time earth imaging. Table 1 lists over 30 representative missions that will or may use formations in the mission design. [4,5]

But many technical issues need to be resolved to enable formation flight, especially for complex formations involving several spacecraft and significant formation reconfiguration. Such issues include space-navigation technology (sensors, etc.); component miniaturization; high-efficiency solar cells; and constellation orbit-design techniques, for example, and research in these areas is ongoing. Receiving less attention, perhaps, are the issues of generic classification of formations; development of convenient

---

[*] Flight Dynamics and Control Laboratory, University of Colorado, Colorado Springs, Colorado 80918-7864.
E-mail: dschmidt@mail.uccs.edu.

coordinate systems for relative navigation; and command-and-control architectures appropriate for complex, highly reconfigurable formations.

Projected Launch Year	Mission	Mission Type
99	New Millennium Program (NMP) EO-1	Earth Science
01	Gravity and Climate Recovery (GRACE)	Earth Science
02	University Nanosats/AFRL	Tech Demo
02	Auroral Multiscale Mission (AMM)/APL (MIDEX)	Space Science
02	Lightweight Synthetic Apeture Radar (LightSAR)	Earth Science
03	New Millennium Program (NMP) ST-3	Space Science
03	New Millennium Program (NMP) ST-5	Space Science
03	Techsat-21/AFRL	Tech Demo
04	Constellation-X	Space Science
05	Magnetospheric Multiscale	Space Science
05	Space Interferometry Mission (SIM)	Space Science
07	Global Precipitation Mission (EOS-9)	Earth Science
07	Global Electrodynamics	Space Science
08	Magnetospheric Constellation	Space Science
08	Laser Interferometric Space Antenna (LISA)	Space Science
	Digital Elevation Mapping (DEM)/EAS	Earth Science
	Earth Radiation Mission (ERM)/ESA	Earth Science
	High X-Ray Energy Satellite Mission (HEXS)/ESA	Space Science
09	DARWIN Space Infrared Interferometer/ESA	Space Science
11	Terrestrial Planet Finder	Space Science
	Astronomical Low-Frequency Array (ALFA)/Explorers	Space Science
	MAXIM X-Ray Interferometry Mission	Space Science
05+	Leonardo (GSFC)	Earth Science
05+	Soil Moisture and Ocean Salinity Observ. Mission (EX-4)	Earth Science
05+	Time-Dependent Gravity Field Mapping Mission (EX-5)	Earth Science
05+	Vegetation Recovery Mission (EX-6)	Earth Science
05+	Cold Land Processes Research Mission (EX-7)	Earth Science
05+	Submillimeter Probe of Evolution of Cosmic Struct. (SPECS)	Space Science
15+	GSFC Earth Sciences Vision	Earth Science
15+	NASA Institute of Advanced Concepts/Very Large Optics for the Study of Extrasolar Terrestrial Planets	Space Science
15+	NASA Institute of Advanced Concepts/Ultra-high Throughput X-Ray Astronomy Observatory	Space Science
15+	NASA Institute of Advanced Concepts/Structureless Extremely Large Lightweight Swarm Array Space Telescope	Space Science

Table 1, Missions Pertinent to Formation Flying

In this paper, we will discuss some of these technical issues from a fairly high level, discuss generic formation requirements derived from example mission requirements, and describe a generic, prototype mission and two nanosatellites designed for formation-flight experiments. This prototypical mission may be utilized as an engineering design case study for performing design tradeoffs, such that insight can be gained regarding some of the issues raised. This case-study mission is being developed by a student group at CU-Colorado Springs, along with faculty and industry advisors, and collaboration with others is of interest. Further, should this mission proceed to launch and on-orbit operations, as planned, it will present an opportunity for technology demonstrations and proof of concepts, with regard to the methods and systems investigated, as well as an on-orbit test bed for new hardware.

Definitions

The first two example missions noted above (interferometry and stereo imaging) require *congruent observations* of the target, while the third mission(short target revisits) require *sequential observations* of the target. Congruent observations must be made within a few seconds, or less, of each other, and sequential observations are perhaps taken minutes apart [6,7].

To define a spacecraft formation, at least four descriptors are required: 1) the *geometric configuration*, 2) the *command-and-control architecture*, 3) the *degree of spacecraft autonomy*, and 4) the *communications architecture* of the constellation. The geometric configuration of the formation describes the spacecraft relative positions and relative attitude orientations. For example, the spacecraft may by in trail, as shown in Figure 1, or in different orbits, ans shown in Figure 2. The formation in Figure 1 is illustrative of a simple geometric configuration for congruent observations of target. P, while the formation in Figure 2 is an example of a geometric formation for sequential observations of the target P. Further, the geometric configuration can be fixed or dynamic. For example, if large on-orbit spatial re-positioning and/or reorientation are involved, the geometric configuration is dynamic.

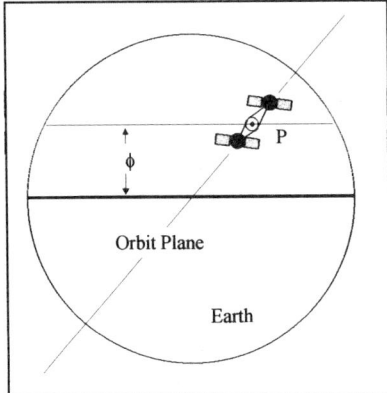
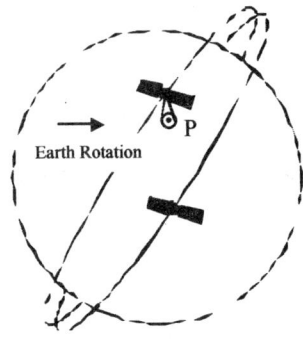

Figure 1. Example Geometric Configuration for Congruent Observations

Figure 2, Example Geometric Configuration for Short-Revisit Observations

A formation's command-and-control (C&C) architecture describes how the constellation maintains and possibly changes its geometric configuration and/or communication architecture (defined below). One example of C&C architecture might include a single lead spacecraft performing all necessary position and attitude determination, and commanding any necessary maneuvers of the other spacecraft. This control architecture is often referred to as a "leader-follower" architecture. Another control architecture is a "cooperative" control architecture, in which all spacecraft interact cooperatively in some fashion to maintain the geometric configuration and/or communication architecture.

Another key component of the C&C architecture is the degree of spacecraft autonomy in the formation. For example, each spacecraft could be independently controlled from the ground, in contrast to total spacecraft autonomy where the constellation spacecraft operate completely independent from ground control.

And finally, the communication architecture of the constellation describes how the spacecraft communicate with the ground and with each other, what information is transferred or crosslinked between the spacecraft, and how that is achieved. One potential communication configuration might involve each spacecraft communicating with every other spacecraft, as well as the ground, while an alternative communication configuration would involve each spacecraft reporting to a single lead spacecraft responsible for all inter-formation as well as ground communications.

Formation Requirements

Attention is now turned to the subject of some probable requirements on any arbitrary constellation formation, derived from mission goals. Rather than delineate just mission-specific requirements, it is also useful to note some rather generic requirements.

First, it is argued that almost any formation will require *relative* position and possibly *relative* orientation stationkeping. Under two-body orbit theory, once the formation geometric configuration has been established no further stationkeeping maneuvering may be necessary. But in reality, maneuvering capability of some sort is almost always required, even if for a "static" geometric configuration. Due to environmental perturbations on the member spacecraft, or any required thrusting for attitude maneuvers, for example, relative-position station keeping and relative orientation station keeping must be performed.

Additionally, in many missions, the geometric configuration must be significantly altered. Examples include any requirement for (rapid) retargeting, or a spacecraft failure requiring adjustment of the formation geometry. As yet another example, suppose the temporal resolution between observations of a target is required of the two-spacecraft constellation depicted in Figure 2. Maneuvering will be necessary to adjust the phasing of the relative longitudes of the ascending nodes and/or relative anomalies, to achieve a new desired temporal resolution over the target. As a result, gross relative maneuvering (rather than finer station-keeping maneuvering) may also be required of many formations.

The above requirements for maneuvering and stationkeeping, along with other mission goals, generate further requirements, including relative navigation. The relative position of the spacecraft within the constellation as well as the relative orientation of the spacecraft within the constellation must somehow be determined. Additionally, efficient command-and-control approaches must be developed, especially for managing complex, highly reconfigurable constellation formations.

All these requirements generate several difficult technical questions, such as how to best accomplish relative navigation to the required precision; whether the information processing for navigation and command and control should be entirely ground based, space based, or some combination; and the degree of spacecraft autonomy within the constellation. Certainly all spacecraft could conceivably be independently commanded from the ground, as with almost all current spacecraft. But if a four-spacecraft formation is to be rapidly retargeted, for example, it is unlikely that individual commanding from the ground is the optimum approach. Quite possibly a better approach would be to command a certain constellation geometric configuration, and perhaps an allowable time frame, and allow the spacecraft formation to autonomously perform the necessary maneuvers, in terms of the required relative positions and orientations. This is reminiscent of controlling platoons of road vehicles, or teams of robots.

If such a reconfiguration is to be autonomously performed, key open issues include the required navigation accuracy, the techniques for relative maneuvering and station-keeping, and the optimum command and control structure. For example, should a leader-follower or a cooperative approach be employed?

Regarding required relative navigation accuracy, a "close formation" is defined as, "...two spacecraft kept to a relative separation ranging from tens of meters to less than one km...."[8]. Further, some mission, such as the HEXS mission listed in Table 1 requires relative navigation accuracy of tens of millimeters [5]. If separation distances and navigation accuracies of this magnitude are required, certainly ground based orbit determination methods are not appropriate for obtaining relative navigation solutions. Even space-based determination techniques such as TDRSS can only deliver an accuracy of hundreds of meters in LEO. Furthermore, these methods are a postori. Hence, they are not appropriate for *real-time* navigation solutions required for autonomous formation reconfiguration.

High-precision ground-based relative navigation solutions can be accomplished with the aide of high precision GPS navigation information, [9,10]. These methods can result in sub-centimeter relative position accuracy. However, the delay in obtaining these solutions is on the order of weeks and requires the use of GPS receivers capable of acquiring both the L1 and L2 frequencies from at least 6 spacecraft simultaneously [9]. Additionally, a network of 10-12 globally distributed ground receivers is required. The significant delay in obtaining high-precision relative navigation solutions in this manner will not meet the almost-real-time requirements for autonomous reconfigurations.

Hence GPS, enhanced in some fashion, perhaps with additional ranging measurements will be necessary to meet a requirement of autonomous relative navigation, with accuracies of centimeters or less.

NavGold Mission Objectives and Constraints

A generic prototype mission and two nanosatellites that can be used for formation-flight experiments and technology demonstrations will now be described. It is suggested that this mission and basic spacecraft design may be used in engineering design case studies for performing design tradeoffs (i.e., a "design challenge") such that insight can be gained into the above open issues. This prototype mission and spacecraft is referred to as the NavGold Mission.

Program Objectives To define a formation-flying case study, for performing engineering design trades, and to establish an on-orbit test bed and technology demonstration for validation of autonomous relative-navigation techniques and guidance algorithms aimed at meeting the critical relative-navigation, maneuvering, and station-keeping requirements of autonomous constellation operations.

Mission Objectives-

- To demonstrate autonomous relative-navigation accuracies of less than 100m, with a goal of 10 m accuracy or better.
- To demonstrate autonomous relative-position maneuvering and station-keeping capability with relative position accuracy equal or better than 100m.
- To demonstrate autonomous relative-attitude maneuvering and station-keeping capability with relative attitude accuracy equal or better than two degrees.
- To limit total mass of the spacecraft constellation to 30 kg or less.
- To accomplish all primary and secondary objectives at the absolute minimum cost.
- To document critical technological accomplishments in developing autonomous navigation, maneuvering and station-keeping capabilities.
- To educate the student team members in all facets of spacecraft mission development, from initial concept to final launch and on-orbit operations.

Spacecraft Design Requirements and Constraints

In order to meet these mission objectives, a minimum of two spacecraft will be required, and at least one spacecraft must have autonomous navigation and maneuvering capability - both position and attitude.

Autonomous navigation capability will require one or more GPS receivers and on-board processing. Maneuvering and station keeping will require maneuvering-propulsion capability plus three-axis attitude determination and control. This in turn will require attitude sensors and torque actuation. Requirements on the remaining functions, such as power generation, environment control, data processing, and structure, will naturally flow from the above requirements and constraints.

Navigation Approaches

To meet a goal of 10-meter relative navigation accuracy, two approaches are to be considered: Differential GPS, and Augmented GPS. For differential GPS, given that the two receivers are in close proximity, the timing errors and atmospheric distortions experience by the two receivers are very similar. Hence, the determination of <u>relative</u> position is more accurate that the determination of absolute position.

By augmented GPS, we mean the addition of another measurement such as relative range, obtained, for example, via tonal or optical method. The tonal method may use a pseudo-random code transmitted on an R.F. cross-link between spacecraft, and a highly accurate timing device.

Two possible benefits are accrued by augmenting GPS with auxiliary range measurements. First, an additional measurement aids in Integer Ambiguity Resolution (IAR), which arises in Carrier-Phase GPS navigation due to the integer differences in the number of wavelengths involved. With an initial range estimate, faster convergence of the navigation solution is achieved. The second benefit arises from blending the auxiliary range measure with the GPS navigation solution. By optimally blending these measurement, additional accuracy may be obtained.

With regard to GPS units, a variety of systems are under consideration. Such systems vary from a complete COTS "black-box" system; to a custom built, in-house system; to a hybrid system. Examples of GPS systems being considered include

1. A Trimble TANS Vector receiver,
2. a unit developed by the University of Surrey in partnership with GEC Plessy,

3. JPL's microGPS unit, with on-board processing
4. a unit being developed by NASA Goddard or
5. a hybrid GPS system including a TIDGET™ GPS front end from NAVSYS Corp., coupled with custom estimation algorithms.

Regarding additional sensors for augmenting the GPS information, several concepts are being considered. The first is the use of a pseudorandom code sequence referred to a high stability clock on board each spacecraft. With one spacecraft transmitting the sequence at a known rate and beginning at a known time the second spacecraft can measure the difference between the received sequence and its internal clock generated sequence thus deriving a "time of flight" measurement hence distance information. This PR code must be of sufficient length to unambiguously accommodate large (e.g. tens of kilometers) separation distances. The PR code may be transmitted at regular intervals or provided on a crosslink subcarrier frequency.

Another concept being explored is the use of transponders. Inter-spacecraft ranges between 10 kilometers and 10 meters are planned for the NavGold experiments, and given this range envelope, on-board timing circuits may prove stable enough to permit measurements of either radio frequency or optical transponder-like beacon signals passed between the spacecraft.

In either case, this additional range measurement can be optimally combined with standard GPS position estimates, to obtain much more accurate estimates of relative position.

Experimental Mission Maneuvers

For discussion purposes here, all maneuvers will be performed "in-plane," although out-of-plane maneuvers are under study. The class of maneuvers to be described are Relative Re-Positioning, Relative Position Station Keeping, Relative Re-Orientation, and Relative Orientation Station Keeping.

Relative Re-Positioning Two types of relative re-positioning maneuvers will be described - a relative phasing maneuver, and a relative altitude maneuver. Any in-plane re-positioning maneuver can be derived from a combination of these two generic maneuvers.

Maneuver 1: Increase/decrease the relative phase angle (or time separation of δt) between the two spacecraft in the same orbit, as in Figure 3. This maneuver essentially involves a Hohmann transfer back to the original orbit, as depicted in Figure 4. The initial relative position of the two spacecraft, S_1 and S_2, is indicated by the filled circles, and the final relative position is indicated by the open circle. After the maneuver, S_2 is closer (indicated by the empty circle) to S_1 by:

$\delta t = T_E - T_C$.

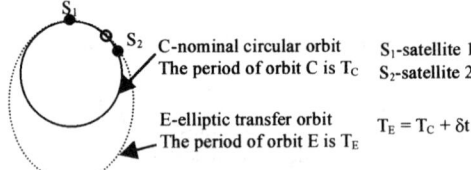

Figure 3. Hohmann Transfer for Relative Phasing Maneuver.

The ΔV required to adjust the phasing is shown parametrically in Figure 4. The y-axis shows the total ΔV required for the maneuver in meters per second, for a given phase angle adjustment. Each line represents different time allotment to perform the maneuver, in orbits.

Maneuver 2: Increase/decrease the relative altitude separation h between the two spacecraft, as denoted in Figure 5. The initial relative position of the two spacecraft, S_1 and S_2, is indicated by the filled circles. After the two-impulse maneuver, S_1 has increased its altitude relative to S_2 by h (indicated by the empty circles).

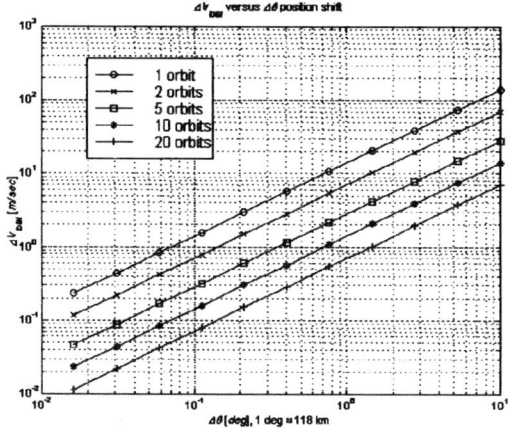

Figure 4. ΔV Trade Study.

Relative Position Station Keeping An example of relative-position station keeping involves the requirement that a relative separation, d, and altitude, h, is to be maintained within some tolerance, over a specified time period, as indicated in Figure 6.

Relative Re-Orientation In Figure 7, θ_1 and θ_2 represent the inertial orientation of spacecraft S_1 and S_2, respectively, and an example re-orientation maneuver or station-keeping maneuver is depicted.

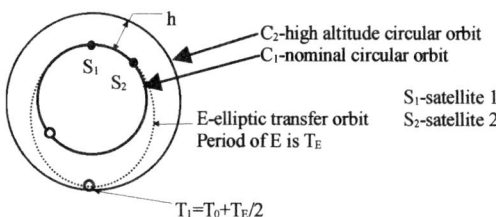

Figure 5. Hohmann Transfer for Relative Altitude Maneuver.

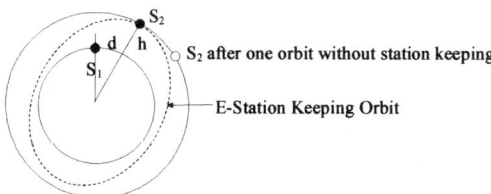

Figure 6. Hohmann Orbit Transfer for Relative Position Station Keeping Maneuver.

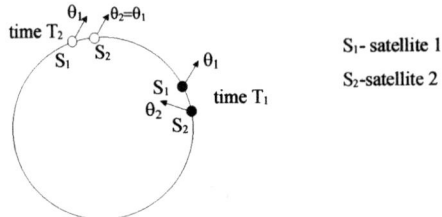

Figure 7 - Example Relative Reorientation.

Spacecraft Bus

Two identical spacecraft will be used, both to provide redundancy, and ease of design. Each spacecraft will have total mass of 14 kg. The spacecraft geometry is shown in Figure 8. Rather than partitioning the spacecraft into traditional subsystems, the spacecraft is broken into functions. These functions will share hardware and software wherever possible to maximize efficiency. The functional areas are Navigation, Guidance and Control, Flight Computer, Electrical Power, Communication, and Structure and Thermal Control.

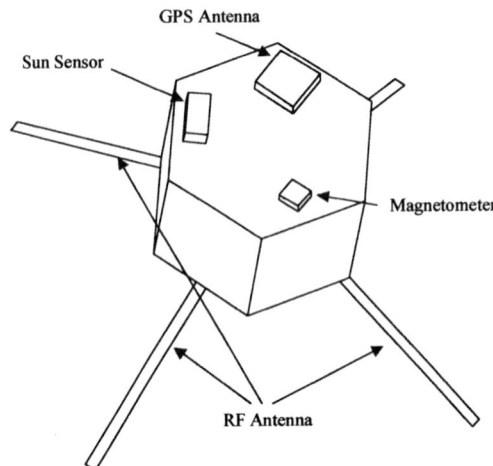

Figure 8. Spacecraft Geometry.

Navigation, Guidance, and Control

The primary payload on the NAVGOLD spacecraft will be the Navigation, Guidance and Control (NGC) package, including the sensor suite, and actuator/propulsion components. The design requirements for NGC package fall under three main categories. First, the spacecraft must be able to perform high-accuracy <u>relative</u> navigation. Second, the NGC system must provide three-axis attitude stabilization and control of the spacecraft, both during maneuvers and coast. Third, the spacecraft must be capable of performing autonomous re-positioning and relative attitude maneuvers and station keeping. Finally, the total mass of the entire NGC package, not including hardware shared with other systems (such as the flight computer and communications antennas), will be approximately 4 kg.

The guidance component of the NGC function consists of the algorithms used for either maneuvering or station keeping. The maneuvering algorithms include discrete-burn timing logic, along with the associated spacecraft pointing algorithms. The station keeping guidance will be based on algorithms derived from Hill's equations, governing the relative position and velocity of orbiting spacecraft.

Baseline Design The primary measures of merit used to evaluate the various design approaches are mass, power consumption, cost, computational complexity, and maturity of the technology. For the propulsion system, emphasis is placed on maturity, while the most risk will be accepted in the navigation and guidance functions (e.g., sensors and algorithms).

Regarding the NGC software, the development environment will be MATLAB/SIMULINK®, include maximum utilization of the Real-Time Workshop to enable rapid prototyping and real-time code generation.

The baseline navigation concept utilizes GPS, augmented with ranging augmentation using radio-frequency (r.f.) signals. The ranging system requires very precise clocks on both spacecraft, and candidates have been identified. The ranging system uses the same hardware as the communications cross-link function, and was a key reason why the r.f.-based method for ranging was considered first.

The baseline propulsion system used in the spacecraft design will be cold gas thrusters. This was chosen primarily due to its proven reliability and low risk. The cold gas thruster will include the propellant tank, propellant (N_2), valves, plumbing and six individual thrusters. The mass of the propulsion system will be approximately 1.5 kg, require approximately 10 watts of power for continuous burn, and supply approximately 25 m/s ΔV total.

The sensors used in the attitude control system will be digital sun sensors, a three-axis magnetometer, and a horizon sensor, all developed in-house at CU-Colorado Springs. (See Figures 9 and 10.) The three-axis magnetometer design is based on a commercially available magneto-sensitive IC. The attitude sensor package will be approximately 0.5 kg in mass and consume very little power. The overall accuracy of each sensor is on the order of 0.5 degrees, one σ.

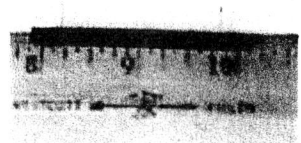

Figure 9. Fine Sun Sensor (Scale in Inches).   Figure 10. Three Axis Magnetometer (Scale in Inches).

Using the digital sun sensors and the three-axis magnetometers described above in conjunction with the Triad and QUEST [11] algorithms for attitude determination, estimates of the attitude-determination accuracy have been obtained. The Triad method for estimating the attitude matrix (the coordinate transformation relating the known inertial coordinate system to the desired body-fixed coordinate system) is outlined below, for example

Let, $\vec{S}_{Inertial}$ be the sun vector from the spacecraft, expressed in inertial coordinates; $\vec{S}_{Body}$ be the same vector expressed in the body-fixed coordinates; $\vec{B}_{Inertial}$ be the local Earth magnetic field vector expressed in inertial coordinates; $\vec{B}_{Body}$ be same vector expressed in the body fixed coordinates. Calculate the following unit vectors,

$$\vec{e}_{1_{Body}} = \frac{\vec{S}_{Body}}{|\vec{S}_{Body}|} \qquad \vec{e}_{1_{Inertial}} = \frac{\vec{S}_{Inertial}}{|\vec{S}_{Inertial}|}$$

$$\vec{e}_{2_{Body}} = \frac{\vec{S}_{Body} \times \vec{B}_{Body}}{|\vec{S}_{Body} \times \vec{B}_{Body}|} \qquad \vec{e}_{2_{Inertial}} = \frac{\vec{S}_{Inertial} \times \vec{B}_{Inertial}}{|\vec{S}_{Inertial} \times \vec{B}_{Inertial}|}$$

$$\vec{e}_{3_{Body}} = \vec{e}_{1_{Body}} \times \vec{e}_{2_{Body}} \qquad \vec{e}_{3_{Inertial}} = \vec{e}_{1_{Inertial}} \times \vec{e}_{2_{Inertial}}$$

and form the matrices,

$$M_{Body} = \left[e_{1_{Body}}, e_{1_{Body}}, e_{1_{Body}} \right]$$

$$M_{Inertial} = \left[e_{1_{Inertial}}, e_{1_{Inertial}}, e_{1_{Inertial}} \right]$$

The attitude matrix, A, is found from $A = M_{Body} M_{Inertial}^T$

Figure 11 shows the attitude-determination accuracy, based on Monte-Carlo analysis, as a function of digital sun sensor and the three-axis magnetometer measurement errors. In the figure, delta$_S$ refers to the 1-σ digital sun sensors errors, and like values for the magnetometer are plotted along the abscissa. For digital sun sensor and three-axis magnetometer accuracy of 0.5 deg, the attitude determination accuracy is about 0.35 deg 1-σ. The error analysis of the QUEST algorithm yields slightly better results.

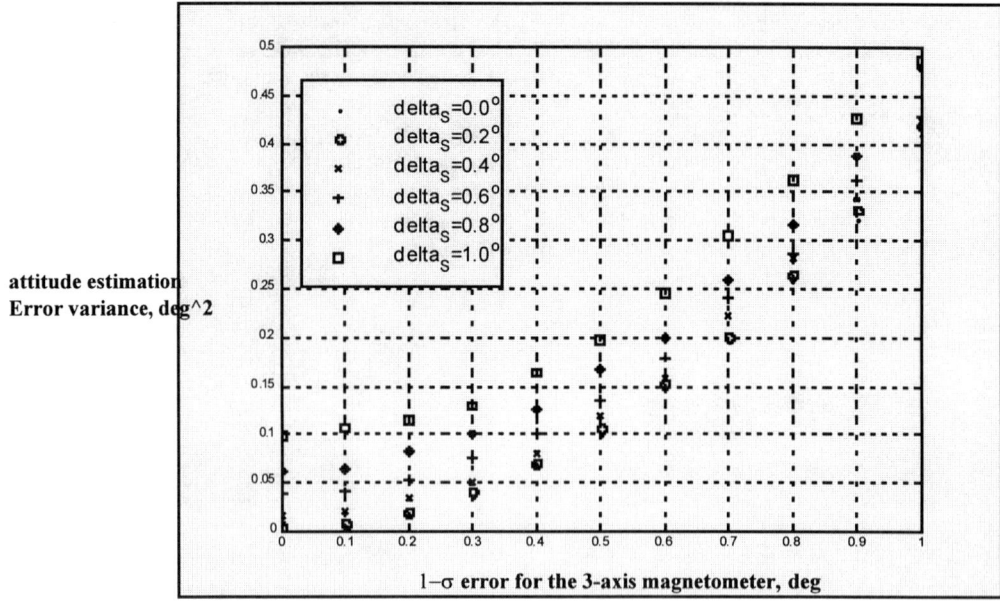

Figure 11. Attitude-Determination Accuracy

There will be three attitude control modes, one for each main operating mode of the spacecraft. These are safe-hold mode, sun-pointing mode, and maneuvering mode. Safe-hold mode, implemented both analog and digitally within the Command and Telemetry Unit (CTU), is designed to keep the spacecraft under positive attitude control after initial orbit insertion to counteract any initial attitude rates, and to provide a simple, reliable reversion mode.

The safe-hold mode, for example, employs the "B-dot" control algorithm. The spacecraft's rotational equation of motion is given as $\frac{d\vec{H}}{dt} = \vec{T}$, where $\vec{H}$ is the spacecraft's angular momentum vector and $\vec{T}$ is the vector of external torques acting on the spacecraft. An external torque can be generated through the interaction of the magnetic field generated by the torque rings, $\vec{m}$, and the local Earth magnetic field, $\vec{B}$. The expression for this torque is $\vec{T} = \vec{m} \times \vec{B}$. In the B-dot control algorithm, the commanded torque ring magnetic moment is given by $\vec{m} = k \frac{d\vec{B}}{dt}$, where k is the control gain, and $\frac{d\vec{B}}{dt}$ is the time rate of change of the local Earth magnetic field.

The actuators to be used by the attitude-control system will be magnetic torque coils and the cold-gas thrusters. Four torque coils will be located around the hexagonal prism structure on the outer panels of each spacecraft, as shown in Figure 12. The coils will be manufactured in-house, will be approximately 20 cm outer diameter by 2.5 cm cross section diameter, will have an approximate mass of 2 kg total, and be capable of generating a magnetic moment of approximately 1.25 A-m^2.

These sizing results are based on the information presented in Figure 13, showing maximum torque per unit inertia plotted versus desired settling time. Using the safe-hold mode, the initial spin rates must be damped with a settling time of at most 3 orbits. From Figure 13, this requires a required maximum torque per unit inertia of $4*10^{-4}$ sec^{-2}. For a body with a moment of inertia of 0.5 kg-m^2, this requires a torque of 0.2 mN-m. With an initial angular rate of 1 deg/sec, the safe-hold mode is capable of reducing the tumble rate to less than 0.1 degrees per second within 3 orbits, as shown in Figure 14. The time response of the torque generated by the torque coils during this maneuver is shown in Figure 15.

Sun-pointing mode will be the mode that is used most frequently. It is designed to orient the spacecraft towards the sun for maximum power generation. Torque coils will also be used for actuation in this control mode.

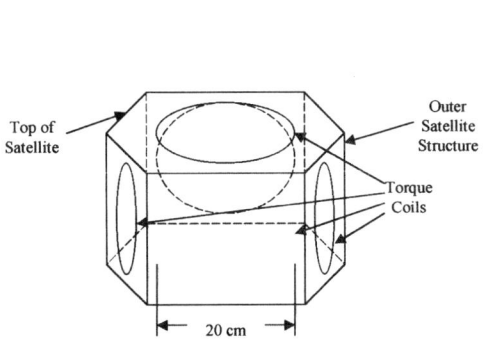

Figure 12. Torque Coil Location.   Figure 13. Settling-Time Torque Requirement.

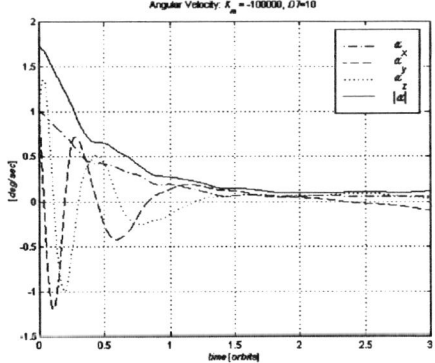

Figure 14. Angular Velocity Time Response.   Figure 15. Torque Time Response.

Maneuvering mode will be employed only during propulsive maneuvering. This mode must be capable of performing large slewing maneuvers, and especially orienting the thrusters properly for a maneuver. In this mode, the cold-gas thrusters will be used for actuation.

Flight Computer

The autonomy requirements of navigation and attitude control mandate the use of an on-board processor. Also, the need to control the spacecraft and communicate with the ground dictates the adoption of a traditional Command and Data Handling (C&DH) function. The flight computer provides the resources for these functions as well as all other telemetry, data handling, and on-board processing, depicted in Figure 16.

The flight computer is divided into two main parts, a Command and Telemetry Unit (CTU) and a main processor. The two parts are logically divided. Physically, they may either be separate or integrated on the same electronic circuit board. Logical separation facilitates parallel development of each function. Keeping the two parts physically separate improves the flexibility of development. Integrating the two physically can make the design smaller and more power efficient.

The main processor will be a COTS unit, presumably having never been flown in space, which makes that part a high risk element from a space environment standpoint. To mitigate against that situation the CTU, which is based on a radiation hardened field programmable gate array (FPGA), takes care of the basic and critical command
and data handling. Since it is radiation hardened, it provides a more reliable control unit.

Command and Telemetry Unit The CTU provides five basic functions:
1. Critical command decoding and execution.
2. Failure detection and spacecraft mode control.
3. Critical telemetry.
4. Electrical power and communication function controls.
5. Timing.

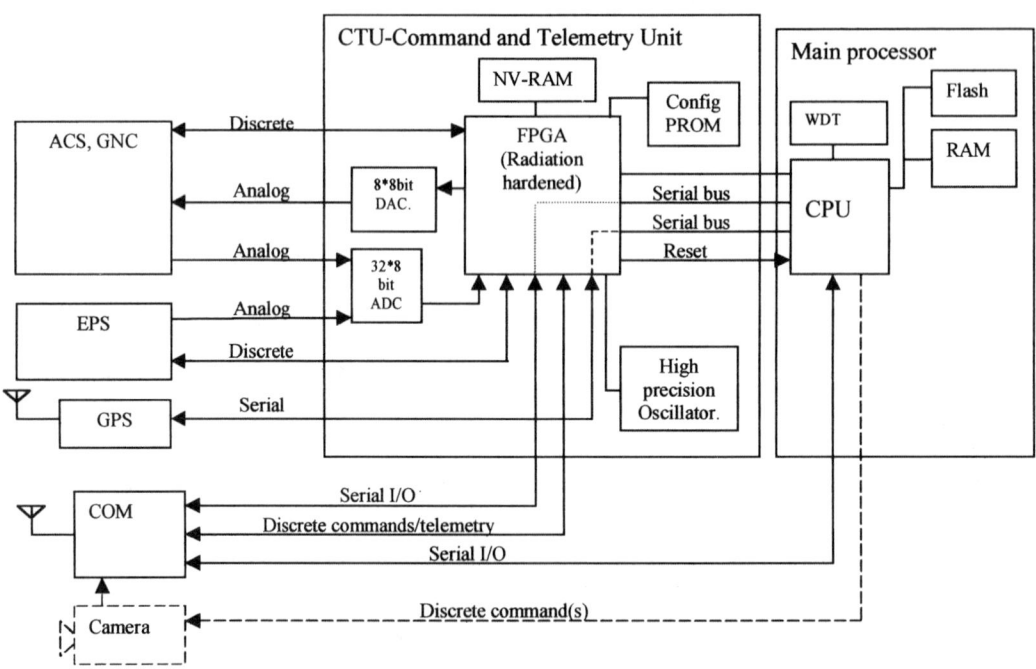

Figure 16 - Flight Computer Block Diagram

All serial commands pass through the CTU. The CTU decodes the command and determines if it should be executed directly by the CTU. There are a limited number of critical commands that will be executed in the CTU.

A separate state machine in the FPGA will monitor select signals and will store the spacecraft mode. This provides the CTU failure detection capability as well as keeping the spacecraft operating mode in the rad-hard portion of the circuitry.

The functionality of the FPGA is stored in a configuration PROM that cannot be reprogrammed after the spacecraft is launched. To provide non-volatile memory for storing spacecraft status information etc., the CTU is equipped with a small non-volatile random access memory (NVRAM). The CTU is equipped with 32 channels of 8-bit analog to digital converter and 8 channels of 8 bit digital to analog converters. This gives the CTU access to critical telemetry and to critical control functions. Access to telemetry points and control outputs for the main processor is provided through the CTU. These control outputs include electrical power controls for critical battery charging and load control.

To be able to provide accurate timing for relative navigation, the CTU is equipped with a high precision oscillator.

Main Processor

The main processor provides five functions:
1. Non critical command decoding
2. Attitude control main processing.
3. Navigation, guidance and control processing
4. Test and diagnostic algorithms.
5.

Commands for controlling payload devices and non-bus critical commands will be executed by the main processor. Attitude control and NGC processing will take the bulk of the main processor's capability. Other processing will include optimal battery charging and power management, spacecraft cross-link internetworking, and processing of any other payload devices in the spacecraft. Test and diagnostic algorithms, primarily used during development, integration, and test will also reside and be run on this processor.

The main processor will use a real-time operating system (RTOS). There are two operating systems being considered: Windows CE and VxWorks. The RTOS based on Windows CE makes it easy to develop software, as development can be performed on most any PC compatible computer. It also allows complete customization of the operating system functions. Since it is a modular operating system, it provides flexibility in design as components are added or removed, i.e. solid-state disks, PCMCIA devices, etc. VxWorks on the other hand is widely used for powerful embedded applications and in addition to being a modular OS, has development advantages. In this project, most of the software development will be in the NGC function. The bulk of the algorithms developed there will be done with MATLAB. The availability of cross compiler utilities enabling conversion and download of MATLAB code directly to VxWorks makes it attractive to shorten integration and test.The main processor must be able to communicate through a serial line and store the code in flash memory, making it possible to update the software during flight. The interface between the CTU and the main processor is *via* serial lines, a parallel bus, and a reset signal that allows the CTU to reset the main processor.

Other Requiremets To allow support functions to size the needs of the flight computer, requirements were assumed. These are listed below.

Electrical -
- +3.3 V 100 mA peak, 30 mA typical
- +5 V 250 mA peak, 20 mA typical, <1 mA standby
- A/D channels: 8-bit, 0-4.096VDC input range
- D/A channels: 8-bit, 0-5VDC output range

Thermal and Other Requirements - All heat needs to be removed to keep the operating temperature within the commercial range: $0<T_0<50°C$. The parts are to be conformal coated to prevent arcing in the event of spacecraft charging.

Electrical Power

The Electrical Power Subsystem function (EPS) provides power and power management for the spacecraft. This function takes power from solar arrays or a ground power source, and uses it to power spacecraft devices and charge spacecraft batteries. The EPS must supply enough power to run the spacecraft throughout all modes of operation including during eclipse.

The EPS provides power to all other electrical dependent functions throughout the life of the spacecraft. The EPS accomplishes this by generating or accepting raw electrical power, converting the raw power into the useable regulated power, storing excess power for use during eclipses, and managing the power distribution throughout the spacecraft. For risk mitigation, the EPS is designed as simple, robust, and reliable as possible.

Power Requirements The worse case power requirement for the spacecraft has been determined to be 7.7 Watts orbital average with a 28.8 Watts peak demand. This requires that the solar arrays generate 18 Watts while in sunlight to support the spacecraft's power requirements and recharge the battery for the next eclipse. These requirements were developed with the following assumptions:
1. Orbit altitude, h_a = 400km (period ~92 min)
2. The spacecraft is in solar eclipse for 40% of the orbital period
3. Ground station visibility duration, t_{vis} = 8 mins. max (transmit time to download data)
4. The ACS is active during the orbit

Design The EPS has three modes of operation: Charge, Discharge, and Maintenance. Charging occurs when the flight computer determines that the batteries are not at 100% charge and there is sufficient power available from a power source to charge them. Discharge occurs when power requirements exceed available power. Maintenance occurs when the batteries are fully charged and there is sufficient power to operate the spacecraft. In this mode, the batteries are trickle charged to maintain full charge. Based on solar array and battery parameters, the flight computer autonomously selects the operating mode. In addition, the mode can be over-ridden by ground command.

The EPS is partitioned into four main functional areas:
1. Generation
2. Regulation
3. Control, Distribution, and Telemetry
4. Storage

Together, these functional blocks work to receive power from the sun (or external power supply during development and testing), convert it to a usable range of voltages, charge the batteries, and distribute power to the various functions.

Generation Power comes from one of the two sources - solar arrays or external power from ground support equipment (GSE). Three types of solar cells were considered. Silicon (Si), Gallium Arsenide (GaAs) and high efficiency cells. GaAs was chosen because Si uses almost twice as many cells to produce the same amount of power, and the high efficiency cells are not available to us at this time. The solar cells will be assembled into arrays, and these will be mounted to and insulated from the body of the spacecraft. This arrangement will provide the necessary power to the EPS without driving unreasonable Sun-pointing attitude requirements.

The GSE power input will provide power during development, integration, and test. This will be supplied by an external power supply through the GSE interface, which will also contain signal lines for communicating with the flight computer.

Regulation This function takes a raw voltage output from the solar arrays or GSE power and converts it for powering the spacecraft and charging the battery DC-to-DC converters are used in each case.

Solar array output will feed a battery charge regulator and the DC-to-DC converters, which generate the regulated power required by the spacecraft components. Battery charge regulation is determined by available power and loading A peak power tracker (PPT) algorithm running in the flight computer will control the battery charge regulator to extract maximum power from the illuminated solar arrays during normal operations. During anomalous operations, the CTU will control basic battery charging. During eclipse and low solar incidence conditions, the battery will feed the DC-to-DC converters.

Regulated power will be provided in two forms: low power, high regulation and high power low regulation. Low power, high regulation will provide voltage ranges of +/- 12VDC, +5VDC, and +3.3VDC. These sources will power the sensors and flight computer, and other low power devices requiring tightly regulated supplies. The high power, low regulation converter will provide nominally +12VDC and will be used to run the transmitters, magnetic torque rods, thruster solenoids, deployment mechanisms, and any other high current devices.

Control, Distribution, and Telemetry The CTU controls spacecraft activation from stand by mode, provides control of power sources, over current protection, and power switching for load shedding. The flight computer controls certain devices that need to be powered on and off through commands to the CTU. The EPS internal busses will be protected with fuses and isolated with diodes. There is also a ground support equipment (GSE) interface that supplies power during development and pre-flight testing. The spacecraft will be equipped with a launch vehicle interface that will supply trickle charge power and inhibit signals to maintain the spacecraft in standby mod. Battery status (voltage, charge, and temperature) will be acquired and relayed as part of the housekeeping telemetry.

Storage The spacecraft power requirements and maximum eclipse duration of approximately 40% of the orbit drive the storage requirements. Due to the 28.8W peak power demand, battery choice is driven by the electrical current demand rather than allowable depth of discharge. A battery constructed from rechargeable nickel-cadmium (NiCd) C size cells meets the requirements. To match the voltage and current requirements of the generation section, a single battery pack of 15 cells in series will be used.

Communications

Ground Station TT&C The ground station will utilize a Yaesu FT-736R Amateur band transceiver capable of providing 25 watts of output r.f. power. The frequencies used for TT&C communications will be in the 430 MHz amateur radio band for both uplink and downlink. Simplex operation will be used. According to communications link budget analysis [12], an existing 15 dB gain 430 MHz cross element Yagi antenna with the previously mentioned uplink r.f. power should be adequate for communications to the spacecraft. The antenna is mounted on a 20 foot tower atop the CU-Colorado Springs Engineering building.

The antenna tracking mechanism consists of Yaesu G-5400B® azimuth and elevation rotors driven by Northern Lights Software Associates NOVA® software. This software accommodates spacecraft passes in all quadrants by incorporating a "flip" mode to overcome the mechanical stops at a south bearing. The tracking software also provides a scripting function by which selected future passes may be programmed for unattended operation.

The ground station TT&C facility will utilize United States radio amateur assigned frequencies in the UHF range to enable the acquisition of low cost TT&C equipment and insure that NavGold operation can be observed by radio amateurs around the world, thus providing an "extended" tracking network.

Spacecraft to Ground Station TT&C The NavGold spacecraft is extremely power limited due to its size and use of body mounted solar photovoltaic cells, therefore the TT&C communications function must operate with minimal power consumption. Link budget calculations, given the ground station output power and ground station receiver sensitivity, indicates that the onboard TT&C transmitter could be limited to approximately 2 watts of output r.f. power. Commercial r.f. equipment will be used onboard the spacecraft to minimize development and test requirements.

Current designs call for the use of 9600 bps TT&C data transfer rate to be compatible with available low cost spacecraft r.f. components. The data will be transmitted using FSK at the UHF operating frequency. A Terminal Node Controller (TNC) will decode data received at the ground station. A similar configuration will be onboard the spacecraft. Data recovered at the ground station will be archived in a local computer. The data will also be transmitted over the CU-Colorado Springs LAN to the Mission Operations Center on the campus, where health and status data will be decoded using LabView® software. This data will be displayed in real-time during a downlink session.

Spacecraft to Spacecraft The inter-spacecraft data link will perform two major functions. One is to relay status information, and GPS data, between spacecraft, as well as record it for later transmission to the ground. The second major function of the communications link is to provide relative spacecraft ranging measurements, discussed previously.

Inter-spacecraft communications will occur on a UHF frequency, but non-concurrent with ground transmissions to avoid radio frequency interference issues. The inter-spacecraft link may also utilize a very low power r.f. transmitter due to power limitations and the relative proximity of the two vehicles. The inter-spacecraft data will include GPS information, payload status, bus status and, potentially, range beacon data. Relatively simple and low power output transceivers are planned for this operation and will be controlled, as the TT&C links, by the C&DH computers. Separate antennas will be used for inter-spacecraft communication to reduce complexity.

Summary

A generic prototype mission and two nanosatellites for formation-flight experiments have been described, and proposed as an engineering design case study for performing design tradeoffs, such that insight can be gained regarding the research and engineering-design issues outlined in this paper. These issues include approaches for precision relative navigation, algorithms and methods for relative maneuvering and station-keeping, and command-and-control techniques appropriate for complex formation reconfigurations. Should this mission proceed to launch and on-orbit operations, as planned, it will also present an opportunity for technology demonstrations and proof of concepts, with regard to the methods and systems investigated, as well as an on-orbit test bed for new hardware.

Acknowledgements

The authors would like to acknowledge the other members of the NavGold team for the definition of the NavGold mission, and spacecraft design, especially Dave Sipple, Jim Torley, Jon Velapoldi, Dave Hanni, Dave Gustavsson, Brett Allard, Tony DeLeon, and Dr. Luigi Morelli.

References
1. Stachnik, Ashlin, and Hamilton, "Space-Station-SAMSI: A Spacecraft Array for Michelson Spatial Interferometry," *Bulletin of American Astronomical Soc.*, Vol. 16, No. 3, 1984.
2. Johnson, and Nock, "Multiple Spacecraft Optical Interferometry Trajectory Analysis," Workshop on Technologies for Space Interferometry JPL, April, 1990.
3. DeCou, "Multiple Spacecraft Optical Interferometry, Preliminary Feasibility Assessment," Internal Rept. D-8811, JPL, August 1991.
4. Hartman, Weidow, and Hadaegh, "Management of Guidance, Navigation, and Control Technologies for Spacecraft Formations Under the NASA Cross-Enterprise Technology Development Program," 1999 Flight Mechanics Conference, NASA Goddard Spaceflight Center, May 1999.
5. Ortega, "Formation Flying Activities at the European Space Agency," IFAC/ION/AAS/ESTEC International Workshop on Aerospace Applications of GPS, Breckenridge, CO, Feb. 2000.
6. Folta, D.C. and Newman, L., "Foundations of Formation Flying for Mission to Planet Earth and New Millennium", AIAA 96-3645, AIAA Guidance, Navigation, and Control Conference and Exhibit, San Diego, CA.
7. Folta D., Bordi, F., and Scolese, C., "Considerations on Formation Flying Separations for Earth Observing Spacecraft Missions," Proceddings of the AIAA/AAS Spaceflight Mechanics Meeting, February 1992, Colorado Springs, CO.
8. Folta, D.C. and Quinn, D., "A 3-D Method for Autonomously Controlling Multiple Spacecraft Orbits", 1998 IEEE Aerospace Conference Proceedings, Vol. 1, March 21-28, 1998, Aspen Co.
9. Yunck, T.P., "TOPSAT Orbit Determination and Spacecraft-Spacecraft Vector Measurement with GPS," Jet Propulsion Laboratory, California Institute of Technology, June 18, 1993.
10. Yunck, T.P., Melbourne, W. G., and Thronton, C. L., "GPS-Based Spacecraft Tracking System for Precise Positioning," IEEE Trans. Geosci. & Remote Sensing, 23, 450-457, 1985.
11. Wertz, "Spacecraft Attitude Determination and Control," Reidel Pub., 1978.
12. Fosha, C. and Torley, J., NavGold Communications Link Budget Analysis, 15 August, 1999; http://mae.uccs.edu/fdcl/navgold/home.html

AAS 00-062

MEMS RATE SENSORS FOR SPACE

Joel Gambino[*]

Micromachined Electro Mechanical System (MEMS) Rate Sensors are an enabling technology for Nanosatellites. The recent award of a Nanosatellite program to the Goddard Space Flight Center (GSFC) underscores the urgency of the development of these systems for space use. The Guidance Navigation and Control Center (GNCC) at the GSFC is involved in several efforts to develop this technology. The GNCC seeks to improve the performance of these sensors and develop flight ready systems for spacecraft use by partnering with industry leaders in MEMS Rate Sensor development. This paper introduces Microgyros and discusses the efforts in progress at the GNCC to improve the performance of these units and develop MEMS Rate Sensors for space use.

INTRODUCTION

Micromachined Electro Mechanical System (MEMS) Rate Sensors offer many advantages that make them attractive for space use. They are smaller, consume less power, and cost less than the systems currently available. MEMS Rate Sensors however, have not been optimized for use on spacecraft. This paper summarizes the state of the art of these devices and describes work being done by the Goddard Space Flight Center and its partners in developing MEMS Rate Sensors systems for space use.

Spacecraft designers are striving to replace traditional heavy, power-hungry mechanical gyro systems with lightweight and extremely reliable solid state rate sensing systems. One such system is the micromachined

---

[*] Guidance, Navigation and Control Center, NASA Goddard Space Flight Center, Code 573, Greenbelt, Maryland 20771.

gyroscope. These systems offer extremely small size (about 8 in^3), low weight (less than 3 lbs.), low power (3 watts), and offer the potential of much lower cost then the existing technology gyros used in space applications.

Many satellites of the future will be much smaller than those currently being produced. Hundreds of "nano-satellites" launched and released into various orbits for space and earth science missions have been proposed. The Nanosat mass and power budgets for an entire Attitude Control System will be less than that currently allocated for a single ACS component. The potential for small size and low cost of the MEMS rate sensor make it an attractive option for these future satellites.

Currently, MEMS Rate Sensor systems are being produced for terrestrial and missile applications. Therefore they have relatively coarse performance, a large rate range (up to 1000 deg/sec), and are not radiation hardened. Before this technology can be used for attitude control on spacecraft, design issues must be addressed.

To address these design issues, the Guidance Navigation and Control Center (GNCC) at the Goddard Space Flight Center (GSFC) and its partners are involved in work that seeks to optimize the MEMS Rate Sensor and develop a MEMS based Inertial Reference Unit (IRU) for space use. This paper will describe this work and future work planned in this area.

MICROMACHINED ELECTROMECHANICAL SYSTEM (MEMS) RATE SENSORS

The MEMS Rate Sensor (also known as Microgyro) uses an oscillating mass to measure rotational rate. Typically, the mass oscillates in one axis and as the body is rotated, coriolis forces cause a change in the line of action of the oscillation of the mass. This change is measured and is proportional to the rotational rate of the body.

Microgyro configurations vary greatly, but in most cases, the fundamental operation of the sensor is typical. A mass is forced into oscillation by electrostatic means. This mass is supported by spring members that are attached at one end to the mass and at the other end to a substrate. When the body rotates about its sensitive axis, coriolis forces cause the mass to oscillate with a component normal to the drive axis. This normal component is sensed by capacitive means and is proportional to the rotational rate of the body.

MICROGYRO DESIGNS

There are many different microgyro configurations in existence however the high performance units seem to display similar performance which is in the 1-10 deg/hr range (over constant temperature). Several examples of various sensor element designs are shown. Despite the significant differences in sensor design, these units all display similar performance.

JPL Microgyro

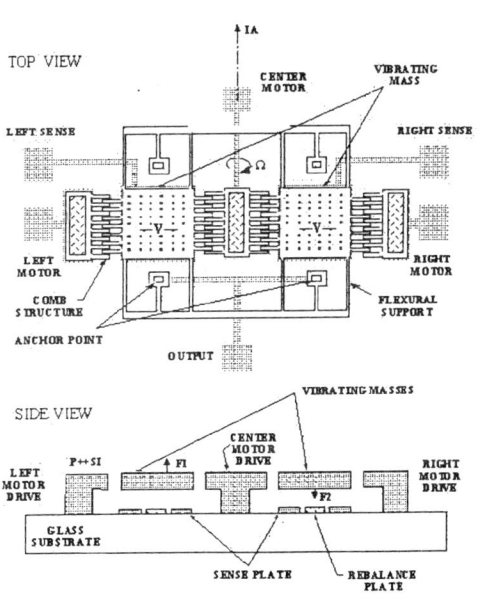

Draper Laboratory Gyro

MICROGYROS FOR NANOSAT

The microgyro is the only existing inertial sensing technology that appears to be capable of meeting the requirements for Nanosatellites. Flight ready Microgyro systems need to be developed in the very near future to support Nanosatellite programs.

Goddard's Nanosat Constellation, targeted for 2003 launch, consists of three spacecraft to validate methods of operating several spacecraft in the Earth's magnetic field. Each Trailblazer spacecraft will be an octagon 16 inches across and 8 inches high. The mission will cost $28 million and will be launched as a secondary payload on an expendable launch vehicle.

Results from the Trailblazer mission will be used to design future missions using constellations of lightweight (about 44 pounds), highly miniaturized autonomous spacecraft. One proposed constellation of up to 100 spacecraft positioned around the Earth will monitor the effects of solar activity that can affect spacecraft, electrical power and communications systems. Others will study global precipitation and the atmospheres of other planets.

Nanosatellites have arrived, and Microgyros will be needed to support their missions. To this end, it is estimated that Microgyro performance will need to improve at least one order of magnitude. This is based on a cursory survey of NASA missions showing the coarsest pointing requirements to be about 0.2 degrees. Assuming a low earth orbit satellite control system that provides a once per orbit attitude update, this roughly correlates to a gyro drift requirement of 0.1 deg/hr.

MICROGYRO PERFORMANCE IMPROVEMENTS

Microgyro performance has improved significantly over the past few years. However these systems still exhibit greater than 1 deg/hr drift at room temperature and performance over varying temperature in the 10 to 40 deg/hr range.

To close the gap between current microgyro performance and spacecraft requirements, efforts are being made to improve the performance of the microgyro. Two areas of focus have been electronics design and, to a lesser extent, sensor element design.

Improvements in microgyro performance can be gained through improved electronics design by reducing electronics noise and parasitic capacitance, and carefully selecting components for low noise performance.

The reduction of both bandwidth and dynamic range is being investigated as a method to improve performance. Most current Microgyro systems have dynamic ranges of about +/- 1000 deg/sec and bandwidths of over 100 hz. Typical spacecraft requirements for dynamic range are +/- 20 deg/sec (or less) and for bandwidth are typically less than 10 Hz. Draper Laboratory, under contract to the GSFC, is investigating how changes in dynamic range and bandwidth affect performance.

Microgyro performance is also a function of the configuration of the sensing element. Work is just beginning at the GNCC in this area with the recent arrival of MEMS design and analysis software. While it is expected that performance can be improved, the magnitude of improvement has not been quantified.

Significant performance gains are expected to be realized with high fidelity temperature compensation. Thermal effects on the output of the gyro are large. This is evident from the order of magnitude degradation in drift performance over varying temperature.

A short-term performance goal of 0.1 deg/hr is expected to be reached with the optimization of electronics design and the implementation of temperature compensation and/or temperature control. This performance level is consistent with the requirements for some spacecraft. A mid-term goal would be to demonstrate 0.01 deg/hr drift. This level of performance would support a significant portion of future missions. The long term performance goal has no bound; work will continue to demonstrate the maximum level of performance of these sensors.

CURRENT GNCC INVOLVEMENT

The Microgyro work being done at the GNCC encompasses several separate efforts. Work is being done with JPL, APL, Kearfott Corp. and Draper Laboratory to develop MEMS rate sensors for space.

Goddard is working with JPL and APL to develop a Micronavigator. The micronavigator will be a 2 inch cube that provides attitude and position information. It will contain three microgyros, three micro accelerometers, GPS and processing capability. The unit will use the JPL Microgyros which continue to be developed. This exciting and challenging program is just getting underway and is planned to last about 4 years.

Work is being done with Kearfott Corp. to develop a three axis Micro Inertial Measurement Unit (IMU) for a test flight in space. This unit will be delivered in fy '01.

Draper Laboratory has been contracted to study Microgyro performance as a function of bandwidth and dynamic range. Microgyro systems currently have bandwidths of over 100 Hz and dynamic ranges of +/- 100 deg/sec or more. Spacecraft requirements for bandwidth are typically less than 10 Hz and requirements for dynamic range are usually about +/- 20 deg/sec. Improvements in performance are expected to be realized by rescaling these parameters.

CONCLUSIONS

The need for very small, low power inertial sensors has driven a large effort to develop microgyros for space use. The short term performance goal of 0.1 deg/hr is expected to be met in fy'01. In parallel, multi axis units are being developed and will be test flown. For the long term, an ultra small Micronavigator is being developed which will perform navigation and attitude functions currently performed by units many times its size. The reality of Nanosatellites is evidenced by the recent award of a Nanosat mission to the GSFC and MEMS rate sensors will be an enabling technology for future Nanosat Missions.

BIBLIOGRAPHY

R. P. Jaffe "DQI Technical Description and Performance Capability", AIAA 96-3711, July 29-31 1996.

J. Bernstein, S. Cho, A. T. King, A. Kourepenis, P. Maciel, and M. Weinberg, "A Micromachined Comb-Drive Tuning Fork Rate Gyroscope", IEEE 0-7803-0957-2/93.

N. Barbour, J. Connelly, J. Gilmore, P. Greiff, A. Kourepenis, M. Weinberg, "Micromechanical Silicon Instrument and Systems Development at Draper Laboratory", AIAA 96-3709, July 29-31, 1996.

J. Connelly, A Kourepenis, D. Larsen, T. Marinis, "Inertial MEMS Development for Space" April, 1999

W.A. Clark, R.T. Howe, and R. Horowitz, "Surface Micromachined, Z-Axis, Vibratory Rate Gyroscope," Tech. Dig. Solid-State Sensor and Actuator Workshop (Hilton Head `96), Hilton Head Island, SC (June 2-6, 1996), pp. 283-287.

M. Kranz, "Design, Simulation and Implementation of Two Novel Micromechanical Vibratory-Rate Gyroscopes", M.S. Thesis, Department of Electrical and Computer Engineering, Carnegie Mellon University, May, 1998.

B. E. Boser, "Electronics for Micromachined Inertial Sensors", Berkeley Sensor & Actuator Center, University of California, Berkeley, 1997.

AAS 00-063

DYNAMICS AND CONTROL OF NANOSATELLITE ASUSAT1

Brian K. Underhill,[*] Assi Friedman[†] and Helen L. Reed[‡]

This paper provides a brief description of the systems and components of the student-designed and -built ASUSat1 nanosatellite, with further detail on the dynamics and controls design. Pictures of final assembly of these components, as well as integration of ASUSat1 with the other payloads and the launch vehicle, are included. Information regarding the January 26 launch and current status is also provided. A brief description of the program's current plans is also given.

INTRODUCTION

The ASUSat Student Satellite Program[§] is managed entirely by undergraduate and graduate students with oversight by a faculty advisor. Industry engineers and additional faculty are available for consultation and periodic evaluations of student progress. There are over half-a-dozen projects currently under way, ranging from soda-can-sized 'satellites' launched from amateur rockets to as high as 40,000 feet before descending under chute, to a constellation of three satellites performing stereoscopic imaging to be launched from the Space Shuttle in 2002.

ASUSat1 is the original project that has made it possible for the program to grow to its current state of multiple in-progress projects. Over 400 students (85+% undergraduate) have participated in the numerous iterations of ASUSat1 from initial concept, through design and development, integration and testing, and flight and ground operations. These students have gained valuable hands-on experience in the design and application of nanosatellite technologies, and today many of them are practicing engineers in the space industry.

---

[*] Program Manager, ASUSat Student Satellite Program. NASA Space Grant Fellow. MS Candidate, Mechanical Engineering, Arizona State University, Box 87-6106, Tempe, Arizona 85287-6106. E-mail: brian.underhill@asu.edu.

[†] Project Leader, ASUSat1. Received BS in Electrical Engineering in May 1999. MBA Candidate, Arizona State University, Box 87-6106, Tempe, Arizona 85287-6106. E-mail: assi.friedman@asu.edu.

[‡] Director, ASUSat Lab. Associate Director, ASU National Space Grant College and Fellowship (NASA Space Grant) Program. Professor, Mechanical and Aerospace Engineering, Arizona State University, Box 87-6106, Tempe, Arizona 85287-6106. E-mail: helen.reed@asu.edu. Web Site: http://nasa.asu.edu/asusat.

ASUSAT1

Miniature satellites are considered to be those under 200 kg[1], microsatellites as between 10 and 50 kg, and a nanosatellite is between 1 and 10 kg. Figure 1 is of ASUSat1, weighing in at a svelte 12.9 pounds (<6 kg), easily being held by one person. This is a prime example that real capability can be achieved in a nanosatellite-size spacecraft.

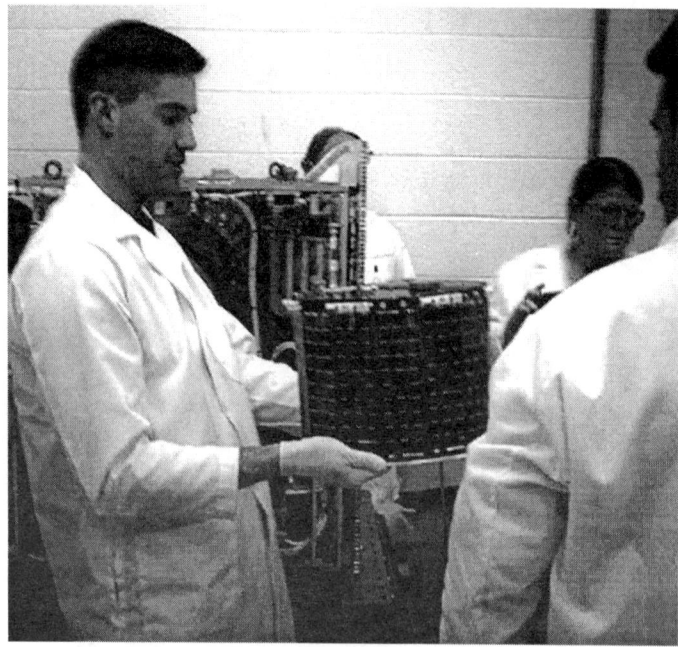

Figure 1. ASUSat1, Weighing Only 12.9 Pounds

The original goal of the ASUSat1 project was to show capability in a 10-pound-class package and provide technology demonstration in flight to enable other nanosatellite missions. The strict mass, volume, and power constraints associated with nanosatellites eliminate the use of many common off-the-shelf components and require innovative rethinking of many commonly used techniques such as active attitude control, radiation shielding, large battery packs, structures, thermal control, and many complex mechanisms. Also with the minimal power that can be generated from the small surface areas, only the lowest power consuming devices could be used.

Overview

A brief view of the systems on the satellite is provided here. Dynamics and Control components are discussed in more detail in the next section.

The Commands, Power, Modem, and Dynamics boards were designed, populated, integrated, and tested by the students. A picture of the Commands board is shown in Figure 2.

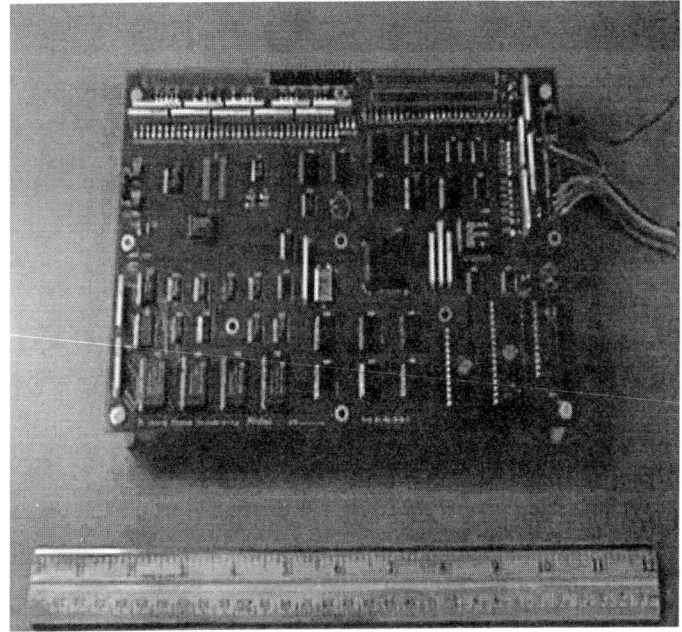

Figure 2. Commands Board

The students also designed the power system, and even performed the actual stringing of the solar arrays. A picture of a student inspecting the GaAs cells is shown in Figure 3.

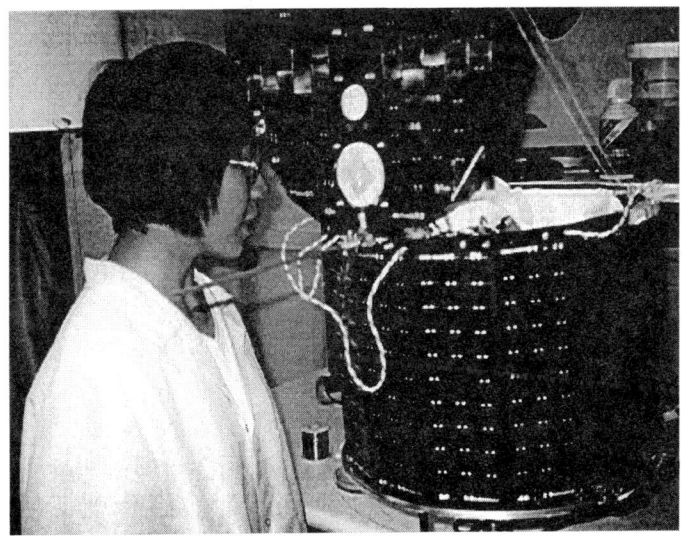

Figure 3. Student Joyce Wong Inspects GaAs Solar Cells

The students, with a lot of guidance from the AMSAT community, also designed the communications system. The satellite has on-board a low-power transmitter and two receivers along with transmit and receive antennas. These boards were not designed from scratch; the students obtained terrestrial COTS hardware and then modified these for space use. Amateur radio operators around the world will also be able to use the satellite as an analog voice repeater as well as to download telemetry.

The software of ASUSat1 was also written by students, again with some help from the AMSAT community. The bootloader keeps the satellite in a power-safe mode and enables basic hardware diagnostics and code upload. Once the satellite is determined to be healthy, the operating system is uploaded to the satellite.

To achieve such a low spacecraft weight, a carbon-composite structure was employed. Students performed the actual layup of the material. Figure 4 is a picture of the qualification structure, affixed with silicon cells and one panel of the GaAs cells.

Figure 4. ASUSat1 Qualification Structure (Carbon-Composite)

To ensure that the structure and all of its components survive the launch environment, testing and finite-element analyses were performed. Static, shock, and vibration simulations were applied to the development structure to ensure the integrity of the satellite. Figure 5 is a graphic from one of the early thermal analyses.

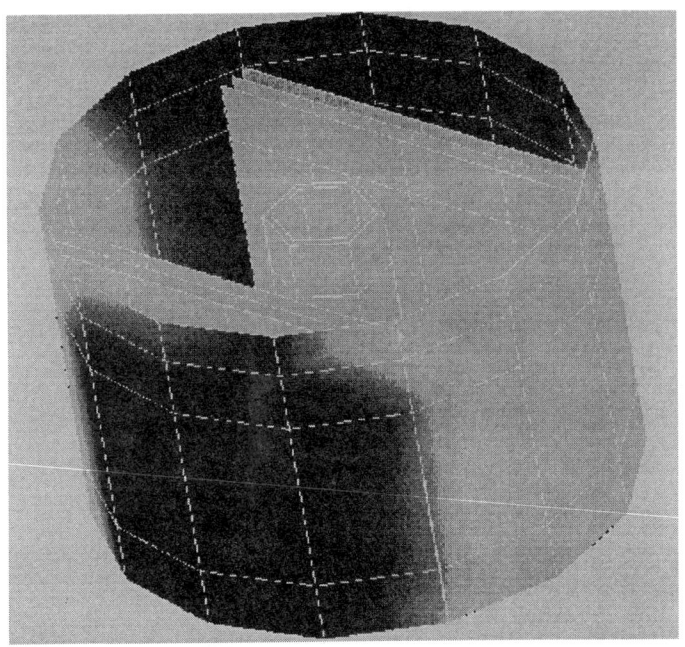

Figure 5. Example of Thermal Modeling of ASUSat1

Figure 6 is a picture of a flight-version of the deployment system. This hardware supports the payload during ascent, and then deploys it safely away from the launch vehicle. The plate is 3/8 inch aluminum that has been pocketed out from the backside to reduce weight. The deployment system, including cabling and ordinance, weighs less than five pounds.

Figure 6. Marmon-Clamp Deployment System

The formal science mission of the spacecraft is earth imaging and vegetation indexing, via two on-board cameras. The cameras are a Dycam product, one of the few components on the satellite that were provided to the team with minimal student involvement. Figure 7 shows a picture of the camera assembly; both cameras are contained within the one housing. Figures 8 and 9 are pictures taken with the camera from the roof of our lab building. The first image is from the camera with a visible blue filter; the second from the camera with an IR / near-IR filter.

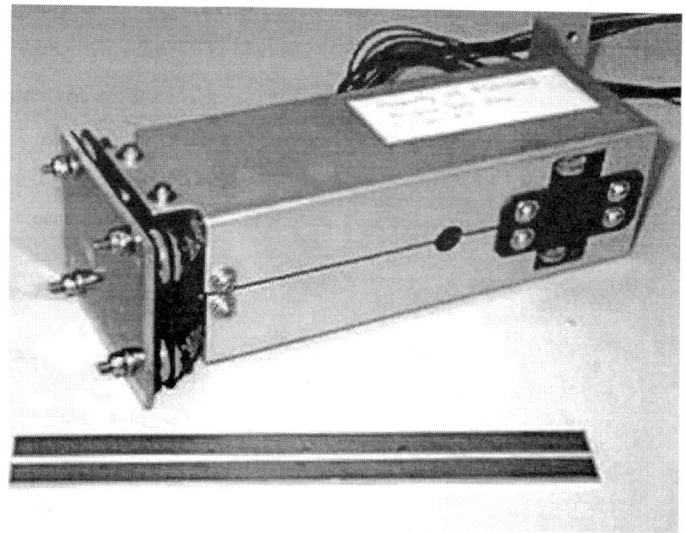

Figure 7. Camera Assembly

Figure 8. Image Taken with Visible Blue Filter

Figure 9. Image Taken with IR / Near-IR Filter

This completes the overview of the satellite subsystems, except for the Dynamics and Control components, which are detailed next.

Dynamics and Control

Stabilizing a nanosatellite is not a trivial task. Due to the strict power, cost, and weight constraints, the dynamics team could not use standard devices such as off-the-shelf torque rods, magnetometers, thrusters, and sensors. However, for earth imaging and for communications optimization, a stable earth-pointing orientation is needed. The ASUSat1 team has developed an innovative passive stabilization and damping collaboration incorporating many student-designed components. One of these components is a passive gravity-gradient fluid damper. This damper coupled with the gravity-gradient boom will provide 3-axis stabilization. Honeywell has provided extensive technical support in this area.

The main stabilization system is the gravity-gradient boom, a cylindrical 2-meter beryllium copper element with a 135-gram tip mass. The boom is deployed from a student-designed release mechanism that is 3.8 x 3.8 x 6.6-cm and weighs less than 130 grams. The release mechanism is an offshoot of current industry designs, but is much smaller and lighter. One electrical signal is required from the launch vehicle at the beginning of the mission to release the element, stabilizing the satellite for the duration of the mission.

Figure 10 shows two students performing the final winding of the boom element into the deployment mechanism.

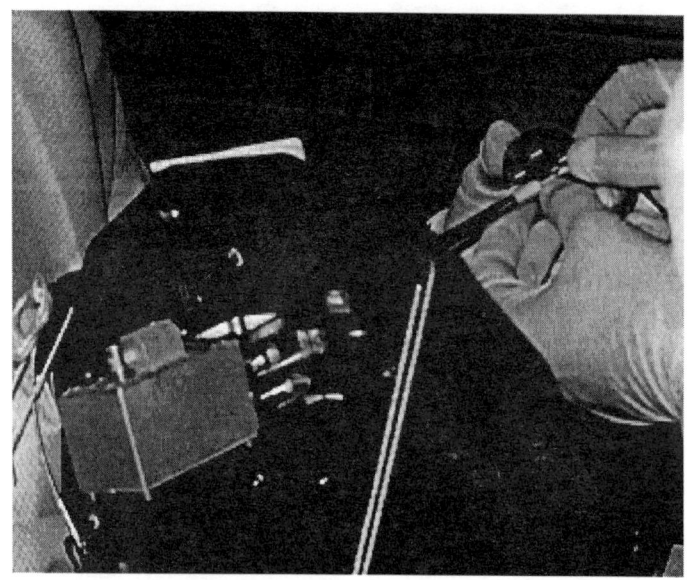

Figure 10. Final Winding of the Boom Element

Figure 11 is a close-up of the 135-gram tip mass, in its stowed configuration.

Figure 11. Tip Mass, GG Boom is Stowed

Figure 12 is a graphic of the spacecraft in its deployed configuration, showing the extended gravity-gradient boom and transmit and receive antennas. The extended tip mass provides the relative difference in principal moments of inertia that enable a gravity-gradient stabilization scheme.

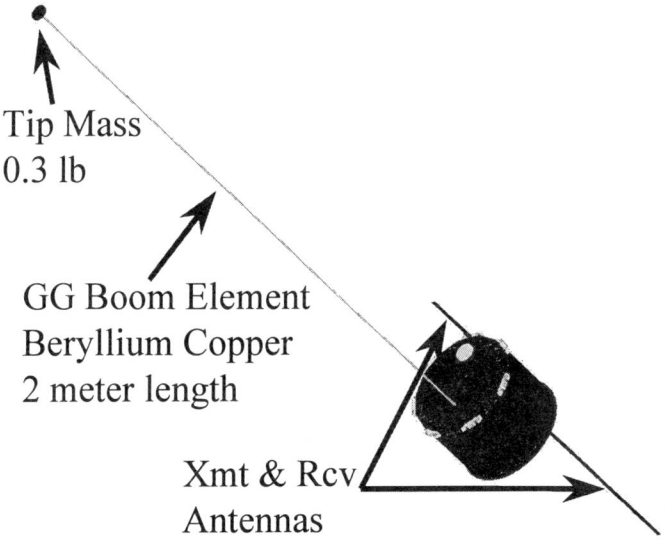

Figure 12. Graphic of ASUSat1 in Deployed Configuration

By utilizing the pitch and roll/yaw decoupled equations of motion of a gravity-gradient stabilized spacecraft in low-earth orbit, the stability of the craft can be easily determined from the two parameters k_1 and k_3, functions of the principal moments of inertia. ASUSat1's principal MOI are [3351, 3289, 234] (lb*in^2). These values give k_1 and k_3 parameters of 0.9 and 0.3, respectively. These parameters are plotted in Figure 13; ASUSat1 is clearly in the stable region of the plot.

One advantage of the extremely light-weight spacecraft is that to obtain the 0.9 parameter value, the required mass for the boom tip (an otherwise un-utilized cost to the mass budget) is only 0.3 pounds. This is an example of the phenomenon that in spacecraft design, and especially in nanosatellite design, mass begets additional mass and the inverse is also true. That is, as component mass increases/decreases, so does structural mass to support it and attitude control hardware to control it.

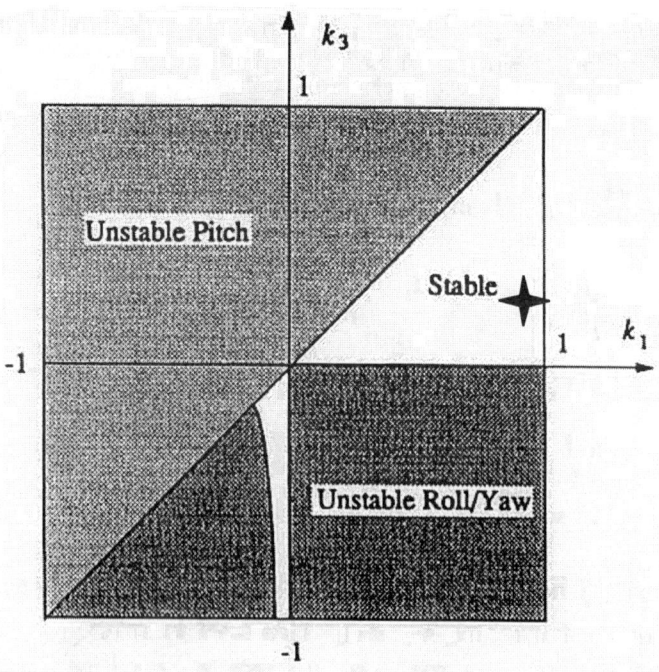

Figure 13. Gravity Gradient Stability Plot, Showing GG Boom Stability for ASUSat1[2]

A gravity-gradient boom (about +/-5 degrees[3]) cannot provide fine stabilization as the satellite is expected to wobble around its equilibrium point. This could cause the images to miss the targeted areas of interest so a finer stabilizing system was added.

The fine attitude control is also a passive system, called the gravity-gradient fluid damper. The system is built around a ball with four different mass concentrations that floats in a viscous liquid inside a larger shell that is attached to the satellite body. The physical principle behind it is that the inner ball should be aligned with both the earth's gravity vector and with the velocity vector of the satellite's orbital motion. Since the satellite will wobble around its equilibrium point, the wobble energy should be dissipated with time in the viscous liquid between the inner ball and the outer shell. This method is based on a new concept, and has not been space proven. If successful, it is expected that the satellite will reach a steady state in about 600 orbits. Both of the methods (gravity-gradient boom and gravity-gradient fluid damper) are completely passive, thus being an ideal solution for a satellite with a low power budget. Figure 14 is an image of the damper housing and the interior ball. The holes in the ball accommodate tungsten caps to provide the different moments of inertia. Honeywell has provided extensive technical support in the development of this system.[4,5]

Figure 14. Gravity-Gradient Fluid Damper

Gravity-gradient stabilization schemes have two stable orientations; one pointing nadir and one pointing zenith. If an uncontrollable event causes the satellite to flip or if the satellite is deployed in the wrong direction from the launch vehicle, many of the satellite's functions would cease to work. The only active means of attitude control added on the satellite was one small, lightweight, student-designed Z-axis magnetorquer. This z-coil will be used to flip the satellite over in case of an upside-down orientation. Because of the large current draw of the coil, it is limited to emergency situations only. Figure 15 is a close-up view of the coil. Honeywell provided use of their facilities to the students, and advisement, to do the winding.

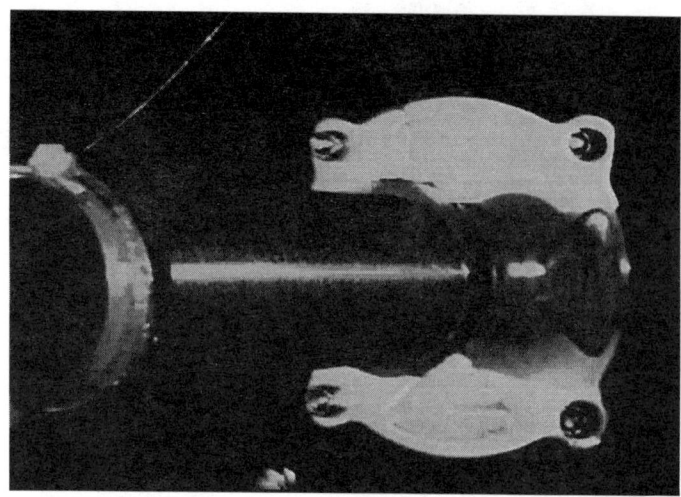

Figure 15. Z-axis Magnetorquer

For attitude determination, various commercial sun/earth horizon sensors were evaluated, but due to the large cost of these units the students reverted to designing their own sun/earth sensors. The emphasis in the design was to build a low-weight, low-cost sensor array for determining the satellite's orientation to within +/-10 degrees. The sensor array was built using twenty-three photosensors mounted on the circumference of the satellite and sampling at three different angles. Such an array reduced costs to under $1000 and minimized required internal volume. The data gathered from the sensors coupled with the camera images will provide the information to refine the attitude determination algorithm. Figure 16 is a close-up of the sensor blocks; the long blocks are mounted on the side of the spacecraft, the short blocks are mounted on the top and bottom bulkheads.

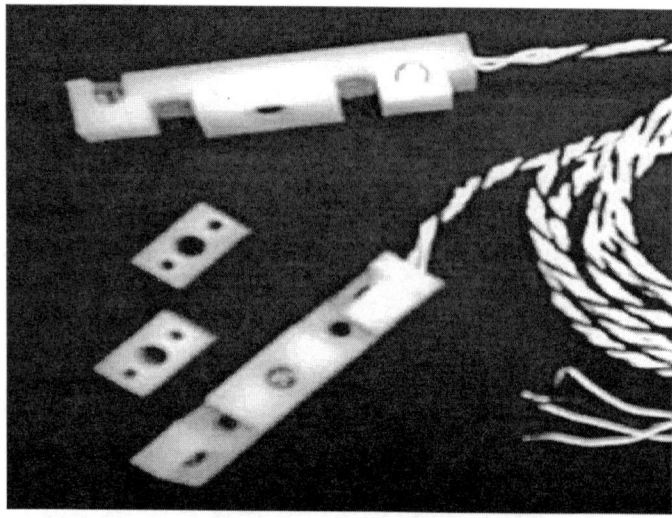

Figure 16. Attitude-Determination Sensors

Ephemeris determination is by GPS. GPS has been introduced to satellites only during the past several years, but is rapidly becoming a standard in spacecraft design. ASUSat1 uses a Trimble SVee-Six GPS receiver. The unit consumes only 1.5 watts and will be used to periodically collect orbital data points that will be stored in the satellite's computer and transmitted to the ground station for analysis. On-board ephemeris determination is not expected at this point due to the fact that the spacecraft computer does not have floating-point capability. This is a terrestrial (non-space-rated) unit that has been conditioned for space by the use of epoxies and shielding and is expected to give position accuracy within 10 km and similar accuracy for velocity measurements. Imagery of the GPS board, EMI box, and patch antenna are provided in Figure 17.

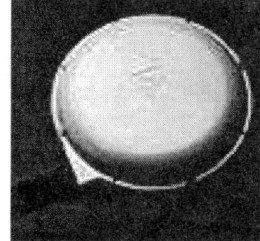

Figure 17. ASUSat1's GPS Components

Using the techniques described above, it is expected that the entire attitude/orbital determination and control system will have a mean power consumption of less then 0.75 watts.

Integration Pictures

Figure 18 shows the final assembly of the ASUSat1 spacecraft. In the first picture can be seen the boom deployer housing, the camera housing, the shrink-wrapped z-coil, and the attitude sensor blocks. The second picture shows the insertion of the commands and communication panel; the visible side is the comm system (receivers, modem, transmitter, and GPS), the commands and dynamics data acquisition boards are on the other side.

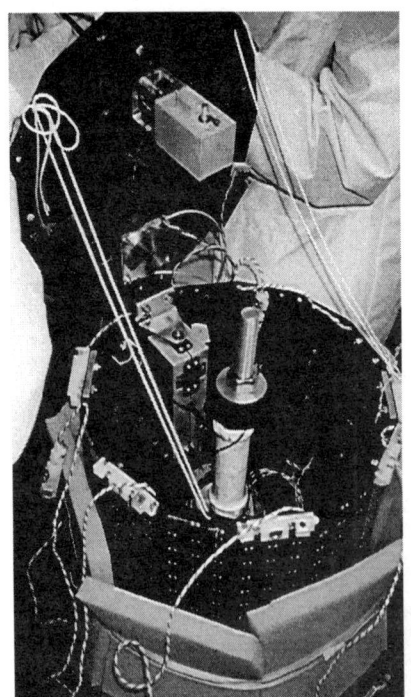 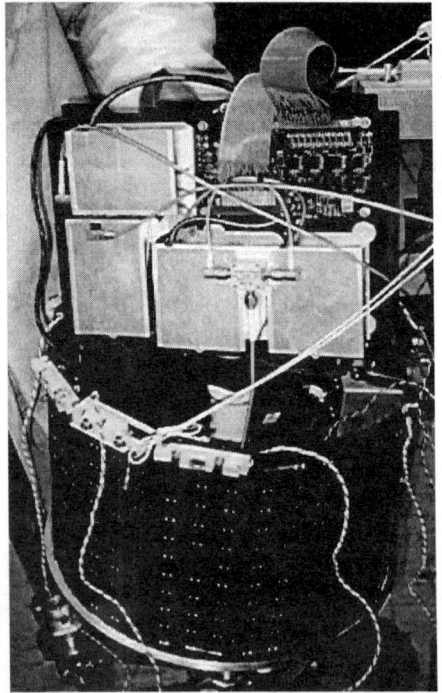

Figure 18. ASUSat1 During Final Assembly

Figure 19 is a picture of ASUSat1 taken on June 23, 1999, after the satellite passed acceptance and functionality tests and was integrated to the Multiple Payload Adapter.

Figure 19. Integration of Fully Functional ASUSat1 to MPA

Launch and Operations

Orbital Science's maiden launch of the OSP Minotaur Space Launch Vehicle was on January 26 at 19:03:03 PST. It was a beautiful launch, and a significant milestone for our program as our first satellite was put into orbit. The ASUSat Program was very fortunate indeed to have such a strong supporter in Orbital Sciences Corporation, who not only donated the launch to us, but also gave us extensive use of their testing facilities for qualification and acceptance of the payload.

ASUSat1 was first in the deployment sequence. We had previously made arrangements with some amateur radio operators in South Africa to listen for our craft. The earliest we could hear something would be 50 minutes after launch, which meant that we were to be the first indication of whether or not the payloads were successfully deployed. We were overjoyed when Paul Roos (ZS6HQ, South Africa) sent us an email that he heard two transmissions, two minutes apart, on our frequency. This verified that the rocket had indeed deployed the payload successfully and that our satellite had survived launch and was now operational on orbit. This was a huge confirmation regarding the viability of our spacecraft and our program.

Current Status

There have been at least eleven transmissions received around the world from the spacecraft so far. From these we were able to glean two frames of telemetry. The first frame had a disturbing data point – that the battery pack was not receiving any charge from the solar array. We had problems with this sensor during testing, so remained hopeful that the pack was being charged and that it was just a bad sensor.

Our second telemetry frame confirmed the bad news for us, as the battery pack voltage had dropped to 7.0 volts, just above the battery curve drop-off. Although our satellite was transmitting a nice strong signal, responding to our commands, and otherwise operating nominally, as one student put it, ASUSat1 was 'singing her death song".

The last we heard from the satellite was fourteen hours after liftoff. Estimated life on batteries alone was fifteen hours. We still remain hopeful that the wound will heal itself, but at this point it doesn't appear likely.

CONCLUSION

Even with such a limited lifetime, thanks to our very capable bootloader we were able to make some very significant conclusions. First of all, since our spacecraft survived launch and was operational on orbit, indeed operating quite strongly, we feel that we have proven the overall system viability – that indeed you can build a nanosatellite that truly has real on-orbit capabilities. We could also show that most of the student-designed and student-built components were performing very well:

computer	carbon-composite structure
boot-loader software	thermal sensors
receivers	boom deployer
modem	GG stabilization
transmitter	sun / earth sensors (?)

The sun / earth sensors did provide readings, but from only two telemetry frames it would be a bit premature to draw any firm conclusions regarding their performance.

Unfortunately, the following components / functions were not due to be tested until a couple of weeks into the mission – after the basic spacecraft had completed it's component and systems checkouts:

cameras	fluid damper
GPS	voice repeater

The intent was to present data pertaining to the attitude determination and control systems at this conference. Firstly with the late launch date of January 26 (originally scheduled for last September, then early December), but definitely with the loss of battery charge, this will obviously not be possible.

As mentioned previously, the team will continue to listen for the next couple of weeks, just in case. After that, we will perform a detailed failure investigation to attempt to find the root cause for the premature end of mission, and to see what we can learn from this event.

Although it is of course disappointing to have our mission lifetime cut short like this, the team's spirits remain high. We proved a lot regarding our capabilities, and have also learned a lot for future missions. And future missions we have!

We are now offering a CanSat class, where new students learn the entire systems design process, hands-on, in one semester. At the end of the term they will have built a functioning, soda-can-sized satellite. These cans will then be launched, as high as 40,000 ft, aboard an amateur high-powered rocket, to then descend under chute while taking sensor readings and transmitting that telemetry back to their mobile ground station.

We also have a joint project, named QUEST, with Kyushu University in Japan and Santa Clara University in California to build a spacecraft to carry a tether experiment. This payload is to be launched aboard an H-IIA rocket in 2002.

We have another joint project, named Three Corner Sat, with the University of Colorado at Boulder and New Mexico State University. This project, under the University Nanosatellite Program, is a three-satellite constellation designed to perform stereoscopic imaging and demonstrate distributed operational capabilities. Three Corner Sat will be launched aboard the Space Shuttle in 2002.

Although ASUSat1 utilizes relatively simple control techniques, both the tether experiment aboard QUEST and the coordinated targeting required for Three Corner Sat present some obvious technical controls challenges and opportunities. Some of the CanSat concepts have involved autonomous return of the payload – so even they, perhaps surprisingly, may present a controls discussion for the future.

ACKNOWLEDGEMENTS

The authors thank the original Program Manager Mr. Joel Rademacher presently at Jet Propulsion Laboratory having completed his MS with Dr. Reed in July 1996, former Program Manager Mr. Shea Ferring presently at Analex Corporation (contracting to Spectrum Astro) having completed his MS in December 1998, former Deputy Program Manager Mr. Christian Lenz presently at Broad Reach Engineering having completed his BS in December 1998, and ALL the students involved with the ASUSat1 project since its inception in October 1993. Many thanks also go to Mr. Scott Webster and Mr. Scott Schoneman (Orbital Sciences Corporation), Mr. Rich Van Riper (Honeywell Space Systems Group), and all the other faculty and industrial sponsors who have provided assistance along the way. Without their help this project would not have been possible.

Support for ASUSat1 has been provided by: Orbital Sciences Corporation, National Space Grant College & Fellowship Program (NASA Space Grant), AMSAT Organization, Honeywell Space Systems Group, Lockheed Martin Tactical Defense Systems, National Science Foundation Faculty Awards for Women in Science and Engineering, Cogitec, Space Quest, KinetX, Hughes Missile Systems, ORCAD, Solid Works Corporation, Zilog, Microchip, Dycam, Motorola (Satcom & University Support), PhotoComm, Inc., Eagle Picher Industries, Intel (University Support), Maxon, Universal Propulsion Company, Inc., ICI Fiberite Composites, DynAir Tech of Arizona (SabreTech), National Technical Systems, SpectrumAstro, Trimble Navigation, Bell Atlantic Cable, Lee Spring Company, Astro Aerospace, BekTek, Jet Propulsion Laboratory, Rockwell, Sinclabs, Inc., Applied Solar Energy Corporation, Gordon Minns and Associates, Communication Specialist, Advanced Foam and Packaging, XL Specialty Percussion, Inc., Simula, Inc., Equipment Reliability Group, and Arizona State University Center for Solid State Electronics Research.

REFERENCES

[1] J. R. Wertz and W. J. Larsen, editors, "Space Mission Analysis and Design", 3rd Edition, 1999, p. 853.
[2] B. Wie, "Space Vehicle Dynamics and Control", *AIAA Education Series*, 1998, pp. 365-371.
[3] J. R. Wertz and W. J. Larsen, editors, "Space Mission Analysis and Design", 3rd Edition, 1999, p. 359.
[4] R. R. Kumar, "Gravity Anchoring for Passive Spacecraft Damping", *AIAA J. Spacecraft*, Vol. 32, No. 5: Engineering Notes, 1995.
[5] E. J. Buerger and R. S. Oxenreider, "Passive Dampers for Gravity Gradient Stabilization", 1968.

ATTITUDE CONTROL – AN AFTERTHOUGHT
The MightySat I Experience

Jeff Benton[*] and Thomas Itchkawich[†]

MightySat I is a technology demonstration satellite built by OSC (Orbital Sciences Corporation) for AFRL (Air Force Research Laboratory). MightySat I was deployed from the first space station assembly shuttle mission in December of 1998. The ejection placed the satellite into a 390 km altitude, 51.6° inclination orbit with a predicted mission life of approximately one year. MightySat I weighs approximately 63.5 kg. The MightySat I bus structure is a regular hexagonal prism, 48.5 cm across the vertices and 69.4 cm overall from separation ring to antenna tips.

Initially designed with no ACS (Attitude Control System), power concerns led to the addition of an ACS late in the program. MightySat I is a major axis spinner designed to orient its spin axis along orbit normal. A simple control scheme was implemented with minimal resources. For example, the ACS algorithms, most of them from the FAST (Fast Auroral SnapshoT) satellite, comprise less than 60 lines of FORTRAN code to ease loading and help reduce system test time. The reduced ACS test program inspired the placement of end-to-end, physical functional testing into the magnetic calibration procedure.

Despite the simple design, satellite testing and verification proved challenging. Difficulties were overcome in many stages of the ACS I&T (Integration and Test). On-orbit, thermal trends, instead of power concerns, motivated the use of the ACS. When MightySat I left the canister, the HES gave it a considerable rate about the Z axis, the minor moment of inertia axis, which lasted until the ground activated the ACS two days after ejection. Since these early operations, the ground has successfully used the ACS. The mission has been a complete success.

---

[*] Engineer, ACS Group, Orbital Sciences Corp., 1521 Westbranch Drive, McLean, Virginia 22102.
[†] Program Manager, Orbital Sciences Corp., 1521 Westbranch Drive, McLean, Virgina 22102.

INTRODUCTION TO THE MIGHTYSAT I SPACECRAFT AND ACS

The purpose of MightySat I, built for AFRL (the Air Force Research Laboratory, formerly Phillips Laboratory) is two-fold: to fly advanced technology experiments and to introduce laboratory personnel to satellite design, integration, and testing. The technology experiments fall into two classes: those that are an integral part of the spacecraft bus and those that are stand-alone payloads in the bus. The integrated experiments are the composite bus structure and the ASCE (Advanced Solar Cell Experiment, cascade gallium-indium-phosphate). The stand-alone payload experiments are MAPLE (Micro-systems And Packaging for Low-power Electronics), SMARD (Shape Memory Actuated Release Device), and MPID (Micro-Particle Impact Detector). MightySat I was deployed from the HES (Hitchhiker Ejection System) on Shuttle mission STS-88 (the first space station assembly mission) in December of 1998. This ejection placed the satellite into a 390-km (210.6-nmi) altitude, 51.6° inclination orbit with a predicted mission life of close to one year. MightySat I weighs 64.5 kg (142 pounds). The MightySat I bus structure is a regular hexagonal prism of 64.5 cm (19.1 inches) across the vertices. This structure is 52.1 cm (20.5 inches) tall and 69.4 cm (27.3 inches) overall from separation ring to antenna tips. See Figure 1.

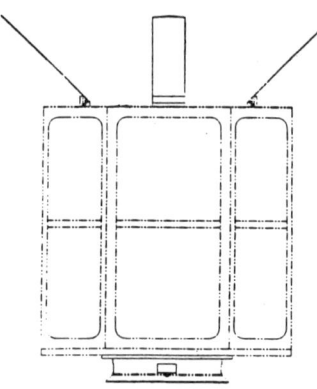

Figure 1. Elevation view of MightySat I sacecraft. Note: all antenna are on top plate.

Uncertainties about whether MightySat I could remain power positive in any orientation or tumble motivated the addition of an ACS. (Although 7 of the 8 sides of MightySat I had solar cells, a fair percentage of this coverage was from the experimental cells.) MightySat I is a major axis spinner (~1.10 inertia ratio) designed to orient its spin axis along the orbit normal. See Table 3, further below. The major inertia moment is not along the axis of symmetry (the Z body axis) but rather along a transverse axis (the Y body axis) through the vertices of the hexagonal. The impromptu nature of the ACS, and very tight cost restrictions dictated a very spare system. The ACS uses one TAM (Three-Axis Magnetometer) from Nanotesla Inc. for its entire sensor suite. The actuators are three orthogonal air coils wound at Orbital Sciences (formerly CTA Space Systems) in McLean, Virginia. The attitude control logic resides in the C&DH (Command and Data Handler), the only spacecraft bus processor, with an enhanced 80C86 CPU (Central Processing Unit). See Figure 2. The ACS algorithms, most of them from the FAST satellite, comprise less than 60 lines of equivalent FORTRAN code to ease processor and memory loading, and to help reduce system test time. The ACS is activated from the ground only and runs for a commanded duration before ending.

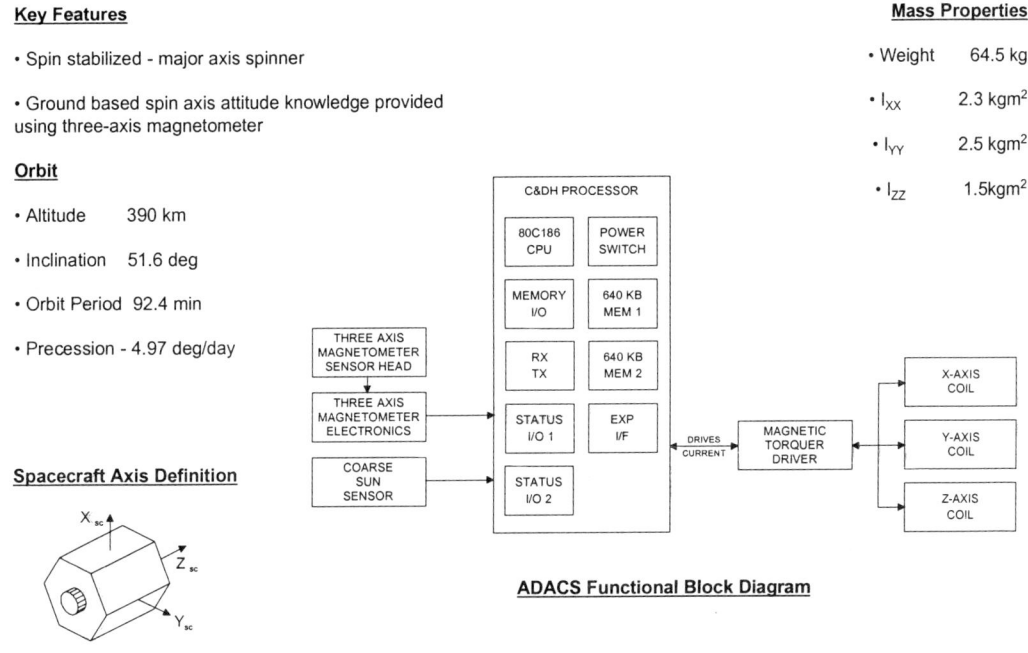

Figure 2. Diagram of MightySat I ACS components. Note axis definitions in lower left.

THE MIGHTYSAT I ACS ALGORITHMS

There are two ACS modes and two control tactics. The two modes are Open Loop and Closed Loop while the two tactics are spin and precession control. Open Loop Mode controls only the spacecraft spin while Closed Loop Mode controls both the spin and the precession. Upon ground activation, the typical ACS command parameters cause the attitude controller to execute Open Loop Mode, then transition to Closed Loop Mode, and then to stop. See the flight software flow diagram, Figure 3 below. The ACS function ends when it exceeds its duration parameter. Engineers choose this time parameter based on a simulation of the ACS performance in the actual flight orbit and on past experience, in accordance with their desire to minimize power usage.

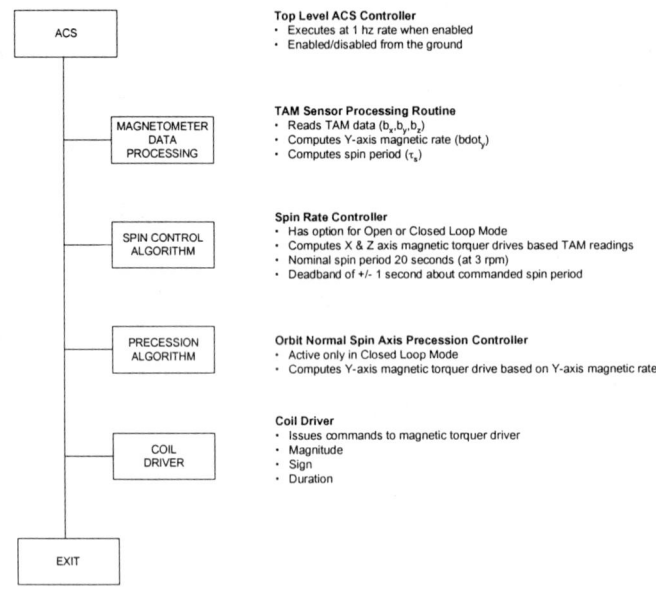

Figure 3. ACS Flight Software Flow Diagram.

The ACS has 6 associated commands. One command controls TAM power, two more commands control automated and manual collection of TAM data, while a fourth and fifth command allow manual torque coil control and algorithm parameter adjustment respectively. The final command activates the ACS. Before activating the controller, satellite operators build the ACS command scenario, usually composed of the ACS parameter adjustment and ACS activation commands. The most frequently changed parameters are: the duration of the ACS activation, the Open Loop Mode time limit, and the target spin period and deadband . After setting these parameters, operators issue the ACS commands to the spacecraft. After MightySat I receives the commands it stores them in a queue and executes them at their scheduled times. The time to begin any command is built into the command itself and is therefore not considered a "parameter" of the command. When the ACS activation begins, the Mode parameter is typically set to start it in Open Loop Mode. After the first ACS activation, however, the duration of the Open Loop Mode is usually quite short before transitioning to Closed Loop Mode. This is to prevent excessive spin from acceleration.

Open Loop Mode Algorithms

In Open Loop Mode, the ACS only tries to spin the spacecraft. It does not attempt to precess the spacecraft. It applies an open loop torque about the spacecraft Y axis using only the X & Z coils. Note that the TORQUE is open loop, not the coils' commanded magnetic moments. To keep the torque in a positive sense about the Y axis, it commutates the coil polarities in phase with the surrounding Earth magnetic field.

Commutation is how the ACS determines which polarity of current to apply to the torque coils in order to accelerate or decelerate the spacecraft's spin about the Y axis. Compared to the spacecraft's spin rate, the rate of change of the Earth's magnetic field local to the spacecraft is quite slow. The local field changes slowly enough (due to Earth rotation and Orbital rate) as to be considered static with respect to the spacecraft's spin. From the spacecraft's point of view, the field rotates about the spacecraft instead of the spacecraft rotating in a static field. The coil polarities must then change to consistently push against the latest ambient field direction. This polarity changing to adapt to the (apparent) changes in the ambient field is analogous to the commutation in a DC (Direct Current) motor with an electromagnetic rotor (coils) and a permanent magnet stator.

The ACS turns the torque coils off during each TAM reading so as to prevent the coils from polluting these readings. In the Open Loop Mode, the controller doesn't check to see how fast the spacecraft is spinning; it just keeps torqueing the spacecraft to get it going steadily in one direction. It applies this torque until the number of ACS execution cycles exceeds the Open Loop Mode time limit, nominally after 1000 seconds. This duration is commanded from the ground and is estimated based on ACS simulation runs and on-orbit experience. After the Open Loop Mode has gotten the satellite spinning, the ACS typically goes into Closed Loop Mode.

Closed Loop Mode Algorithms

In Closed Loop Mode, the MightySat I ACS controls both the spin and the precession of the spacecraft. It uses only the X and Z coils to control the spin and only the Y coil to control the precession. The difference between the Open Loop Mode and Closed Loop Mode spin tactic is that the Closed Loop Mode spin is closed loop control. The ACS still applies a commutation to maintain a consistent Y torque as in Open Loop Mode. Every cycle, though, it also checks to see if it has spun into the deadband about the commanded spin period. If it is within the deadband, it stops spin torqueing by commanding the X and Z coils to zero magnetic moments.

The precession tactic, unlike the closed loop spin tactic, is open loop (despite the fact that it occurs in the Closed Loop Mode). The ACS does not check to see that it has precessed the Y (spin) axis sufficiently close to the orbit normal. In Closed Loop Mode, the ACS is repeatedly taking the difference between current and previous Y-axis magnetic field readings. When this difference is non-zero, it sets the Y coil to the opposite sign of the difference. When this difference is zero, it maintains the coil at its previous polarity. The effect of this algorithm is to try to keep the body frame's Y component of the Earth's magnetic field constant.

As the satellite travels in its orbit, the local magnetic field vector rotates at twice the orbital rate, with respect to the fixed stars. This rotation is roughly about the orbit normal for sufficiently high inclination orbits. Since the controller tries to keep the Y field component constant, by opposing the change in this component, the resulting torque is about the orbit normal as the Y axis tries to follow the local magnetic field vector. The law of gyroscopes then forces the spin axis into the torque axis, that is, the orbit normal. Thus the precession control automatically drives the spin axis to the orbit normal. The reason that the ACS restricts the (open loop) precession control to the Closed Loop Mode is that it relies on the law of gyroscopes, which requires a spinning body. Being open loop, the precession stops when the ACS function ends.

Runaway Spin Acceleration

All of the antennas are on the cartwheeling top plate. If MightySat I had spun much faster than the nominal rate of 3 rpm, the ground could potentially have lost all communications due to the possible dropout in the RF (Radio Frequency) link. The basic control algorithms lack a spin sense determination and thus there exists the possibility for runaway spin acceleration. Although the risk of this occurring was small, the scenario was analyzed.

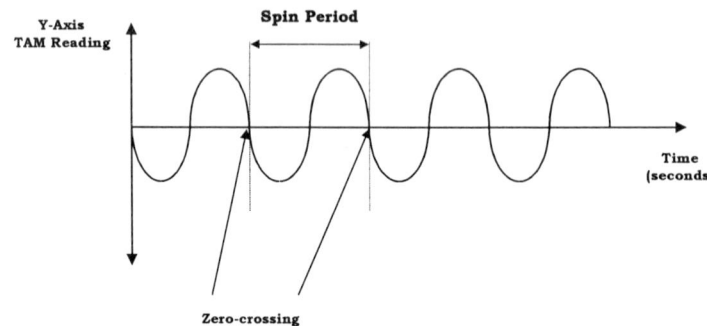

Figure 4. Estimating the spin period from consecutive TAM zero crossing.

The root cause of the potential for runaway acceleration is the use of the spin period to control the spin rate, without knowledge of the spin sense. The spin period is estimated by the number of control cycles between consecutive, negative-going zero-crossings on the TAM's Z axis. See Figure 4. Every one-second cycle through the controller, an algorithm increments an ACS cycle counter. Then the algorithm checks to see if the current TAM Z component is less than zero at the same time that the previous TAM Z component was greater than zero. If (and only if) these two conditions are met, the controller stores the counter value in the measured spin period variable and then clears the counter to zero. To reduce processor loading, this spin period is not converted into a spin rate. The spin control algorithms work only with the spin period.

The measured spin period is compared to the commanded spin period and the difference is taken as a spin error. If the magnitude of this error is less than the ground-commanded deadband about the commanded spin period, the ACS makes no attempt to adjust the spin period. If it is outside the commanded deadband, the ACS attempts to reduce the error to be within the deadband. If the error is larger than the deadband and is positive, the algorithm tries to accelerate the spin. Conversely, if the error is larger than the deadband and is negative, the algorithm tries to decelerate the spin.

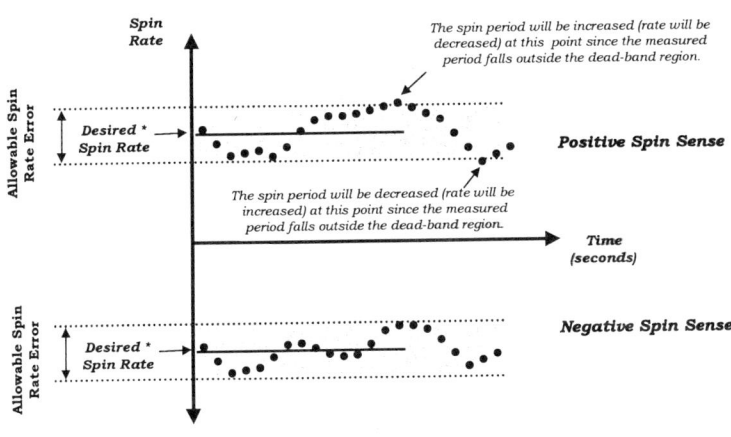

Figure 5. Spin rate deadbands equivalent to commanded spin period deadband.

The potential for trouble arises because the spin period is always positive while the spin rate can be positive or negative. See Figure 5. The algorithm nominally assumes the spin sense (positive or negative rate) to be positive. It assumes this because the Open Loop mode is hard coded to produce positive spin acceleration, and this mode has usually preceded the Closed Loop Mode. The spin sense is represented by a variable that can be either positive or negative "1". By commanding a change in this spin sense variable, the ground can command a reversal in the spin acceleration, should the spin be negative. This negative spin could come from ejection tip-off rates, for example. There is no way, however, for the ACS to autonomously determine the spin sense. Thus, if the spacecraft is spinning negatively and the ACS is laboring under the assumption of positive spin, then attempts to accelerate the spin will have the effect of decelerating the spin and vice versa. If this occurs while the measured spin period is on the large-period side of the deadband (the spin period is large and thus the absolute value of spin rate is small), then the spacecraft will ultimately reverse its spin sense (direction). See Figure 6. If this occurs while the measured spin period is on the small-period side of the deadband (the spin period is small and thus the absolute value of spin rate is large), then the spacecraft spin period will dwindle to two seconds. This corresponds to the spin rate running away to the Nyquist rate of the controller at 180° per second, Figure 7.

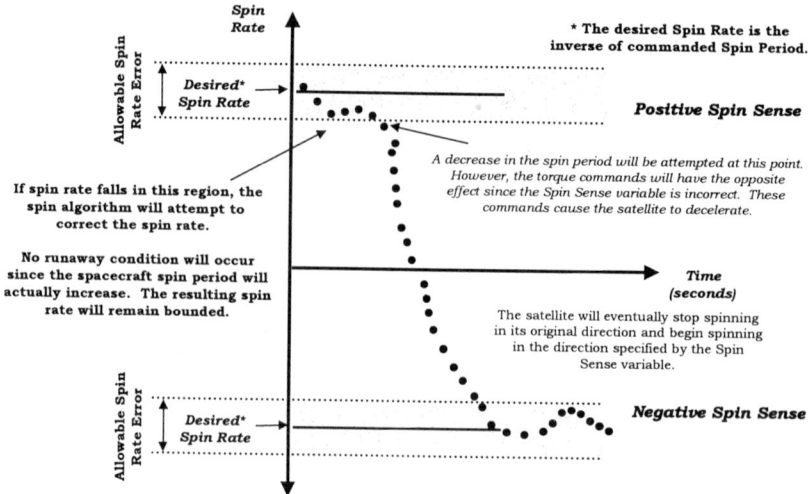

Figure 6. Incorrect spin sense results in reversing direction of spin. Period remains bounded.

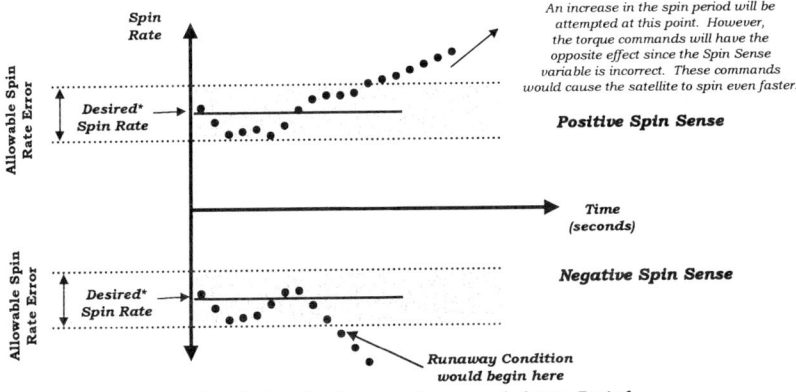

Figure 7. Incorrect spin sense results in runaway spin acceleration.

MightySat I encountered no negative spin or runaway spin conditions on orbit. Close attention to telemetry and discipline in ACS commanding prevented the spin sense variable from ever being commanded to the opposite of the actual spacecraft spin sense. This condition is discussed to indicate the potential hazards of flying with a minimal set of ACS algorithms. The austere set of algorithms reduced both test time and overall dollar cost. The price of this simplicity, though, is the more rigorous operational discipline required of the ground crew.

DESIGNING, BUILDING, AND TESTING THE MIGHTYSAT I ACS

Simulation of ACS Algorithms

Simulation results showed MightySat I precessing from about 130° off to within about 3° of orbit normal after approximately 18 hours, Figure 8. (Note that in the comments in Figure 8, Open Loop Mode is denoted as Mode 1 and Closed Loop Mode is denoted as Mode 2.) The spin rate during this simulated maneuver was approximately 17.2°/second and the spin axis inertia was approximately 2.55 kg*m^2. The Y coil magnitude was 4.0 a*m^2 and the simulation included residual magnetic, aerodynamic, and gravity gradient disturbances. This precession performance is comparable to the precession rate observed on orbit.

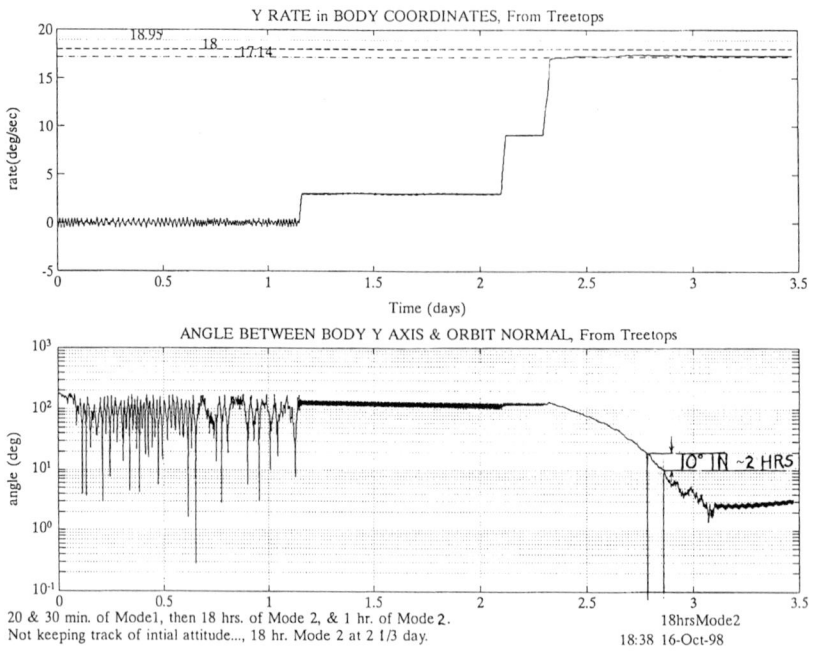

Figure 8. Simulated Spin and Precession Control After Approximately a Day of Tumble.

MightySat I was scheduled to launch on any one of 4 or 5 shuttle missions, and perhaps even the OSP (Orbital/Sub-Orbital launch vehicle) or Lockheed Launch Vehicles. Orbit altitudes varying from 306 km (165 nmi) to 394 km (213 nmi) and inclinations varied from 28° to 52°. The ACS was simulated at a variety of altitudes and inclinations to cover any contingencies and to set acceptable limits on the orbit. These Treetops™ simulations showed that the controller is fairly robust to parameter variations. Various initial attitudes and initial rates up to 3° per second, per axis were simulated. Magnetometer and coil errors and misalignments also went into the simulation. Off-axis inertia products of 0.0000, 0.0005, and 0.0050 kg*m^2 were simulated. Each time a certain orbit looked likely, the ACS simulation was rerun with the new orbit parameters to verify acceptable performance. When MightySat I was finally manifested on the first ISS (International Space Station) assembly shuttle mission, engineers increased the ballistic coefficient to achieve a one-year mission life at the 390 km (210.5 nmi) altitude.

As stated earlier, the ACS was designed for minimal testing. Besides component-level testing and simple integration testing of the sensor and actuators, the main pillars of the ACS test program were the ACS closed-loop testing (not to be confused with Closed-Loop Mode) and Magnetic Calibration Testing.

ACS Closed-Loop Testing

The ACS and software engineers conducted the closed-loop testing using the CATS (Closed-loop ACS Test Simulator) on one PC and the flight software on another. The CATS closes the loop for the flight software. It takes the actuator commands from the flight software and applies them to a model of the spacecraft dynamics. It propagates the simulated orbit and applies disturbances. Finally, it returns sensor readings to the flight software in time for it to calculate another set of actuator commands. The CATS simulation is totally independent from the Treetops™ models. The CATS not only served to test the flight software, but also to independently verify the control algorithms.

Because of the differing models, one cannot expect the simulation results to agree precisely. For example, when precessing from anti-orbit normal, a Treetops™ simulation showed precession to within approximately 3° after about 3.9 hours. The corresponding CATS simulation showed similar precession taking approximately 5.6 hours. These times are from simulations before the ballistic coefficient was increased. Since there was never any requirement levied on how fast the spacecraft precessed to the orbit normal, precession times from 2 to 20 hours, for precessions of 130° or more, were acceptable.

Another place where the simulations did not precisely agree was in the spin rate performance. Both simulations spun up to approximately 17°/second in approximately 33 minutes. The Treetops™ simulation, however, spun into the ~1-second deadband around 18°/second while the CATS stopped at about 16.5°/seconds. This difference can be attributed to delays in the X-Modem communications between the flight software PC and the CATS PC. A delay of 38 msec per ACS cycle adds up to a net shift of approximately 0.8 seconds in the total (21-ACS-cycle) revolution of the local magnetic field vector around the spinning spacecraft. Since the spin period measurement is quantized to the ACS controller period, one second, a 0.8-second lag drags a 21-second period out to 22 seconds. From the CATS point of view, it keeps on propagating the local magnetic field vector around the spacecraft, whether the rest of the spacecraft loop can keep up or not. The flight software got through its calculations in time, but due to the extra delay of the loop-closing communications, ACS cycles took a little longer than 1 second each to complete. Each cycle took about 1.038 seconds. Thus, after about 21 cycles, the ACS had another Z TAM channel zero crossing but those 21 cycles took approximately 22 seconds! The lesson here is that non-ideal loop-closing performance can lead to apparent simulation performance variations.

Evolution of MightySat I Inertia Tensor

In preparation for the spin balancing of MightySat I, several simulations were run to characterize the acceptable limits on the inertia products. In August of 1996, the simulations indicated that the Ixy and Iyz POI (Products Of Inertia) must remain fairly tight. The Ixz product of inertia, however, could be two orders of magnitude greater than the other products while still maintaining nominal precession performance. The magnitudes of Ixy and Iyz should be held to

0.0008 kg*m² while the magnitude of Ixz could be as large as 0.08 kg*m². With these POIs, a simulation confirmed that the satellite spun into the deadband in approximately 4000 seconds (67 minutes) and precessed from 90° to within 10° of the orbit normal within approximately 20,000 seconds (5.6 hours). See Table 1 for the inertia tensor. The inertia moments in this table are estimates while the products are requirements.

Table 1.
ORIGINAL INERTIA REQUIREMENTS IN INERTIA TENSOR FORMAT

(kg*m²)

1.915	0.0008	0.0800
0.0008	2.143	0.0008
0.0800	0.0008	1.1660

In February of 1997, MightySat I underwent the first spin balance. The initial spin balance was done on a new portable table proposed by the SDL (Spacecraft Dynamics Laboratory) at Utah State University. The accuracy of this table did not meet the intended standards, and it was determined that the balancing would have to be repeated at a later time. The inertia matrix came back with products of 0.013 ± 0.005 kg*m². This original balancing violated the inertia product specification by an order of magnitude. See Figure 2. (This violation was mainly due to a failure to put the hard requirements into the test procedure.) Simulations the following month (March 1997) indicated that precession with the 0.013 kg*m² products took 90,000 seconds (approximately one day) to go from 90° to within 10° of the orbit normal. With the upper limit of 0.018 kg*m², the precession got no closer than about 50° from the orbit normal.

Table 2.
INERTIA TENSOR AFTER FIRST SPIN BALLANCE

(kg*m²)

2.1450	0.0130	0.0130
0.0130	2.790	0.0130
0.0130	0.0130	1.1900

The customer was initially resigned to these large inertia products because of tight budgets and schedules. By May, however, the decision was taken to investigate the envelope further, and ACS engineers resumed simulations of the effects of various inertia products. Because other testing was being done in the meantime, several months of occasional simulations were run with various results and recommendations. During this testing, the AF (Air Force) engineers learned a valuable lesson in material selection. After vibration testing, they put the satellite into thermal vacuum testing without much intermediate testing. There was an apparent change in performance of the TAM during this testing. After thermal vacuum testing they discovered, that they had inadvertently used highly permeable stainless steel in a few of the balance masses, rather than the originally procured low permeability stainless steel. The process of vibration testing had magnetized the steel. It shows that you really need to know what you are buying, and also that a systems approach must be adhered to throughout a program.

Table 3.
FINAL INERTIA TENSOR RESULTS

$(kg*m^2)$

2.3239	0.00014	-0.003202
0.00014	2.5466	-0.002344
-0.003202	-0.002344	1.4966

The final balance of the satellite (in September, 1997) was performed at Sandia National Labs, on the Kirtland AFB (Air Force Base). See Table 3. This balance was done to very exacting standards, and the products of inertia were reduced to low levels (<0.003 kg-m2 in the sensitive axes). However, it was during this testing that it was discovered that the inertia ratio was not quite what the team expected. The measured inertia ratio was below 1.10. The decrease in the ratio was investigated closely to evaluate the impact on mission success, and it was determined that the increased nutation and precession time to orbit normal could be tolerated. The inertia ratio and products still violated the original recommendations by about a factor of three, but the simulated precession performance was deemed acceptable. The spacecraft precessed from about 150° to within 5° of the orbit normal after about 20 hours. The nutation angle was approaching 3 to 4°. A major driver in the decision was the schedule, but the lack of good places to add mass also weighed heavily in the choice not to balance any tighter. MightySat I had already been allowed to violate the static envelope in the radial direction by 0.25 cm, and in order to balance to these tight requirements, masses had to be added to the upper and lower deck edges, further violating the HES keep-out zones.

There were two lessons from this balancing experience. First, the moments and products of inertia have an overwhelming impact on the control performance of a spinner. Second, if you have hard requirements to meet, you better know the requirements and how critical they are before going into test.

Magnetic Calibration Testing

The minimal test plan of component-level testing, hardware integration, closed-loop testing, and magnetic calibration inspired the placement of end-to-end, physical functional testing into the magnetic calibration procedure. This was a less than optimal place for such testing due to its late position in the schedule. This testing was performed at the Goddard Space Flight Center's Spacecraft Magnetic Test Facility after the satellite had completed all other environmental testing. It used the 42-foot circular coils of the Braunbek three-axis coil system.

After nulling the Earth's magnetic field, the spacecraft's residual magnetic moment was measured. Afterwards, the spacecraft was degaussed to reduce this residual magnetic moment. Typically, the spacecraft was degaussed by application of a 60 Hz, 60 Gauss field that decayed gradually. Concern over potential instrument damage from exposure to these levels resulted in degaussing at lower amplitude fields. The goal of the degaussing was to reduce the residual magnetic moment to less than 0.05 a*m^2. Degaussing at the lower field levels resulted in a residual of 0.33 a*m^2, an order of magnitude higher than desired. Analysis showed that this residual dipole was not detrimental to ACS performance. The simulated spacecraft still precessed to within 5° of the orbit normal within 24 hours. The angular momentum, 0.331 r/s * 2.55 kg*m^2 = 0.844 kg*m^2/s (N*m*s), provided enough gyroscopic stiffness for MightySat I to easily weather this increased magnetic disturbance.

The TAM was calibrated to determine its scale factor and zero offset in response to known facility-generated fields. For each axis, TAM output in counts was plotted against the field readings in nano-Tesla (nT) to derive the scale factor (in nT/count) and offset (in nT). The TAM scale factors and offsets (one each for each channel) were acceptably nominal. Proper magnetic torquer operation was verified by maintaining a null field, manually commanding each coil in each direction and verifying the measured moment. Following this, end-to-end physical functional tests were performed, referred to as the three phasing tests.

The first phasing test was to confirm the proper measurement of the spacecraft spin period. The facility generated a rotating field about the spacecraft Y axis and test personnel recorded the ACS measured period. This test went smoothly and MightySat I correctly determined the periods of the rotating fields. The second phasing test, to verify proper commutation, began with the facility generating a field vector between the spacecraft spin coil axes (in the spin plane). For this vector, test personnel measured the resulting coil moments with a hand-held test magnetometer. Engineers also recorded the coil currents as a check on the hand-held measurements. This test revealed significant discrepancies from nominal performance. The third (and final) phasing test was the precession verification. The facility generated slowly rotating fields about the X and Z axes. For each of these rotating fields, the resulting Y coil moment and current were recorded to check proper operation of the precession. This test also revealed discrepancies.

ACS Functional Troubleshooting

Troubleshooting indicated five problems. Three were performance problems and two more were operational. The three performance problems were that the Z coil frequently had the wrong sign, the precession coil activation was intermittent, and the closed-loop commutation was always half wrong (one of two coils). The operational stumbling blocks were that the coil driver enable command was not being issued before initializing the ACS and that there was no way to power off any coil independently from the others. The operational stumbling blocks were met with operational fixes. Operators would always send a command to enable the coils before starting the ACS. The all or none coil powering was deemed a feature. The performance problems, however required further troubleshooting.

It was quickly discovered that the Z coil had been measured inconsistently. Sometimes engineers had measured the coil moment near the center of the coil while at other times the magnetic calibration facility TAM had measured the field far from the coil axis. This is the difference between measuring the flux coming out of a magnetic pole versus measuring the flux far from the pole as it loops back around to the opposite pole. See Figure 9. For a given magnet coil polarity, these two flux lines are in opposite directions. Consistent measurements indicated that the Z coil was wired backwards. An adjustment to the flight software fixed this problem since it was undesirable to open the satellite for any reason at this point in the testing.

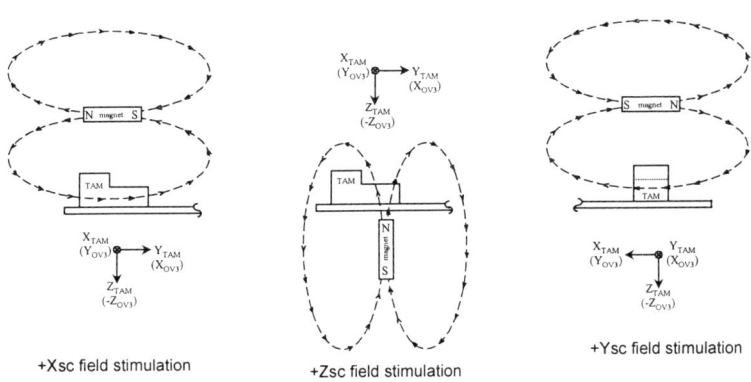

Figure 9. Hand-held Magnetometer measurements of Z coil were along Z axis, similar to the TAM in the center drawing. Facility measurements of Z coil were far from axis, similar to TAM in the left drawing.

The fluctuating precession coil (Y axis) was traced to a fluctuating TAM Y channel. The precession algorithm couldn't work with these Y-channel fluctuations. A review of the measured coil properties showed that the time constants, T = L / R (Henrys/Ohm), were all less than 4 msec. That was before their installation in the spacecraft, however. Measurements at the spacecraft level indicated that the coil moment took 36 ms to decay. The software had allotted only 25 msec. Again, the problem was fixed in software.

The Closed Loop Mode commutation anomaly was traced to a software error. While the Open Loop Mode commutation had never failed during test, the Closed Loop Mode commutation always did. The root cause of the commutation anomaly lay in the process by which the ACS algorithms became flight code. The algorithms had gone from the algorithm simulation code, through pseudo-code, to the flight code. In the translation, some of the algebraic operations had been combined in a mathematically acceptable fashion. The flight code running on the PC had no problem with this equation. To the 80C86, however, it was ambiguous. The PC to PC closed-loop testing had not caught a subtlety of the 80C86's implementation of the flight code.

THE MIGHTYSAT I FLIGHT EXPERIENCE

Although slight, the possibility of runaway spin acceleration, the limited ACS testing, the restriction to scheduled commands, and the lack of real time telemetry combined to provide for an exciting early orbit operations. Thankfully everything functioned as designed and no anomalies were observed. MightySat I has solar arrays mounted on all sides except for the bottom with the separation ring. Because of this coverage, MightySat I was very power positive without the need for ACS intervention. Thermal trends, however, motivated the early use of the ACS. When MightySat I left the canister, the HES gave it a considerable rate about the Z axis, the minor moment of inertia axis. Because the composite bus is very rigid, with little mechanical energy dissipation, this large Z rate component remained until the ground activated the ACS two days after ejection.

This Z-axis spin was confirmed by use of the solar array power and TAM data, as well as the video footage from the Shuttle. The panel voltages peaked in order around the circumference of the satellite. This data enabled the team to determine that the satellite was coning about the Z-axis. This orientation, along with rapidly increasing battery temperatures, which were on the Sun side of the bus, engendered a desire to activate the ACS. A prolonged ACS activation (1 hour) gave a clean Y-axis spin at a rate of about 2.6 rpm. This mitigated the battery temperature trend to a small extent and evened out all of the other bus temperatures to predicted values. However it took a change in the charge regulator settings to fix the battery temperature problem. Since these early operations, the ground has successfully used the ACS to keep the satellite spinning at the commanded period of 20 seconds (3 rpm).

During the mission, the ACS was also used to orient the satellite along certain vectors by timing a short precession to leave it at the right point while it was moving towards orbit normal. Operators also used it to precess to orbit normal on several other occasions. The customer estimated the pointing of the spin axis using the TAM data and solar panel currents. See Figure 10 for a plot of the solar array power and the corresponding TAM data. The solar array currents provided a sanity check on this TAM attitude. For example, looking again at Figure 10, the relative output of the top and side panels indicates crudely what the angle to the Sun must be. The spin rate can also be determined from the cyclic nature of the panel power generation. For instance in this plot, the Top Panel comes around every ~33 seconds, as do the other panels. The power output of the Top Panel is ~190 mA, or ~73% of the peak panel output. This is equivalent to a combined angle of about 43° from the cosine law, i.e., arccos(0.73) = 43°. The other two panels that are generating power at the same time are about 110 mA each, or 42% of maximum output. This equates to an angle of about 65°, which is roughly 47° plus the 30° slant of each panel (60° total angle between the panels for the hexagonal shape). This is also borne out by the ground based, Kalman filter attitude estimator, which shows an angle around 40 degrees off the orbit normal, which at this time of year and the orbit was about -5 degrees, in the same time frame as the solar panel data was taken. Figure 11 shows this plot.

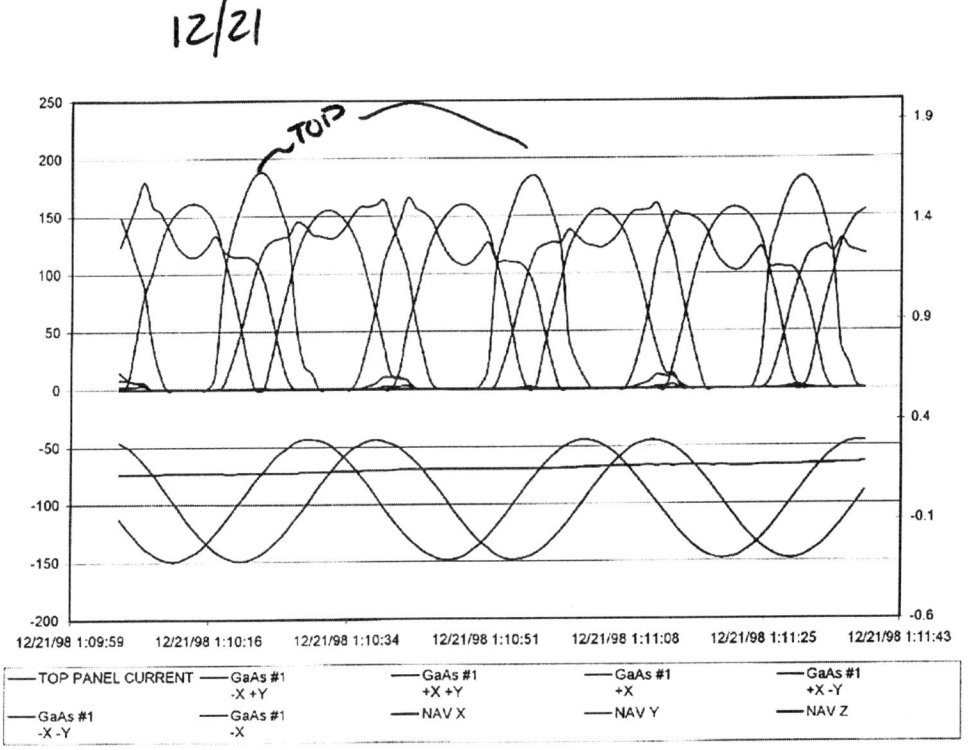

Figure 10. Solar array ouputs (in mA) versus time.

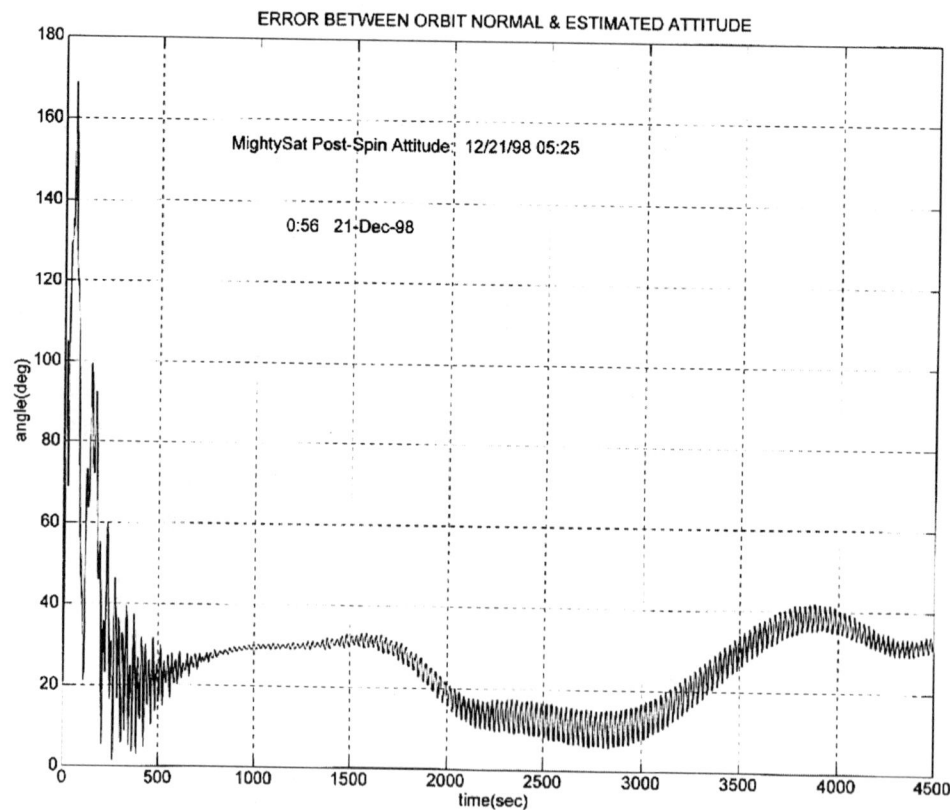

Figure 11. TAM-based attitude estimator output.

CONCLUSION

The MightySat I mission was a complete success. While it is possible to interpolate some ACS algorithms from a previous program in about a week, following through with the actual analysis, flight software validation, spacecraft level I&T and mission operations of such a system is still challenging.

ACKNOWLEDGEMENTS

The authors wish to express their appreciation to all those who participated in the success of the MightySat I ACS:

> Capt. Barbara Braun (AFRL Test Engineer and Program Manager)
> Robert Davis (The Aerospace Corporation)
> Mark Edison (Orbital Sciences Corporation)
> Doug Freesland (Orbital Sciences Corporation)
> Don Gibbons (Orbital Sciences Corporation)
> Rich Kuha (Orbital Sciences Corporation)
> Jon Thurneysen (Orbital Sciences Corporation)
> and others...

The authors also wish to express their appreciation to those who assisted the preparation of this paper:

> Elisa Ambrose -- spin period, deadband, and runaway acceleration plots (Figures 4 to 7)
> Doug Freesland -- ACS block diagram and summary (Figure 2)
> Mark Ponton -- magnetic flux measurement drawing (Figure 9).

AAS 00-065

MEMS-BASED GN&C SENSORS AND ACTUATORS FOR MICRO/NANO SATELLITES

J. Connelly, N. Dennehy, P. Hattis, W. Johnson, D. Sargent and M. Socha[*]

A new generation of miniature, low mass/power, high performance GN&C sensors and actuators will be required for envisioned Micro Satellite and Nano Satellite missions. As satellite volume decreases correspondingly small sensors and actuators are needed to perform satellite operational functions. Therefore technologies which enable very small devices are quite important for miniature satellite development. Draper Laboratory, a world leader in MicroElectroMechanical System (MEMS) inertial sensing, is currently developing attitude determination and attitude control devices to meet the emerging needs of these new missions. This paper describes designs for both a miniature stellar-inertial attitude sensor package and a MEMS wafer wheel momentum device.

The Draper Attitude Sensor Suite (DASS) integrates complementary technologies for Electron Bombarded Charge Coupled Device (EBCCD) stellar cameras and MEMS gyroscopes into one miniature package. Both the EBCCD sensor and MEMS gyro technologies have been continuously matured at Draper over the past several years with all major hardware components having been successfully tested in both laboratory and operational environments. The EBCCD stellar camera provides an order of magnitude improvement in S/N ratio (as compared to standard CCD technology) at high bandwidth allowing detection of faint stars with a relatively small optical aperture. Star detection capability can be further improved by implementing the Draper-developed "Drag-Back" technique for integrating the star images over periods of several seconds. The backthinned EBCCD detector is inherently radiation hard. Performance analysis indicates that arc-second accuracy can be accomplished with the DASS. This EBCCD star camera performance is uniquely suited to detect faint stars on highly dynamic Micro/Nano spacecraft or spin stabilized satellites. Laboratory testing indicates the ability of the unit to detect stars as faint as Magnitude 8 while spinning @ 20 rpm. Second generation MEMS Tuning Fork Gyros (TFG's) provide low power/mass/volume three-axis attitude rate measurements for high bandwidth updates between stellar fixes.

The Draper MEMS Wafer Momentum Device (MWMD) has been designed as a feasible near-term Micro/Nano Satellite attitude control actuator. The wafer rotor in this device is planar and is fabricated as a wafer thick (500 micron) silicon part. In the current implementation, spin axis rotational freedom is furnished by a very small, but standard ball bearing cartridge supplied by a commercial vendor. This MEMS wheel has been designed to operate at a speed of 50,000 rpm and will provide an angular

---

[*] The Charles Stark Draper Laboratory, Inc., 555 Technology Square, Cambridge, Massachusetts 02139.

momentum of about 150 dyne-cm-s. This device can be configured to serve either as a momentum wheel, a reaction wheel, or a control moment gyro. As a long-term alternative to the conventional cartridge bearing Draper is currently investigating a novel ball bearing in which 11 mil steel balls are captured within raceways fabricated into the silicon rotor. This silicon-steel bearing combination is unique and provides a potentially elegant and compact low cost design. Limited testing has shown the viability of the silicon-steel approach. This very promising investigation showed that silicon mated to steel was potentially a good bearing combination. Looking even further into the future, another suspension concept has been developed for a wafer control moment gyro device with a hemispherical gas bearing. Here the gas bearing provides the radial and axial stiffness from the boundary layer dynamic forces.

1.0 Introduction

1.1 Draper Laboratory - MEMS Attitude Determination Subsystems

1.1.1 ST5 Mission Motivation

Draper Laboratory is a member of the New Millennium Programs (NMP) project concept definition team focused on the Space Technology 5 (ST5) small satellite constellation program. For NASA's New Millennium Program, Draper Laboratory is proposing new sensor technology that has been developed at Draper Laboratory over the last several years. In particular, a new technology of optical sensors and a new technology of inertial sensors have been developed that relate directly to NASA's NMP objectives of small, low cost, space qualified and autonomous sensors. These sensors would be key components in the Attitude Sensing and Control and the overall Guidance, Navigation and Control (GN&C) for the Space Technology 5 Constellation of Small Satellites. The design, development, integration, and operation of a full service 20KG class spacecraft is titled the NanoSat Constellation Trailblazer (NCT) program.

The specific new families of sensors proposed for the NCT program include the Electron Bombarded Charge Coupled Device (EBCCD) low light level sensor and the Micro Electro Mechanical System (MEMS) inertial sensors (in particular, the gyro). The advantage of the EBCCD is its excellent performance characteristics (high sensitivity, high bandwidth, high resolution, small size, and radiation tolerant). The advantages of the MEMS sensors are very low size, weight and power.

The concept for the combined EBCCD and MEMS technology-based sensor for the NCT spin-stabilized satellite is shown in Figure 1. The combination of the EBCCD star tracker and the MEMS inertial rate sensors (gyros) will provide a smoothed attitude output measurement that can be used as a reference for NASA science measurements and for spacecraft attitude control.

The operation of the EBCCD star tracker for the NCT mission will take advantage of the limited star field that is scanned as shown in Figure 1. Depending on the limited field of view of the EBCCD star tracker (nominally 2 degrees), only the stars near the spin axis need to be used for the attitude update.

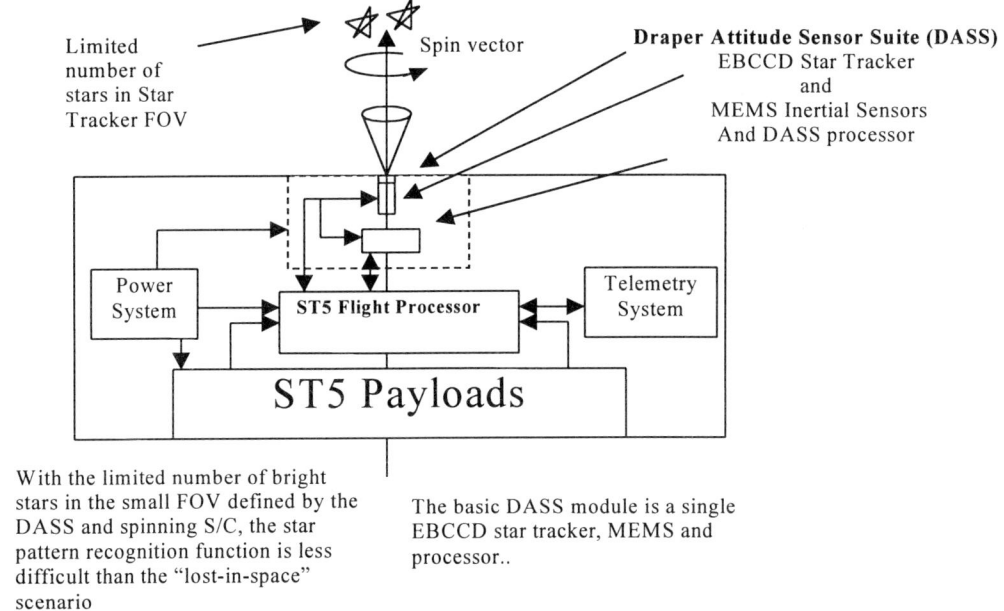

Figure 1. DASS subsystem on NCT spacecraft

1.1.2 Technical Challenge

Draper Laboratory has extensive experience in star tracker design, development, test & evaluation including the current Navy Fleet Ballistic Missile Program. It was recognized that the sensitivity of the EBCCD star tracker allows relatively dim stars in the field of view to be candidates for the attitude update measurement. The current required sensitivity for the NCT mission is to detect stars brighter than visual magnitude 8 stars.

The radiation tolerance of the EBCCD star tracker is an inherent characteristic of the backside-illuminated and thinned CCD sensor for the EBCCD. The thinning to 10-15 microns reduces the radiation sensitivity while the optical sensitivity is maintained by the high EBCCD gain (~200).

The MEMS inertial rate sensors complement the EBCCD star tracker. The MEMS gyros together with the constant spacecraft spin rate provide attitude memory for the NCT satellite. The star measurements can be filtered and the gyro biases can be updated using conventional techniques in the flight processor. The inertial data can either be processed in real time for attitude control or down-linked to the ground with the science data for post-flight processing. The size, weight and power of the MEMS components are low enough that redundancy could be considered to further improve the system reliability.

In summary, a new generation of optical and inertial sensors is available that can be applied to the NASA NMP to meet the desired requirements for small, low cost, space qualified and autonomous systems for the small satellite focus technology. The EBCCD and MEMS sensor technologies address the spin-stabilized NCT mission and are also applicable to other non-spinning satellite missions.

1.2 Draper Laboratory - MEMS Attitude Control Devices

1.2.1 Attitude Control Motivation

Micro- and nano-satellites will require correspondingly small actuators to perform satellite operational functions. This discussion examines the viability of using micro-electro-mechanical systems (MEMS) to build miniaturized momentum wheel assembly (MWA) devices. Implementation concepts for such devices are illustrated, and near term development activities that might be undertaken to further develop this technology are described. Related MEMS development, upon which the space actuator work would be based; now being conducted at the Charles Stark Draper Laboratory is also briefly noted.

1.2.2 Technical Challenge

The micromechanical work at Draper includes diverse areas of research and capabilities including inertial sensors, chem-bio sensors, microphones and hydrophones. To date the approach to these devices is to fabricate flat planar masses in silicon that are suspended using miniature flexures and electro-static comb drivers. Recently, an effort was initiated to investigate performance of MEMS fabricated spinning mass configurations; much like a traditional gyro is constructed.

Spin axis rotational freedom is furnished by a very small, but standard ball bearing cartridge supplied by a commercial vendor. This MEMS wheel will be tested for operational speeds of 50,000 rpm and an angular momentum, H, of about 150 dyne-cm-s. A computer, cut-away drawing of the assembled device is shown in Figure 2.

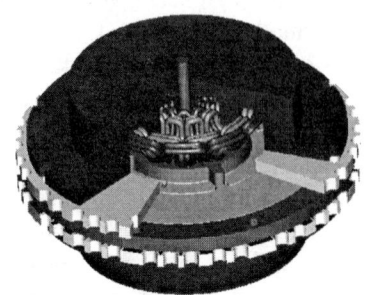

Figure 2. Wafer Momentum Wheel with Conventional Bearing

Extending this idea further, the concept that a spinning mass inertial sensor could be used as an actuator for envisioned nanosatellites became apparent. The concept of using a conventional spinning-wheel instrument gyro (sensor) as a control moment gyro (actuator) has been developed before at Draper Laboratory. The current technical challenge is to implement this concept using MEMS technology. (See Section 3.0 for more detailed discussion)

2.0 Draper Attitude Sensor Suite (DASS)

2.1 General Description

The use of stellar sensors to enable satellite attitude determination has been well established – especially for inertially stabilized platforms. The use of stellar sensors on spin stabilized vehicles has been much more limited due to the high rates imposed on a spinning scanner combined with the previously low star tracker frame rates that were possible. The detection and identification of stars is further complicated if the stellar sensor must also be miniaturized.

The recently developed Draper Laboratory EBCCD camera technology provides an inherently high signal-to-noise ratio in combination with excellent frame rate characteristics. This technology offers a true breakthrough for accomplishing accurate and reliable star sightings for a spacecraft with high spin rates At rotations of 20 rpm for vehicles such as the NCT satellites, the sensor can detect stars down to magnitude 8 in brightness.

The EBCCD camera has undergone prototype development and qualification in two major DoD programs (the Navy FBM star tracker Advanced Technology Demonstration System and the Airborne Laser Program). The Navy program requires the EBCCD accuracy and radiation tolerance. The Airborne Laser Program also requires the EBCCD signal-to-noise ratio and the high sensor frame rate. The NCT application of the DASS star sensor applies this newly-developed technology to micro satellites and will miniaturize the star sensor package to achieve the program objectives.

The Draper Laboratory development of MEMS gyro technology provides an inertial rate sensor capability to complete the miniature stellar-inertial system. The MEMS gyros have extremely low size, weight and power, and can provide the inertial attitude memory function between stellar updates and can also maintain attitude knowledge during rapid programmed slewing. The stellar updates can, in turn, be used to determine the MEMS gyro inertial drift rates. The current 2nd generation of the single-axis MEMS gyro instrument is 0.35 cubic inches in volume, consumes only 50 mW of power and has a mass of less than 20 grams. The 3rd generation of MEMS gyro improves these numbers even further by putting MEMS gyro sensors for measurements of all 3 rate axes on a single board. (see Figure 13)

The layout of the EBCCD star tracker and MEMS gyro sensors is shown in Figure 3. The DASS system package size is driven by the aperture that is required for detecting visual magnitude 8 stars with high probability.

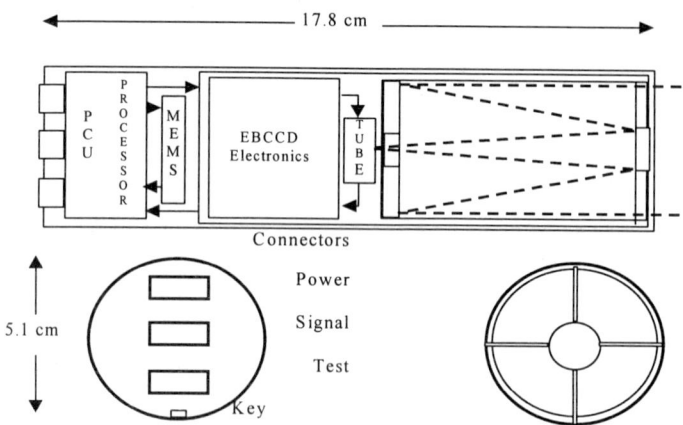

Figure 3. Layout of DASS subsystem

2.2   EBCCD Technology and Operation of DASS

A comparison of EBCCD design and performance characteristics with other photosensor technologies is provided in Figures 4 and 5. Figure 4 shows that the EBCCD sensor has the desired characteristics of high gain from electron multiplication (like the photomultiplier tube) and good resolution (like the multi-pixel CCD) in contrast to the capabilities of other relevant photosensor technologies. The EBCCD uses a different technology than the intensified CCD (ICCD) – resulting in fewer sources of image and bandwidth degradation when operating at low light levels.

The signal-to-noise performance of the EBCCD is compared to a conventional CCD sensor (with no electronic gain) in Figure 5. At low light levels and high frame rates (> 60 Hz), the EBCCD has an order of magnitude improvement in the signal-to-noise ratio.

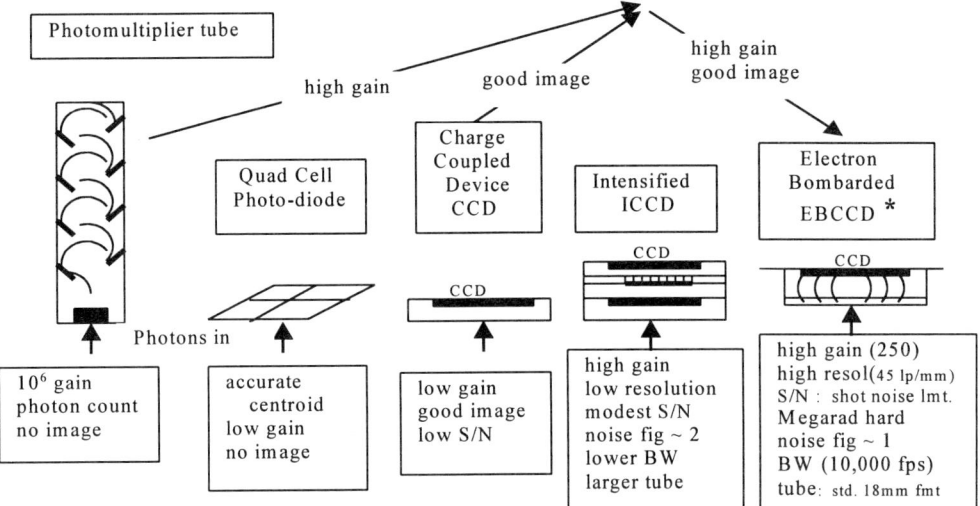

The EBCCD has the high gain resulting from electron multiplication (like the photomultiplier tube) and has the good imaging capability of a multi-pixel CCD – without the image degradation caused by ICCD multiple interfaces or the noise introduced by the MCP or the persistence of the phosphor component of the ICCD.

Figure 4. Comparison of EBCCD generic capabilities to other photosensors

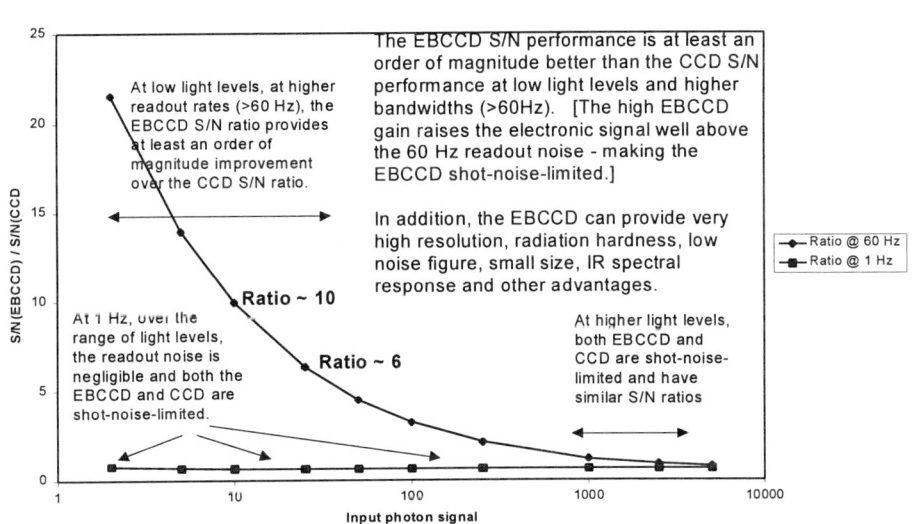

Figure 5. EBCCD signal-to-noise ratio comparison

The expected operation characteristics of the DASS in the NCT spacecraft is shown in Figures 6 and 7. The DASS sensor design parameters have been selected to be consistent with the NCT spacecraft and mission requirements. The requirement to detect visual magnitude 8 stars comes from the orientation of the 2 degree field-of-view optical axis along the spacecraft 20 rpm spin vector. The expected sensor field of view from the NCT satellites is near the north ecliptic pole which has very few bright stars. The selected DASS frame rate of 60 Hz limits smear on the CCD pixels at 20 rpm while also limiting the star image integration time per pixel per frame. Figure 6 summarizes the response of the EBCCD to dim stars in terms of electrons/pixel/frame.

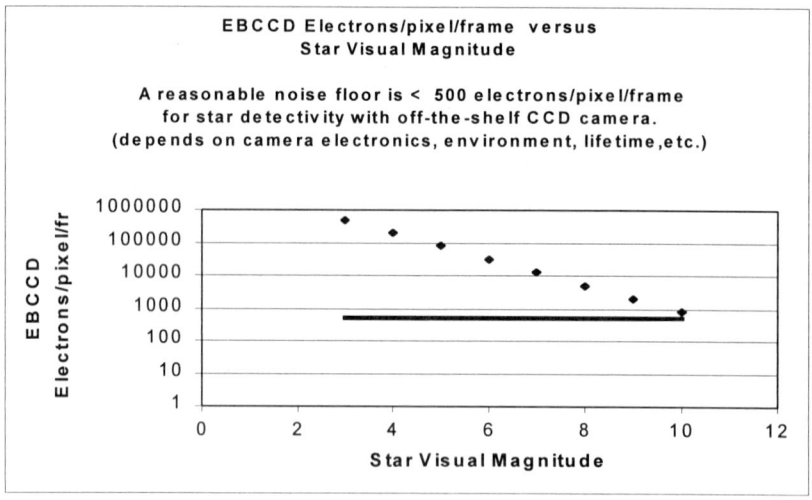

Figure 6. EBCCD sensitivity to dim stars

In the DASS application on the NCT satellites, the spin axis orients the DASS optical axis along the normal to the ecliptic plane at a 20 RPM spin rate. In this geometry, the star(s) that are in the FOV will appear as nearly circular traces every 3 second as seen in Figure 7 (with alignment errors resulting in some out-of-round characteristics of the star traces).

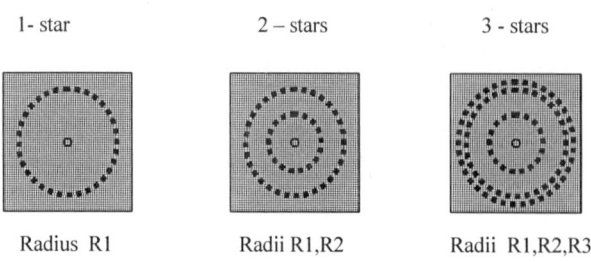

Figure 7. Near-circular traces of DASS-detected stars over a 3-second period

Identification of the satellite spin axis location in right ascension. and declination coordinates from EBCCD star trace data is graphically shown in Figure 8, using locations of several stars in the limited celestial sphere region around the normal to the ecliptic. This process of attitude determination from the star traces can be easily implemented on a processor – either on-board the satellite(real-time) or on the ground (post-flight).

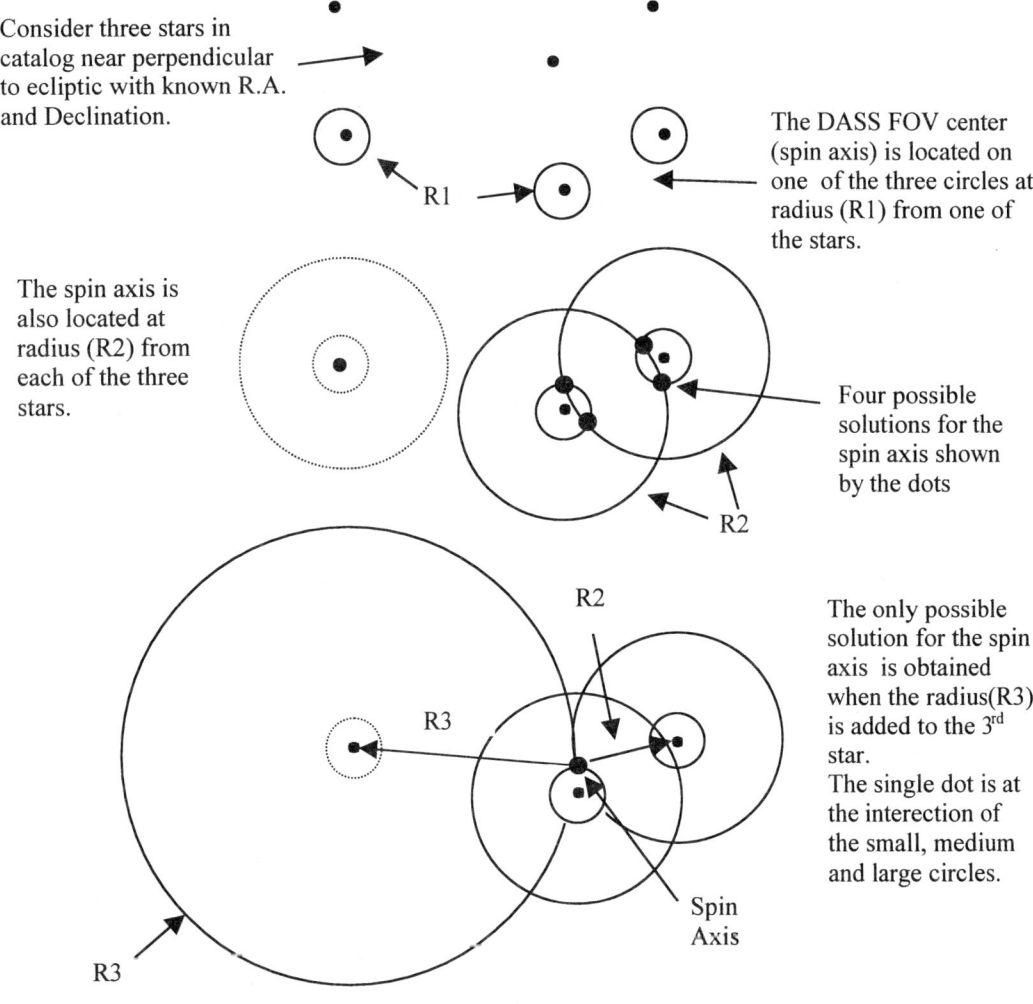

Figure 8. DASS identification of the satellite spin axis location

2.3   MEMS Gyro Technology and Application to DASS

Micromachined silicon inertial sensors offer revolutionary improvements in cost, size, and reliability for sensors used by guidance, navigation, and control systems.

Inertial sensors represent an important application of MEMS technology, which combines semiconductor materials and processing to create integrated mechanical and electrical systems. Batch manufacturing techniques produce thousands of virtually identical MEMS devices, each a few square millimeters in size, enabling inertial systems at a fraction of the cost, size, and power of any previous technology.

Draper has been developing micromachined gyroscopes and accelerometers for over 10 years, and was the first to demonstrate rate sensing with a micromachined silicon sensor. Micromachining borrows processes from the semiconductor industry to fabricate tiny sensors and actuators on silicon chips. Many new applications are enabled by the reduced size and lower cost inherent to sensors developed with this process.

A recent Draper research focus has been the development of MEMS inertial sensor technology for military and space systems. Improved sensors and electronics combined with advancements in MEMS packaging have enabled new applications, the first being the addition of guidance to low-end, previously unguided artillery shells. New MEMS integration processes are now being developed to create complete 6-DOF inertial sensor sets on a single chip. When available, these microsystems can be packaged as separate molded components or as multi-chip modules for chip-scale integration with higher-level systems. They are ideal for advanced space system concepts. It is expected that MEMS inertial technology will be tolerant to space radiation environments, with testing now underway to verify this expectation.

Figure 9 is a scanning electron microscope (SEM) photograph of a MEMS gyroscope. The sensor structure that is shown is approximately 1 mm on a side and a few microns thick. Figure 10 is a SEM photograph of a MEMS accelerometer with dimensions similar to the gyroscope. Proof mass and flexure design variations yield devices with full-scale ranges from 1 to 100,000 g's. Figure 11 shows the features and fully integrated scale of an inertial system that applies to the 2nd generation MEMS single-axis inertial sensor designs. Figure 12 shows the progression of the MEMS instrument scale through the 3rd generation sensor design.

Figure 9. Micromachined comb drive tuning-fork gyro

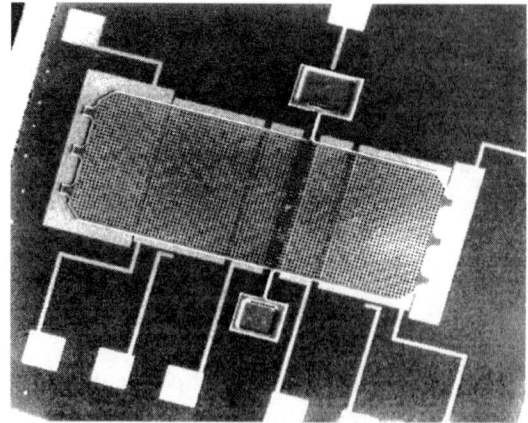

Figure 10. 100-g accelerometer sensor

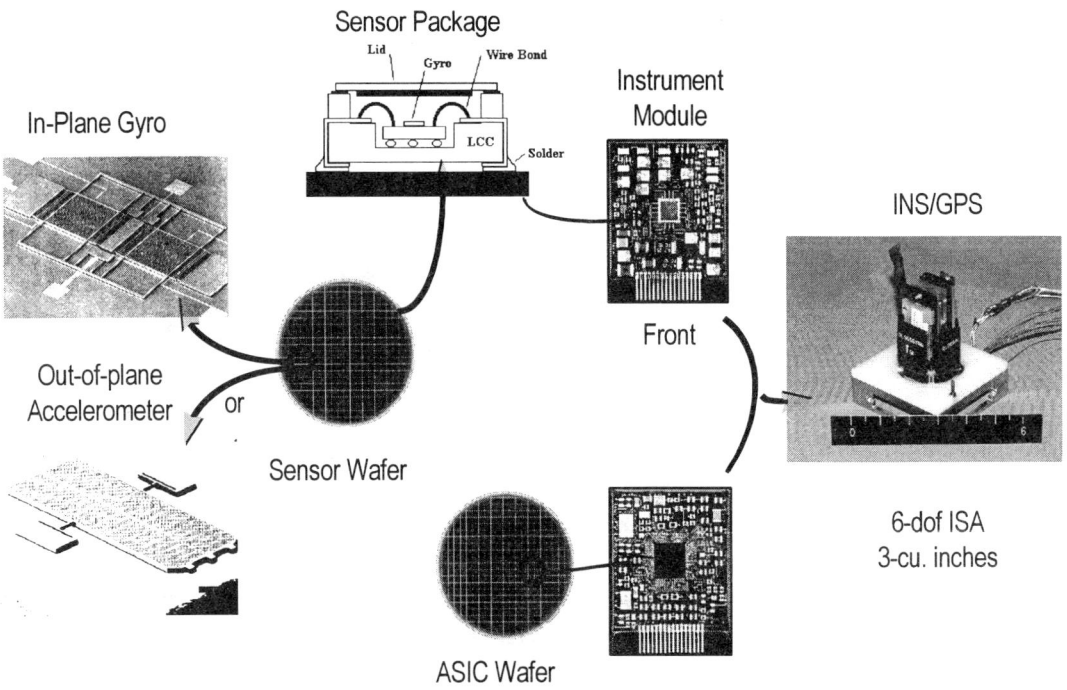

Figure 11. MEMS 2nd generation inertial system components and integated package

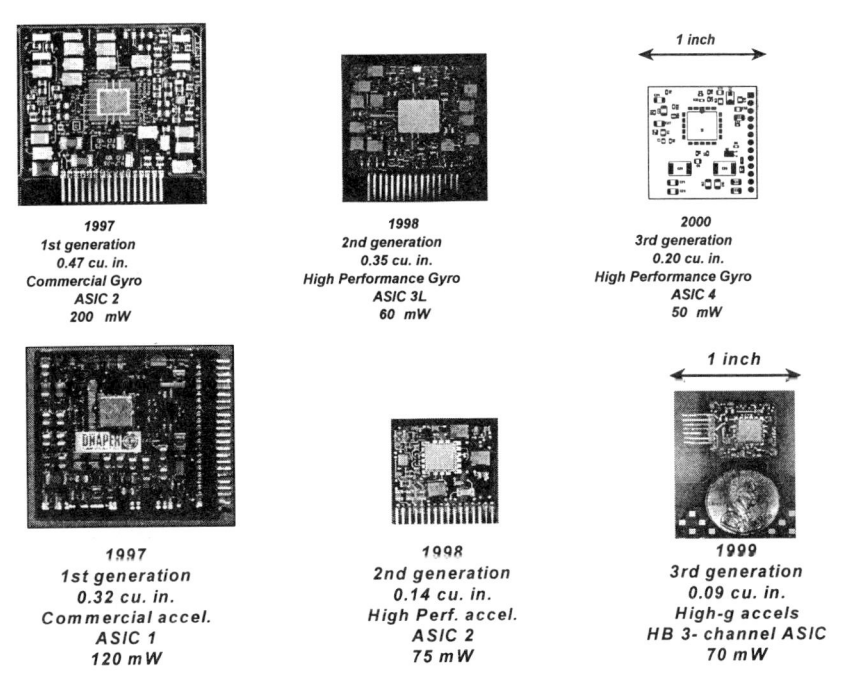

Figure 12. MEMS single-axis instrument scale progression

571

A 3rd generation three-axis accelerometer set that was built with two in-plane and one out-of-plane sensors is shown in Figure 13. Its performance is currently being evaluated, while design of a three-axis gyro set of similar construction is also proceeding.

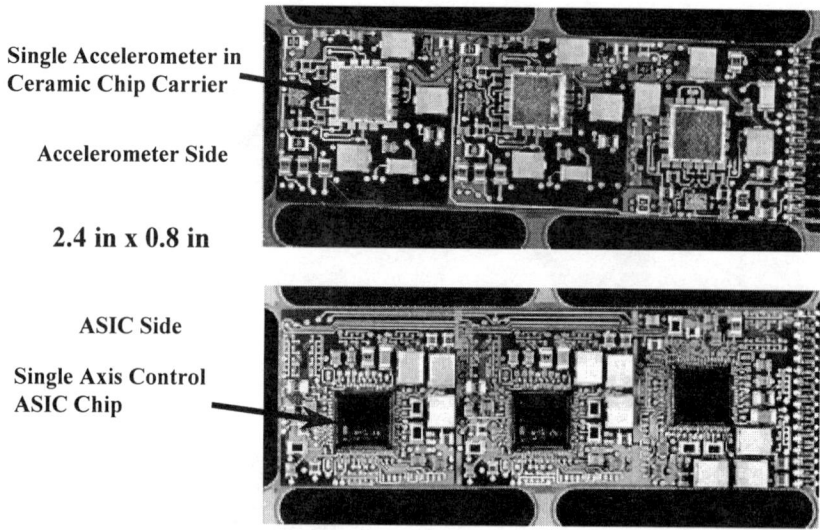

Figure 13. MEMS 3rd generation three axis accelerometer module

2.4   Orbital Operations for the NCT Program

The objective of the NCT program is to demonstrate breakthrough technology to support revolutionary cost and size reduction as well as enhanced operational performance required to enable future constellation satellite operations. The goal is deployment of many autonomous, small spacecraft that can perform planned science missions. A generic need of Micro Satellite and Nano Satellite missions is low mass, low power, high performance, GN&C sensors.

The DASS stellar/inertial sensor system in the NCT application is designed to flight demonstrate critical technology features while limiting the integration and support costs. This is achieved by recording the near-circular stellar measurments (shown in Figure 7) and down-linking them to the ground for post-flight processing to determine the spacecraft spin axis and attitude (using means such as those shown in Figure 8). The NCT science data would be processed post-flight using the DASS-determined attitude data. In the NCT application, the MEMS gyro data would also be down-linked directly to ground to be post-processed as necessary to provide the necessary attitude memory function. The extent of ground processing of DASS data in the NCT application is the result of an effort to limit NCT-flight-specific DASS software development costs for a technology demonstration mission with very stringent cost limitations.

Subsequent flights of the DASS concept could implement the attitude determination algorithms in real-time on-board the spacecraft using the NCT-flight-proven DASS star tracker and MEMS sensors along with NCT-ground-proven attitude determination algorithms. This approach to cost and schedule risk mitigation for the NCT DASS experiment enables the demonstration of the revolutionary DASS sensors in the space environment without risk of losing the entire DASS experiment due to a flight attitude determination software error. This NCT DASS orbital operations plan provides a proactive risk management approach and a cost control approach in line with the NMP Space Technology Program objectives and constraints.

3.0 Wafer Reaction Wheel/ Momentum Wheel

3.1 General description

Reaction wheels are rotary motors that torque against high-inertia rotors (the wheel). When the motor in the reaction wheel exerts a torque on the wheel, an equal and opposite torque is transmitted to the spacecraft structure. Reaction wheels are operated bi-directionally to provide torques about a single axis.

Momentum wheels are designed to rotate in one direction and depend on the momentum of the spinning wheel, $\mathbf{H} = I\omega$, where I is the moment of inertia about the rotation axis and ω is the angular velocity. Momentum wheels can be fixed to the structure or gimbaled so that the axis can be moved with respect to the spacecraft. When gimbaled, it is commonly referred to as a Control Moment Gyro. Because the spin axis is gimbaled, rotation of the satellite is caused by a torque along the output axis due to a change in the direction of the angular momentum of the spinning wheel. The magnitude of this torque depends upon the speed of the rotor and the rate about the gimbal input axis.

3.2 New technology description

Using the specifications of the current fabricated design (See Figure 14), a control system actuator was considered for a concept nanosatellite. Assuming the nanosatellite to be spherical in shape, 1 kg mass, 10 cm diameter, with an articulation requirement of 180° in 10 minutes, a calculation was performed to determine the required angular momentum operating in a reaction wheel mode.

Starting from an inertially fixed position, angular momentum is accumulated in the wheel during the first half of the slew. During the second half of the slew, a reverse torque is applied to stop the satellite by the end of the 180 degrees. The required angular momentum is determined from the max angular rate, which the wheel will achieve at the midpoint of the slew. The peak angular momentum required to accomplish this maneuver was determined to be 102 dyne-cm-sec. The device currently being developed will be tested at 50,000 rpm, achieving an angular momentum of 150 dyne-cm-sec. This is well within the current capabilities of the existing design.

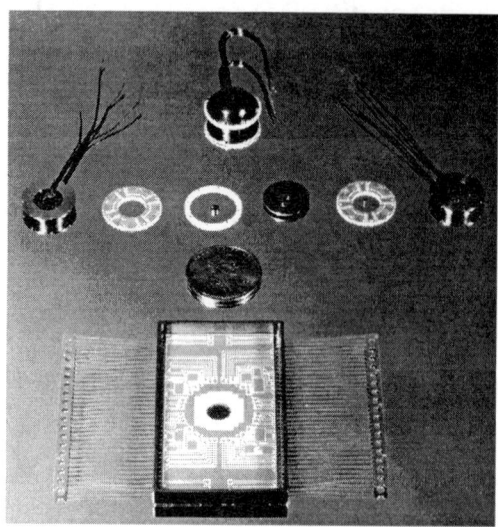

Figure 14. Fabricated Parts For Wafer Spinning-Mass Gyro

The other important design parameter to consider is the required torque. In a reaction wheel mode of operation for a max acceleration in the above slew maneuver, the required torque was determined to be 0.35 dyne-cm from,

$$\frac{\theta}{2} = \frac{1}{2}\frac{T}{I}\left(\frac{t}{2}\right)^2$$

where   T = torque
         I = satellite inertia
         θ = slew angle

For high torque applications, the device could be designed to operate as a single gimbal control-moment-gyro (CMG). The magnitude of the output torque of a CMG is the cross product of the gimbal input rate and the angular momentum of the rotor, $\vec{T}_{out} = \vec{w} \times \vec{H}$. To achieve the slew profile, the above required torque would be applied for 300 sec, for the expected angular momentum of 150 dyne-cm-sec. The gimballed wheel would move through an angle of approximately 40 degrees at a rate of 2.33E-3 rad/sec, before reversing direction.

3.3 Future Development for MEMS Actuators

Although the current configuration adequately satisfies the design requirements, it can be seen that this is only a point design. Many other mission scenarios can be envisioned and would require comparable component design.

The silicon fabricated raceway bearing is a concept that would be considered in future designs. In support of the concept, silicon steel tribological compatibility was evaluated by machining macro-sized R6 bearing raceways from thick silicon stock. The parts were mated to appropriately sized steel balls and the assembly lubricated with Nye 177 bearing oil. The assembled bearing was run in a fixture initially for 190 hours at 700 rpm. No observable wear or measurable loss of preload resulted from this test. The assembly was later run for a few days at 1,500 rpm and then for 160 hours at 6,000 rpm, again with excellent results. This very promising investigation showed that silicon mated to steel was potentially a good bearing combination.

Looking further out to the future, Figure 15 shows a concept for a wafer control-moment-gyro concept with hemispherical gas bearing. Here the gas bearing provides the radial and axial stiffness from the boundary layer dynamic forces. This has not been built and remains at the concept stage.

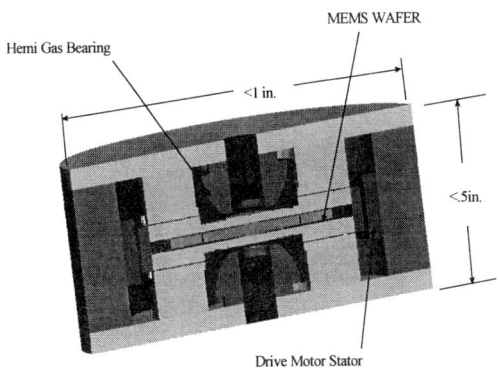

Figure 15. Wafer Momentum Wheel Concept with Hemispherical Gas Bearing

4.0 Summary of Sensors and Actuators for Micro/Nano Satellites

The concept of a miniature stellar-inertial sensor for a Micro/Nano Satellite has been developed in response to the needs of NASA's New Millennium Program. This program, through the Space Technology 5 (ST-5) mission, is focussed on small satellite constellations. Future NASA science programs will require the use of multiple, low-cost, autonomous satellites to collect the desired spatially distributed science data. Inertial attitude data and data regarding the relative position between the multiple satellites will be needed to meet the system goals. MEMS attitude control actuators will be required to complement the MEMS sensing devices.

This paper has described some of the breakthrough technologies that will enable the Micro/Nano Satellite missions to meet their goals. In particular, a new optical attitude sensor technology and a new inertial sensor technology have been developed that relate directly to NASA's NMP objectives of small, low cost, space sensors. The EBCCD star tracker can detect visual magnitude 8 stars and the MEMS gyros provide very low weight, low volume, low power inertial attitude memory. Also the MEMS Wafer Momentum Wheel has been developed as a Micro/Nano Satellite attitude control actuator. This device can function in any of the standard attitude control configurations (momentum wheel, reaction wheel or control moment gyro) providing an angular momentum of about 150 dyne-cm-s.

In addition to supporting the attitude measurement/control function, these small sensors can also support other Micro/Nano Satellite functions such as pointing/tracking, station-keeping, surveillance, communication, interception, inspection, re-supply, and redundancy. The implementation of all of these functions depends on the ability to provide attitude measurement and control. The ability to make these sensors small and accurate with low power will enable the Micro/Nano Satellite missions to meet their mission objectives within their expected stringent design limits on satellite weight, volume, and power.

References

1. "Inertial MEMS Development for Space," J. Connelly, A. Kourepenis, D. Larsen, T. Marinis, Draper Laboratory Report, CSDL-P-3726, April, 1999.

2. "Gallium Arsenide Electron Bombarded CCD Technology," John J. Boyle and William L. Robbins, Draper Laboratory et al., presented at SPIE Annual Conference in San Diego, July 1997.

AAS 00-066

DIGITAL REACTION WHEEL ASSEMBLY RSI 01-5 FOR SMALL AND LOW-COST SPACECRAFT

Armin Landes and Stephen Böttcher-Arff[*]

The RSI 01-5 is a ball bearing momentum/reaction wheel system with digital electronics completely integrated in the wheel housing. It was developed and manufactured for Orbital Sciences Corporation, USA to be used in the ORBCOMM data communication system (a constellation of up to 36 satellites in low Earth orbit). All wheels are in orbit representing a cummulated lifetime of approx. 40 years.
An extended version was designed and manufactured for PROBA, an European Space Agency mission to be launched in 2000.

INTRODUCTION

For more than 20 years TELDIX has been very successful in the field of gyroscopic actuators like momentum and reaction wheels for satellites.
Our momentum wheels DRALLRAD® are utilised for the stabilisation of communication satellites. In 1990 TELDIX started the development of a experimental low cost reaction/momentum wheel with integrated Drive Electronics for the use in small satellites in cooperation with the Technical University of Berlin. In order to achieve the cost goal, commercial parts were used. The wheels of this type were delivered in 1993 for the satellites TUBSAT B and TUBSAT C, built by the Technical University of Berlin, MAROCSAT and KITSAT-3 and were furthermore selected and delivered for a small spacecraft to be built by DASA, Bremen named INSPECTOR.

The more stringent requirements of small commercially used satellites necessitate an adaptation of the experimental design to a high reliable, medium-life space-qualified Reaction Wheel.
The technical discussion whether to use a conventional analog electronics or to design a state-of-the-art digital electronics led to the decision for a fully digital system. The main arguments for the decision to use a digital system were:
- Only 1 Central Processing Unit, universally adaptable by software to any kind of wheel (large/small reaction/momentum wheel). Hardware development to be carried out only once. The central processing unit co-operates with any kind of DC-motor (typically 3-phase, 8 poles) by use of an appropriate universal commutation logic.

---

[*] TELDIX GmbH, Postfach 105608, D-69046 Heidelberg, Germany.

- Implementation of a Digital Controller permits versatile adaptation of control laws and extension of parameter ranges. Control parameters are software parameters. There is no need for hardware modification to adapt parameters to special wheel requirements (e.g. different moment of inertia).
- Simple Upgrading for New Features by loading new software version via (serial) customer interface.
- Parameters (typically time constants) are free of ageing and temperature drift effects.
- Very high noise immunity.

1. TECHNICAL DESCRIPTION MECHANICS

The RSI 01-5 is a ball bearing momentum/reaction wheel system with digital electronics completely integrated in the wheel housing. The complete wheel is protected by an airtight housing, which is sealed by bonding and filled with a special Helium/Nitrogen mix at ambient pressure. It is designed for a in-orbit lifetime (at 775 km) of 5 years. The total mass is less than 0,6 kg; the dimensions are $\varnothing$ 95 mm and 102 mm height.

The wheel RSI 01-5/15 provides an angular momentum storage capacity of 0,04 Nms and a reaction torque of 10 mNm at start-up and 5 mNm at 1500 rpm. The serial interface (RS-485) handles commands/data with a rate of 9600 baud. The bus capability allows to address up to four small wheels connected to one serial data bus.

An extended version RSI 01-5/28 with a momentum storage capacity of 0,12 Nms and a reaction torque of 10 mNm at start-up and 5 mNm at 2800 rpm is available.

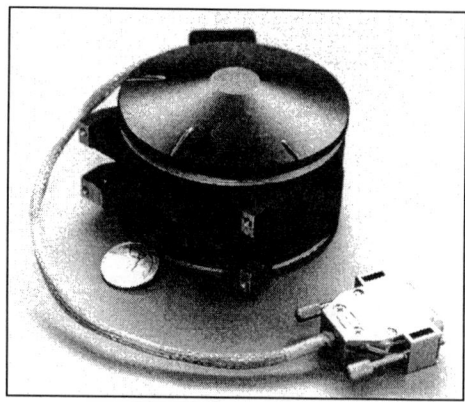

Figure 1: TELDIX Digital Reaction Wheel Assembly RSI 01-5/15

Figure 2 shows a cross-sectional view of the Reaction Wheel Assembly RSI 01-5/15.

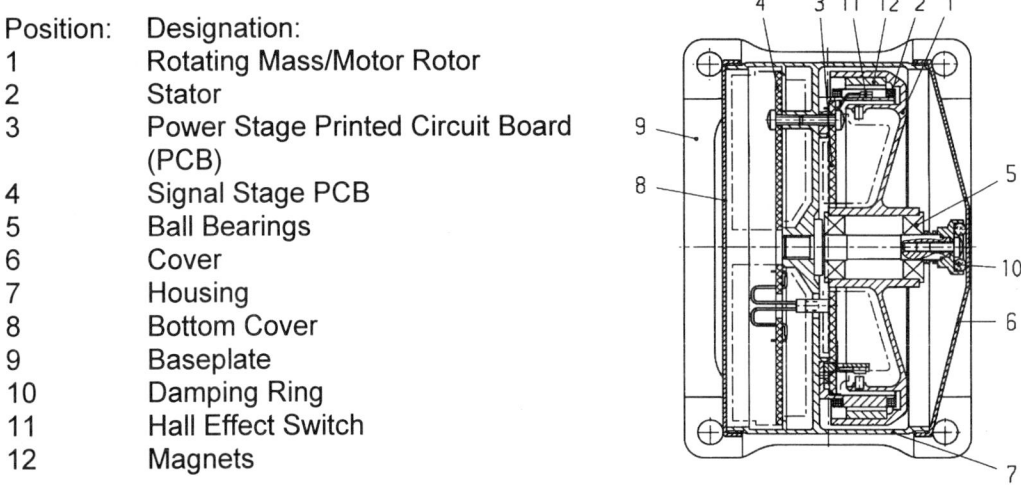

Position:	Designation:
1	Rotating Mass/Motor Rotor
2	Stator
3	Power Stage Printed Circuit Board (PCB)
4	Signal Stage PCB
5	Ball Bearings
6	Cover
7	Housing
8	Bottom Cover
9	Baseplate
10	Damping Ring
11	Hall Effect Switch
12	Magnets

Figure 2: Cross-sectional view of RSI 01-5/15

Main parts and subassemblies as described above are shown in Figure 3.

The Rotating Mass with Motor Rotor (1)
The rotating mass and the motor rotor are essentially one unit made of steel alloy. The simple turned part features a substantial hub construction. The rotating mass is screwed to the shaft of the DC Motor. It is suspended by two solidly preloaded ball bearings (5). They are located on the axis which again is elastically fixed into the Cover (6) by means of a damping ring (10). In this way, radial oscillations of the rotating mass are suppressed. The rotating mass-motor-unit will be balanced as an assembly. Correction of the static and dynamic unbalance is carried out by means of adding balancing screws to the rim in two separate planes.

The Housing
The complete wheel is protected by an airtight housing, which is hermetically sealed and filled with a special Helium/Nitrogen Mix at ambient pressure. This procedure/material is qualified and approved in various military applications (mainly gyroscopes). The housing consists of three major parts, the housing (7) with an upper cover (6) and a bottom cover (8). The mechanical interface is defined by the baseplate (9), which is mounted to four feet of the housing resulting in a high vibration load capability.

The Motor
The motor of the wheel can be subdivided into three subassemblies:
- Motor Rotor/Rotating Mass
- Stator
- Power Stage PCB including Commutation and Speed Sensor System

The motor of the RSI 01-5/15 is a brushless 3 phase DC motor (8 pole pairs). The motor consists of an ironless stator or coil-carrier and a permanent magnet rotor. Commutation is performed electronically, using Hall sensors responding to the permanent magnets of the rotor.

The motor rotor contains the permanent magnets and an integrated feedback ring. The stator coils form the three motor phases which are electrically 120° apart. The electronic commutation triggered by Hall sensors avoids the possibility of life restrictions caused by wear since physical contact is avoided completely. Twelve position sensors are mounted to the motor coils which generate 192 pulses/revolution for high resolution and extremly precise speed/torque loop characteristics also for very low speeds (< 10 rpm).

The Bearings
The ball bearings used are preloaded and in accordance with the dimensional and running accuracy requirements of the Quality Standard ABEC 9P. The retainer is made from Meldin 9000 ROD and had been developed by TELDIX especially for this type of bearing and was used in TELDIX gyros with speeds of up to 25000 rpm. In order to improve the lubrication and running characteristics the balls are TiC coated. Life calculations show that the bearing life L_0 (non failure lifetime) is appr. 90 years for reaching the max. number of revolutions.

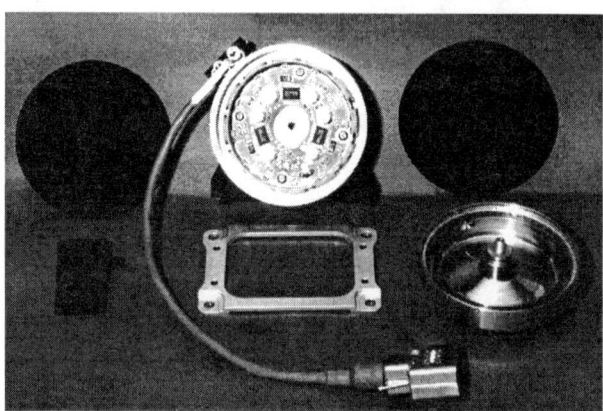

Figure 3: Main Subassemblies of RSI 01-5/15

TECHNICAL DESCRIPTION ELECTRONICS

General Design

The electronics for the momentum wheel consist of two sections which are physically located on two separate Printed Circuit Boards (PCB). The first PCB accommodates the Signal Electronics (Signal PCB), the second PCB the Drive Electronics (Power PCB). Both PCBs are designed and manufactured in accordance with MIL-STD-275. Conformal coating is used to prevent microcircuits from moisture.

The Signal Electronics is dominated by a high speed fix point Digital Signal Processor (DSP) with 16-bit architecture which operates the main functions:
- Command and Data I/O via serial interface (RS-485)
- Speed/Torque calculation for loop-modes and telemetry
- PWM management (for motor current regulation)
- Telemetry-Data acquisition management
- Monitoring and limitation of motor current and speed
- Watchdog management
- Correction of non-linear behaviour of electromechanical and thermal influences to the system to guarantee high precision performance

The commutation electronics for operating the brushless 3-phase 8-pole DC motor is programmed into a Field Programmable Gate Array (FPGA) as well as digital subunits to support the DSP. Motor management is also integrated in the FPGA and is realised through Pulse Width Modulation (PWM) with a fixed frequency of 40 kHz. The digital input for the commutation section is generated by twelve Hall sensors which are located on the Power PCB; one additional Hall sensor generates an additional index pulse (1 pulse/rev) for synchronisation. This signal is directly available at the interface connector. This magnetic sensor system leads to a resolution of 192 pulses/rev.

The embedded Software is re-programmable (via serial interface) and stored in a EEPROM to permit an update/adaptation of software without any hardware modification. After Power-on the application software is loaded automatically from the external EEPROM into the DSP on-chip program memory. This allows high speed performance by use of a standard EEPROM as an external boot memory.

Using the serial interface, the wheel is capable to operate in a RS-485 bus system. Wheel address is determined by two hardware pins available on the wheel interface connector. So up to four wheels can be operated on one bus system.

The Drive Electronics interfaces the commutation signals from the DSP and is supporting digital logic to the motor windings. It mainly consists of 6 power FETs in 3 H-Bridge configuration with its integrated drivers and the above described

Hall Sensors for speed control. The driver supply voltage is decoupled from the power bus by a low dropout linear voltage regulator.

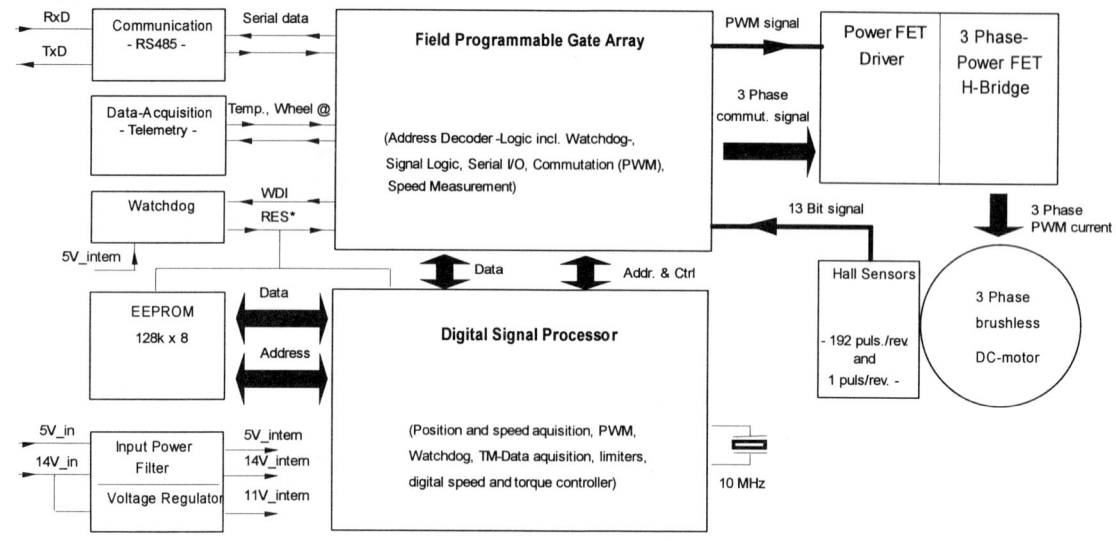

Figure 4: RSI 01-5/15 Functional Block Diagram

Detailed Design

The Microprocessor

The microprocessor used is a single chip high speed and high reliable 16-bit fixed point Digital Signal Processor with on-chip memory in CMOS technology with low power consumption. This processor is optimised for digital signal processing and other high speed numeric processing applications. Other features are integrated I/O peripherals e.g. serial ports and timer.

The FPGA

The commutation, most of the signal logic and the watchdog generator are implemented in a Field Programmable Gate Array (FPGA). This device is user programmable and non-volatile and offers gate array flexibility, high performance reliability under minimum need of space. It is fabricated in 1.0 micron CMOS technology for low power applications and replaces up to 50 TTL packages.

The EEPROM

The operational software is programmed into an electrically erasable and programmable memory (EEPROM). It is organised as 131072-word x 8 bit and realises high speed, low power consumption and a high level of reliability, employing advanced MNOS memory technology and CMOS process and circuitry technology. During operation the software is write protected.

The Watchdog
For monitoring the supply voltage and the correct microprocessor activity respectively a microprocessor supervisor circuit is used. The watchdog service routine is triggered by the DSP internal 1 kHz Timer-Interrupt and resets the watchdog regularly. If the watchdog times out a „System Reset" is generated. In order to recover the system in case of radiation single event upsets (SEU) autonomously the software implemented is verifying the program memory integrity online („Checksum Calculation"). In case of a bad checksum, a „System-Reset" will be generated and the application software is loaded from the EEPROM into DSP program memory once again.

Communication
ACS communication is possible via a full duplex serial differential RS-485 interface in asynchronous mode. Commanding and message response operates at a rate of 9600 Baud, no parity, eight data bit and one stop bit (9600,n,8,1). The used interface circuit is full compatible with EIA Standard RS-485.
To operate the wheels in a RS-485 bus system wheel address (@00 to @03) is determined by two hardware pins available on the interface connector.

The Internal Power Distributor
The wheel is operating with an non-regulated power source ranging from 11 to 17 VDC and a regulated power source of 5V ± 0.25 VDC with an transient in-rush ripple of 50 mV. For EMI purposes both input lines are filtered. A low dropout linear voltage regulator generates the decoupled 11 V supply voltage for the motor driver section.

The Data Acquisition (Telemetry)
The data acquisition block prepares all peripheral data for telemetry response. Thus the wheel address, wheel status, motor current, readback of set speed and torque, wheel speed and wheel inner temperature are available. They are reported via the Reaction Wheel software interface.

The Motor Driver Section
Wheel motor is driven by 6 Power FET in a bipolar H-Bridge configuration with 3 integrated drivers. The Power FETs are operated from a 11 VDC supply decoupled from the motor supply. Commutation signals are provided by 3 of the 12 Hall Sensors, electrically 120° apart. During active braking the power is delivered back into the power capacitor respectively into the satellite bus system.

The Hall Sensor System
The Hall Sensor System is part of the stator and consists of 12 position sensors plus 1 additional sensor for synchronising. The motor magnets located in the motor rotor generate 24 leading and trailing edges, electrically 15° apart. Choosing 8 pairs of motor poles thus 192 pulses per revolutions are generated. Motor position is identified using the additional sensor as a reference signal („zero sensor") for synchronising.

Operation Philosophy

The wheel can be switched either to speed or to torque mode by an 8-bit command via the software interface. In speed mode the resolution of 192 pulses/rev guarantees a high precision speed control down to 4 rpm with an accuracy of better than 1 rpm for the whole speed range of ± 1500 rpm. In torque mode the DSP calculates the set speed in dependence of the moment of inertia of the rotating mass, which is an individual programmed parameter, and the desired torque value. This feature avoids the influence of loss torque which is always present in traditional analogue motor current setting electronics.

To achieve very high precision measurement between two Hall sensor signals and to overcome non-avoidable tolerances during rotating mass assembly the wheel is self-calibrating. These correction data are gained during wheel final assembly and are permanently stored in the EEPROM.

The wheel provides three operational modes:
- Stand-by
- Speed Control
- Torque Control

Control Loop Design

Global Structure
Figure 5 presents the global structure of the wheel control loop including the major I/O signals.

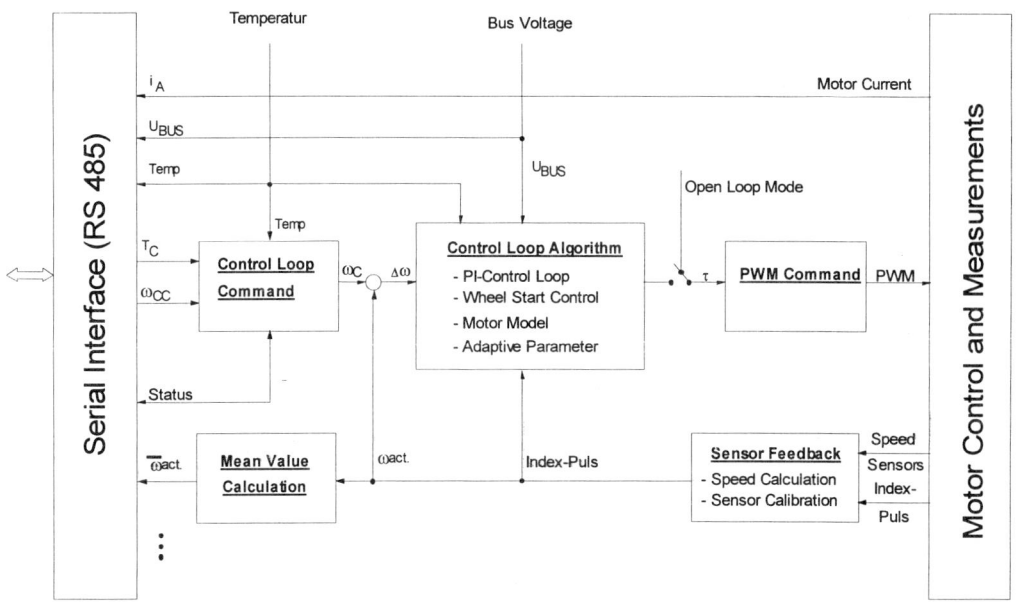

Figure 5:   Control Loop Structure including major I/O Signals

A Torque Command (T_c) respectively Speed Command (ω_{cc}) via the serial interface triggers the calculations in the Control Loop Command block. The output in both cases is the speed command ω_c. This values is compared continuously with the actual speed value ω_{act} which is the result of the sensor feedback derived from the 12 Speed Sensors. Using Hall Effect switches the passed angle in a definite time period is measured. Noise minimisation is achieved by sensor calibration during manufacturing process. An additional sensor generates an Index Pulse (1 pulse/rev.) for synchronising. The difference of commanded and actual value ($\Delta\omega$) is the input for the Control Loop Algorithm block. This stage includes the PI-Control Loop itself, the Wheel Start Control, the Motor Model and the Adaptive Parameter Set. A multiple polynom describes the motor model and is programmed in the wheel operation software. Wheel Start Control as well as operation through zero speed is realised as a pulse mode.

The PWM Command block generates the motor input current calculated in the control loop algorithm stage for each operational range. For measurement purposes the control loop can be switched to an Open Loop Mode (during manufacturing process only).

The mean value of the actual speed (ω_{act}) together with other output signals e.g. $i_{Mot.}$, U_{Bus} and Wheel Status is available via the serial interface.

Control Loop simplified Model
Figure 6 shows a simplified model of the Reaction Wheel Control Loop. This model is valid for the linear operation mode only. The existing limits are not schematised.

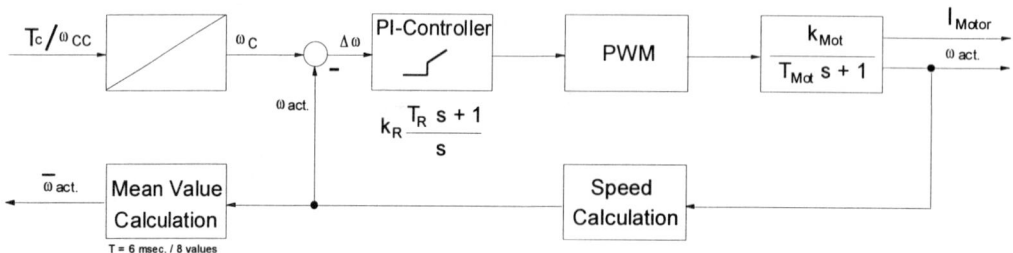

Figure 6: Simplified model of the Reaction Wheel Control Loop

The abbreviations T_R, k_R respectively T_{Mot} and k_{Mot} are the constants of the control loop and are variable depending on the wheel actual operation range.

Closed Loop System Performance Data (extracts)

- Constant Torque Spinup for Torque Command = +1 µNm

Reported is current speed [rpm] versus time [sec.] in the critical speed range through zero speed in acceleration mode. Wheel Reaction Torque Output is very linear.

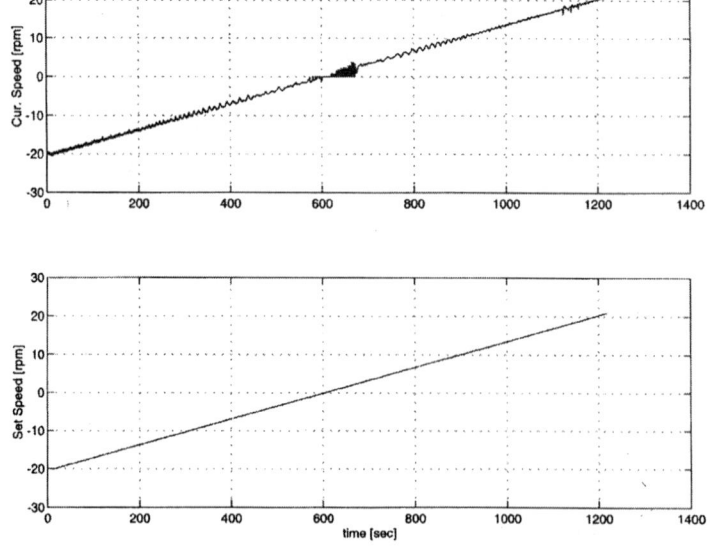

- Constant Torque Spinup for Torque Command = +5 mNm

Reported is current speed [rpm] versus time [sec.] for constant torque in acceleration mode. Risetime is less than 9 sec (starting from zero to max. speed) with an overshoot
< 10 % for t < 3 sec

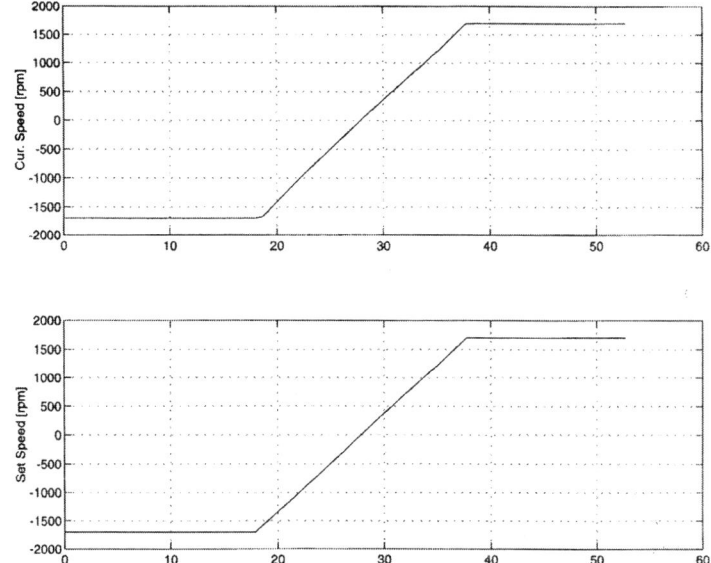

- Stair Stepped RPM Output

Reported is current speed [rpm] versus time [sec.] for constant speed.
Speed accuracy is better than 1 rpm over complete operation range.

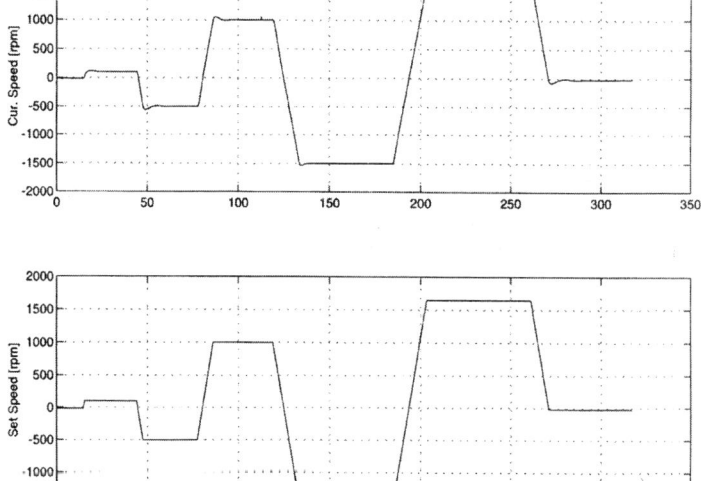

PERFORMANCE SUMMARY RSI 01-5/15 *(RSI 01-5/28)*

Angular Momentum at Nominal Speed	0,04 Nms *(0,12 Nms)*
Operational Speed Range	± 1500 rpm *(± 2800 rpm)*
Software Speed Limiter	< 1700 rpm *(3800 rpm)*
Reaction Torque (1500 rpm)	5 mNm
Torque Mode Accuracy	1 µNm averaged over 30 sec.
Speed Mode Accuracy	< 1 rpm (operation range ± 20 to ± 1500 rpm)
Operational Modes	Standby • Speed-Loop • Torque-Loop

Mechanical:
- Dimensions: Diameter x Height — 95 mm x 102 mm
- Mass — < 0,6 kg *(0,7 kg)*
- Static Imbalance — < 0,050 gcm
- Dynamic Imbalance — < 0,025 gcm^2
- Alignment Spin Axis to Mounting Plane — < ± 0.4°

Electrical:
- Power Consumption:
- Steady State at nom. Speed — < 1,5 W *(< 1,8 W)*
- Max. Torque at nom. Speed — < 3 W *(< 3,8 W)*
- Supply Voltage — 14V ± 3V, 5V ± 0.25V *(20V - 0,2V, 5V ± 0.25V)*
- Input Current (14&20V / 5V-line) — < 0.30 A / < 0.15 A
- Serial I/O — RS-485, full duplex (9600, n, 8, 1)
- Connector — 15 pin high density (GFSC type)
- Telemetry Data — Speed • Torque • Motor current • Inner Temperature

Environmental Conditions:
- Operating Temperature — - 20 °C ... + 60 °C
- Non-operating Temperature — - 35 °C ... + 70 °C
- Random Vibration — 7 grms (max. flight level)
- Life Time — 5 years (in-orbit) 2 years (storage)

AAS 00-067

AN ATTITUDE CONTROL SYSTEM FOR AN ASAP5-LAUNCHED INTERPLANETARY SPACECRAFT

Tobin Anthony and Bhavesh Patel[*]

The Ariane 5 Structure for Auxiliary Payloads (ASAP5) platform provides a low-cost means of launching interplanetary missions using the Ariane 5 launch vehicle. However, use of the ASAP5 platform presents several key constraints on the spacecraft system design. In particular, the ASAP5 places a tight constraint on spacecraft mass which, coupled with the need to carry a large amount of fuel for interplanetary propulsive maneuvers, leaves very little flexibility for subsystem design. Orbital Sciences Corporation has conceived a spacecraft bus that accommodates a variety of interplanetary mission payloads such as probes, remote sensing instruments, and communication transponders. The attitude control system is comprised of heritage components and software to provide high performance with low mass, power, and cost. This single-string platform provides arcminute-level pointing accuracy, in either inertial or nadir-pointing orientations, using a star tracker, a fiber-optic gyro, and six coarse sun sensors. Furthermore, the attitude control system manages spacecraft attitude during trajectory correction maneuvers as well as dynamic events such as probe deployment using a monopropellant-based reaction control system. This paper quantitatively discusses the difficult tradeoff decisions made in the design of this control system. In addition, performance, and possible applications of this powerful new spacecraft bus are presented.

INTRODUCTION

Deep space missions to date have typically required relatively large launch vehicles to provide the necessary launch energy to provide direct insertion into interplanetary trajectories. The use of these launch vehicles results in high overall mission costs. In an effort to enable lower cost deep space missions, alternative mission design options have been explored. Using the ASAP5 option on the Ariane 5, the utilization of multiple lunar swingby trajectories provides a low cost means to launch deep space missions.

The Ariane 5 can launch a spacecraft in the ASAP5 configuration of up to 220 kg to Geostationary Transfer Orbit (GTO) as a secondary payload on a commercial Ariane 5 launch; the payload volume is shown in Figure 1. The available volume is in the form of a 80° segment of a toroidal volume surrounding the primary payloads. This uniquely shaped spacecraft is inserted and deployed in GTO by the Ariane 5. The spacecraft performs a burn at perigee to raise the orbit apogee to a nearly lunar altitude. After a wait of a few tens of days, the spacecraft performs an unpowered swingby of the moon. A couple days later, it performs a powered swingby of the Earth and injects into an interplanetary transfer trajectory to a deep space location such as Mars, Venus, or one of the Earth-Sun Lagrangian points. Multiple lunar gravity assists may be required to attain the correct alignment of the spacecraft velocity vector with the required Earth escape asymptote.

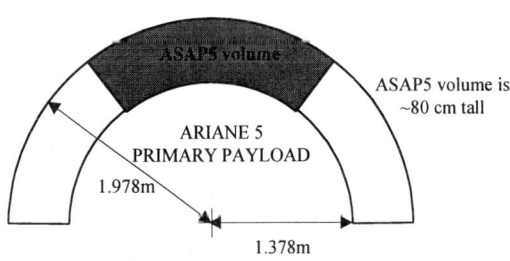

Figure 1 – The ASAP5 volume encompasses roughly 80° of the volume surrounding an Ariane 5 primary payload

The utilization of an ASAP5 launch and lunar gravity assists does indeed result in significant cost savings on the launch vehicle

---

[*] Orbital Sciences Corporation, 21700 Atlantic Blvd., Dulles, Virginia 20166.

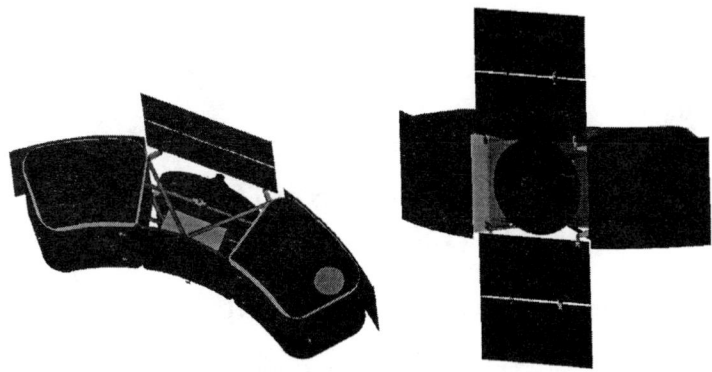

Figure 2 – Different Views of the DeepStar Spacecraft

albeit, at the expense of a more complicated spacecraft design. The ASAP5 launch vehicle places stringent constraints on center of mass location, spacecraft volume, and structural stiffness. Extended time in both the GTO and trans-lunar orbit requires spacecraft components that are significantly radiation-hardened. Further complications in the spacecraft design result from the need to provide a high propulsive delta-V capability simultaneously with a capability to aerobrake into planetary atmospheres.

Leveraging its expertise and experience in innovative design and low cost production of small spacecraft, Orbital has developed an ASAP5-compatible deep space bus that can support a wide variety of deep space missions. Orbital has developed such a deep space bus concept resulting from participation in a study sponsored by JPL to review options for sending an ASAP5-based microspacecraft to Mars [1]. Views of this bus are shown in Figure 2. The spacecraft design is based on Orbital's flight-proven MicroStar bus which epitomizes high efficiency in terms of mass, power, and volume. Centered around a MicroStar avionics suite, the spacecraft provides a high performance ACS capable of arc-minute-level pointing accuracy. In addition, a dual mode propulsion system provides a delta-V capability up to 3000 m/s enabling a wide variety of deep space applications. A standard X-band communications subsystem provides compatibility with the Deep Space Network (DSN). Tentatively called the DeepStar bus, this spacecraft bus can support a number of missions to a number of different targets such as Earth's Moon, Mars, Venus, and the asteroids. Among other missions, the DeepStar bus can support orbiter, probe carrier, flyby and rendezvous missions with either science, communications, or technology demonstration payloads. This wide-ranging capability is facilitated by a highly capable ACS that constitutes the focus of this paper.

ACS OVERVIEW

The DeepStar attitude control system (ACS) is designed to provide a low mass, high-performance means of controlling the payload pointing. For this reason, a high premium was placed on the selection of hardware components that require the least mass, power, and special accommodations.

The variety of missions for which the DeepStar spacecraft bus can be used requires that the ACS provide an accurate means of determining attitude regardless of spacecraft position or orientation. For this reason, the ACS is built around a single star tracker. In recent years, the "quaternion-out" tracker has become available in which the sensor processes the star locations and executes the determination algorithms using self-contained hardware and software. While these trackers are more massive and require more power than earlier versions, that required the ACS to maintain the star catalog and determination algorithms, they provide a "plug-and-play" ability allowing ease of integration with the ACS. A unit such as the Ball CT-633 provides sub-arc-minute accuracy in a low-mass package. Payloads that require higher accuracy could incorporate units such as the Sodern SED-16 or Raytheon HD 1003 albeit at the cost of slightly extra mass. At the time of this writing, the Ball unit has undergone a full qualification process while the Sodern and Raytheon units are in the process of qualification.

The star tracker provides a quaternion output which can be used to develop pointing errors; however, in mission phases where the tracker is not available, attitude position and rate information are developed using an inertial reference unit (IRU). Orbital has extensive experience with the Litton LN-200S on the MicroStar platform upon which the Orbcomm spacecraft is based. This unit provides a low-mass, low-cost rate feedback solution. The short-term noise present in the LN-200S output is significantly higher than common flight-qualified gyros; however, for reasons discussed later, this gyro provides adequate fidelity for use in the DeepStar ACS.

The classic star tracker/gyro combination provides proper attitude determination for a variety of missions. However, the type of mission specified by the DeepStar payload could enable substitution of the star tracker with a sun sensor/horizon sensor combination. A digital sun sensor, such as the Adcole or Barnes units flown on a variety of Orbital LEO missions, could be complemented by a horizon sensor to provide a proper attitude basis. As was done for the Lunar Prospector mission[2], the horizon sensor could be based on an Earth-based IR sensor adapted to sense the limb of whichever planet the spacecraft is orbiting. However, the startup costs of developing and testing such a horizon sensor might outweigh the cost of using an existing star tracker.

The set of six Coarse Sun Sensors (CSS) positioned about the spacecraft provide a coarse sun reference. The information gleaned from these sensors is used to develop attitude errors in the Safehold mode, as discussed later in this paper.

In order to provide precision pointing for telecommunications or science payloads, the DeepStar ACS uses a set of three reaction wheels provided by Teldix. These low-cost, low-volume wheels are derived from the reaction wheels used on the ORBCOMM constellation. The electronics and motor of these wheels have over 30 spacecraft-years of heritage in flight.

Some DeepStar payloads may not need the precise pointing of the reaction wheels and may opt to remove them in favor of the coarser pointing provided by the reaction control system (RCS) thrusters. These six thrusters are positioned about the spacecraft and can provide attitude control using a deadband pulsing scheme. The tradeoff between reaction wheels and RCS must include a lifetime mass calculation of required propellant.

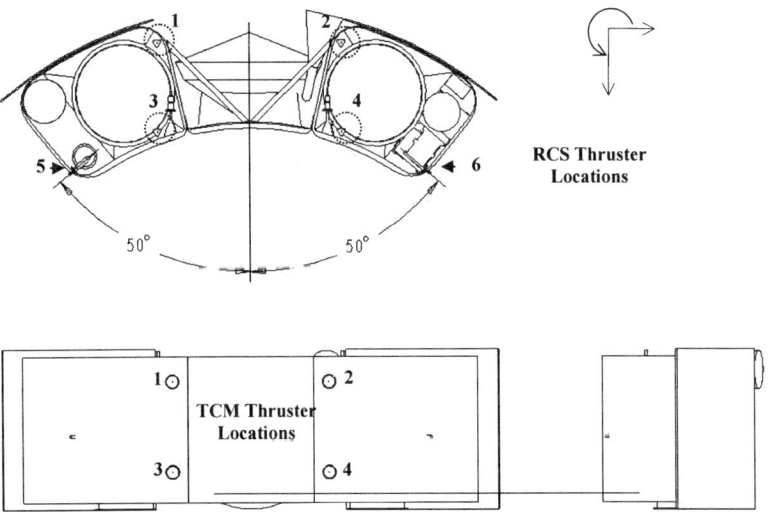

Figure 3 – DeepStar Reaction Control System and Trajectory Control Maneuver Thrusters

For nominal payloads that require reaction wheels, the primary function of the RCS is to manage the momentum of the DeepStar reaction wheels. Using the monopropellant part of the DeepStar propulsion system, these thrusters provide an efficient means of dumping the reaction wheel momentum accumulated through external torques.

In addition, the spacecraft will employ the RCS during such periods when disturbance torques exceed the capabilities of the reaction wheels such as during aerobraking and probe release events. As seen in Figure 3, the six RCS thrusters are positioned and operated so as to provide three-axis control authority. However, operation of the RCS will induce spacecraft translation as well as rotation. On a typical deep space mission, a set of 12 RCS thrusters would be used that provide pure torques without translation. However, this extra set of six RCS thrusters was not included on the DeepStar bus because of mass constraints. As a result, the maneuver errors accumulated through the RCS thruster firings will need to be tracked during mission operations.

The positioning of the 5-6 thruster pair was designed so that jet exhaust would not impact directly into the relative wind during aerobraking operations. During the Mars Global Surveyor aerobraking operations, it was determined that the reaction jet firings into the atmosphere occasionally provided the opposite dynamic effect [3]. In the free molecular flow regime, exhaust from the RCS firings actually evacuated the sparse atmosphere immediately forward of the thruster. The rapid change in local atmospheric density in front of these particular thrusters attenuated the disturbance torques on the spacecraft and at times, altered the effective polarity of the thruster firing. Hence, this thruster pair is positioned so as to prevent the RCS thruster exhaust from directly interacting with the atmosphere. Although not necessarily a part of the Deep Star ACS, the trajectory control maneuver (TCM) thrusters provide three-axis control during TCM operations.

ACS OPERATIONAL MODES

An operational mode diagram for the DeepStar bus is shown in Figure 4. The DeepStar operational modes are largely derived from heritage Microstar software. The initial power-up sequence is used to initialize spacecraft operations from the dormant Launch mode. One of the ASAP5 requirements is that the spacecraft launch in an unpowered state. The spacecraft is initialized through the use of a microswitch attached to the launch vehicle that powers up the DeepStar bus upon separation. This Launch mode also doubles as a test mode used for ground testing of the integrated spacecraft.

Safehold is used to provide a thermally-safe and power-positive means of pointing the spacecraft. Furthermore, this mode is used to stabilize the spacecraft following separation from the Ariane-5 third stage. Safehold uses a minimal amount of hardware components ensuring a highly reliable spacecraft. Using software developed for Orbital's GALEX and SORCE programs, a Safehold mode is used that requires only reaction wheels and the CSS. In the case of payloads that do not require wheels, the RCS is the primary Safehold actuator.

The Sun Point mode is the primary mode from which all other modes are entered. In this mode, the spacecraft is oriented in a three-axis inertially fixed attitude with the x axis towards the Sun to provide maximum power. The spacecraft resides in this mode during operations near Earth as well as during deep space cruise to the desired destination. Normally, such an ACS mode would use a digital sun sensor to develop a Sun-pointing error; however, mass constraints preclude adding extra hardware to the system. Using onboard ephemeris that calculates the Sun's position in a heliocentric coordinate system, the Sun pointing information is provided using the star tracker.

On many Orbital spacecraft, the star tracker and gyro are used in tandem to provide high precision pointing. However, for certain deep-space applications, the DeepStar bus may not provide sufficient power to operate the gyro and star tracker simultaneously. In this case, the Sun Point mode is operated solely using the star tracker for both position and rate feedback. This is accomplished by filtering the star tracker output using a gyroless rate estimation scheme [4]. During periods when the star tracker is unavailable, such as

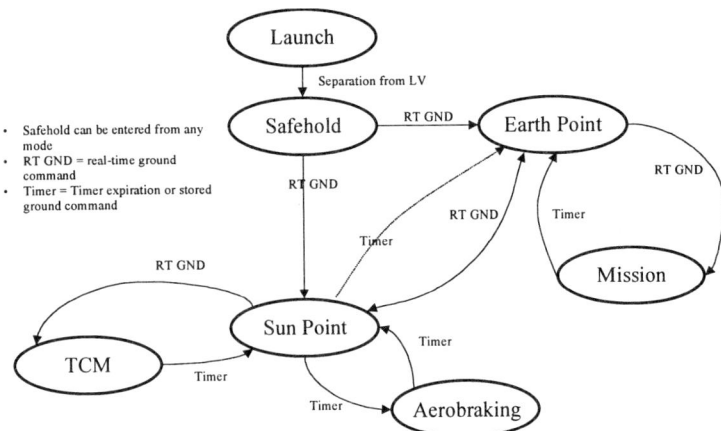

Figure 4 – DeepStar ACS Operational Modes

Aerobraking and TCM, the tracker is turned off to conserve power while the gyro provides rate and position feedback.

Periodically, the spacecraft will need to point to the Earth for the purpose of communication. The Earth Point mode, similar to the Sun Point mode, provides that capability. While antenna beamwidth dictates the pointing requirements it is expected that pointing will be required within a 0.5-1° deadband for the spacecraft within Venus/Mars/L1/L2 distances. This can be easily accomplished with the reaction wheels or even the RCS. Similar to the Sun Point mode, the spacecraft will use a heliocentric Earth ephemeris and a star tracker to provide a proper pointing reference.

As described earlier in the paper, the TCM mode is used to provide three-axis control during delta-V burns. Prior to entering this mode, the spacecraft receives an updated gyro bias from a standard six-state Kalman filter that operates in Sun Point mode. During TCM mode, the spacecraft operates the four 22 N thrusters in parallel providing close to 88 N thrust capability. Using gyro feedback, the DeepStar bus provides rate and position feedback to maintain a prescribed trajectory.

ACS PERFORMANCE

The primary attitude control mode for the majority of DeepStar payloads is the Sun Point mode. Pointing and knowledge budgets for the DeepStar ACS in Sun Point mode are provided in Table 1. These tables assume that the bus includes reaction wheels for actuation and uses the star tracker as a primary attitude reference.

The largest pointing error comes from the high-frequency jitter noise associated with rate filtering the star tracker attitude. This error would be greatly attenuated if rate feedback were provided using a gyro; this is an area of future work. The control error is a result of the accuracy of the Teldix reaction wheels that provide a high-bandwidth internal tachometer control loop for precise pointing. The wheels are also statically and dynamically balanced to such a degree that wheel-induced jitter is negligible. Furthermore, most environmental torques are small in deep space; while the spacecraft flies in Sun Point, the primary disturbance torque is solar pressure. As seen in Figure 2, the DeepStar geometry is symmetric and the magnitude of the disturbance torque is small and varies secularly, presenting a negligible source of pointing error. The primary effect of solar pressure on the spacecraft is in the accumulation of disturbance torque over a long period of time that requires periodic RCS burns to manage momentum.

The knowledge budget shows that the primary source of error is in the attitude estimation algorithm. The filter used in Sun Point propagates the spacecraft dynamics, using assumptions of the spacecraft

Table 1 – DeepStar ACS Pointing and Knowledge Budget

Attitude Knowledge Error Source (3σ)		Roll (arc-min)	Pitch (arc-min)	Yaw (arc-min)
Star Tracker Aberration		0.01	0.01	0.01
Star Tracker Thermal Misalignment		0.15	0.34	0.34
Kalman Filter Estimation		0.84	0.45	0.45
	RSS	0.85	0.56	0.56
Attitude Control Error Source (3σ)				
Jitter		0.55	0.19	0.19
Reaction Wheel Imbalance		0.01	0.01	0.01
Controller Lag		0.17	0.17	0.17
	RSS	0.58	0.25	0.25
Accuracy (RSS Know. & Ctrl.)		**1.03**	**0.62**	**0.62**

disturbance torques, to complement the rate estimates provided by the algorithm. Again, one trade that will be worked in the future is the possibility of refining the algorithm to include gyro inputs in order to reduce jitter. Following filter error, star tracker alignment and aberration errors resulting from spacecraft motion, present the next greatest sources of knowledge error.

Pointing and knowledge errors will differ in other operational modes. For example, if the spacecraft is designed to operate without reaction wheels, primary actuation will be handled with the RCS. The resulting pointing will be much coarser without the highly accurate Teldix wheels. The pointing deadband can therefore be determined a priori with a tradeoff between RCS fuel usage and deadband size.

For planetary operations, the spacecraft mission mode may require a nadir-pointing configuration. In this environment, the spacecraft will experience some degree of atmospheric torque disturbance depending on the primary planet. The effect on a reaction wheel-enabled DeepStar bus will be a slight pointing error depending on the atmospheric density and the bandwidth of the attitude controller; the more important effect will be the increase in RCS propellant usage to manage momentum. The attitude knowledge will differ slightly in a nadir-pointing environment in that the star tracker attitude knowledge degrades as the spacecraft slowly spins to maintain a non-inertially-fixed attitude.

APPLICATIONS

In this section, several applications of the DeepStar ACS are presented. These involve 1) attitude control during trajectory control maneuvers, 2) attitude control during aerobraking maneuvers, and 3) attitude control during probe release. As described above, all three of these scenarios use the gyro for rate and position feedback and propulsion for actuation.

Trajectory Correction Maneuver Control

The DeepStar spacecraft is designed to provide roughly 3000 m/sec of delta-V in the form of trajectory correction maneuvers. These TCMs can be for several minutes of duration depending on the application. During TCMs, the spacecraft will experience disturbances in the form of attitude determination error, thruster misalignment, thruster miscalibration, or offsets in the center of mass. Each of the four 22 N TCM thrusters shown in Figure 3 are oriented so as to provide a thrust component in all three axes; the thrusters are primarily oriented along the x-axis but have slight components along the y and z axes. By off-pulsing the thruster pairs shown in Figure 3, the DeepStar ACS exerts three-axis control over the spacecraft using the TCM during maneuvers. This eliminates the need to operate the RCS during TCMs; disturbance torques arising from thruster misalignment would be large enough to scale the design of the RCS or limit the duration of the TCM firings. The control system would be designed to attenuate the effects of slosh dynamics. The disadvantage of this system is that thruster canting decreases the effective thrust and specific impulse of the bipropellant TCM propulsion system thereby requiring more propellant to accomplish a given TCM. This thrust-canting arrangement, used on the OrbView-2 spacecraft [5], it involves a complicated LQR-based thrusting design based on the multivariable nature of the controller.

Aerobraking

One potential feature of the DeepStar bus is the ability to reduce the need for propellant through the use of aerobraking or aerocapture. In an aerobraking maneuver, the spacecraft kinetic energy is reduced by atmospheric drag by repeated passes through the planetary atmosphere. During aerocapture, the spacecraft makes a single pass low into the atmosphere to reduce the propellant required for a planetary insertion burn. During an aerobraking trajectory, the spacecraft is in a highly elliptical orbit with the periapsis deep into the planetary atmosphere. Aerodynamic drag works on the spacecraft throughout the passage through the atmosphere but with particular intensity near the periapsis. As a result of this drag, the spacecraft orbital angular momentum decreases with each periapsis passage resulting in a lower apoapsis height. After consequent periapsis passages, the spacecraft orbit is circularized to a degree where payload operations can commence.

The aerobraking environment presents unusually harsh conditions on the spacecraft - especially during periapsis passage. As the spacecraft passes through the planetary atmosphere, the same aerodynamic drag that alters the vehicle's orbit also acts to heat the spacecraft (proportional to the cube of the velocity) as well as exert static loads on the spacecraft (proportional to the square of the velocity). Of these two effects, the ACS engineer is obviously more concerned with countering the effects of the static loading of the spacecraft during aerobraking.

Most LEO satellites encounter the atmosphere in a regime known as free molecular flow where molecules transfer momentum to the spacecraft surface in the form of elastic collisions. During an aerobraking maneuver, the spacecraft largely resides in this regime. However, in the dense atmosphere near periapsis, the spacecraft enters a transition region where the air molecules interact with one another in a less predictable manner. Aerodynamic principles are applied to examine of the stability of spacecraft in this regime.

The DeepStar bus uses the RCS to control its attitude during periapsis maneuvers. However, RCS propellant usage greatly decreases when the spacecraft is aerodynamically stable. When the aerobraking body is aerodynamically stable, the aerodynamic center of pressure lies aft of the center of mass. In this manner, perturbations to the spacecraft attitude are counteracted with a proportional restoring torque; unlike terrestrial applications, the transition flow does not provide a damping torque. This aerodynamic control torque provides position control whereas the RCS provides rate damping using thrusters and the gyro for rate feedback.

The geometry, center of mass location, and volume of the DeepStar platform is dictated by the constraints of the ASAP5 payload envelope. Figure 5 shows this basic volume with the center of mass and center of pressure locations in two different configurations. In Figure 5-a, the inner radial surface is used as the aerobraking surface; however, the center of pressure is still foreword of the center of mass. This configuration is aerodynamically unstable and would require substantial amounts of RCS fuel to control the spacecraft attitude during periapsis passage. In Figure 5-b, the outer radial surface is used as the aerobraking surface. In this case, the aerodynamic center of pressure is also foreword of the spacecraft center of mass. As a result, both configurations shown in Figure 5 are aerodynamically unstable.

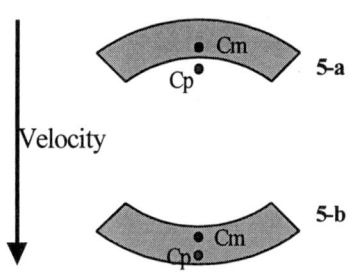

Figure 5 – Center of Mass vs. Center of Pressure Locations for Nominal ASAP5 Geometry

In order to passively stabilize the spacecraft, the configuration needs to be modified in order to move the center of pressure aft of the center of mass. Adding spacecraft cross-sectional area as far behind the center of mass as possible can do this. Several means of modifying the center of pressure location were evaluated including ballutes, parachutes, and deployable flaps. An inflatable ballute would greatly increase the stability of the spacecraft by increasing cross-sectional area. However, the aerobraking operations could require repeated passes over several weeks or months time. Maintaining the structural integrity of the ballute over that period of time would complicate the design of the system. A parachute would be similarly difficult to deploy and use after repeated periapsis passes.

For this reason, the DeepStar bus maintains two deployable flaps as shown in Figure 2. These panels nearly double the amount of surface area exposed to the atmosphere providing additional braking that requires less time to complete aerobraking operations. The inner radial surface is being used, as smaller and less complicated deployable panels are required to provide static margin for the aerobraking design. If the outer radial surface were used for aerobraking, larger panels and complicated release hinges would be required to provide adequate static margin.

The aerobraking control system is actuated entirely by the RCS. Attitude position and rate feedback are calculated using the gyros, as the star tracker is turned off during this phase of the mission. The passively stable design will maintain a stable attitude, the control thrusters are only required to correct attitude errors outside of a 15° deadband. The aerodynamic restoring torque provides position control but no rate damping. For this reason, the DeepStar ACS provides rate damping using the RCS. The thrusters will react only when rates exceed a 3 °/sec threshold. The rate and position deadbands can be reduced in order to provide tighter attitude control at the expense of additional RCS fuel.

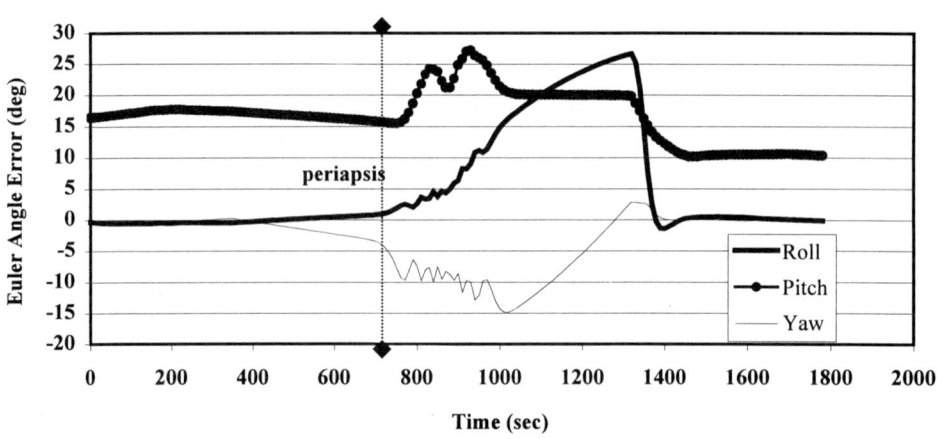

Figure 6 – Aerobraking Attitude History

Figure 6 shows the Euler angle time history of one of the first periapsis passes during an aerobraking sequence in the Martian atmosphere. Note that, for the sake of simulation, the spacecraft has a 15° offset in pitch at the onset of aerobraking which simulates a non-zero angle-of-attack condition. In this simulation, the spacecraft is in a highly elliptical orbit with a 115 km periapsis. At around 400 km altitude, the spacecraft attitude begins to drift as a result from the larger aerodynamic torque in the increasingly dense atmosphere. At about 240 km altitude, the spacecraft attitude begins to oscillate at a frequency corresponding to the magnitude of the aerodynamic restoring torque. While this oscillation about all three axes ensues throughout periapsis, note that the frequency of the oscillation decreases as the spacecraft reaches higher altitude. Eventually, the restoring torque dissipates with the decreasing atmospheric density to the point where the spacecraft is left with residual rates from the aerobraking oscillations. At a 400 km altitude, the DeepStar RCS acts to null rate and position errors to predetermined levels.

Probe Release

As part of the original MMSB concept developed at JPL, the bus is designed to carry small probes for entry into planetary atmospheres or to meet other scientific objectives. The spacecraft can release multiple small probes in a sequential fashion or a single larger probe. The release of these probes is a significant dynamic event for which the DeepStar ACS needs to recover to a stable attitude. The bus is designed to release the spacecraft, null the ensuing attitude rates, and act as a communications relay between the probes and Earth for a predetermined amount of time. Because of the thruster arrangement discussed above and highlighted in Figure 2, a multivariable controller similar to what is designed for TCM attitude control is used for RCS attitude control as well.

Prior to probe release, the gyro biases are corrected through use of filter star tracker information. After the event, the spacecraft encounters attitude errors resulting from the reaction torque and force of the probe separation. Using thrusters, the control system acts to null these errors to return the spacecraft to an inertially pointing attitude. Figure 7 shows the attitude history of the spacecraft recovering from a probe release event. In this scenario, the spacecraft encountered a 5°/sec angular rate along the x-axis resulting from release of the probe. The spacecraft recovers from the probe release event within 20 seconds and limit cycles in a wide deadband after this time. Trading off cycle time with RCS fuel usage can refine this limit cycle amplitude. Figure 8 shows the time history for the six RCS thrusters for this maneuver. Note that thrusters 5 and 6 do not operate during this event while the remaining thrusters work in pairs to recover the roll attitude.

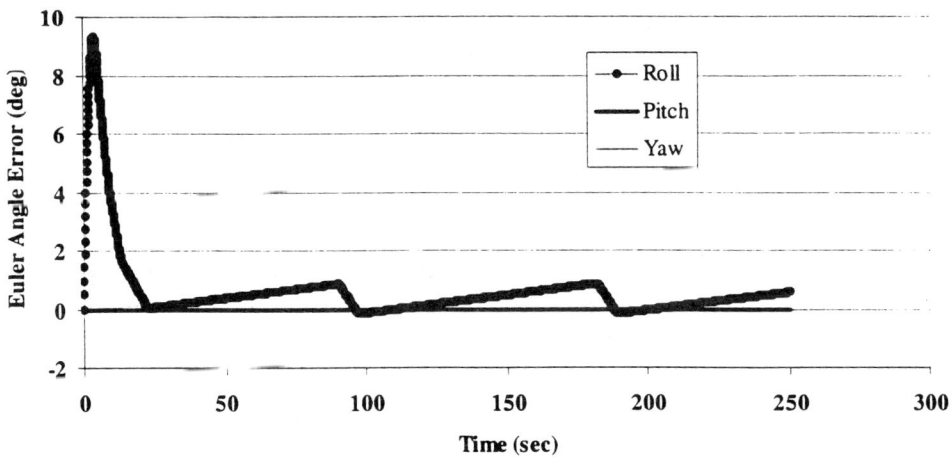

Figure 7 – Attitude History Following Probe Release

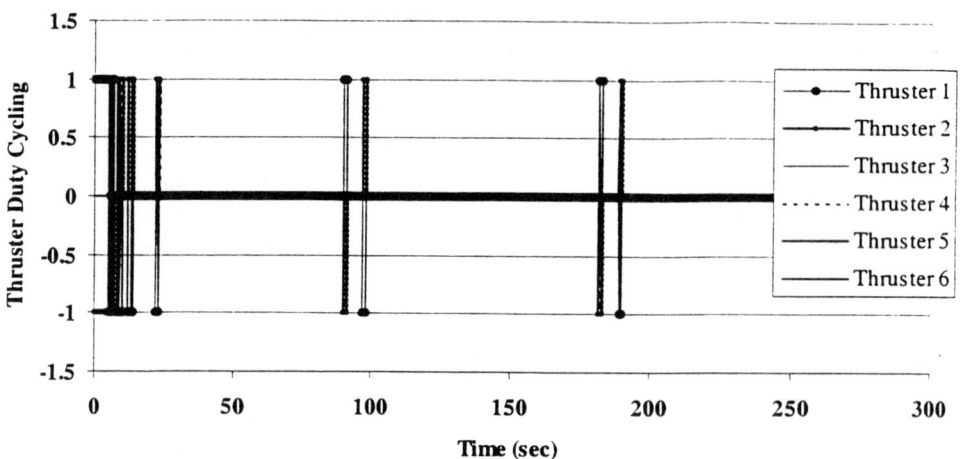

Figure 8 – Probe Release Thruster Cycling

FUTURE WORK

Orbital is currently undertaking a number of efforts to improve the performance of the DeepStar bus. These include the incorporation of advanced technology currently under development. The DeepStar's MicroStar-based open avionics architecture is ideally suited to accommodating advanced technology such as star trackers, gyros, communications transponders, and so forth. In addition, the common spacecraft design is being customized for specific applications to obtain further improvements in performance without compromising the commonality and heritage of the core spacecraft bus. Specifically, Orbital is investigating more accurate low-mass gyros yielding more precise attitude control during TCM and aerobraking events. There is also opportunity for low-power low-mass reaction wheels that provide greater precision and torque authority than the current baseline.

CONCLUSIONS

The DeepStar spacecraft bus offers low cost access to deep space by providing a synergistic combination of innovative Earth escape methods and flight proven low cost avionics. Leveraging the MicroStar's open architecture, the DeepStar bus provides high performance using a combination of flight-proven avionics with state-of-the-art components developed for deep space missions. The high performance provided by the spacecraft enables a wide variety of missions that range from science orbiters, probe carriers, to flyby platforms. The DeepStar bus provides the space community with a low cost bus that provides high performance at low risk.

ACKNOWLEDGEMENTS

The authors would like to acknowledge the contribution of Dr. Robert H. Tolson of George Washington University to the understanding of MGS aerobraking operational experience. Similarly, the authors would like to acknowledge the contribution of the Orbital MMSB study team, specifically Ann Mauritz, Chris Costello, in the formulation of this paper.

REFERENCES

1. "Mars Micromissions", Matousek, M., Leschley, K., Gershman, R., and Reimer J., 13th Annual AIAA/USU Conference on Small Satellites, Logan UT, August 23-26, 1999.
2. *Lunar Prospector Mission Handbook*, Andolz, F.J., Report # LMMS/P458481, Lockheed-Martin Missiles & Space Co, April 10, 1998.
3. Tolson, R.H, Personal communication. May 18, 1999.
4. "Satellite Angular Rate Estimation From Vector Measurements", Azor, R., Bar-Itzhack, I.Y., Harman, R.R., 1996 NASA Flight Mechanics/Estimation Theory Symposium, Greenbelt MD.
5. "Satellite Orbit-Raising Using LQR Control with Fixed Thrusters", Stoltz, P.M, Sivapiragasam, S., Anthony T.C., 21st Annual AAS Guidance and Control Conference, AAS 98-007, Feb 4-8, 1998.

Section VI
RECENT EXPERIENCES IN GUIDANCE AND CONTROL

SESSION VI

Joint Session Chairperson: Marv Levenson
Naval Research Laboratory

Joint Session Charperson: Angela Bukley
The Aerospace Corporation

Local Session Chairperson: Dr. William Frazier
Ball Aerospace & Technologies Corp.

The following paper was not available for publication:

AAS 00-075 "Attitude Determination for the STEX Spacecraft Using Virtual Gyros," by C. Heatwole, L. Herman, G. Manke, The Aerospace Corp.; C. Voth, Lockheed Martin Astronautics

The following paper numbers were not assigned:
AAS 00-078 to -099

AAS 00-071

AUTONOMOUS ORBIT CONTROL: INITIAL FLIGHT RESULTS FROM UoSAT-12

James R. Wertz,[*] Jeffrey L. Cloots,[*] John T. Collins,[*] Simon D. Dawson,[*] Gwynne Gurevich,[*] Brian K. Sato[*] and L. Jane Hansen[†]

Microcosm, under funding from the Air Force Research Laboratory Space Vehicles Directorate, has developed the first on-orbit demonstration of autonomous, on-board in-track and cross-track orbit control. The flight demonstration was conducted on the Surrey Satellite Technology Laboratory's (SSTL) UoSAT-12 mission. The satellite is in a 650 km, near circular orbit at 65 deg. Navigation data is provided by an experimental GPS receiver built by SSTL. The first full orbit control test was initiated Sept. 23, 1999, and lasted for 29 days until other propulsion system experiments were initiated. During the test period, thrust calibration was initially off by a factor of 2 and was corrected midway in the testing. Due to active work on the GPS software, GPS data outages occurred that were as long as 8 hours, with as much as 11 hours out during a 24 hour period. In spite of these unexpected anomalies and lack of data, OCK maintained a time late at the ascending node to within a standard deviation of 0.12 sec, equivalent to ± 0.93 km in-track over the entire 29 day test run. This is in contrast to an in-track slippage of approximately 4,500 km over the preceding 4 month period. During the test period, OCK applied a total ΔV of 73.3 mm/sec in 53 burns. All burns were in the positive direction (i.e., provided drag make-up). The total ΔV was equivalent to or slightly less than that which would have been required to return the spacecraft to its initial altitude had all of the thrust been applied at the end of the test period.

INTRODUCTION

The Microcosm Orbit Control Kit (OCK)[‡] is currently being flight-tested on board SSTL's ongoing UoSAT-12 mission. This demonstration validates the use of autonomous orbit control to maintain a spacecraft's long-term orbital position. The spacecraft orbit position, in terms of orbital phase and "longitudinal phase," is controlled individually or within a larger constellation. OCK[**]technology opens the door to significant simplification of day-to-day operations of constellations, large and small, and the associated cost benefits. In addition, OCK can greatly simplify mission planning and operations for single satellite missions.

OCK continues Microcosm's long-term commitment to transfer those operations that can be properly carried out by an onboard system from the ground to the spacecraft. This approach leads to efficiencies in ground operations, lower cost, and reduced system risk and will allow resources to be utilized on problems at which human operators excel, such as on-orbit anomalies and one-of-a-kind operations such as check-out. For further discussion of the history of OCK and its applications to constellation management and orbit control see Wertz [1996], Königsmann et al. [1996], Collins et al. [1996], and Wertz et al. [1998].

---

[*]   Microcosm, Inc., 401 Coral Circle, El Segundo, California 90245.
[†]   HRP Systems, Inc.
[‡]   Microcosm holds US and European patents for doing autonomous onboard orbit control.
[**] OCK consists of software developed under Phase 1 and 2 Small Business Innovative Research (SBIR) contracts sponsored by the Air Force Research Laboratory, Albuquerque, New Mexico.

Figure 1 UoSAT-12

ARCHITECTURE

The Microcosm OCK software is flying on the Surrey Satellite Technology Limited (SSTL) UoSAT-12 spacecraft, where it co-resides on a customized 386 onboard computer, developed by SSTL, with their attitude determination and control system software. The inputs for OCK are generated by the SSTL-built 12-channel L1-code GPS receiver (SSTL model SGAR 20) with an output frequency of 1 Hz. GPS is a relatively low-cost solution to acquiring the needed navigational input to the Microcosm OCK controller. This approach enables absolute orbit control instead of having to do relative orbit control for a constellation of Earth-orbiting satellites.

The Microcosm OCK approach employs absolute control rather than relative control. Absolute orbit control maintains each satellite in a "box" which will keep a predictable, regular position relative to users on the ground and will keep the satellite orbit from decaying. (For a discussion of absolute vs. relative orbit control, see Wertz [1999, 2000].) In the Microcosm OCK implementation, each satellite maintains its own position and velocity relative to a pre-defined stationkeeping box which has a very predictable periodic motion in an inertial reference frame. In this way, the position of one satellite in the constellation does not determine the fate of the rest of the satellites. If one satellite "dies" prematurely, this has no detrimental effect on the other satellites in the constellation, since each maintains its own state relative to its uniquely defined orbit control box. Another benefit of absolute control is that the satellites will have a longer life than those in systems employing relative control. In a relative control implementation, the entire constellation typically decays as quickly as the slowest falling member, to maintain the desired uniform distribution of satellites.

ALGORITHMIC OVERVIEW

The Microcosm OCK project is validating two different high-accuracy in-track orbit controllers as well as a single cross-track controller. Additionally, the use of a general filter for GPS inputs assures that a consistent and continuous data set is provided to the control software and provides improved accuracy of node crossing detection when possible. Nominally, UoSAT-12's attitude determination and control system provides a fresh GPS state vector, with GPS time, every ten seconds. The filter's role is to aid OCK through "back-filling" of skipped vectors and, as best it can, aid in reducing the vector-to-vector noise that is encountered.

The filter has four operational modes:

1) *Pass Through Mode*, where GPS solutions are converted to ECI and passed on to Orbit Control,

2) *Sample Mode*, where every n^{th} GPS solution is converted to ECI and passed to Orbit Control (the sample rate is defined by the user),

3) *Short Term Average Mode*, where n GPS solutions, taken over a $n \times 10$ sec time period, are propagated to a common epoch and averaged to produce a state estimate for Orbit Control, and

4) *Long Term Average Mode,* where, in addition to the short term average, a long term average based on the last *m* short term average estimates is produced. Long term averages are only used for testing the filtering technique and are not used by Orbit Control.

The filter has been tested using data from a high precision orbit propagator with noise added to simulate GPS selective availability and with on-orbit GPS data. Preliminary results show that the filter can reduce position noise by a factor of about 2. Results with simulated data show that the filter provides accurate data during short GPS unavailability to allow orbit control processing to continue.

In the implementation of both of the in-track controllers, the basic measurement to be controlled is the deviation in the time of crossing from South to North of the Earth's equator from the desired or expected value. Thus, the system controls a pre-defined period (and, specifically, the orbit phase) and longitudinal phase, rather than any particular set of orbital elements. In essence, the end-user is uninterested in whether a spacecraft can hold its orbital elements x, y, z to some prescribed level of accuracy; they are more interested in whether or not a telephone signal will get through. By use of OCK, this problem is now reduced to a timing problem.

This approach allows the controller freedom from particular force models' differences with reality and negates the need for onboard propagation. Therefore, great savings in onboard processing requirements are accrued without impinging on accuracy. This data is supplied to the OCK software in the form of ECEF GPS state vectors and their associated epochs, which are then processed within OCK to extract the relevant information.

The in-track controllers differ in their filter implementation. Both filter types reduce the effects of high-order terms in the Earth's gravitational field by removing the majority of the effects of the tesseral and sectoral terms, that is, the non-axially symmetric terms in the expansion. Each technique provides a different level of fidelity based on the complexity of the implementation. Since different filtering techniques are employed, the nature of the signal fed to the in-track controller is different between the two.

In addition, onboard targeting of frozen orbit conditions is used to better control the "orbit average" performance of the controller. Frozen orbit conditions are those where the natural rotation of the argument of perigee and the oscillations in the orbital eccentricity are essentially frozen out through the playing off of J_2-derived perturbations against perturbations from higher odd-numbered zonal coefficients. A proprietary method is used to continually move the orbit toward frozen orbit conditions and, once achieved, hold it there. Orbit-averaged mean elements are calculated on board and are used to facilitate this feature.

In terms of cross-track control, an analogous process to in-track control has been implemented. The longitudinal phase of the orbit is controlled and not the orbital inclination. This means that any secular drifts in the placement of a constellation's orbital plane are removed slowly over time until the desired longitudinal position and, importantly, "longitudinal speed" are maintained. From a constellation operator's point of view, the control of the longitude of a spacecraft with respect to other planes and within a plane can be as important to coverage as in-track slippage and altitude decay.

FLIGHT SOFTWARE

The Microcosm OCK software, while originating from two different SBIR programs, is a single computer software configuration item (CSCI) consisting of four computer software components (CSC) as shown in Figure 2.

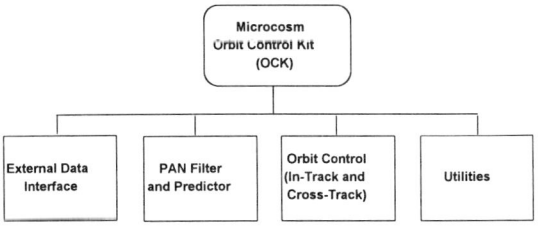

Figure 2 Microcosm OCK CSCI and CSCs

The general input, processing, and output flow, shown in Figure 3, demonstrates how compact the software is. The OCK software executes in under 100 msec when called at a 10 sec time interval.

Originally, the Microcosm OCK software was to reside on an Intel 186 processor with no math co-processor. Much of the accuracy associated with the OCK implementation was dependent on double precision arithmetic that would not be available without a co-processor. Low level testing was performed to see how single precision would effect accuracy, timing and code size. The results confirmed that the accuracy of the overall system would be substantially affected by the lack of double precision arithmetic.

In addition to the initial arithmetic constraints, the Intel 186 onboard computer (OBC) would only allow an executable of less than 64 Kbytes. The size of the code needed to be minimized, implying that using software solutions for improved arithmetic would not be acceptable. A version of the OCK software is available for an Intel 186 processor without a math co-processor, with slightly reduced accuracy.

When the Intel 386-based OBC became available, the OCK software could now make use of a math co-processor and/or expand to provide double precision arithmetic in software. Both of these implementations are also available. The OCK software based on the 386 OBC with math co-processor is ~65 Kbytes and the OCK software with its own software based arithmetic libraries is ~75 Kbytes. The use of the math co-processor provides higher fidelity output.

INTEGRATION

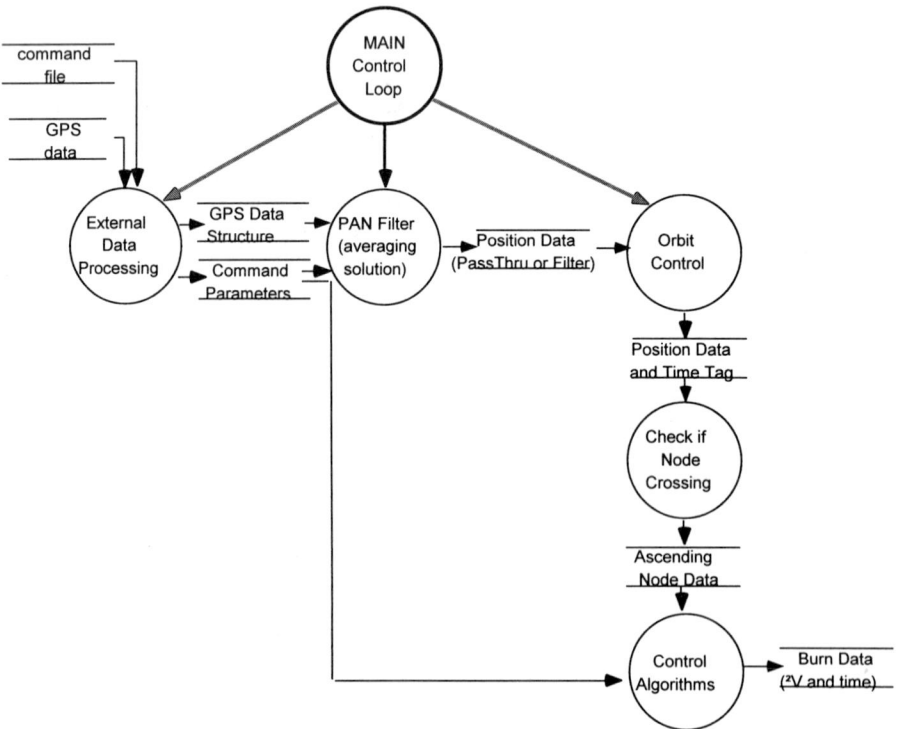

Figure 3 Microcosm OCK Data Flow

The Microcosm OCK software was developed for implementation on prototype hardware being developed at SSTL in Guildford, England (see Figure 4). Early in the development process an interface control document (ICD) was created between the SSTL shell software which calls OCK and the OCK software itself. This ICD was necessarily adhered to by both sets of team members. When changes were needed by one team member, the implications were discussed and both team members agreed upon a solution. Often this interaction took several days, and even weeks, since the groups were 8 time zones apart.

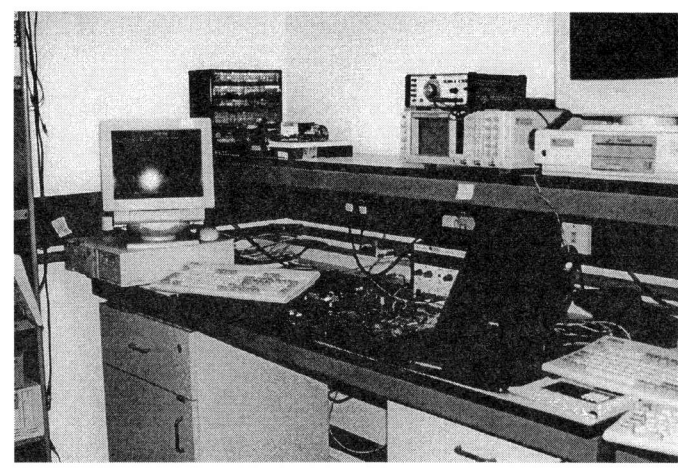

Figure 4 Integration on Engineering Model at SSTL's Facilities.

When the OCK software was ready for final integration on the hardware, a trip was made to SSTL. Prior to any "hands-on" work, the ICD was re-addressed, questions were resolved, and details were firmed up. Each team member made changes to the application software as required. The OCK software was loaded onto the OBC and execution began. The interface between the OCK software and the SSTL shell was proven early in the integration process.

The 386-based OBC is a new product for SSTL and the kernel functions that support the hardware were worked concurrently with the hardware development. Thus, the kernel support for the floating-point math co-processor was not available during early integration. The Microcosm OCK software was tested using the software implementation of the double precision functions and compared to similar baseline test cases. This integration was successful (see Figure 5).

Prior to uploading the Microcosm software to the spacecraft, the software was re-tested on the SSTL OBC using the math co-processor. The integration process for the Microcosm OCK software was quite successful based on strong adherence to an established ICD and regular communications between the team members. The total effort for hardware/software integration in the laboratory was less than a man-month performed over a two-week calendar period.

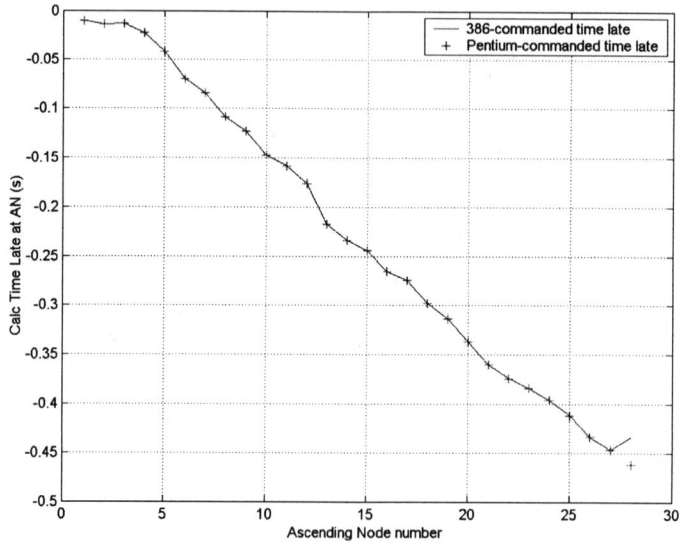

Figure 5 Sample Integration Results.
Note strong correlation between reference and 386-derived results.

EARLY RESULTS: PAN FILTER

The precision autonomous navigation (PAN) element pan was uploaded to UoSAT-12 and commenced operations on September 22. Results from the complete run are given below. These represent PAN's best estimates of the orbit's mean elements and show the level of noise to be expected from the proprietary method used. The need for orbital elements on board the spacecraft is as a check value for orbit control burns. They also provide information on the orbit's relative orientation with respect to the desired orientation. The actual values sometimes differ strongly from NORAD-derived data – the only "truth" model available to the project. This is to be expected given the differences in averaging models. NORAD apparently uses an amalgam of Brouwer's and Kozai's theories when it creates Two-Line Element (TLE) sets whereas OCK uses a time-averaged (orbital frequency) osculating to mean equinoctial conversion before the final "mean" Keplerian elements are created. This difference in techniques has significant implications for the absolute values of the calculated elements. It is the broad agreement with trends in both data sets that is important, not the disagreement in absolute values. OCK requires stability in the answers it receives on eccentricity and argument of perigee and not on their absolute values – these can always be offset from the "real" (read NORAD) values in the telecommanded-desired values.

Finally, Figure 11 shows the NORAD-derived estimate of UoSAT-12's ballistic coefficient for the duration of the run. This shows some alarming effects as NORAD's filtering technique attempts to track the – apparently – continuously maneuvering spacecraft. NORAD's solution suggests that the spacecraft has become alternately draggy, drag-free and "negatively-draggy", i.e., gaining altitude – and all within the space of days! Note that the OCK run began on Day of Year 267 and ended on Day of Year 296.

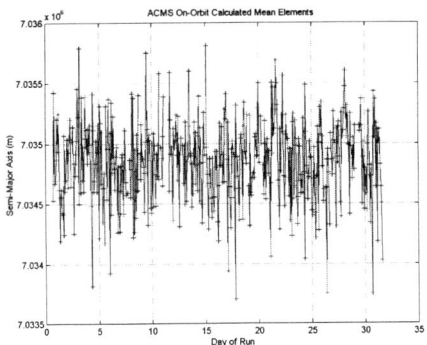

Figure 6 Semimajor Axis Measured On Orbit

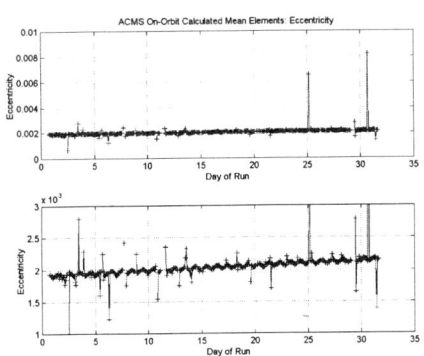

Figure 7 Eccentricity Measured On Orbit with Enlargement of Detail

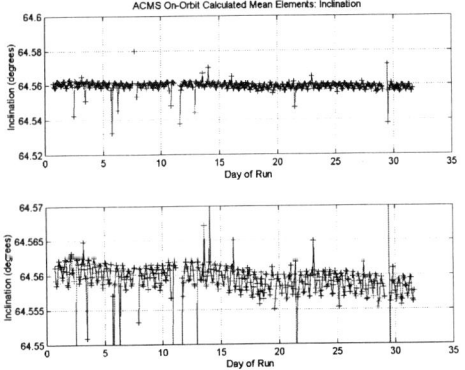

Figure 8 Inclination Measured On Orbit with Enlargement of Detail

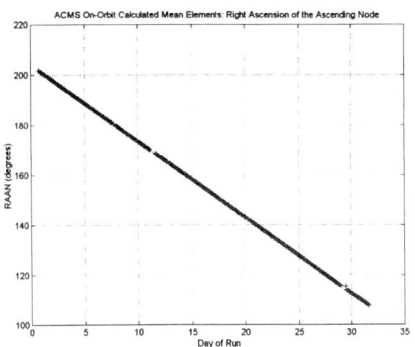

Figure 9 Right Ascension of the Ascending Node Inclination Measured On Orbit

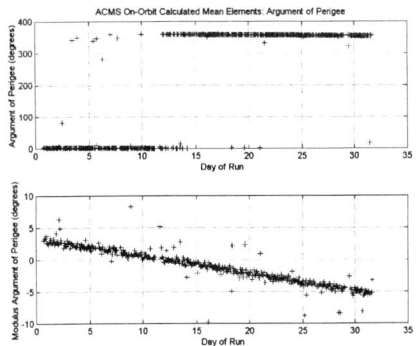

Figure 10 Argument of Perigee Measured On Orbit with Enlargement of Detail (Modulo 360 Degrees)

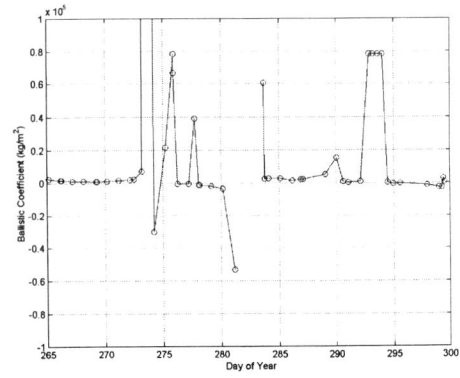

Figure 11 NORAD-Estimated Ballistic Coefficient

EARLY RESULTS: OCK SWITCH-ON AND RESULTS

Fully autonomous, onboard orbit control was activated on September 23 at 14:29:41 UT. The controller first commanded a burn of 1.386 mm/s some 8 hours later. This burn was executed as planned by the

UoSAT-12 ADCS system. OCK then commanded 52 subsequent burns over the next 29 days totaling 73.3 mm/s of expended ΔV. The results are shown in Figure 12 and Figure 13.

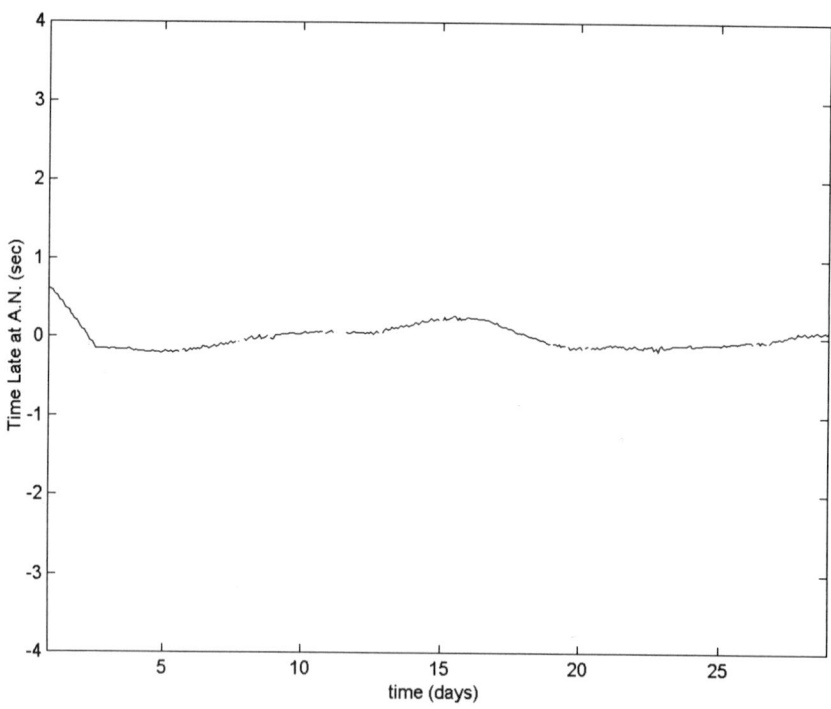

Figure 12 On-Orbit Time Late at The Ascending Node
The steep drop off for the first few days represents the time late prior to OCK being switched on at 14:29:41 UT, September 23, 1999.

Examination of this chart shows the slow 'sinking" of the "time late at the ascending node," or deviation from expected crossing time, over the first five days. This was expected and represents the controller reacting to the building of the time late due to the effects of drag. As can be seen up to around day 12, the magnitude of the burns increases before starting to level off as the time late ceases to increase and returns to close to the desired zero level. This zero level conforms to the spacecraft being exactly on time to the level of control.

Near day 12, SSTL staff identified a problem with one of the two thrusters that were being used to effect the velocity change. Essentially, the software was preventing the second thruster from activating. This imbalance was detected in anomalous attitude motions. OCK compensated for the lack of restorative impulse by upping its demanded ΔV. Hence, the 'sink" upon initialization was a little further than one would have expected from simulation results alone. The second thruster's non-performance was rectified on approximately day 12, but by then OCK's internal integral term had built up to a significant level. Thus, when, from the perspective of OCK, the ΔV was effectively doubled – or, alternatively, drag halved – we proceeded to overshoot into positive time late and subsequently rebounded into negative territory over the next ten days or so.

Examining a little further the ramifications of the controller's behavior is instructive:

1. The controller dealt with an effective doubling of the ballistic coefficient upon initialization and controlled the time late to an acceptable level.

2. Upon "re-instatement" of the second thruster the controller "hiccoughed" as to be expected but never lost control.

Thus, the controller has demonstrated its ability to overcome pre-launch mis-modeling of the spacecraft as well as unforeseen variations in the atmosphere.

OCK is expected to show a variation in the time late signal of < 0.2 sec (3 σ) after the controller has had time to settle. Short runs, as depicted in the figures, will have larger standard deviations. If allowed to run for an extended duration then these data would improve over time towards the value of < 0.2 sec.

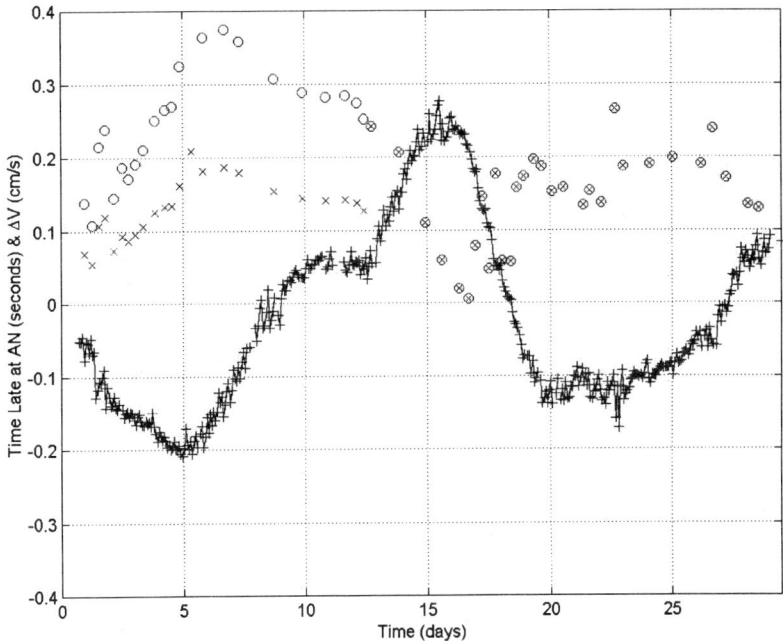

Figure 13 On-Orbit Time Late at The Ascending Node and Commanded and Actual In-Track Burn Magnitudes
Solid line and plus signs depict measured time late; empty circles depict commanded ΔV; crosses depict actual ΔV.

NAVIGATION PERFORMANCE

The performance of the SSTL GPS receiver was excellent while good data was flowing to OCK. However, for a significant fraction of the run data was either unavailable from the receiver or was deemed by PAN to be unreliable and was therefore discarded. The SSTL GPS receiver is a prototype unit and the positioning and communications software was in a state of flux throughout the run. In addition, the UoSAT-12 receiver did not benefit from having EDAC protection that will be available on future versions. Later data runs showed much greater reliability as the software matured. Consequently, we feel certain that there will be a more stable data stream from similar runs in the future. However, this data was the data available to OCK and hence provided a much needed "work-out" of the data checking capabilities resident in our code. Figure 14 shows the number of frames that were dropped in any single day but does not differentiate between long-term outages and singly dropped frames. Since OCK received data at 10 sec intervals multiplying the number of frames that were dropped translates this into fractions of a day. As one can see, on two occasions almost a half-day was lost to outages.

Despite these outages, OCK continued to work through them, inhibiting burns when the age of the last good GPS vector exceeded a pre-set limit. This limit was set empirically from simulation work before the run and put a limit on the accuracy or usefulness of a vector from the simple onboard predictor. This tool was only meant to take OCK through short duration outages (few minutes) and therefore, while very helpful in that area, was unsuccessful in taking OCK through many hour outages.

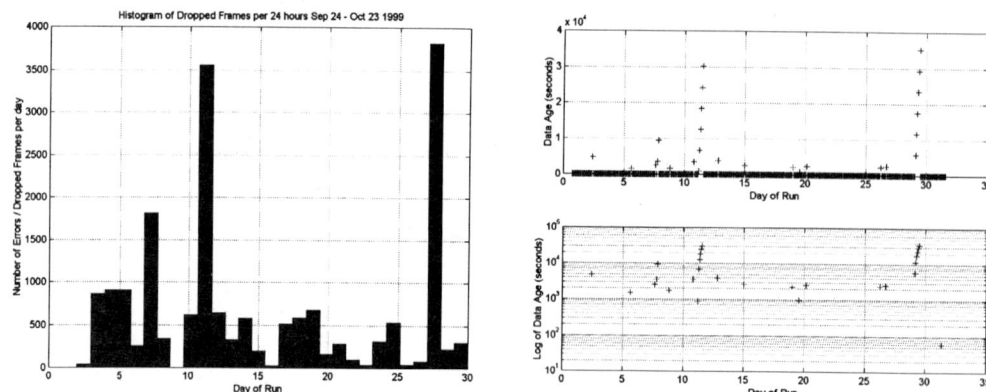

Figure 14 Histogram of Dropped Data or Bad Data Frames and
Age of Last Good Vector Prior to Ascending Node.

SUMMARY OF PRELIMINARY RESULTS

OCK operated continuously for 29 days and demonstrated accurate autonomous in-track orbit control under the adverse conditions of long GPS outages and an initial halving of thrust followed by a return to full thrust midway into the run. Due to navigation drop-outs 21 ascending nodes were missed out of a total of 418 (i.e., 5% were dropped). The standard deviation of the time late was 0.1237 sec representing a 3σ value of 0.3711 sec. Multiplying these time late data by the orbital velocity of 7.531 km/s gives an estimate of the in-track slippage over the entire length of run.

Length of Run	29 days
Number of Burns	53
Maximum burn size	2.7 mm/s
Minimum burn size	0.053 mm/s
Mean burn size	1.4 mm/s
Standard deviation of burn size	0.564 mm/s (1σ)
Sum of burns	73.3 mm/s

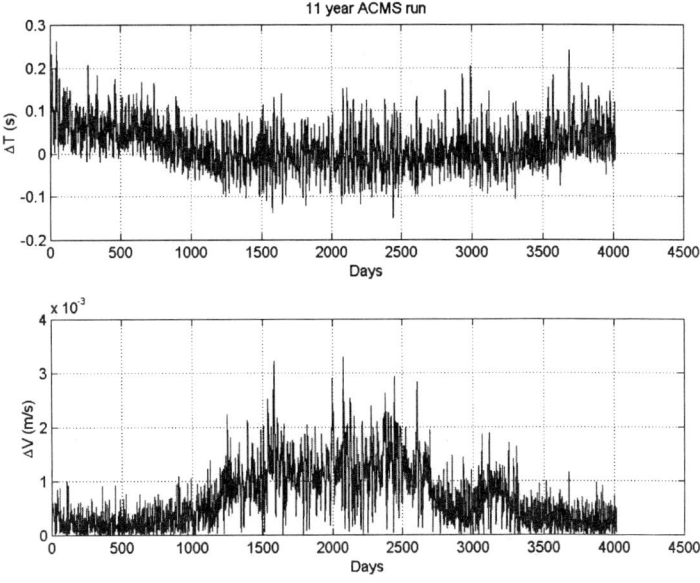

Figure 15 Simulated OCK Performance (Integrated Time Late at the ascending Node and Associated Propellant Usage) over a Full 11-year Solar Cycle
UoSAT-12 Configuration shown.

Ground-based simulations of both in-track control implementations, run for durations of 11 years, suggest a 3 σ time late at the ascending node for a nominal LEO (650 km altitude close circular) spacecraft of ~0.12 sec and ~0.08 sec respectively. These simulations were run using Microcosm's High Precision Orbit Propagator (HPOP) using the JGM-3 gravity field (truncated to 21 × 21), MSIS-86 atmospheric model using historical F10.7 plus random noise solar flux, solar radiation perturbations, and third body lunar and solar perturbation from the standard JPL ephemeris. Given the excellent agreement between expected and achieved results, it is expected that forthcoming on-orbit implementations can produce similar levels of performance.

CONCLUSION

The Microcosm OCK project, consisting of two SBIRs sponsored by USAF AFRL (Albuquerque), has produced flight code demonstrably capable of *autonomously* controlling constellation members" in-track and cross-track positions indefinitely to approximately 1 km. Therefore, a ground user can predict the position of the same spacecraft any time into the future with 1 km accuracy. The tasks of orbit prediction, re-prediction and re-re-prediction, and all the associated burden of planning and re-planning of activities, can be automated to the level of a simple spreadsheet operation.

ACKNOWLEDGEMENTS

The authors would like to take this opportunity to thank staff at USAF AFRL for their support. This work was funded by SBIR funds originating from AFRL and from internal R&D funds. We would like to thank SSTL for allowing this demonstration on UoSAT-12 and commend the excellent work and support of the whole SSTL team, without whose help this work would not have been possible.

REFERENCES

Collins, J. T., S. D. Dawson, and J. R. Wertz. 1996. "Autonomous Constellation Maintenance System." Presented at the 10th Annual AIAA/Utah State University Conference on Small Satellites, Logan UT, September 15-19.

Königsmann, Hans, J. T. Collins, S. Dawson, and J. R. Wertz. 1996. "Autonomous Orbit Maintenance System ." Presented at the IAA Symposium on Small Satellites for Earth Observation, Berlin, Germany, November 4–8.

Wertz, J. R. 1996. "Implementing Autonomous Orbit Control." AAS paper 96-004, presented at the 19th Annual Guidance and Control Conference, Breckenridge, CO, Feb. 7–11.

Wertz, J. R. 2000. *Spacecraft Orbit and Attitude Systems. Volume 1: Mission Geometry; Orbit and Constellation Design and Management.* Torrance, CA and Dordrecht, The Netherlands: Microcosm, Inc. and Kluwer Academic Publishers.

Wertz, J. R., John T. Collins, Simon Dawson, and Curtis Potterveld. 1998. "Autonomous Constellation Maintenance." Presented at the IAF Workshop on Satellite Constellations, Toulouse, France, November 18–19.

Wertz, J.R. 1999. "Guidance and Navigation," Sec. 11.7 in *Space Mission Analysis and Design*, Third Edition, edited by J.R. Wertz and W.J. Larson. Torrance, CA and Dordrecht, The Netherlands: Microcosm, Inc. and Kluwer Academic Publishers.

AAS 00-072

FUEL OPTIMIZATION DURING MARS GLOBAL SURVEYOR AEROBRAKING

Stuart R. Spath[*] and Dave F. Eckart[†]

Due to a solar array on the –y-axis side of the Mars Global Surveyor (MGS) spacecraft that was severely damaged during deployment, the MGS aerobraking mission was significantly replanned. The aerobraking mission was extended from about 400 aeropasses to about 893 aeropasses. At 0.013 kg/aeropass, the required increase in fuel use due to the additional 493 aeropasses would be 6.4 kg. There was insufficient margin to support this increase in fuel use.

As a result, several techniques were analyzed to reduce the fuel use on each aeropass. Four of the techniques were employed on the spacecraft, each with favorable results. Each of these techniques is discussed in detail. The combined result was a reduction from 0.013 kg/aeropass to about 0.004 kg/aeropass. With these fuel savings, the extended aerobraking mission was accomplished with less fuel than was in the original budget, and the MGS spacecraft was able to successfully achieve the objectives of its mapping mission.

INTRODUCTION

Mars Global Surveyor (MGS) is a three-axis stabilized spacecraft that was launched on November 7, 1996. The spacecraft was designed, assembled, and tested by Lockheed Martin Astronautics (LMA) for NASA's Jet Propulsion Laboratory (JPL). See Figure 1 for an illustration of the spacecraft configuration. The +x-axis is defined to be out of the high gain antenna (HGA) boresight when stowed, the +z-axis is defined to be out the payload deck, and the +y-axis completes the triad. The spacecraft has two symmetric solar arrays used to generate electric power that are driven by two-axis gimbals. The primary actuators for attitude control are three reaction wheel assemblies (RWAs) mounted along the spacecraft orthogonal axes. The spacecraft also contains 12 thrusters nominally rated at 4.45 N. The purpose of the thrusters is to provide lateral delta-v, to desaturate the RWAs, and to provide attitude control during periods when more control torque is required than RWAs can provide.

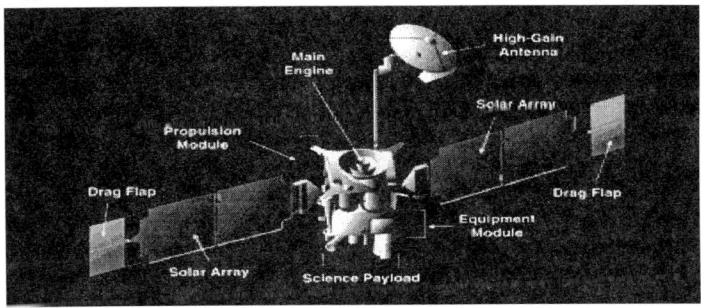

Figure 1. MGS Spacecraft Configuration

---

[*] Mission Operations Spacecraft Team Chief, Lockheed Martin Astronautics, P.O. Box 179, Denver, Colorado 80201.
[†] Mission Operations Attitude Control Engineer, Lockheed Martin Astronautics, P.O. Box 179, Denver, Colorado 80201.

The MGS mission plan called for a ten-month cruise to the red planet followed by large main engine delta-v maneuver to capture the spacecraft into orbit around Mars. Immediately after capture into a highly elliptical orbit with a period of 45 hours, the spacecraft was to embark on a four-month aerobraking phase to circularize the orbit. Using delta-v maneuvers, the spacecraft was to reduce its periapsis altitude to a level that was under the influence of the Mars atmosphere. Aerodynamic drag experienced during each periapsis passage would result in a loss of velocity at periapsis, which in turn would reduce the apoapsis altitude of the next orbit. Repeated periapsis passes through the atmosphere, called *aeropasses,* were to slowly change the highly elliptical capture orbit into a nearly circular mapping orbit with a period of 1.96 hours. Upon achieving the desired orbit period, the spacecraft was to raise its periapsis above the atmosphere to the nominal mapping altitude, deploy its HGA, and begin its two-year mapping mission. Figure 2 shows the gradual circularization of the orbit using aerobraking.

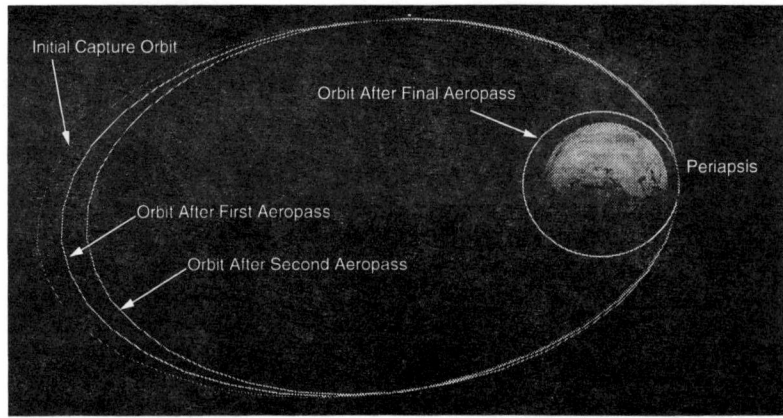

Figure 2. Orbit Circularization Using Aerobraking

ORIGINAL AEROBRAKING MISSION

The basic goal during aerobraking is to place the spacecraft center of pressure (CP) aft of the spacecraft center of mass (CM). This results in aerodynamic torques that tend to stabilize the spacecraft like a weather vane in the wind. The CP is moved aft by re-positioning the two solar arrays back from the flow by 30° as shown in Figure 3. Gimbal hard stops were manufactured at these -30° points to prevent the solar arrays from slipping during the large aerodynamic torques. The –z-axis of the spacecraft leads into the flow during an aeropass, and the +x-axis is pointed at nadir. Thus, the pitchover axis is the –y-axis.

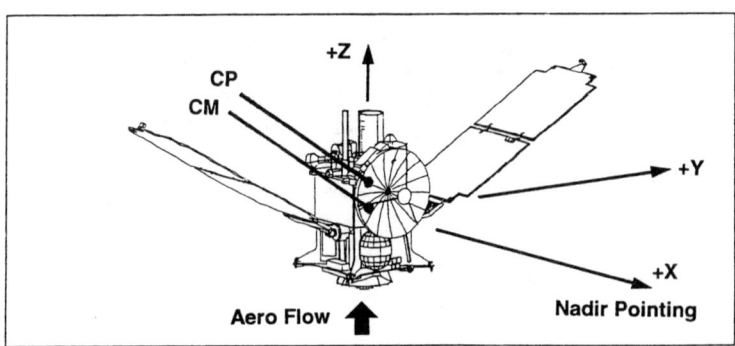

Figure 3. Aerobraking Configuration

All nadir pointing target attitudes are derived from a Mars ephemeris file provided by the JPL navigation team. Accurate predictions of the actual periapsis times are critical in computing the correct target attitudes. However, these predictions are difficult because (1) the Martian atmosphere has a large orbit-to-orbit variability, and (2) the periapsis times must be predicted by navigation several hours prior to the event. The requirement for the navigation team is to predict periapsis times within 3.75 minutes. In addition, 1.25 minutes is added for atmospheric uncertainty. The total time of five minutes of timing uncertainty is the key driver for the design of the attitude control law and the aerobraking command sequence. See Figure 4 for an illustration of the timing uncertainty.

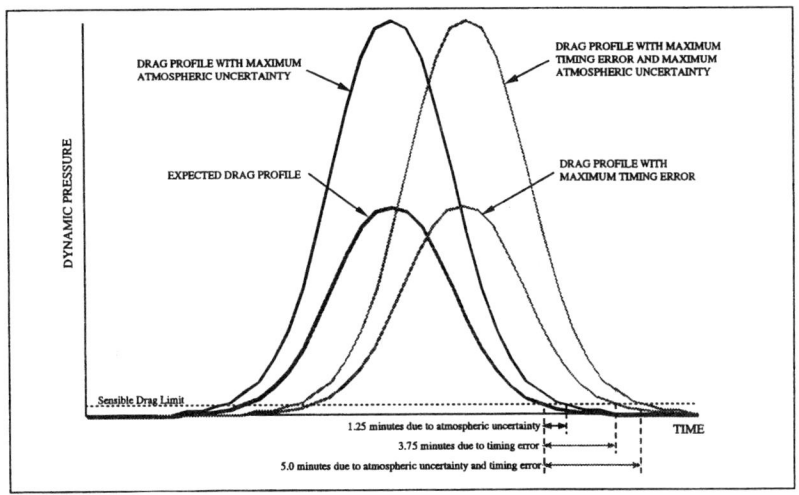

Figure 4. Timing Uncertainty

The aerobraking attitude control law is actually two independent control laws, both using thruster control only. The first, known as the *aeropass controller*, is used during the aeropass. The second, called the *post-aeropass controller*, is used immediately after the aeropass prior to passing control back to the RWA control law. The purpose of the aeropass controller is to serve as a very loose safety net in x and y, and to let the aerodynamic torques provide the control. The size of the deadbands in x and y is governed by the five-minute timing uncertainty. The average rate about the –y-axis, known as the *pitchover rate*, during an aeropass is approximately 0.067 °/sec. For a five-minute error in timing, the attitude error would be about 20°. Thus, the deadbands in x and y were selected to be 20°. In z, the aeropass controller simply provides a slow limit cycle response using a smaller attitude deadband of 7.5°.

After the aeropass has ended, the post-aeropass controller is selected which damps out large rates and attitude errors. The design strategy for the post-aeropass controller was to find a balance between complete damping and partial damping. The goal was to save fuel without imparting momentum to the RWAs. See the fuel savings section for a more detailed discussion of the post-aeropass controller.

The aeropass command sequence issues the instructions to spacecraft to perform the following major activities: slew to the aeropass attitude, switch to thruster control using the aeropass controller, switch to the post-aeropass controller, switch back to RWA control, slew back to the nominal Earth-pointing attitude. Due to the possibility of timing errors, the sequence must be padded before and after the expected drag period. During these pre-drag and post-drag delay segments, the spacecraft must be on thruster control in case the drag occurs early or late. Initially, 1.5 minutes and 5.0 minutes were used for the pre-drag and post-drag delays, respectively. See Figure 5 for a description of the command sequence and the spacecraft attitudes during the aeropass.

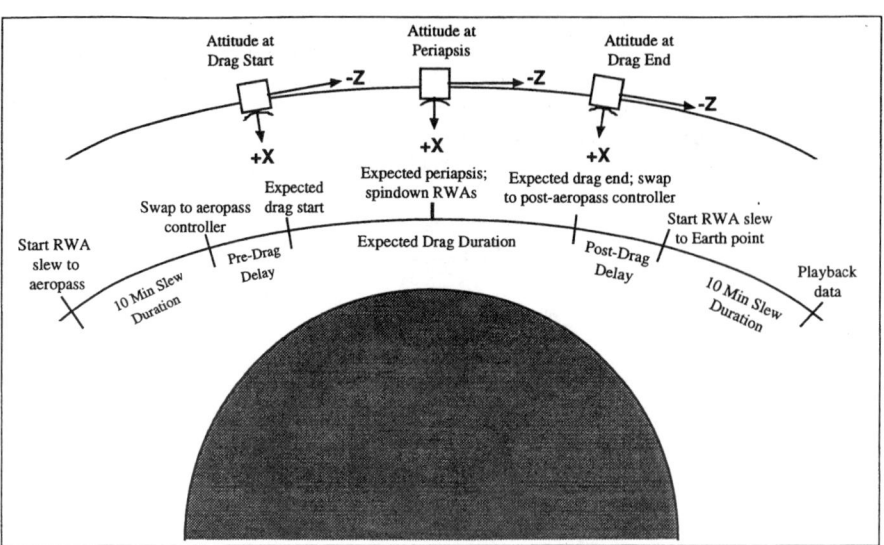

Figure 5. Aeropass Commands and Attitudes

Fuel Savings During the Original Aerobraking Plan

Due to mass constraints imposed by the launch vehicle, the fuel budget was extremely tight. The entire aerobraking phase was designed to require approximately 400 aeropasses with a total fuel budget of 8 kg, or 0.020 kg/aeropass. For rough comparison, the Magellan spacecraft, which completed aerobraking around Venus a few years earlier, used 0.031 kg/aeropass. A separate entry in the fuel budget covered various contingencies such as safe mode entries during aerobraking. As the aerobraking implementation was being designed, three fuel savings techniques were included to reduce the fuel use per aeropass.

Single String Thruster Control. All thruster control laws on MGS used dual string control for control authority reasons. In other words, the thrusters always fired in redundant pairs with identical pulse durations for each thruster string. However, for the aerobraking control laws, special flight software was added to support single string control. This allowed for smaller minimum impulses and an overall reduction in fuel use. The tradeoff was lack of redundancy. If a thruster were to fail during single string operations, no control torque would be generated. To mitigate this risk, new fault protection was added to the flight software to monitor for a failed thruster. Upon detection of this fault, the flight software would autonomously switch to the redundant thruster string.

RWA Spindown At Periapsis. Performing RWA desaturations during the vacuum portion of aerobraking orbits was undesirable for two reasons. First, each desaturation requires about 0.005 kg of fuel, which would equate to 2 kg over the duration of aerobraking if one were to occur every orbit. Second, small accelerations during the vacuum portion of the orbit can significantly affect the accuracy of the predicted periapsis times, resulting in greater fuel use. To prevent desaturations during aerobraking, it was decided to hold the RWAs at their current speeds at the start of each aeropass. Then, at the predicted periapsis time, the RWAs would execute a spindown to parameterized values. During the peak dynamic pressure, the large aerodynamic torques dwarf the small torques imparted by the RWA spindown, and therefore no additional fuel would be required for the desaturation.

Rate Damping During Post-Drag Delay. After an aeropass, the spacecraft must transition back to RWA control and then slew to an Earth-pointed attitude to playback the recorded data from that aeropass. Any excessive body rates at the point where RWA control is re-established will effectively load momentum into the RWAs, partially defeating the advantages of the periapsis spindown. Therefore, body rates must be fully damped. Attitude errors, on the other hand, do not need to be damped out. There is no momentum penalty for re-establishing RWA control with large attitude errors. However, excessively large attitude errors can cause payload sun avoidance violations. Using a control law during the post-drag delay segment that primarily damps body rates but has some slight attitude damping, additional fuel savings were realized. The attitude deadbands selected were 3°.

Simulation Description

The main simulation used to validate the control laws was a Mars Observer heritage three degree-of-freedom FORTRAN simulation. A Magellan heritage drag model was added to the simulation to perform aerobraking. A Free Molecular Aerodynamics Code (FreeMAC) program was used to generate the moment coefficients in each axis for all possible angles-of-attack. These moment coefficients were added to the simulation using table lookup. Fuel use was estimated by executing a number of different cases and averaging them. The average fuel use was 0.013 kg/aeropass. Over the course of 400 aeropasses, this would equate to 5.2 kg. The fuel budget was changed from 8 kg to 5.2 kg to reflect the simulation results.

AEROBRAKING REPLAN #1

Immediately after launch, telemetry was received from MGS indicating that the Solar Array on the Minus-y side of the spacecraft, known as *SAM*, did not fully deploy. Using the SAM-mounted sun sensor, it was estimated that the deployment shortfall was about 21°. Power level differences between the two solar arrays confirmed this estimate. A series of eight "wiggle" tests were commanded on the spacecraft during early cruise to assess the nature of the damage, and to determine how the aerobraking mission would be affected. The results of these in-flight tests indicated that the solar array was not locked into its latched position, but was resting against an obstruction. This obstruction was shown to be somewhat rigid when a force was applied to the active side of the solar array. However, when even the smallest force was applied to the SAM backside, the solar array would buckle. Since the original aerobraking configuration called for the aerodynamic drag to hit both solar arrays on the backside, the aerobraking mission would have to be re-designed.

Powered Hold Mode

The only chance of performing a successful aerobraking mission now required that the SAM array be re-positioned with the active side into the drag force. This was accomplished by rotating the inner gimbal from 180° to 0°, and re-targeting the outer gimbal from -30° to +30° (plus the 21° deployment shortfall). This new configuration retained the same drag area, the same center of pressure, the same center of mass, and the same symmetry on each side of the spacecraft. However, since the outer gimbal on SAM was no longer against the gimbal hard stop at -30°, there was a strong possibility of back-driving the gimbal motor in the presence of large aerodynamic drag. A special mode that simultaneously applies power to all three motor windings, known as *powered hold mode*, was available on the gimbal motors but not supported by the original MGS flight software. The holding torque of this powered hold mode was enough to prevent the drag force from back-driving the motor. A flight software patch was created and uplinked to implement powered hold mode, and the command sequences were updated to use this new capability on SAM.

Hinge Angle Telemetry

As learned during the wiggle tests and verified during subsequent ground tests, the obstructed area would "crush" or deform under the influence of large forces. The amount of this deflection, referred to as the hinge angle, varies with dynamic pressure and creates a bell-shaped attitude error profile about the x-axis. See Figure 6. Since it is difficult to determine the hinge angle from dynamics data alone, a derived telemetry channel was created to use the raw sun sensor data to directly measure the hinge angle. This hinge angle telemetry was used on every aeropass to provide a health assessment of the damaged array.

Under the assumptions of the SAM failure theory, if the hinge angle ever got to 0°, the solar array would latch into place and become rigid in both directions.

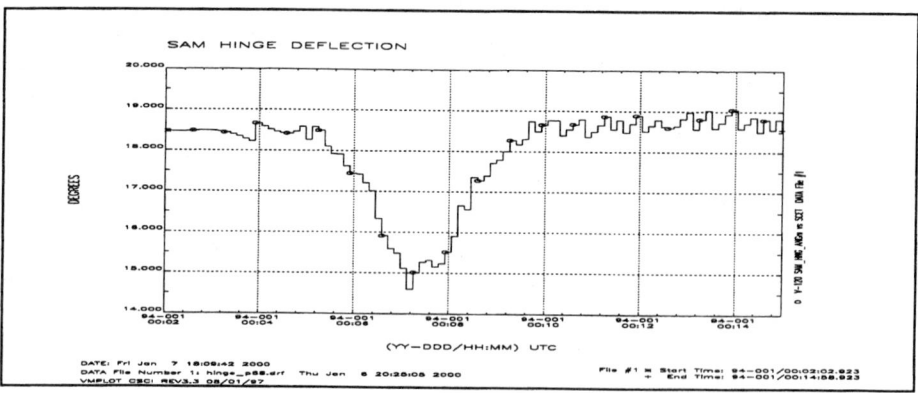

Figure 6. Typical Hinge Angle Response

Fast Fourier Transform Telemetry

In addition to hinge angle, 10 Hz gyro data was used to monitor the health of the SAM. Passing the gyro data through a fast Fourier transform (FFT), the first frequency of the damaged array could be determined. During the wiggle tests, the first frequency was about 0.175 Hz. On every aeropass, the first frequency would be extracted from gyro telemetry. If this frequency ever dropped below a red alarm limit of 0.14 Hz, aerobraking would be discontinued. Figure 7 shows a typical FFT plot.

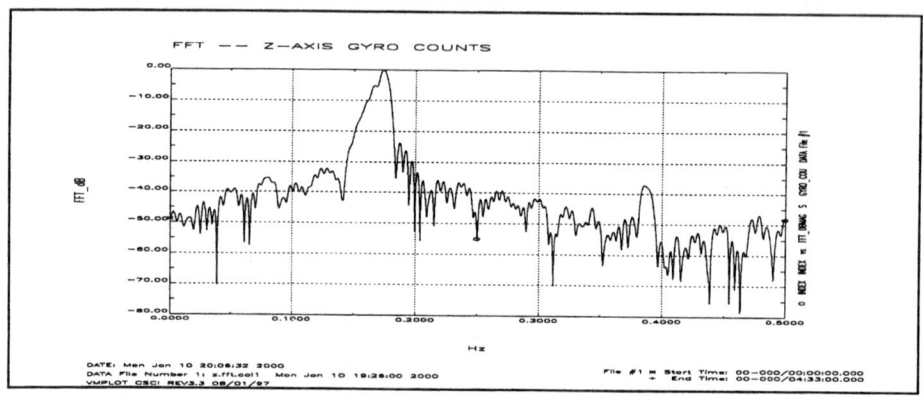

Figure 7. Typical FFT Response

Flight Results for Early Passes

The first 12 aeropasses (#3 through #14) performed fairly well. The powered hold mode held the gimbal motor fixed in the presence of large drag forces. The fuel use was within the 0.013 kg/aeropass budget, and the FFT frequency appeared to be reasonably constant. However, the hinge angle showed some strange behavior. During each aeropass, the SAM array was being deflected as expected, but was not returning to its original position after the aeropass. Furthermore, on a few aeropasses, it appeared as though the array had been deflected enough to latch.

Then, during aeropass #15, the SAM array appeared to deflect about 17° past the latch point. Using the original failure theory, the array should have latched into place at 0° hinge angle and experience no further deflection. Aeropass #15 clearly showed this assumption to be erroneous. The data was extensively reviewed, and the conclusion was that the damage to the SAM array was more severe than originally anticipated. Therefore, after aeropass #18, the periapsis was raised to an altitude completely out of the atmosphere for an evaluation period and a short science campaign.

AEROBRAKING REPLAN #2

During the evaluation period, numerous ground tests were conducted. The qualification solar array was taken to failure and produced a new failure theory, one which entailed substantial damage to the composite face sheets at the yoke attachment points. The first frequency of the damaged qualification array was about 0.17 Hz, very closely matching the FFT of the flight telemetry. Additional tests were performed with the failed qualification array to determine what dynamic pressure it could withstand and how many aeropasses were possible.

Reduction of Dynamic Pressure

At the end of the evaluation period, the conclusion was the SAM array was no longer capable of withstanding the dynamic pressures required for the nominal mission. It was recommended that the mission be replanned to fly at a higher altitude where the dynamic pressure would not exceed 0.45 N/m2. This new mission would require 493 more aeropasses and four additional months of time. Simply extending aerobraking by four months would result in a sun-synchronous orbit at a local mean solar time (LMST) of about 10:30 AM at the descending node. Since the science instruments were optimized for a LMST of 2:00 PM at the descending node, the new aerobraking plan would not accomplish the science objectives. Thus, the final plan was to divide aerobraking into two phases of five months and four months, respectively. Between the two aerobraking phases would be a six-month science mission named Science Phasing Orbit. If successful, the results of this final plan would be the achievement of a 2:00 AM LMST at the descending node, which was identical to the original orbit from the science perspective. In addition, the Science Phasing Orbit would provide a science campaign at periapsis altitudes lower than in the mapping phase (170 km vs. 375 km), thereby producing unique science data unavailable in the original mission.

Fuel Budget

Although the new mission plan was acceptable to the science teams, there were still engineering hurdles to be cleared. The most significant one was the fuel use. The original fuel budget for aerobraking was 400 aeropasses at 0.013 kg/aeropass, or 5.2 kg. With an additional 493 aeropasses at 0.013 kg/aeropass, the fuel penalty would be 6.4 kg. This penalty would be unacceptably large, and likely would result in shortening the primary mapping mission. As a result, the attitude control group was given the challenge of reducing the fuel use per aeropass such that the new two-phase aerobraking mission could be performed within the original budget of 5.2 kg. The following paragraphs describe the techniques that were analyzed to accomplish this new goal.

Simulation Enhancements

The first step was to compare the simulation results with the actual flight data that had been obtained. The simulation matched reasonably well, but some simulation updates were required. One major discrepancy was due to the fact that the RWA spindown was not modeled in the simulation. This was added to the simulation with a parameterized commanded speed for each RWA, and the correlation to the flight data was much better. Another difference was due to the fact that the simulation assumed zero initial RWA speeds at the start of the aeropass. In reality, the flight data showed substantial rates were present on most aeropasses. Therefore, the simulation was updated to allow parameterized initial conditions for RWA speeds. Again, the correlation with the flight data improved. Finally, the z-axis thruster activity differed between flight and simulation. The flight data showed a greater number of z-axis firings, and always in the same direction. In was concluded that a small, inner gimbal error on the solar arrays was causing a constant z-axis torque, known as a *windmill torque*. This windmill torque was modeled as a constant disturbance torque with a parameterized magnitude and was added to the simulation. After this final change, the flight data and simulation agreed very well as shown in Figures 8 and 9.

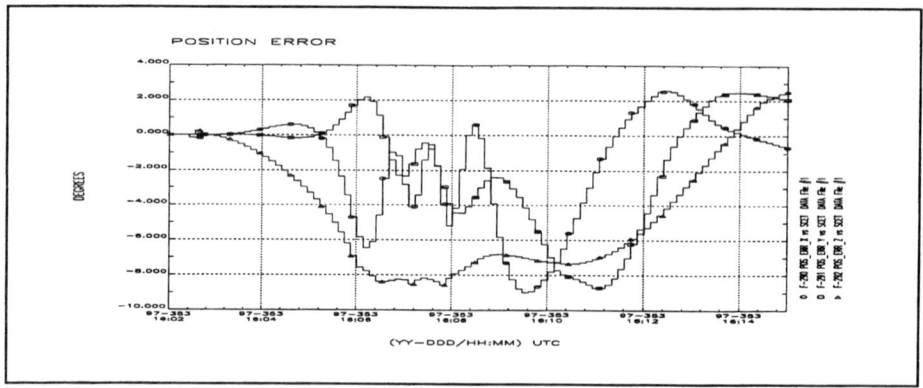

Figure 8. Flight Data

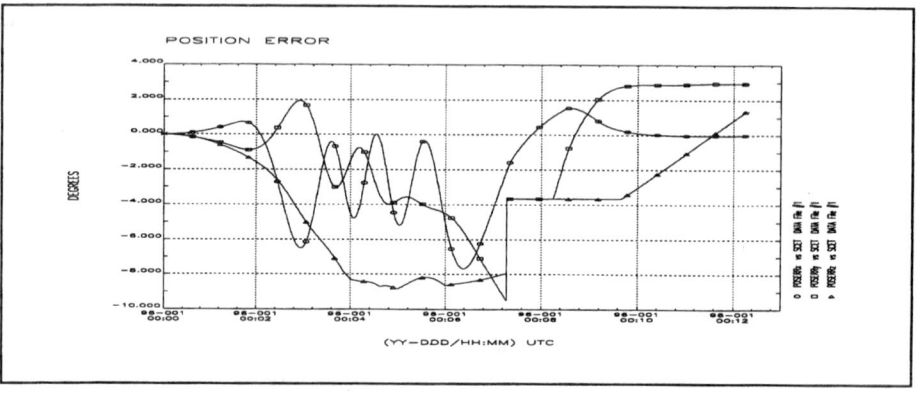

Figure 9. Simulation Data

Minimizing Gyroscopic Torques

The spindown of the RWAs to 0 RPM at periapsis had the desired effect of eliminating the need for desaturations during the vacuum part of the orbit. However, after the spindown, the environmental torques accumulated during the rest of the orbit. This caused the x and z-axis RWA speeds to be large when the spacecraft completed its slew to the aeropass attitude. At the same time, the y-axis body rate was being controlled to 0.067 °/sec to match the average pitchover rate during the aeropass. As a result, during the pre-drag delay, significant gyroscopic torques occurred that caused large initial attitude errors in x and z. When the drag finally occurred, the magnitude of the oscillations was much larger than necessary, which resulted in excess fuel use. The solution was to select spindown target speeds for the RWAs that would result in near-zero RWA speeds on x and z at start of the next aeropass. A tool that would predict the momentum buildup during the vacuum part of the orbit was created. This tool also calculated the transfer of momentum in the body frame during the slew to the aeropass attitude. After predicting the RWA speeds at drag start, the tool would calculate new parameters for the spindown target speeds that would minimize the gyroscopic torques for the next aeropass. The technique was verified by simulation to produce meaningful fuel savings, and the flight showed similar behavior with a reduction of about 0.0025 kg/aeropass. The lower attitude errors during the pre-drag delay segment are shown in Figure 10.

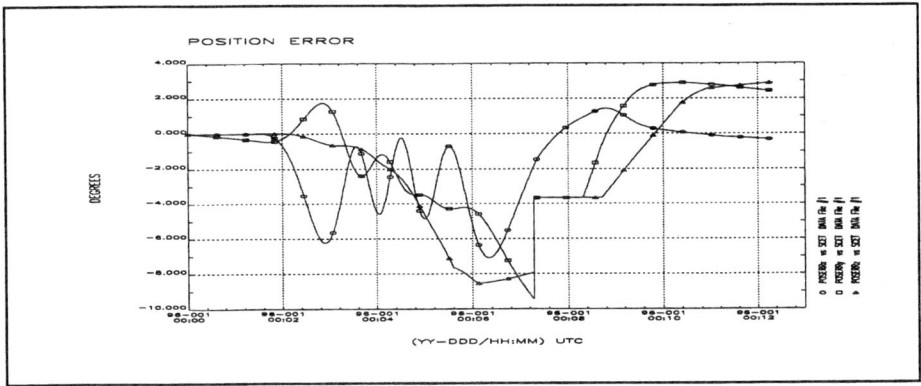

Figure 10. Minimization of Gyroscopic Torques

Eliminating Windmill Torques

Using the enhanced simulation with the windmill torque tuned to match the flight data, the impact on the fuel use was analyzed. It was estimated to account for 0.0013 kg/aeropass. Reducing this excess fuel use would be significant. Using the FreeMAC code, it was determined that the inner gimbal of each solar array would have to be off by 0.2° to produce the magnitude of windmill torque being observed. The solar arrays were commanded on the spacecraft to positions that were 0.2° from the original targets in a manner that would intentionally produce windmill torques in the opposite direction. The spacecraft results matched those predicted by the simulation. Figure 11 shows the lower z-axis attitude error during the aeropass.

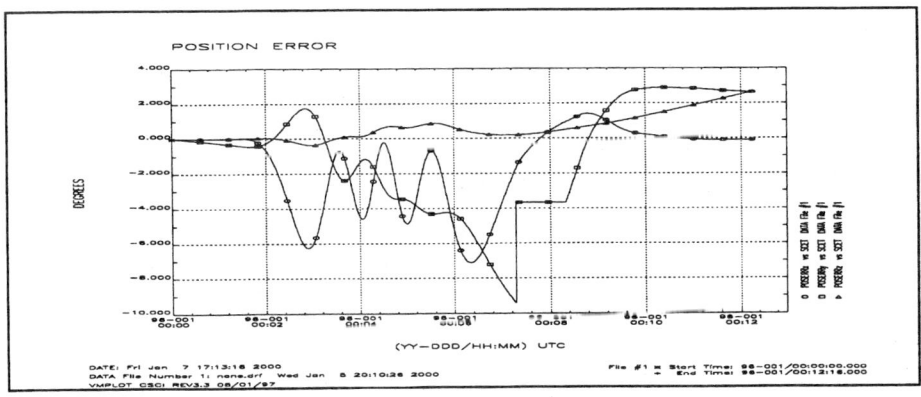

Figure 11. Elimination of Windmill Torques

Reducing the Attitude Error Limiter

As stated previously, the post-aeropass controller was primarily a rate damping control law with slight attitude damping. Damping the attitude errors too strongly would cause excess fuel use, while damping too lightly would increase the probability of violating the payload sun avoidance constraint. The original design value for the attitude deadband was 3°. However, there was also an attitude error limiter of 3.7° that truncated any large attitude errors down to 3.7°. Since this limited value was greater than the 3° attitude deadband, the post-aeropass controller would fire thrusters until the attitude errors were damped to less than 3°. The enhanced simulation showed that the attitude error limiter could be reduced to 2.5° without violating the sun avoidance requirement. Large attitude errors would be limited to 2.5°, and since this was below the 3° deadband, no firing would occur. In essence, the net effect would be to eliminate attitude error damping except in conjunction with rate damping. The decision was made to reduce the attitude error limiter to 2.5° on the spacecraft. The corresponding fuel savings were about 0.0012 kg/aeropass. The effect of this strategy can be observed in the post-aeropass damping of Figure 12.

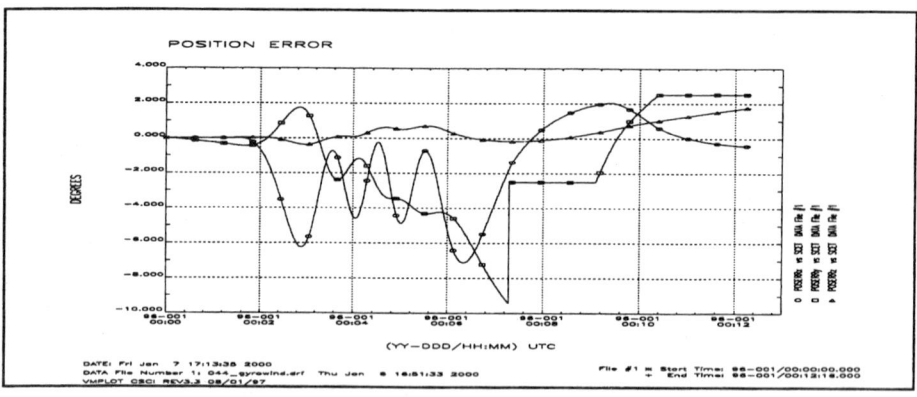

Figure 12. Reduction of Attitude Error Limiter

Increasing Pre-Drag and Post-Drag Delay Segments

Using the three techniques above, the average fuel use was substantially reduced. However, it was observed that during instances when large timing errors were present, the fuel use increased dramatically. This occurred in cases where the post-aeropass controller was being selected prior to the end of the actual aeropass. The post-aeropass controller was attempting rate damping during the presence of large aerodynamic torques. Figure 13 illustrates this for a case with a two-minute timing error.

The solution was to increase the delay segments to accommodate larger timing errors. The pre-drag delay segment was raised to five minutes to account for the maximum five minutes of timing error. This would prevent an instance where the atmospheric drag occurred before the switch to thruster control. The post-drag delay segment was raised to ten minutes to account for five minutes of timing error, plus an additional five minutes for performing rate damping with the post-aeropass controller. The amount of fuel savings from this technique depends on the timing error. In cases where the timing error was near-zero, the fuel savings were negligible. In cases where the timing error was large, the savings were about 0.002 kg/aeropass. On the average, modest savings of 0.0003 kg/aeropass were realized. The improvement shows up in the post-aeropass damping as seen in Figure 14.

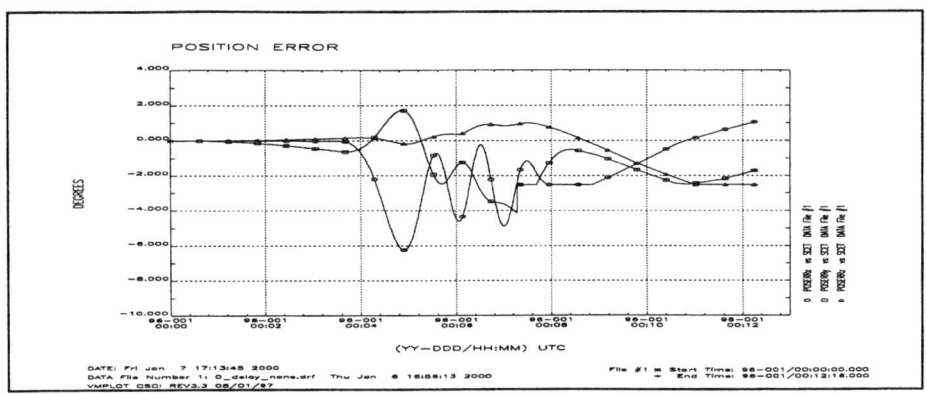

Figure 13. Effect of Large Timing Errors

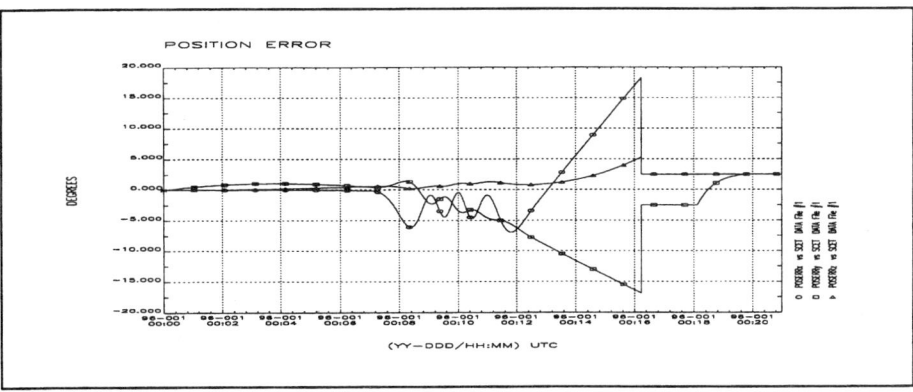

Figure 14. Increasing the Delay Segments

CONCLUSION

The final fuel use results for the entire aerobraking mission are shown in Appendix A, along with supporting plots of dynamic pressure, hinge angle, and FFT frequency. Through the techniques described above, the total fuel use for aeropass control was 3.7 kg over the 893 aeropasses, or 0.0041 kg/aeropass. Even with 493 additional aeropasses, the total fuel use was 1.5 kg below the original budget. MGS was able to successfully achieve its desired mapping orbit, and had saved enough fuel for a possible extended mission. As of January 2000, MGS had completed ten months and over 4000 orbits of mapping operations.

ACKNOWLEDGEMENTS

The authors would like to recognize several individuals for their contributions to this paper, or for their development of the concepts described within. Included in this list are Pat Esposito, Tim Gasparrini, Al Herzl, Dan Johnston, Jon Nichols, Norm Pence, Ralph Roncoli, David Shafter, Owen Short, Wayne Sidney, Bill Willcockson, the MGS design team, and all the members of the mission operations team.

APPENDIX-A

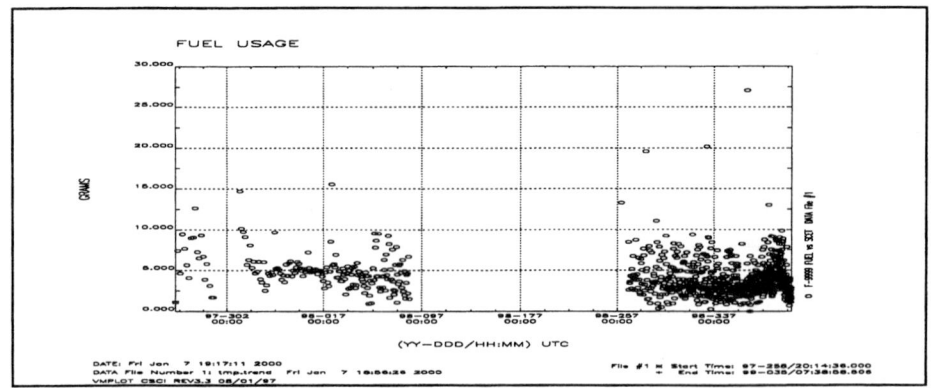

Figure A-1. Fuel Use Over Entire Aerobraking Mission

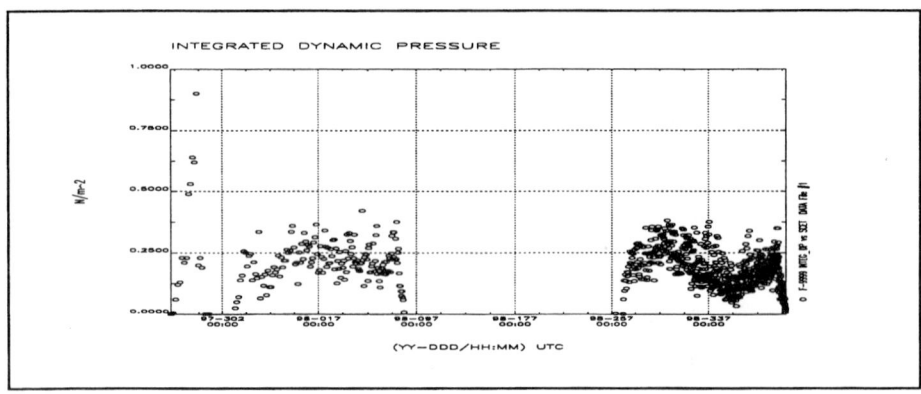

Figure A-2. Dynamic Pressure Over Entire Aerobraking Mission

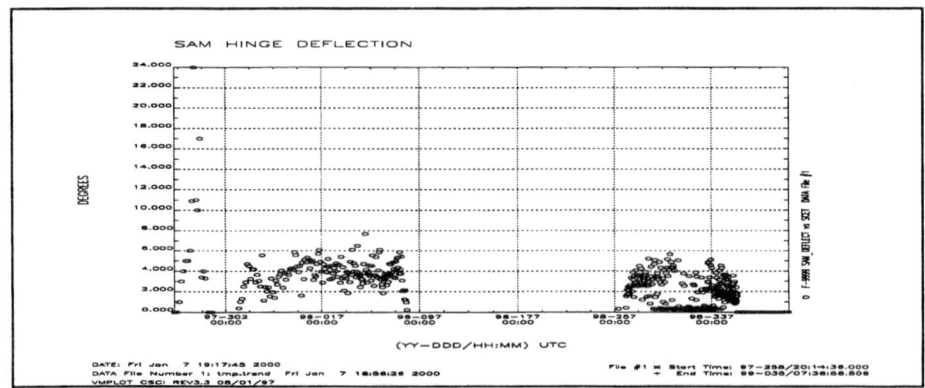

Figure A-3. Hinge Angle Over Entire Aerobraking Mission

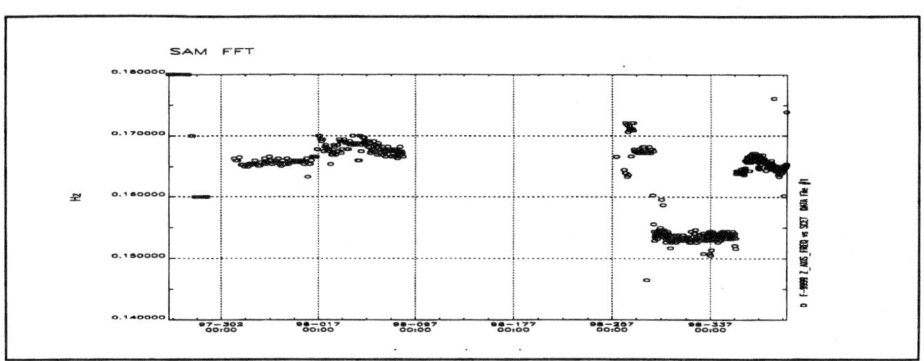

Figure A-4. FFT Frequency Over Entire Aerobraking Mission

AAS 00-073

QuikSCAT ATTITUDE CONTROL SYSTEM INITIALIZATION AND EARLY ON-ORBIT OPERATIONS

Dan Hegel and Scott Mitchell[*]

The QuikSCAT spacecraft was the first spacecraft procured through NASA's Rapid Spacecraft Acquisition initiative. A wind field scatterometer provided by JPL was integrated onto a commercial remote sensing bus that was already in development. QuikSCAT was launched on June 20, 1999 from VAFB using a Titan booster into a nominal 800x260 km orbit. Spacecraft initialization activities proceeded rapidly and nominally. All ADCS initialization was completed, and the first orbit raising burn performed less than four days after launch. The spacecraft was placed into its final 800 km frozen orbit about 18 days after launch using a series of 24 burns, totaling approximately 4 hours of thrusting with 1-lb thrusters. During this time, the ADCS performed nominally, except for a couple flight software/star tracker timing bugs which were discovered, diagnosed, and repaired via software loads. Pointing performance is well within the 0.05-degree (3-sigma) knowledge requirement.

INTRODUCTION

Mission Description

QuikSCAT is a NASA mission to measure the wind speed and direction over the world's oceans. The system produces a complete radar map of the oceans every two days. These data are used by climatologists, meteorologists, and oceanographers for a variety of purposes, including:

- improving weather forecasting;
- detecting and monitoring severe marine storms;
- studying world climate and global warming, including phenomena such as El Nino;
- monsoon monitoring;
- ship routing;
- oil spill cleanup.

The QuikSCAT spacecraft hosts the SeaWinds scatterometer. This instrument operates by transmitting 13.4 GHz microwave pulses to the ocean surface and measuring the returned or "backscattered" radar pulses. The returned pulses are changed by the small (< 3 cm) wind-caused ripples on the ocean surface. Algorithms extract wind speed and direction from the raw data with accuracies of 2 m/s in speed and 20 degrees in direction. SeaWinds uses a 1 m rotating dish antenna to collect data over an 1800 km swath, making approximately 400,000 measurements and covering 90% of the earth's surface each day. This remote sensing system provides orders of magnitude more observations of the ocean winds vector than can ships and buoys, and can operate in all weather. The SeaWinds scatterometer was built by JPL.

The QuikSCAT spacecraft bus was the first obtained under NASA's Indefinite Delivery/Indefinite Quantity program for rapid acquisition of spacecraft. This procurement method allows NASA to purchase satellite systems through a "catalog," providing significant time-savings from mission conception to launch. The QuikSCAT spacecraft was ready for launch less than one year after contract award. Such rapid completion of the spacecraft was possible due to the availability of a partially constructed spacecraft bus at Ball Aerospace and Technologies Corp.

---

[*] Ball Aerospace & Technologies Corporation, P.O. Box 1062, Boulder, Colorado 80306.

The QuikSCAT mission was speeded into operation to replace NASA Scatterometer (NSCAT), which flew on Japan's ADEOS spacecraft. ADEOS failed only 9 months after launch due to loss of power. QuikSCAT is intended to "fill the data gap" for scatterometer data between the loss of NSCAT and the launch of a scatterometer on the ADEOS II satellite.

Mission operations are performed by the Laboratory for Atmospheric and Space Physics (LASP) at the University of Colorado. LASP is staffed by a mix of professionals and students who plan and schedule operations, monitor and control the satellite, and analyze the engineering data to monitor health and status of the satellite. NASA's Earth Polar Ground Station (EPGS) network is used to communicate with the satellite through sites in Poker Flats Alaska, Svalbard Norway, Wallops Island Virginia, and McMurdo Antarctica. Contacts are scheduled each orbit to downlink the data to optimize the timeliness of the science data.

Science data from QuikSCAT are processed at NOAA in Suitland Maryland and forwarded to meteorological agencies around the world and the general public. The data are also processed and archived at JPL.

Spacecraft Description

The QuikSCAT spacecraft was designed and built by Ball Aerospace and Technologies Corp. QuikSCAT is a version of the Ball Commercial Platform (BCP) 2000 bus. Figure 1 shows two views of the spacecraft, with major components labeled.

The main structure is a 2.2 by 1.7 by 1.4 m box made of aluminum honeycomb panels connected by the aluminum cornerposts. Components are mounted both inside and outside the box, with antennas, star trackers, magnetometers, and sun sensors mounted outside. The bus is fully redundant, with a design life of five years. The pre-launch mass of the fueled spacecraft was 870 kg.

Communication is via S-band. Housekeeping data can be transmitted at 4, 16, or 256 kb, and science data at 2 mb per second. The satellite receives commands from the ground at 2 kb per sec. There are two communication antennas; one near the scatterometer dish is nominally pointed at nadir and is used for most communications. The second points in the opposite direction to provide complete antenna coverage for contingency operations.

Power is provided by two deployable single-axis articulating solar panels. Each panel is 71 by 168 cm, and covered with gallium arsenide solar cells. Each solar panel produces 642 Watts of power at the beginning of life. Power in the earth's shadow is provided by a 40 amp-hour nickel hydrogen battery.

Spacecraft thermal control is mostly passive. Multi-Layer Insulation (MLI) blankets cover most spacecraft surfaces. Twenty-seven heaters can be used if certain components become too cold, while four radiators and panel fins dissipate excess heat.

QuikSCAT has a monopropellant hydrazine blowdown propulsion system. Each of the four thrusters produces nominally 1 lb (4.4 N) of thrust. At launch the propellant tank contained 76 kg of fuel, pressurized to 415 psi.

At the heart of the command and data handling system is a RHC3001 processor. This is a radiation-hardened microprocessor built by Harris and used on many spacecraft. Data are stored on an 8 Gb solid-state recorder, which is capable of simultaneous recording and playback. The flight software was written by Ball, and is modular, table-driven, and written in the Ada programming language.

The BCP-2000 attitude control system (ACS) is a three-axis stabilized, zero-momentum system. Redundant star trackers, IRUs, and reaction wheels provide precise attitude determination and control. Sun sensors and magnetometers are used for coarse attitude determination, and torque rods are used for momentum control. Four thrusters provide three-axis closed-loop control for orbit adjust. In addition, the satellite has two GPS receivers, which produce single-frequency data for orbit determination. Figure 2 shows a functional block diagram of the ACS equipment. The BCP-2000 ACS is described in more detail in ref. (1).

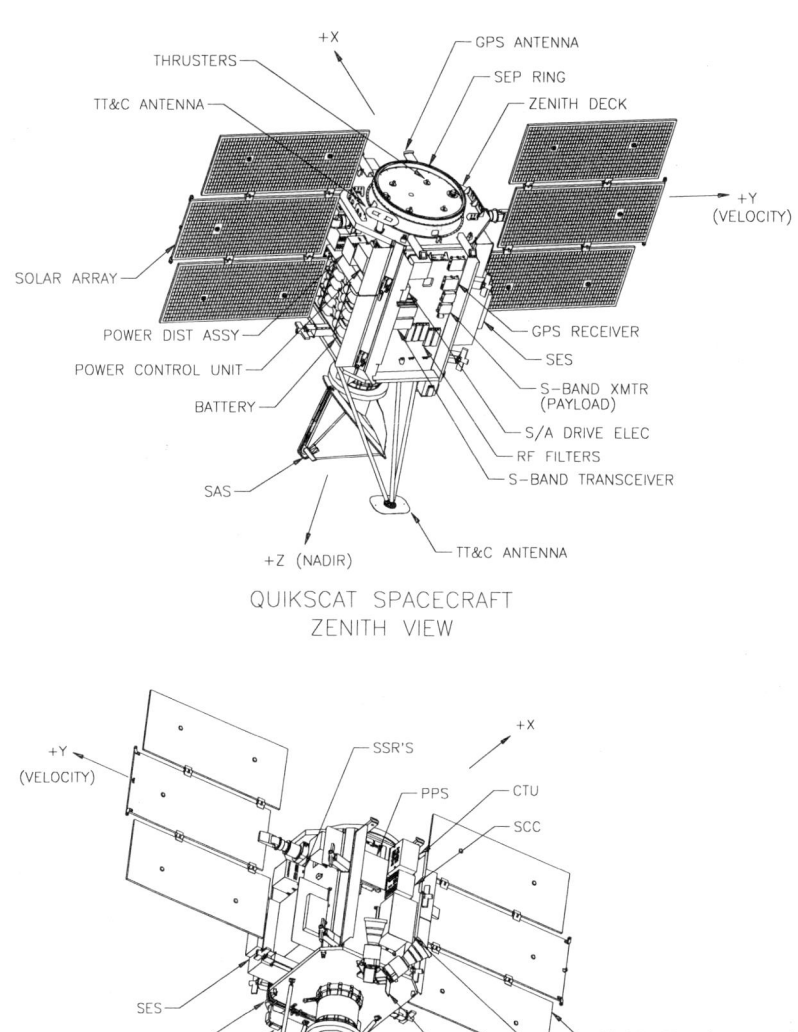

Figure 1. The QuikSCAT spacecraft layout

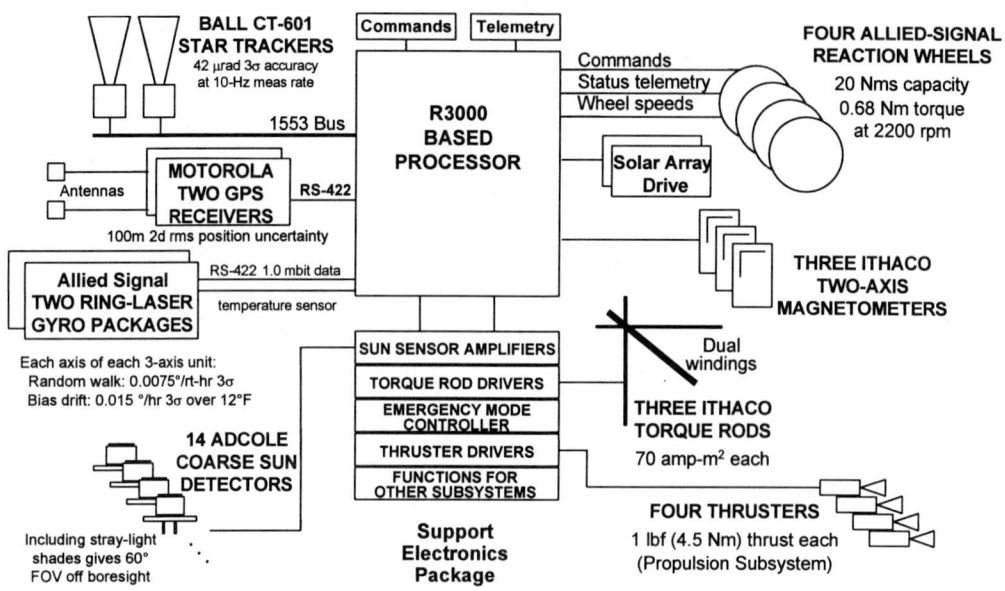

Figure 2. ACS Functional Block Diagram

EARLY OPERATIONS

Launch

The QuikSCAT spacecraft was launched on a Lockheed Martin Titan II rocket on June 20, 1999, from Vandenburg AFB. The Titan II is a decommissioned intercontinental ballistic missile that has been refurbished as a space launch vehicle. Launch occurred at 0215z, at the beginning of the 10-minute launch window. The spacecraft separated from the launch vehicle over the Malindi (Kenya) tracking station at 0316z. After separation from the launch vehicle the Spacecraft Control Computer automatically booted up and began processing the initialization sequence of commands from EEPROM. Approximately 15 minutes later, after solar arrays had been deployed, the reaction wheels were turned on and the spacecraft maneuvered to its Acquire Sun safemode orientation as planned.

The default Acquire Sun mode uses only sun sensors and magnetometers (combined with a fixed-gain Kalman filter) for attitude and rate determination, reaction wheels for control, and torque rods for momentum dumping. Figures 3 and 4 show estimates of attitude and rate error determined by the spacecraft, and how quickly errors settled out after the wheels were turned on. Attitude error is always zero about the sun-pointing axis (the -X axis) --- only rate is controlled. Because several sun sensors are located on the solar panels, attitude and rate estimates are not accurate until the panels are deployed (at approximately 700 seconds). Since no ephemeris (and therefore no magnetic field reference) or gyros are used during initial sun pointing, the control system perceives a rate about the sun-pointing axis as the magnetic field moves during the orbit. As the control system responds to the perceived rate, the on-board estimate is maintained near zero; however the spacecraft actually rotates at twice-orbit rate as it follows the moving magnetic field. Figure 5 shows the rapid change in reaction wheel speeds during the maneuver to sun pointing, and then a gradual decay as the torque rods dump the excess momentum. The tip-off rates were less than 1 deg/s --- well within the combined specification levied on the launch vehicle and separation system.

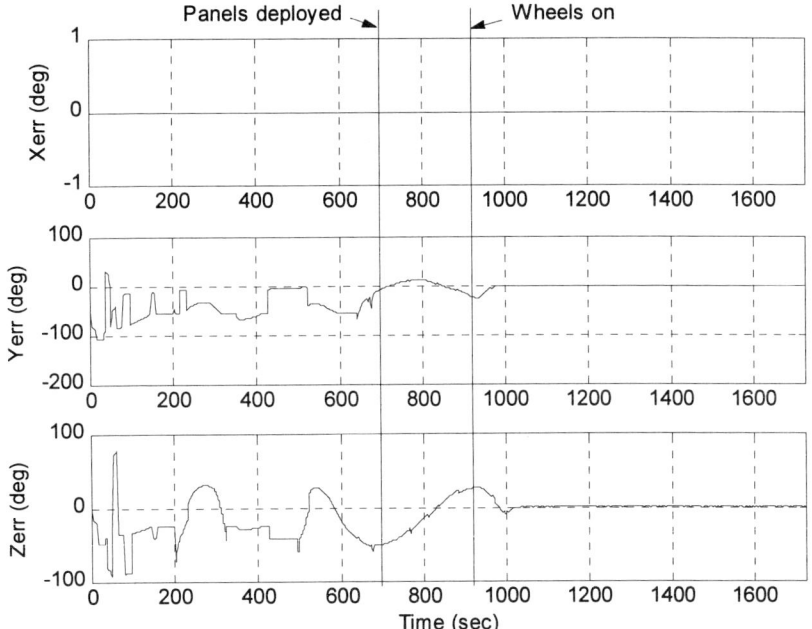

Figure 3. Post-Separation Attitude Error

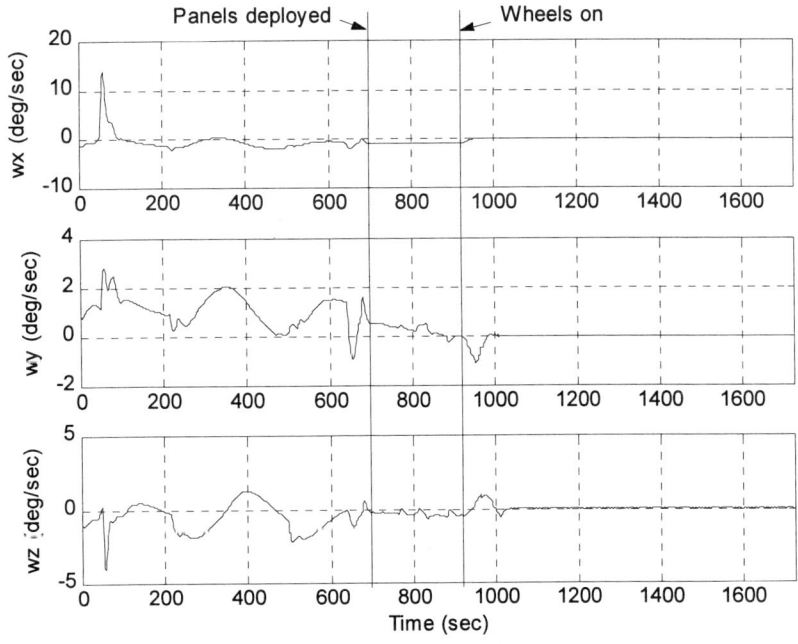

Figure 4. Post-Separation Rate Error

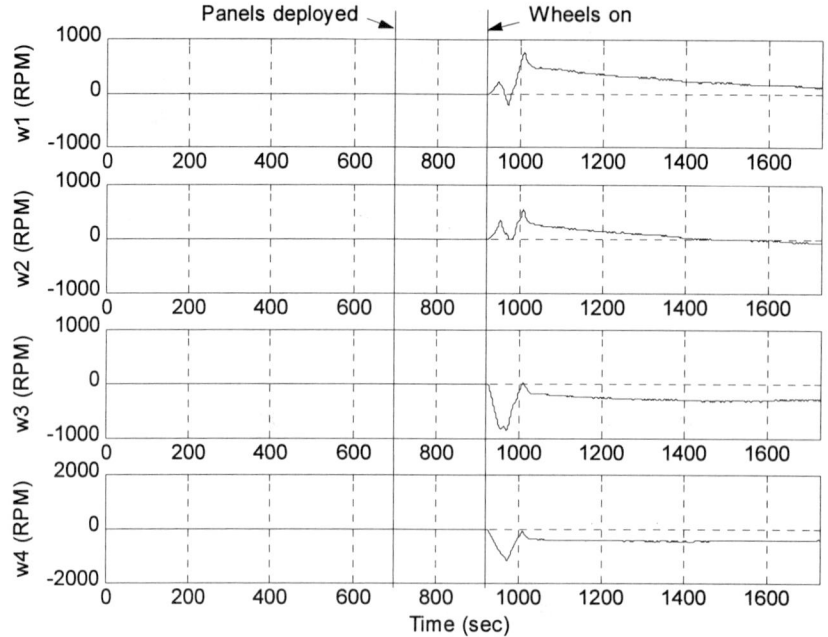

Figure 5. Post-Separation Reaction Wheel Speeds

The controllers on the ground then began commanding the spacecraft. Within a few orbits, the basetime and on-board orbit propagator were initialized, IRU #1 was turned on, and the spacecraft was commanded to its nominal Point mode. GPS receiver #1 was turned on and locked up on four satellites in about 2.5 minutes. GPS data are used both on-board by the orbit propagator/filter, and on the ground to estimate the satellite orbit for planning.

The two star trackers were the last ACS hardware to be powered on. Once data requests had been enabled, the trackers each began tracking five stars within 30 seconds. The flight software (FSW) quickly identified the stars and calculated an attitude, at which time the trackers stopped tracking stars. This cycle repeated a few times until an attitude error fault placed the spacecraft in Acquire Sun mode. Data requests to the trackers were disabled and the spacecraft was commanded back to Point mode. At this time, safe and efficient orbit raising was the highest priority. Therefore, it was decided to start the orbit-raising process, without trackers, and to trouble-shoot the problem only if it did not interfere. Such a contingency had already been considered prior to launch; ample analysis and testing had been performed to verify that orbit-raising could still be performed without the star trackers. The problems were fixed shortly after orbit-raising was completed. The causes and fixes are described later in this paper, along with final pointing performance.

Pre-Orbit-Raising Check-Out

Before beginning the extended orbit-raising campaign, the spacecraft maneuver and burn sequences were checked out in a series of steps designed to incrementally build confidence in the hardware and software. First a maneuver to re-orient to the burn attitude was performed to verify ACS functionality. Next an attitude maneuver with a 30-second test burn was performed to verify functionality of the propulsion and ACS subsystems together. (Wheels are used during the maneuvers, and thrusters alone are used during the burn.) Then a command sequence with a 5-minute calibration burn was performed. Figure 6 shows the angular rates (measured by the IRU) during the maneuver to the burn attitude (instrument in velocity direction), the 5-minute calibration burn, and the maneuver back to the nominal mission attitude (instrument nadir). The "noisiness" is actual motion due to control system response to the relatively coarse sun sensor and magnetometer signals. The IRU is the primary sensor in this mode, with corrections from sun/mag to keep from drifting. However, the mixing gain was higher than it needed to be, resulting in unnecessary response from the control system. One of the mission operations rules was to not change any table values unless it was necessary; therefore, it was decided to live with the noise. There are also some "spikes" visible in the data which were found to be caused by a FSW table that had an error in the boresight direction of sun sensor 4, and a table threshold that allowed sun sensor signals to be used when the sun was just beyond the useful limit of the sensor FOV. Figure 7 shows the sensors that were used during the sequence; the change in sensors is correlated with the spikes in rate. (A table was eventually uploaded that corrected the sensor boresight direction and reduced the useable signal ranges. These changes were made to avoid unnecessary attitude error faults that might interrupt the orbit-raising process.) Figures 8 and 9 show the attitude error and the resulting duty cycle of each of the thrusters during the closed-loop burn. The attitude spikes are visible, as well as a 0.5-degree offset error just after the start of the burn (the integrator gradually removed the error). Predicted thruster alignment and performance values were used to generate the default parameters in the thruster control table. The offset in the Y-axis was due to a slight difference between the predicted and actual values for the thrusters. Using the actual data from the calibration burn, a new thruster control table was generated and uploaded to the flight software, which removed any perceptible offsets on subsequent burns.

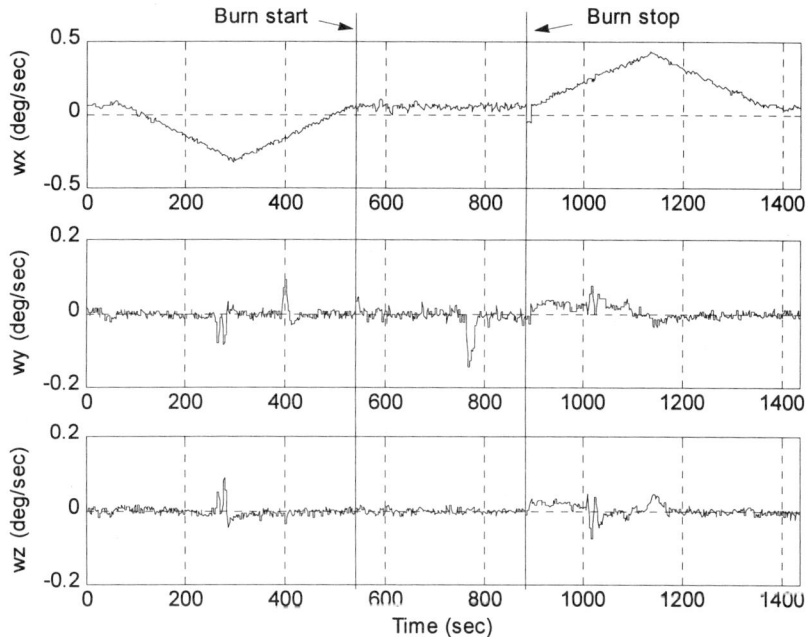

Figure 6. Angular Rates During the 5-minute Calibration Sequence

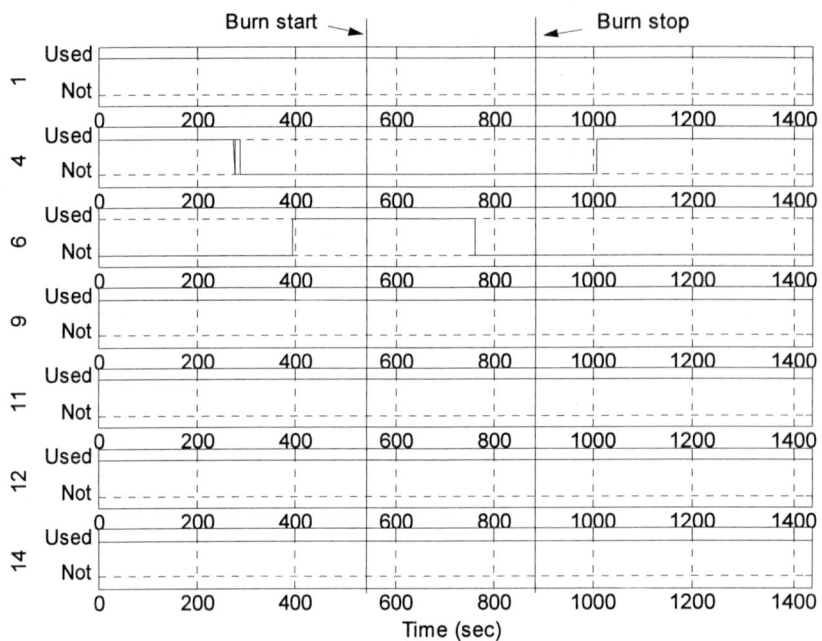

Figure 7. Sun Sensors Used During the 5-minute Calibration Sequence

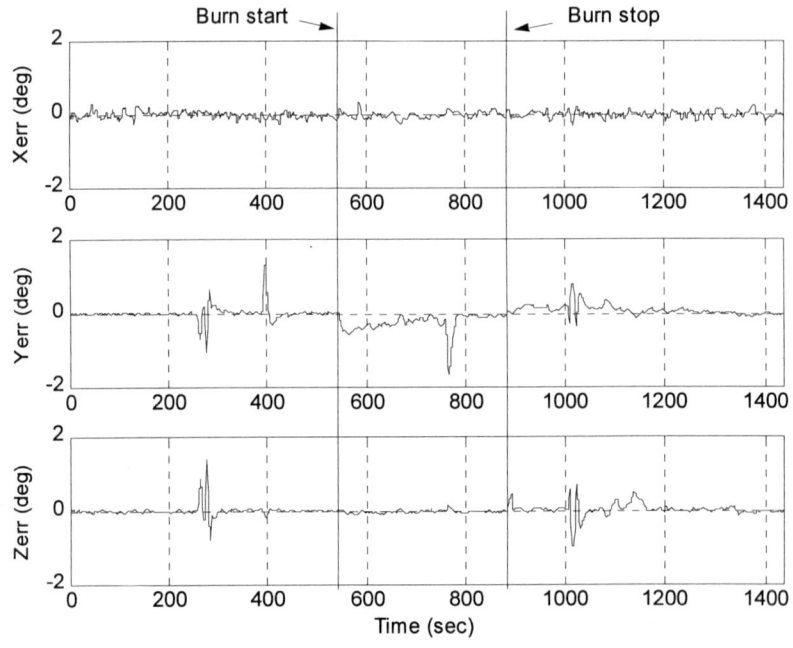

Figure 8. Attitude Error During the 5-minute Calibration Sequence

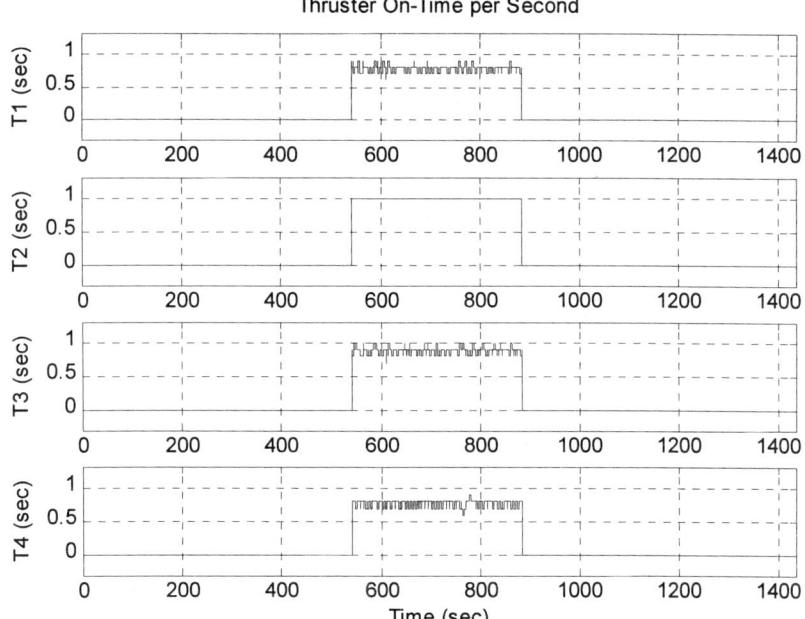

Figure 9. Thruster Duty Cycle During the 5-minute Calibration Sequence

ORBIT RAISING

Preparation

Planning and development of the QuikSCAT orbit raising process began long before launch with the formation of an Orbit Raising Working Group (ORWG). The ORWG consisted of appropriate engineering and operations personnel from Ball, LASP (Laboratory for Atmospheric and Space Physics of the University of Colorado), CCAR (Colorado Center for Astrodynamics Research of the University of Colorado), and JPL (Jet Propulsion Laboratory). Two of the overriding goals for the orbit raising process were to have it as automated as possible (i.e. minimize the necessity for human intervention) and to verify the process in the most realistic flight-like scenario possible. The key was to get engineering and operations personnel working together early in the development so that there would be no surprises on either side regarding what could or couldn't be done during the hectic and exciting launch and early orbit operations period. The ORWG worked out details such as the Sequence of Events (SOE), data flows, orbit determination accuracy requirements (for both burn planning and tracking), and ground operations constraints. Once details were worked out by the ORWG, an Interface Control Document (ICD) was written to capture the details. The ICD contained details such as: who was to provide what to whom and when; where data would be located; data formats; reference frames; units; etc. The various groups were then able to develop the necessary tools and procedures for each to do their part of the orbit raising process.

A high-level diagram of the QuikSCAT orbit raising process is shown in Figure 10. Spacecraft housekeeping data (including GPS position and velocity) is telemetered to LASP via the ground stations. The GPS navigation solution data are extracted from the housekeeping files and given to the orbit determination system. This system was built and optimized by CCAR, and is based on the MicroCosm orbit determination software. One of the products of this system is a file of the orbit state at all apogees for the next 24 hours. Ball selects one of the states for the start of the next cluster of burns; this selection is influenced by the contact times provided by LASP --- i.e. the first burn of cluster is performed while in real-time contact with ground station, if possible. Software written by Ball then provides LASP and JPL a file of burn times, durations, and directions for each burn in the next cluster. Also produced is file of expected accelerations, which are used by the MicroCosm software to propagate the trajectory through the all burns of the cluster for communication antenna pointing. JPL provides independent validation of the burn cluster design. LASP uses a procedure template to generate the complete time-tagged sequence of commands to upload to the spacecraft Command Storage Memory (CSM) --- including all spacecraft attitude maneuvers, state changes, enabling and disabling thruster power bus, and telemetry format changes. The Ball ACS engineer provides updated control system tables, based on previous burn performance and expected blow-down to LASP, which then formats them for upload to the spacecraft. Ball reviews the command sequence and table uploads and provides the go/no-go decision for upload to the spacecraft. The commands are then executed autonomously by the spacecraft from CSM.

Each piece of the process was tested individually before testing as a whole. The ORWG determined what tests and criteria were acceptable to declare each piece "verified". The whole process was then verified by using the QuikSCAT Software Test Bench (SWTB --- a closed-loop spacecraft simulation environment with an engineering unit Spacecraft Control Computer and Command and Telemetry Unit) and a T1 link to LASP. Only a few minor formatting errors were uncovered during the integrated test. More testing was performed during formal mission operations rehearsals where simulated anomalies were applied to the spacecraft and personnel to verify robustness of the process (e.g. key personnel got the "flu" so that backup personnel had to step in and perform the necessary tasks).

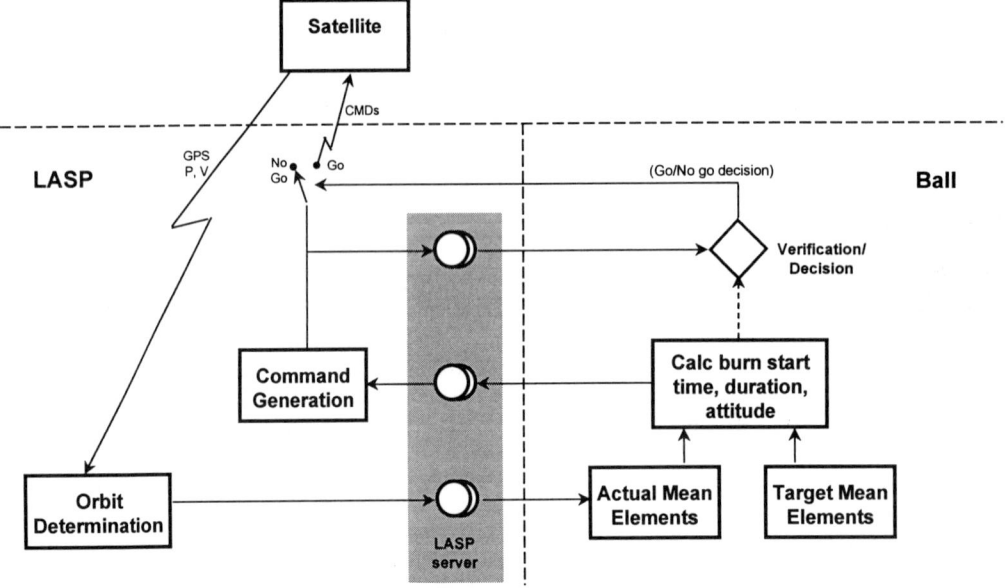

Figure 10. Orbit Raising Process Data Flow

Orbit Requirements

The QuikSCAT orbit requirements are:
- the spacecraft shall be in a 803 km altitude near-circular orbit;
- the ascending node shall remain within +/- 30 minutes of 6 AM over the three year mission;
- the altitude shall remain constant to within 1 km over the three year mission;
- the orbit shall remain frozen.

(A frozen orbit is one in which the eccentricity and argument of perigee are chosen such that the geopotential perturbations on perigee and eccentricity cancel, resulting in nearly stationary values of these mean elements. Frozen orbits provide nearly constant altitude history from orbit to orbit, and so are desirable for many remote-sensing missions.)

The second requirement implies a sun synchronous orbit, with the launch time and orbit raising carefully planned. The third requirement may dictate the performance of drag make-up maneuvers later in the mission. The last requirement implies careful adjustment of eccentricity and argument of perigee at the start of the mission.

The mean orbit elements for the mission are required to be:
- Semi-major axis   7181 +/- 1 km
- Eccentricity   0.00115 +/- 0.00015
- Inclination   98.615 degrees
- Argument of perigee   90 +/- 15 degrees

Injection Orbit

The injection orbit was estimated using GPS data from the on-board receiver, and is shown in Table 1. The actual and expected 3-σ launch dispersions are shown in Table 2. All values shown in Tables 1 and 2 are mean orbit elements.

Table 1. QuikSCAT Injection Orbit

Parameter (mean values)	Value
Semi-major axis	6921.877 km
Eccentricity	.03899
Inclination	98.650 deg
Longitude of the ascending node	355.914 deg
Argument of perigee	147.361 deg
Apogee altitude	813.6 km
Perigee altitude	273.8 km

Table 2. QuikSCAT Launch Dispersions

Parameter	Target	Actual	Dispersions	
			Expected (3σ)	Actual
Apogee altitude	797 km	814 km	22.4 km	16.6 km
Perigee altitude	as high as possible	274 km	0.4 km	n/a
Inclination	98.615 deg	98.650 deg	0.06 deg	0.035 deg

Burn Parameters

In order to perform its science mission, the QuikSCAT spacecraft was changed from the eccentric injection orbit to the near-circular final orbit. This change consisted mainly of raising the orbit perigee from 274 km to 795 km. The other orbit elements were also adjusted to their final values.

Approximately four hours of thruster burning was required to move QuikSCAT into its operational orbit. The spacecraft has relatively small (1-lb) thrusters, which can change spacecraft velocity by approximately 1 m/s per minute of burning (near the start of the mission). The propellant tank was loaded with 76 kg of hydrazine, which can produce about 190 m/s of velocity change for QuikSCAT. The propellant load was selected to maximize the total amount of velocity change available.

Orbit raising was performed using clusters (or batches) of five burns. Each burn was centered on apogee. The first burn of a cluster (as well as the test and calibration burns) was performed while in contact with a ground station; subsequent burns were performed where they were most efficient without regard to ground station visibility. The clusters were performed two days apart to allow time for tracking, orbit determination, burn cluster design, building the command load, and upload to the spacecraft. The commands to perform the entire cluster were built and uploaded in a single block. The burns of a cluster were separated by two orbits, to allow for monitoring and evaluation of performance. During the contacts between burns of a cluster, real-time telemetry was examined to evaluate performance. This evaluation was based on: propellant tank pressure decrease; thruster, valve, and tank temperatures; and orbit element changes from the GPS data and on-board filter. If these items were within 10% of the expected values, then the burn was assumed to have executed nominally, and the next burn of the cluster was allowed to execute. If any of these items had been significantly different than expected, the CSM would have been canceled to abort the remainder of the cluster.

Each burn was ten minutes long. This introduced some inefficiency, as perigee is most efficiently changed by an impulsive burn performed at apogee. A Matlab program was written to determine the efficiency of long burns. This program divided the maneuver into small segments and calculated the change in orbit elements due to an impulsive maneuver applied at each time. Efficiency is defined in this context as a combination of two factors: as the amount of actual perigee increase produced by the burn as compared to an impulsive (instantaneous) burn, and as the amount of (unwanted) apogee increase. The inefficiency of ten minute long burns was estimated in this way to be approximately 1.5%, which was judged to be acceptable.

Propulsion System

Table 3 shows the propulsion tank pressure and mass of fuel remaining after each burn, the velocity change produced, and the resulting perigee altitude.

Table 3. QuikSCAT Fuel Pressure and Mass

Event (burn)	Tank pressure (psi)	Fuel mass (kg)	Delta-V (m/s)	Perigee altitude (km)
Launch	402	76.0	n/a	273.8
Test (30 sec)	399	75.7	0.42	274.2
Calibration (5 min)	350	73.1	7.17	294.6
Cluster 1, burn 1	286	68.9	11.86	336.8
Cluster 1, burn 2	247	65.3	10.46	373.6
Cluster 1, burn 3	219	61.9	9.49	407.8
Cluster 1, burn 4	199	58.9	8.76	439.1
Cluster 1, burn 5	184	56.2	8.20	468.8
Cluster 2, burn 1	171	53.5	7.77	495.3
Cluster 2, burn 2	160	50.9	7.39	522.3
Cluster 2, burn 3	151	48.4	7.06	548.3
Cluster 2, burn 4	143	46.0	6.79	573.3
Cluster 2, burn 5	136	43.6	6.54	597.6
Cluster 3, burn 1	130	41.4	6.33	620.5
Cluster 3, burn 2	125	39.3	6.14	643.5
Cluster 3, burn 3	120	37.2	5.97	665.9
Cluster 3, burn 4	116	35.2	5.82	687.8
Cluster 3, burn 5	112	33.2	5.68	709.3
Cluster 4, burn 1	108	31.4	5.56	728.2
Cluster 4, burn 2	105	29.5	5.44	749.0
Cluster 4, burn 3	102	27.6	5.34	769.4
Cluster 4, burn 4 (8 min)	99	25.8	3.93	784.4
Cluster 4, burn 5 (inc)	98	25.0	4.81	784.4
Trim 1	96	23.8	3.33	794.7
Trim 2	95	22.6	3.57	794.7

Figures 11 and 12 show the propellant tank pressure and delta-v produced as a function of time. All burns are labeled (e.g. "1-5" means the 5th burn of cluster 1). The observed behavior of the tank pressure and realized delta-v decreasing rapidly at first and leveling out later is as expected.

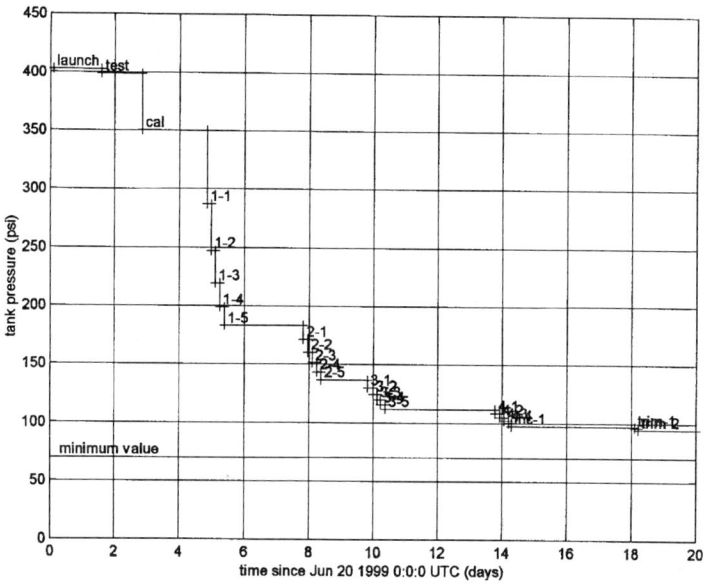

Figure 11. Propellant tank pressure versus time during QuikSCAT orbit raising

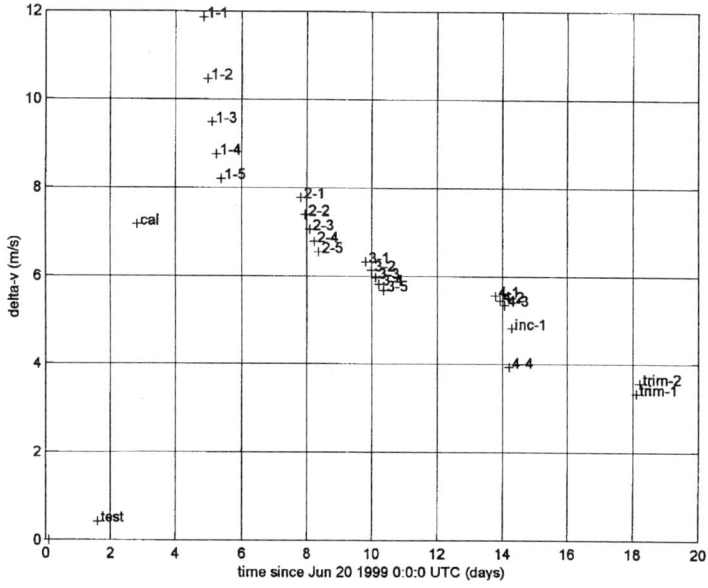

Figure 12. Delta-v produced by each burn versus time during QuikSCAT orbit raising

Orbit Adjustments

Figures 13 through 16 show the changes in the orbit during perigee raising.

The steps in perigee altitude shown in Figure 13 decrease in size due to the reduction in tank pressure and thrust. An increase in apogee altitude during perigee raising, as shown in Figure 14, was expected due to the finite (10 minute) burn duration. However, the total increase in apogee altitude of 10 km is about twice that expected. This is mostly because the burns were longer than assumed in pre-launch analysis – 10 minutes of cumulative thruster on-time were accomplished in approximately 11.75 minutes of clock time, due to off-modulation of thrusters for attitude control. The perigee raising burns circularized the orbit and so reduced eccentricity, as shown in Figure 15.

Figure 16 shows that eccentricity and argument of perigee were driven to their required frozen orbit values by a combination of perigee raising (which reduces eccentricity) and natural drift of the argument of perigee. By the time the final trim burns were performed, the argument of perigee had drifted into its required range.

The trim burns occurred about four days after cluster 4 for two reasons: a longer tracking arc was used to produce a more accurate orbit solution; and problems were encountered with one component of the burn design software (described below). The first trim burn mainly raised perigee by about 13 km. The second trim burn mainly lowered apogee by about 14 km.

Thruster efficiency calibration was typically performed before each cluster of burns based on the change in orbit from the previous set of burns. However, calibration was not possible prior to cluster 1 because the orbit solution following the 5 minute calibration burn (required to assess performance) was available only after the design of cluster 1 started.. Consequently, calibration was guessed at for cluster 1; performance was significantly different from that predicted as evident by the large and growing (up to 1 minute) time biases observed by the tracking stations. Subsequent calibrations were much more accurate; predicting very well the observed behavior and orbit changes due to each burn. Following cluster 1, time biases observed at the tracking stations were never more than a few seconds, which is in the noise.

Analysis of pre- and post-burn telemetry and orbit solutions indicated that propulsion system performance was about as expected. The mass flow rate was a few percent lower than expected, based on the observed tank pressure changes. In addition, the thrust produced was a few percent higher than expected, based on pre- and post-burn orbit solutions. These deviations are due to thruster, fuel, and other propulsion system variations, and were compensated for by including calibration factors in the burn design software.

The software used to design the burns (calculate burn times, burn magnitudes, propellant tank pressure blowdown, thrust decrease with tank pressure, etc.) worked with mean rather than osculating orbit elements. Mean elements must be used for targeting because of the large, natural variations in the osculating elements. The orbit elements were converted from osculating to mean using the Osmean software from JPL (ref. 2). The Osmean component of the burn design software was observed to produce erroneous results before and after the execution of cluster 4. The mean orbit eccentricity calculated with Osmean after cluster 4 was more than a factor of two too low. The actual eccentricity was observed by calculating and plotting the eccentricity from the GPS nav solutions, which are produced every 10 seconds and downlinked from the spacecraft. Subsequent analysis showed that this problem was present from the beginning of orbit raising, although it was less severe as well as less important for more eccentric orbits early in the orbit raising process.

The problem was finally traced to one of the inputs to Osmean (this was suggested by George Born of the University of Colorado). Previously Osmean was allowed to select the 1000 day default value for the input parameter MAXPER, which is the maximum period of the perturbations to be considered. Some of the geopotential coefficients, particularly the odd zonals, induce long period perturbations on the orbit elements. Osmean was then calculating the mean elements over approximately a 3 year period, while the desired values were the mean elements over a few days. When MAXPER was changed to 5 days, the problem disappeared; the mean elements produced by Osmean were now close to those calculated using only J2, and match those from the GPS nav and orbit determination solutions.

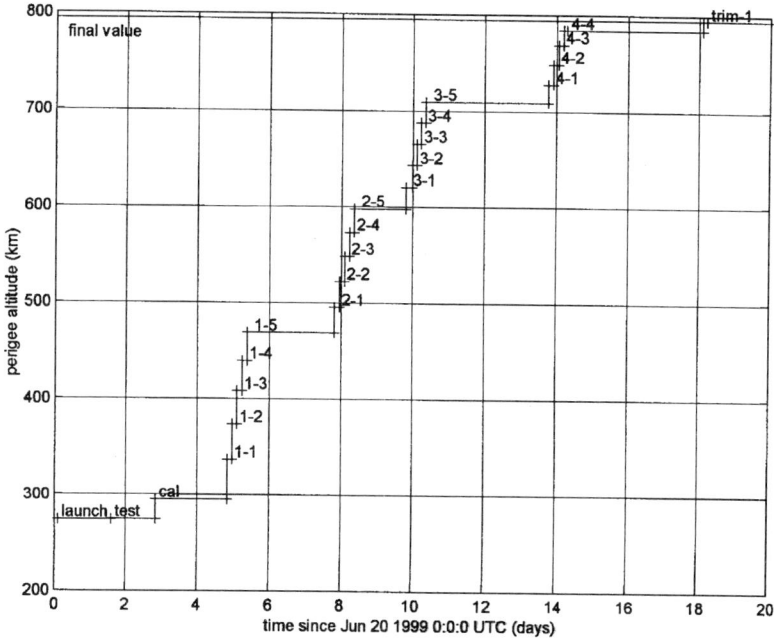

Figure 13. Perigee altitude versus time during QuikSCAT orbit raising

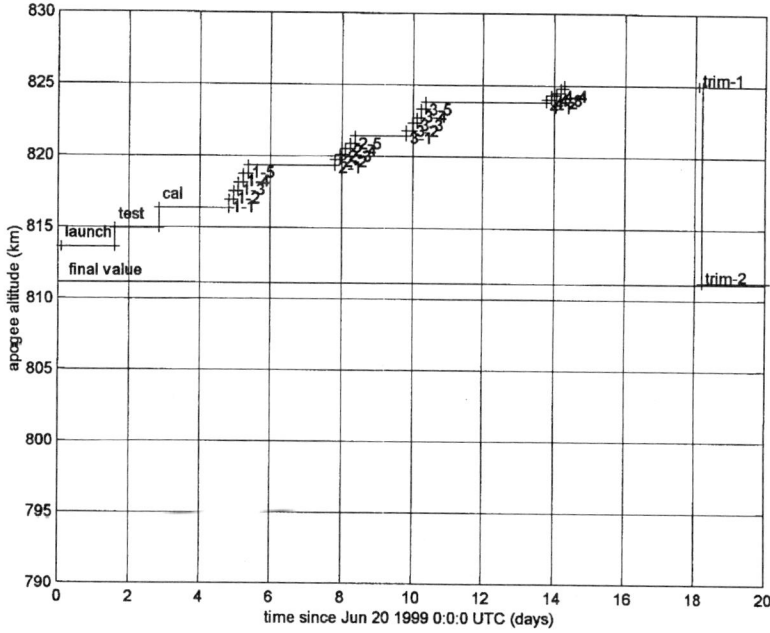

Figure 14. Apogee altitude versus time during QuikSCAT orbit raising

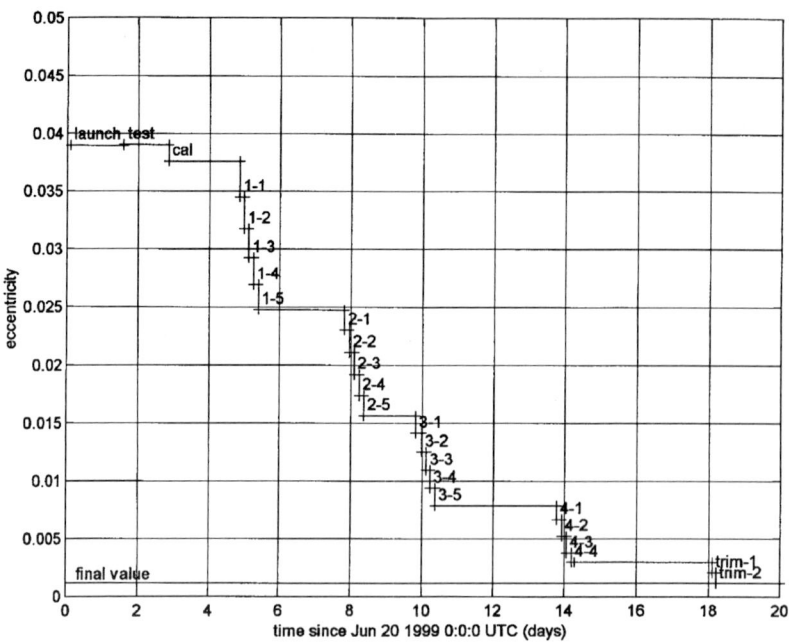

Figure 15. Eccentricity versus time during QuikSCAT orbit raising

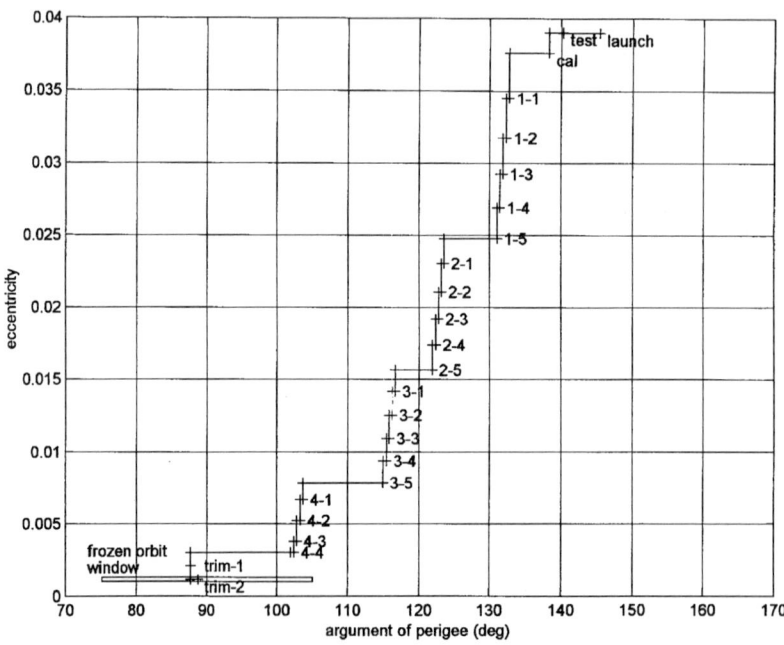

Figure 16. Argument of perigee versus eccentricity during QuikSCAT orbit raising

Observed and Expected Orbit Evolution

Figure 17 shows the actual and expected semi-major axis decrease for the first seven months of the mission. The orbit decays due to atmospheric drag; the amount of decay depends largely on solar activity. The atmospheric density was calculated using the Marshall Engineering Thermosphere (MET) model, using solar and geomagnetic activity predictions published by the Marshall Space Flight Center. Predictions assuming the +2 sigma, nominal, and -2 sigma solar and geomagnetic activity predictions, as published by Marshall, were used. The solar and geomagnetic activity prediction data were used with the LOP orbit propagation program (ref. 3). This plot indicates that the actual amount of orbit decay is within the expected range, and that based on the observed decay, the solar activity has been about 1-σ higher than nominal. When projected out for the three-year life of the mission, with the worst case (plus 2 sigma) solar activity, the orbit altitude may decay by up to approximately 2 km. In this case, an orbit maintenance maneuver will be necessary to maintain orbit altitude to within the range required. This maneuver, to increase semi-major axis by 1 km, performed approximately 1 ½ years after launch, will consume approximately 0.2 kg of fuel and increase spacecraft velocity by 0.5 m/s.

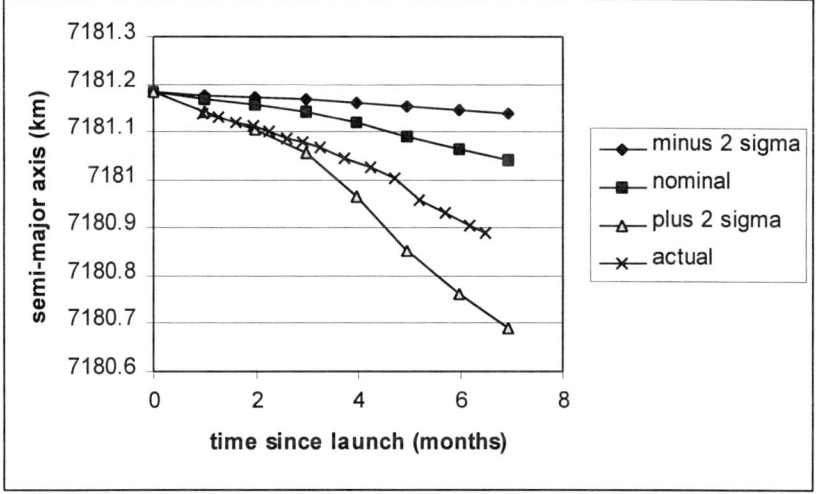

Figure 17. Predicted and actual semi-major axis decay to date

Figure 18 shows the actual and predicted eccentricity versus argument of perigee for the first seven months of the mission. To remain in a frozen orbit, these two values must oscillate within fairly narrow bounds, as shown in this figure. Longer-term orbit propagation show that the orbit will remain frozen over the 3 year mission. Since orbit decay affects the orbit elements, this analysis was performed for +2 sigma, nominal, and −2 sigma solar activity, assuming no altitude maintenance maneuvers. Results for all three cases are similar; the maximum excursions in eccentricity are from 0.0011 to 0.00125 (requirement is 0.0010 to 0.0013), while those in argument of perigee are 86 to 94 degrees (requirement is 75 to 105 degrees).

The QuikSCAT ascending node is required to be within 30 minutes of 6 AM. The node is currently about 5.5 minutes earlier than 6 AM, and drifting slowly towards earlier times. However, the drift rate is sufficiently slow that after 3 years the ascending node will be only about 11 minutes earlier than 6 AM, well within the 30-minute requirement. The target injection ascending node was biased a few minutes early. If the perigee raising process had taken longer, the ascending node would currently be closer to 6 AM. But because perigee raising was executed quickly and on schedule, the current ascending node is a few minutes earlier than 6 AM.

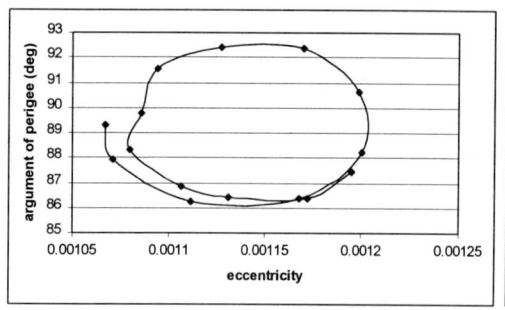 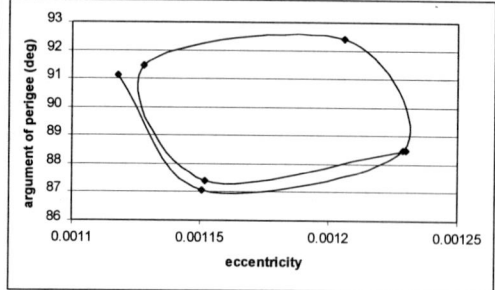

Figure 18. Actual (left) and predicted (right) eccentricity versus argument of perigee to date

ATTITUDE SYSTEM CONTINUING PERFORMANCE

Flight Software-Star Tracker Problems and Fixes

After QuikSCAT was placed in its final orbit a series of tests were performed to troubleshoot the flight software star tracker anomaly. It was found that there were two errors in the way that the FSW communicated with the star trackers.

The first problem was that the FSW was not properly commanding the trackers to perform Directed Searches (a method of telling the trackers where to look for selected stars so as to optimize spatial distribution of stars in the field of view, thereby improving the attitude solution). The second problem was that interrupts in the FSW were causing data requests to one of the trackers to occur at irregular intervals. Modifications were uploaded to the spacecraft and the problems were resolved.

Mission Performance

Once the flight software star tracker anomalies were resolved, the instrument went through a calibration phase to determine any misalignments. A new attitude quaternion was then uploaded to optimize instrument pointing. The pointing knowledge requirement for QuikSCAT is 0.05 deg 3-sigma. Analysis of on-orbit data shows that QuikSCAT is pointing to better than 0.005 deg, 3-sigma.

References

1. Wiemer, D. "Attitude Determination and Control for the Global Imaging System 2000," AAS 98-012, 21st Annual AAS Guidance and Control Conference, Breckenridge, CO, Feb 4-8, 1998.
2. Guinn, J. R. "Periodic Gravitational Perturbations for Conversion Between Osculating and Mean Orbit Elements" paper AAS 91-430, AAS/AIAA Astrodynamics Specialist Conference, Durango, CO, August 19-22, 1991.
3. Kwok, J. W. "The Long-Term Orbit Predictor (LOP)," JPL Report EM 312/86-151, 30 June, 1986.

For a look at the scientific results of QuikSCAT, go to winds.jpl.nasa.gov.

AAS 00-074

CHANDRA X-RAY OBSERVATORY POINTING CONTROL SYSTEM PERFORMANCE DURING TRANSFER ORBIT AND INITIAL ON-ORBIT OPERATIONS

Peter Quast,[*] Frank Tung,[*] John Wider[†] and Mark West[*]

The Chandra X-ray Observatory (CXO, formerly AXAF) is the third of the four NASA great observatories. It was launched from Kennedy Space Flight Center on 23 July 1999 aboard the Space Shuttle Columbia and was successfully inserted in a 330 x 72,000 km orbit by the Inertial Upper Stage (IUS). Through a series of five Integral Propulsion System burns, CXO was placed in a 9,650 x 139,200 km orbit. After initial on-orbit checkout, Chandra's first light images were unveiled to the public on 26 August 1999.

The CXO Pointing Control and Aspect Determination (PCAD) subsystem is designed to perform attitude control and determination functions in support of transfer orbit operations and on-orbit science mission. After a brief description of the PCAD subsystem, the paper highlights the PCAD activities during the transfer orbit and initial on-orbit operations. These activities include: CXO/IUS separation, attitude and gyro bias estimation with earth sensor and sun sensor, attitude control and disturbance torque estimation for delta-v burns, momentum build-up due to gravity gradient and solar pressure, momentum unloading with thrusters, attitude initialization with star measurements, gyro alignment calibration, maneuvering and transition to normal pointing, and PCAD pointing and stability performance.

INTRODUCTION

The Chandra X-ray Observatory (CXO) is the third of the four NASA great observatories. Its predecessors include the Hubble Space Telescope and the Compton Gamma Ray Observatory. Chandra was formerly known as AXAF, the Advanced X-ray Astrophysics Facility, and was renamed in December 1998 to honor Subrahmanyan Chandrasekhar, the late Indian-American Nobel laureate.

The CXO is a space astronomy mission for the observation of celestial objects radiating at X-ray wavelengths between 1.15 angstroms and 115.7 angstroms (~ 0.1 to 10 keV energy range). The CXO was designed and built by TRW with team members Ball and Kodak. The program is sponsored by NASA Marshall Space Flight Center.

A drawing of the CXO design is shown in Figure 1 (Reference [1]). CXO is composed of four elements: the spacecraft system, the telescope system, the science instrument module, and the science instruments. The flight sheet of each element can be found in the Appendix. The spacecraft system constitutes the core vehicle and houses the avionics and power generation and distribution functions.

---

[*] TRW Space & Electronics Group, One Space Park, Redondo Beach, California 90278.
[†] NASA Marshall Space Flight Center, Huntsville, Alabama 35812.

The telescope system consists of the high resolution mirror assembly (HRMA) and the optical bench assembly. The science instrument module (SIM) is placed at the aft end of the telescope. In addition to support for the two focal plane science instruments, the SIM provides the mechanism for interchanging the instruments and adjusting their position at the focal plane of the HRMA. There are two objective transmission gratings mounted at the aft end of the HRMA.

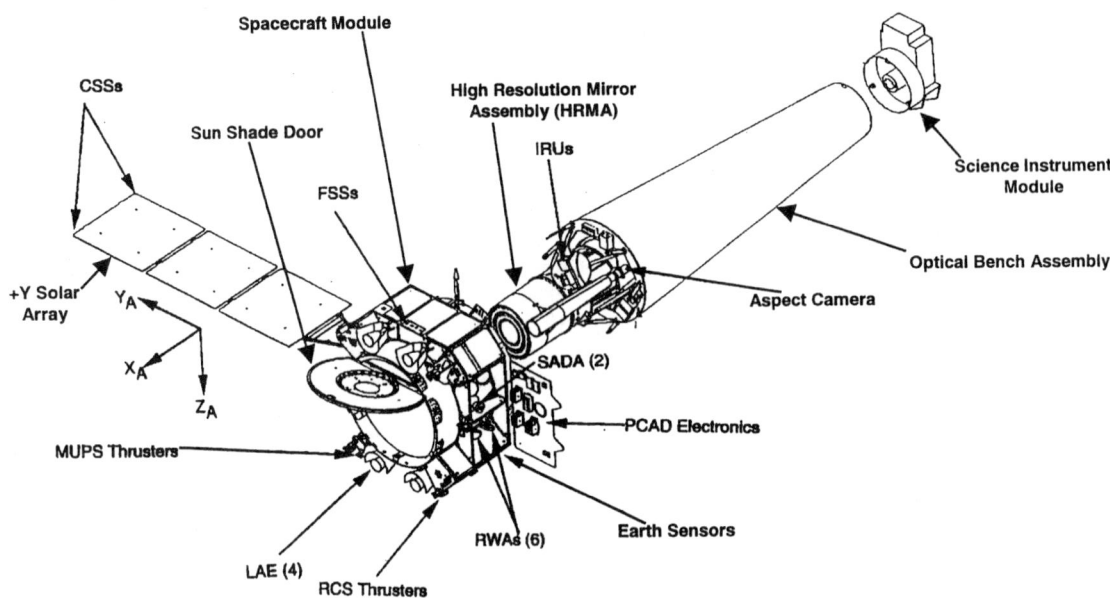

Figure 1. The Chandra X-ray Observatory

POINTING CONTROL AND ASPECT DETERMINATION SUBSYSTEM

The Pointing Control and Aspect determination (PCAD) subsystem points the CXO at desired science targets, slews it to new targets, supplies data and algorithms for post-facto image reconstruction, and provides safe modes in response to detected failures. PCAD also performs the attitude control and determination functions during both powered flight and coast phases of the transfer orbit.

Subsystem Configuration

The PCAD hardware configuration is shown in Figure 2. The algorithms for the normal operational modes are implemented in the on-board computer (OBC). Attitude control during safemode contingency operations is performed by the Control Processing Electronics (CPE) of the Control Electronics Assembly (CEA). The PCAD equipment list is presented in Table 1.

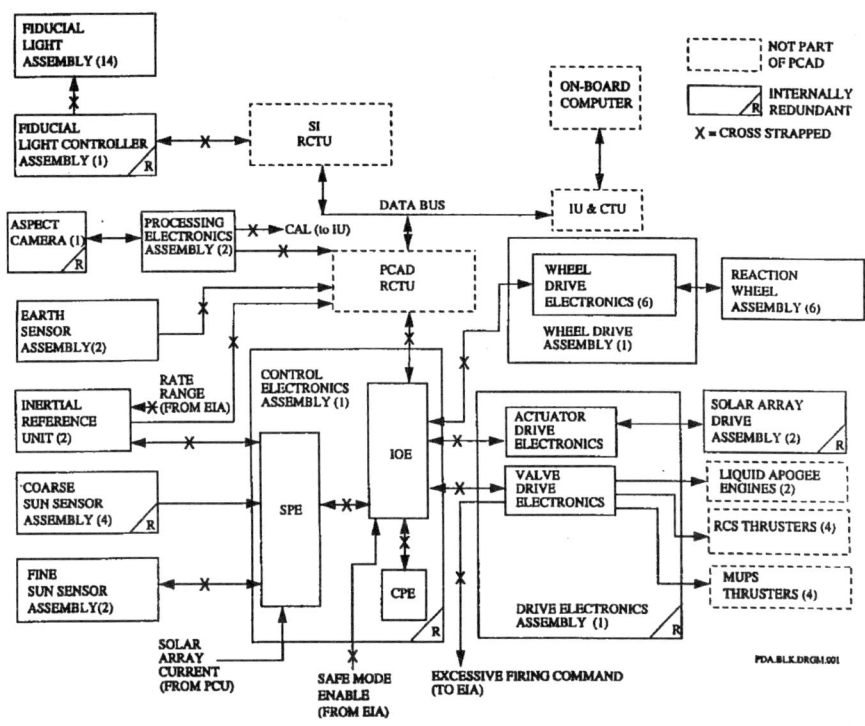

Figure 2. PCAD Hardware Configuration

Table 1 PCAD Equipment List

EQUIPMENT	TOTAL WEIGHT (LB)	TOTAL POWER (W)	QUANTITY	SOURCE	HERITAGE
Inertial Reference Unit (IRU)	28.8	28.5	2	Kearfott	Classified
Aspect Camera Assembly (ACA)	68.1	14.1	1	BASD	CT-601
Fiducial Light Assembly (FLA)	1.3	0.1	14	TRW	New
Fiducial Light Controller Assembly (FLCA)	6.3	4.8	1	TRW	New
Fine Sun Sensor Assembly (FSSA)	8.8	2.2	2	Adcole	TRMM
Coarse Sun Sensor Assembly (CSSA)	0.2	0	4	TRW	GRO
Reaction Wheel Assembly (RWA)	116.4	130.6	6	Teldix	GOES
Wheel Drive Assembly (WDA)	23.6	98.4	1	TRW	IRAD
Control Electronics Assembly (CEA)	16.8	19.6	1	TRW	IRAD
Drive Electronics Assembly (DEA)	13.8	14.5	1	TRW	IRAD
Solar Array Drive Assembly (SADA)	34.0	42.6	2	TRW	WCP
Earth Sensor Assembly (ESA)	11.3	7.0	2	Ithaco	New
Reaction Wheel Isolator Assembly (RWIA)	24.0	0	6	TRW	New
Total	353.4				

The aspect camera and two inertial reference units (IRUs) are mounted on the HRMA support structure. Each IRU has two 2 degrees of freedom dry-tuned gyros. The IRUs are skewed such that any two gyros can provide three-axis rate measurements. The IRU supports two rate ranges: low rate with a resolution of 0.02 arcsec/pulse and high rate with a resolution of 0.75 arcsec/pulse. High rate supports a maximum rate of 4 deg/sec. The aspect camera has two (primary and redundant) 1024x1024 pixel CCD detectors with a 1.4°x1.4° field of view. The IRU with star updates from the aspect camera provides the primary attitude reference.

Six reaction wheels, arranged in a pyramidal configuration, provide torquing and momentum storage. Each wheel is mounted on a reaction wheel isolator assembly (RWIA) to reduce the transmission of RW disturbances to the telescope and science instruments. A brief description of the RWIA is presented in Reference [2]. Each wheel is capable of generating 0.104 ft-lbf of torque and storing 50 ft-lb-sec of momentum. Secular momentum unloading is accomplished using thrusters. Four solar array mounted coarse sun sensors with overlapping field of views provide full sky coverage. A fine sun sensor assembly provides more accurate sun position data and also performs the bright object detection function (determination of improper sun attitudes). The earth sensors and fine sun sensor together provided the ground with the necessary telemetry required for ground attitude determination during transfer orbits.

Subsystem Modes

PCAD provides 6 normal modes and 3 safe modes for various phases of the CXO mission. The normal modes implemented in the OBC, support all planned operations in both transfer orbit and on-orbit phases. The safe modes, implemented in the CPE, provide contingency operation when an on-board anomaly is detected. All PCAD operational modes are summarized in Table 2. Figure 3 depicts the PCAD mode transition for normal operations.

Table 2. Summary of PCAD Operational Modes

Mode	Sensors/Actuators	Processor	Function
Standby (SBM)	IRU, FSSA, CSSA	OBC	• Process sensor data • No OBC command to actuators
Normal Pointing (NPM)	IRU, ACA, RWA, MUPS, FLA, FLCA	OBC	• Points to science target • Provide star and fiducial pixel data for post-facto image reconstruction • Provide dither function • Acquire guide stars
Normal Maneuver (NMM)	IRU, RWA, SADA, MUPS	OBC	• Slew to new science target • Slew for ESA data collection • Provide dither function
Normal Sun (NSM)	IRU, CSSA, FSSA, SADA, RWA, MUPS, RCS	OBC	• Acquire sun to solar arrays • Position solar arrays to –Z • Point –Z to sun • Hold attitude during eclipse • Rotate about sun line
Powered Flight (PFM)	IRU, RCS, LAE	OBC	• ΔV
RCS Maneuver (RMM)	IRU, RCS	OBC	• Slew to ΔV attitude

Mode	Sensors/Actuators	Processor	Function
Safe Sun (SSM)	IRU, CSSA, SADA, FSSA, RWA, MUPS	CPE	• Acquire sun to solar arrays • Position solar arrays to –Z • Point –Z to sun • Hold attitude during eclipse • Rotate about sun line
RCS Safe Sun (RSM)	IRU, CSSA, SADA, FSSA, RCS	CPE	Same as above
Derived Rate Safe Sun (DSM)	IRU (1 gyro), CSSA, SADA, FSSA, RWA, MUPS	CPE	• Acquire sun to solar arrays • Position solar arrays to –Z • Point –Z to sun • Hold attitude during eclipse

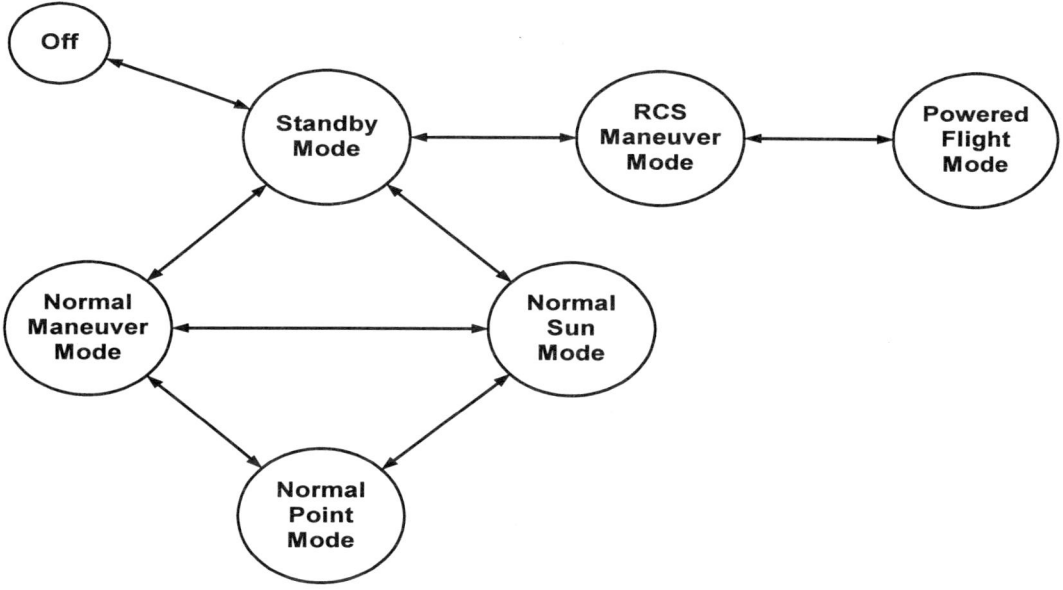

Figure 3. PCAD Normal Mode Transitions

Operational Constraints

Although there are many sun exclusion constraints for the observatory, the constraints can be generalized to the following:

1. Keep the sun within +/-20° roll of a nominal roll attitude. (Nominal roll is defined as orientations for which the center of the sun is on the –Z half of the XZ plane). This constraint is determined primarily by the geometry of the sun shades which are used to shield science instruments.
2. Maintain sun shade door shadowing of the High Resolution Mirror Assembly (HRMA). For nominal roll orientations, this corresponds to not allowing the sun to get within 45° of the +X axis.

Key PCAD Performance Requirements

Key system level performance requirements and their allocations to PCAD are summarized in Table 3.

Table 3 Key PCAD Requirement

Requirements	System Level Requirements	PCAD Allocation
Absolute LOS Pointing	Within a radius of 30 arcsec 99% of the time	4 arcsec (1σ) per axis
Pointing Stability	Less than 0.25 arcsec (rms) half-cone angle 95% of all 10 second periods	Less than 0.12 arcsec (rms) per axis 95% of all 10 second period
Maneuver Time	Maneuver 90 deg in 45 minutes with no wheel failure	Maneuver 90 deg in 45 minutes with no wheel failure

TRANSFER ORBIT OPERATIONS

PCAD was activated 13.5 minutes prior to separation from IUS. One minute after separation from IUS, PCAD entered normal sun mode to damp out separation rates and maintain solar arrays at a sun pointed attitude. PCAD stayed in the normal sun mode for most of the transfer orbit except for earth viewing maneuvers and Delta-V burns. The Chandra transfer orbit history is presented in Table 4 and depicted in Figure 4.

Table 4 Significant Events and Timeline, 1999

Date	Time	Event
23 July	04:31	Liftoff of STS 93
23 July	11:47	Chandra/IUS Deploy from Space Shuttle (268 km x 295 km)
23 July	12:54	IUS Burns Complete (330 km x 72,000 km)
23 July	13:49	Separation of Chandra from IUS
25 July	01:15	IPS 1 (1190 km x 72,000 km)
26 July	01:52	IPS 2 (3464 km x 72,000 km)
31 July	22:51	IPS 3 (3480 km x 139,200 km)
4 August	16:35	IPS 4 (5650 km x 139,200km)
7 August	05:43	IPS 5 (9,650 km x 139,200km)
12 August	17:59	Sun Shade Door Open
12 August	21:00	Fine Attitude Initialization Complete

Delta-V Burns

After deployment by the Space Shuttle Columbia on 23 July 1999 at 7:30 MET, a two stage Inertia Upper Stage (IUS manufactured by Boeing) was used to transfer the Chandra Observatory from the shuttle parking orbit (268 x 295 km) to an intermediate orbit of 330 x 72,000 km with a period of 24.3 hours. A series of 5 burns were then performed over the following 14 days using Chandra's integral propulsion system (IPS) to transfer the observatory to an orbit of approximately 9,650 x139,200 km and period of 63.5 hours. The burns were each performed at inertially fixed attitudes with the first two performed at apogee followed by one performed at perigee and two more performed at apogee.

The four liquid apogee engines (LAEs) on Chandra are grouped into sets of two diagonally opposed primary engines and two diagonally opposed redundant engines. Each engine produced approximately 106 lbs of thrust. During LAE burns, only the primary or redundant set of LAEs were used at one time. Attitude control was maintained using the four 20 pound thrusters of the reaction control system (RCS).

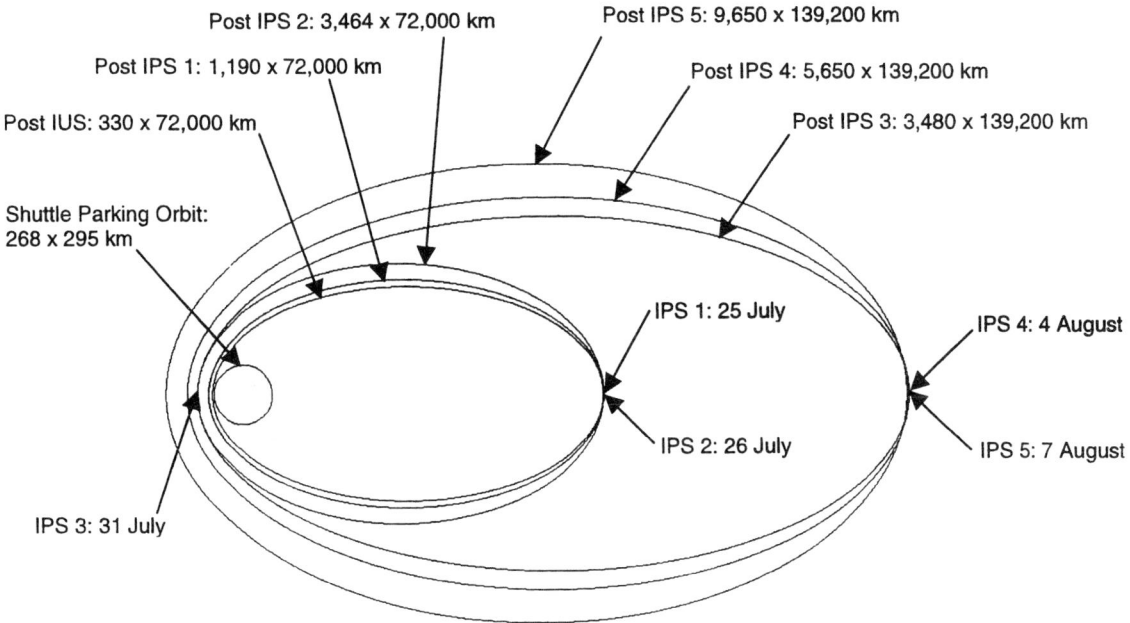

Orbital Data at Mission Start:
Perigee altitude: 9,650 km
Apogee altitude: 139,200 km
Inclination: 28.5 degrees
Right Ascension of the Ascending node: 200 degrees
Argument of perigee: 270 degrees

Figure 4. Chandra Transfer Orbit History

After each burn, detailed analysis was conducted to assess the performance of the IPS and attitude control subsystem. Following the IPS Burn 3, detailed review of the IPS and PCAD telemetry data as well as the DSN state vector (as processed by JPL) revealed indications of anomalous behavior during the burn. The three main indications were as follows:

1. The state vector showed a shortfall in effective thrust against that predicted by approximately 0.9%.
2. The disturbance torque, as calculated by RCS firings commanded during the burn, did not match those predicted by the center of mass migration alone. The asymmetric thrust calculated to account for the full disturbance torque corresponded to a difference in thrust between the engines at the end of IPS 3 of approximately 0.7%.
3. The temperature signature on the LAEs was different between IPS burns 1 and 2 and IPS burn 3.

Based on the observed performance and analysis, the decision was made to perform the remaining burns on the redundant engines. This conclusion was based on the belief that there was lower risk by performing the final burns on the redundant engines. The subsequent burns were performed satisfactorily. The true cause and effect of the anomaly is not yet completely understood.

During the burns, the sun-orbit configuration on the date of the burns resulted in the sun position being in the XZ plane of the observatory and within 20° of the –Z axis. This sun position in the body frame resulted in the relatively accurate inertial attitude knowledge provided by the fine sun sensors translating to very accurate attitude knowledge predominately about roll and pitch during the burns. The less accurate inertial attitude knowledge about the sun line

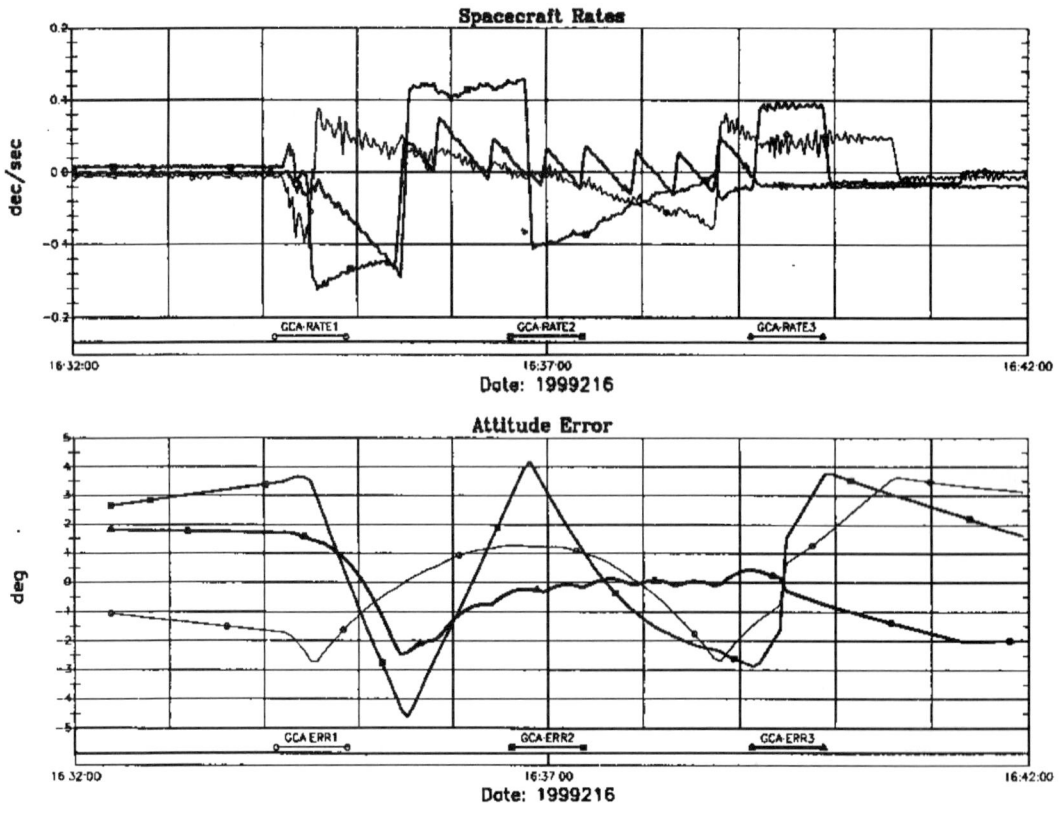

Figure 5. Rate and Attitude Error During IPS 4

provided by the earth sensors resulted in slightly less accurate attitude knowledge about the yaw axis during the burns.

The rate and attitude error history during IPS 4 is presented in Figure 5. The performance illustrated is typical of that observed during all IPS burns. Note that although the deadband behavior produces attitude errors greater than 1°, the overall averaged effective attitude error over the entire burn is less 1°. Indeed, based on the assessment of the delta-V vector after each burn provided by the JPL, the resultant true delta-V vector generated during each burn was within on the order of 1° of that desired. This was well below the tolerance needed for suitable attainment of the final orbit desired and indicated very good attitude control performance as well as very accurate onboard attitude knowledge during all burns.

Attitude and Gyro Bias Estimation

Attitude knowledge is maintained by propagation of spacecraft rates as measured by Kearfott SKIRU V gyroscopes. During transfer orbit, updates to the onboard estimated attitude and gyro bias were made periodically by the ground. The onboard attitude was initially updated based on the attitude reported by the IUS at time of separation. Throughout the duration of transfer orbit, the attitude and gyro bias updates were calculated by the ground using data from the fine sun sensors and the earth sensors. Since the sun shade door was closed during transfer orbit, aspect camera information was not available for updating attitude and gyro bias knowledge.

Using fine sun sensor data, the inertial attitude knowledge of the observatory about axes orthogonal to the sun line was determinable to within approximately $0.01°$. However, the sun sensors are in general unable to provide accurate inertial attitude knowledge about the direction of the sun line. Attitude knowledge about the sun line was updated by determination of earth nadir vector locations calculated using measurements made with the earth sensors during earth scan maneuvers. These maneuvers were performed on both the ascending and descending legs of each orbit during transfer orbit between optimal altitudes of approximately 15,000 km and 60,000 km. Based on the inherent accuracy of the earth sensors combined with the degree of precision recoverable by calibration, the observatory attitude knowledge about the sun line, as determined by the earth sensor measurements, was accurate to within approximately $0.5°$.

The earth sensors were manufactured by Ithaco Corporation and calibrated by TRW. The instantaneous field of view of each earth sensor subtends an arc of approximately $1.5°$. While scanning, this field of view prescribes a cone which has an angle between the symmetric axis of the cone and the cone surface of approximately $45°$. Each scan cone was electronically blanked over an arc of approximately $200°$. The earth sensors were oriented on the observatory such that the center of the remaining unblanked arc is centered near the +Z axis of the observatory. The earth viewing attitudes were constructed so that the earth would pass across the scan cones while the observatory maintained sun exclusion constraints.

ON-ORBIT ACTIVATION AND CHECKOUT

Fine Attitude Initialization

For the initialization of attitude knowledge after sun shade door opening, the attitude uncertainty resulting from the limited accuracy of the fine sun sensor and the earth sensors prohibited the identification of the first star acquisition using the on-board NPM software. Ground software was developed which utilized Aspect Camera telemetry of acquired star positions to generate a fine attitude update, which could then be uploaded to the spacecraft. Due to concerns with the first use of the nominal fine attitude initialization ground software and the initial Aspect Camera star acquisition, a backup manual attitude determination ground software capability, which relied on visual pattern recognition, was developed based upon a similar system developed for the NASA Spacelab Astro-2 mission. The backup fine attitude determination system was successfully utilized to remove the initial fine attitude error of approximately 7 degrees about the sun line. Figure 6 shows the pattern recognition display after the successful attitude initialization. The rectangles indicate the ACA real-time field of view position telemetry for the eight successfully acquired stars.

Hardware failures or ground commanding problems leading to safemode or failure of the autonomous on-board star acquisition will require subsequent fine attitude initialization procedures. The nominal ground software attitude initialization software was successfully utilized to provide an attitude update to the CXO after a safemode recovery on 19 August 1999.

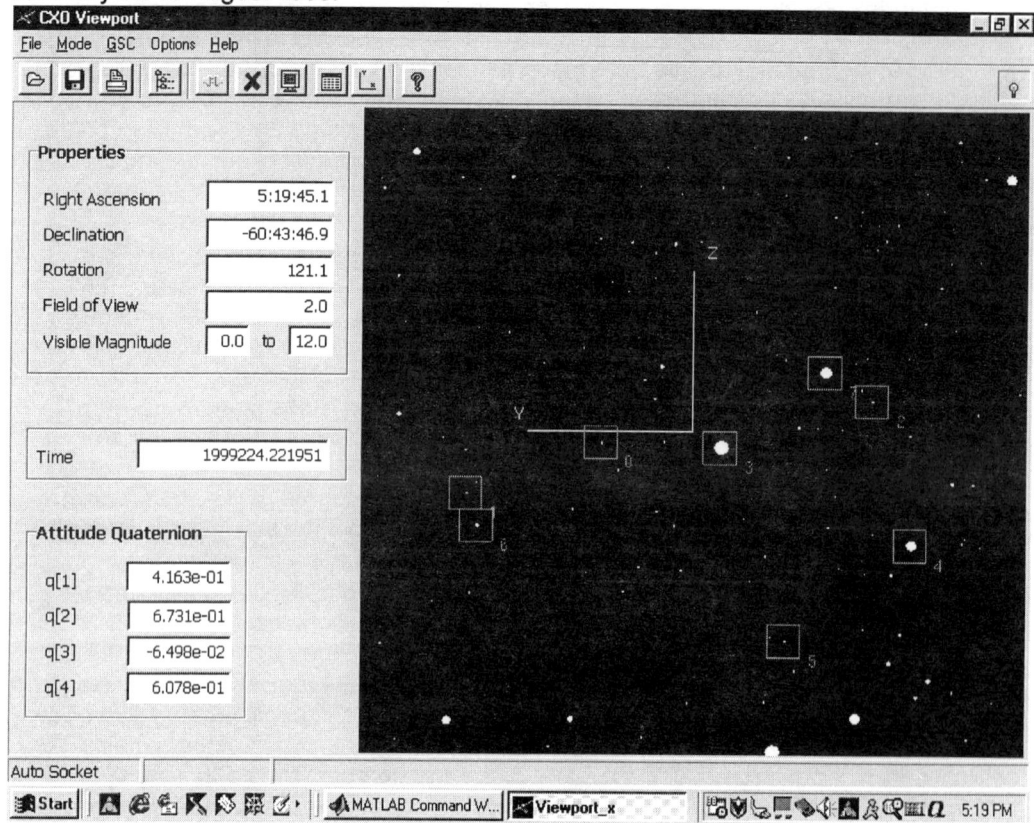

Figure 6. Star Pattern Matching for Attitude Initialization

Pointing and Stability Performance

Because the photon flux from most x-ray sources arrives at relatively low rates, long exposure times ranging from several minutes to hours will be required for most observations. Photon arrivals are time-tagged and, with the aid of time-tagged attitude and fiducial light data from the IRU and aspect camera respectively, the image is reconstructed on the ground. Rather than requiring the observatory to maintain precision target pointing over long observation intervals, CXO has a relatively relaxed absolute pointing requirement but a very tight pointing stability requirement. These requirements apply to Normal Pointing Mode, including during dither motion, but excluding the transition time after a maneuver. These requirements also apply for 15 minutes after completion of momentum unloading.

PCAD Absolute Pointing Performance. The CXO is required to point the telescope line-of-sight (LOS) to within a radius of 30 arcsec of the commanded direction 99% of the time. PCAD's share of the error budget is 4 arcsec (1σ, per axis). Figure 7 shows pitch and yaw pointing error in a 4 hour period after a maneuver. Dither mode with dither magnitude of 8 arcsec is active during this period (see later section for a description of dither capability). The pointing errors are less than 0.2 arcsec for most of the time.

PCAD Pointing Stability Performance. The top level relative LOS stability is required to be less than 0.25 arcsec (rms) half-cone angle with respect to the commanded direction over 95% of all 10 second periods. The error budget allocation to PCAD is 0.12 arcsec (rms) per axis for 95% of all 10 second periods. The PCAD pointing stability for the same 4 hour period is presented in Figure 8. The plots represent rms values of pointing errors, computed in a 10 second moving window. The plots show pointing stability is less than 0.05 arcsec with the most of the errors below 0.03 arcsec.

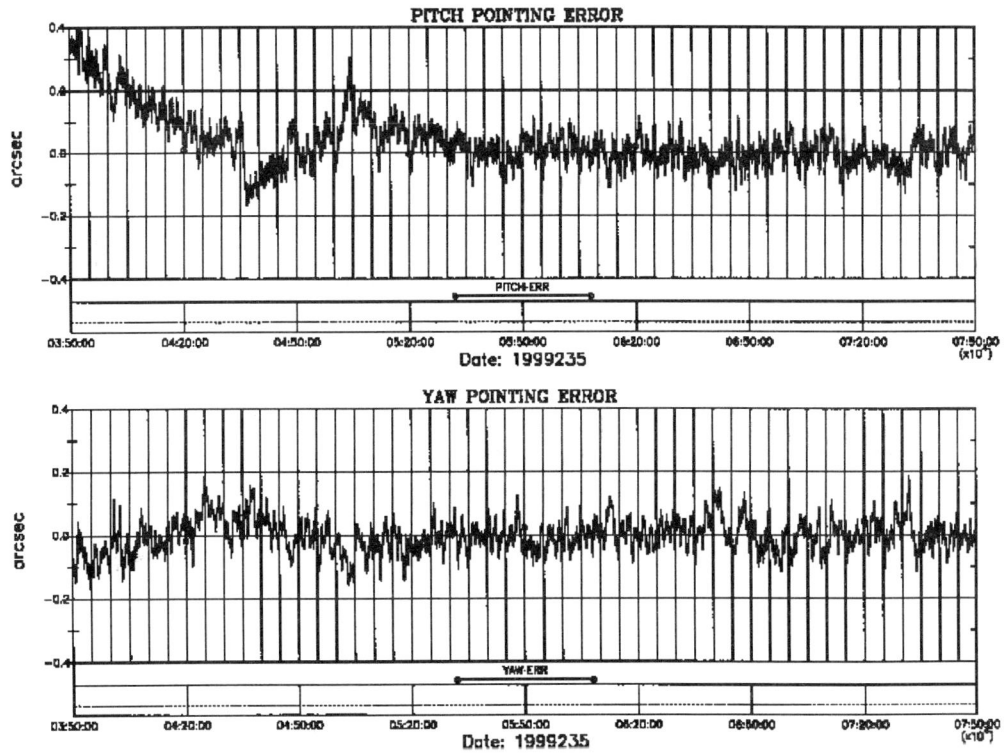

Figure 7. Pitch and Yaw Pointing Error

Momentum Unloading

Reaction wheel momentum accumulated due to disturbance torques is dumped using the Momentum Unloading Propulsion System (MUPS) thrusters. Fed by a blowdown system, the nominal beginning of life thrust is 0.26 lbf. A maximum of 3 thrusters are used at any given time, with each thruster individually pulsewidth modulated so as to align the unloading torque as closely as possible opposite the direction of momentum.

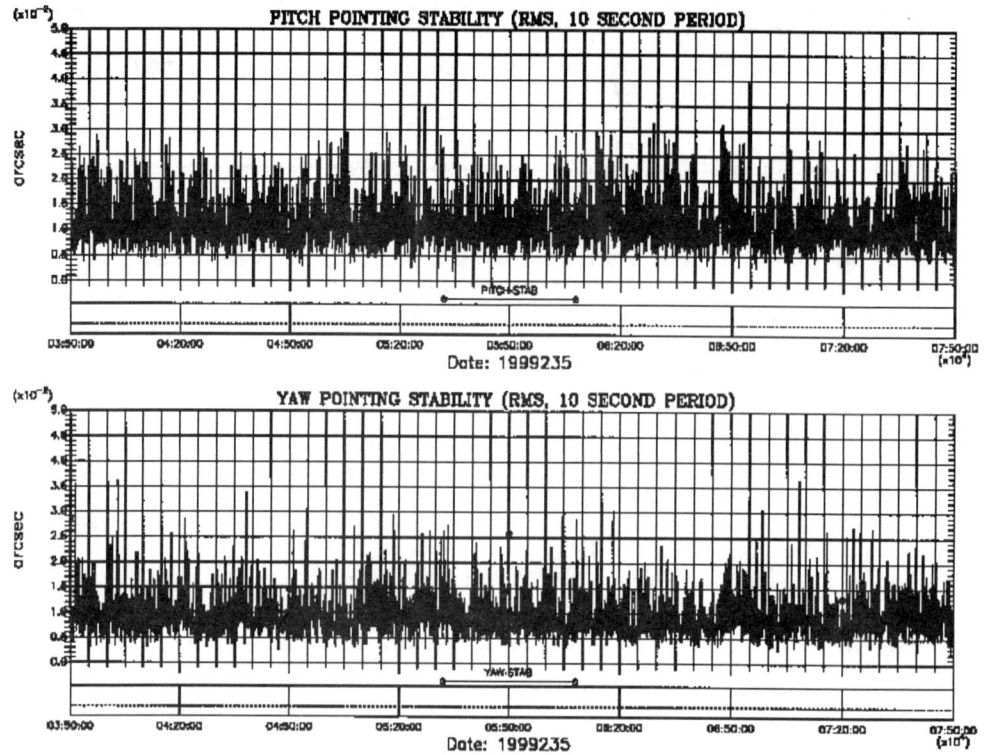

Figure 8. Pitch and Yaw Stability

The time to commence unloading and the desired final momentum state are commanded by the ground and the selection and firing of thrusters are performed on board. For minimum disruption to science data gathering, the unloading can be performed at the beginning of a long maneuver. If the accumulated momentum exceeds a pre-set value, the unloading function is performed autonomously. Figure 9 shows pointing errors during a typical momentum unloading which are less than 0.015 deg.

Prior to launch, the largest secular disturbance torque was predicted to be solar. However, the actual solar torque (see Figure 10) is only about one-half that predicted. Instead, gravity gradient torque tends to dominate the timing of momentum unloads. The less than expected solar disturbance torques have resulted in MUPS unloading fuel being expended at a rate approximately ¼ that planned for the nominal 5 year mission. The cause of the less than expected solar torque is not known, but may be due to modeling errors in the surface properties of the observatory.

Thus far in the mission, unloading performance has been nominal with the unloading control law selecting appropriate on-times and thrusters producing expected torque. PCAD's allocation for pointing (4 arc-sec) and stability (0.12 arc-sec) apply 15 minutes after completion of momentum unloading. However, the above requirements are actually met within 2 minutes following unloading. Further, pointing to within 0.4 arc-sec is met within 3 minutes following unloading.

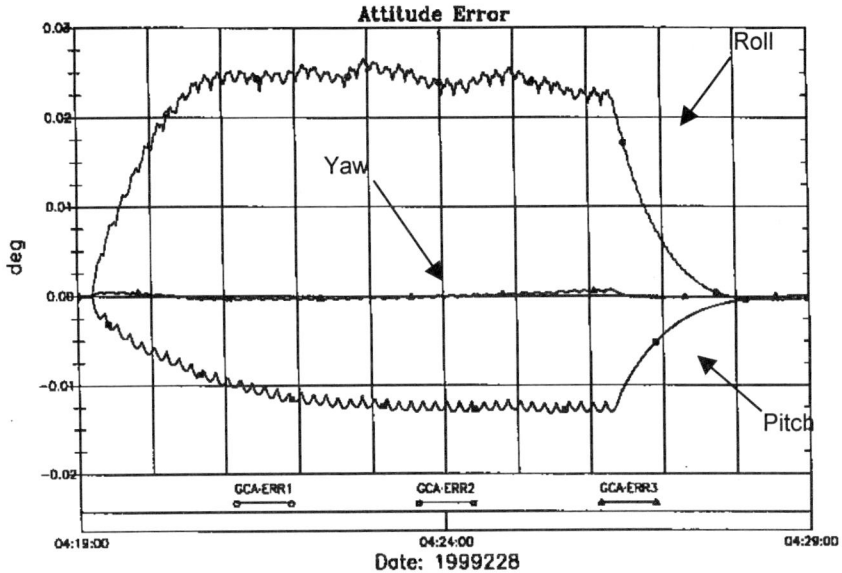

Figure 9. Attitude Errors During a Typical Momentum Unloading

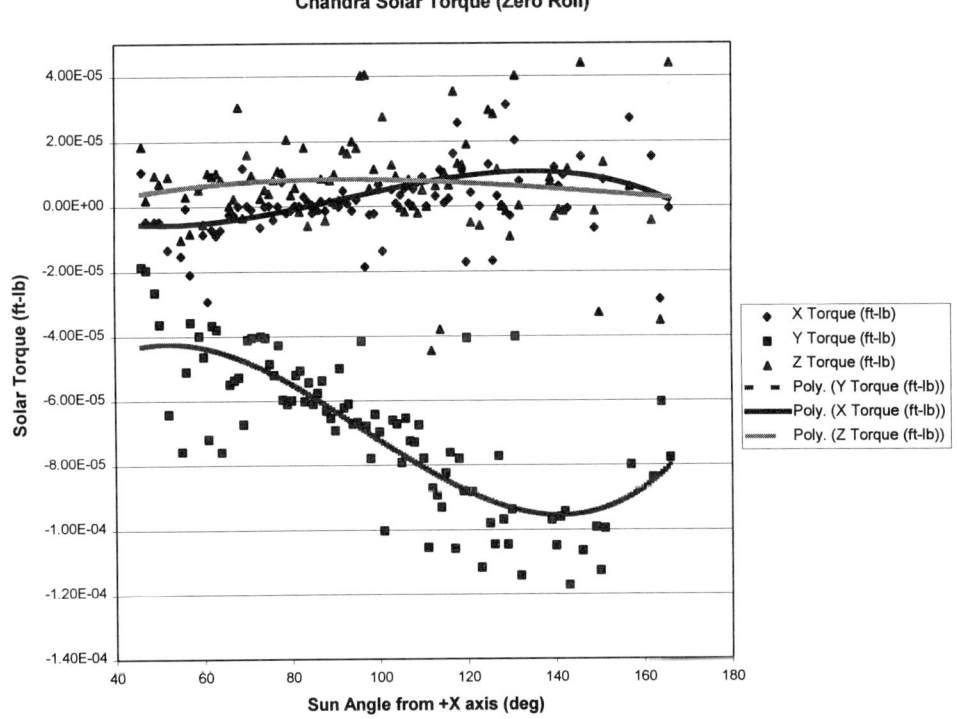

Figure 10. Actual Solar Torque Derived from Flight Data

Slew Maneuvers

The observatory maneuvers by direct eigen axis slews from a given commanded attitude to the next target attitude. The trajectory of the slew angle, angular rate and acceleration about the eigen axis is generated consistent with the maneuver parameters given in Table 5.

Table 5 On Orbit Maneuver Parameters

Parameter	Value
Jerk Time (time over which the acceleration increases to a maximum)	60 seconds
Maximum Angular Acceleration	1.25e-4 deg/sec^2
Maximum Angular Rate	0.075 deg/sec

The resulting maneuver times over all maneuver angles are depicted in Figure 11. Throughout all Chandra flight operations, maneuver performance has been nominal.

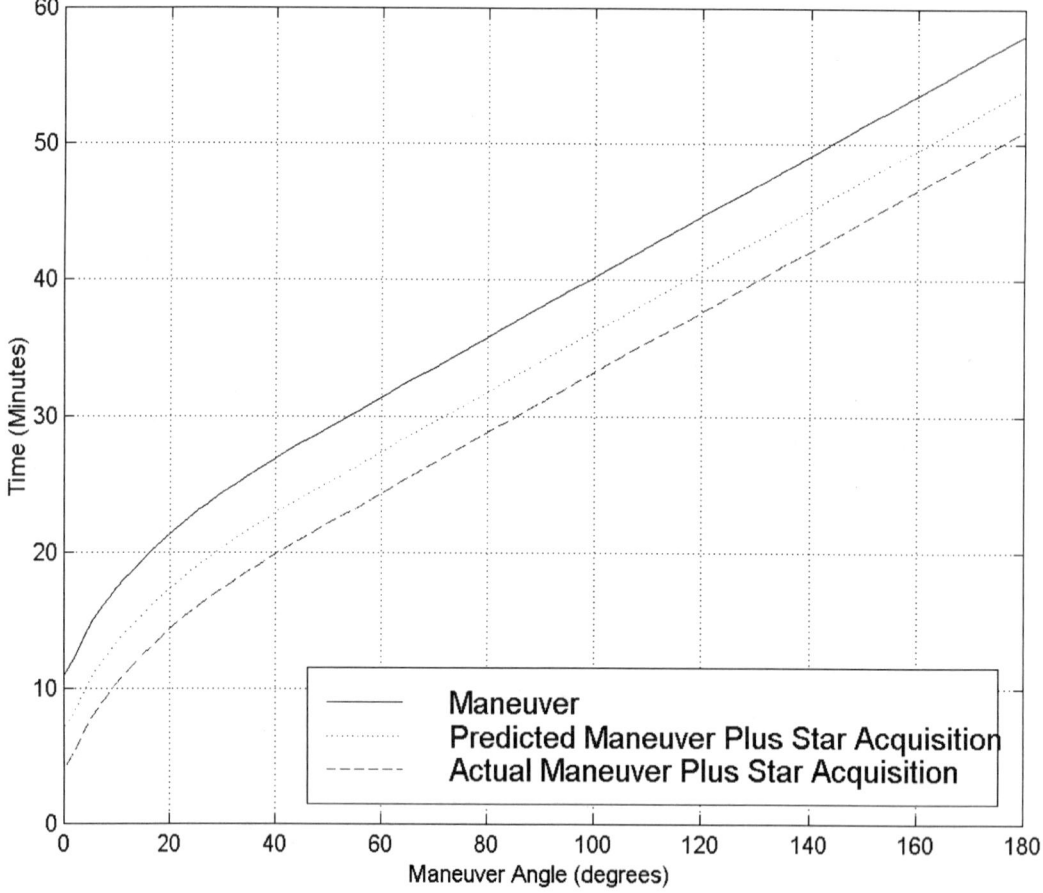

Figure 11. Maneuver Time versus Maneuver Angle

Spacecraft Rate Determination and Gyro Scale Factor and Misalignment Compensation

The estimation of the gyro axis rates is performed by the flight software each sampling period using a back difference of the gyro counts along with nominal gyro scale factors. The calculation of the estimated body axis rates using these gyro axis rates is then performed by the flight software using the calculation $\omega_{body} = [\,I+M\,]G\,\omega_{gyro}$ where I is the identity matrix, M is a misalignment matrix and G is the pseudoinverse matrix which maps gyro rates in each of the 4 gyro axis in use to the 3 body axis. The pseudoinverse matrix includes known misalignments of the IRUs as measured in calibration on the ground.

In order to account for further uncompensated gyro misalignment and scale factor differences from those already accounted for, the flight software uses the misalignment matrix M to modify the calculation of estimated body axis rates. Determination of this matrix was made during a calibration performed soon after sun shade door opening. The calibration consisted of a series of large maneuvers for which each eigenaxis was near the observatory roll, pitch or yaw axis and for which sun exclusion constraints were maintained. Before and after each maneuver, a full field search of the aspect camera was commanded and stars were acquired. Using the ground software, the precise attitude of the observatory at each star acquisition was then determined. By knowing the precise attitude before and after each maneuver as well as the time history of the gyro counts during each maneuver, the best fit estimation of the misalignment matrix M was made. Based upon on orbit performance results, the calibration was very successful. The resulting attitude knowledge errors, even after very large maneuvers, have only been on the order of tens of arc seconds.

Star Acquisition and Kalman Filter Performance

Since the sun shade door was opened, the spacecraft attitude and gyro bias estimation updates have subsequently been determined by a Kalman filter using inertial attitude data provided by the aspect camera. The aspect camera has the capability to track eight images simultaneously. For typical science operations, five stars and three fiducial lights are tracked by the aspect camera. The fiducial lights are located on the translation table (SIM Table) to which the science instruments are attached within the science module at the –X end of the observatory. The fiducial light images are reflected into the aspect camera by a series of mirrors and used for a posteriori reconstruction of the true position of the science images relative to the inertial reference of the tracked stars. Star catalogs are uploaded for each observation and logic in the flight software is used to command the aspect camera to acquire stars. After stars are acquired by the aspect camera, the position and magnitude information of the stars which were found is evaluated by the flight software logic. The data is screened to ensure that only stars which were correctly acquired by the aspect camera are actually used. A Kalman Filter implemented in flight software uses the aspect camera star measurement data to determine updates to the onboard estimates of attitude and gyro bias. Attitude and bias updates are performed by the flight software each time new data becomes available from the aspect camera. The period of updates is driven by the integration period selected for the aspect camera, but typically is between one and five seconds.

Inertial Reference Unit Performance

Both IRUs were used in high rate range prior to IUS separation, with near zero null measurement vectors demonstrating proper operation by each unit. Science operation uses IRU-1 in low rate range. IRU-2 is left off, except when powered on by autonomous

safing action. IRU acceleration insensitive drift rate (AIDR) performance specifications include:
- 7.2 arc-sec/sec absolute
- 1 arc-sec/sec life time variation
- 0.1 arc-sec/sec/month variation

All operations have been within the above specifications.

Safe Mode Performance

For safing configuration, the observatory has a backup string of sensors hardware, Communication, Command and Data Management (CCDM) hardware and On Board Computer (OBC). In safemode, there is also a dedicated processor called the Control Processing Electronics (CPE) used for attitude control. A series of monitors implemented in flight software continuously evaluate spacecraft performance and will command a failover to the redundant set of hardware as well as attitude control by the CPE in the event anomalous behavior is detected.

There have been two transitions to safe mode in the Chandra X-ray Observatory. Both have been related to ground processing errors and did not involve hardware failures. The first transition to safemode occurred on 17 August and resulted from a timing problem in the sequencing of maneuver commands. Each maneuver sequence has a series of commands associated with it, namely updating the target quaternion and commanding the start of the maneuver. If the maneuver is to be followed by a star acquisition, the star catalog is also updated at the beginning of the maneuver and a flag is set to enable the autonomous transition from NMM to NPM at the end of the maneuver. This first safemode transition occurred on the occasion of the first ever execution of a segmented maneuver by the ground software. The maneuver was segmented for sun exclusion purposes and was to involve a maneuver to an intermediate attitude after which stars were not to be acquired, followed immediately by a second maneuver to the desired target attitude after which stars were to be acquired. Not enough time was allocated by the ground system after the completion of the first maneuver sequence prior to the beginning of the second maneuver. In this case, the command to enable autonomous transition from NMM to NPM at the end of the second maneuver was incorrectly issued just seconds prior to the end of the first maneuver. At the time the target quaternion was updated for the second maneuver, the observatory was already in NPM. This updating of the target quaternion while in NPM had the effect of creating a very large attitude error resulting in the tripping of the Attitude Error / Rate Error Monitor and subsequent transition to safemode. The observatory was successfully recovered from safemode in 30 hours which included a thorough reevaluation and retesting of existing safemode recovery procedures.

The second transition to safemode occurred on 26 September and resulted from a incorrect issuance of commands during a recovery from a Bright Star Hold event. Bright Star Hold occurs when the stars which were to be acquired for an observation cannot be found. In this case, the camera is commanded to search for the brightest stars it can find and then the flight software "holds" on these stars until recovery by the ground. As a result of a slightly incorrect star catalog computed by the ground system, an attitude knowledge error of approximately 400 arc seconds was systematically introduced into the flight software during an observation. At the next observation, the desired stars were not acquired as a result of this attitude error and the observatory went to Bright Star Hold. In the process of recovery by the ground during the next DSN contact with the observatory, the process of fine attitude initialization was performed to correct the attitude knowledge error. However, the onboard attitude was inadvertently updated while in NPM. While not resulting in enough attitude error to trip the Attitude Error / Rate Error Monitor, there was enough of an attitude error for the Kalman Filter logic to conclude that the bright stars that it was currently tracking were not valid. In this case, the observatory autonomously transitioned to NSM as a safing action. As part of this safing action, the SIM table is moved to a safe location. The angular momentum discrepancy associated with moving the SIM table would ordinarily be enough to trip the Spacecraft Momentum Monitor, and therefore this monitor is disabled during all SIM table moves during normal science operations. However, this disabling of the Spacecraft Momentum Monitor had not yet been implemented as part of the onboard sequence to move the SIM table during transition to NSM. Therefore, in this case, the Spacecraft Momentum Monitor tripped and a transition to safemode resulted. The observatory was successfully recovered from safemode in 12 hours and science observations were resumed within 24 hours.

IMPROVEMENTS

Continuous Dither

Dither is the capability to superimpose a dither pattern onto the commanded attitude while maneuvering as well as while pointing "steadily" at a target. The typical dither profile is a Lissajous Pattern with maximum amplitude of 20 arc seconds.

The originally intended purposes of the dither capability was to ensure that X-ray sources do not steadily irradiate individual pores of the Micro Channel Plate (MCP) in the High Resolution Camera (HRC) during observations. Steady irradiation of the same pores while HRC voltage is ramped up could cause permanent degradation of the HRC instrument. In addition to this originally intended purpose of dither, the dither capability is now used for all observations (ACIS and HRC) and is considered particularly important for ACIS bias calibrations which occur regularly.

The Chandra X-ray Observatory was launched with the capability to dither during observations only and would not allow HRC voltage to ramp up until after stars had been acquired at the observation attitude. Further, once a maneuver began, dither stopped and HRC voltage was ramped down. In order to enhance science efficiency and reduce mission planning complexity, the flight software was altered to provide dither capability continuously at all times, including during maneuvers. This capability now allows ACIS and HRC to perform science at all times, as well as allow ACIS to perform bias calculations at any time.

CONCLUSION

The Chandra X-ray observatory was successfully activated and configured for normal operation during transfer orbit with no significant difficulties. The observatory has since performed its mission in all respects in an exemplary manner, equaling or surpassing all specifications and expectations for a successful mission to date.

REFERENCES

1. M. Hamidi, R. P. Iwens, J. F. Donaghy, J. A. Spina, J. A. Wynn, D. D. Johnston, G. F. Flanagan, and L. D. Hill, "Design of the Advanced X-Ray Astrophyscial Facility – Imaging," AIAA 93-4199, AIAA Space Programs and Technology Conference and Exhibit, September 1993.
2. C. Schauwecker, S. Shawger, F. Tung, and G. Nurre, "Imaging Pointing Control and Aspect Determination System for the NASA Advanced X-Ray Astrophysics Facility," AAS 97-064, 20th Annual AAS Guidance and Control Conference, February 1997.

APPENDIX
NASA'S ADVANCED X-RAY ASTROPHYSICS FACILITY (AXAF)

Overall Specifications

Size:	45.3 ft x 64.0 ft (solar arrays depolyed)
Weight:	10,560 pounds
Orbit:	6,200 x 86,000 miles 28.5 deg. inclination
Ascending node:	200 degrees
Argument of perigee:	270 degrees
Life:	minimum 5 years

Spacecraft Specifications

Power:	two 3-panel silicon solar arrays (2350W); three 40 amp-hour nickel hydrogen batteries
Antennas:	two low-gain, conical log spiral antennas
Frequencies:	transmit 2250 MHz, receive 2071.8 MHz
Command Link:	2 kilobits per second (kbps)
Data Recording:	solid state recorder; 1.8 gigabits (16.8 hours) recording capability
Downlink options:	selectable rates from 32 to 1024 kbps
Downlink Operations:	downloaded typically every 8 hours
Contigency Mode:	32kbps
Safing:	autonomous operation

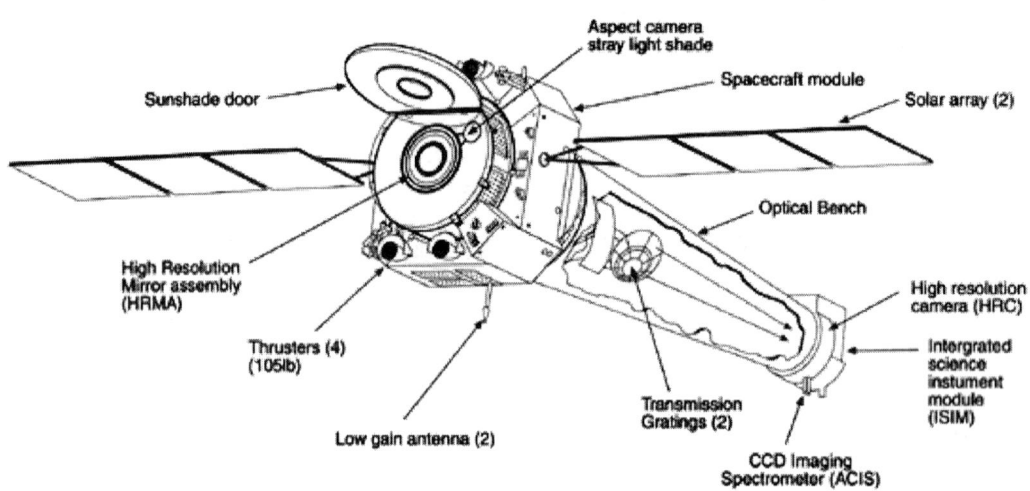

Telescope System

High Resolution Mirror Assembly:	4 sets of nested, grazed incidence mirror pairs
Length:	each 83.3 cm long
Weight:	2104 pounds
Focal Length:	10 meters
Outer diameter:	1.2 meters
Field of view:	1.0 degree diameter
Ang. resolution:	0.5 arc sec
Altitude Control:	6 reaction wheel control 2 inertial reference units
Aspect Camera:	1.40deg x 1.40deg field of view
Pointing Stability:	0.25 arc-sec (RMS) radius over 95% of all 10 second periods
Pointing Accuracy:	30 arc-sec 99% of viewing time
Remarks:	Mirrors have an effective area of 400 sq. cm. @ 1 keV; 330 Å iridium coating

Science Instruments

AXAF Charged Coupled Imaging Spectrometer (ACIS):	Ten CCD chips in 2 arrays provide imaging and spectroscopy; over an energy range 0.2 - 10 keV; sensitivity: 4×10^{15} ergs-cm^{-2} sec^{-1} in 10^5 s
High Resolution Camera (HRC):	Uses large field-of-view micro-channel plates to makes X-ray images; ang. resolution < 0.5 arc-sec over field-of-view 31x31 arc-min; time resolution: 16 micro-sec sensitivity: 4×10^{15} ergs-cm^{-2} sec^{-1} in 10^5 s
High Energy Transmission Grating (HETG):	To be inserted into focused X-ray beam; provides spectral resolution of 60-1000 over energy range 0.4 - 10 keV
Low Energy Transmission Grating (LETG):	To be inserted into focused X-ray beam; provides spectral resolution of 40-2000 over the energy range 0.09 - 3 keV

The AXAF program is managed by the Marshall Center for the Office of Space Science, NASA Headquarters.
TRW is the prime contractor and has assembled and tested the observatory for NASA.

AAS 00-076

THE RECOVERY OF TOMS-EP

Brent Robertson,[*] Phil Sabelhaus,[*] Todd Mendenhall[†] and Lorraine Fesq[‡]

On December 13th 1998, the Total Ozone Mapping Spectrometer – Earth Probe (TOMS-EP) spacecraft experienced a Single Event Upset which caused the system to reconfigure and enter a Safe Mode. This incident occurred two and a half years after the launch of the spacecraft which was designed for a two year life. A combination of factors, including changes in component behavior due to age and extended use, very unfortunate initial conditions and the safe mode processing logic prevented the spacecraft from entering its nominal long term storage mode. The spacecraft remained in a high fuel consumption mode designed for temporary use. By the time the onboard fuel was exhausted, the spacecraft was Sun pointing in a high rate flat spin.

Although the uncontrolled spacecraft was initially in a power and thermal safe orientation, it would not stay in this state indefinitely due to a slow precession of its momentum vector. A recovery team was immediately assembled to determine if there was time to develop a method of de-spinning the vehicle and return it to normal science data collection. A three stage plan was developed that used the onboard magnetic torque rods as actuators. The first stage was designed to reduce the high spin rate to within the linear range of the gyros. The second stage transitioned the spacecraft from sun pointing to orbit reference pointing. The final stage returned the spacecraft to normal science operation. The entire recovery scenario was simulated with a wide range of initial conditions to establish the expected behavior. The recovery sequence was started on December 28th 1998 and completed by December 31st. TOMS-EP was successfully returned to science operations by the beginning of 1999.

This paper describes the TOMS-EP Safe Mode design and the factors which led to the spacecraft anomaly and loss of fuel. The recovery and simulation efforts are described. Flight data are presented which show the performance of the spacecraft during its return to science. Finally, lessons learned are presented.

---

[*] NASA Goddard Space Flight Center, Greenbelt, Maryland 20771.
[†] TRW, One Space Park, Redondo Beach, California 90278.
[‡] Massachusetts Institute of Technology, 77 Massachusetts Avenue, Cambridge, Massachusetts 02139.

INTRODUCTION

The Total Ozone Mapping Spectrometer - Earth Probe (TOMS-EP) is a National Aeronautics and Space Administration (NASA) mission to continue the long-term daily mapping of the global distribution of Earth's atmospheric ozone layer. The satellite was built by TRW for NASA's Goddard Space Flight Center. TOMS-EP collects high resolution measurements of the total column of ozone. The NASA-developed instrument measures ozone directly by mapping ultraviolet light emitted by the Sun to that scattered from the Earth's atmosphere back to the satellite. The TOMS instrument has mapped in detail the global ozone distributions as well as the Antarctic "ozone hole," which forms September through November of each year. In addition, TOMS measures sulfur-dioxide released in volcanic eruptions which may be used to detect volcanic ash clouds that are hazardous to commercial aviation.

TOMS-EP was inserted into orbit by the Pegasus XL booster on July 2, 1996. In the nine days following launch, the spacecraft executed a series of Delta V burns to reach a 500 km circular Sun-synchronous mission orbit with an ascending node mean local time crossing of 11:18 AM. Originally, the data obtained from TOMS-EP were intended to complement data obtained from ADEOS TOMS, which gave complete equatorial coverage due to its higher orbit. However, with the failure of ADEOS in June 1997, the orbit of TOMS-EP was boosted to 740 km and circularized to provide coverage that is almost daily. TOMS-EP is currently the only satellite providing scientific data with an operating TOMS instrument. A QuickTOMS mission is planned for launch in August, 2000 with another TOMS instrument. Figure 1 illustrates the TOMS-EP satellite.

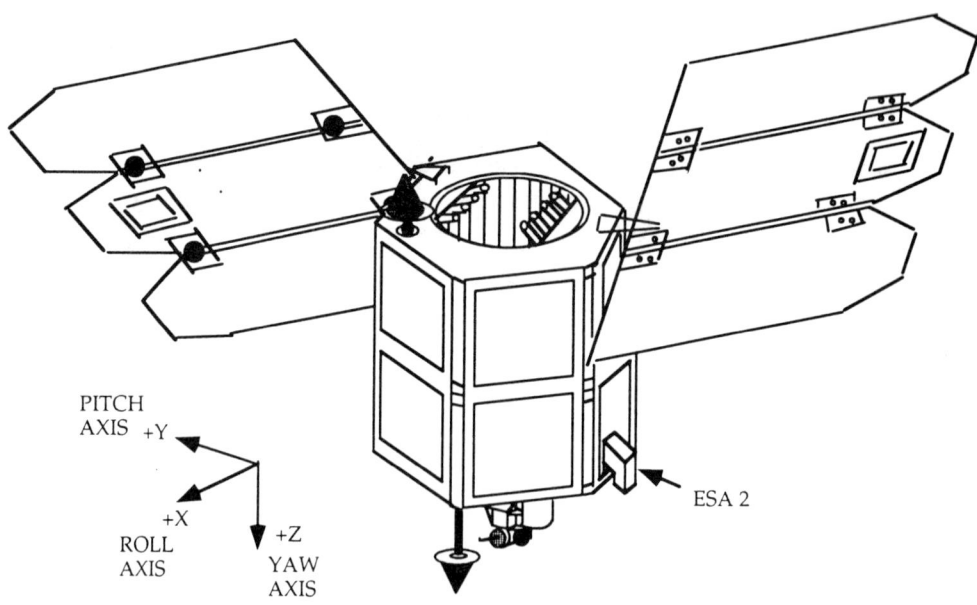

Figure 1 TOMS-EP Satellite

SYSTEM SAFE MODES

To understand the anomaly, it is necessary to understand the system implementation of the active safe modes. The Safe Power Mode uses all standby redundant equipment. It has two submodes, Sun Point Recovery and Long Term Hold, whose functions are defined in Table 1. Both submodes point the +X (roll) spacecraft axis to the Sun. The coarse sun sensor assembly (CSSA) is used for pitch and yaw attitude error and a single two-axis gyro provides rate information about pitch and yaw. The spacecraft undergoes an open loop roll spin-up by two 1 pound hydrazine thrusters prior to entering Long Term Hold.

**Table 1
Safe Power Submodes**

Mode	Submode	Description	Automatic Transitions
Safe Power	Sun Point Recovery	Two axis inertial sun pointing mode. CSSA and gyro are used as sensors. Thrusters used as actuators	Entry from any other mode due to fault condition. Entry from Long Term Hold due to excessive Sun pointing error.
	Long Term Hold	Spin stabilized Sun pointing precession control mode with two axis rate control. CSSA and gyro are used as sensors. Thrusters used as actuators	Entry from Sun Point Recovery only after successful Sun acquisition. Exit from mode if there is excessive Sun pointing error.

ANOMALY OUTLINE

The anomaly began when an event caused the spacecraft to transition from the prime processor to the redundant processor in response to a critical parameter that exceeded an established limit. The spacecraft successfully aligned the +X axis with the sun line using a two axis inertial controller based on processed coarse sun sensor measurements and a single two axis rate gyro. At this point, the flight software should automatically spin up the spacecraft about the roll axis and transition to a very low fuel consumption momentum based controller. At some point in the transition, the flight software failed to complete the transfer to the momentum based controller. Table 2 provides a concise timeline of events starting just before the processor reboot.

Within approximately 6 hours from entering Safe Power Mode, TOMS-EP had used virtually all of the 25 lb of Hydrazine fuel that remained before the anomaly occurred. The spacecraft was pointed at the Sun, but was uncontrolled and spinning at approximately 18 deg/sec about the +X (roll) axis.

The large amount of thruster activity had a small effect on the TOMS-EP orbit. TOMS-EP is required to stay within an ascending node crossing time of between 11:03 and 11:30. Before the anomaly, ascending node crossing time evolution was not a science life-limiting factor. After the anomaly, the rate of change of the ascending node crossing time was increased by about 3.6 min per year. This rate of change still allows more than 4 years of operation before the ascending node crossing time begins to degrade science collection.

Table 2
Anomaly Timeline

Event	Time	Notes
Corrupted Ephemeris Position (ECI) Data In Telemetry.	347/15:11:26	Previous value of position was 2139.6, 4193.99, -5330.01. Position reading at this time was 1042.67, 5484.46, 4412.70.
Large Pitch Error.	347/15:11:30	Error is calculated by subtracting the onboard propagated position quaternion from the commanded quaternion. Since error did not appear in either roll or yaw, suspect variables for SEU are those related to time (onboard clock, software time or epoch time).
First Thruster Firing To Counter Wheel Spin Down.	347/15:13:06	The first thruster activity occurs more than 1.5 minutes after the Redundant Processor boot is finished. This is the required time to configure the ADCS hardware and initialize Sun Point Recovery. Pitch thruster firings seem to be very clean. The system started virtually sun pointed. Correct thruster pair participate in the removal of wheel momentum as it bleeds into the spacecraft.
End of minimum ten minute window required in Sun Point Recover.	347/15:23:07	The safing logic waits a minimum of 10 minutes in Sun Point Recovery to allow the wheels to run down. This should prevent momentum coupling while the spacecraft spins up.
System begins to monitor the five required conditions necessary to begin the transition from two single axis inertial controllers to a spin stabilized momentum controller.	347/15:23:07	The five conditions required to start the transition are: 1. No presence in Fine Sun Sensor #2, 2. Pitch rate within specified threshold, 3. Yaw rate within specified threshold, 4. Pitch angle within specified threshold, 5. Yaw angle within specified threshold. At this time, the processed telemetry showed that all five of the conditions above were satisfied. The flight software changes the flag "runup" from 0 (as initialized) to 1 to denote that the system is ready to be spun up.
Start of Roll Spin-up	347/15:23:07	Immediately after the minimum time window, the roll thrusters begin to spin up the spacecraft. Telemetry from the thruster commands shows the total roll on time to be approximately 19.15 seconds. The expected roll rate with this duration pulse should be 3.9 to 4.5 deg/sec. This matches with the algorithm in the flight software and the tank reading in telemetry of 36 counts (8 bit reading) which represents 85 psi. At the start of the roll spin up, the flight software sets the flag "runup" to 2 to let the system know that the roll spin-up has started.
Completion of roll spin up / transition to spin stabilized controller.	347/15:24:04	The telemetry shows that the roll spin-up completed on time and yet the system failed transition to the spin stabilized controller.
Continuous Firing of Pitch Thrusters.	347/15:24:04	Once there was angular velocity in the roll axis, imperfections in the alignment of the inertial and control axes caused a constant pitch rate to appear on the pitch gyro. The inertial control law continuously fired the pitch thrusters to compensate for this rate. The thrusters were ineffective due to the spinning dynamics. A small torque coupling between pitch and roll resulted in a continuous increase in roll rate as the pitch thrusters were fired.
1st Contact after anomaly. Ground acquires downlink with only 3 min to Horizon LOS.	347/16:01:00	Ground observes spacecraft in Sun Point Recovery. Tank pressure 84 psi.
First expiration of sun acquisition timeout.	347/17:08:11	The failure to reach the spin stabilized mode caused the Redundant Processor to reset after 7000 seconds and attempt to acquire the Sun again in Sun Point Recovery. This was the first of three or four resets due to this trigger. The subsequent attempts to acquire the Sun failed due to the system dynamics.
2nd Contact.	347/17:44:00	Ground observes Sun Point Recovery failure to acquire. Tank pressure 78 psi.
3rd Contact.	347/19:19	Ground evaluating problem.
4th Contact.	347/20:57	Ground turns on GRA 1 & 2. Spacecraft processor reset occurs during pass. Tank pressure 77 psi.
5th Contact.	347/22:38	Tank pressure 9 psi. Spacecraft spinning at 18 deg/sec.

ANOMALY CONTRIBUTING FACTORS

There were several factors that combined to produce the state of the spacecraft at the time all of the fuel was spent. This condition is referred to as the "end condition". These factors were distinguished as belonging to one of two classes: factors that were necessary for the end condition and factors that contributed to the end condition. Those that were necessary are:
1. Initial fail over,
2. Wheel bearing friction,
3. Safe mode transition logic,
4. Safe mode design philosophy,
5. Ground controller response.

Those that were contributors are:
6. Location of the failure in the orbit,
7. Thruster force level.

Each of these factors will be examined in the following section.

Factor #1 Initial Fail Over
The anomaly was started by what appears to be a Single Event Upset (SEU) in the on-board Primary Processor. The telemetry stream recorded a jump in the estimated position of the spacecraft at the UTC time 347/15:11:26. This position is calculated onboard to facilitate the nadir pointing function of the attitude control system. The change in position was calculated to be greater than 9888 km in 32.768 seconds. The nominal change in position should be around 245 km.

After identifying and analyzing all reasonable candidates for this anomaly, it is believed the erroneous change in position was due to an SEU in the calculation of the spacecraft state (contained in the ephemeris routine). This conclusion is supported by the fact that:
1. The magnitude of the orbit position vector is consistent between the two vectors. This significantly narrows the possible locations in code for the SEU to occur; and
2. The angle between the position vectors was about 88 degrees. This error appeared in the pitch angle error telemetry as a value of 81.05 degrees (quaternion "small angle" approximation accounts for the difference). Virtually no error appeared in the roll or yaw angle telemetry. This suggests that the spurious position was in the correct orbit plane. Again, this points to a very limited number of points in the processing.

Factor #2 Wheel Bearing Friction
The initial behavior in Safe Power Mode was very nominal. This event represented the seventh entry into Safe Power Mode since the start of the mission and all other entries successfully safed the spacecraft. What made this occurrence different? The key can be found in the timing of the transition from the two axis controlled sun pointing inertial mode (Sun Point Recovery) to the spin stabilized sun pointing momentum based control (Long Term Hold). Initial examination of the playback data showed that there was an

anomaly in the dynamics of the spacecraft during the transition between Sun Point Recovery and Long Term Hold. Although there is no direct evidence of the cause because both the attitude decoder electronics (ADE) and the motor driver electronics (MDE) are turned off during Safe Power Mode, the circumstantial evidence presented below points to residual momentum in the wheels.

There is a minimum delay period of ten minutes that the system must spend in Sun Point Recovery before it is allowed to transition to Long Term Hold. This delay was designed to allow for wheel rundown. Thruster activity, gyro readings and CSSA data during this ten minute time period give us important clues about the dynamic condition of the spacecraft upon attempted entry into Long Term Hold. Figure 2 shows the thruster usage within the ten minute delay interval. Note that only thrusters number 2 and 3 are firing and that they are firing in perfect unison. Thrusters 2 and 3 provide positive pitch torque which would be expected as the negative pitch momentum bias is transferred from the wheels to the spacecraft body. Figure 3 shows the spacecraft body rates in the pitch and yaw axes (no roll information is available in the backup mode). The shape of the pitch rate curve shows classic saw-tooth behavior associated with a thruster based controller with a fixed minimum pulse width subject to a near constant disturbance torque (due to the wheel run-down).

The total angular impulse provided to the system in this ten minutes adds up to between 2.0 and 2.25 N-m-sec. This is based on the expected force level of about 0.35 lbf per thruster and the telemetry data which showed 586 counts (2.93 sec) of pitch thruster firing. Since the wheels started with 3.0 N-m-sec of momentum at their nominal 2000 rpm, there was 0.75 to 1.0 N-m-sec of residual momentum in the system when the spacecraft attempted to spin up about the roll axis. This residual momentum would certainly cause the "wobble" observed as the spacecraft began to spin up in roll. This is an unusual case where *lower* than expected wheel bearing drag caused the problem.

Figure 4 is generated from on orbit data and shows a plot of the average voltage needed to keep the TOMS-EP wheels at 2000 rpm over the life of the spacecraft. Based on a linear estimate of the voltage to torque ratio, the drag seems to have leveled off at around 2 mN-m. Figure 5 show the results of a type A scan wheel life test performed at Ithaco over the course of three years. This test was performed under flight like conditions (in vacuum). The data shows that the drag varied from 4.25 mN-m at near beginning of life to around 3.25 mN-m at the end of three years. The lower limit was actually established 16 months into the test.

The shapes in Figures 4 and 5 are very similar. The data suggest that the wheels have reached a steady state and there is no reason for concern over the health of the wheels. The difference is the magnitude of the drop in drag torque. The test wheel showed less then a 25% drop in drag over a 3 year interval. The on orbit wheels show greater than a 60% drop in torque in less than 2 years. The analysis below will show how the unexpectedly low drag torque caused the system to fail.

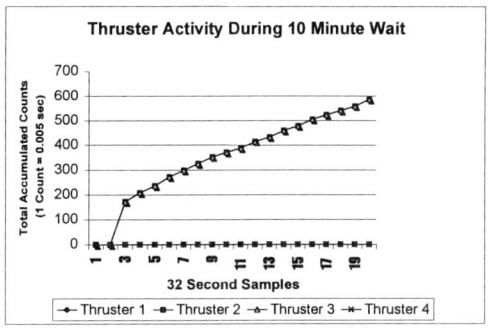

Figure 2 Thruster Firing

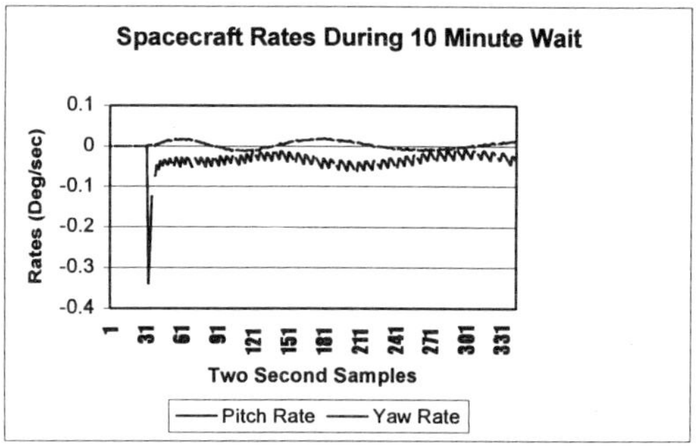

Figure 3 Spacecraft Body Rates

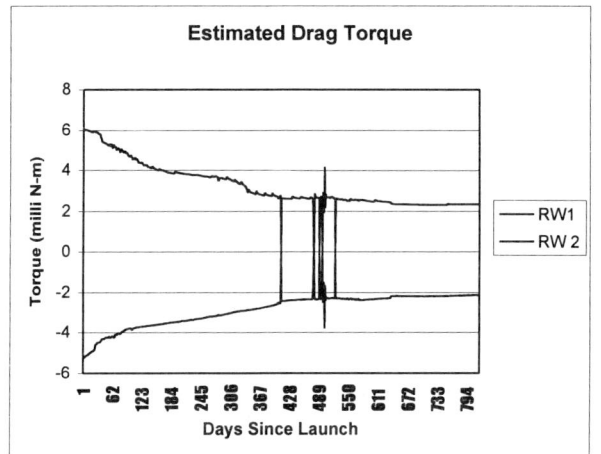

Figure 4 Lifetime Drag Torque (Estimated From Voltage)

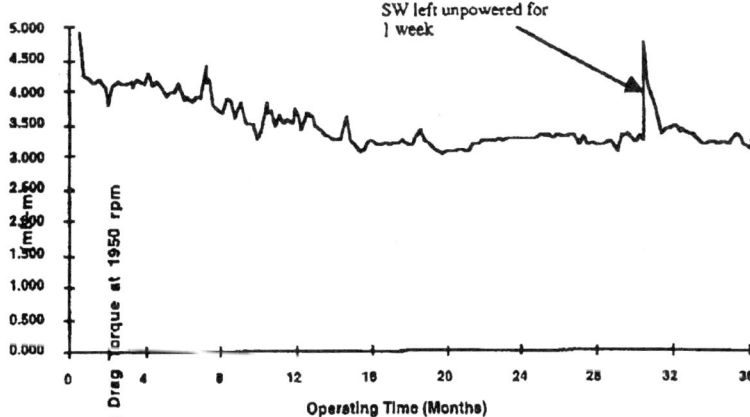

Figure 5 Ithaco SCANWHEEL Drag Torque Life Test Data

Anomaly Simulations

Simulations were run in an attempt to match the behavior of the anomaly. The attitude control and determination subsystem verification simulation (TOMSIM) was used to try and duplicate the behavior of the spacecraft at the time of the failure. Using initial conditions similar to the state of the spacecraft at the time of the failure, the transition from Normal Science Mode to Safe Power Mode was repeated for different levels of wheel bearing drag. The drag value was decreased until the system failed the transition from Sun Point Recovery to Long Term Hold. For reference, the top line in Figure 6 shows the drag torque requirement imposed on Ithaco during the procurement of the wheels.

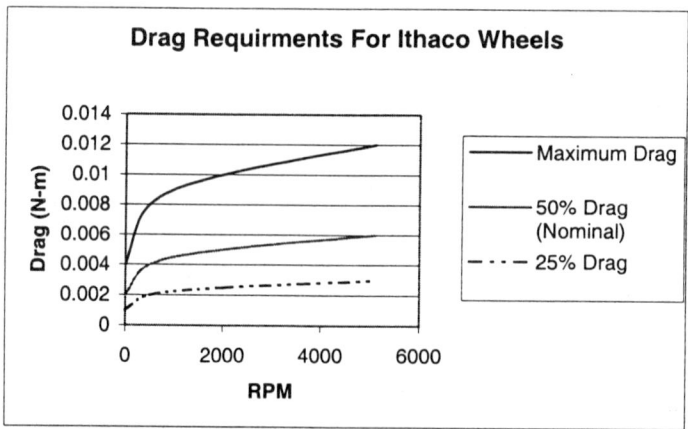

Figure 6 Drag Data Used in Simulations is Derived From Max Drag Requirement

Wheel Drag at 50% of the Maximum Allowed

This simulation shows the expected end of life performance of the Normal Science Mode to Safe Power Mode transition. In this case, the wheel model used the 50% line from Figure 6. Figures 7-10 show the behavior of a system that has the same initial conditions as the anomaly. Figure 7 shows the wheel speeds. The wheel that starts near –2000 rpm is the +Y wheel and the wheel that starts near +2000 rpm is the –Y wheel. At 50% of the maximum specified friction, the wheels are run down before the 10 minute waiting period is finished. Figure 8 shows the spacecraft body rates. A saw-tooth pattern that is similar to the actual anomaly data can be seen. There is a small rate transient when the spacecraft is spun up in roll at around 775 seconds. Figure 9 shows the processed CSSA data which gives sun angles for pitch and yaw. At the time of spin-up, the pitch and yaw error do not exceed 5 degrees. Figure 10 shows the thruster command "on" flags. There is near continuous thruster activity during the spin-up but after the spin-up is completed, thruster usage drops to zero.

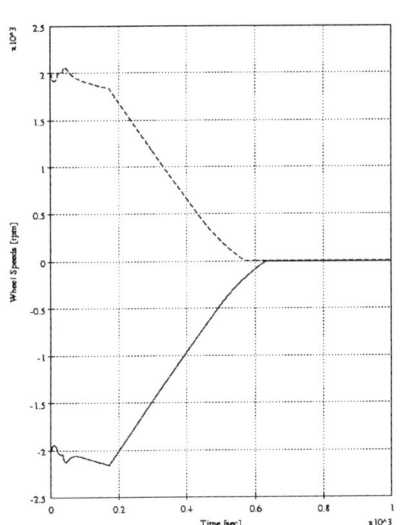

Figure 7 Wheel Spin Down

Figure 8 Spacecraft Body Rates

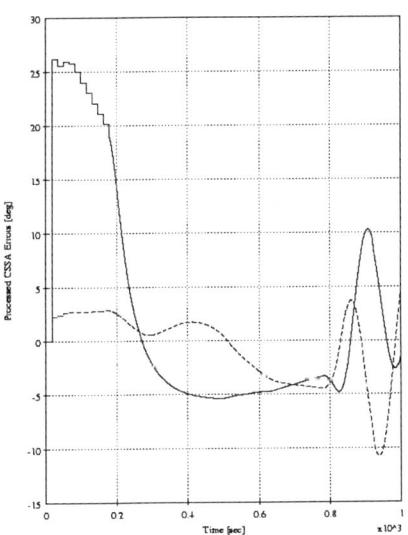

Figure 9 Processed Sun Sensor Angle

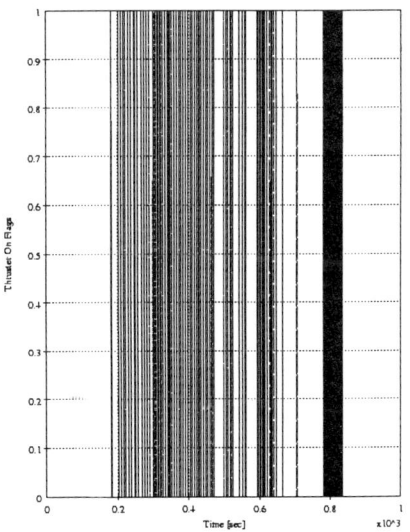

Figure 10 Thruster Commands

Wheel Drag at 20% of Maximum Allowed

The second simulation case presented here shows what happens when there is too much momentum in the system at the time of roll spin-up. Figures 11-14 show the behavior of a system that has the same initial conditions as the anomaly but wheel drag is scaled to 20% of maximum. Figure 11 shows the wheel speeds. At 20% drag, there is still 800 rpm (1.2 N-m-sec) remaining in the wheels when the spacecraft begins to spin up about roll. Figure 12 shows the spacecraft body rates. Coning and nutation are now apparent in the motion of the spacecraft. Figure 13 shows the processed CSSA data which gives sun angles for pitch and yaw. The system is unable to complete the transition from Sun Point Recovery to Long Term Hold because the processed sun angle error is too large. Figure 14 shows the thruster command "on" flags. Since the spacecraft was unable to complete the transition to the momentum based controller, the system is now using a two axis inertial sun pointing control law (Sun Point Recovery) with a high roll rate. This controller is unsuited for systems with a large momentum bias and the pitch thrusters begin to fire continuously in a futile attempt to reduce the observed pitch rate (caused by misalignment of control and inertial axes and the presence of a significant roll rate). The combination of very small misalignments in the thrusters and CG migration over the life of the spacecraft caused a slight pitch/roll torque coupling. As the pitch thrusters continued to fire, the roll rate slowly increased to 18 deg/sec at which point the 25 lb of hydrazine was exhausted.

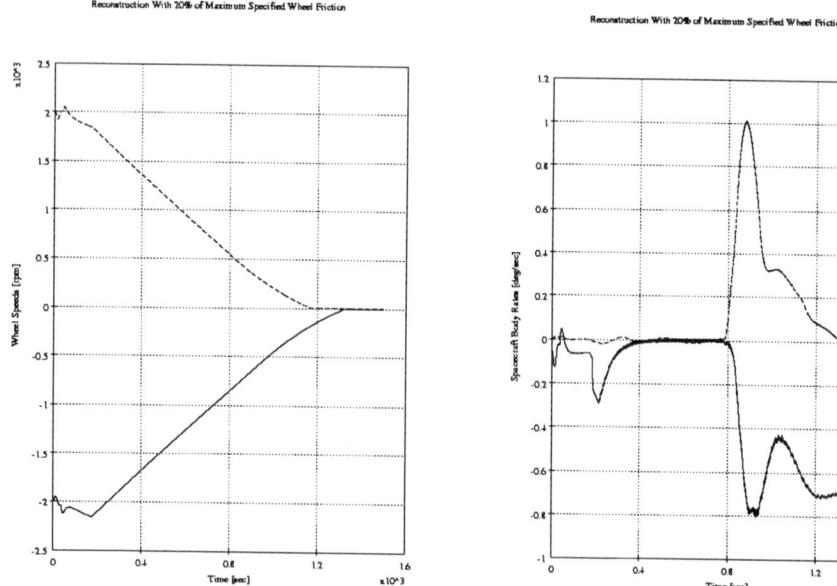

Figure 11 Wheel Spin Down   **Figure 12 Spacecraft Body Rates**

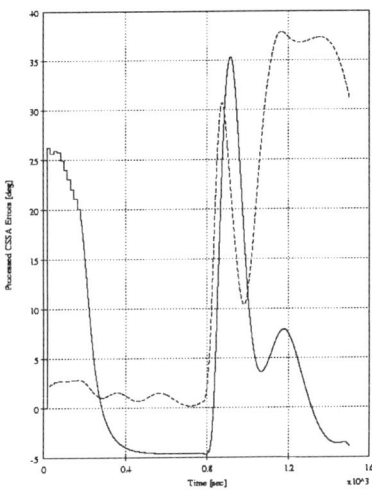
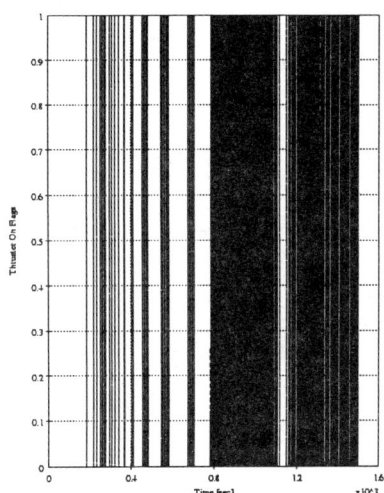

Figure 13 Processed Sun Sensor Angle   **Figure 14 Thruster Commands**

Factor #3 Safe Mode Transition Logic
The National Transportation Safety Board (NTSB) approaches investigations with the motto "If any link in the chain of events were broken, the accident would not occur." Of all the contributors, the safe mode transition logic would have been the easiest link to break. The crux of the problem is this: a control flag was used for two purposes, both to turn on and off the momentum controller and to signal the end of the roll spin-up maneuver. The Sun Point Recovery logic interfered with the function of the roll spin-up logic, thus preventing the transition to the momentum control mode.

There are two flags of interest that control the transition to Long Term Hold. These two flags are named "runup" and "isunon" (integer sun control on/off flag). In a nominal scenario, the flight software should go through the procedure outlined below:
1. Sun Point Recovery is entered, "runup" is initialized to 0 and "isunon" is initialized to 1.
2. The spacecraft tries to satisfy the five conditions listed in Table 2 by acquiring the sun and becoming quiescent.
3. When the five conditions are satisfied, "runup" is set to 1.
4. If ten or more minutes have passed, the necessary roll thrust is calculated and the roll thrusters begin to spin up the spacecraft. "runup" is set to 2.
5. When the spin-up is complete, "isunon" is set to 0 to tell the mode transition logic to transition to Long Term Hold.
6. Long Term Hold is initialized with the controller off ("isunon" = 0) and it is usually a day or two before the controller needs to be turned on.

The potential flaw in this logic comes from the fact that the mode transition logic runs at 1.024 second intervals and the Sun Point Recovery controller runs at 0.256 second intervals. There is potential delay between when the "isunon" flag is set to 0 and the mode transition logic reads it. That delay may be anywhere from 30 msec to 798 msec. In that time, the Sun Point Recovery controller may be run either 0,1,2 or 3 times. If the sun angle is outside the 12 degree outer deadzone, the control logic will set the "isunon" flag back to 1 before the mode transition logic can read it.

This logic was tested extensively in both simulation and fixed based test without discovering this potential flaw. That is because the spin-up process does not start unless the angle error is within the 5 degree deadzone and the spacecraft is under active position and rate control during the spin-up. The transition was simulated using worst case thruster misalignments, force mismatch, CG offsets, force vector rotations and nozzle exit location errors. In all cases, the transition to Long Term Hold was achieved.

Factor #4 Safe Mode Design Philosophy
In order to insure that the cause of an anomaly is removed from the system and to eliminate software health checks for equipment, it was decided to use all standby redundant components for the thruster based safe modes. Because of budget constraints, it was impossible to meet the criteria for using all standby redundant equipment in Safe Power and have rate information for all three axes. The choice was made to use a single gyro in the backup mode and maximize the control stability in other ways.

As many "smart" decisions as possible were made to mitigate the lack of roll rate information:
1. Point major moment of inertia at Sun.
2. Use a two-stage safing procedure. The first mode (Sun Point Recovery) is temporary and once the Sun is acquired, the system transitions to a spin stabilized mode that used a momentum controller.
3. The momentum controller was designed to be stable over a wide range of roll rates. Simulation has shown this controller to be stable from 0.75 to > 20 deg/sec.
4. Bias rejection filters were added to remove DC signals associated with roll rate.
5. Control only executed at orbit location where "Earth shine" is minimum.
6. A failsafe check will return to two axis inertial control if momentum controller failed to hold Sun.

It was known that the two axis inertial mode would use a large amount of fuel if significant roll rate accumulated. This was an acceptable risk to the program since the system was designed to pass through this mode in a short period of time.

Factor #5 Ground Controller Response
Once the anomaly occurred, TOMS-EP consumed fuel at a very high rate for a period of approximately 6 hours before the fuel was depleted. Ground controllers had only 4 contacts during this time, the first of which was only 3 minutes long by the time the ground acquired a signal. The other 3 passes were on the order of 10 minutes each.

During the 4 passes, the ground had the ability to disable the thrusters, which in hind sight would have saved fuel and prevented such a high spin-up.

Although ground controllers could have prevented or minimized the effects of the anomaly, it is understandable why they were unable to do so. Although tank pressure was dropping and the spacecraft was spinning up about roll, TOMS-EP remained in the proper Sun-pointing attitude at each contact. Furthermore, while in Safe Power Mode, there is not a direct measurement of roll rate available in telemetry. Further complicating understanding at the time was the fact that when ground controllers turned on other gyros to look at the roll rate, the spacecraft processor reset during the same pass. It was later determined that this was just a coincidental 7000 second timer reset which had nothing to do with turning on gyros.

Factor #6 Location of Failure in the Orbit
The location of the failure had a significant role in the behavior of the system for two reasons. First, after the Redundant Processor had booted up, the spacecraft was almost exactly sun pointed. Figure 15 shows the processed CSSA angles over the entire 10 minute wait period. The maximum angle observed was just above 3 degrees. At the end of the wait period, the spacecraft immediately began to spin up. Second, the presence of Earth shine fooled the spacecraft into thinking it was still sun pointed even after rotating more than 20 degrees. If the spacecraft pitch rate shown in Figure 8 is integrated, it should produce a change in pitch angle as shown in Figure 16. Figure 17 shows the processed CSSA angles with the effects of Earth shine removed. These data match the integrated gyro data much more closely. It is quite possible that in the absence of Earth shine, the system would not have satisfied the five conditions for spin-up immediately and the reaction wheels would have had more time to spin down.

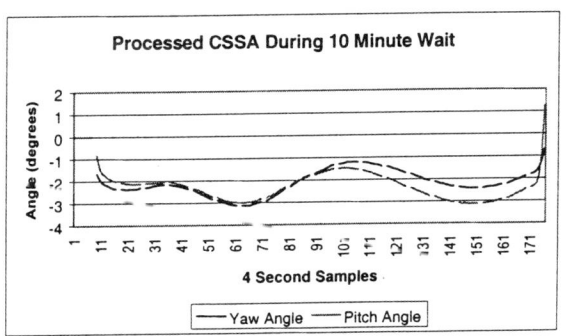

Figure 15 Processed CSSA Sun Angles

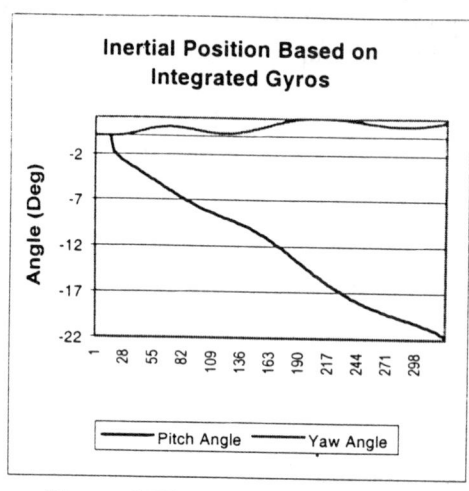
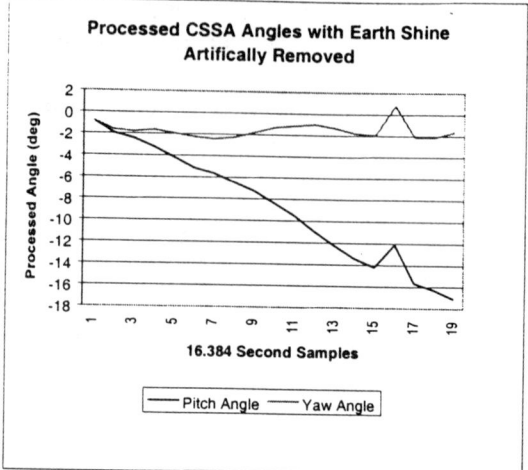

Figure 16 Integrated Rates   **Figure 17 Improved Estimate of CSSA Angle**

Factor #7 Thruster Force Level

The last contributor identified was the level of force available at the time of the anomaly. Due to of the initial orbit insertion burns and subsequent orbit change, over 90% of the hydrazine fuel had been exhausted from the blow down propulsion system and the force level was around 35% of the force level available at the beginning of life. The lower force level made the system less capable of countering the $\omega \times H$ torques generated by the residual wheel momentum.

RECOVERY EFFORT

The focus of the recovery effort was to generate a scenario that was easily implemented and that maximized the probability of spacecraft recovery. To be successful, the recovery must first and foremost maintain the health of the power subsystem.

Power Subsystem Considerations

The original design of the power subsystem makes it very robust to attitude anomalies. The fixed arrays are arranged in a cruciform orientation off of the –Z body axis as shown in Figure 1. Figure 18 shows the power output of the arrays as a function of the solar normal vector (neglecting shadowing effects) scaled to the output of a single sun pointed array. The power system produces enough power to run the spacecraft in all orientations except when the sun is within about 45 degrees of the plus or minus Y spacecraft axis. Since the Y axis is the intermediate axis of inertia, the Sun should not dwell near the axis if there is any significant angular rate in the spacecraft body.

The TOMS-EP battery has 9 amp-hours of capacity. With normal loads, the battery can sustain the spacecraft for about 3 hours without solar array power. It should be noted that there is no provision for "jump starting" the power system after the battery is discharged since the solar array regulators (SARs) are powered from the battery.

Normalized Solar Array Power Potential

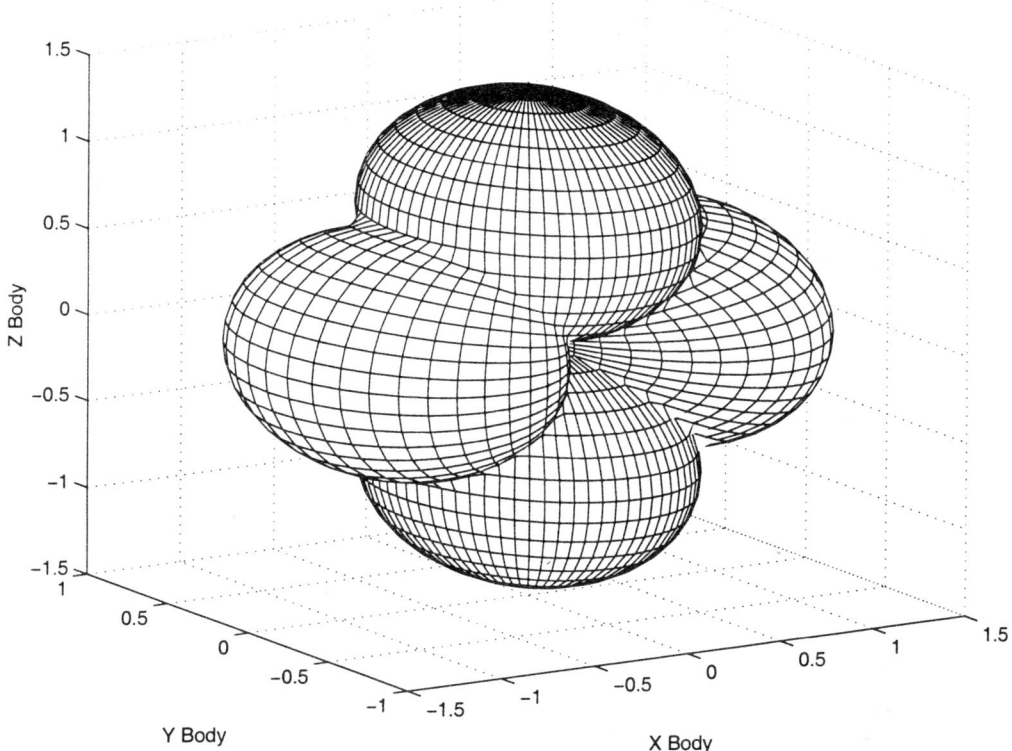

Figure 18 Normalized Power Output For Different Sun Vector Orientations

Attitude Control and Determination System Considerations

The Attitude Control and Determination System (ACDS) was essentially disabled after the fuel was exhausted. A task was immediately undertaken to ascertain the current state of the spacecraft and predict the future orientation with respect to the sun.

The gyros could not be used for on-board roll rate determination because the spacecraft was spinning at 18 deg/sec and the gyros lose polarity at 7 deg/sec. At the beginning of the anomaly, the Sun was within the field of view of one of the FSS; the other FSS was pointed anti-Sun. From the CSSA, it was known that the spacecraft +X axis was pointed within about 5 degrees of the Sun and moving away at a rate of approximately 2-3 degrees per day. Although data from the magnetometer was available, the absolute inertial attitude was very difficult to determine due to the interaction of the high roll rate and processing and telemetry delays. It was certain that if the precession rate observed on the CSSAs continued, the power system would see significant reductions in available power within 2 weeks.

With so little time available, the recovery procedure had to be designed and tested within 10 days. This requirement drove the team toward trying to use the onboard algorithms with minimum modifications. Looking at a block diagram of the TOMS-EP ACDS hardware shown in Figure 19, it can be seen that wheels and torque rods are the only actuators available for maneuvering the spacecraft once fuel is depleted. It was not possible to use the wheels for a large angle maneuver since each was only capable of containing 4 N-m-sec of momentum and the spacecraft body contained about 36 N-m-sec of momentum. The torque rods could be used to slowly maneuver the spacecraft if there were sufficient time. Fortunately, there are two onboard magnetic control algorithms onboard to choose from. The first is a cross product law used for momentum unloading and the second is a B-dot law used for a magnetic Safe Hold Mode. This was the extent of the tools to be used in the recovery.

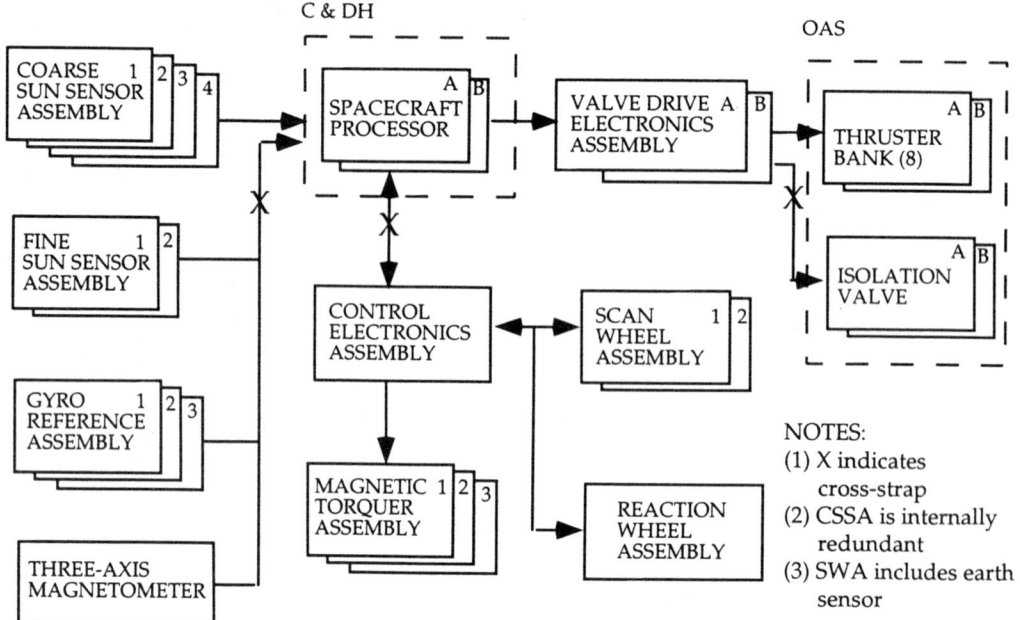

Figure 19 ACDS Equipment Block Diagram

Thermal Subsystem Considerations
The thermal subsystem was designed to radiate most of the heat generated in the spacecraft out the panels on the –Y spacecraft axis. The constraint this placed on the recovery scenario was essentially enveloped by the power system requirements.

Propulsion Subsystem Considerations
Although it was thought that all the fuel in the tank was exhausted, there was known to be fuel in the prime side thruster lines. In addition, there may have been some fuel trapped by the tank bladder against the side of the tank. Immediately after the thruster valves on the redundant side were closed, the pressure in the tank began to rise slowly. Although the recovery could not count on using the impulse from the trapped fuel, an attempt could be made to use it in the most constructive manner possible. When the spacecraft was returned to the prime processor, the Long Term Hold momentum control was selected and the residual propellant precessed the spin vector nearly 10 degrees towards the Sun.

This "last gasp" contribution from the propulsion subsystem gave the recovery team several more days to plan the recovery procedure. It also allowed the team to turn off all propulsion heaters and save power at critical points in the recovery.

RECOVERY PLAN

Of the two onboard magnetic control laws, the B-dot law would need the least amount of modification because it is primarily a minimum energy based design. In the absence of internal momentum, a spacecraft in a polar orbit using a B-dot control law will eventually end up with the maximum moment of inertia (roll) perpendicular to the orbit normal and the roll body rate would approach 2 revolutions per orbit (RPO). If the wheels were spun up to produce momentum in the –Y spacecraft axis, this momentum would end up perpendicular to the orbit plane with 2 RPO rate about the pitch axis.

In this instance, knowing the start and end conditions did not answer the question of whether the spacecraft would pass through an unfavorable power condition somewhere in between. If the wheels were running, the Y axis could act as a pseudo maximum moment of inertia and it is possible for the Sun vector to remain near the Y axis long enough to discharge the battery. The best way to minimize this possibility was to break the recovery into three stages:
1. B-Dot magnetic despin without internal momentum,
2. Wheel capture into the nominal Safe Hold Mode,
3. Science Return into nadir pointing.

Simulations
The entire recovery scenario was extensively simulated using TOMSIM prior to the start of the spacecraft recovery attempt. These simulations calculated both attitude and power potential. Direct measurements from the CSSA provided data on the angle between the spacecraft X axis and the Sun vector. That narrowed the uncertainty in the spin axis attitude to the surface of a cone about the Sun. After the recovery scenario was established, the robustness of the recovery approach was examined by simulating the process using four different sets of initial conditions that resided on the surface of the uncertainty cone. The simulated recovery was successful in all four cases.

Figure 20 shows the spacecraft body rates for a simulated recovery from a position 30 degrees east of the Sun. The simulation predicted a stage 1 duration of approximately 2 days. Although the roll rate could not be directly measure during the actual recovery, Figure 21 shows the rate estimated from the DC bias on the pitch gyro which should be proportional to the roll rate if the system is in a flat spin and has roll-pitch cross products of inertia.

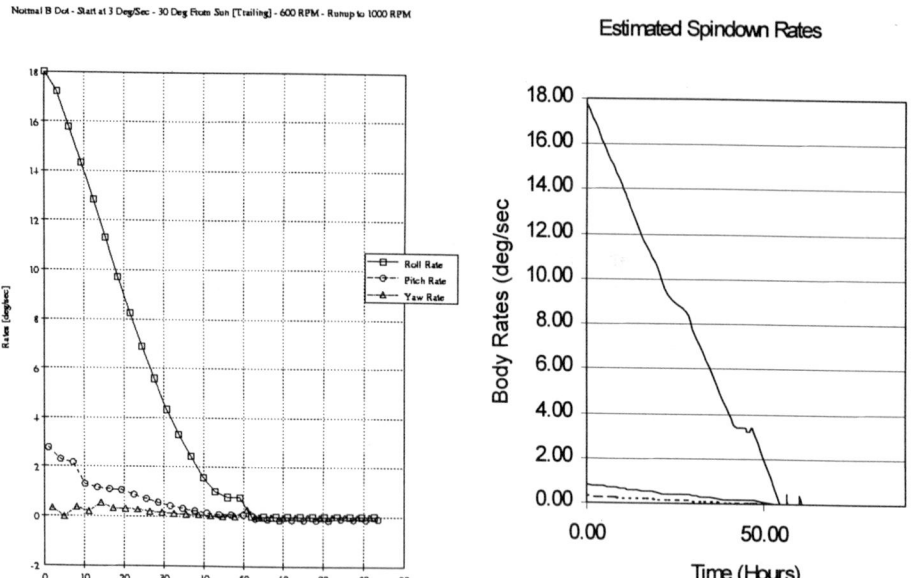

Figure 20 Simulated Recovery Rates   **Figure 21 Estimated Recovery Rates**

Stage 1

The goal of the first stage was to reduce the spin rate from 18 deg/sec to within the capture range of the Safe Hold Mode (between 2-3 deg/sec). The normal B-dot algorithm processing is executed every 16.384 seconds which allows multiple magnetometer samples to be averaged for noise reduction. In addition, the magnetic field rate is calculated with a differential filter that has a time constant longer than 16.384 seconds. Clearly, the algorithm could not be used successfully with the spacecraft spinning at 18 deg/sec. Changes in the database allowed us to successfully reduce the number of magnetometer samples to 1 and change the characteristics of the differential filter. The next step involved changing the flight software executive to call the B-dot algorithm every 2.048 seconds. Fortunately, this was accomplished by replacing a single byte in an inequality statement. The final "high speed" B-dot algorithm had 1.024 seconds allocated to magnetometer sampling and 1.024 seconds allocated to torque rod firing.

The question still remained whether the magnetic despin would put the spacecraft in an unfavorable power attitude. If you base your guess on the known end condition, you might assume that the maximum moment of inertia would be pushed perpendicular to the orbit plane. In fact, just the opposite is true. Since the body rates are much higher than orbit rate, the B-dot algorithm simply sheds energy wherever it can. The key to understanding its behavior is in the available torque. In a near polar orbit, the magnetic field remains close to the orbit plane (rotating twice per orbit). If you break the total system momentum into the portion projected into the orbit plane and the portion perpendicular to the orbit plane, there is *always* magnetic torque available to reduce the momentum perpendicular to the orbit plane and four points in the orbit where it is very difficult to affect the momentum in the orbit plane. Thus as long as the body rates remain high relative to orbit rate, the maximum moment of inertia will remain close to the orbit plane.

Since TOMS-EP resided in a sun synchronous orbit with an 11:00 to 11:30 ascending node, pulling the maximum moment of inertia toward the orbit plane should not degrade the power potential.

Stage 2
Stage 2 was the riskiest portion of the recovery scenario. At some point, the spacecraft had to transition from spinning about the maximum moment of inertia to spinning about the Y axis. The transition had the potential of pointing the spacecraft in a low power attitude for an extended period of time.

Once TOMS-EP was despun to 3 deg/sec, the nominal B-Dot Mode software parameters were restored. The wheels were set to their minimum rotation rate to minimize the time in transition between sun pointing and normal B-dot pointing. The spacecraft was prepared for the low power attitude by shedding *all* loads not directly involved in the recovery. These loads included prime and redundant platform heaters, prime and redundant propulsion heaters, gyros, and all transmitters. In this configuration, the average load was reduced to 45 watts (1/3 of orbit average). Essentially, it was up to the physics of the B-dot controller to complete the transition. Interference from the ground would only reduce the chance of a successful recovery.

Two orbits after starting stage 2, contact was re-established with the spacecraft. The system had settled with the pitch momentum bias perpendicular to the orbit plane. The wheel speeds were slowly increased until they matched the normal Science Mode speeds.

Stage 3
In the final Science return stage, TOMS-EP was commanded into its originally designed Science Return Mode. Although never used prior on-orbit, the mode was well tested and simulated prior to delivery of TOMS-EP. This mode allowed automatic transition into Science Mode within one orbit.

POST RECOVERY FAULT MANAGEMENT IMPLICATIONS

The loss of the propulsion subsystem left the Safe Power Mode incapable of active control. If the spacecraft switches to an uncontrolled mode, it is known that the momentum stored in the pitch momentum bias will eventually end up as a roll rate of +/- 1.9 deg/sec. This should be sufficient to prevent complete battery discharge.

It is still preferred to keep under active control if possible. For this reason, the onboard fault detection software was modified to minimize the number of faults that send the system to Safe Power Mode. Only those faults that require reconfiguration of power system or processor faults send the system to Safe Power Mode. All other faults (pointing anomalies, wheel speed delta, etc.) cause a transition to the B-dot Safe Hold without switching processors.

LESSONS LEARNED

1. SEUs happen. Be ready.

2. Where it is possible, directly measure states that are driving decisions. In our case, it would have been preferable to measure the wheel speeds directly and start sun acquisition after they had completed their run down. Unfortunately, that was not possible because tachometer data is unavailable when the motor drivers or the attitude decoder electronics are off. Additional fault management risk would have been assumed if the wheel electronics were left on in Safe Power Mode.

3. Designers often concentrate on accommodating behavior associated with "worst case" conditions (CG locations, misalignments, friction, structural flexibility etc.). Sometimes, an ideal CG location, perfect alignments, better than expected friction or higher than expected stiffness can cause problems. These should be considered also.

4. Do not use flags for multiple purposes no matter how closely related they are. Carefully check the logic of flags that are set and read asynchronously.

5. The emphasis in fault management at TRW has shifted since the design of TOMS-EP. In subsequent programs, the inherent robustness of the safe mode was considered to be more important than using standby redundant components. Even in light of the TOMS-EP on orbit experience, this is not a clear cut decision. To address the issue of using a single gyro for safe modes on the new programs, flight software chooses which pair of gyros to turn on based on a number of comparison tests. It has been demonstrated in fixed based test that these tests can be fooled under certain circumstances. If the software chooses a failed gyro, the consequences will be worse than the TOMS-EP anomaly.

6. Murphy works smarter than you do. Rely more on general principles to prove system robustness rather than attempting to find degenerative cases.

7. A number of spacecraft have been lost or nearly lost due to anomalous autonomous thruster operation. If a spacecraft has the capability to use wheels rather than thrusters to acquire its safe mode orientation, then it is usually prudent to use wheels over thrusters. Although thruster hardware and the associated electronics are very reliable, a system using expendables always introduces a risk of imparting unwanted high momentum to the spacecraft. This high momentum input can be caused by spacecraft hardware, logic or software anomalies. A wheel-based safe mode limits the amount of spin-up while in the mode, should an anomaly occur.

8. The canted, double-sided solar array orientation on TOMS-EP is very forgiving of spacecraft attitude anomalies. Adequate power can be generated from most spinning or tumbling conditions. This was of great relief while TOMS-EP was spinning uncontrolled

for 18 days. Multiple solar array viewing angles increase the robustness of a spacecraft to anomalies.

CONCLUSIONS

The TOMS-EP spacecraft was successfully recovered in less than 3 weeks from a severe anomaly that depleted all fuel and left the spacecraft uncontrolled with a high spin rate. A team of engineers and spacecraft operators quickly determined the cause of the anomaly and implemented a recovery effort. The TOMS-EP satellite continues to successfully perform its mission well beyond its design life of mapping the global distribution of Earth's atmospheric ozone layer.

OPERATIONAL EXPERIENCES WITH GLOBALSTAR CONSTELLATION CONTROL AND STATIONKEEPING USING SOLAR RADIATION PRESSURE

Lee Barker and Benjamin Lange[*]

Globalstar satellites operate in yaw steering mode approximately 80-90 percent of their on orbit life. The remainder of the time is primarily spent in orbit normal when the magnitude of the sun angle with respect to the orbit plane (beta) is less than 12 degrees. By operating with a small yaw bias applied to the yaw steering profile, the external force resulting from the reflective component of the solar radiation pressure can be used to raise or lower mean semi-major axis and provide a 'thrusterless' station keeping capability. Globalstar spacecraft operating this way have been observed to gain (or lose) orbital energy as a proportional function of the beta angle and the yaw bias angle. This energy change, manifesting itself as a change in absolute semi-major axis, can be controlled to overcome atmospheric drag, and perform station keeping maneuvers. It is possible to measure such small maneuver effects thanks to the Globalstar's filtered GPS navigation solution available for orbit determination. Globalstar on orbit flight data will be presented demonstrating this 'thrusterless' station keeping technique.

---

[*] Space Systems/Loral, 3825 Fabian Way, Palo Alto, California 94303.

Introduction

Globalstar operates a constellation of low Earth orbiting (LEO) satellites in a Walker constellation at 1414.0 km altitude, 52.0° inclination, in frozen orbits. Globalstar operational attitude control is performed primarily with a wheel momentum bias system and the attitude is generally defined as orbit normal or yaw steering depending on the sun-orbit plane geometry.

Globalstar satellites operate in yaw steering mode approximately 80-90 percent of their on orbit life. The remainder of the time is primarily spent in orbit normal when the magnitude of the sun angle with respect to the orbit plane (beta) is less than 12 degrees. During the first year of on-orbit operations, those Globalstar spacecraft operating with a yaw bias in the yaw steering mode exhibited orbital behavior different from those spacecraft not operating with a bias in yaw steering. Specifically, while operating with a yaw bias of 165 degrees, the spacecraft were observed to gain or lose orbital energy as a direct proportional function of the beta angle. This energy change manifested itself as a change in absolute semi-major axis and the maximum rate of change approached plus or minus two meters per day. It was possible to measure this change primarily thanks to the Globalstar's filtered GPS navigation solution available for orbit determination.

An example of this behavior is illustrated in Figure 1. Figure 1 shows the spacecraft semi-major axis over time for a spacecraft (FM04) operating in unbiased yaw steering and orbit normal operations and two spacecraft (FM01 and FM02) operating in yaw steering with a 165 degree yaw bias. The time span is two station keeping cycles. All three spacecraft are in the same orbital plane. Beta traverses from zero to approximately -66 degrees then back through zero to +55 degrees, then back to zero over this time span. The discontinuity in the yaw biased spacecraft represents station keeping thruster firings conducted during orbit normal operations when beta is near zero. FM04 shows the expected behavior due to atmospheric drag over the same period.

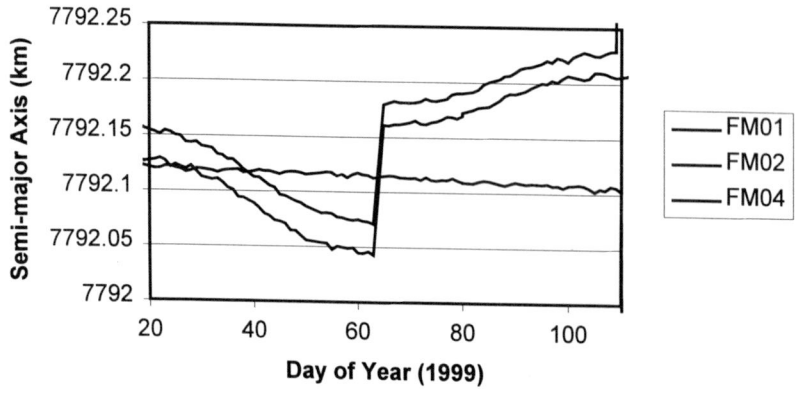

Figure 1

In observing the orbital behavior of Globalstar spacecraft operating in yaw steering with a 165 degree yaw bias, it has been discovered that solar radiation pressure (SRP) can apply a force parallel or anti-parallel to the orbital velocity vector. This force has an average sign around the entire orbit which is the same as the sign of the orbit plane beta angle. When beta is positive the force is posigrade, and when beta is negative the force is retrograde. By using this effect, **it is possible to station keep the spacecraft without using thrusters**. Instead, a yaw bias of plus or minus 10 to 15 degrees would be used. This makes the Globalstar spacecraft less sensitive to thruster failure.

This paper explains the unusual orbital behavior observed in the Globalstar satellites currently configured to operate with a yaw bias while in yaw steering. These spacecraft, all located in plane 'A' of the Globalstar constellation, have been observed to gain or lose orbit energy in the form of absolute semi-major axis. This energy source/sink **exceeds the atmospheric drag force by up to an order of magnitude**. This energy exchange is observable, repeatable, and has been found to be a directly proportional function of the orbit plane 'beta' angle. Careful analysis has shown the source of the force to be a solar radiation pressure differential due to reflected solar energy resulting from operations with yaw biases. The resulting orbital behavior has implications for both the station keeping strategy of the current Globalstar satellites as well as design implications for future spacecraft that will operate with yaw steering. With these considerations in mind, future yaw steered spacecraft operating in the LEO regime could take advantage of differential solar radiation reflection to augment thrusters for orbit maintenance.

Analysis

The observed data shows the rate of semi-major axis change to be directly proportional to the current beta angle for the satellite. For the case where beta is positive and orbit energy is increasing, a typical station keeping cycle of approximately 50 days exhibited about 50 meters increase in altitude, or an average gain of about 1 meter per day. The peak rate of change occurred when beta was near its maximum value. During this period, semi-major axis was observed to increase 28 meters in 14 days, or about 2.0 meters per day.

To calculate the force required to change the semi-major axis of a spacecraft in near circular orbit using a low thrust source, in reference (1), Battin derives the new orbit radius, r, as a time dependent function of the original orbit radius, r_0, and the tangential acceleration:

$$r = r_0 / (1 - (2 \cdot dt \cdot a_T / V_0))$$

Solving for the tangential acceleration, a_T, for the Globalstar orbit, one gets a required acceleration of approximately $1.06e{-}11$ Km/s^2 in order to produce a 2.0 meter per day change in radius. Assuming nominal on orbit mass values for the spacecraft (mass = ~430 kg), one can now solve for the required force using $F=ma$. The resulting required force for a Globalstar satellite is approximately 4.6 μN applied in the tangential direction averaged throughout the orbit.

A spacecraft in yaw steering mode, by definition, holds its Y axis (solar array axis) normal to the plane containing the sun vector and the nadir vector. For these spacecraft, the tangential force due to solar radiation pressure, both absorptive and reflective, is sinusoidal over the orbit and sums to zero. While there may be some small effects on eccentricity, like those experienced by geosynchronous spacecraft, the overall energy gain to the system should be zero. Figures 2 and 3 illustrate the geometry of the solar radiation pressure force on the satellite at the 6 AM and 6 PM satellite local time points in the orbit.

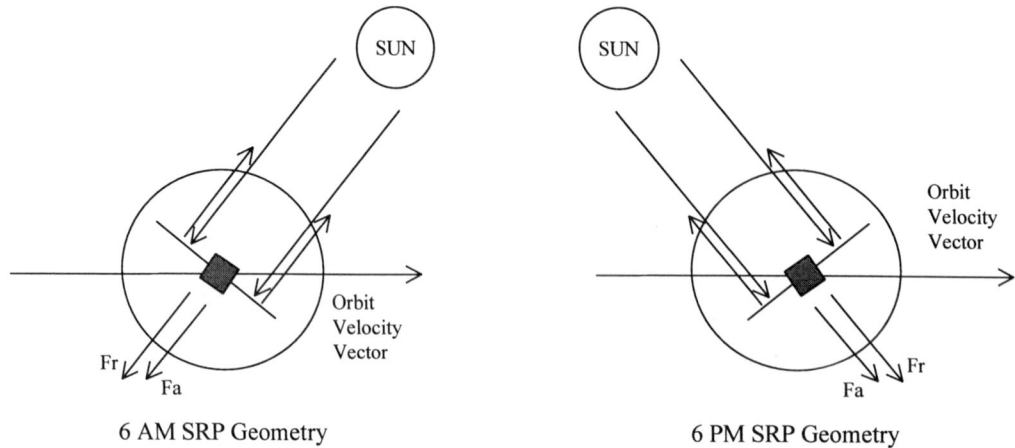

Figures 2 and 3

While the absorptive component of solar radiation pressure continues to sum to zero over the orbit for a spacecraft in yaw steering mode with yaw bias, the reflective component does not. Figures 4 and 5 illustrate the geometry of the solar radiation pressure force on the satellite with a 165 degree yaw bias at the 6 AM and 6 PM satellite local time points in the orbit.

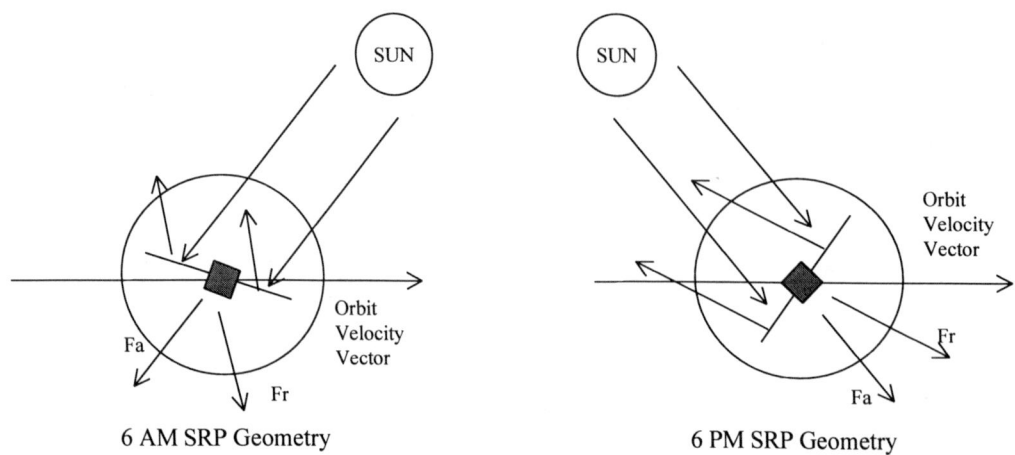

Figures 4 and 5

Note that, at both 6 AM and 6 PM, the reflected sun light may have force components in the direction of orbital motion. This results in an increase in orbit velocity and thus energy and semi-major axis.

Some texts assume that the reflected component of solar radiation pressure is two times the momentum change of the reflected light in the direction normal to the reflecting surface. This is the resulting force from the inbound component of the photon momentum, which lies in the same path as the absorbed component of solar radiation pressure force, and the reflected component of force, which is opposite the direction of the reflected light. For our purposes, the inbound portion of the momentum can be treated in the same way the absorbed photons are treated, averaging to zero over the orbit. This leaves a reflected component not perpendicular to the reflecting surface, but at an equal and opposite angle to the normal from the absorbed force. Since the incoming momentum change averages to zero around the orbit, the only component left is the reflected momentum change. This reflected component does not average to zero around the orbit. Thus, this component should not be multiplied by two.

It is difficult to visualize the geometry at other points around the orbit. For this a simulation was run to observe the intrack components of the absorptive and reflective components of solar radiation pressure force on the yaw biased spacecraft. Figures 6 and 7 illustrate these for high positive and negative beta angles.

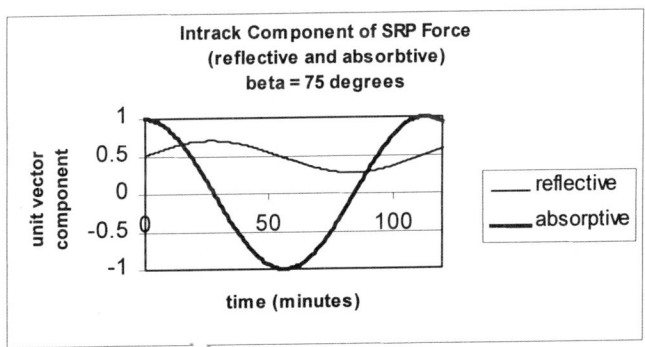

Figure 6

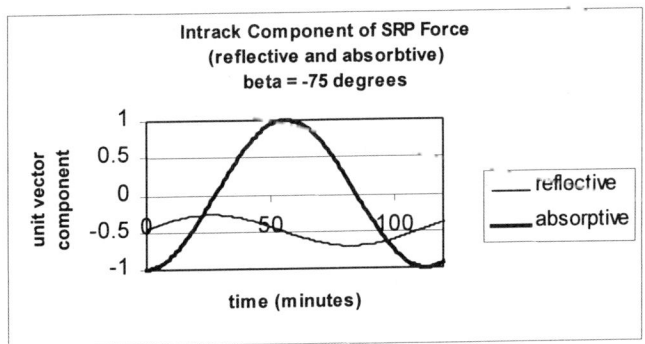

Figure 7

As beta progresses from positive down through zero and increases in the negative direction, the average intrack reflected solar radiation pressure force component moves from positive to negative. The average absorptive component remains zero.

The unit vector component of the reflected solar radiation pressure force in the intrack direction vs orbit position by beta angle is plotted in figures 8 and 9. The average over the orbit for each beta angle is also shown. These plots apply for yaw biases of 15, -15, and –165 degrees as well.

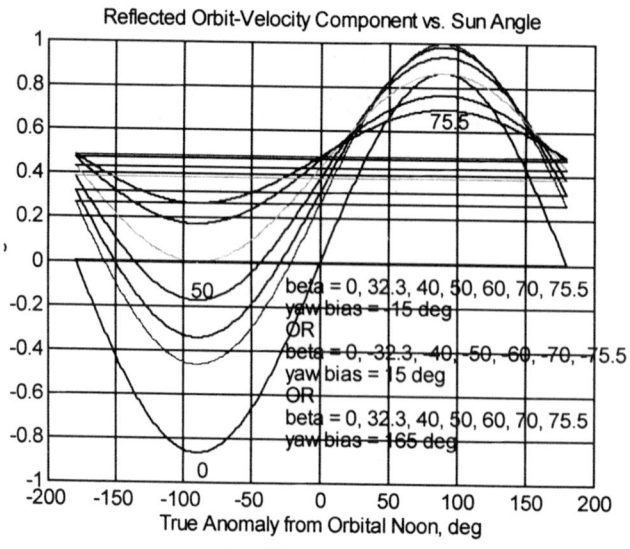

Figure 8

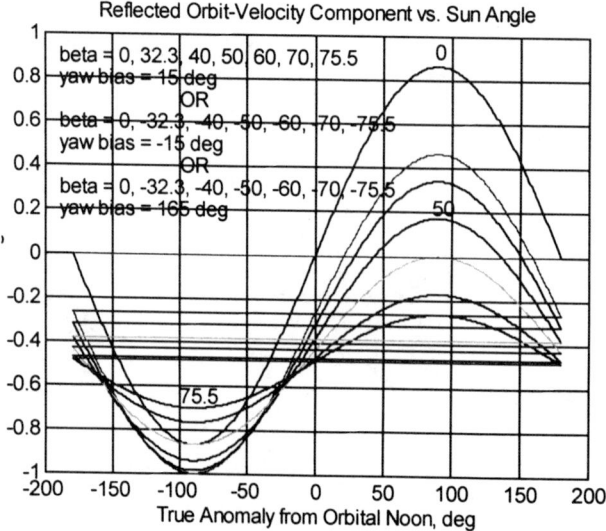

Figure 9

Over an entire orbit, the average value of the unit vector in the direction of the reflected solar radiation pressure projected onto the orbit velocity vector is given by

$$-\sin(2\Delta\varphi)\sin\beta$$

where $\Delta\varphi$ is the yaw bias angle and β is the beta angle. Figure 10 shows the average value of the reflected component of the solar radiation pressure in the direction of spacecraft motion as a function of beta angle for spacecraft in yaw steering with 15, -15, 165 or -165 degrees yaw bias.

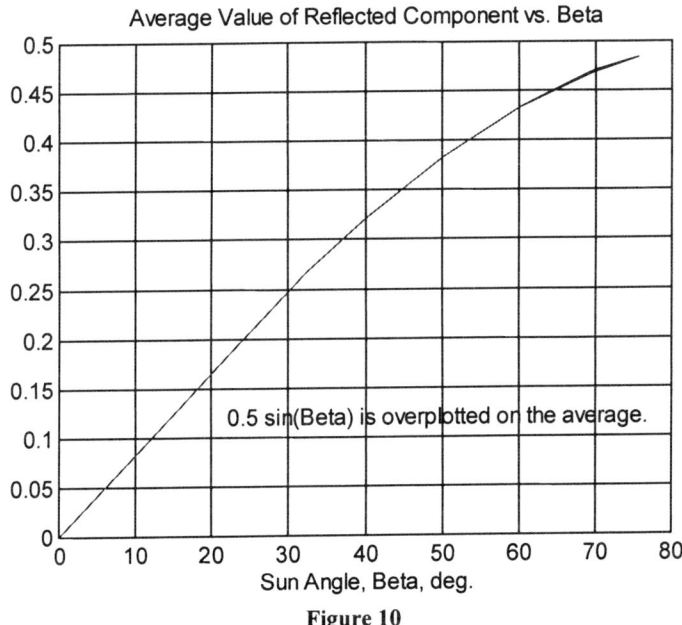

Figure 10

Having now identified a potential source for the external force acting on the spacecraft, we next set about to show that the force provided by the reflected solar radiation pressure is sufficient and reasonable to result in the 2.0 meter per day change in semi-major axis observed.

The following equation defines the intrack force due to the reflected component of solar radiation:

Force = SRP reflected force vector component in velocity direction • $C_{reflectivity}$

where $C_{reflectivity}$, the coefficient of reflectivity for the solar array, is approximately 0.26 for sun light striking the array.

The incoming solar radiation pressure near Earth is approximately 4.5e-6 N/m². Multiplying by the area of the spacecraft solar array, the force due to solar radiation pressure is calculated. For Globalstar, this force is approximately 60 µN. If we assume that 1) the coefficient of absorption for the solar array is 0.74 (coefficient of reflection

0.26), 2) the beta angle is 55 degrees, which provides an average component of the normalized reflection force vector over the orbit of 0.4 (from figure 8), and 3) our 165 degree yaw bias, the resulting force is 6 µN. Some of the reflected light is diffused. For the diffused portion of reflected light, the resulting force is perpendicular to the reflecting surface. It is difficult to estimate the diffused vs. specularly reflected light ratio, however this effect will reduce the force by some amount. Diffusion of 20 to 30 percent leads to a resulting average intrack force very close to the 4.6 µN estimated from orbital measurements. This force corresponds to the measured peak rate of change in semi-major axis.

Obviously, there are also perturbations in the radial and crosstrack directions due to the solar radiation pressure, however, these are less observable due to the higher energies required to make out-of-plane orbit changes and other perturbations affecting the system.

Results/Conclusion

Applying the above described method to varying yaw biases, the component of the reflected SRP vector in the intrack direction as a function of beta can be predicted and an estimate of the daily change in semi-major axis can be plotted. This is shown in figures 11 and 12 and provides operationally useful information for mission planning.

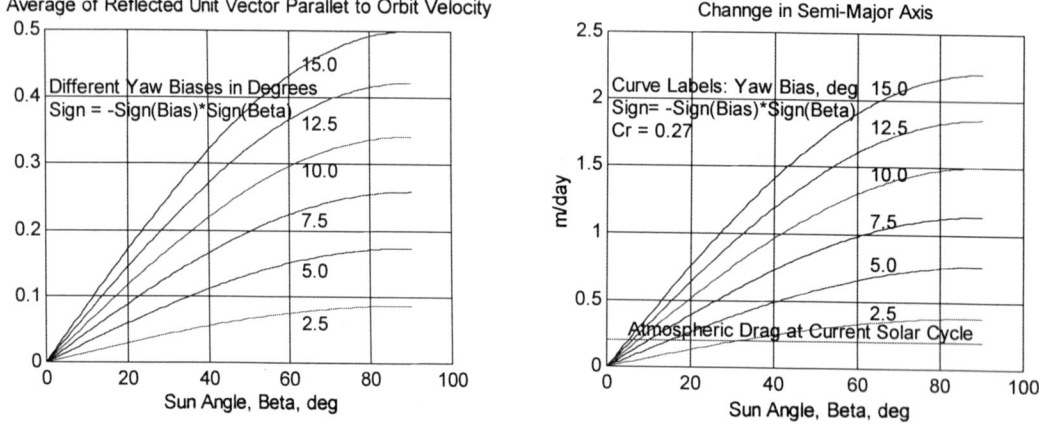

Figures 11 and 12

Having satisfied ourselves as to the cause and effects of the phenomena in question, we can now use this knowledge to augment the current station keeping strategy and even conceive of new station keeping strategies that take advantage of controlled use of the reflected solar radiation pressure force.

The ability to station keep without thrusters leads to a number of possibilities:

1) The spacecraft are now much less sensitive to thruster failure.
2) It may be operationally simpler and less risky to station keep by yaw biasing, since the only commanding requirement is a yaw bias angle as needed.
3) The life time of fuel critical spacecraft could be considerably extended.

Some disadvantages to consider:

1) System software must be able to accommodate the changes resulting from the bias.
2) There is a small power loss for operating with small yaw biases. For the 15 degree (or 165 degree) yaw bias, the power available is reduced to 96.6%. This will not present a serious problem until near end-of-life for the spacecraft. At smaller biases (5 degrees or less) this is almost negligible.

References

1) An Introduction to the Mathematics and Methods of Astrodynamics, Richard H. Battin, American Institute of Aeronautics and Astronautics, New York, 1987.

APPENDICES

PUBLICATIONS OF THE AMERICAN ASTRONAUTICAL SOCIETY

Following are the principal publications of the American Astronautical Society:

JOURNAL OF THE ASTRONAUTICAL SCIENCES (1954 -   )
Published quarterly and distributed by AAS Business Office, 6352 Rolling Mill Place, Suite #102, Springfield, Virginia 22152. Back issues available from Univelt, Inc., P.O. Box 28130, San Diego, California 92198.

SPACE TIMES (1986 -   )
Published bi-monthly and distributed by AAS Business Office, 6352 Rolling Mill Place, Suite #102, Springfield, Virginia 22152.

AAS NEWSLETTER (1962 - 1985)
Incorporated in *Space Times*. Back issues available from AAS Business Office, 6352 Rolling Mill Place, Suite #102, Springfield, Virginia 22152.

ASTRONAUTICAL SCIENCES REVIEW (1959 -1962)
Incorporated in *Space Times*. Back issues still available from Univelt, Inc., P.O. Box 28130, San Diego, California 92198.

ADVANCES IN THE ASTRONAUTICAL SCIENCES (1957 -   )
Proceedings of major AAS technical meetings. Published and distributed for the American Astronautical Society by Univelt, Inc., P.O. Box 28130, San Diego, California 92198.

SCIENCE AND TECHNOLOGY SERIES (1964 -   )
Supplement to *Advances in the Astronautical Sciences*. Proceedings and monographs, most of them based on AAS technical meetings. Published and distributed for the American Astronautical Society by Univelt, Inc., P.O. Box 28130, San Diego, California 92198.

AAS HISTORY SERIES (1977 -   )
Supplement to *Advances in the Astronautical Sciences*. Selected works in the field of aerospace history under the editorship of R. Cargill Hall. Published and distributed for the American Astronautical Society by Univelt, Inc., P.O. Box 28130, San Diego, California 92198.

AAS MICROFICHE SERIES (1968 -   )
Supplement to *Advances in the Astronautical Sciences*. Consists principally of technical papers not included in the hard-copy volume. Published and distributed for the American Astronautical Society by Univelt, Inc., P.O. Box 28130, San Diego, California 92198.

Subscriptions to the *Journal of the Astronautical Sciences* and the *Space Times* should be ordered from the AAS Business Office. Back issues of the *Journal* and all books and microfiche should be ordered from Univelt, Incorporated.

ADVANCES IN THE ASTRONAUTICAL SCIENCES SERIES
(1957-   )

ISSN 0065-3438,

LIBRARY OF CONGRESS CARD NO. 57-43769

Proceedings of Major AAS Technical Meetings

Vol. 1 Third Annual AAS Meeting, Dec. 6-7, 1956, New York, NY, 1957, 184p., ed. Norman V. Petersen, Microfiche only, $20 (ISBN 0-87703-002-2)

Vol. 2 Fourth Annual AAS Meeting, Jan. 29-31, 1958, New York, NY, 1958, 440p., eds. Norman V. Petersen, Horace Jacobs, Microfiche only, $20 (ISBN 0-87703-003-0)

Vol. 3 First Western National AAS Meeting, Aug. 18-19, 1958 530p., eds. Norman V. Petersen, Horace Jacobs, Microfiche only, $20 (ISBN 0-87703-004-9)

Vol. 4 Fifth Annual AAS Meeting, Dec. 27-31, 1958, Washington, D.C., 1959, 462p., ed. Horace Jacobs, Microfiche only, $20 (ISBN 0-87703-005-7)

Vol. 5 Second Western National AAS Meeting, Aug. 4-5, 1959, Los Angeles, CA, 1960, 364p., ed. Horace Jacobs, Microfiche only, $20 (ISBN 0-87703-006-3)

Vol. 6 Sixth Annual AAS Meeting, Jan. 18-21, 1960, New York, NY, 1961, 968p., eds. Horace Jacobs and Eric Burgess, Hard Cover $45 (ISBN 0-87703-007-3)

Vol. 7 Third Western National AAS Meeting, Aug. 4-5, 1960, Seattle, WA, 1961, 464p., eds. Horace Jacobs and Eric Burgess, Microfiche only, $20 (ISBN 0-87703-008-1)

Vol. 8 Seventh Annual AAS Meeting, Jan. 16-18, 1961, Dallas, TX, 1963, 602p., ed. Horace Jacobs, Microfiche only, $20 (ISBN 0-87703-009-X)

Vol. 9 Fourth Western Regional AAS Meeting, Aug. 1-3, 1961, San Francisco, CA, 1963, 910p., ed. Eric Burgess, Hard Cover $45 (ISBN 0-87703-010-3)

Vol. 10 Manned Lunar Flight (AAS/AAAS Symposium) Dec. 19, 1961, Denver, CO, 1963, 310p., eds. George W. Morgenthaler and Horace Jacobs, Hard Cover $35 (ISBN 0-87703-011-1)

Vol. 11 Eighth Annual AAS Meeting, Jan. 16-18, 1962, Washington, D.C., 1963, 808p., ed. Horace Jacobs, Hard Cover $45 (ISBN 0-87703-012-X)

Vol. 12 Scientific Satellites - Mission and Design (AAS/AAAS Symposium), Dec. 27, 1962, Philadelphia, PA, 1963, 262p., ed. Irving E. Jeter, Hard Cover $25 (ISBN 0-87703-013-8)

Vol. 13 Interplanetary Missions, 9th Annual AAS Meeting, Jan. 15-17, 1963, Los Angeles, CA, 1963, 690p., ed. Eric Burgess, Hard Cover $45 (ISBN 0-87703-014-6)

Vol. 14 Second AAS Symposium on Physical and Biological Phenomena under Zero G Conditions, Jan. 18, 1963, Los Angeles, CA, 1963, 382p., eds. Elliot T. Benedikt and Robert W. Halliburton, Hard Cover $30 (ISBN 0-87703-015-4)

Vol. 15 Exploration of Mars Symposium, Jun. 6-7, 1963, Denver, CO, 1963, 634p., ed. George W. Morgenthaler, Hard Cover $45 (ISBN 0-87703-016-2)

Vol. 16 Space Rendezvous, Rescue, and Recovery Symposium, Sept. 10-12, 1963, Edwards, CA 1963, 1408p., ed. Norman V. Petersen, Hard Cover, Part 1, 1028p., $45 (ISBN 0-87703-017-0); Part 2, 380p., $30 (ISBN 0-87703-018-9)

Vol. 17 Bioastronautics - Fundamental and Practical Problems (AAS/AAAS Symposium), Dec. 30, 1963, Cleveland, OH, 1964, 128p., ed. William C. Kaufman, Microfiche only, $10 (ISBN 0-87703-019-7)

Vol. 18 Lunar Flight Programs, 10th Annual AAS Meeting, May 4-7, 1964, New York, NY, 1964, 630p., ed. Ross Fleisig, Hard Cover $45 (ISBN 0-87703-020-0)

Vol. 19 Unmanned Exploration of the Solar System Symposium, Feb. 8-10, 1965, Denver, CO, 1965, 1000p., eds. George W. Morgenthaler, Robert G. Morra, Hard Cover $45 (ISBN 0-87703-021-9)

Vol. 20 Post Apollo Exploration, 11th Annual AAS Meeting, May 3-6, 1965, Chicago, IL, 1966, 1220p., ed. Francis Narin, Microfiche only, Part I, 572p., $30 (ISBN 0-87703-022-7); Part 2, 648p., $35 (ISBN 0-87703-023-5)

Vol. 21 Practical Space Applications Symposium, Feb. 21-23, 1966, San Diego, CA, 1967, 508p., ed. Lawrence L. Kavanau, Microfiche Only $40 (ISBN 0-87703-024-3)

Vol. 22 The Search for Extraterrestrial Life, 12th Annual AAS Meeting, May 23-25, 1966, Anaheim, CA, 1967, 388p., ed. James S. Hanrahan, Microfiche only $30 (ISBN 0-87703-025-1); Microfiche Suppl. (Vol. 1 AAS Microfiche Series) $12 (ISBN 0-87703-132-0)

Vol. 23 Commercial Utilization of Space, 13th Annual AAS Meeting, May 1-3, 1967, Dallas, TX, 1968, 512p., eds. J. Ray Gilmer, Alfred M. Mayo, Ross C. Peavey, Hard Cover (ISBN 0-87703-026-X); plus Microfiche Suppl. (Vol. 3 AAS Microfiche Series) $60 (ISBN 0-87703-216-5)

Vol. 24 Exploitation of Space for Experimental Research, 14th Annual AAS Meeting, May 13-15, 1968, Dedham, MA, 1968, 363p., ed. Harry Zuckerberg, Hard Cover $30 (ISBN 0-87703-027-8)

Vol. 25 Advanced Space Experiments, Sept. 16-18, 1968, Ann Arbor, MI, 1969, 530p., eds. O. Lyle Tiffany and Eugene M. Zaitzeff, Hard Cover $40 (ISBN 0-87703-028-6)

Vol. 26 Planning Challenges of the 70"s in Space, 15th Annual AAS Meeting, Jun. 17-20, 1969, Denver, CO, 1970, 470p., eds. George W. Morgenthaler and Robert G. Morra, Hard Cover $35 (ISBN 0-87703-053-7); Microfiche Suppl. (Vol. 14 AAS Microfiche Series) $20 (ISBN 0-87703-130-4)

Vol. 27/28 Space Stations (v27) and Space Shuttles and Interplanetary Missions (v28), 16th Annual AAS Meeting, Jun. 8-10, 1970, Anaheim, CA, 1970, Vol. 27, eds. Lewis Larmore and Robert L. Gervais, 606p., Hard Cover $45 (ISBN 0-87703-054-5); Vol. 28, eds. Lewis Larmore and Robert L. Gervais, 488p., Hard Cover $35 (ISBN 0-87703-055-3)

Vol. 29 The Outer Solar System, 17th Annual AAS Meeting, Jun. 28-30, 1971, Seattle, WA, 1971, ed. Juris Vagners, Part 1 (on Microfiche Only), 618p., $40 (ISBN 0-87703-059-6); Part 2, Hard Cover, 740p., $45 (ISBN 0-87703-060-X)

Vol. 30 International Congress of Space Benefits, 19th Annual AAS Meeting, Jun. 19-21, 1973, Dallas, TX, 1974, 528p., ed. Francis S. Johnson, Hard Cover $40 (ISBN 0-87703-065-0)

Vol. 31 The Skylab Results, 20th Annual AAS Meeting, Aug. 20-22, 1974, Los Angeles, CA, 1975, 1174p., eds. William C. Schneider and Thomas E. Hanes, Microfiche only (ISBN 0-87703-072-3); Plus Microfiche Suppl. (Vol. 22 AAS Microfiche Series) $60 (ISBN 0-87703-143-6)

Vol. 32 Space Shuttle Missions of the 80's, 21st Annual AAS Meeting, Aug. 26-28, 1975, Denver, CO, 1977, 1364p., eds. William J. Bursnall, George W. Morgenthaler, Gerald E. Simonson, Hard Cover, Part 1, 598p., $40 (ISBN 0-87703-078-2); Hard Cover, Part 2, 766p., $55 (ISBN 0-87703-087-1); Microfiche Suppl. (Vol. 25 AAS Microfiche Series $65 (ISBN 0-87703-133-9)

Vol. 33 AAS/AIAA Astrodynamics Conference, July 28-30, 1975, Nassau, Bahamas, 1976, 390p., eds. William F. Powers, Herbert E. Rauch, Byron D. Tapley, Carmelo E. Velez, Hard Cover $35 (ISBN 0-87703-079-0); Microfiche Suppl. (Vol. 26 AAS Microfiche Series) $40 (ISBN 0-87703-142-8)

Vol. 34 Apollo Soyuz Mission Report, 1977, 336p., ed. Chester M. Lee, Hard Cover $35 (ISBN 0-87703-089-8)

Vol. 35 The Bicentennial Space Symposium - New Themes for Space: Mankind"s Future Needs and Aspirations, 22nd AAS Meeting, Oct. 6-8, 1976, Washington, D.C., 1977, 242p., ed. William C. Schneider, Hard Cover $25 (ISBN 0-87703-090-1)

Vol. 36 The Industrialization of Space, 23rd Annual AAS Meeting, Oct. 18-20, 1977, San Francisco, CA, 1978, 1160p., eds. Richard A. Van Patten, Paul Siegler, Edward V.B. Stearns, Hard Cover, Part 1, 610p., $55 (ISBN 0-87703-094-4); Hard Cover, Part 2, 550p., $45 (ISBN 0-87703-095-2); Microfiche Suppl. (Vol. 28 AAS Microfiche Series) $15 (ISBN 0-87703-121-5)

Vol. 37 Space Shuttle and Spacelab Utilization, What are the Near-Term and Long-Term Benefits for Mankind?, 16th Goddard Memorial Symposium, 24th Annual AAS Meeting, March 8-10, 1978, Washington, D.C., 1978, 865p., eds. George W. Morgenthaler and Manfred Hollstein, Hard Cover, Part 1, 400p., $40 (ISBN 0-87703-096-0); Hard Cover, Part 2, 465p., $45 (ISBN 0-87703-097-9)

Vol. 38 The Future U.S. Space Program, 25th Anniversary Conference, Oct. 20 - Nov. 2, 1978, Houston, TX, 1979, 880p., eds. Richard S. Johnston, Albert Naumann, Jr., Clay W. G. Fulcher, Hard Cover, Part 1, 444p., $45 (ISBN 0-87703-098-7); Hard Cover, Part 2, 436p., $40 (ISBN 0-87703-099-5); Microfiche Suppl. (Vol. 30 AAS Microfiche Series) $15 (ISBN 0-87703-129-0)

Vol. 39 Guidance and Control 1979, Feb. 24-28, 1979, Keystone, CO, 1979, 492p., ed. Robert D. Culp, Hard Cover $45 (ISBN 0-87703-100-2); Microfiche Suppl. (Vol. 31 AAS Microfiche Series) $10 (ISBN 0-87703-128-2)

Vol. 40 AAS/AIAA Astrodynamics Conference, Jun. 25-27, 1979, Provincetown, MA, 1980, 996p., eds. Paul A. Penzo, Bernard Kaufman, Louis Friedman, Richard Battin, Hard Cover, Part 1, 494p., $45 (ISBN 0-87703-107-X); Soft Cover $35 (ISBN 0-87703-108-8); Hard Cover, Part 2, 502p., $45 (ISBN 0-87703-109-6); Soft Cover $35 (ISBN 0-87703-110-X); Microfiche Suppl. (Vol. 32 AAS Microfiche Series) $20 (ISBN 0-87703-139-8)

Vol. 41 Space Shuttle: Dawn of an Era, 26th Annual AAS Meeting, Oct. 29-Nov. 1, 1979, Los Angeles, CA, 1980, 980p., eds. William F. Rector, III and Paul A. Penzo, Hard Cover, Part 1, 452p., $45 (ISBN 0-87703-111-8); Soft Cover $35 (ISBN 0-87703-112-6); Hard Cover, Part 2, 528p., $55 (ISBN 0-87703-113-4); Soft Cover $40 (ISBN 0-87703-114-2); Microfiche Suppl. (Vol. 33 AAS Microfiche Series) $10 (ISBN 0-87703-136-3)

Vol. 42 Guidance and Control 1980, Feb. 17-21, 1980, Keystone, CO, 1980, 738p., ed. Louis A. Morine, Hard Cover $60 (ISBN 0-87703-137-1); Soft Cover $45 (ISBN 0-87703-138-X)

Vol. 43 Shuttle/Spacelab - The New Transportation System and its Utilization, (3rd DGLR/AAS Symposium), Apr. 28-30, 1980, Hannover, Germany, 1981, 342p., eds. Dietrich E. Koelle and George V. Butler, Hard Cover $45 (ISBN 0-87703-144-4); Soft Cover $35 (ISBN 0-87703-146-0)

Vol. 44 Space-Enhancing Technological Leadership, 27th Annual AAS Meeting, Oct. 20-23, 1980, Boston, MA, 1981, 580p., ed. Lawrence P. Greene, Hard Cover $65 (ISBN 0-87703-147-9); Soft Cover $50 (ISBN 0-87703-148-7); Microfiche Suppl. (Vol. 35 AAS Microfiche Series) $10 (ISBN 0-87703-164-9)

Vol. 45 Guidance and Control 1981, Jan. 31-Feb. 4, 1981, Keystone, CO, 1981, 506p., ed. Edward J. Bauman, Hard Cover $60 (ISBN 0-87703-150-9); Soft Cover $50 (ISBN 0-87703-151-7); Microfiche Suppl. (Vol. 36 AAS Microfiche Series) $15 (ISBN 0-87703-156-8)

Vol. 46 AAS/AIAA Astrodynamics Conference, Aug. 3-5, 1981, North Lake Tahoe, NV, 1982, 1124p., eds. Alan L. Friedlander, Paul J. Cefola, Bernard Kaufman, Walt Williamson, G.T. Tseng, Hard Cover, Part 1, 552p., $55 (ISBN 0-87703-159-2); Soft Cover $45 (ISBN 0-87703-160-6); Hard Cover, Part 2, 572p., $55 (ISBN 0-87703-161-4); Soft Cover $45 (ISBN 0-87703-162-2); Microfiche Suppl. (Vol. 37 AAS Microfiche Series) $40 (ISBN 0-87703-163-0)

Vol. 47 Leadership in Space - For Benefits on Earth, 28th Annual AAS Meeting, Oct. 26-29, 1981, San Diego, CA, 1982, 310p., ed. William F. Rector, III, Hard Cover $45 (ISBN 0-87703-168-1); Soft Cover $35 (ISBN 0-87703-169-X)

Vol. 48 Guidance And Control 1982, Jan. 30 -Feb. 3, 1982, Keystone, CO, 1982, 558p., eds. Robert D. Culp, Edward J. Bauman, W. E. Dorroh, Jr., Hard Cover $65 (ISBN 0-87703-170-3); Soft Cover $50 (ISBN 0-87703-171-1); Microfiche Suppl. (Vol. 38 AAS Michrofiche Series) $10 (ISBN 0-87703-180-0)

Vol. 49 Spacelab, Space Platforms, and the Future, Fourth AAS/DGLR Symposium and 20th Goddard Memorial Symposium, Mar. 17-19, 1982, Greenbelt, MD, 1982, 502p., eds. Peter M. Bainum, Dietrich E. Koelle, Hard Cover $55 (ISBN 0-87703-174-6); Soft Cover $45 (ISBN 0-87703-175-4); Microfiche Suppl. (Vol. 42 AAS Microfiche Series) $15 (ISBN 0-87703-181-9)

Vol. 50 Proceedings on an International Symposium on Engineering Sciences and Mechanics, Dec. 29-31, Tainan, Taiwan, 1983, two parts, 1570p., eds. Han-Min Hsia, Richard W. Longman, You-Li Chou, Hard Cover $120 (ISBN 0-87703-176-2); Microfiche Suppl. (Vol. 43 AAS Microfiche Series) $10 (ISBN 0-87703-215-7)

Vol. 51 Guidance and Control 1983, Feb. 5-9, 1983, Keystone, CO, 1983, 494p., eds. Edward J. Bauman, Zubin W. Emsley, Hard Cover $60 (ISBN 0-87703-182-7); Soft Cover $50 (ISBN 0-87703-183-5); Microfiche Suppl. (Vol. 44 AAS Microfiche Series) $10 (ISBN 0-87703-214-9)

Vol. 52 Developing the Space Frontier, 29th Annual AAS Meeting, Oct. 25-27, 1982, Houston, TX, 1983, 436p., eds. Albert Naumann, Grover Alexander, Microfiche Only $45 (ISBN 0-87703-189-4)

Vol. 53 Space Manufacturing 1983, May 9-12, 1983, Princeton, NJ, 1983, 496p., eds. James D. Burke, April S. Whitt, On Microfiche Only $50 (ISBN 0-87703-188-6)

Vol. 54 AAS/AIAA Astrodynamics Conference, Aug. 22-25, 1983, Lake Placid, NY, 1984, two parts, 1370p., eds. G.T. Tseng, Paul J. Cefola, Peter M. Bainum, David A. Levinson, Hard Cover $120 (ISBN 0-87703-190-8); Soft Cover $90 (ISBN 0-87703-191-6); Microfiche Suppl. (Vol. 45 AAS Microfiche Series) $40 (ISBN 0-87703-192-4)

Vol. 55 Guidance and Control 1984, Feb. 4-8, 1984, Keystone, CO, 1984, 500p., eds. Robert D. Culp, Parker S. Stafford, Hard Cover $60 (ISBN 0-87703-199-1); Soft Cover $50 (ISBN 0-87703-200-9); Microfiche Suppl. (Vol. 48 AAS Microfiche Series $15 (ISBN 0-87703-201-7)

Vol. 56 From Spacelab to Space Station, Fifth DGLR/AAS Symposium, Oct. 3-5, 1984, Hamburg, Germany, 1985, 270p., eds. H. Stoewer, Peter M. Bainum, Microfiche Only $30 (ISBN 0-87703-209-2)

Vol. 57 Guidance and Control 1985, Feb. 2-6, 1985, Keystone, CO, 1985, 618p., eds. Robert D. Culp, Edward J. Bauman, Charles A. Cullian, Hard Cover $65 (ISBN 0-87703-211-4); Soft Cover $50 (ISBN 0-87703-212-2); Microfiche Suppl. (Vol. 50 AAS Microfiche Series) $15 (ISBN 0-87703-213-0)

Vol. 58 AAS/AIAA Astrodynamics Conference, Aug. 12-15, 1985, Vail, CO, 1986, two parts, 1556p., eds. Bernard Kaufman, Joseph J.F. Liu, Robert A. Calico, Felix R. Hoots, Hard Cover $140 (ISBN 0-87703-245-9); Soft Cover $110 (ISBN 0-87703-246-7); Microfiche Suppl. (Vol. 51 AAS Microfiche Series); $60 (ISBN 0-87703-247-5)

Vol. 59 Space Station Beyond IOC, 32nd Annual AAS Meeting, Nov 6-7, 1985, Los Angeles, CA, 1986, 188p., ed. M. Jack Friedenthal, Hard Cover $40 (ISBN 0-87703-252-1); Soft Cover $30 (ISBN 0-87703-253-X)

Vol. 60 Space Exploitation and Utilization, First AAS/JRS Symposium, Dec. 15-19, 1985, Honolulu, HI, 1986, 740p., eds. Gayle L. May, Peter M. Bainum, Kenji Ikeda, Tamiya Nomura, Tatsuo Yamanaka, Ryojiro Akiba, Hard Cover $70 (ISBN 0-87703-254-8); Soft Cover $55 (ISBn 0-87703-255-6); Microfiche Suppl. (Vol. 52 AAS Microfiche Series) $10 (ISBN 0-87703-256-4)

Vol. 61 Guidance and Control 1986, Feb. 1-5, 1986, Keystone, CO, 1986, 460p., eds. Robert D. Culp, John C. Durrett, Hard Cover $60 (ISBN 0-87703-257-2); Soft Cover $50 (ISBN 0-87703-258-0); Microfiche Suppl. (Vol. 53 AAS Microfiche Series) $10 (ISBN 0-87703-259-9)

Vol. 62 Tethers in Space, Proceedings of First International Conference on Tethers in Space (NASA & PSN Sponsors; AIAA, AAS, & AIDAA Co-Sponsors), Sept. 17-19, 1986, Arlington, VA, 1987, 784p., eds. Peter M. Bainum, Ivan Bekey, Luciano Guerriero, Paul A. Penzo, Hard Cover $80 (ISBN 0-87703-264-5); Soft Cover $70 (ISBN 0-87703-265-3)

Vol. 63 Guidance and Control 1987, Jan. 31 -Feb. 4, 1987, Keystone, CO, 1987, 638p., eds. Robert D. Culp, Terry J. Kelly, Hard Cover $75 (ISBN 0-87703-268-8); Soft Cover $60 (ISBN 0-87703-269-6)

Vol. 64 Aerospace Century XXI, 33rd AAS Annual Meeting, Oct. 26-29, 1986, Boulder, CO, 1987, all three parts, Hard Cover $225 (ISBN 0-87703-276-9); Soft Cover $180 (ISBN 0-87703-277-7); Part I, Space Missions and Policy, 686p., eds. George W. Morgenthaler, Gayle L. May, Hard Cover $75 (ISBN 0-87703-279-3); Soft Cover $60 (ISBN 0-87703-282-3); Part II, Space Flight Technologies, 608p., eds. George W. Morgenthaler, W. Kent Tobiska, Hard Cover $75 (ISBN 0-87703-280-7); Soft Cover $60 (ISBN 0-87703-283-1); Part III, Space Sciences, Applications, and Commercial Developments, 724p., eds. George W. Morgenthaler, Jean N. Koster, Hard Cover $75 (ISBN 0-87703-281-5); Soft Cover $60 (ISBN 0-87703-284-X); Microfiche Suppl. (Vol. 54 AAS Microfiche Series) $25 (ISBN 0-87703-278-5)

Vol. 65 AAS/AIAA Astrodynamics Conference, Aug. 10-13, 1987, Kalispell, MT, 1988, two parts, 1774p., eds. John K. Soldner, Arun K. Misra, Robert E. Lindberg, Walton Williamson, Hard Cover $180 (ISBN 0-87703-285-8); Soft Cover $150 (ISBN 0-87703-286-6); Microfiche Suppl. (Vol. 55 AAS Microfiche Series); $70 (ISBN 0-87703-287-4)

Vol. 66 Guidance and Control 1988, Jan. 30 -Feb. 3, 1988, Keystone, CO, 1988, 576p., eds. Robert D. Culp, Paul L. Shattuck, Hard Cover $75 (ISBN 0-87703-288-2); Soft Cover $60 (ISBN 0-87703-289-0); Microfiche Suppl. (Vol. 56 AAS Microfiche Series) $10 (ISBN 0-87703-290-4)

Vol. 67 Space - A New Community of Opportunity, 34th AAS Annual Meeting, Nov. 3-5, 1987, Houston, TX, 1989, 472p., eds. William G. Straight, Henry N. Bowes, Hard Cover $70 (ISBN 0-87703-297-1); Soft Cover $55 (ISBN 0-87703-298-X)

Vol. 68 Guidance and Control 1989, Feb. 4-8, 1989, Keystone, CO, 1989, 708p., eds. Robert D. Culp, Robert A. Lewis, Hard Cover $85 (ISBN 0-87703-299-8); Soft Cover $70 (ISBN 0-87703-300-5)

Vol. 69 Orbital Mechanics and Mission Design, Apr. 24-27, 1989, Greenbelt, MD, 1989, 862p., ed. Jerome Teles, Hard Cover $95 (ISBN 0-87703-311-0); Soft Cover $80 (ISBN 0-87703-312-9); Microfiche Suppl. (Vol. 57 AAS Microfiche Series) $10 (ISBN 0-87703-313-7)

Vol. 70 The 21st Century in Space, 35th AAS Annual Meeting, Oct. 24-26, 1988, St. Louis, MO, 1990, 446p., ed. George V. Butler, Hard Cover $90 (ISBN 0-87703-314-5); Soft Cover $75 (ISBN 0-87703-315-3); Microfiche Suppl. (Vol. 58 AAS Microfiche Series) $10 (ISBN 0-87703-316-1)

Vol. 71 AAS/AIAA Astrodynamics Conference, Aug. 7-10, 1989, Stowe, VT, 1990, two parts, 1472p., eds. Catherine L. Thornton, Ronald J. Proulx, John E. Prussing, Felix R. Hoots, Hard Cover $200 (ISBN 0-87703-317-X); Soft Cover $170 (ISBN 0-87703-318-8); Microfiche Suppl. (Vol. 59 AAS Microfiche Series) $50 (ISBN 0-87703-319-6)

Vol. 72 Guidance and Control 1990, Feb. 3-7, 1990, Keystone, CO, 1990, 676p., eds. Robert D. Culp, Arlo D. Gravseth, Hard Cover $95 (ISBN 0-87703-320-X); Soft Cover $80 (ISBN 0-87703-321-8)

Vol. 73 Space Utilization and Applications in the Pacific, Third (PISSTA) AAS/JRS/CSA Symposium, Nov. 6-8, 1989, Los Angeles, CA, 1990, 764p., eds. Peter M. Bainum, Gayle L. May, Tatsuo Yamanaka, Yang Jiachi, Hard Cover $95 (ISBN 0-87703-325-0); Soft Cover $80 (ISBN 0-87703-326-9)

Vol. 74 Guidance and Control 1991, Feb. 2-6, 1991, Keystone, CO, 1991, 730p., eds. Robert D. Culp, James P. McQuerry, Hard Cover $120 (ISBN 0-87703-334-X); Soft Cover $90 (ISBN 0-87703-335-8)

Vol. 75 AAS/AIAA Spaceflight Mechanics Meeting, Feb. 11-13, 1991, Houston, TX, 1991, two parts, 1354p., eds. John K. Soldner, Arun K. Misra, Lester L. Sackett, Richard Holdaway, Hard Cover $220 (ISBN 0-87703-338-2); Soft Cover $190 (ISBN 0-87703-339-0); Microfiche Suppl. (Vol. 62 AAS Microfiche Series) $50 (ISBN 0-87703-340-4)

Vol. 76 AAS/AIAA Astrodynamics Conference, Aug. 19-22, 1991, Durango, CO, 1992, three parts, 2590p., eds. Bernard Kaufman, Kyle T. Alfriend, Ronald L. Roehrich, Robert R. Dasenbrock, Hard Cover $390 (ISBN 0-87703-347-1); Microfiche Suppl. (Vol. 63 AAS Microfiche Series) $75 (ISBN 0-87703-348-X)

Vol. 77 International Space Year (ISY) in the Pacific Basin, Fourth ISCOPS (formerly PISSTA) AAS/JRS/CSA Symposium, Nov. 17-20, 1991, Kyoto, Japan, 1992, 798p., eds. Peter M. Bainum, Gayle L. May, Makoto Nagatomo, Yoshiaki Ohkami, Yang Jiachi, Hard Cover $120 (ISBN 0-87703-351-X); Soft Cover $90 (ISBN 0-87703-352-8)

Vol. 78 Guidance and Control 1992, Feb. 2-6, 1992, Keystone, CO, 1992, 754p., eds. Robert D. Culp, Richard P. Zietz, Hard Cover $120 (ISBN 0-87703-353-6); Soft Cover $90 (ISBN 0-87703-354-4); Microfiche Suppl. (Vol. 64 AAS Microfiche Series) $20 (ISBN 0-87703-355-2)

Vol. 79 AAS/AIAA Spaceflight Mechanics Meeting, Feb. 24-26, 1992, Colorado Springs, CO, 1992, two parts, 1312p., eds. Roger E. Diehl, Ralph G. Schinnerer, Walton E. Williamson, Daryl G. Boden, Hard Cover $240 (ISBN 0-87703-358-7); Microfiche Suppl. (Vol. 65 AAS Microfiche Series) $40 (ISBN 0-87703-359-5)

Vol. 80 Space Business Opportunities, 37th and 38th AAS Annual Meetings, Nov. 5-7, 1990, Dec. 3-5, 1991, Los Angeles, CA, 1992, 380p., eds. Wayne J. Esser, Don K. Tomajan, Hard Cover $90 (ISBN 0-87703-360-9); Soft Cover $70 (ISBN 0-87703-361-7); Microfiche Suppl. (Vol. 61 AAS Microfiche Series) $50 (ISBN 0-87703-331-5); (Vol. 66 AAS Microfiche Series) $60 (ISBN 0-87703-362-5)

Vol. 81 Guidance and Control 1993, Feb. 6-10, 1993, Keystone, CO, 1993, 648p., eds. Robert D. Culp, George Bickley, Hard Cover $120 (ISBN 0-87703-365-X); Soft Cover $90 (ISBN 0-87703-366-8); Microfiche Suppl. (Vol. 67 AAS Microfiche Series) $20 (ISBN 0-87703-367-6)

Vol. 82 AAS/AIAA Spaceflight Mechanics Meeting, Feb. 22-24, 1993, Pasadena, CA, 1993, two parts, 1454p., eds. Robert G. Melton, Lincoln J. Wood, Roger C. Thompson, Stuart J. Kerridge, Hard Cover $240 (ISBN 0-87703-368-4); Microfiche Suppl. (Vol. 68 AAS Microfiche Series) $15 (ISBN 0-87703-369-2)

Vol. 83 Dynamics of Space Tether Systems (English Language Edition), 1993, 508p., by Vladimir V. Beletsky, Evgenii M. Levin, Hard Cover $120 (ISBN 0-87703-370-6); Soft Cover $90 (ISBN 0-87703-371-4)

Vol. 84 AAS/GSFC International Symposium on Spaceflight Dynamics, Apr. 26-30, 1993, Greenbelt, MD, 1993, two parts, 1450p., eds. Jerome Teles, Mina V. Samii, Hard Cover $240 (ISBN 0-87703-378-1); Microfiche Suppl. (Vol. 69 AAS Microfiche Series) $10 (ISBN 0-87703-379-X)

Vol. 85 AAS/AIAA Astrodynamics Conference, Aug. 16-19, 1993, Victoria, British Columbia, Canada, 1994, three parts, 2750p., eds. Arun K. Misra, Vinod J. Modi, Richard Holdaway, Peter M. Bainum, Hard Cover $390 (ISBN 0-87703-380-3); Microfiche Suppl. (Vol. 70 AAS Microfiche Series) $30 (ISBN 0-87703-381-1)

Vol. 86 Guidance and Control 1994, Feb. 2-6, 1994, Keystone, CO, 1994, 700p., eds. Robert D. Culp, Ronald D. Rausch, Hard Cover $120 (ISBN 0-87703-384-6); Soft Cover $90 (ISBN 0-87703-385-4)

Vol. 87 AAS/AIAA Spaceflight Mechanics Meeting, Feb. 14-16, 1994, Cocoa Beach, FL, 1994, two parts, 1272p., eds. John E. Cochran, Jr., Charles D. Edwards, Jr., Stephen J. Hoffman, Richard Holdaway, Hard Cover $240 (ISBN 0-87703-386-2)

Vol. 88 Guidance and Control 1995, Feb. 1-5, 1995, Keystone, CO, 1995, 600p. eds. Robert D. Culp, James D. Medbery, Hard Cover $120 (ISBN 0-87703-399-4); Soft Cover $90 (ISBN 0-87703-400-1)

Vol. 89 AAS/AIAA Spaceflight Mechanics Meeting, Feb. 13-16, 1995, Albuquerque, NM, 1995, two parts, 1774p., eds. Ronald J. Proulx, Joseph J. F. Liu, P. Kenneth Seidelmann, Salvatore Alfano, Hard Cover $280 (ISBN 0-87703-401-X); Microfiche Suppl. (Vol. 71 AAS Microfiche Series) $15 (ISBN 0-87703-402-8)

Vol. 90 AAS/AIAA Astrodynamics Conference, Aug. 14-17, 1995, Halifax, Nova Scotia, Canada, 1996, two parts, 2270p., eds. K. Terry Alfriend, I. Michael Ross, Arun K. Misra, C. Fred Peters, Hard Cover $290 (ISBN 0-87703-407-9); Microfiche Suppl. (Vol. 72 AAS Microfiche Series) $15 (ISBN 0-87703-408-7)

Vol. 91 Strengthening Cooperation in the 21st Century, Sixth ISCOPS (formerly PISSTA) AAS/JRS/CSA Symposium, Dec. 6-8, 1995, Marina Del Rey, CA, 1996, 1154p., eds. Peter M. Bainum, Gayle L. May, Yoshiaki Ohkami, Kuninori Uesugi, Qi Faren, Li Furong, Hard Cover $145 (ISBN 0-87703-409-5)

Vol. 92 Guidance and Control 1996, Feb. 7-11, 1996, Breckenridge, CO, 1996, 744p. Eds. Robert D. Culp, Marv Odefey, Hard Cover $120 (ISBN 0-87703-412-5); Soft Cover $90 (ISBN 0-87703-413-3)

Vol. 93 AAS/AIAA Spaceflight Mechanics Meeting, Feb. 12-15, 1996, Austin, TX, 1996, two parts, 1776p., eds. G. Edward Powell, Robert H. Bishop, John B. Lundberg, Robert H. Smith, Hard Cover $280 (ISBN 0-87703-414-1); Microfiche Suppl. (Vol. 73 AAS Microfiche Series) $10 (ISBN 0-87703-415-X)

Vol. 94 Guidance and Control 1997, Feb. 5-9, 1997, Breckenridge, CO, 1997, 458p. Eds. Robert D. Culp, Stuart B. Wiens, Hard Cover $120 (ISBN 0-87703-430-3); Soft Cover $90 (ISBN 0-87703-431-1)

Vol. 95 AAS/AIAA Spaceflight Mechanics Meeting, Feb. 10-12, 1997, Huntsville, AL, 1997, two parts, 1168p., eds. Kathleen C. Howell, David A. Cicci, John E. Cochran, Jr., Thomas S. Kelso, Hard Cover $240 (ISBN 0-87703-432-X)

Vol. 96 Space Cooperation into the 21st Century, Seventh ISCOPS (formerly PISSTA) JRS/AAS/CSA Symposium, Jul. 15-18, 1997, Nagasaki, Japan, 1997, 1089p., eds. Peter M. Bainum, Gayle L. May, Makoto Nagatomo, Kuninori T. Uesugi, Fu Bing-chen, Zhang Hui, Hard Cover $145.00 (ISBN 0-87703-438-9)

Vol. 97 AAS/AIAA Astrodynamics Conference, Aug. 4-7, 1997, Sun Valley, ID, 1998, two parts, 2184p., eds. Felix R. Hoots, Bernard Kaufman, Paul J. Cefola, David B. Spencer, Hard Cover $310 (ISBN 0-87703-441-9)

Vol. 98 Guidance and Control 1998, Feb. 4-8, 1998, Breckenridge, CO, 1998, 706p. Eds. Robert D. Culp, David Igli, Hard Cover $120 (ISBN 0-87703-448-6); Soft Cover $90 (ISBN 0-87703-449-4)

Vol. 99 AAS/AIAA Spaceflight Mechanics Meeting, Feb. 9-11, 1998, Monterey, CA, 1998, two parts, 1638p., eds. Jay W. Middour, Lester L. Sackett, Louis A. D"Amario, Dennis V. Byrnes, Hard Cover $280 (ISBN 0-87703-450-8); Microfiche Suppl. (Vol. 78 AAS Microfiche Series) $10 (ISBN 0-87703-452-4)

Vol. 100 AAS/GSFC International Symposium on Space Flight Dynamics, May 11-15, 1998, Greenbelt, MD, 1998, two parts, 1092p., ed. Thomas H. Stengle, Hard Cover $250 (ISBN 0-87703-453-2)

Vol. 101 Guidance and Control 1999, Feb. 3-7, 1999, Breckenridge, CO, 1999, 528p. Eds. Robert D. Culp, Douglas Wiemer, Hard Cover $120 (ISBN 0-87703-456-7); Soft Cover $90 (ISBN 0-87703-457-5)

Vol. 102 AAS/AIAA Spaceflight Mechanics Meeting, Feb. 7-10, 1999, Breckenridge, CO, 1999, two parts, 1600p., eds. Robert H. Bishop, Robert D. Culp, Donald L. Macklson and Maria Evans, Hard Cover $280 (ISBN 0-87703-458-3)

Vol. 103 AAS/AIAA Astrodynamics Conference, Aug. 16-19, 1999, Girdwood, AK, 2000, three parts, 2724p., eds. Kathleen C. Howell, Felix R. Hoots, Bernard Kaufman, K. Terry Alfriend, Hard Cover $450 (ISBN 0-87703-467-2)

Vol. 104 Guidance and Control 2000, Feb. 2-6, 2000, Breckenridge, CO, 2000, 738p. Eds. Robert D. Culp, Eileen M. Dukes, Hard Cover $130 (ISBN 0-87703-468-0); Soft Cover $95 (ISBN 0-87703-469-9)

Order from Univelt, Inc., P.O. Box 28130, San Diego, California 92198
STANDING ORDERS ACCEPTED

SCIENCE AND TECHNOLOGY SERIES (1964-   )

ISSN 0278-4017

A Supplement to *Advances in the Astronautical Sciences*. Proceedings and monographs, most of them based on AAS technical meetings.

Vol. 1 Manned Space Reliability Symposium, Jun. 9, 1964, Anaheim, CA, 1964, 112p., ed. Paul Horowitz, Hard Cover $20 (ISBN 0-87703-029-4)

Vol. 2 Towards Deeper Space Penetration (AAS/AAAS Symposium), Dec. 29, 1964, Montreal, Canada, 1964, 182p., ed. Edward R. Van Driest, Hard cover $20 (ISBN 0-87703-030-8)

Vol. 3 Orbital Hodograph Analysis, 1965, 150p., ed. Samuel P. Altman, Hard Cover $20 (ISBN 0-87703-031-6)

Vol. 4 Scientific Experiments for Manned Orbital Flight, 3rd Goddard Memorial Symposium, Mar. 18-19, 1965, Washington, D.C., 1965, 372p., ed. Peter C. Badgley, Hard Cover $30 (ISBN 0-87703-032-4)

Vol. 5 Physiological and Performance Determinants in Manned Space Systems (AAS/HFS Symposium, Apr. 14-15, 1965, Northridge, CA, 1965, 220p., ed. Paul Horowitz, Microfiche Only $20 (ISBN 0-87703-033-2)

Vol. 6 Space Electronics Symposium (AAS/AES Meeting), May 25-27, 1965, Los Angeles, CA, 1965, 404p., ed. Chung-Ming Wong, Hard Cover $30 (ISBN 0-87703-034-0)

Vol. 7 Theodore von Karman Memorial Seminar, May 12, 1965, Los Angeles, CA, 1966, 140p., ed. Shirley Thomas, Hard Cover $30 (ISBN 0-87703-035-9)

Vol. 8 Impact of Space Exploration on Society, Aug. 18-20, 1965, San Francisco, CA, 1966, 382p., ed. William E. Frye, Hard Cover $30 (ISBN 0-87703-036-7)

Vol. 9 Recent Developments in Space Flight Mechanics, (AAS/AAAS Symposium), Dec. 29, 1965, Berkeley, CA, 1966, 280p., ed. Paul B. Richards, Hard Cover $25 (ISBN 0-87703-037-5)

Vol. 10 Space in the Fiscal Year 2001, 4th Goddard Memorial Symposium, Mar. 15-16, 1966, Washington, D.C., 1967, 458p., eds. Eugene B. Konecci, Maxwell W. Hunter, II, Robert F. Trapp, Hard Cover $35 (ISBN 0-87703-038-3)

Vol. 11 Space Flight Specialist Conference, Jul. 6-8, 1966, Denver, CO, 1967, 618p., ed. Maurice L. Anthony, Microfiche Only (ISBN 0-87703-039-1); Plus Microfiche Suppl. (Vol. 2 AAS Microfiche Series) $60 (ISBN 0-87703-221-1)

Vol. 12 Management of Aerospace Programs Conference, Nov. 16-18, 1966, Columbia, MO, 1967, 392p., ed. Walter K. Johnson, Hard Cover $30 (ISBN 0-87703-040-5)

Vol. 13 Physics of the Moon (AAS/AAAS Symposium), Dec. 29, 1966, Washington, D.C., 1967, 260p., ed. S. Fred Singer, Hard Cover $25 (ISBN 0-87703-041-3)

Vol. 14 Interpretation of Lunar Probe Data, Sept. 17, 1966, Huntington Beach, CA, 1967, 270p., ed. Jack Green, Hard Cover $25 (ISBN 0-87703-042-1)

Vol. 15 Future Space Program and Impact on Range and Network Development Symposium, Mar. 22-24, 1967, Las Cruces, NM, 1967, 588p., ed. George W. Morgenthaler, Hard Cover $40 (ISBN 0-87703-043-X)

Vol. 16 Voyage to the Planets, 5th Goddard Memorial Symposium, Mar. 14-15, 1967, Washington, D.C., 1968, 184p., ed. S. Fred Singer, Hard Cover $20 (0-87703-044-8)

Vol. 17 Use of Space Systems for Planetary Geology and Geophysics Symposium, May 25-27, 1967, Boston, MA, 1968, 623p., ed. Robert D. Enzmann, Hard Cover $45 (ISBN 0-87703-045-6); Microfiche Suppl. (Vol. 5 AAS Microfiche Series) $15 (ISBN 0-87703-135-5)

Vol. 18 Technology and Social Progress, 6th Goddard Memorial Symposium, Mar. 12-13, 1968, Washington, D.C., 1969, 170p., ed. Philip K. Eckman, Hard Cover $20 (ISBN 0-87703-046-4)

Vol. 19 Exobiology - The Search for Extraterrestrial Life (AAS/AAAS Symposium) Dec. 30, 1967, New York, NY, 1969, 184p., eds. Martin M. Freundlich, Bernard W. Wagner, Hard Cover $20 (ISBN 0-87703-047-2)

Vol. 20 Bioengineering and Cabin Ecology (AAS/AAAS Symposium) Dec. 30, 1968, Dallas, TX, 1969, 162p., ed. William Cassidy, Hard Cover $20 (ISBN 0-87703-048-0)

Vol. 21 Reducing the Cost of Space Transportation, 7th Goddard Memorial Symposium, Mar. 4-5, 1969, Washington, D.C., 1969, 264p., ed. George K. Chacko, Microfiche only $25 (ISBN 0-87703-049-9)

Vol. 22 Planning Challenges of the 70"s in the Public Domain, 15th Annual AAS Meeting, Jun. 17-20, 1969, Denver, CO, 1970, 504p., eds. William J. Burnsnall, George K. Chacko, George W. Morgenthaler, Hard Cover $40 (ISBN 0-87703-050-2); Microfiche Suppl. (Vol. 13 AAS Microfiche Series) $20 (ISBN 0-87703-131-2); See also Vols. 15-17, AAS Microfiche Series

Vol. 23 Space Technology and Earth Problems Symposium, Oct. 23-25, 1969, Las Cruces, NM, 1970, 418p., ed. C. Quentin Ford, Hard Cover $35 (ISBN 0-87703-051-0); Microfiche Suppl. (Vol. 12 AAS Microfiche Series) $20 (ISBN 0-87703-134-7)

Vol. 24 Aerospace Research and Development, Jul. 14, 1966, Holloman AFB, NM, 1970, 500p., ed. Ernst A. Steinhoff, Microfiche Only $40 (ISBN 0-87703-052-9)

Vol. 25 Geological Problems in Lunar and Planetary Research, Feb. 17-18, 1969, Huntington Beach, CA, 1971, 750p., ed. Jack Green, Hard Cover $45 (ISBN 0-87703-056-1)

Vol. 26 Technology Utilization Ideas for the 70s and Beyond, Oct. 30, 1970, Winrock, AR, 1971, 312p., eds. Fred W. Forbes, Paul Dergarabedian, Microfiche only $30 (ISBN 0-87703-057-X)

Vol. 27 International Cooperation in Space Operations and Exploration, 9th Goddard Memorial Symposium, Mar. 11, 1971, Washington, D.C. 1971, 194p., ed. Michael Cutler, Hard Cover $20 (ISBN 0-88703-058-8)

Vol. 28 Astronomy from a Space Platform (AAS/AAAS Symposium) Dec. 27-28, 1971, Philadelphia, PA, 1972, 416p., eds. George W. Morgenthaler, Howard D. Greyber, Hard Cover $35 (ISBN 0-87703-061-8)

Vol. 29 Space Technology Transfer to Community and Industry, 10th Goddard Memorial Symposium, 18th Annual AAS Meeting, Mar. 13-14, 1972, Washington, D.C., 1972, 196p., eds. Ralph H. Tripp, John K. Stotz, Jr., Hard Cover $20 (ISBN 0-87703-062-6); on Microfiche $15

Vol. 30 Space Shuttle Payloads (AAS/AAAS Symposium) Dec. 27-28, 1972, Washington, D.C., 1973, 532p., eds. George W. Morgenthaler, William J. Bursnall, Hard Cover $40 (ISBN 0-87703-063-4)

Vol. 31 The Second Fifteen Years in Space, 11th Goddard Memorial Symposium, Mar. 8-9, 1973, Washington, D.C., 1973, 212p., ed. Saul Ferdman, Hard Cover $25 (ISBN 0-87703-064-2)

Vol. 32 Health Care Systems Conference, Nov. 21-22, 1972, Dallas, TX, 1974, 265p., ed. Eugene B. Konecci, Hard Cover $25 (ISBN 0-87703-067-7)

Vol. 33 Orbital International Laboratory, 3rd and 4th IAF/OIL Symposia, Oct. 5-6, 1970, Constance, Germany, Sept. 24-25, 1971, Brussels, Belgium, 1974, 322p., ed. Ernst A. Steinhoff, Hard Cover $30 (ISBN 0-87703-068-5)

Vol. 34 Management and Design of Long-Life Systems, Apr. 24-26, 1973, Denver, CO, 1974, 198p., ed. Harris M. Schurmeier, Hard Cover $20 (ISBN 0-87703-069-3)

Vol. 35 Energy Delta, Supply vs. Demand, (AAS/AAAS Symposium) Feb. 25-27, 1974, San Francisco, CA, 1975, 2nd Printing 1976, 604p., eds. George W. Morgenthaler, Aaron N. Silver, Hard Cover $35 (ISBN 0-87703-070-7); Soft Cover $25 (ISBN 0-87703-082-0); on Microfiche $20

Vol. 36 Skylab and Pioneer Report, 12th Goddard Memorial Symposium, Mar. 8, 1974, Washington, D.C., 1975, 160p., eds. Philip H. Bolger, Paul B. Richards, Hard Cover $20 (ISBN 0-87703-071-5)

Vol. 37 Space Rescue and Safety 1974, 7th International IAA Symposium, Sept. 30 - Oct. 5, 1974, Amsterdam, Netherlands, 1975, 294p., ed. Philip H. Bolger, Microfiche Only $25 (ISBN 0-87703-073-1)

Vol. 38 Skylab Science Experiments, (AAS/AAAS Symposium) Feb. 28, 1974, San Francisco, CA, 1976, 274p., eds. George W. Morgenthaler, Gerald E. Simonson, Microfiche only $20 (ISBN 0-87703-074-X)

Vol. 39 Environmental Control and Agri-Technology, 1976, 346p., ed. Eugene B. Konecci, Microfiche only $20 (ISBN 0-87703-075-8)

Vol. 40 Future Space Activities, 13th Goddard Memorial Symposium, Apr. 11, 1975, Washington, D.C., 1976, 182p., ed. Carl H. Tross, Microfiche only $20 (ISBN 0-87703-076-6)

Vol. 41 Space Rescue and Safety 1975, 8th International IAA Symposium, Sept. 21-27, 1975, Lisbon, Portugal, 1976, 230p., ed. Philip H. Bolger, Hard Cover $25 (ISBN 0-87703-077-4)

Vol. 42 The End of an Era in Space Exploration, From International Rivalry to International Cooperation, 1976, 216p., by J.C.D. Blaine, Hard Cover $25 (ISBN 0-87703-084-7); without volume number (ISBN 0-87703-080-4)

Vol. 43 The Eagle Has Returned, Part I, International Space Hall of Fame Dedication Conference, Oct. 5-9, 1976, Alamogordo, NM, 1976, 370p., ed. Ernst. A. Steinhoff, Hard Cover $30 (ISBN 0-87703-086-3)

Vol. 44 Satellite Communications in the Next Decade, 14th Goddard Memorial Symposium, Mar. 12, 1976, Washington, D.C., 1977, 188p., ed. Leonard Jaffe, Hard Cover $20 (ISBN 0-87703-088-X)

Vol. 45 The Eagle Has Returned, Part 2, International Space Hall of Fame Dedication Conference, Oct. 5-9, 1976, Alamogordo, NM, 1977, 454p., ed. Ernst A. Steinhoff, Hard Cover $35 (ISBN 0-87703-092-8)

Vol. 46 Export of Aerospace Technology, 15th Goddard Memorial Symposium, Mar. 31 - Apr. 1, 1977, Washington, D.C., 1978, 174p., ed. Carl H. Tross, Hard Cover $20 (ISBN 0-87703-093-6)

Vol. 47 Handbook of Soviet Lunar and Planetary Exploration, 1979, 276p., by Nicholas L. Johnson, Microfiche Only $25 (ISBN 0-87703-105-3)

Vol. 48 Handbook of Soviet Manned Space Flight, 2nd Edition, 1988, 474p., by Nicholas L. Johnson, Hard Cover $60 (ISBN 0-87703-115-0); Soft Cover $45 (ISBN 0-87703-116-9)

Vol. 49 Space - New Opportunities for International Ventures, 17th Goddard Memorial Symposium, Mar. 28-30, 1979, Washington, D.C., 1980, 300p., ed. William C. Hayes, Jr., Hard Cover $35 (ISBN 0-87703-124-X); Soft Cover $25 (ISBN 0-87703-125-8); see also Vol. 2 AAS History Series

Vol. 50 Remember the Future - The Apollo Legacy, Jul. 20-21, 1979, San Francisco, CA, 1980, 218p., ed. Stan Kent, Hard Cover $25 (ISBN 0-87703-126-6); Soft Cover $15 (ISBN 0-87703-127-4)

Vol. 51 Commercial Operations in Space 1980-2000, 18th Goddard Memorial Symposium, Mar. 27-28, 1980, Washington, D.C., 1981, 214p., eds. John L. McLucas, Charles Sheffield, Hard Cover $30 (ISBN 0-87703-140-1); Soft Cover $20 (ISBN 0-87703-141-X); Microfiche Suppl. (Vol. 34 AAS Microfiche Series) $10 (ISBN 0-87703-165-7); see also Vols. 2 and 3, AAS History Series

Vol. 52 International Space Technical Applications, 19th Goddard Memorial Symposium, Mar. 26-27, 1981, Washington, D.C., 1981, 186p., eds. Andrew Adelman, Peter M. Bainum, Hard Cover $30 (ISBN 0-87703-152-5); Soft Cover $20 (ISBN 0-87703-153-3); see also Vol. 5, AAS History Series

Vol. 53 Space in the 1980"s and Beyond, 17th European Space Symposium, Jun. 4-6, 1980, London, England, 1981, 302p., ed. Peter M. Bainum, Hard Cover $40 (ISBN 0-87703-154-1); Soft Cover $30 (ISBN 0-87703-155-X)

Vol. 54 Space Safety and Rescue 1979-1981 (with abstracts 1976-1978), Proceedings of symposia of the International Academy of Astronautics held in conjunction with the 30th, 31st, and 32nd International Astronautical Federation Congresses, Munich, Germany, 1979, Tokyo, Japan, 1980, and Rome, Italy, 1981, 1983, 456p., ed. Jeri W. Brown, Hard Cover $45 (ISBN 0-87703-177-0); Soft Cover $35 (ISBN 0-87703-178-9); Microfiche Suppl. (Vols. 39-41 AAS Microfiche Series) $39 (ISBN 0-87703-222-X); (ISBN 0-87703-223-8); (ISBN 0-87703-224-6)

Vol. 55 Space Applications at the Crossroads, 21st Goddard Memorial Symposium, Mar. 24-25, 1983, Greenbelt, MD, 1983, 308p., eds. John H. McElroy, E. Larry Heacock, Hard Cover $45 (ISBN 0-87703-186-X); Soft Cover $35 (ISBN 0-87703-187-8)

Vol. 56 Space: A Developing Role for Europe, 18th European Space Symposium, Jun. 6-9, 1983, London, England, 1984, 278p., eds. Len J. Carter, Peter M. Bainum, Hard Cover $45 (ISBN 0-87703-193-2); Soft Cover $35 (ISBN 0-87703-194-0); Microfiche Suppl. (Vol. 46 AAS Microfiche Series) $15 (ISBN 0-87703-195-9)

Vol. 57 The Case for Mars, Apr. 29 - May 2, 1981, Boulder, CO, 1984, Second Printing 1987, 348p., ed. Penelope J. Boston, on Microfiche only $25 (ISBN 0-87703-198-3)

Vol. 58 Space Safety and Rescue 1982-1983, Proceedings of the International Academy of Astronautics held in conjunction with the 33rd and 34th International Astronautical Congresses, Paris, France, Sept. 27 - Oct. 2, 1982, and Budapest, Hungary, Oct. 10-15, 1983, 1984, 378p., ed. Gloria W. Heath, Hard Cover $50 (ISBN 0-87703-202-5); Soft Cover $40 (ISBN 0-87703-203-3)

Vol. 59 Space and Society - Challenges and Choices, April 14-16, 1982, University of Texas at Austin, 1984, 442p., eds. Paul Anaejionu, Nathan C. Goldman, Philip J. Meeks, Hard Cover $55 (ISBN 0-87703-204-1); Soft Cover $35 (ISBN 0-87703-205-X)

Vol. 60 Permanent Presence - Making It Work, 22nd Goddard Memorial Symposium, Mar. 15-16, 1984, Greenbelt, MD, 1985, 190p., ed. Ivan Bekey, Hard Cover $40 (ISBN 0-87703-207-6); Soft Cover $30 (ISBN 0-87703-208-4)

Vol. 61 Europe/United States Space Activities - With a Space Propulsion Supplement, 23rd Goddard Memorial Symposium/19th European Space Symposium, Mar. 27-29, 1985, Greenbelt, MD, 31st Annual AAS Meeting, Oct. 22-24, 1984, Palo Alto, CA, 1985, 442p., eds. Peter M. Bainum, Friedrich von Bun, Hard Cover $55 (ISBN 0-87703-217-3); Soft Cover $45 (ISBN 0-87703-218-1)

Vol. 62 The Case for Mars II, July 10-14, 1984, Boulder, CO, 1985, 730p., ed. Christopher P. McKay, Hard Cover $60 (ISBN 0-87703-219-1); Soft Cover $40 (ISBN 0-87703-220-3)

Vol. 63 Proceedings of 4th International Conference on Applied Numerical Modeling, Dec. 27-29, 1984, Tainan, Taiwan, 1986, 800p., ed. Han-Min Hsia, You-Li Chou, Shu-Yi Wang, Sheng Jii Hsieh, Hard Cover $70 (ISBN 0-87703-242-4)

Vol. 64 Space Safety and Rescue 1984-1985, Proceedings of the International Academy of Astronautics held in conjunction with the 35th and 36th International Astronautical Congresses, Lausanne, Switzerland, Oct. 7-13, 1984, and Stockholm, Sweden, Oct. 7-12, 1985, 1986, 400p., ed. Gloria W. Heath, Hard Cover $55 (ISBN 0-87703-248-3); Soft Cover $45 (ISBN 0-87703-249-1)

Vol. 65 The Human Quest in Space, 24th Goddard Memorial Symposium, Mar. 20-21, 1986, Greenbelt, MD, 1987, 312p., ed. Gerald L. Burdett, Gerald A. Soffen, Hard Cover $55 (ISBN 0-87703-262-9); Soft Cover $45 (ISBN 0-87703-263-7)

Vol. 66 Soviet Space Programs 1980-1985, 1987, 298p., by Nicholas L. Johnson, Microfiche Only $45 (ISBN 0-87703-266-1)

Vol. 67 Low-Gravity Sciences, Seminar Series 1986, University of Colorado at Boulder, 290p., ed. Jean N. Koster, Hard Cover $55 (ISBN 0-87703-270-X); Soft Cover $45 (ISBN 0-87703-271-8)

Vol. 68 Proceedings of the Fourth Annual L5 Space Development Conference, Apr. 25-28, 1985, Washington, D.C., 1987, 268p., ed. Frank Hecker, Hard Cover $50 (ISBN 0-87703-272-6); Soft Cover $35 (ISBN 0-87703-273-4)

Vol. 69 Visions of Tomorrow: A Focus on National Space Transportation Issues, 25th Goddard Memorial Symposium, Mar. 18-20, 1987, Greenbelt, MD, 1987, 338p., ed. Gerald A. Soffen, Hard Cover $55 (ISBN 0-87703-274-2); Soft Cover $45 (ISBN 0-87703-275-0)

Vol. 70 Space Safety and Rescue 1986-1987, Proceedings of the International Academy of Astronautics held in conjunction with the 37th and 38th International Astronautical Congresses, Innsbruck, Austria, Oct. 4-11, 1986, and Brighton, England, Oct. 11-16, 1987, 1988, 360p., ed. Gloria W. Heath, Hard Cover $55 (ISBN 0-87703-291-2); Soft Cover $45 (ISBN 0-87703-292-0)

Vol. 71 The NASA Mars Conference, Jul. 21-23, 1986, Washington, D.C., 1988, 570p., ed. Duke B. Reiber, Hard Cover $50 (ISBN 0-87703-293-9); Soft Cover $30 (ISBN 0-87703-294-7)

Vol. 72 Working in Orbit and Beyond: The Challenges for Space Medicine, Jun. 20-21, 1987, Washington, D.C., 1989, 188p., ed. David Lorr, Victoria Garshnek, Hard Cover $45 (ISBN 0-87703-295-5); Soft Cover $35 (ISBN 0-87703-296-3)

Vol. 73 Technology and the Civil Future in Space, 26th Goddard Memorial Symposium, Mar. 16-18, 1988, Greenbelt, MD, 1989, 246p., ed. Leonard A. Harris, Hard Cover $50 (ISBN 0-87703-301-3); Soft Cover $35 (ISBN 0-87703-302-1)

Vol. 74 The Case for Mars III: Strategies for Exploration - General Interest and Overview, July 18-22, 1987, Boulder, CO, 1989, 744p., ed. Carol Stoker, Hard Cover $75 (ISBN 0-87703-303-X); Soft Cover $55 (0-87703-304-8)

Vol. 75 The Case for Mars III: Strategies for Exploration - Technical, July 18-22, 1987, Boulder, CO, 1989, 646p., ed. Carol Stoker, Hard Cover $70 (ISBN 0-87703-305-6); Soft Cover $50 (ISBN 0-87703-306-4)

Vol. 76 Global Environmental Change: The Role of Space in Understanding Earth, 27th Goddard Memorial Symposium, Mar. 8-10, 1989, Washington, D.C., 1990, 178p., ed. Richard G. Johnson, Hard Cover $50 (ISBN 0-87703-322-6); Soft Cover $40 (ISBN 0-87703-323-4); Microfiche Suppl. (Vol. 60 AAS Microfiche Series) $10 (ISBN 0-87703-324-2)

Vol. 77 Space Safety and Rescue 1988 - 1989, Proceedings of the International Academy of Astronautics held in conjunction with the 39th and 40th International Astronautical Congresses, Bangalore, India, Oct. 8-15, 1988, and Málaga, Spain, Oct. 7-12, 1989, 1990, 500p., ed. Gloria W. Heath, Hard Cover $70 (ISBN 0-87703-327-7); Soft Cover $55 (ISBN 0-87703-328-5)

Vol. 78 Leaving the Cradle: Human Exploration of Space in the 21st Century, 28th Goddard Memorial Symposium, Mar. 14-16, 1990, Washington, D.C., 1991, 348p., ed. Thomas O. Paine, Hard Cover $70 (ISBN 0-87703-336-6); Soft Cover $55 (ISBN 0-87703-337-4)

Vol. 79 Space Safety and Rescue 1990, Proceedings of the International Academy of Astronautics held in conjunction with the 41st International Astronautical Congress, Dresden, Germany, Oct. 6-12, 1990, 1991, 232p., ed. Gloria W. Heath, Hard Cover $65 (ISBN 0-87703-341-2); Soft Cover $50 (ISBN 0-87703-342-0)

Vol. 80 Prospects for Interstellar Travel, 1992, 390p., by John H. Mauldin, Hard Cover $50 (ISBN 0-87703-344-7); Soft Cover $27 (ISBN 0-87703-345-5)

Vol. 81 Humans and Machines in Space: The Vision, The Challenge, The Payoff, 29th Goddard Memorial Symposium, Mar. 14-15, 1991, Washington, D.C., 1992, 204p., ed. Bradley Johnson, Gayle L. May, Paula Korn, Hard Cover $50 (ISBN 0-87703-356-0); Soft Cover $35 (ISBN 0-87703-357-9)

Vol. 82 Space Safety and Rescue 1991, Proceedings of the International Academy of Astronautics held in conjunction with the 42nd International Astronautical Congress, Montreal, Canada, Oct. 5-11, 1991, 1993, 270p., ed. Gloria W. Heath, Hard Cover $65 (ISBN 0-87703-372-2); Soft Cover $50 (ISBN 0-87703-373-0)

Vol. 83 Space: A Vital Stimulus to Our National Well-Being, 31st Goddard Memorial Symposium, March 9-10, 1993, Arlington, Virginia, and World Space Programs and Fiscal Reality, 30th Goddard Memorial Symposium, April 9-10, 1992, Alexandria, Virginia, 1994, 334p., ed. Gayle L. May, Saunders B. Kramer, Paula Korn, Leonard David, Barbara Sprungman, Hard Cover $70 (ISBN 0-87703-389-7); Soft Cover $50 (ISBN 0-87703-390-0)

Vol. 84 Space Safety and Rescue 1992, Proceedings of the International Academy of Astronautics held in conjunction with the World Space Congress, Washington, D.C., Aug. 28 to Sept. 5, 1992, 1994, 372p., ed. Gloria W. Heath, Hard Cover $70 (ISBN 0-87703-391-9); Soft Cover $55 (ISBN 0-87703-392-7)

Vol. 85 Civil Space in the Clinton Era, 32nd Goddard Memorial Symposium, March 1-2, 1994, Crystal City, Virginia, and Partners in Space . . . 2001, 41st Annual Meeting, November 14-16, 1994, Crystal City, Virginia, 1995, 292p., ed. Donald R. McConathy, Paula Korn, Hard Cover $70 (ISBN 0-87703-397-8); Soft Cover $50 (ISBN 0-87703-398-6)

Vol. 86 Strategies for Mars: A Guide to Human Exploration, 1996, 644p, ed. Carol R. Stoker, Carter Emmart, Hard Cover $70 (ISBN 0-87703-405-2); Soft Cover $45 (ISBN 0-87703-406-0)

Vol. 87 Space Safety and Rescue 1993, Proceedings of the International Academy of Astronautics held in conjunction with the 44th International Astronautical Congress, Graz, Austria, Oct. 16-22, 1993, 1996, 344p., ed. Gloria W. Heath, Hard Cover $70 (ISBN 0-87703-410-9); Soft Cover $50 (ISBN 0-87703-411-7)

Vol. 88 Space Safety and Rescue 1994, Proceedings of the International Academy of Astronautics held in conjunction with the 45th International Astronautical Congress, Jerusalem, Israel, Oct. 9-14, 1994, 1996, 326p., ed. Gloria W. Heath, Hard Cover $70 (ISBN0-87703-416-8); Soft Cover $50 (ISBN 0-87703-417-6)

Vol. 89 The Case for Mars IV: The International Exploration of Mars"Mission Strategy and Architectures, June 4-8, 1990, Boulder, CO, 1997, 790p., ed. Thomas R. Meyer, Hard Cover $80 (ISBN 0-87703-418-4); Soft Cover $55 (ISBN 0-87703-419-2)

Vol. 90 The Case for Mars IV: The International Exploration of Mars"Considerations for Sending Humans, June 4-8, 1990, Boulder, CO, 1997, 502p., ed. Thomas R. Meyer, Hard Cover $70 (ISBN 0-87703-420-6); Soft Cover $55 (ISBN 0-87703-421-4)

Vol. 91 From Imagination to Reality: Mars Exploration Studies of the Journal of the British Interplanetary Society (Part I, Precursors and Early Piloted Missions), 1997, 388p., ed. Robert M. Zubrin, Hard Cover $70 (ISBN 0-87703-426-5); Soft Cover $45 (ISBN 0-87703-427-3)

Vol. 92 From Imagination to Reality: Mars Exploration Studies of the Journal of the British Interplanetary Society (Part II, Base Building, Colonization and Terraformation), 1997, 376p., ed. Robert M. Zubrin, Hard Cover $70 (ISBN 0-87703-428-1); Soft Cover $45 (ISBN 0-87703-429-X)

Vol. 93 Space Safety and Rescue 1995, Proceedings of the International Academy of Astronautics held in conjunction with the 46th International Astronautical Congress, Jerusalem, Israel, Oct. 2-6, 1995, 1997, 482p., ed. Gloria W. Heath, Hard Cover $80 (ISBN 0-87703-416-8); Soft Cover $55 (ISBN 0-87703-417-6)

Vol. 94 Fluid and Electrolyte Regulation in Spaceflight, 1998, 238p., Carolyn S. Leach Huntoon, Anatoliy I. Grigoriev, Yuri V. Natochin, Hard Cover $60 (ISBN 0-87703-442-7); Soft Cover $40 (ISBN 0-87703-443-5)

Vol. 95 Space Safety and Rescue 1996, Proceedings of the International Academy of Astronautics held in conjunction with the 47th International Astronautical Congress, Beijing, China, Oct. 7-11, 1996, 1998, 362p., ed. Gloria W. Heath, Hard Cover $75 (ISBN 0-87703-446-X); Soft Cover $50 (ISBN 0-87703-447-8)

Vol. 96 Space Safety and Rescue 1997, Proceedings of the International Academy of Astronautics held in conjunction with the 48th International Astronautical Congress, Turin, Italy, Oct. 6-10, 1997, 1999, 412p., ed. Gloria W. Heath, Hard Cover $80 (ISBN 0-87703-454-0); Soft Cover $55 (ISBN 0-87703-455-9)

Vol. 97 The Case for Mars V, May 26-29, 1993, Boulder, CO, 2000, 564p., ed. Penelope J. Boston, Hard Cover $80 (ISBN 0-87703-459-1); Soft Cover $55 (ISBN 0-87703-460-5)

Vol. 98 The Case for Mars VI: Making Mars an Affordable Destination, July 17-20, 1996, Boulder, CO, 2000, 578p., ed. Kelly R. McMillen, Hard Cover $80 (ISBN 0-87703-461-3); Soft Cover $55 (ISBN 0-87703-462-1)

Vol. 99 Space Safety and Rescue 1998, Proceedings of the International Academy of Astronautics held in conjunction with the 49th International Astronautical Congress, Melbourne, Australia, Sept. 28 - Oct. 2, 1998, 2000, 410p., ed. Macgregor S. Reid and Walter Flury, Hard Cover $80 (ISBN0-87703-463-X); Soft Cover $55 (ISBN 0-87703-464-8)

Order from Univelt, Inc., P.O. Box 28130, San Diego, California 92198
STANDING ORDERS ACCEPTED

AAS HISTORY SERIES

Vol. 1 Two Hundred Years of Flight in America: A Bicentennial Survey, Edited by Eugene M. Emme, 1977, 326p, Third Printing 1981, Hard Cover $35 (ISBN 0-87703-091-X); Soft Cover $25 (ISBN 0-87703-101-0); special price for classroom text or bulk purchase.

Vol. 2 Twenty-Five Years of the American Astronautical Society: Historical Reflections and Projections, 1954-1979, Edited by Eugene M. Emme, 1980, 248p, Hard Cover $25 (ISBN 0-87703-117-7); Soft Cover $15 (ISBN 0-87703-118-5).

Vol. 3 Between Sputnik and the Shuttle: New Perspectives on American Astronautics, 1957-1980, Edited by Frederick C. Durant, III, 1981, 350p, Hard Cover $40 (ISBN 0-87703-145-2); Soft Cover $30 (ISBN 0-87703-149-9).

Vol. 4 The Endless Space Frontier: A History of the House Committee on Science and Astronautics, By Ken Hechler, Abridged and edited by Albert E. Eastman, 1982, 460p, Hard Cover $45 (ISBN 0-87703-157-6); Soft Cover $35 (ISBN 0-87703-158-4).

Vol. 5 Science Fiction and Space Futures: Past and Present, Edited by Eugene M. Emme, 1982, 278p, Hard Cover $35 (ISBN 0-87703-172-X); Soft Cover $25 (ISBN 0-87703-173-8).

Vol. 6 First Steps Toward Space, Edited by Frederick C. Durant, III and George S. James, 1986, 318p, Hard Cover $45 (ISBN 0-87703-243-2); Soft Cover $35 (ISBN 0-87703-244-0).

Vol. 7 History of Rocketry and Astronautics, Edited by R. Cargill Hall, 1986, Part I, 250p, Part II, 502p, sold as a set, Hard Cover $100 (ISBN 0-87703-260-2); Soft Cover $80 (ISBN 0-87703-261-0).

Vol. 8 History of Rocketry and Astronautics, Edited by Kristan R. Lattu, 1989, 368p, Hard Cover $50 (ISBN 0-87703-307-2); Soft Cover $35 (ISBN 0-87703-308-0).

Vol. 9 History of Rocketry and Astronautics, Edited by Frederick I. Ordway, III, 1989, 330p, Hard Cover $50 (ISBN 0-87703-309-9); Soft Cover $35 (ISBN 0-87703-310-2).

Vol. 10 History of Rocketry and Astronautics, Edited by Å. Ingemar Skoog, 1990, 330p, Hard Cover $60 (ISBN 0-87703-329-3); Soft Cover $40 (ISBN 0-87703-330-7).

Vol. 11 History of Rocketry and Astronautics, Edited by Roger D. Launius, 1994, 236p, Hard Cover $60 (ISBN 0-87703-382-X); Soft Cover $40 (ISBN 0-87703-383-8).

Vol. 12 History of Rocketry and Astronautics, Edited by John L. Sloop, 1991, 252p, Hard Cover $60 (ISBN 0-87703-332-3); Soft Cover $40 (ISBN 0-87703-333-1).

Vol. 13 History of Liquid Rocket Engine Development in the United States 1955-1980, Edited by Stephen E. Doyle, 1992, 176p, Hard Cover $50 (ISBN 0-87703-349-8); Soft Cover $35 (ISBN 0-87703-350-1).

Vol. 14 History of Rocketry and Astronautics, Edited by Tom D. Crouch, Alex M. Spencer, 1993, 222p, Hard Cover $50 (ISBN 0-87703-374-9); Soft Cover $35 (ISBN 0-87703-375-7).

Vol. 15 History of Rocketry and Astronautics, Edited by Lloyd H. Cornett, Jr., 1993, 452p, Hard Cover $60 (ISBN 0-87703-376-5); Soft Cover $40 (ISBN 0-87703-377-3).

Vol. 16 Out From Behind the Eight-Ball: A History of Project Echo, by Donald C. Elder, 1995, 176p, Hard Cover $50 (ISBN 0-87703-387-0); Soft Cover $30 (ISBN 0-87703-388-9).

Vol. 17 History of Rocketry and Astronautics, Edited by John Becklake, 1995, 480p, Hard Cover $60 (ISBN 0-87703-395-1); Soft Cover $40 (ISBN 0-87703-396-X).

Vol. 18 Organizing for the Use of Space: Historical Perspectives on a Persistent Issue, Edited by Roger D. Launius, 1995, 234p, Hard Cover $60; Soft Cover $40

Vol. 19 History of Rocketry and Astronautics, Edited by J. D. Hunley, 1997, 318p, Hard Cover $60 (ISBN 0-87703-422-2); Soft Cover $40 (ISBN 0-87703-423-0).

Vol. 20 History of Rocketry and Astronautics, Edited by J. D. Hunley, 1997, 344p, Hard Cover $60 (ISBN 0-87703-424-9); Soft Cover $40 (ISBN 0-87703-425-7).

Vol. 21 History of Rocketry and Astronautics, Edited by Philippe Jung, 1997, 368p, Hard Cover $60 (ISBN 0-87703-439-7); Soft Cover $40 (ISBN 0-87703-440-0).

Vol. 22 History of Rocketry and Astronautics, Edited by Philippe Jung, 1998, 418p, Hard Cover $60 (ISBN 0-87703-444-3); Soft Cover $40 (ISBN 0-87703-445-1).

Order from Univelt, Incorporated, P.O. Box 28130, San Diego, California 92198
STANDING ORDERS ACCEPTED

INDEX

INDEX TO ALL AMERICAN ASTRONAUTICAL SOCIETY PAPERS AND ARTICLES 1954 - 1992

This index is a numerical/chronological index (which also serves as a citation index) and an author index. (A subject index volume will be forthcoming.)

It covers all articles that appear in the following:
Advances in the Astronautical Sciences (1957 - 1992)
Science and Technology Series (1964 -1992)
AAS History Series (1977 - 1992)
AAS Microfiche Series (1968 - 1992)
Journal of the Astronautical Sciences (1954 -September 1992)
Astronautical Sciences Review (1959 - 1962)

If you are in aerospace you will want this excellent reference tool which covers the first 35 years of the Space Age.

Numerical/Chronological/Author Index in three volumes,

Ordered as a set:
Library Binding (all three volumes) $120.00;
Soft Cover (all three volumes) $90.00.
Ordered by individual volume:
Volume I (1954 - 1978) Library Binding $40.00; Soft Cover $30.00;
Volume II (1979 - 1985/86) Library Binding $60.00; Soft Cover $45.00;
Volume III (1986 - 1992) Library Binding $70.00; Soft Cover $50.00.

Order from Univelt, Inc., P.O. Box 28130, San Diego, California 92198.

NUMERICAL INDEX

VOLUME 104 ADVANCES IN THE ASTRONAUTICAL SCIENCES,
GUIDANCE AND CONTROL 2000 (2000)

(AAS Annual Rocky Mountain Guidance and Control Conference, February 2-6, 2000, Breckenridge, Colorado)

AAS 00-001 Enhanced Fault-Tolerant Attitude Control for Lockheed Martin's A2100 Spacecraft, N. Goodzeit, M. Patel and H. Weigl

AAS 00-002 The Two-Step Optimal Estimator and Example Applications, N. Jeremy Kasdin

AAS 00-003 Test Results for the Automated Rendezvous and Capture System, Craig A. Cruzen, James J. Lomas and Richard W. Dabney

AAS 00-004 Lost-in-Space: A Star Pattern Recognition and Attitude Estimation Approach for the Case of No Prior Attitude Information, Gwanghyoek Ju, Hye-Young Kim, Thomas C. Pollock, John L. Junkins, Jer-Nan Juang and Daniele Mortari

AAS 00-005 Design, Implementation, and Flight Results for All-Stellar Attitude Determination, Jim D. Chapel, Stephen M. Micciche and Richard Kiessig

AAS 00-006 Attitude Dynamics of the Genesis Spacecraft, Carl Hubert

AAS 00-007 Fire Control Design for High Energy (Laser) Systems, Timothy J. Schneeberger, S. M. Seltzer and Robert Van Allen

AAS 00-008 to -010 Not Assigned

AAS 00-011 A Projection Approach to Spacecraft Formation Attitude Control, Jonathan Lawton, Randal W. Beard and Fred Y. Hadaegh

AAS 00-012 Gravitational Perturbations, Nonlinearity and Circular Orbit Assumption Effects on Formation Flying Control Strategies, Kyle T. Alfriend, Hanspeter Schaub and Dong-Woo Gim

AAS 00-013 Nonlinear Dynamics, Trajectory Generation, and Adaptive Control of Multiple Spacecraft in Periodic Relative Orbits, Qiguo Yan, Guang Yang, Vikram Kapila and Marcio S. de Queiroz

AAS 00-014 Validating a Formation Flying Control System Design: The GRACE Project Experience, H. D. Stevens, Jack Rodden, Phil Morton and Matthias Fehrenbach

AAS 00-015 A Tethered Formation Flying Concept for the SPECS Mission, David A. Quinn and David C. Folta

AAS 00-016 Project Orion: Carrier Phase Differential GPS Navigation for Formation Flying, Franz D. Busse, Gokhan Inalhan and Jonathan P. How

AAS 00-017	Mode and Logic-Based Switching for the Formation Flying Control of Multiple Spacecraft, Mehran Mesbahi and Fred Y. Hadaegh
AAS 00-018 to -020	Not Assigned
AAS 00-021	Evolution of International Space Station GN&C System Across ISS Assembly Stages, Roscoe Lee
AAS 00-022	International Space Station Assembly and Operation Control Challenges, Nazareth Bedrossian
AAS 00-023	Not Available (Withdrawn)
AAS 00-024	Not Available (Withdrawn)
AAS 00-025	Studies on the Attitude Control System Design for the Crew Return Vehicle (X38), K. Abdel-Motagaly, O. Rombout, R. Gonzalez, K. Berrier, D. Hasan, D. Strack and B. Rishikof
AAS 00-026	Thermal Radiator Pointing for International Space Station, Scott A. Green
AAS 00-027	Command Level Maneuver Optimization for the International Space Station, Gregory E. Chamitoff, Adam L. Dershowitz and Amy L. Bryson
AAS 00-028 to -030	Not Assigned
AAS 00-031	Adaptation of a Spaceborne Geolocation System to Airborne Experiments, Alan S. Hope, Henry M. Pickard and Jay W. Middour
AAS 00-032	Nanosol – A Next Generation Sun Sensor, John Glaberson
AAS 00-033	Not Assigned
AAS 00-034	The TERMA Star Tracker for the NEMO Satellite, L. Maresi, T. Paulsen, R. Noteborn, O. Mikkelsen and R. Nielsen
AAS 00-035	A Cost Effective, High Reliable RWA Solution to a Commercial Market Demand, Terance Marshall, Mitchell Fletcher and Joseph Zuckerbrow
AAS 00-036	Redundant Launch Vehicle Guidance, R. Joe Wright
AAS 00-037	Radiation Hardened Power PC 603e^{TM} Space Processor, Robert D. Campbell, Richard F. Elmhurst and Gary R. Brown
AAS 00-038	ARU Architectural Solutions for Long Life Satellite Missions, Edward C. Moulton, Robert H. Fall and Thomas G. Stottlar
AAS 00-039 to -040	Not Available
AAS 00-041	Turbo-Charged Torqwheels^{TM}, Bill Bialke
AAS 00-042	Not Available
AAS 00-043	Progress of Fiber Optic Gyroscope Development for Space Applications, James Goodwin, Pei-hwa Lo, Mark Mariak and Ming Yu
AAS 00-044	SOHO: Loss and Recovery 1998, Gyroless 1999, A. van Overbeek
AAS 00-045	A Novel New GEO Earth Sensor Provides High Accuracy, George Rullman, Richard Burton and Len Anderson
AAS 00-046	The Design and Operation of a COTS Space GPS Receiver, Martin J. Unwin and Michael K. Oldfield

AAS 00-047 to -048 Not Available

AAS 00-049 Rocket Sounding Balloon Experimental Section 2000, Kenneth Dalton, Gretchen England, Justin Eisenach, Chris Kilzer, Thanh Tran and Echezona Onwuatuegwu

AAS 00-050 In Flight Performance of the ZARM Magnetic Torquers MT80-1/MT140-2 Flown on the ABRIXAS Mission, Matthias Wiegand, Oliver Matthews, Peter Offterdinger and H. J. Rath

AAS 00-051 to -060 Not Assigned

AAS 00-061 Formation Flying and Relative Navigation – A Nanosatellite Research Mission, Frank R. Chavez and David K. Schmidt

AAS 00-062 MEMS Rate Sensors for Space, Joel Gambino

AAS 00-063 Dynamics and Control of Nanosatellite ASUSat1, Brian K. Underhill, Assi Friedman and Helen L. Reed

AAS 00-064 Attitude Control – An Afterthought: The MightySat I Experience, Jeff Benton and Thomas Itchkawich

AAS 00-065 MEMS-Based GN&C Sensors and Actuators for Micro/Nano Satellites, J. Connelly, N. Dennehy, P. Hattis, W. Johnson, D. Sargent and M. Socha

AAS 00-066 Digital Reaction Wheel Assembly RSI 01-5 for Small and Low-Cost Spacecraft, Armin Landes and Stephen Böttcher-Arff

AAS 00-067 An Attitude Control System for an ASAP5-Launched Interplanetary Spacecraft, Tobin Anthony and Bhavesh Patel

AAS 00-068 to -070 Not Assigned

AAS 00-071 Autonomous Orbit Control: Initial Flight Results From UoSAT-12, James R. Wertz, Jeffrey L. Cloots, John T. Collins, Simon D. Dawson, Gwynne Gurevich, Brian K. Sato and L. Jane Hansen

AAS 00-072 Fuel Optimization During Mars Global Surveyor Aerobraking, Stuart R. Spath and Dave F. Eckart

AAS 00-073 QuikSCAT Attitude Control System Initialization and Early On-Orbit Operations, Dan Hegel and Scott Mitchell

AAS 00-074 Chandra X-Ray Observatory Pointing Control System Performance During Transfer Orbit and Initial On-Orbit Operations, Peter Quast, Frank Tung, John Wider and Mark West

AAS 00-075 Not Available (Withdrawn)

AAS 00-076 The Recovery of TOMS-EP, Brent Robertson, Phil Sabelhaus, Todd Mendenhall and Lorraine Fesq

AAS 00-077 Operational Experiences With Globalstar Constellation Control and Stationkeeping Using Solar Radiation Pressure, Lee Barker and Benjamin Lange

AAS 00-078 to -099 Not Assigned

AUTHOR INDEX*

Abdel-Motagaly, K., AAS 00-025, Adv v104, pp281-296

Alfriend, K. T., AAS 00-012, Adv v104, pp139-158

Anderson, L., AAS 00-045, Adv v104, pp447-464

Anthony, T., AAS 00-067, Adv v104, pp589-599

Barker, L., AAS 00-077, Adv v104, pp687-695

Beard, R. W., AAS 00-011, Adv v104, pp119-137

Bedrossian, N., AAS 00-022, Adv v104, pp259-279

Benton, J., AAS 00-064, Adv v104, pp541-559

Berrier, K., AAS 00-025, Adv v104, pp281-296

Bialke, B., AAS 00-041, Adv v104, pp415-423

Böttcher-Arff, S., AAS 00-066, Adv v104, pp577-588

Brown, G. R., AAS 00-037, Adv v104, pp387-401

Bryson, A. L., AAS 00-027, Adv v104, pp311-326

Burton, R., AAS 00-045, Adv v104, pp447-464

Busse, F. D., AAS 00-016, Adv v104, pp197-212

Campbell, R. D., AAS 00-037, Adv v104, pp387-401

Chamitoff, G. E., AAS 00-027, Adv v104, pp311-326

Chapel, J. D., AAS 00-005, Adv v104, pp73-90

Chavez, F. R., AAS 00-061, Adv v104, pp499-514

Cloots, J. L., AAS 00-071, Adv v104, pp603-614

Collins, J. T., AAS 00-071, Adv v104, pp603-614

Connelly, J., AAS 00-065, Adv v104, pp561-576

Cruzen, C. A., AAS 00-003, Adv v104, pp35-56

Dabney, R. W., AAS 00-003, Adv v104, pp35-56

Dalton, K., AAS 00-049, Adv v104, pp475-482

Dawson, S. D., AAS 00-071, Adv v104, pp603-614

Dennehy, N., AAS 00-065, Adv v104, pp561-576

de Queiroz, M. S., AAS 00-013, Adv v104, pp159-174

Dershowitz, A. L., AAS 00-027, Adv v104, pp311-326

Eckart, D. F., AAS 00-072, Adv v104, pp615-627

Eisenach, J., AAS 00-049, Adv v104, pp475-482

Elmhurst, R. F., AAS 00-037, Adv v104, pp387-401

England, G., AAS 00-049, Adv v104, pp475-482

Fall, R. H., AAS 00-038, Adv v104, pp403-414

Fehrenbach, M., AAS 00-014, Adv v104, pp175-182

Fesq, L., AAS 00-076, Adv v104, pp665-685

* For each author the paper number is given. The page numbers refer to Volume 104, *Advances in the Astronautical Sciences*.

Fletcher, M., AAS 00-035, Adv v104, pp365-378

Folta, D. C., AAS 00-015, Adv v104, pp183-196

Friedman, A., AAS 00-063, Adv v104, pp523-540

Gambino, J., AAS 00-062, Adv v104, pp515-521

Gim, D.-W., AAS 00-012, Adv v104, pp139-158

Glaberson, J., AAS 00-032, Adv v104, pp347-354

Gonzalez, R., AAS 00-025, Adv v104, pp281-296

Goodwin, J., AAS 00-043, Adv v104, pp425-432

Goodzeit, N., AAS 00-001, Adv v104, pp3-14

Green, S. A., AAS 00-026, Adv v104, pp297-310

Gurevich, G., AAS 00-071, Adv v104, pp603-614

Hadaegh, F. Y., AAS 00-011, Adv v104, pp119-137; AAS 00-017, Adv v104, pp213-240

Hansen, L. J., AAS 00-071, Adv v104, pp603-614

Hasan, D., AAS 00-025, Adv v104, pp281-296

Hattis, P., AAS 00-065, Adv v104, pp561-576

Hegel, D., AAS 00-073, Adv v104, pp629-646

Hope, A. S., AAS 00-031, Adv v104, pp329-346

How, J. P., AAS 00-016, Adv v104, pp197-212

Hubert, C., AAS 00-006, Adv v104, pp91-104

Inalhan, G., AAS 00-016, Adv v104, pp197-212

Itchkawich, T., AAS 00-064, Adv v104, pp541-559

Johnson, W., AAS 00-065, Adv v104, pp561-576

Ju, G., AAS 00-004, Adv v104, pp57-72

Juang, J.-N., AAS 00-004, Adv v104, pp57-72

Junkins, J. L., AAS 00-004, Adv v104, pp57-72

Kapila, V., AAS 00-013, Adv v104, pp159-174

Kasdin, N. J., AAS 00-002, Adv v104, pp15-34

Kiessig, R., AAS 00-005, Adv v104, pp73-90

Kilzer, C., AAS 00-049, Adv v104, pp475-482

Kim, H.-Y., AAS 00-004, Adv v104, pp57-72

Landes, A., AAS 00-066, Adv v104, pp577-588

Lange, B., AAS 00-077, Adv v104, pp687-695

Lawton, J., AAS 00-011, Adv v104, pp119-137

Lee, R., AAS 00-021, Adv v104, pp243-257

Lo, P.-h., AAS 00-043, Adv v104, pp425-432

Lomas, J. J., AAS 00-003, Adv v104, pp35-56

Maresi, L., AAS 00-034, Adv v104, pp355-364

Mariak, M., AAS 00-043, Adv v104, pp425-432

Marshall, T., AAS 00-035, Adv v104, pp365-378

Matthews, O., AAS 00-050, Adv v104, pp483-495

Mendenhall, T., AAS 00-076, Adv v104, pp665-685

Mesbahi, M., AAS 00-017, Adv v104, pp213-240

Micclche, S. M., AAS 00-005, Adv v104, pp73-90

Middour, J. W., AAS 00-031, Adv v104, pp329-346

Mikkelsen, O., AAS 00-034, Adv v104, pp355-364

Mitchell, S., AAS 00-073, Adv v104, pp629-646

Mortari, D., AAS 00-004, Adv v104, pp57-72

Morton, P., AAS 00-014, Adv v104, pp175-182

Moulton, E. C., AAS 00-038, Adv v104, pp403-414

Nielsen, R., AAS 00-034, Adv v104, pp355-364

Noteborn, R., AAS 00-034, Adv v104, pp355-364

Offterdinger, P., AAS 00-050, Adv v104, pp483-495

Oldfield, M. K., AAS 00-046, Adv v104, pp465-473

Onwuatuegwu, E., AAS 00-049, Adv v104, pp475-482

Patel, B., AAS 00-067, Adv v104, pp589-599

Patel, M., AAS 00-001, Adv v104, pp3-14

Paulsen, T., AAS 00-034, Adv v104, pp355-364

Pickard, H. M., AAS 00-031, Adv v104, pp329-346

Pollock, T. C., AAS 00-004, Adv v104, pp57-72

Quast, P., AAS 00-074, Adv v104, pp647-664

Quinn, D. A., AAS 00-015, Adv v104, pp183-196

Rath, H. J., AAS 00-050, Adv v104, pp483-495

Reed, H. L., AAS 00-063, Adv v104, pp523-540

Rishikof, B., AAS 00-025, Adv v104, pp281-296

Robertson, B., AAS 00-076, Adv v104, pp665-685

Rodden, J., AAS 00-014, Adv v104, pp175-182

Rombout, O., AAS 00-025, Adv v104, pp281-296

Rullman, G., AAS 00-045, Adv v104, pp447-464

Sabelhaus, P., AAS 00-076, Adv v104, pp665-685

Sargent, D., AAS 00-065, Adv v104, pp561-576

Sato, B. K., AAS 00-071, Adv v104, pp603-614

Schaub, H., AAS 00-012, Adv v104, pp139-158

Schmidt, D. K., AAS 00-061, Adv v104, pp499-514

Schneeberger, T. J., AAS 00-007, Adv v104, pp105-116

Seltzer, S. M., AAS 00-007, Adv v104, pp105-116

Socha, M., AAS 00-065, Adv v104, pp561-576

Spath, S. R., AAS 00-072, Adv v104, pp615-627

Stevens, H. D., AAS 00-014, Adv v104, pp175-182

Stottlar, T. G., AAS 00-038, Adv v104, pp403-414

Strack, D., AAS 00-025, Adv v104, pp281-296

Tran, T., AAS 00-049, Adv v104, pp475-482

Tung, F., AAS 00-074, Adv v104, pp647-664

Underhill, B. K., AAS 00-063, Adv v104, pp523-540

Unwin, M. J., AAS 00-046, Adv v104, pp465-473

Van Allen, R., AAS 00-007, Adv v104, pp105-116

van Overbeek, A., AAS 00-044, Adv v104, pp433-445

Weigl, H., AAS 00-001, Adv v104, pp3-14

Wertz, J. R., AAS 00-071, Adv v104, pp603-614

West, M., AAS 00-074, Adv v104, pp647-664

Wider, J., AAS 00-074, Adv v104, pp647-664

Wiegand, M., AAS 00-050, Adv v104, pp483-495

Wright, R. J., AAS 00-036, Adv v104, pp379-386

Yan, Q., AAS 00-013, Adv v104, pp159-174

Yang, G., AAS 00-013, Adv v104, pp159-174

Yu, M., AAS 00-043, Adv v104, pp425-432

Zuckerbrow, J., AAS 00-035, Adv v104, pp365-378